Asimov's
New Guide to Science

ISAAC ASIMOV

PENGUIN BOOKS

PENGUIN BOOKS

Published by the Penguin Group
Penguin Books Ltd, 27 Wrights Lane, London W8 5TZ, England
Penguin Books USA Inc., 375 Hudson Street, New York, New York 10014, USA
Penguin Books Australia Ltd, Ringwood, Victoria, Australia
Penguin Books Canada Ltd, 2801 John Street, Markham, Ontario, Canada L3R 1B4
Penguin Books (NZ) Ltd, 182–190 Wairau Road, Auckland 10, New Zealand

Penguin Books Ltd, Registered Offices: Harmondsworth, Middlesex, England

First published in the USA by Basic Books, Inc. 1984
First published in Great Britain by Viking 1985
Published in Penguin Books 1987
5 7 9 10 8 6

Printed in England by Clays Ltd, St Ives plc

Penguin Books

Asimov's New Guide to Science

Dr Isaac Asimov was born in Russia in 1920 and emigrated
with his family to the United States when he was three. In
1949 he received a Ph.D. from Columbia University and joined
the School of Medicine, Boston University, as Professor of
Biochemistry. He has published over two hundred books, both
fiction and non-fiction, and won a Nebula Award and three
Hugo Awards, including one in 1966 for *The Foundation Trilogy*
(1963). One of Asimov's main contributions to science-fiction
lore has been his formulation (with John W. Campbell) of the
Three Robotic Laws, which he used as material in his short
stories (*I, Robot*, 1956; and *The Rest of the Robots*, 1964) and
novels (*The Caves of Steel*, 1954; and *The Naked Sun*, 1957). He
has written for children under the pseudonym Paul French.
Other works include *Nine Tomorrows* (1959); a history of our
attitudes and discoveries about the cosmos, *The Universe*
(Pelican); *Asimov's Guide to Science* in two volumes (Pelican);
Exploring the Earth and Cosmos (Pelican); two volumes of
autobiography, *In Memory Yet Green* and *In Joy Still Felt*; *In the
Beginning* and *Foundation's Edge*.

Contents

ILLUSTRATIONS   xiv   PREFACE   xv

Chapter 1   MATHEMATICS   8
                                           THE DEDUCTIVE PROCESS   9
What Is Science?   3   THE RENAISSANCE AND
                                           COPERNICUS   10
    THE DESIRE TO KNOW   4   EXPERIMENTATION AND
    THE GREEKS   6   INDUCTION   12
    GEOMETRY AND   MODERN SCIENCE   14

PART I
The Physical Sciences

Chapter 2   SPIRAL GALAXIES   34
                                           CLUSTERS OF GALAXIES   36
The Universe   19   The Birth of the Universe   37
                                               THE AGE OF THE EARTH   38
    The Size of the Universe 19   THE SUN AND THE SOLAR
    EARLY MEASUREMENTS   20   SYSTEM   39
    MEASURING THE SOLAR   THE BIG BANG   43
        SYSTEM   22   The Death of the Sun   45
    THE FARTHER STARS   24   NOVAE AND SUPERNOVAE   46
    MEASURING A STAR'S   THE EVOLUTION OF THE
        BRIGHTNESS   27   STARS   51
    DETERMINING THE   The Windows to the
        GALAXY'S SIZE   30   Universe   56
    ENLARGING THE UNIVERSE   32

THE TELESCOPE 57
THE SPECTROSCOPE 58
PHOTOGRAPHY 60
RADIO ASTRONOMY 61
LOOKING BEYOND OUR
  GALAXY 64
The New Objects 65
QUASARS 66
NEUTRON STARS 69
BLACK HOLES 73
"EMPTY" SPACE 75

Chapter 3

The Solar System 79

Birth of the Solar System 79
The Sun 85
The Moon 88
MEASURING THE MOON 90
GOING TO THE MOON 91
ROCKETRY 91
EXPLORING THE MOON 93
ASTRONAUTS AND THE
  MOON 96
Venus and Mercury 98
MEASURING THE PLANETS 99
THE VENUS PROBES 100
THE MERCURY PROBES 103
Mars 104
MAPPING MARS 106
THE MARS PROBES 108
THE MARTIAN SATELLITES 110
Jupiter 111
JOVIAN SATELLITES 111
JUPITER'S SHAPE AND
  SURFACE 114
JUPITER'S SUBSTANCE 116
THE JUPITER PROBES 117
Saturn 119
SATURN'S RINGS 119
THE SATURNIAN SATELLITES 121
The Outermost Planets 124
URANUS 124
NEPTUNE 126
PLUTO 128
Asteroids 130

ASTEROIDS BEYOND MARS'S
  ORBIT 130
EARTH GRAZERS AND
  APOLLO OBJECTS 133
Comets 135

Chapter 4

The Earth 138

Of Shape and Size 138
THE EARTH AS SPHERE 138
MEASURING THE GEOID 142
WEIGHING THE EARTH 144
Earth's Layers 146
EARTHQUAKES 146
VOLCANOES 149
FORMATION OF EARTH'S
  CRUST 153
THE LIQUID CORE 155
EARTH'S MANTLE 156
THE ORIGIN OF THE MOON 159
THE EARTH AS LIQUID 161
The Ocean 162
THE CURRENTS 163
THE OCEAN'S RESOURCES 166
THE OCEAN DEPTHS AND
  CONTINENTAL CHANGES 167
LIFE IN THE DEEP 175
DEEP-SEA DIVING 177
The Icecaps 178
THE NORTH POLE 178
THE SOUTH POLE—
  ANTARCTICA 180
THE INTERNATIONAL
  GEOPHYSICAL YEAR 182
GLACIERS 183
CAUSES OF ICE AGES 186

Chapter 5

The Atmosphere 191

The Shells of Air 191
MEASURING AIR 191
AIR TRAVEL 194
The Gases in Air 199
THE LOWER ATMOSPHERE 199
THE STRATOSPHERE 201
THE IONOSPHERE 204

CONTENTS

Magnets   207

  MAGNETISM AND
    ELECTRICITY   207
  EARTH'S MAGNETIC FIELD   210
  THE SOLAR WIND   214
  THE MAGNETOSPHERE   215
  PLANETARY
    MAGNETOSPHERES   218

Meteors and Meteorites   220

  METEORS   220
  METEORITES   222

*Air: Keeping It and
Getting It*   226

  ESCAPE VELOCITY   226
  THE ORIGINAL
    ATMOSPHERE   229

Chapter 6

The Elements   232

The Periodic Table   232

  EARLY THEORIES   232
  ATOMIC THEORY   234
  MENDELEEV'S PERIODIC
    TABLE   237
  ATOMIC NUMBERS   240

Radioactive Elements   242

  IDENTIFYING THE
    ELEMENTS   242
  FINDING THE MISSING
    ELEMENTS   245
  TRANSURANIUM ELEMENTS   246
  SUPER-HEAVY ELEMENTS   248

Electrons   249

  THE PERIODICITY OF THE
    PERIODIC TABLE   250
  THE NOBLE, OR INERT,
    GASES   251
  THE RARE-EARTH
    ELEMENTS   255
  THE TRANSITIONAL
    ELEMENTS   257
  THE ACTINIDES   261

Gases   262

  LIQUEFACTION   262
  ROCKET FUEL   267

  SUPERCONDUCTORS AND
    SUPERFLUIDS   268
  CRYOGENICS   271
  HIGH PRESSURES   273

Metals   276

  IRON AND STEEL   278
  NEW METALS   281

Chapter 7

The Particles   285

The Nuclear Atom   285

  IDENTIFYING THE
    PARTICLES   285
  THE ATOMIC NUCLEUS   287

Isotopes   289

  UNIFORM BUILDING BLOCKS   289
  TRACKING PARTICLES   292
  TRANSMUTATION OF
    ELEMENTS   295

New Particles   295

  THE NEUTRON   296
  THE POSITRON   298
  RADIOACTIVE ELEMENTS   302
  PARTICLE ACCELERATORS   304
  PARTICLE SPIN   310
  COSMIC RAYS   311
  THE STRUCTURE OF THE
    NUCLEUS   313

Leptons   315

  NEUTRINOS AND
    ANTINEUTRINOS   316
  TRACKING DOWN THE
    NEUTRINO   318
  NUCLEAR INTERACTION   320
  THE MUON   322
  THE TAUON   324
  THE NEUTRINO'S MASS   324

Hadrons and Quarks   326

  PIONS AND MESONS   326
  BARYONS   327
  THE QUARK THEORY   329

Fields   333

  ELECTROMAGNETIC
    INTERACTION   334
  THE CONSERVATION LAWS   335
  A UNIFIED FIELD THEORY   338

Chapter 8

The Waves   341

Light   341

THE NATURE OF LIGHT   341
THE SPEED OF LIGHT   344
RADAR   347
LIGHT-WAVES THROUGH
  SPACE   348
THE MAGNETIC
  MONOPOLES   349
ABSOLUTE MOTION   350

Relativity   352

THE LORENTZ-FITZGERALD
  EQUATIONS   352
RADIATION AND PLANCK'S
  QUANTUM THEORY   353
EINSTEIN'S PARTICLE-WAVE
  THEORY   355
THE THEORY OF
  RELATIVITY   357
SPACE-TIME AND THE
  CLOCK PARADOX   358
GRAVITY AND EINSTEIN'S
  GENERAL THEORY   359
TESTING THE GENERAL
  THEORY   361

Heat   362

MEASURING TEMPERATURE   362
TWO THEORIES OF HEAT   364
HEAT AS ENERGY   365
HEAT AND MOLECULAR
  MOTION   367

Mass to Energy   368

Particles and Waves   371

ELECTRON MICROSCOPY   372
ELECTRONS AS WAVES   374
THE UNCERTAINTY
  PRINCIPLE   375

Chapter 9

The Machine   379

Fire and Steam   379

EARLY TECHNOLOGY   379
THE STEAM ENGINE   382

Electricity   386

STATIC ELECTRICITY   386
DYNAMIC ELECTRICITY   389
GENERATING ELECTRICITY   390
EARLY APPLICATION OF
  ELECTRICITY TO
  TECHNOLOGY   392

Electrical Technology   394

THE TELEPHONE   394
RECORDING SOUND   395
ARTIFICIAL LIGHT BEFORE
  ELECTRICITY   396
ELECTRIC LIGHT   397
PHOTOGRAPHY   400

Internal Combustion
  Engines   402

THE AUTOMOBILE   402
THE AIRPLANE   407

Electronics   410

THE RADIO   410
TELEVISION   414
THE TRANSISTOR   415

Masers and Lasers   419

MASERS   419
LASERS   421

Chapter 10

The Reactor   425

Energy   425

COAL AND OIL: FOSSIL
  FUELS   426
SOLAR ENERGY   431

The Nucleus in War   432

THE DISCOVERY OF FISSION   432
THE CHAIN REACTION   435
THE FIRST ATOMIC PILE   437
THE NUCLEAR AGE   440
THE THERMONUCLEAR
  REACTION   442

The Nucleus in Peace   444

NUCLEAR-POWERED
  VESSELS   446
NUCLEAR REACTORS FOR
  ELECTRIC POWER   446

BREEDER REACTORS   447   FALLOUT   452
THE DANGERS OF
  RADIATION   448   *Controlled Nuclear*
USING FISSION PRODUCTS   451   *Fusion*   456

PART II
The Biological Sciences

Chapter 11

The Molecule   463

Organic Matter   463
  CHEMICAL STRUCTURE   465
The Details of Structure   468
  OPTICAL ACTIVITY   469
  THE PARADOX OF THE
    BENZENE RING   473
Organic Synthesis   477
  THE FIRST SYNTHESIS   477
  ALKALOIDS AND PAIN
    DEADENERS   480
  THE PROTOPORPHYRINS   485
  NEW PROCESSES   487
Polymers and Plastics   488
  CONDENSATION AND
    GLUCOSE   488
  CRYSTALLINE AND
    AMORPHOUS POLYMERS   491
  CELLULOSE AND
    EXPLOSIVES   494
  PLASTICS AND CELLULOID   496
  HIGH POLYMERS   497
  GLASS AND SILICONE   500
Synthetic Fibers   501
Synthetic Rubber   504

Chapter 12

The Proteins   507

Amino Acids   507
  THE COLLOIDS   510

THE POLYPEPTIDE CHAINS   514
PROTEINS IN SOLUTION   515
BREAKING DOWN A
  PROTEIN MOLECULE   516
ANALYZING THE PEPTIDE
  CHAIN   519
SYNTHETIC PROTEINS   522
THE SHAPE OF THE
  PROTEIN MOLECULE   523
Enzymes   523
  CATALYSIS   525
  FERMENTATION   526
  PROTEIN CATALYSTS   528
  ENZYME ACTION   529
Metabolism   534
  THE CONVERSION OF SUGAR
    TO ETHYL ALCOHOL   534
  METABOLIC ENERGY   535
  THE METABOLISM OF FATS   538
Tracers   539
  CHOLESTEROL   542
  THE PORPHYRIN RING OF
    HEME   543
Photosynthesis   544
  THE PROCESS   545
  CHLOROPHYLL   546

Chapter 13

The Cell   551

Chromosomes   551
  CELL THEORY   554
  ASEXUAL REPRODUCTION   560

Genes 561

MENDELIAN THEORY 561
GENETIC INHERITANCE 564
CROSSING OVER 566
THE GENETIC LOAD 568
BLOOD TYPES 569
EUGENICS 570
CHEMICAL GENETICS 572
ABNORMAL HEMOGLOBIN 574
METABOLIC ABNORMALITY 577

Nucleic Acids 578

GENERAL STRUCTURE 578
DNA 581
THE DOUBLE HELIX 582
GENE ACTIVITY 586

The Origin of Life 592

EARLY THEORIES 592
CHEMICAL EVOLUTION 595
THE FIRST CELLS 598
ANIMAL CELLS 599

Life in Other Worlds 600

Chapter 14

The Microorganisms 606

Bacteria 606

MAGNIFYING DEVICES 606
NAMING THE BACTERIA 608
THE GERM THEORY OF
  DISEASE 610
IDENTIFYING BACTERIA 611

Chemotherapy 613

SULFA DRUGS 614
THE ANTIBIOTICS 616
RESISTANT BACTERIA 617
PESTICIDES 618
HOW CHEMOTHERAPY
  WORKS 620
BENEFICENT BACTERIA 621

Viruses 623

NONBACTERIAL DISEASE 623
SUBBACTERIA 625
THE ROLE OF NUCLEIC
  ACID 630

Immune Reactions 633

SMALLPOX 634

VACCINES 635
ANTIBODIES 639

Cancer 644

THE EFFECTS OF
  RADIATION 646
MUTAGENS AND
  ONCOGENES 647
THE VIRUS THEORY 648
POSSIBLE CURES 649

Chapter 15

The Body 652

Food 652

THE ORGANIC FOODS 654
PROTEINS 655
FATS 657

Vitamins 657

DEFICIENCY DISEASES 658
ISOLATING VITAMINS 660
CHEMICAL COMPOSITION
  AND STRUCTURE 661
VITAMIN THERAPY 664
VITAMINS AS ENZYMES 666
VITAMIN A 669

Minerals 670

COBALT 673
IODINE 675
FLUORIDES 675

Hormones 676

INSULIN AND DIABETES 679
THE STEROID HORMONES 683
THE PITUITARY AND THE
  PINEAL GLANDS 685
THE ROLE OF THE BRAIN 687
THE PROSTAGLANDINS 688
HORMONE ACTION 688

Death 690

ATHEROSCLEROSIS 691
OLD AGE 692

Chapter 16

The Species 696

Varieties of Life 696

CLASSIFICATION 697
THE VERTEBRATES 701

CONTENTS

Evolution   706

  EARLY THEORIES   707
  DARWIN'S THEORY   710
  OPPOSITION TO THE
    THEORY   713
  EVIDENCE FOR THE
    THEORY   715

The Course of Evolution   716

  ERAS AND AGES   717
  BIOCHEMICAL CHANGES   721
  RATE OF EVOLUTION   723

The Descent of Man   724

  EARLY CIVILIZATIONS   724
  THE STONE AGE   727
  HOMINIDS   731
  PILTDOWN MAN   734
  RACIAL DIFFERENCES   735
  BLOOD GROUPS AND RACE   736

Humanity's Future   738

  THE POPULATION
    EXPLOSION   738
  LIVING IN THE SEA   745
  SETTLING IN SPACE   746

Chapter 17

The Mind   751

The Nervous System   751

NERVE CELLS   751
BRAIN DEVELOPMENT   754
THE HUMAN BRAIN   756
INTELLIGENCE TESTING   757
THE SPECIALIZATION OF
  FUNCTIONS   758
THE SPINAL CORD   763

Nerve Action   764

REFLEX ACTION   765
ELECTRICAL IMPULSES   768

Human Behavior   771

CONDITIONED RESPONSES   772
THE BIOLOGICAL CLOCK   777
PROBING HUMAN BEHAVIOR   778
DRUG USE   782
MEMORY   783

Automatons   784

FEEDBACK   786
EARLY AUTOMATION   788
ARITHMETICAL
  CALCULATIONS   789
CALCULATING MACHINES   792

Artificial Intelligence   795

ELECTRONIC COMPUTERS   795
ROBOTS   798

Appendix

Mathematics in
Science   803

Gravitation   803

  THE FIRST LAW OF
    MOTION   804

THE SECOND AND THIRD
  LAWS   808
Relativity   811

THE MICHELSON-MORLEY
  EXPERIMENT   811
THE FITZGERALD EQUATION   814
THE LORENTZ EQUATION   815
EINSTEIN'S EQUATION   818

BIBLIOGRAPHY   820

NAME INDEX   829

SUBJECT INDEX   847

Illustrations

I. The Solar System and Earth and
   Space Exploration   *following page 224*

II. Aspects of Technology and Aspects
    of Evolution   *following page 660*

Individual illustrations may be located by consulting the index.

Preface

The rapid advance of science is exciting and exhilarating to anyone who is fascinated by the unconquerability of the human spirit and by the continuing efficacy of the scientific method as a tool for penetrating the complexities of the universe.

But what if one is also dedicated to keeping up with every phase of scientific advance for the deliberate purpose of interpreting that advance for the general public? For that person, the excitement and exhilaration is tempered by a kind of despair.

Science will not stand still. It is a panorama that subtly dissolves and changes even while we watch. It cannot be caught in its every detail at any one moment of time without leaving us behind at once.

In 1960, *The Intelligent Man's Guide to Science* was published; and at once, the advance of science flowed past it. In order to consider quasars and lasers, for instance (which were unknown in 1960 and household words a couple of years later), *The New Intelligent Man's Guide to Science* was published in 1965.

But still science drove on inexorably. Now there came the question of pulsars, of black holes, of continental drift, men on the moon, REM sleep, gravitational waves, holography, cyclic-AMP, and so forth—all post-1965.

So it was time for a new edition, the third. And what did we call it? *The New New Intelligent Man's Guide to Science?* Obviously not. The third edition was named, straightforwardly, *Asimov's Guide to Science* and was published in 1972.

And *still* science refused to stop. Enough was learned of the solar system, thanks to our probes, to require an entire chapter. And now we have the new inflationary universe, new theories on the end of the dinosaurs, on quarks, gluons, as well as unified field theories, magnetic monopoles, the energy crisis, home computers, robots, punctuated evolution, oncogenes, and on, and on, and on.

So it is time for another new edition, the fourth; and since for each edition, I always change the name, I shall do so again. It is now *Asimov's New Guide to Science.*

ISAAC ASIMOV

New York
1984

Asimov's
New Guide
to Science

Chapter 1

What Is Science?

Almost in the beginning was curiosity.

Curiosity, the overwhelming desire to know, is not characteristic of dead matter. Nor does it seem to be characteristic of some forms of living organism, which, for that very reason, we can scarcely bring ourselves to consider alive.

A tree does not display curiosity about its environment in any way we can recognize; nor does a sponge or an oyster. The wind, the rain, the ocean currents bring them what is needful, and from it they take what they can. If the chance of events is such as to bring them fire, poison, predators, or parasites, they die as stoically and as undemonstratively as they lived.

Early in the scheme of life, however, independent motion was developed by some organisms. It meant a tremendous advance in their control of the environment. A moving organism no longer had to wait in stolid rigidity for food to come its way, but went out after it.

Thus, adventure entered the world—and curiosity. The individual that hesitated in the competitive hunt for food, that was overly conservative in its investigation, starved. Early on, curiosity concerning the environment was enforced as the price of survival.

The one-celled paramecium, moving about in a searching way, cannot have conscious volitions and desires in the sense that we do, but it has a drive, even if only a "simple" physical-chemical one, which causes it to behave as if it were investigating its surroundings for food or safety, or both. And this "act of curiosity" is what we most easily recognize as being inseparable from the kind of life that is most akin to ours.

As organisms grew more intricate, their sense organs multiplied and became both more complex and more delicate. More messages of greater variety were received from and about the external environment. At the same time, there developed (whether as cause or effect we cannot tell), an increasing complexity of the nervous system, the living instrument that interprets and stores the data collected by the sense organs.

THE DESIRE TO KNOW

There comes a point where the capacity to receive, store, and interpret messages from the outside world may outrun sheer necessity. An organism may be sated with food, and there may, at the moment, be no danger in sight. What does it do then?

It might lapse into an oysterlike stupor. But the higher organisms at least still show a strong instinct to explore the environment. Idle curiosity, we may call it. Yet, though we may sneer at it, we judge intelligence by it. The dog, in moments of leisure, will sniff idly here and there, pricking up its ears at sounds we cannot hear; and so we judge it to be more intelligent than the cat, which in its moments of leisure grooms itself or quietly and luxuriously stretches out and falls asleep. The more advanced the brain, the greater the drive to explore, the greater the "curiosity surplus." The monkey is a byword for curiosity. Its busy little brain must and will be kept going on whatever is handy. And in this respect, as in many others, man is a supermonkey.

The human brain is the most magnificently organized lump of matter in the known universe, and its capacity to receive, organize, and store data is far in excess of the ordinary requirements of life. It has been estimated that, in a lifetime, a human being can learn up to 15 trillion items of information.

It is to this excess that we owe our ability to be afflicted by that supremely painful disease, boredom. A human being, forced into a situation where one has no opportunity to utilize one's brain except for minimal survival, will gradually experience a variety of unpleasant symptoms, up to and including serious mental disorganization. The fact is that the normal human being has an intense and overwhelming curiosity. If one lacks the opportunity to satisfy it in immediately useful ways, one will satisfy it in other ways—even regrettable ways to which we have attached admonitions such as "Curiosity killed the cat," and "Mind your own business."

The overriding power of curiosity, even with harm as the penalty, is reflected in the myths and legends of the human race. The Greeks had the tale of Pandora and her box. Pandora, the first woman, was given a box that she was forbidden to open. Quickly and naturally enough she opened it and found it full of the spirits of disease, famine, hate, and all kinds of evil—which escaped and have plagued the world ever since.

In the Biblical story of the temptation of Eve, it seems fairly certain (to me, at any rate) that the serpent had the world's easiest job and might have saved his words: Eve's curiosity would have driven her to taste the forbidden fruit even without external temptation. If you are of a mind to interpret the Bible allegorically, you may think of the serpent as simply the representation of this inner compulsion. In the conventional cartoon picturing Eve standing under the tree with the forbidden fruit in her hand, the serpent coiled around the branch might be labeled "Curiosity."

If curiosity can, like any other human drive, be put to ignoble use—the prying invasion of privacy that has given the word its cheap and unpleasant connotation—it nevertheless remains one of the noblest properties of the human mind. For its simplest definition is "the desire to know."

This desire finds its first expression in answers to the practical needs of human life: how best to plant and cultivate crops, how best to fashion bows and arrows, how best to weave clothing—in short, the "applied arts." But after these comparatively limited skills have been mastered, or the practical needs fulfilled, what then? Inevitably the desire to know leads on to less limited and more complex activities.

It seems clear that the "fine arts" (designed to satisfy inchoate and boundless and spiritual needs) were born in the agony of boredom. To be sure, one can easily find more mundane uses and excuses for the fine arts. Paintings and statuettes were used as fertility charms and as religious symbols, for instance. But one cannot help suspecting that the objects existed first and the use second.

To say that the fine arts arose out of a sense of the beautiful may also be putting the cart before the horse. Once the fine arts were developed, their extension and refinement in the direction of beauty would have followed inevitably, but even if this had not happened, the fine arts would have developed nevertheless. Surely the fine arts antedate any possible need or use for them, other than the elementary need to occupy the mind as fully as possible.

Not only does the production of a work of fine art occupy the mind satisfactorily; the contemplation or appreciation of the work supplies a similar service to the audience. A great work of art is great precisely because it offers a stimulation that cannot readily be found elsewhere. It contains enough data of sufficient complexity to cajole the brain into exerting itself past the usual needs; and, unless a person is hopelessly ruined by routine or stultification, that exertion is pleasant.

But if the practice of the fine arts is a satisfactory solution to the problem of leisure, it has this disadvantage: it requires, in addition to an active and creative mind, physical dexterity. It is just as interesting to pursue mental activities that involve only the mind, without the supplement of manual skill. And, of course, such activity is available. It is the pursuit of knowledge itself, not in order to do something with it but for its own sake.

Thus, the desire to know seems to lead into successive realms of greater etherealization and more efficient occupation of the mind—from knowledge of accomplishing the useful, to knowledge of accomplishing the esthetic, to "pure" knowledge.

Knowledge for itself alone seeks answers to such questions as How high is the sky? or, Why does a stone fall? This is sheer curiosity—curiosity at its idlest and therefore perhaps at its most peremptory. After all, it serves no apparent purpose to know how high the sky is or why the stone falls. The lofty sky does not interfere with the ordinary business of life; and, as for the stone, knowing why it falls does not help us to dodge it more skillfully or soften the blow if it happens to hit us. Yet there have always been people who ask such apparently useless questions and try to answer them out of the sheer desire to know —out of the absolute necessity of keeping the brain working.

The obvious method of dealing with such questions is to make up an esthetically satisfying answer: one that has sufficient analogies to what is already known to be comprehensible and plausible. The expression "to make up" is rather bald and unromantic. The ancients liked to think of the process of discovery as the inspiration of the muses or as a revelation from heaven. In any case, whether it was inspiration, revelation, or the kind of creative thinking that goes into storytelling, the explanations depended heavily on analogy. The lightning bolt is destructive and terrifying but appears, after all, to be hurled like a weapon and does the damage of a hurled weapon—a fantastically violent one. Such a weapon must have a wielder similarly enlarged in scale, and so the thunderbolt becomes the hammer of Thor or the flashing spear of Zeus. The more-than-normal weapon is wielded by a more-than-normal man.

Thus a myth is born. The forces of nature are personified and become gods. The myths react on one another, are built up and improved by generations of mythtellers until the original point may be obscured. Some myths may degenerate into pretty stories (or ribald ones), whereas others may gain an ethical content important enough to make them meaningful within the framework of a major religion.

Just as art may be fine or applied, so may mythology. Myths may be maintained for their esthetic charm or bent to the physical uses of human beings. For instance, the earliest farmers were intensely concerned with the phenomenon of rain and why it fell capriciously. The fertilizing rain falling from the heavens on the earth presented an obvious analogy to the sex act; and, by personifying both heaven and earth, human beings found an easy explanation of the release or the withholding of the rains. The earth goddess, or the sky god, was either pleased or offended, as the case might be. Once this myth was accepted, farmers had a plausible basis for the art of bringing rain —namely, appeasing the god by appropriate rites. These rites might well be orgiastic in nature—an attempt to influence heaven and earth by example.

THE GREEKS

The Greek myths are among the prettiest and most sophisticated in our Western literary and cultural heritage. But it was the Greeks also who, in due course, introduced the opposite way of looking at the universe—that is, as something impersonal and inanimate. To the mythmakers, every aspect of nature was essentially human in its unpredictability. However mighty and majestic the personification, however superhuman the powers of Zeus, or

Ishtar or Isis or Marduk or Odin, they were also—like mere humans—frivo-
lous, whimsical, emotional, capable of outrageous behavior for petty reasons,
susceptible to childish bribes. As long as the universe was in the control of such
arbitrary and unpredictable deities, there was no hope of understanding it,
only the shallow hope of appeasing it. But in the new view of the later Greek
thinkers, the universe was a machine governed by inflexible laws. The Greek
philosophers now devoted themselves to the exciting intellectual exercise of
trying to discover just what the laws of nature might be.

The first to do so, according to Greek tradition, was Thales of Miletus,
about 600 B.C. He was saddled with an almost impossible number of discoveries by later Greek writers, and it may be that he first brought the gathered
Babylonian knowledge to the Greek world. His most spectacular achievement
is supposed to have been predicting an eclipse for 585 B.C.—which actually
occurred.

In engaging in this intellectual exercise, the Greeks assumed, of course, that
nature would play fair; that, if attacked in the proper manner, it would yield
its secrets and would not change position or attitude in midplay. (Over two
thousand years later, Albert Einstein expressed this feeling when he said, "God
may be subtle, but He is not malicious.") There was also the feeling that the
natural laws, when found, would be comprehensible. This Greek optimism has
never entirely left the human race.

With confidence in the fair play of nature, human beings needed to work
out an orderly system for learning how to determine the underlying laws from
the observed data. To progress from one point to another by established rules
of argument is to use "reason." A reasoner may use "intuition" to guide the
search for answers, but must rely on sound logic to test particular theories. To
take a simple example: if brandy and water, whiskey and water, vodka and
water, and rum and water are all intoxicating beverages, one may jump to the
conclusion that the intoxicating factor must be the ingredient these drinks
hold in common—namely, water. There is something wrong with this reasoning, but the fault in the logic is not immediately obvious; and in more subtle
cases, the error may be hard indeed to discover.

The tracking down of errors or fallacies in reasoning has amused thinkers
from Greek times to the present. And we owe the earliest foundations of
systematic logic to Aristotle of Stagira who in the fourth century B.C. first
summarized the rules of rigorous reasoning.

The essentials of the intellectual game of man-against-nature are three.
First, you must collect observations about some facet of nature. Second, you
must organize these observations into an orderly array. (The organization does
not alter them but merely makes them easier to handle. This is plain in the
game of bridge, for instance, where arranging the hand in suits and order of
value does not change the cards or show the best course of play, but makes
it easier to arrive at the logical plays.) Third, you must derive from your orderly
array of observations some principle that summarizes the observations.

For instance, we may observe that marble sinks in water, wood floats, iron sinks, a feather floats, mercury sinks, olive oil floats, and so on. If we put all the sinkable objects in one list and all the floatable ones in another and look for a characteristic that differentiates all the objects in one group from all in the other, we will conclude: Objects denser than water sink in water, and objects less dense than water, float.

The Greeks named their new manner of studying the universe *philosophia* ("philosophy"), meaning "love of knowledge" or, in free translation, "the desire to know."

GEOMETRY AND MATHEMATICS

The Greeks achieved their most brilliant successes in geometry. These successes can be attributed mainly to the development of two techniques: abstraction and generalization.

Here is an example. Egyptian land surveyors had found a practical way to form a right angle: they divided a rope into twelve equal parts and made a triangle in which three parts formed one side, four parts another, and five parts the third side—the right angle lay where the three-unit side joined the four-unit side. There is no record of how the Egyptians discovered this method, and apparently their interest went no further than to make use of it. But the curious Greeks went on to investigate why such a triangle should contain a right angle. In the course of their analysis, they grasped the point that the physical construction itself was only incidental; it did not matter whether the triangle was made of rope or linen or wooden slats. It was simply a property of "straight lines" meeting at angles. In conceiving of ideal straight lines, which are independent of any physical visualization and can exist only in imagination, the Greeks originated the method called abstraction—stripping away nonessentials and considering only those properties necessary to the solution of the problem.

The Greek geometers made another advance by seeking general solutions for classes of problems, instead of treating individual problems separately. For instance, one might have discovered by trial that a right angle appeared in triangles, not only with sides 3, 4, and 5 feet long, but also in triangles of 5, 12, and 13 feet and of 7, 24, and 25 feet. But these were merely numbers without meaning. Could some common property be found that would describe all right triangles? By careful reasoning, the Greeks showed that a triangle is a right triangle if, and only if, the lengths of the sides have the relation $x^2 + y^2 = z^2$, z being the length of the longest side. The right angle lies where the sides of length x and y meet. Thus for the triangle with sides of 3, 4, and 5 feet, squaring the sides gives $9 + 16 = 25$; similarly, squaring the sides of 5, 12, and 13 gives $25 + 144 = 169$; and squaring 7, 24, and 25 gives $49 + 576 = 625$. These are only three cases out of an infinity of possible ones and, as such, trivial. What intrigued the Greeks was the discovery of a proof that the relation must hold in all cases. And they pursued geometry as an elegant means of discovering and formulating such generalizations.

Various Greek mathematicians contributed proofs of relationships existing among the lines and points of geometric figures. The one involving the right triangle was reputedly worked out by Pythagoras of Samos about 525 B.C. and is still called the Pythagorean theorem in his honor.

About 300 B.C., Euclid gathered the mathematical theorems known in his time and arranged them in a reasonable order, such that each theorem could be proved through the use of theorems proved previously. Naturally, this system eventually worked back to something unprovable: if each theorem had to be proved with the help of one already proved, how could one prove theorem no. 1? The solution was to begin with a statement of truths so obvious and acceptable to all as to need no proof. Such a statement is called an "axiom." Euclid managed to reduce the accepted axioms of the day to a few simple statements. From these axioms alone, he built an intricate and majestic system of "Euclidean geometry." Never before was so much constructed so well from so little, and Euclid's reward is that his textbook has remained in use, with but minor modification, for more than 2,000 years.

THE DEDUCTIVE PROCESS

Working out a body of knowledge as the inevitable consequence of a set of axioms ("deduction") is an attractive game. The Greeks fell in love with it, thanks to the success of their geometry—sufficiently in love with it to commit two serious errors.

First, they came to consider deduction as the only respectable means of attaining knowledge. They were well aware that, for some kinds of knowledge, deduction was inadequate; for instance, the distance from Corinth to Athens could not be deduced from abstract principles but had to be measured. The Greeks were willing to look at nature when necessary; however, they were always ashamed of the necessity and considered that the highest type of knowledge was that arrived at by cerebration. They tended to undervalue knowledge directly involved with everyday life. There is a story that a student of Plato, receiving mathematical instruction from the master, finally asked impatiently, "But what is the use of all this?" Plato, deeply offended, called a slave and, ordering him to give the student a coin, said, "Now you need not feel your instruction has been entirely to no purpose." With that, the student was expelled.

There is a well-worn belief that this lofty view arose from the Greek's slave-based culture, in which all practical matters were relegated to the slaves. Perhaps so, but I incline to the view that the Greeks felt that philosophy was a sport, an intellectual game. Many people regard the amateur in sports as a gentleman socially superior to the professional who makes his living at it. In line with this concept of purity, we take almost ridiculous precautions to make sure that the contestants in the Olympic games are free of any taint of professionalism. The Greek rationalization for the "cult of uselessness" may similarly have been based on a feeling that to allow mundane knowledge (such as the distance from Athens to Corinth) to intrude on abstract thought was to allow imperfection to enter the Eden of true philosophy. Whatever the rationalization, the Greek thinkers were severely limited by their attitude. Greece was not barren of practical contributions to civilization, but even its great engineer, Archimedes of Syracuse, refused to write about his practical inventions and discoveries; to maintain his amateur status, he broadcast only his achievements in pure mathematics. And lack of interest in earthly things —in invention, in experiment, in the study of nature—was but one of the factors that put bounds on Greek thought. The Greeks' emphasis on purely abstract and formal study—indeed, their very success in geometry—led them into a second great error and, eventually, to a dead end.

Seduced by the success of the axioms in developing a system of geometry, the Greeks came to think of the axioms as "absolute truths" and to suppose that other branches of knowledge could be developed from similar "absolute truths." Thus in astronomy they eventually took as self-evident axioms the notions that (1) the earth was motionless and the center of the universe, and (2) whereas the earth was corrupt and imperfect, the heavens were eternal, changeless, and perfect. Since the Greeks considered the circle the perfect curve, and since the heavens were perfect, it followed that all the heavenly bodies must move in circles around the earth. In time, their observations (arising from navigation and calendar making) showed that the planets do not move in perfectly simple circles, and so the Greeks were forced to allow planets to move in ever more complicated combinations of circles, which, about 150 A.D., were formulated as an uncomfortably complex system by Claudius Ptolemaeus (Ptolemy) at Alexandria. Similarly, Aristotle worked up fanciful theories of motion from "self-evident" axioms, such as the proposition that the speed of an object's fall was proportional to its weight. (Anyone could see that a stone fell faster than a feather.)

Now this worship of deduction from self-evident axioms was bound to wind up at the edge of a precipice, with no place to go. After the Greeks had worked out all the implications of the axioms, further important discoveries in mathematics or astronomy seemed out of the question. Philosophic knowledge appeared complete and perfect; and for nearly 2,000 years after the Golden Age of Greece, when questions involving the material universe arose, there was a tendency to settle matters to the satisfaction of all by saying, "Aristotle says . . . ," or, "Euclid says. . . ."

THE RENAISSANCE AND COPERNICUS

Having solved the problems of mathematics and astronomy, the Greeks turned to more subtle and challenging fields of knowledge. One was the human soul.

Plato was far more interested in such questions as What is justice? or, What is virtue? than in why rain falls or how the planets move. As the supreme moral philosopher of Greece, he superseded Aristotle, the supreme natural philosopher. The Greek thinkers of the Roman period found themselves drawn more and more to the subtle delights of moral philosophy and away from the apparent sterility of natural philosophy. The last development in ancient philosophy was an exceedingly mystical "neo-Platonism" formulated by Plotinus about 250 A.D.

Christianity, with its emphasis on the nature of God and His relation to man, introduced an entirely new dimension into the subject matter of moral philosophy that increased its apparent superiority as an intellectual pursuit over natural philosophy. From 200 A.D. to 1200 A.D., Europeans concerned themselves almost exclusively with moral philosophy, in particular with theology. Natural philosophy was nearly forgotten.

The Arabs, however, managed to preserve Aristotle and Ptolemy through the Middle Ages; and, from them, Greek natural philosophy eventually filtered back to Western Europe. By 1200, Aristotle had been rediscovered. Further infusions came from the dying Byzantine empire, which was the last area in Europe to maintain a continuous cultural tradition from the great days of Greece.

The first and most natural consequence of the rediscovery of Aristotle was the application of his system of logic and reason to theology. About 1250, the Italian theologian Thomas Aquinas established the system called "Thomism," based on Aristotelian principles, which still represents the basic theology of the Roman Catholic Church. But Europeans soon began to apply the revival of Greek thought to secular fields as well.

Because the leaders of the Renaissance shifted emphasis from matters concerning God to the works of humanity, they were called "humanists," and the study of literature, art, and history is still referred to as the "humanities."

To the Greek natural philosophy, the Renaissance thinkers brought a fresh outlook, for the old views no longer entirely satisfied. In 1543, the Polish astronomer Nicolaus Copernicus published a book that went so far as to reject a basic axiom of astronomy: he proposed that the sun, not the earth, be considered the center of the universe. (He retained the notion of circular orbits for the earth and other planets, however.) This new axiom allowed a much simpler explanation of the observed motions of heavenly bodies. Yet the Copernican axiom of a moving earth was far less "self-evident" than the Greek axiom of a motionless earth, and so it is not surprising that it took more than half a century for the Copernican theory to be accepted.

In a sense, the Copernican system itself was not a crucial change. Copernicus had merely switched axioms; and Aristarchus of Samos had already anticipated this switch to the sun as the center 2,000 years earlier. I do not mean to say that the changing of an axiom is a minor matter. When mathematicians of the nineteenth century challenged Euclid's axioms and developed "non-Euclidean geometries" based on other assumptions, they influenced thought on many matters in a most profound way: today the very history and form of the universe are thought to conform to a non-Euclidean geometry rather than the "commonsense" geometry of Euclid. But the revolution initiated by Copernicus entailed not just a shift in axioms but eventually involved a whole new approach to nature. This revolution was carried through in the person of the Italian Galileo Galilei toward the end of the sixteenth century.

EXPERIMENTATION AND INDUCTION

The Greeks, by and large, had been satisfied to accept the "obvious" facts of nature as starting points for their reasoning. It is not on record that Aristotle ever dropped two stones of different weight to test his assumption that the speed of fall is proportional to an object's weight. To the Greeks, experimentation seemed irrelevant. It interfered with and detracted from the beauty of pure deduction. Besides, if an experiment disagreed with a deduction, could one be certain that the experiment was correct? Was it likely that the imperfect world of reality would agree completely with the perfect world of abstract ideas; and if it did not, ought one to adjust the perfect to the demands of the imperfect? To test a perfect theory with imperfect instruments did not impress the Greek philosophers as a valid way to gain knowledge.

Experimentation began to become philosophically respectable in Europe with the support of such philosophers as Roger Bacon (a contemporary of Thomas Aquinas) and his later namesake Francis Bacon. But it was Galileo who overthrew the Greek view and effected the revolution. He was a convincing logician and a genius as a publicist. He described his experiments and his point of view so clearly and so dramatically that he won over the European learned community. And they accepted his methods along with his results.

According to the best-known story about him, Galileo tested Aristotle's theories of falling bodies by asking the question of nature in such a way that all Europe could hear the answer. He is supposed to have climbed to the top of the Leaning Tower of Pisa and dropped a 10-pound sphere and a 1-pound sphere simultaneously; the thump of the two balls hitting the ground in the same split second killed Aristotelian physics.

Actually Galileo probably did not perform this particular experiment, but the story is so typical of his dramatic methods that it is no wonder it has been widely believed through the centuries.

Galileo undeniably did roll balls down inclined planes and measured the distance that they traveled in given times. He was the first to conduct time experiments and to use measurement in a systematic way.

His revolution consisted in elevating "induction" above deduction as the logical method of science. Instead of building conclusions on an assumed set of generalizations, the inductive method starts with observations and derives generalizations (axioms, if you will) from them. Of course, even the Greeks obtained their axioms from observation; Euclid's axiom that a straight line is the shortest distance between two points was an intuitive judgment based on experience. But whereas the Greek philosopher minimized the role played by induction, the modern scientist looks on induction as the essential process of gaining knowledge, the only way of justifying generalizations. Moreover, the scientist realizes that no generalization can be allowed to stand unless it is repeatedly tested by newer and still newer experiments—the continuing test of further induction.

The present general viewpoint is just the reverse of the Greeks. Far from considering the real world an imperfect representation of ideal truth, we consider generalizations to be only imperfect representatives of the real world. No amount of inductive testing can render a generalization completely and absolutely valid. Even though billions of observations tend to bear out a generalization, a single observation that contradicts or is inconsistent with it must force its modification. And no matter how many times a theory meets its tests successfully, there can be no certainty that it will not be overthrown by the next observation.

This, then, is a cornerstone of modern natural philosophy. It makes no claim of attaining ultimate truth. In fact, the phrase "ultimate truth" becomes meaningless, because there is no way in which enough observations can be made to make truth certain and, therefore, "ultimate." The Greek philosophers recognized no such limitation. Moreover, they saw no difficulty in applying exactly the same method of reasoning to the question What is justice? as to the question What is matter? Modern science, on the other hand, makes a sharp distinction between the two types of question. The inductive method cannot make generalizations about what it cannot observe; and, since the nature of the human soul, for example, is not observable by any direct means yet known, this subject lies outside the realm of the inductive method.

The victory of modern science did not become complete until it established one more essential principle—namely, free and cooperative communication among all scientists. Although this necessity seems obvious now, it was not obvious to the philosophers of ancient and medieval times. The Pythagoreans of ancient Greece were a secret society who kept their mathematical discoveries to themselves. The alchemists of the Middle Ages deliberately obscured their writings to keep their so-called findings within as small an inner circle as possible. In the sixteenth century, the Italian mathematician Niccolò Tartaglia, who discovered a method of solving cubic equations, saw nothing wrong in attempting to keep it a secret. When Geronimo Cardano, a fellow mathematician, wormed the secret out of Tartaglia on the promise of confidentiality and published it, Tartaglia naturally was outraged; but aside from Cardano's trickery in breaking his promise, he was certainly correct in his reply that such a discovery had to be published.

Nowadays no scientific discovery is reckoned a discovery if it is kept secret. The English chemist Robert Boyle, a century after Tartaglia and Cardano, stressed the importance of publishing all scientific observations in full detail. A new observation or discovery, moreover, is no longer considered valid, even after publication, until at least one other investigator has repeated the observation and "confirmed" it. Science is the product not of individuals but of a "scientific community."

One of the first groups (and certainly the most famous) to represent such a scientific community was the Royal Society of London for Improving Natural Knowledge, usually called simply the "Royal Society." It grew out of informal meetings, beginning about 1645, of a group of gentlemen interested in the new scientific methods originated by Galileo. In 1660, the society was formally chartered by King Charles II.

The members of the Royal Society met and discussed their findings openly, wrote letters describing them in English rather than Latin, and pursued their experiments with vigor and vivacity. Nevertheless, through most of the seventeenth century, they remained in a defensive position. The attitude of many of their learned contemporaries might be expressed by a cartoon, after the modern fashion, showing the lofty shades of Pythagoras, Euclid, and Aristotle staring down haughtily at children playing with marbles and labeled "Royal Society."

All this was changed by the work of Isaac Newton, who became a member of the society. From the observations and conclusions of Galileo, of the Danish astronomer Tycho Brahe, and of the German astronomer Johannes Kepler, who figured out the elliptical nature of the orbits of the planets, Newton arrived by induction at his three simple laws of motion and his great fundamental generalization—the law of universal gravitation. (Nevertheless, when he published his findings, he used geometry and the Greek method of deductive explanation.) The educated world was so impressed with this discovery that Newton was idolized, almost deified, in his own lifetime. This majestic new universe, built upon a few simple assumptions derived from inductive processes, now made the Greek philosophers look like boys playing with marbles. The revolution that Galileo had initiated at the beginning of the seventeenth century was triumphantly completed by Newton at the century's end.

MODERN SCIENCE

It would be pleasant to be able to say that science and human beings have lived happily ever since. But the truth is that the real difficulties of both were only beginning. As long as science remained deductive, natural philosophy could be part of the general culture of all educated men (women, alas, being rarely educated until recent times). But inductive science became an immense labor—of observation, learning, and analysis. It was no longer a game for amateurs. And the complexity of science grew with each decade. During the century after Newton, it was still possible for a man of unusual attainments to master all fields of scientific knowledge. But, by 1800, this had become entirely impracticable. As time went on, it was increasingly necessary for a scientist to limit himself to a portion of the field with which he was intensively concerned. Specialization was forced on science by its own inexorable growth. And with each generation of scientists, specialization has grown more and more intense.

The publications of scientists concerning their individual work have never been so copious—and so unreadable for anyone but their fellow specialists. This has been a great handicap to science itself, for basic advances in scientific knowledge often spring from the cross-fertilization of knowledge from different specialties. Even more ominous, science has increasingly lost touch with nonscientists. Under such circumstances, scientists come to be regarded almost as magicians—feared rather than admired. And the impression that science is incomprehensible magic, to be understood only by a chosen few who are suspiciously different from ordinary mankind, is bound to turn many youngsters away from science.

Since the Second World War, strong feelings of outright hostility toward science were to be found among the young—even among the educated young in the colleges. Our industrialized society is based on the scientific discoveries of the last two centuries, and our society finds it is plagued by undesirable side effects of its very success.

Improved medical techniques have brought about a runaway increase in population; chemical industries and the internal-combustion engine are fouling our water and our air; the demand for materials and for energy is depleting and destroying the earth's crust. And this is all too easily blamed on "science" and "scientists" by those who do not quite understand that while knowledge can create problems, it is not through ignorance that we can solve them.

Yet modern science need not be so complete a mystery to nonscientists. Much could be accomplished toward bridging the gap if scientists accepted the responsibility of communication—explaining their own fields of work as simply and to as many as possible—and if nonscientists, for their part, accepted the responsibility of listening. To gain a satisfactory appreciation of the developments in a field of science, it is not essential to have a total understanding of the science. After all, no one feels that one must be capable of writing a great work of literature in order to appreciate Shakespeare. To listen to a Beethoven symphony with pleasure does not require the listener to be capable of composing an equivalent symphony. By the same token, one can appreciate and take pleasure in the achievements of science even though one does not oneself have a bent for creative work in science.

But what, you may ask, would be accomplished? The first answer is that no one can really feel at home in the modern world and judge the nature of its problems—and the possible solutions to those problems—unless one has some intelligent notion of what science is up to. Furthermore, initiation into the magnificent world of science brings great esthetic satisfaction, inspiration to youth, fulfillment of the desire to know, and a deeper appreciation of the wonderful potentialities and achievements of the human mind.

It is to provide such initiation that I have undertaken to write this book.

PART I

The Physical Sciences

Chapter 2

The Universe

The Size of the Universe

There is nothing about the sky that makes it look particularly distant to a casual observer. Young children have no great trouble in accepting the fantasy that "the cow jumped over the moon"—or "he jumped so high, he touched the sky." The ancient Greeks, in their myth telling stage, saw nothing ludicrous in allowing the sky to rest on the shoulders of Atlas. Of course, Atlas might have been astronomically tall, but another myth suggests otherwise. Atlas was enlisted by Hercules to help him with the eleventh of his famous twelve labors —fetching the golden apples (oranges?) of the Hesperides ("the far west"— Spain?). While Atlas went off to fetch the apples, Hercules stood on a mountain and held up the sky. Granted that Hercules was a large specimen, he was nevertheless not a giant. It follows then that the early Greeks took quite calmly to the notion that the sky cleared the mountaintops by only a few feet.

It is natural to suppose, to begin with, that the sky is simply a hard canopy in which the shining heavenly bodies are set like diamonds. (Thus the Bible refers to the sky as the "firmament," from the same Latin root as the word *firm.*) As early as the sixth to the fourth centuries B.C., Greek astronomers realized that there must be more than one canopy. For while the "fixed" stars moved around Earth in a body, apparently without changing their relative positions, this was not true of the sun, the moon, and five bright starlike objects (Mercury, Venus, Mars, Jupiter, and Saturn): in fact, each moved in a separate path. These seven bodies were called *planets* (from a Greek word meaning "wanderer"), and it seemed obvious that they could not be attached to the vault of the stars.

The Greeks assumed that each planet was set in an invisible spherical vault of its own, and that the vaults were nested one above the other, the nearest belonging to the planet that moved fastest. The quickest motion belonged to the moon, which circled the sky in about twenty-seven and a third days. Beyond it lay in order (so thought the Greeks) Mercury, Venus, our sun, Mars, Jupiter, and Saturn.

EARLY MEASUREMENTS

The first scientific measurement of any cosmic distance came about 240 B.C. Eratosthenes of Cyrene, the head of the Library at Alexandria, then the most advanced scientific institution in the world, pondered the fact that on 21 June, when the noonday sun was exactly overhead at the city of Syene in Egypt, it was not quite at the zenith at noon in Alexandria, 500 miles north of Syene. Eratosthenes decided that the explanation must be that the surface of the earth curved away from the sun. From the length of the shadow in Alexandria at noon on the solstice, straightforward geometry could yield the amount by which the earth's surface curved in the 500-mile distance from Syene to Alexandria. From that one could calculate the circumference and the diameter of the earth, assuming it to be spherical in shape—a fact Greek astronomers of the day were ready to accept (figure 2.1).

Eratosthenes worked out the answer (in Greek units), and, as nearly as we can judge, his figures in our units came out at about 8,000 miles for the diameter and 25,000 miles for the circumference of the earth. These figures, as it happens, are just about right. Unfortunately, this accurate value for the size of the earth did not prevail. About 100 B.C. another Greek astronomer, Posidonius of Apamea, repeated Eratosthenes' work but reached the conclusion that the earth was but 18,000 miles in circumference.

It was the smaller figure that was accepted throughout ancient and medieval times. Columbus accepted the smaller figure and thought that a 3,000-mile westward voyage would take him to Asia. Had he known the earth's true size, he might not have ventured. It was not until 1521–23, when Magellan's fleet (or rather the one remaining ship of the fleet) finally circumnavigated the earth, that Eratosthenes' correct value was finally established.

In terms of the earth's diameter, Hipparchus of Nicaea, about 150 B.C., worked out the distance to the moon. He used a method that had been suggested a century earlier by Aristarchus of Samos, the most daring of all Greek astronomers. The Greeks had already surmised that eclipses of the moon were caused by the earth coming between the sun and the moon. Aristarchus saw that the curve of the earth's shadow as it crossed the moon should indicate the relative sizes of the earth and the moon. On this basis, geometric methods offered a way to calculate how far distant the moon was

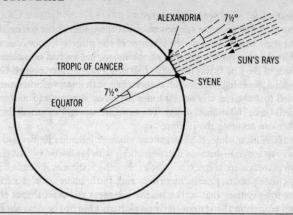

Figure 2.1. Eratosthenes measured the size of the earth from its curvature. At noon, on 21 June, the sun is directly overhead at Syene, which lies on the Tropic of Cancer. But, at the same time, the sun's rays, seen from farther north in Alexandria, fall at an angle of 7.5 degrees to the vertical and therefore cast a shadow. Knowing the distance between the two cities and the length of the shadow in Alexandria, Eratosthenes made his calculations.

in terms of the diameter of the earth. Hipparchus, repeating this work, calculated that the moon's distance from the earth was 30 times the earth's diameter. If Eratosthenes' figure of 8,000 miles for the earth's diameter was correct, the moon must be about 240,000 miles from the earth. This figure again happens to be about correct.

But finding the moon's distance was as far as Greek astronomy managed to carry the problem of the size of the universe—at least correctly. Aristarchus had made a heroic attempt to determine the distance to the sun. The geometric method he used was absolutely correct in theory, but it involved measuring such small differences in angles that, without the use of modern instruments, he was unable to get a good value. He decided that the sun was about 20 times as far as the moon (actually it is about 400 times). Although his figures were wrong, Aristarchus nevertheless did deduce from them that the sun must be at least 7 times larger than the earth. Pointing out the illogic of supposing that the large sun circled the small earth, he decided that the earth must be revolving around the sun.

Unfortunately, no one listened to him. Later astronomers, beginning with Hipparchus and ending with Claudius Ptolemy, worked out all the heavenly movements on the basis of a motionless earth at the center of the universe, with the moon 240,000 miles away and other objects an undetermined distance farther. This scheme held sway until 1543, when Nicolaus Copernicus published his book, which returned to the viewpoint of Aristarchus and forever dethroned Earth's position as the center of the universe.

MEASURING THE SOLAR SYSTEM

The mere fact that the sun was placed at the center of the solar system did not in itself help determine the distance of the planets. Copernicus adopted the Greek value for the distance of the moon, but he had no notion of the distance of the sun. It was not until 1650 that a Belgian astronomer, Godefroy Wendelin, repeated Aristarchus' observations with improved instruments and decided that the sun was not 20 times the moon's distance (5 million miles) but 240 times (60 million miles). The estimate was still too small, but it was much more accurate than before.

In 1609, meanwhile, the German astronomer Johannes Kepler had opened the way to accurate distance determinations with his discovery that the orbits of the planets were ellipses, not circles. For the first time, it became possible to calculate planetary orbits accurately and, furthermore, to plot a scale map of the solar system: that is, the relative distances and orbit shapes of all the known planets in the system could be plotted. Thus, if the distance between any two planets in the system could be determined in miles, all the other distances could be calculated at once. The distance to the sun, therefore, need not be calculated directly, as Aristarchus and Wendelin had attempted to do. The determination of the distance of any nearer body, such as Mars or Venus, outside the Earth-moon system would do.

One method by which cosmic distances can be calculated involves the use of *parallax*. It is easy to illustrate what this term means. Hold your finger about 3 inches before your eyes and look at it first with just the left eye and then with just the right. Your finger will shift position against the background, because you have changed your point of view. Now if you repeat this procedure with your finger farther away—say, at arm's length—the finger again will shift against the background, but this time not so much. The amount of shift can be used to determine the distance of the finger from your eye.

Of course, for an object 50 feet away, the shift in position from one eye to the other begins to be too small to measure; we need a wider "baseline" than just the distance between our two eyes. But all we have to do to widen the change in point of view is to look at the object from one spot, then move 20 feet to the right and look at it again. Now the parallax is large enough to be measured easily, and the distance can be determined. Surveyors make use of just this method for determining the distance across a stream or ravine.

The same method, precisely, can be used to measure the distance to the moon, with the stars playing the role of background. Viewed from an observatory in California, for instance, the moon will be in one position against the stars. Viewed at the same instant from an observatory in England, it will be in a slightly different position. From this change in position, and the known distance between the two observatories (in a straight line through the earth), the distance of the moon can be calculated. Of course, we can, in theory, enlarge the baseline by making observations from observatories at directly opposite sides of the earth; the length of the baseline is then 8,000 miles. The resulting angle of parallax, divided by two, is the *geocentric parallax*.

The shift in position of a heavenly body is measured in degrees or in subunits of a degree—minutes and seconds. One degree is 1/360 of the circuit around the sky; each degree is split into 60 minutes of arc, and each minute into 60 seconds of arc. A minute of arc is therefore 1/(360 × 60) or 1/21,600 of the circuit of the sky, while a second of arc is 1/(21,600 × 60) or 1/1,296,000 of the circuit of the sky.

Using trigonometry (the interrelationship of the sides and angles of triangles), Claudius Ptolemy was able to measure the distance of the moon from its parallax, and his result agreed with the earlier figure of Hipparchus. It turned out that the geocentric parallax of the moon is 57 minutes of arc (nearly a full degree). The shift is about equal to the width of a twenty-five-cent piece as seen at a distance of five feet. This is easy enough to measure even with the naked eye. But when it came to measuring the parallax of the sun or a planet, the angles involved were too small. The only conclusion that could be reached was that the other bodies were much farther than the moon. How much farther, no one could tell.

Trigonometry alone, in spite of its refinement by the Arabs during the Middle Ages and by European mathematicians of the sixteenth century, could not give the answer. But measurement of small angles of parallax became possible with the invention of the telescope (which Galileo first built and turned to the sky in 1609, after hearing of a magnifying tube that had been made some months earlier by a Dutch spectaclemaker).

The method of parallax passed beyond the moon in 1673, when the Italian-born French astronomer Jean Dominique Cassini determined the parallax of Mars. He determined the position of Mars against the stars while, on the same evenings, the French astronomer Jean Richer, in French Guiana, was making the same observation. Combining the two, Cassini obtained his parallax and calculated the scale of the solar system. He arrived at a figure of 86 million miles for the distance of the sun from the earth—a figure only 7 percent less than the actual one.

Since then, various parallaxes in the solar system have been measured with increasing accuracy. In 1931, a vast international project was made out of the determination of the parallax of a small planetoid named Eros, which at that time approached the earth more closely than any heavenly body except the moon. Eros on this occasion showed a large parallax that could be measured with considerable precision, and the scale of the solar system was determined more accurately than ever before. From these calculations and by the use of methods still more accurate than those involving parallax, the distance of the sun from the earth is now known to average approximately 92,965,000 miles, give or take a thousand miles or so. (Because the earth's orbit is elliptical, the actual distance varies from 91,400,000 to 94,600,000 miles.)

This average distance is called an *astronomical unit* (A.U.), and other distances in the solar system are given in this unit. Saturn, for instance, turned out to be, on the average, 887 million miles from the sun, or 9.54 A.U. As the outer planets—Uranus, Neptune, and Pluto—were discovered, the boundaries of the solar system were successively enlarged. The extreme diameter of Pluto's orbit is 7,300 million miles, or 79 A.U. And some comets are known to recede to even greater distances from the sun.

By 1830, the solar system was known to stretch across billions of miles of space, but obviously this was by no means the full size of the universe. There were still the stars.

THE FARTHER STARS

The stars might, of course, still exist as tiny objects set into the solid vault of the sky that formed the boundary of the universe just outside the extreme limits of the solar system. Until about 1700, this remained a rather respectable view, although there were some scholars who did not agree.

As early as 1440, a German scholar, Nicholas of Cusa, maintained that space was infinite, and that the stars were suns stretching outward in all directions without limit, each with a retinue of inhabited planets. That the stars did not look like suns but appeared as tiny specks of light, he attributed to their great distance. Unfortunately Nicholas had no evidence for these views but advanced them merely as opinion. The opinion seemed a wild one, and he was ignored.

In 1718, however, the English astronomer Edmund Halley, who was working hard to make accurate telescopic determinations of the position of various stars in the sky, found that three of the brightest stars—Sirius, Procyon, and Arcturus—were not in the positions recorded by the Greek astronomers. The change was too great to be an error, even allowing for the fact that the Greeks were forced to make naked-eye observations. Halley concluded that the stars are not fixed to the firmament after all, but that they move independently, like bees in a swarm. The movement is very slow and so unnoticeable until the telescope was available that they *seemed* fixed.

The reason this *proper motion* is so small is that the stars are so distant from us. Sirius, Procyon, and Arcturus are among the nearer stars, and their proper motion eventually became detectable. Their relative proximity to us makes them seem so bright. Dimmer stars are, in general, farther away, and their proper motion remained undetectable even over the time that elapsed between the Greeks and ourselves.

The proper motion itself, while testifying to the distance of the stars, did not actually give us the distance. Of course, the nearer stars should show a parallax when compared with the more distant ones. However, no such parallax could be detected. Even when the astronomers used as their baseline the full diameter of the earth's orbit around the sun (186 million miles), looking at the stars from the opposite ends of the orbit at half-year intervals, they still could observe no parallax. Hence, even the nearest stars must be extremely distant. As better and better telescopes failed to show a stellar parallax, the estimated distance of the stars had to be increased more and more. That they were visible at all at the vast distances to which they had to be pushed made it plain that they must be tremendous balls of flame like our own sun. Nicholas of Cusa was right.

But telescopes and other instruments continued to improve. In the 1830s, the German astronomer Friedrich Wilhelm Bessel made use of a newly invented device, called the *heliometer* ("sun measure") because it was originally intended to measure the diameter of the sun with great precision. It could be used equally well to measure other distances in the heavens, and Bessel used it to measure the distance between two stars. By noticing the change in this distance from month to month, he finally succeeded in measuring the parallax of a star (figure 2.2). He chose a small star in the constellation Cygnus, called

Figure 2.2. Parallax of a star measured from opposite points on the earth's orbit around the sun.

61 Cygni. His reason for choosing it was that it showed an unusually large proper motion from year to year against the background of the other stars and thus must be nearer than the others. (This steady proper motion should not be confused with the back-and-forth shift against the background that indicates parallax.) Bessel pinpointed the successive positions of 61 Cygni against the "fixed" neighboring stars (presumably much more distant) and continued his observations for more than a year. Then, in 1838, he reported that 61 Cygni had a parallax of 0.31 second of arc—the width of a twenty-five-cent piece as seen from a distance of 10 miles! This parallax, observed with the diameter of the earth's orbit as the baseline, meant that 61 Cygni was about 64 trillion (64,000,000,000,000) miles away—9,000 times the width of our solar system. Thus, compared with the distance of even the nearest stars, the solar system shrinks to an insignificant dot in space.

Because distances in trillions of miles are inconvenient to handle, astronomers shrink them by giving them in terms of the speed of light—186,282 miles per second. In a year, light travels 5,880,000,000,000 (nearly 6 trillion) miles. That distance is therefore called a *light-year*. In terms of this unit, 61 Cygni is about 11 light-years away.

Two months after Bessel's success (so narrow a margin by which to lose the honor of being the first!), the British astronomer Thomas Henderson reported the distance of the star Alpha Centauri. This star, located low in the southern skies and not visible north of the latitude of Tampa, Florida, is the third brightest in the heavens. It turned out that Alpha Centauri has a parallax of 0.75 second of arc, more than twice that of 61 Cygni. Alpha Centauri is therefore correspondingly closer. In fact, it is only 4.3 light-years from the solar system and is our nearest stellar neighbor. Actually it is not a single star, but a cluster of three.

In 1840, the German-born Russian astronomer Friedrich Wilhelm von Struve announced the parallax of Vega, the fourth brightest star in the sky. He was a little off in his determination as it turned out, but understandably, because Vega's parallax is very small and it is much farther away—27 light-years.

By 1900, the distances of about seventy stars had been determined by the parallax method (and by the 1980s, many thousands). One hundred light-years is about the limit of the distance that can be measured with any accuracy, even with the best instruments. And beyond are countless stars at much greater distances.

With the naked eye, we can see about 6,000 stars. The invention of the telescope at once made plain that these were only a fragment of the universe. When Galileo raised his telescope to the heavens in 1609, he not only found new stars previously invisible but, on turning to the Milky Way, received an even more profound shock. To the naked eye, the Milky Way is merely a luminous band of foggy light. Galileo's telescope broke down this foggy light into myriads of stars, as numerous as the grains in talcum powder.

The first man to try to make sense out of this was the German-born English astronomer William Herschel. In 1785, Herschel suggested that the stars of the heavens were arranged in a lens shape. If we look toward the Milky Way, we see a vast number of stars, but when we look out to the sky at right angles to this wheel, we see relatively few stars. Herschel deduced that the heavenly bodies formed a flattened system, with the long axis in the direction of the Milky Way. We now know that, within limits, this picture is correct, and we call our star system the *galaxy*, which is actually another term for Milky Way, because *galaxy* comes from the Greek word for milk.

Herschel tried to estimate the size of the galaxy. He assumed that all the stars had about the same intrinsic brightness, so that one could tell the relative distance of a star by its brightness. (By a well-known law brightness decreases as the square of the distance, so if star A is one-ninth the brightness of star B, it should be three times as far as star B.)

By counting samples of stars in various spots of the Milky Way, Herschel estimated that there were about 100 million stars in the galaxy altogether. From the levels of their brightness, he decided that the diameter of the galaxy was 850 times the distance to the bright star Sirius, and that the galaxy's thickness was 155 times that distance.

We now know that the distance to Sirius is 8.8 light-years, so Herschel's estimate was equivalent to a galaxy about 7,500 light-years in diameter and 1,300 light-years thick. This estimate turned out to be far too conservative.

But like Aristarchus' overconservative measure of the distance to the sun, it was a step in the right direction.

It was easy to believe that the stars in the galaxy move about (as I said before) like bees in a swarm, and Herschel showed that the sun itself also moves in this manner.

By 1805, after he had spent twenty years determining the proper motions of as many stars as possible, he found that, in one part of the sky, the stars generally seemed to be moving outward from a particular center (the *apex*). In a place in the sky directly opposite to the first, the stars generally seemed to be moving inward toward a particular center (the *anti-apex*).

The easiest way of explaining this phenomenon was to suppose that the sun was moving away from the anti-apex and toward the apex, and that the clustered stars seemed to be moving apart as the sun approached, and to be closing in behind. (This is a common effect of perspective. We would see it if we were walking through a grove of trees, and would be so accustomed to the effect that we would scarcely notice it.)

The sun is not, therefore, the immovable center of the universe as Copernicus had thought, but moves—yet not in the way the Greeks had thought. It does not move about the earth but carries the earth and all the planets along with it as it moves through the galaxy. Modern measurements show that the sun is moving (relative to the nearer stars) toward a point in the constellation of Lyra at a speed of 12 miles a second.

Beginning in 1906, the Dutch astronomer Jacobus Cornelis Kapteyn conducted another survey of the Milky Way. As he had photography at his disposal and knew the true distance of the nearer stars, he was able to make a better estimate than Herschel had. Kapteyn decided that the dimensions of the galaxy were 23,000 light-years by 6,000. Thus Kapteyn's model of the galaxy was four times as wide and five times as thick as Herschel's; but it was still overconservative.

To sum up, by 1900 the situation with respect to stellar distances was the same as that with respect to planetary distances in 1700. In 1700, the moon's distance was known, but the distance of the farther planets could only be guessed. In 1900, the distance of the nearer stars was known, but that of the more distant stars could only be guessed.

MEASURING A STAR'S BRIGHTNESS

The next major step forward was the discovery of a new measuring rod—certain variable stars that fluctuate in brightness. This part of the story begins with a fairly bright star called Delta Cephei, in the constellation Cepheus. On close study, the star was found to have a cycle of varying brightness: from its dimmest stage, it rather quickly doubled in brightness, then slowly faded to its dim point again. It did this over and over with great regularity. Astronomers found a number of other stars that varied in the same regular way; and in honor of Delta Cephei, all were named *cepheid variables* or, simply, *cepheids*.

The cepheids' periods (the time from dim point to dim point) vary from less than a day to as long as nearly two months. Those nearest our sun seem to have a period in the neighborhood of a week. The period of Delta Cephei itself is 5.3 days, while the nearest cepheid of all (the Pole Star, no less) has a period of 4 days. (The Pole Star, however, varies only slightly in luminosity —not enough to be noticeable to the unaided eye.)

The importance of the cepheids to astronomers involves their brightness— a subject that requires a small digression.

Ever since Hipparchus, the brightness of stars has been measured by the term *magnitude* according to a system he invented. The brighter the star, the lower the magnitude. The twenty brightest stars he called *first magnitude*. Somewhat dimmer stars are *second magnitude*. Then, third, fourth, and fifth, until the dimmest, those just barely visible, are of the *sixth magnitude*.

In modern times—1856, to be exact—Hipparchus' notion was made quantitative by the English astronomer Norman Robert Pogson. He showed that the average first-magnitude star was about 100 times brighter than the average sixth-magnitude star. Allowing this interval of five magnitudes to represent a ratio of 100 in brightness, the ratio for 1 magnitude must be 2.512. A star of magnitude 4 is 2.512 times as bright as a star of magnitude 5, and 2.512 × 2.512, or about 6.3 times as bright as a star of magnitude 6.

Among the stars, 61 Cygni is a dim star with a magnitude of 5.0 (modern astronomical methods allow magnitudes to be fixed to the nearest tenth and even to the nearest hundredth in some cases). Capella is a bright star, with a magnitude of 0.9; Alpha Centauri still brighter, with a magnitude of 0.1. And the measure goes on to still greater brightnesses which are designated by magnitude 0 and beyond this by negative numbers. Sirius, the brightest star in the sky, has a magnitude of −1.42. The planet Venus attains a magnitude of −4.2; the full moon, −12.7; the sun, −26.9.

These are the *apparent magnitudes* of the stars as we see them—not their absolute *luminosities* independent of distance. But if we know the distance of a star and its apparent magnitude, we can calculate its actual luminosity. Astronomers base the scale of *absolute magnitudes* on the brightness at a standard distance, which has been established at ten parsecs, or 32.6 light-years. (The *parsec* is the distance at which a star would show a parallax of 1 second of arc; it is equal to a little more than 19 trillion miles, or 3.26 light-years.)

Although Capella looks dimmer than Alpha Centauri and Sirius, actually it is a far more powerful emitter of light than either of them. It merely happens to be a great deal farther away. If all were at the standard distance, Capella would be much the brightest of the three. Capella has an absolute magnitude of −0.1; Sirius, 1.3; and Alpha Centauri, 4.8. Our own sun is just about as bright as Alpha Centauri, with an absolute magnitude of 4.86. It is an ordinary, medium-sized star.

Now to get back to the cepheids. In 1912, Henrietta Leavitt, an astronomer at the Harvard Observatory, was studying the smaller of the Magellanic

Clouds—two huge star systems in the Southern Hemisphere named after Ferdinand Magellan, because they were first observed during his voyage around the globe. Among the stars of the Small Magellanic Cloud, Miss Leavitt detected twenty-five cepheids. She recorded the period of variation of each and, to her surprise, found that the longer the period, the brighter the star.

As this is not true of the cepheid variables in our own neighborhood, why should it be true of the small Magellanic cloud? In our own neighborhood, we know only the apparent magnitudes of the cepheids; not knowing their distances or absolute brightnesses, we have no scale for relating the period of a star to its brightness. But in the Small Magellanic Cloud, all the stars are effectively at about the same distance from us, because the cloud itself is so far away. It is as though a man in New York were trying to calculate his distance from each person in Chicago. He would conclude that all the Chicagoans were about equally distant from himself—what is a difference of a few miles in a total distance of a thousand? Similarly, a star at the far end of the Cloud is not significantly farther away than one at the near end.

With the stars in the Small Magellanic Cloud at about the same distance from us, their apparent magnitude could be taken as a measure of their comparative absolute magnitude. So Leavitt could consider the relationship she saw a true one: that is, the period of the cepheid variables increases smoothly with the absolute magnitude. She was thus able to establish a *period-luminosity curve*—a graph that shows what period a cepheid of any absolute magnitude must have; and, conversely, what absolute magnitude a cepheid of a given period must have.

If cepheids everywhere in the universe behaved as they did in the Small Magellanic Cloud (a reasonable assumption), then astronomers had a *relative* scale for measuring distances, as far out as cepheids could be detected in the best telescopes. If they spotted two cepheids with equal periods, they could assume that both were equal in absolute magnitude. If cepheid A seemed four times as bright as cepheid B, cepheid B must be twice as distant from us. In this way, the relative distances of all the observable cepheids could be plotted on a scale map. Now if the actual distance of just one of the cepheids could be determined, so could the distances of all the rest.

Unfortunately, even the nearest cepheid, the Pole Star, is hundreds of light-years away, much too far to measure its distance by parallax. Astronomers had to use less direct methods. One usable clue was proper motion: on the average, the more distant a star is, the smaller its proper motion. (Recall that Bessel decided 61 Cygni was relatively close because it had a large proper motion.) A number of devices were used to determine the proper motions of groups of stars, and statistical methods were brought to bear. The procedure was complicated, but the results gave the approximate distances of various groups of stars which contained cepheids. From the distances and the apparent magnitudes of those cepheids, their absolute magnitudes could be determined, and these could be compared with the periods.

In 1913, the Danish astronomer Ejnar Hertzsprung found that a cepheid of absolute magnitude −2.3 had a period of 6.6 days. From that finding, and using Leavitt's period-luminosity curve, he could determine the absolute magnitude of any cepheid. (It turned out, incidentally, that cepheids generally are large, bright stars, much more luminous than our sun. Their variations in brightness are probably the result of pulsations. The stars seem to expand and contract steadily, as though they are ponderously breathing in and out.)

A few years later, the American astronomer Harlow Shapley repeated the work and decided that a cepheid of absolute magnitude −2.3 had a period of 5.96 days. The agreement was close enough to allow astronomers to go ahead. They had their yardstick.

DETERMINING THE GALAXY'S SIZE

In 1918, Shapley began observing the cepheids of our own galaxy in an attempt to determine the galaxy's size by this new method. He concentrated on the cepheids found in groups of stars called *globular clusters*—densely packed spherical aggregates of tens of thousands to tens of millions of stars, with diameters of the order of 100 light-years.

These clusters (whose nature had first been observed by Herschel a century earlier) present an astronomical environment quite different from that prevailing in our own neighborhood in space. At the center of the larger clusters, stars are packed together with a density of 500 per 10 cubic parsecs, as compared with 1 star per 10 cubic parsecs in our own neighborhood. Starlight under such conditions would be far brighter than moonlight on Earth, and a hypothetical planet situated near the center of such a cluster would know no true night.

There are about 100 known globular clusters in our galaxy and probably as many again that have not yet been detected. Shapley calculated the distance of the various globular clusters at from 20,000 to 200,000 light-years from us. (The nearest cluster, like the nearest star, is in the constellation Centaurus and is visible to the naked eye as a starlike object, Omega Centauri. The most distant, NGC 2419, is so far off as scarcely to be considered a member of the galaxy.)

Shapley found the clusters to be distributed in a large sphere that the plane of the Milky Way cuts in half, and to surround a portion of the main body of the galaxy like a halo. Shapley made the natural assumption that they encircle the center of the galaxy. His calculations placed the central point of this halo of globular clusters within the Milky Way in the direction of the constellation Sagittarius and about 50,000 light-years from us. The implication was that our solar system, far from being at the center of the galaxy, as Herschel and Kapteyn had thought, is far out toward one edge.

Shapley's model pictured the galaxy as a giant lens about 300,000 light-years in diameter. This time its size was overestimated, as another method of measurement soon showed.

From the fact that the galaxy had a disk shape, astronomers from William Herschel on assumed it had to be rotating in space. In 1926, the Dutch astronomer Jan Oort set out to measure this rotation. Since the galaxy is not a solid object, but is composed of numerous individual stars, it is not to be expected to rotate in one piece, as a wheel does. Instead, stars close to the gravitational center of the disk must revolve around it faster than those farther away (just as the planets closest to the sun travel fastest in their orbits). Hence, the stars toward the center of the galaxy (that is, in the direction of Sagittarius) should tend to drift ahead of our sun, whereas those farther from the center (in the direction of the constellation Gemini) should tend to lag behind us in their revolution. And the farther a star is from us, the greater this difference in speed should be.

On these assumptions, it became possible to calculate, from the relative motions of the stars, the rate of rotation around the galactic center. The sun and nearby stars, it turned out, travel at about 150 miles a second relative to the galactic center and make a complete revolution around the center in approximately 200 million years. (The sun travels in a nearly circular orbit, but the orbit of some stars, such as Arcturus, are quite elliptical. The fact that the various stars do not rotate in perfectly parallel orbits accounts for the sun's relative motion toward the constellation Lyra.)

Having estimated a value for the rate of rotation, astronomers were then able to calculate the strength of the gravitational field of the galactic center and, therefore, its mass. The galactic center (which contains most of the mass of the galaxy) turns out to be well over 100 billion times as massive as our sun. Since our sun is a star of greater than average mass, our galaxy therefore contains perhaps 200 to 300 billion stars—up to 3,000 times the number estimated by Herschel.

From the curve of the orbits of the revolving stars, it is also possible to locate the center around which they are revolving. The center of the galaxy in this way has been confirmed to be in the direction of Sagittarius, as Shapley found, but only 27,000 light-years from us, and the total diameter of the galaxy comes to 100,000 light-years, instead of 300,000. In this new model, now believed to be correct, the thickness of the disk is some 20,000 light-years at the center and falls off toward the edge: at the location of our sun, which is two-thirds of the way out toward the extreme edge, the disk is perhaps 3,000 light-years thick (figure 2.3). But these are only rough figures, because the galaxy has no sharply definite boundaries.

If the sun is so close to the edge of the galaxy, why is not the Milky Way much brighter in the direction toward the center than in the opposite direction, where we look toward the edge? Looking toward Sagittarius, we face the main body of the galaxy with some 200 billion stars, whereas out toward the edge there is only a scattering of some millions. Yet in each direction the band of the Milky Way seems of about the same brightness. The answer appears to be that huge clouds of obscuring dust hide much of the center of the galaxy from us. As much as half the mass of the galactic outskirts may be composed of such clouds of dust and gas. Probably we see no more than 1/10,000 (at most) of the light of the galactic center.

Figure 2.3. A model of our galaxy seen edgewise. Globular clusters are arrayed around the central portion of the galaxy. The position of our sun is indicated by +.

Thus it is that Herschel and other early students of the galaxy thought our solar system was at the center; and also, it seems, that Shapley originally overestimated the size of the galaxy. Some of the clusters he studied were dimmed by the intervening dust, so that the cepheids in them seemed dimmer and therefore more distant than they really were.

ENLARGING THE UNIVERSE

Even before the size and mass of the galaxy itself had been determined, the cepheid variables of the Magellanic Clouds (where Leavitt had made the crucial discovery of the period-luminosity curve) were used to determine the distance of the Clouds, which proved to be more than 100,000 light-years away. The best modern figures place the Large Magellanic Cloud at about 150,000 light-years from us and the Small Magellanic Cloud at 170,000 light-years. The Large Cloud is no more than half the size of our galaxy in diameter; the Small Cloud, no more than one-fifth. Besides, they seem to be less densely packed with stars. The Large Magellanic Cloud contains 5 billion stars (only 1/20 or less the number in our galaxy), while the Small Magellanic Cloud has only 1.5 billion.

That was the situation in the early 1920s: the known universe was less than 200,000 light-years in diameter and consisted of our galaxy and its two neighbors. The question then arose whether anything existed outside that.

Suspicion rested upon certain small patches of luminous fog, called *nebulae* (from the Greek word for "cloud"), which astronomers had long noted. The French astronomer Charles Messier had catalogued 103 of them in 1781. (Many are still known by the numbers he gave them, preceded by the letter *M* for Messier.)

Were these nebulosities merely the clouds they seemed? Some, such as the Orion Nebula (first discovered in 1656 by the Dutch astronomer Christian Huygens), seemed to be just that: a cloud of gas and dust, equal in mass to about 500 suns like ours, and illuminated by hot stars within. Other nebulosities, on the other hand, turned out to be globular clusters—huge assemblages of stars.

But there remained patches of luminous cloud that seemed to contain no stars at all. Why, then, were they luminous? In 1845, the British astronomer William Parsons (third Earl of Rosse), using a 72-inch telescope he had spent his life building, had ascertained that some of these patches had a spiral structure, which gave them the name "spiral nebulae" but did not help explain the source of the luminosity.

The most spectacular of these patches, known as M–31, or the Andromeda Nebula (because it is in the constellation Andromeda), was first studied in 1612 by the German astronomer Simon Marius. It is an elongated oval of dim light about half the size of the full moon. Could it be composed of stars so distant that they could not be made out separately even in large telescopes? If so, the Andromeda Nebula must be incredibly far away and incredibly large to be visible at all at such a distance. (As long ago as 1755, the German philosopher Immanuel Kant had speculated on the existence of such far-distant star collections: *island universes*, he called them.)

In the 1910s, there was a strong dispute over the matter. The Dutch-American astronomer Adriaan Van Maanen had reported that the Andromeda Nebula was rotating at a measurable rate. To do so, it had to be fairly close to us. If it were beyond the galaxy, it would be too far away to display any perceptible motion. Shapley, a good friend of Van Maanen, used his results to argue that the Andromeda Nebula was part of the galaxy.

Arguing against this assumption was the American astronomer Heber Doust Curtis. Although no stars were visible in the Andromeda Nebula, every once in a while an exceedingly faint star would make its appearance. Curtis felt this to be a *nova*, a star that suddenly brightens several thousand fold. In our galaxy, such stars end up being quite bright for a short while before fading again; but in the Andromeda Nebula, they were just barely visible, even at their brightest. Curtis reasoned that the novas were exceedingly dim because the Andromeda Nebula was exceedingly far away. Ordinary stars in the Andromeda Nebula were altogether too dim to be made out, but just melted together in a kind of faintly luminous fog.

On 26 April 1920, Curtis and Shapley held a well-publicized debate on the matter. On the whole, it was a standoff, although Curtis turned out to be a surprisingly good speaker and presented an impressive defense of his position.

Within a few years, however, it was clear that Curtis was in the right. For one thing, Van Maanen's figures turned out to be wrong. The reason is uncertain, but even the best can make errors, and Van Maanen had apparently done so.

Then, in 1924, the American astronomer Edwin Powell Hubble turned the new 100-inch telescope at Mount Wilson in California on the Andromeda Nebula. (It was called the *Hooker telescope* after John B. Hooker who had provided the funds for its construction.) This powerful instrument resolved portions of the nebula's outer edge into individual stars, thus showing at once that the Andromeda Nebula, or at least parts of it, resembled the Milky Way and that there might be something to this "island universe" notion.

Among the stars at the edge of the Andromeda Nebula are cepheid variables. Using these measuring rods, Hubble decided that the nebula was nearly a million light-years away! So the Andromeda Nebula was far, far outside our galaxy. Allowing for its distance, its apparent size showed that it must be a huge conglomeration of stars, almost rivaling our own galaxy.

Other nebulosities, too, turned out to be conglomerations of stars, even farther away than the Andromeda Nebula. These *extra-galactic nebulae* all had to be recognized as galaxies—new universes that reduced our own to just one of the many in space. Once again the universe had expanded. It was larger than ever—not merely hundreds of thousands, but perhaps hundreds of millions, of light-years across.

SPIRAL GALAXIES

Through the 1930s, astronomers wrestled with several nagging puzzles about these galaxies. For one thing, on the basis of their assumed distances, all of them were apparently much smaller than our own. It seemed an odd coincidence that we should be inhabiting the largest galaxy in existence. For another thing, globular clusters surrounding the Andromeda galaxy seemed to be only one-half or one-third as luminous as those of our own galaxy. (Andromeda is about as rich in globular clusters as our own galaxy, and its clusters are spherically arranged about Andromeda's center. This finding seems to show that Shapley's assumption that our own clusters are so arranged was reasonable. Some galaxies are amazingly rich in globular clusters. The galaxy M-87, in Virgo, possesses at least 1,000.)

The most serious problem was that the distances of the galaxies seemed to imply that the universe was only about 2 billion years old (for reasons I shall discuss later in this chapter). This was puzzling, for the earth itself was considered by geologists to be older than that, on what was thought to be the very best kind of evidence.

The beginning of an answer came during the Second World War, when the German-born American astronomer Walter Baade discovered that the yardstick by which the galaxies' distances had been measured was wrong.

In 1942, Baade took advantage of the wartime blackout of Los Angeles, which cleared the night sky at Mount Wilson, to make a detailed study of the Andromeda galaxy with the 100-inch telescope. With the improved visibility, he was able to resolve some of the stars in the inner regions of the galaxy. He immediately noted some striking differences between these stars and those in the outskirts of the galaxy. The brightest stars in the interior were reddish, whereas those of the outskirts were bluish. Moreover, the red giants of the interior were not nearly so bright as the blue giants of the outskirts: the latter had up to 100,000 times the luminosity of our sun, whereas the internal red giants had only up to 1,000 times that luminosity. Finally, the outskirts of the Andromeda galaxy, where the bright blue stars were found, was loaded with dust; whereas the interior, with its somewhat less bright red stars, was free of dust.

To Baade, it seemed that here were two sets of stars with different structure and history. He called the bluish stars of the outskirts *Population I* and the reddish stars of the interior, *Population II.* Population I stars, it turns out, are relatively young, with high metal content, and follow nearly circular orbits about the galactic center in the median plane of the galaxy. Population II stars are relatively old, with low metal content, with orbits that are markedly elliptical, and with considerable inclination to the median plane of the galaxy. Both populations have been broken down into finer subgroups since Baade's discovery.

When the new 200-inch Hale telescope (named for the American astronomer, George Ellery Hale, who supervised its construction) was set up on Palomar Mountain after the war, Baade continued his investigations. He found certain regularities in the distribution of the two populations, and these depended on the nature of the galaxies involved. Galaxies of the class called *elliptical* (systems with the shape of an ellipse and with rather uniform internal structure) apparently were made up mainly of Population II stars, as were globular clusters in any galaxy. On the other hand, in *spiral galaxies* (galaxies with arms that make them look like a pinwheel) the spiral arms were composed of Population I, set against a Population II background.

It is estimated that only about 2 percent of the stars in the universe are of the Population I type. But our own sun and the familiar stars in our neighborhood fall into this class. From this fact alone, we can deduce that ours is a spiral galaxy, and that we lie in one of the spiral arms. (Hence, the many dust clouds, both light and dark, in our neighborhood: the spiral arms of a galaxy are clogged with dust.) Photographs show that the Andromeda galaxy also is of the spiral type.

Now to get back to the yardstick. Baade began to compare the cepheid stars found in globular clusters (Population II) with those found in our spiral arm (Population I). It turned out that the cepheids in the two populations were really of two different types, as far as the relation between period and luminosity was concerned. Cepheids of Population II followed the period-luminosity curve set up by Leavitt and Shapley. With this yardstick, Shapley had measured the distances to the globular clusters and the size of our galaxy with reasonable accuracy. But the cepheids of Population I, it now developed, were a different yardstick altogether! A Population-I cepheid was four or five times as luminous as a Population-II cepheid of the same period. Hence, use of the Leavitt scale would result in miscalculation of the absolute magnitude of a Population-I cepheid from its period. And if the absolute magnitude was wrong, the calculation of distance must be wrong: the star would actually be much farther away than the calculation indicated.

Hubble had gauged the distance of the Andromeda galaxy from the cepheids (of Population I) in its outskirts—the only ones that could be resolved at the time. Now, with the revised yardstick, the galaxy proved to be about 2.5 million light-years away, instead of less than a million. And other galaxies had to be moved out in proportion. (The Andromeda galaxy is still a close neighbor, however. The average distance between galaxies is estimated to be something like 20 million light-years.)

At one stroke, the size of the known universe was more than doubled, and the problems that had plagued the 1930s were solved. Our galaxy was no longer larger than all the others; the Andromeda galaxy, for instance, was definitely more massive than ours. Second, it now appeared that the Andromeda galaxy's globular clusters were as luminous as ours; they had seemed less bright only because of the misjudgment of their distance. Finally, for reasons I will explain later, the new scale of distances allowed the universe to be considered much older, bringing it into line with the geologists' estimates of the age of the earth.

CLUSTERS OF GALAXIES

Doubling the distance of the galaxies does not end the problem of size. We must now consider the possibility of still larger systems—of clusters of galaxies and clusters of clusters.

Actually, modern telescopes have shown that clusters of galaxies do exist. For instance, in the constellation of Coma Berenices there is a large, ellipsoidal cluster of galaxies about 8 million light-years in diameter. The Coma cluster contains about 11,000 galaxies, separated by an average distance of only 300,000 light-years (as compared with an average of something like 3 million light-years between galaxies in our own vicinity).

Our own galaxy seems to be part of a *local group* that includes the Magellanic Clouds, the Andromeda galaxy, and three small *satellite galaxies* near it, plus some other galaxies; a total of nineteen members altogether. Two of these, called Maffei One and Maffei Two (for Paolo Maffei, the Italian astronomer, who first reported them), were discovered only in 1971. The lateness of the discovery was owing to the fact that they can only be detected through dust clouds that lie between them and ourselves.

Of the local group, only our own galaxy, Andromeda, and the two Maffeis are giants, whereas the rest are dwarfs. One of the dwarfs, IC 1613, may contain only 60 million stars; hence it is scarcely more than a large globular cluster. Among galaxies, as among stars, dwarfs far outnumber giants.

If galaxies do form clusters and clusters of clusters, does that mean that the universe goes on forever and that space is infinite? Or is there some end, both to the universe and to space? Well, astronomers can make out objects up to

an estimated 10 billion light-years away, and there they seem to be reaching
a limit. To see why, I must now shift the direction of discussion a bit. Having
considered space, let us next consider time.

The Birth of the Universe

Mythmakers have invented many fanciful creations of the universe (usually
concentrating on the earth itself, with all the rest dismissed quickly as the
"sky" or the "heavens"). Generally, the time of creation is set not very far in
the past (although we should remember that, to people in the preliterate stage,
a time of a thousand years was even more impressive than a billion years is
today).

The creation story with which we are most familiar is, of course, that given
in the first chapters of Genesis, which, some people hold, is an adaptation of
Babylonian myths, intensified in poetic beauty and elevated in moral grandeur.

Various attempts have been made to work out the date of the Creation on
the basis of the data given in the Bible (the reigns of the various kings, the
time from the Exodus to the dedication of Solomon's temple, the ages of the
patriarchs both before and after the flood). Medieval Jewish scholars put the
date of the Creation at 3760 B.C., and the Jewish calendar still counts its years
from that date. In 1658 A.D., Archbishop James Ussher of the Anglican
Church calculated the date of the Creation to be 4004 B.C.; while others
following his lead placed it exactly at 8 P.M. on 22 October of that year. Some
theologians of the Greek Orthodox Church put Creation as far back as
5508 B.C.

Even as late as the eighteenth century, the Biblical version was accepted
by the learned world, and the age of the universe was considered to be only
6,000 or 7,000 years at most. This view received its first major blow in 1785
in the form of a book entitled *Theory of the Earth,* by a Scotch naturalist
named James Hutton. Hutton started with the proposition that the slow
natural processes working on the surface of the earth (mountain building and
erosion, the cutting of river channels, and so on) had been working at about
the same rate throughout the earth's history. This *uniformitarian principle* implied that the processes must have been working for a stupendously long
time to produce the observed phenomena. Therefore the earth must be not
thousands but many millions of years old.

Hutton's views were immediately derided. But the ferment worked. In the
early 1830s, the British geologist Charles Lyell reaffirmed Hutton's views and,
in a three-volume work entitled *Principles of Geology,* presented the evidence
with such clarity and force that the world of science was won over. The
modern science of geology can be dated from that work.

THE AGE OF THE EARTH

Attempts were made to calculate the age of the earth on the basis of the uniformitarian principle. For instance, if one knew the amount of sediment laid down by the action of water each year (a modern estimate is 1 foot in 880 years), one could calculate the age of a layer of sedimentary rock from its thickness. It soon became obvious that this approach could not accurately determine the earth's age, because the record of the rocks was obscured by erosion, crumbling, upheavals, and other forces. Nevertheless, even the incomplete evidence indicated that the earth must be at least 500 million years old.

Another way of measuring the age of the earth was to estimate the rate of accumulation of salt by the oceans, a suggestion first advanced by Edmund Halley as long ago as 1715. Rivers steadily washed salt into the sea; since only fresh water left it by evaporation, the salt concentration rose. The assumption was that the ocean had started as fresh water; hence, the time necessary for the rivers to have endowed the oceans with their salt content of over 3 percent could have been as long as a billion years.

This great age was very agreeable to the biologists, who, during the latter half of the nineteenth century, were trying to trace the slow development of living organisms from primitive one-celled creatures to the complex higher animals. They needed long eons for the development to take place, and a billion years gave them sufficient time.

However, by the mid-nineteenth century, astronomical considerations brought sudden complications. For instance, the principle of the *conservation of energy* raised an interesting problem with respect to the sun. The sun was pouring out energy in colossal quantities and had been doing so throughout recorded history. If the earth had existed for countless eons, where had all this energy come from? It could not have come from the usual familiar sources. If the sun had started as solid coal burning in an atmosphere of oxygen, it would have been converted to carbon dioxide (at the rate it was delivering energy) in the space of about 2,500 years.

The German physicist Hermann Ludwig Ferdinand von Helmholtz, one of the first to enunciate the law of conservation of energy, was particularly interested in the problem of the sun. In 1854, he pointed out that if the sun were contracting, its mass would gain energy as it fell toward its center of gravity, just as a rock gains energy when it falls. This energy could be converted into radiation. Helmholtz calculated that a contraction of the sun by a mere 1/10,000 of its radius could provide it with a 2,000-year supply of energy.

The British physicist William Thomson (later Lord Kelvin) did more work on the subject and decided that, on this basis, the earth could not be more than 50 million years old; for at the rate the sun had spent energy, it must have contracted from a gigantic size, originally as large as the earth's orbit around the sun. (This assumption meant, of course, that Venus must be younger than the earth and Mercury still younger.) Lord Kelvin went on to estimate that if the earth itself had started as a molten mass, the time needed to cool to its present temperature, and therefore its age, would be about 20 million years.

By the 1890s, the battlelines were drawn between two apparently invincible armies. The physicists seemed to have shown conclusively that the earth could not have been solid for more than a few million years, while geologists and biologists seemed to have proved just as conclusively that the earth must have been solid for not less than a billion years.

And then something new and completely unexpected turned up, and the physicists' case began to crumble.

In 1896, the discovery of radioactivity made it clear that the earth's uranium and other radioactive substances were liberating large quantities of energy and had been doing so for a very long time. This finding made Kelvin's calculations meaningless, as was pointed out first, in 1904, by the New Zealand-born British physicist Ernest Rutherford in a lecture—with the aged (and disapproving) Kelvin himself in the audience.

There is no point in trying to decide how long it would take the earth to cool if you do not take into account the fact that heat is being constantly supplied by radioactive substances. With this new factor, it might take the earth billions of years, rather than millions, to cool from a molten mass to its present temperature. The earth might even be warming with time.

Actually, radioactivity itself eventually gave the most conclusive evidence of the earth's age (in ways that will be described later in chapter 6) for it allowed geologists and geochemists to calculate the age of rocks directly from the quantity of uranium and lead they contain. By the clock of radioactivity, some of the earth's rocks are now known to be over 3 billion years old, and there is every reason to think that the earth is somewhat older than that. An age of 4.6 billion years for the earth in its present solid form is now accepted as likely. And, indeed, some of the rocks brought back from our neighbor world, the moon, have proven to be nearly that old.

THE SUN AND THE SOLAR SYSTEM

And what of the sun? Radioactivity, together with discoveries concerning the atomic nucleus, introduced a new source of energy, much larger than any previously known. In 1930, the British physicist Sir Arthur Eddington set a train of thought working when he suggested that the temperature and pressure at the center of the sun must be extraordinarily high: the temperature might be as high as 15 million degrees. At such temperatures and pressures, the nuclei of atoms could undergo reactions that could not take place in the bland mildness of the earth's environment. The sun is known to consist largely of hydrogen. If four hydrogen nuclei combined (forming a helium atom), they would liberate large amounts of energy.

Then, in 1938, the German-born American physicist Hans Albrecht Bethe worked out two possible ways in which this combination of hydrogen to helium could take place under the conditions at the center of stars like the sun: one way involved the direct conversion of hydrogen to helium; the other involved a carbon atom as an intermediate in the process. Either set of reactions can occur in stars; in our own sun, the direct hydrogen conversion seems to be the dominant mechanism. Either brought about the conversion of mass to energy. (Einstein, in his special theory of relativity, advanced in 1905, had shown that mass and energy were different aspects of the same thing and could be interconverted; and, furthermore, that a great deal of energy could be liberated by the conversion of a small amount of mass.)

The rate of radiation of energy by the sun requires the disappearance of solar mass at the rate of 4,200,000 tons per second. At first blush, this seems a frightening loss, but the total mass of the sun is 2,200,000,000,000,000,-000,000,000,000 tons, so the sun loses only 0.00000000000000000002 percent of its mass each second. If the sun has been in existence for 5 billion years, as astronomers now believe, and if it has been radiating at its present rate all that time, it would have expended only 1/33,000 of its mass. It is easy to see, then, that the sun can continue to radiate energy at its present rate for billions of years to come.

By 1940, then, an age of nearly 5 billion years for the solar system as a whole seemed reasonable. The whole matter of the age of the universe might have been settled, but astronomers had thrown another monkey wrench into the machinery. Now the universe as a whole seemed too youthful to account for the age of the solar system. The trouble arose from an examination of the distant galaxies by the astronomers and from a phenomenon first discovered in 1842 by an Austrian physicist named Christian Johann Doppler.

The *Doppler effect* is familiar enough; it is most commonly illustrated by the whistle of a passing locomotive, which rises in pitch as it approaches and drops in pitch as it recedes. The change in pitch is due simply to the fact that the number of sound waves striking the eardrum per second changes because of the source's motion.

As Doppler suggested, the Doppler effect applies to light waves as well as to sound. When light from a moving source reaches the eye, there is a shift in frequency—that is, color—when the source is moving fast enough. For instance, if the source is traveling toward us, more light waves are crowded into each second, and the light perceived shifts toward the high-frequency violet end of the visible spectrum. On the other hand, if the source is moving away, fewer waves arrive per second, and the light shifts toward the low-frequency red end of the spectrum.

Astronomers have been studying the spectra of stars for a long time and are well acquainted with the normal picture—a pattern of bright lines against a dark background or dark lines against a bright background showing the emission or the absorption of light by atoms at certain wavelengths, or colors. They have been able to calculate the velocity of stars moving toward or away from us (*radial velocity*) by measuring the displacement of the usual spectral lines toward the violet or red end of the spectrum.

It was the French physicist Armand Hippolyte Louis Fizeau who, in 1848, pointed out that the Doppler effect in light could best be observed by noting

Figure 2.4. The Doppler-Fizeau effect. The lines in the spectrum shift toward the violet end *(left)* when the light source is approaching. When the source recedes, the spectral lines shift toward the red end *(right)*.

the position of the spectral lines. For that reason, the Doppler effect is called the *Doppler-Fizeau effect* where light is concerned (figure 2.4).

The Doppler-Fizeau effect has been used in a variety of ways. Within our solar system, it could be used to demonstrate the rotation of the sun in a new way. The spectral lines originating from that limb of the sun being carried toward us in the course of its vibration would be shifted toward the violet (a *violet shift*). The lines from the other limb would show a *red shift* since it was receding.

To be sure, the motion of sunspots is a better and more obvious way of detecting and measuring solar rotation (which turns out to have a period of about 26 days, relative to the stars). However, the effect can also be used to determine the rotation of featureless objects, such as the rings of Saturn.

The Doppler-Fizeau effect can be used for objects at any distance, so long as those objects can be made to produce a spectrum for study. Its most dramatic victories, therefore, were in connection with the stars.

In 1868, the British astronomer Sir William Huggins measured the radial velocity of Sirius and announced that it was moving away from us at 29 miles per second. (We have better figures now, but he came reasonably close for a first try.) By 1890, the American astronomer James Edward Keeler, using more accurate instruments, was producing reliable results in quantity; he showed, for instance, that Arcturus was approaching us at a rate of 3.75 miles per second.

The effect can even be used to determine the existence of star systems, whose details cannot be made out by telescope. In 1782, for instance, an English astronomer, John Goodricke (a deaf-mute who died at twenty-two— a first-rate brain in a tragically defective body), studied the star Algol, whose brightness increases and decreases regularly. Goodricke explained this effect by supposing that a dark companion circles Algol, periodically passing in front of it, eclipsing it, and dimming its light.

A century passed before this plausible hypothesis was supported by additional evidence. In 1889, the German astronomer Hermann Carl Vogel showed that the lines of Algol's spectrum undergoes alternate red and violet shifts that match its brightening and dimming. First it recedes while the dark companion approaches and then approaches while the dark companion recedes. Algol was seen to be an *eclipsing binary star*.

In 1890, Vogel made a similar and more general discovery. He found that some stars were both advancing and receding: that is, the spectral lines showed both a red shift and a violet shift, appearing to have doubled. Vogel concluded that the star was an eclipsing binary, with the two stars (both bright) so close together that they appeared as a single star even in the best telescopes. Such stars are *spectroscopic binaries*.

But there was no need to restrict the Doppler-Fizeau effect to the stars of our galaxy. Objects beyond the Milky Way could be studied in this way, too. In 1912, the American astronomer Vesto Melvin Slipher found, on measuring the radial velocity of the Andromeda galaxy, that it was moving toward us at approximately 125 miles per second. But when he went on to examine other galaxies, he discovered that most of them were moving away from us. By 1914, Slipher had figures on fifteen galaxies; of these, thirteen were receding, all at the healthy clip of several hundred miles per second.

As research along these lines continued, the situation grew more remarkable. Except for a few of the nearest galaxies, all were fleeing from us. Furthermore, as techniques improved so that fainter, and presumably more distant, galaxies could be tested, the observed red shift increased further.

In 1929, Hubble at Mount Wilson suggested that there was a regular increase in these velocities of recession in proportion to the distance of the particular galaxy. If galaxy A was twice as far from us as galaxy B, then galaxy A receded at twice the velocity of galaxy B. This relationship is sometimes known as *Hubble's law*.

Hubble's law certainly continued to be borne out by observations. Beginning in 1929, Milton La Salle Humason at Mount Wilson used the 100-inch telescope to obtain spectra of ever dimmer galaxies. The most distant galaxies he could test were receding at 25,000 miles per second. When the 200-inch telescope came into use, still more distant galaxies could be studied; and by the 1960s, objects were detected so distant that their recession velocities were as high as 150,000 miles per second.

Why should this be? Well, imagine a balloon with small dots painted on it. When the balloon is inflated, the dots move apart. To a manikin standing on any one of the dots, all the other dots would seem to be receding, and the farther away from him a particular dot was, the faster it would recede. It would not matter on which particular dot he was standing; the effect would be the same.

The galaxies behave as though the universe were expanding like the three-dimensional skin of a four-dimensional balloon. Astronomers have now generally accepted the fact of this expansion, and Einstein's "field equations" in his general theory of relativity can be construed to fit an expanding universe.

THE BIG BANG

If the universe has been expanding constantly, it is logical to suppose that it was smaller in the past than it is now; and that, at some time in the distant past, it began as a dense core of matter.

The first to point out this possibility, in 1922, was the Russian mathematician Alexander Alexandrovich Friedmann. The evidence of the receding galaxies had not yet been presented by Hubble, and Friedmann worked entirely from theory, making use of Einstein's equations. However, Friedmann died of typhoid fever three years later at the age of thirty-seven, and his work was little known.

In 1927, the Belgian astronomer, Georges Lemaître, apparently without knowledge of Friedmann's work, worked out a similar scheme of the expanding universe. Since it was expanding, there was a time in the past when it was very small and as dense as it could be. Lemaître called this state the *cosmic egg.* In accordance with Einstein's equations, the universe could do nothing but expand; and, in view of its enormous density, the expansion had to take place with superexplosive violence. The galaxies of today are the fragments of that cosmic egg; and their recession from each other, the echo of that long-past explosion.

Lemaître's work also went unnoticed until it was called to the attention of scientists generally by the more famous English astronomer Arthur Stanley Eddington.

It was the Russian-American physicist George Gamow, however, who, in the 1930s and 1940s, truly popularized this notion of the explosive start of the universe. He called this initial explosion the *big bang*—the name by which it has been everywhere known ever since.

Not everyone was satisfied with the big bang as a way of starting the expanding universe. In 1948, two Austrian-born astronomers, Hermann Bondi and Thomas Gold, put forward a theory—later extended and popularized by a British astronomer, Fred Hoyle—that accepted the expanding universe but denied a big bang. As the galaxies move apart, new galaxies form between them, with matter being created from nothing at a rate too slow to detect with present-day techniques. The result is that the universe remains essentially the same through all eternity. It has looked as it does now through countless eons in the past and will look as it does now through countless eons in the future, so that there is neither a beginning nor an end. This theory is referred to as *continuous creation* and results in a *steady-state universe.*

For over a decade, the controversy between big bang and continuous creation went on heatedly, but there was no actual evidence to force a decision between the two.

In 1949, Gamow had pointed out that, if the big bang had taken place, the radiation accompanying it should have lost energy as the universe expanded, and should now exist in the form of radio-wave radiation coming from all parts of the sky as a homogeneous background. The radiation should be characteristic of objects at a temperature of about 5° K (that is, 5 degrees above absolute zero, or −268° C). This view was carried farther by the American physicist Robert Henry Dicke.

In May 1964, the German-American physicist Arno Allan Penzias and an American radio astronomer, Robert Woodrow Wilson, following the advice of Dicke, detected a radio-wave background with characteristics much like those predicted by Gamow. It indicated an average temperature of the universe of 3° K.

The discovery of this radio-wave background is considered by most astronomers to be conclusive evidence in favor of the big-bang theory. It is now generally accepted that the big bang did take place, and the notion of continuous creation has been abandoned.

When did the big bang take place?

Thanks to the easily measured red shift, we know with considerable certainty the rate at which the galaxies are receding. We need to know also the distance of the galaxies. The greater the distance, the longer it has taken them to reach their present position as a result of the recession rate. It is not, however, easy to determine the distance.

A figure that is generally accepted as at least approximately correct is 15 billion years. If an *eon* is 1 billion years, then the big bang took place 15 eons ago, although it might just possibly have taken place as recently as 10 eons ago or as long as 20 eons ago.

What happened before the big bang? Where did the cosmic egg come from?

Some astronomers speculate that actually the universe began as a very thin gas that slowly condensed, forming stars and galaxies perhaps, and continued to contract until it formed a cosmic egg in a *big crunch*. The formation of the cosmic egg was followed instantaneously by its explosion in a big bang, forming stars and galaxies again, but now expanding until some day it will be a thin gas again.

It may be that, if we look into the future, the universe will be expanding forever, growing thinner and thinner with a smaller and smaller overall density, approaching nearer and nearer to a vacuum of nothingness. And if we look into the past, beyond the big bang, and imagine time moving backward, again the universe will seem to be expanding forever and approaching a vacuum.

Such a "once in, once out" affair, with ourselves now occupying a place close enough to the big bang for life to be possible (were it not so, we would not be here to observe the universe and attempt to draw conclusions) is called an *open universe*.

There is no way now (and there may never be a way) to obtain any evidence for what happened before the big bang, and some astronomers are reluctant to speculate on the matter. Recently there have been arguments to the effect that the cosmic egg formed out of nothing, so that rather than a "once in, once out" universe, there is simply a "once out" universe—still an open universe.

On this assumption, it may be that, in an infinite sea of nothingness, an infinite number of big bangs may occur at various times, and that ours is therefore but one of an infinite number of universes, each with its own mass, its own point of development, and, for all we know, its own set of natural laws. It may be that only a very rare combination of natural laws make possible stars, galaxies, and life, and that we are in one such unusual situation, only because we cannot be in any other.

Needless to say, there is no evidence yet for the appearance of a cosmic egg out of nothing or for a multiplicity of universes—and there may never be. It would, however, be a harsh world indeed if scientists were not allowed to speculate poetically in the absence of evidence.

For that matter, can we be sure the universe will expand forever? It is expanding against the pull of its own gravity, and the gravity may be sufficient to slow the rate of recession to zero and eventually impose a contraction. The universe may expand and then contract into a big crunch and disappear back into nothingness—or expand again in a bounce and then some day contract again in an endless series of oscillations. Either way we have a *closed universe*.

It may yet be possible to decide whether the universe is closed or open, and I shall return to this matter later, in chapter 7.

The Death of the Sun

The expansion of the universe, even if it continues indefinitely, does not directly affect individual galaxies or clusters of galaxies. Even if all the distant galaxies recede and recede until they are out of range of the best possible instruments, our own galaxy will remain intact, its component stars held firmly within its gravitational field. Nor will the other galaxies of the local group leave us. But changes within our galaxy, not connected with universal expansion and possibly disastrous to our planet and its life, are by no means excluded.

The whole conception of changes in heavenly bodies is modern. The ancient Greek philosophers—Aristotle, in particular—believed the heavens to be perfect and unchangeable. All change, corruption, and decay were confined to the imperfect regions that lay below the nethermost sphere—the moon. This seemed only common sense, for certainly, from generation to generation and from century to century, there was no important change in the heavens. To be sure, mysterious comets occasionally materialized out of nowhere—erratic in their comings and goings, ghostlike as they shrouded stars with a thin veil, baleful in appearance, for the filmy tail looks like the streaming hair of a distraught creature prophesying evil. About twenty-five of these objects are visible to the naked eye each century. (Comets will be discussed in more detail in the next chapter.)

Aristotle tried to reconcile these apparitions with the perfection of the heavens by insisting that they belonged to the atmosphere of the corrupt and changing earth. This view prevailed until late in the sixteenth century. But, in 1577 (before the days of the telescope), the Danish astronomer Tycho Brahe attempted to measure the parallax of a bright comet and discovered that it could not be measured. Since the moon's parallax *was* measurable, Tycho Brahe was forced to conclude that the comet lay far beyond the moon and that there was change and imperfection in the heavens. (The Roman philosopher Seneca had suspected such change in the first century A.D.)

Actually, changes even in the stars had been noticed much earlier but apparently had aroused no great curiosity. For instance, there are the variable stars that change noticeably in brightness from night to night, even to the naked eye. No Greek astronomer made any reference to variations in the brightness of any star. It may be that we have lost the records of such references; on the other hand, perhaps the Greek astronomers simply chose not to see these phenomena. One interesting case in point is Algol, the second brightest star in the constellation Perseus, which loses two-thirds of its brightness, then regains it, and does this regularly every 69 hours. (We know now, thanks to Goodricke and Vogel, that Algol has a dim companion star that eclipses it and diminishes its light at 69-hour intervals.) The Greek astronomers made no mention of the dimming of Algol, nor did the Arab astronomers of the Middle Ages. Nevertheless, the Greeks placed the star in the head of Medusa, the demon who turned men to stone; and the very name Algol, which in Arabic, means "ghoul," is suggestive. Clearly, the ancients felt uneasy about this strange star.

A star in the constellation Cetus, called Omicron Ceti, varies irregularly. Sometimes it is as bright as the Pole Star; sometimes it vanishes from sight. Neither the Greeks nor the Arabs said a word about it, and the first man to report it was a Dutch astronomer, David Fabricius, in 1596. It was later named Mira (Latin for "wonderful"), astronomers having grown less frightened of heavenly change by then.

NOVAE AND SUPERNOVAE

Even more remarkable was the sudden appearance of *new stars* in the heavens, the Greeks could not altogether ignore. Hipparchus is said to have been so impressed by the sighting of such a new star, in the constellation Scorpio in 134 B.C., that he designed the first star map, in order that future new stars might be more easily detected.

In 1054 A.D., in the constellation Taurus, another new star was sighted— a phenomenally bright one. It surpassed Venus in brightness and for weeks was visible in broad daylight. Chinese and Japanese astronomers recorded its position accurately, and their records have come down to us. In the Western world, however, the state of astronomy was so low at the time that no European record of this remarkable occurrence has survived, probably because none was kept.

It was different in 1572, when a new star as bright as that of 1054 appeared in the constellation Cassiopeia. European astronomy was reviving from its long sleep. The young Tycho Brahe carefully observed the new star and wrote a book entitled *De Nova Stella*. It is from the title of that book that the word *nova* was adopted for any new star.

In 1604, still another remarkable nova appeared, in the constellation Serpens. It was not quite as bright as that of 1572, but it was bright enough to outshine Mars. Johannes Kepler observed this one, and he too wrote a book about the subject.

After the invention of the telescope, novae became less mysterious. They were not new stars at all, of course, but faint stars that had suddenly brightened to visibility.

Increasing numbers of novae were discovered with time. They would brighten many thousandfold, sometimes within the space of a few days, and then dim slowly over a period of months to their previous obscurity. Novae showed up at the average rate of twenty per year per galaxy (including our own).

From an investigation of the Doppler-Fizeau shifts that took place during nova formation and from certain other fine details of their spectra, it became plain that the novae were exploding stars. In some cases, the star material blown into space could be seen as a shell of expanding gas, illuminated by the remains of the star.

On the whole, the novae that have appeared in modern times have not been particularly bright. The brightest, Nova Aquilae, appeared in June 1918 in the constellation Aquila. This nova was, at its peak, nearly as bright as the star Sirius, which is itself the brightest in the sky. No novae, however, have appeared to rival the bright planets Jupiter and Venus, as the novae observed by Tycho and by Kepler did.

The most remarkable nova discovered since the beginning of the telescope was not recognized as such. The German astronomer Ernst Hartwig noted it in 1885; but even at its peak, it reached only the seventh magnitude and was never visible to the unaided eye.

It appeared in what was then called the Andromeda nebula and, at its peak, was one-tenth as bright as the nebula. At the time, no one realized how distant the Andromeda nebula was, or understood that it was actually a galaxy made up of several hundred billion stars, so the apparent brightness of the nova occasioned no particular excitement.

After Curtis and Hubble worked out the distance of the Andromeda galaxy (as it then came to be called), the brilliance of that nova of 1885 suddenly staggered astronomers. The dozens of novae discovered in the Andromeda galaxy by Curtis and Hubble were far dimmer than that remarkably (for the distance) bright one.

In 1934, the Swiss astronomer Fritz Zwicky began a systematic search of distant galaxies for novae of unusual brightness. Any nova that blazed up in similar fashion to the one of 1885 in the Andromeda would be visible, for such novae are almost as bright as entire galaxies, so that if the galaxy can be seen, the nova can be as well. By 1938, he had located no fewer than twelve of such galaxy-bright novae. He called these extraordinarily bright novae *super-novae*. As a result, the 1885 nova was named at last—S Andromedae, the S standing for "supernova."

Whereas ordinary novae attain an absolute magnitude of, on the average, −8 (they would be 25 times as bright as Venus, if they were seen at a distance of 10 parsecs), a supernova could have an absolute magnitude of as much as −17. Such a supernova would be 4,000 times as bright as an ordinary nova, or nearly 1,000,000,000 times as bright as the sun. At least, it would be that bright at its temporary peak.

Looking back now, we realize that the novae of 1054, 1572, and 1604 were also supernova. What is more, they must have flared up in our own galaxy, to account for their extreme brightness.

A number of novae recorded by the meticulous Chinese astronomers of ancient and medieval times must also have been supernovae. One such was reported as early as A.D. 185; and a supernova in the far southern constellation of Lupus in 1006 must have been brighter than any that have appeared in historic times. It may, at its peak, have been 200 times as bright as Venus and one-tenth as bright as the full moon.

Astronomers, judging from remnants left behind, suspect that an even brighter supernova (one that may actually have rivaled the full moon) appeared in the far southern constellation Vela 11,000 years ago, when there were no astronomers to watch, and the art of writing had not yet been invented. (It is possible, however, that certain prehistoric pictograms may have been drawn that refer to this nova.)

Supernovae are quite different in physical behavior from ordinary novae, and astronomers are eager to study their spectra in detail. The main difficulty is their rarity. About 1 per 50 years is the average for any one galaxy. Although astronomers have managed to spot more than 50 so far, all these are in distant galaxies and cannot be studied in detail. The 1885 supernova of Andromeda, the closest to us in the last 350 years, appeared a couple of decades before photography in astronomy had been fully developed; consequently, no permanent record of its spectrum exists.

However, the distribution of supernovae in time is random. In one galaxy recently, 3 supernovae were detected in just 17 years. Astronomers on earth may yet prove lucky. Indeed, one particular star is now attracting attention. Eta Carinae is clearly unstable and has been brightening and dimming for quite a while. In 1840, it brightened to the point where, for a time, it was the second brightest star in the sky. There are indications that make it appear as though it may be on the point of exploding into a supernova. One trouble, though, is that, to astronomers, "on the point of" can mean tomorrow or ten thousand years from now.

Besides, the constellation Carina, in which Eta Carinae is found, is like the constellations Vela and Lupus, so far south that the supernova, when and if it occurs, will not be visible from Europe or from most of the United States.

But what causes stars to brighten with explosive violence, and why do some become novae and some supernovae? The answer to this question requires a digression.

As early as 1834, Bessel (the astronomer who was later the first to measure the parallax of a star) noticed that Sirius and Procyon shifted position very slightly from year to year in a manner that did not seem related to the motion of the earth. Their motions were not in a straight line but wavy, and Bessel decided that each must actually be moving in an orbit around something.

From the manner in which Sirius and Procyon were moving in these orbits, the "something" in each case had to have a powerful gravitational attraction that could belong to nothing less than a star. Sirius's companion, in particular, had to be as massive as our own sun to account for the bright star's motions. So the companions were judged to be stars; but since they were invisible in telescopes of the time, they were referred to as *dark companions*. They were believed to be old stars growing dim with time.

Then, in 1862, the American instrument maker Alvan Clark, testing a new telescope, sighted a dim star near Sirius; and, sure enough, on further observation, this turned out to be the companion. Sirius and the dim star circled about a mutual center of gravity in a period of about fifty years. The companion of Sirius (Sirius B, it is now called, with Sirius itself being Sirius A) has an absolute magnitude of only 11.2 and so is only about 1/400 as bright as our sun, though it is just as massive.

Sirius B seemed to be a dying star. But, in 1914, the American astronomer Walter Sydney Adams, after studying the spectrum of Sirius B, decided that the star had to be as hot as Sirius A itself and hotter than our sun. The atomic vibrations that gave rise to the particular absorption lines found in its spectrum could be taking place only at very high temperatures. But if Sirius B was so hot, why was its light so faint? The only possible answer was that it was considerably smaller than our sun. Being hotter, it radiated more light per unit of surface; but to account for the small total amount of light, its total surface had to be small. In fact, we now know that the star cannot be more than 6,900 miles in diameter; it is smaller than the earth in volume, even though it has a mass equal to that of our sun! With all that mass squeezed into so small a volume, the star's average density would have to be about 130,000 times that of platinum.

Here was nothing less than a completely new state of matter. Fortunately, by this time physicists had no trouble in suggesting the answer. They knew that in ordinary matter the atoms are composed of very tiny particles, so tiny that most of the volume of an atom is "empty" space. Under extreme pressure, the subatomic particles can be forced together into a superdense mass. Yet even in superdense Sirius B, the subatomic particles are far enough apart to move about freely so that the far-denser-than-platinum substance still acts as a gas. The English physicist Ralph Howard Fowler suggested in 1925 that this be called a *degenerate* gas, and the Soviet physicist Lev Davidovich Landau pointed out in the 1930s that even ordinary stars such as our own sun ought to consist of degenerate gas at the center.

The companion of Procyon (Procyon B), first detected in 1896 by J. M. Schaberle at Lick Observatory in California, was also found to be a super-dense star although only five-eighths as massive as Sirius B; and, as the years passed, more examples were found. These stars are called *white dwarfs,* because they combine small size with high temperature and white light. White dwarfs are probably numerous and may make up as much as 3 percent of all stars. However, because of their small size and dimness, only those in our own neighborhood are likely to be discovered in the foreseeable future. (There are also *red dwarfs,* considerably smaller than our sun, but not as small as white dwarfs. Red dwarfs are cool and of ordinary density. They are the most common of all stars—making up three-fourths of the total—but, because of their dimness, are as difficult to detect as white dwarfs. A pair of red dwarfs, a mere six light-years distant from us, was only discovered in 1948. Of the thirty-six stars known to be within fourteen light-years of the sun, twenty-one are red dwarfs, and three are white dwarfs. There are no giants among them, and only two, Sirius and Procyon, are distinctly brighter than our sun.)

The year after Sirius B was found to have its astonishing properties, Albert Einstein presented his general theory of relativity, which was mainly concerned with new ways of looking at gravity. Einstein's views of gravity led to the prediction that light emitted by a source possessing a very strong gravitational field should be displaced toward the red (the *Einstein shift*). Adams, fascinated by the white dwarfs he had discovered, carried out careful studies of the spectrum of Sirius B and found that there was indeed the red shift predicted by Einstein. This was a point in favor not only of Einstein's theory but also of the superdensity of Sirius B; for in an ordinary star such as our sun, the red-shift effect would be only one-thirtieth as great. Nevertheless, in the early 1960s this very small Einstein shift produced by our sun was detected, and the general theory of relativity was further confirmed.

But what have white dwarfs to do with supernovae, the subject that prompted this discussion? To work toward an answer, let us go back to the supernova of 1054.

In 1844, the Earl of Rosse, investigating the location in Taurus where the Oriental astronomers had reported finding the 1054 supernova, studied a small cloudy object. Because of its irregularity and its clawlike projections, he named the object the Crab Nebula. Continued observation over decades showed that the patch of gas was slowly expanding. The actual rate of expansion could be calculated from the Doppler-Fizeau effect, which, combined with the apparent rate of expansion, made it possible to compute the distance of the Crab Nebula as 3,500 light-years from us. From the expansion rate it was also determined that the gas had started its expansion from a central explosion point nearly 900 years ago, which agrees well with the date 1054. So there can be little doubt that the Crab Nebula, which now spreads over a volume of space some 5 light-years in diameter, represents the remnants of the 1054 supernova.

No similar region of turbulent gas has been observed at the reported sites of the supernovae of Tycho and Kepler, although small spots of nebulosity have been observed close to each site. There are some 150 planetary nebulae, however, in which doughnut-shaped rings of gas may represent large stellar explosions. A particularly extended and thin gas cloud, the Veil Nebula in Cygnus, may be what is left of a supernova explosion 30,000 years ago. It must have been even closer and brighter than the supernova of 1054—but no civilization existed on earth to record the spectacle.

There are even suggestions that a very faint nebulosity enveloping the constellation Orion may be what is left of a still older supernova.

In all these cases, though, what happened to the stars that exploded? Have they simply vanished in one enormous puff of gas? Is the Crab Nebula, for instance, *all* that is left of the 1054 supernova, and will this simply spread out until all visible sign of the star is forever gone? Or is some remnant left that is still a star but too small and too dim to be detected? Is there, in other words, a white dwarf left behind (or something even more extreme), and are white dwarfs, so to speak, the corpses of stars that were once like our sun? These queries lead us into the problem of the evolution of stars.

THE EVOLUTION OF THE STARS

Of the stars near us, the bright ones seem to be hot and the dim ones cooler, according to a fairly regular brightness-temperature scale. If the surface temperatures of various stars are plotted against their absolute magnitudes, most of the familiar stars fall within a narrow band, increasing steadily from dim coolness to bright hotness. This band is called the *main sequence*. It was first plotted in 1913 by the American astronomer Henry Norris Russell, following work along similar lines by Hertzsprung (the astronomer who first determined the absolute magnitudes of the cepheids). A graph showing the main sequence is therefore called a *Hertzsprung–Russell diagram*, or *H–R diagram* (figure 2.5).

Not all stars belong in the main sequence. There are some red stars that, despite their rather low surface temperature, have large absolute magnitudes, because their substance is spread out in rarefied fashion into tremendous size, and the sparse heat per unit area is multiplied over the enormous surface to a huge total. Among these *red giants,* the best-known are Betelgeuse and Antares. They are so cool (it was discovered in 1964) that many have atmospheres rich in water vapor, which would decompose to hydrogen and oxygen at the higher temperatures of our own sun. The high-temperature white dwarfs also fall outside the main sequence.

In 1924, Eddington pointed out that the interior of any star must be very hot. Because of a star's great mass, its gravitational force is immense. If the star is not to collapse, this huge force must be balanced by an equal internal pressure—from heat and from radiation energy. The more massive the star,

Figure 2.5. The Hertzsprung-Russell diagram. The dotted line indicates the evolution of a star. The relative size of the stars are given only schematically, not according to scale.

the higher the central temperature required to balance the gravitational force. To maintain this high temperature and radiation pressure, the more massive stars must be burning energy faster, and they must be brighter, than less massive ones: this is the *mass-luminosity* law. The relationship is a drastic one, for luminosity varies as the sixth or seventh power of the mass. If the mass is increased by 3 times, then the luminosity increases by a factor of six or seven 3's multiplied together—say, 750-fold.

It follows that the massive stars are spendthrift with their hydrogen fuel and have a shorter life. Our sun has enough hydrogen to last it at its present radiation rate for billions of years. A bright star such as Capella must burn out in about 20 million years, and some of the brightest stars—for example, Rigel —cannot possibly last more than 1 or 2 million years. Hence, the very brightest stars must be very youthful. New stars are perhaps even now being formed in regions where space is dusty enough to supply the raw material.

Indeed, in 1955, the American astronomer George Herbig detected two stars in the dust of the Orion Nebula that were not visible in photographs of the region taken some years before. These may be stars that were actually born in our lifetime.

By 1965, hundreds of stars were located that were so cool, they did not quite shine. They were detected by their infrared radiation and are therefore called *infrared giants* because they are made up of large quantities of rarefied matter. Presumably, these are quantities of dust and gas, gathering together and gradually growing hotter. Eventually, they will become hot enough to shine.

The next advance in the study of the evolution of stars came from analysis of the stars in globular clusters. The stars in a cluster are all about the same distance from us, so their apparent magnitude is proportional to their absolute magnitude (as in the case of the cepheids in the Magellanic Clouds). Therefore, with their magnitude known, an H-R diagram of these stars can be prepared. It is found that the cooler stars (burning their hydrogen slowly) are on the main sequence, but the hotter ones tend to depart from it. In accordance with their high rate of burning, and their rapid aging, they follow a definite line showing various stages of evolution, first toward the red giants and then back, across the main sequence again, and down toward the white dwarfs.

From this and from certain theoretical considerations about the manner in which subatomic particles can combine at certain high temperatures and pressures, Fred Hoyle has drawn a detailed picture of the course of a star's evolution. According to Hoyle, in its early stages, a star changes little in size or temperature. (This is the position our sun is in now and will continue to be in for a long time.) As in its extremely hot interior, a star converts its hydrogen into helium, the helium accumulates at its center. When this helium core reaches a certain size, the star starts to change in size and temperature dramatically. It expands enormously and its surface becomes cooler. In other words, it leaves the main sequence and moves in the red-giant direction. The more massive the star, the more quickly it reaches this point. In the globular clusters, the more massive ones have already progressed varying lengths along the road.

Despite its lower temperature, the expanded giant releases more heat because of its larger surface area. In the far distant future, when the sun leaves the main sequence, or even somewhat before, it will have heated to the point where life will be impossible on the earth. That point, however, is still billions of years in the future.

But what precisely is the change in the helium core that brings about expansion to a red giant? Hoyle suggested that the helium core itself contracts and, as a result, rises to a temperature at which the helium nuclei can fuse to form carbon, with the liberation of additional energy. In 1959, the American physicist David Elmer Alburger showed in the laboratory that this reaction actually can take place. It is a very rare and unlikely sort of reaction, but there are so many helium atoms in a red giant that enough such fusions can occur to supply the necessary quantities of energy.

Hoyle goes further. The new carbon core heats up still more, and still more complicated atoms, such as those of oxygen and neon, begin to form. While this is happening, the star is contracting and getting hotter again; it moves back toward the main sequence. By now the star has begun to acquire a series of layers, like an onion. It has an oxygen-neon core, then a layer of carbon, then one of helium, and the whole is enveloped in a skin of still-unconverted hydrogen.

However, in comparison with its long life as a hydrogen consumer, the star is on a quick toboggan slide through the remaining fuels. Its life cannot continue for long, since the energy produced by helium fusion and beyond is about one-twentieth that produced by hydrogen fusion. In a comparatively short time, the energy required to keep the star expanded against the inexorable pull of its own gravitational field begins to fall short, and the star contracts ever more swiftly. It contracts not only back to what would have been the size of a normal star, but beyond—to a white dwarf.

During the contraction, the outermost layers of the star may be left behind or even blown off because of the heat developed by the contraction. The white dwarf is thus surrounded by an expanding shell of gas, which shows up in our telescopes at the edges where the quantity of gas in the line of sight is thickest and therefore greatest. Such white dwarfs seem to be surrounded by a small "smoke ring" or "doughnut" of gas. These are called *planetary nebulae* because the smoke surrounds the star like a planetary orbit made visible. Eventually, the ring of smoke expands and thins into invisibility, and we have white dwarfs such as Sirius B with no sign of any surrounding nebulosity.

White dwarfs form, in this way, rather quietly; and such a comparatively quiet "death" lies in the future for stars like our sun and smaller ones. What's more, white dwarfs, if undisturbed, have, in prospect, an indefinitely prolonged life—a kind of long rigor mortis—in which they slowly cool until, eventually, they are no longer hot enough to glow (many billions of years in the future) and then continue for further billions and billions of years as *black dwarfs*.

On the other hand, if a white dwarf is part of a binary system, as Sirius B and Procyon B are, and if the other star is main-sequence and very close to the white dwarf, there can be exciting moments. As the main-sequence star expands in its own evolutionary development, some of its matter may drift outward under the pull of the white dwarf's intense gravitational field and move into orbit about the latter. Occasionally, some of the orbiting material will spiral to the white dwarf's surface, where the gravitational pull will compress it and cause it to undergo fusion so that it will emit a burst of energy. If a particularly large gout of matter drops to the white dwarf's surface, the energy emission may be large enough to see from Earth, and astronomers record the existence of a nova. Naturally, this sort of thing can happen more than once, and *recurrent novas* do exist.

But these are still not supernovas. Where do these come in? To answer that, we have to turn to stars that are distinctly more massive than our sun. These are relatively rare (in all classes of astronomical objects, large members are rarer than small ones) so that perhaps only one star in thirty is considerably more massive than our sun. Even so there may be 7 billion such massive stars in our galaxy.

In massive stars, the core is more compressed under a gravitational field pull that is greater than those in smaller stars. The core is therefore hotter, and fusion reactions can continue past the oxygen-neon stage of smaller stars. The neon can combine further to magnesium, which can combine in turn to form silicon, and then, in turn, iron. At a late stage in its life, the star may be built up of more than half a dozen concentric shells, in each of which a different fuel is being consumed. The central temperature may have reached 3 billion to 4 billion degrees by then. Once the star begins to form iron, it has reached a dead end, for iron atoms represent the point of maximum stability and minimum energy content. To alter iron atoms in the direction of more complex or less complex atoms requires, either way, an input of energy.

Furthermore, as central temperatures rise with age, radiation pressure rises, too, and in proportion to the fourth power of the temperature. When the temperature doubles, the radiation pressure increases sixteenfold, and the balance between it and gravitation becomes ever more delicate. Eventually, the central temperatures may rise so high, according to Hoyle's suggestion, that the iron atoms are driven apart into helium. But for this to happen, as I have just said, energy must be poured into the atoms. The only place the star can get this energy from is its gravitational field. When the star shrinks, the energy it gains can be used to convert iron to helium. The amount of energy needed is so great, however, that the star must shrink drastically to a tiny fraction of its former volume, and must do so according to Hoyle, in about a second.

When such a star starts to collapse, its iron core is still surrounded with a voluminous outer mantle of atoms not yet built up to a maximum stability. As the outer regions collapse and their temperature rises, these still combinable substances "take fire" all at once. The result is an explosion that blasts the outer material away from the body of the star. This explosion is a supernova. It was such an explosion that created the Crab Nebula.

The matter blasted into space as a result of a supernova explosion is of enormous importance to the evolution of the universe. At the time of the big bang, only hydrogen and helium atoms were formed. In the core of stars, other atoms, more complex ones, are formed—all the way up to iron. Without supernova explosions, these complex atoms would remain in the cores and, eventually, in white dwarfs. Only trivial amounts would make their way into the universe generally through the halos of planetary nebulas.

In the course of the supernova explosion, material from the inner layers of stars would be ejected forcefully into surrounding space. The vast energy of the explosion would even go into the formation of atoms more complex than those of iron.

The matter blasted into space would be added to the clouds of dust and gas already existing and would serve as raw material for the formation of new, *second-generation* stars, rich in iron and other metallic elements. Our own sun is probably a second-generation star, much younger than the old stars of some of the dust-free globular clusters. Those *first-generation* stars are low in metals and rich in hydrogen. The earth, formed out of the same debris of which the sun was born, is extraordinarily rich in iron—iron that once may have existed at the center of a star that exploded many billions of years ago.

But what happens to the contracting portion of the stars that explode in supernova explosions? Do they form white dwarfs? Do larger, more massive stars simply form larger, more massive white dwarfs?

The first indication that they cannot do so, and that we cannot expect larger and larger white dwarfs, came in 1939 when the Indian astronomer Subrahmanyan Chandrasekhar, working at Yerkes Observatory near Williams Bay, Wisconsin, calculated that no star more than 1.4 times the mass of our sun (now called *Chandrasekhar's limit*) could become a white dwarf by the "normal" process Hoyle described. And, in fact, all the white dwarfs so far observed turn out to be below Chandrasekhar's limit in mass.

The reason for the existence of Chandrasekhar's limit is that white dwarfs are kept from shrinking farther by the mutual repulsion of the electrons (subatomic particles I will discuss later, in chapter 7) contained in its atoms. With increasing mass, gravitational intensity increases; and at 1.4 times the mass of the sun, electron repulsion no longer suffices, and the white dwarf collapses to form a star even tinier and denser, with subatomic particles in virtual contact. The detection of such further extremes had to await new methods of probing the universe, taking advantage of radiations other than those of visible light.

The Windows to the Universe

The greatest weapons in the conquest of knowledge are an understanding mind and the inexorable curiosity that drives it on. And resourceful minds have continually invented new instruments which have opened up horizons beyond the reach of our unaided sense organs.

THE TELESCOPE

The best-known example is the vast surge of new knowledge that followed the invention of the telescope in 1609. The telescope, essentially, is simply an oversized eye. In contrast to the quarter-inch pupil of the human eye, the 200-inch telescope on Palomar Mountain has more than 31,000 square inches of light-gathering area. Its light-collecting power intensifies the brightness of a star about a million times, compared with what the naked eye can see. This telescope, first put into use in 1948, is the largest in use today in the United States; but in 1976, the Soviet Union began observations with a 236.2-inch telescope (that is, one with a mirror that is 600 centimeters in diameter) located in the Caucasus mountains.

This is about as large as telescopes of this kind are likely to get; and, to tell the truth, the Soviet telescope does not work well. There are other ways, however, of improving telescopes than by simply making them larger. During the 1950s Merle A. Ture developed an image tube which electronically magnifies the faint light gathered by a telescope, tripling its power. Clusters of comparatively small telescopes, working in unison, can produce images that are equivalent to those produced by a single telescope larger than any of the components; and plans are in progress both in the United States and the Soviet Union to build clusters that will far outstrip the 200-inch and 236.2-inch telescopes. Then, too, a large telescope put into orbit about the earth would be able to scan the skies without atmospheric interference and to see more clearly than any telescope likely to be built on Earth. That, too, is in the planning stage.

But mere magnification and light-intensification are not the full measure of the telescope's gift to human beings. The first step toward making it something more than a mere light collector came in 1666 when Isaac Newton discovered that light could be separated into what he called a *spectrum* of colors. He passed a beam of sunlight through a triangularly shaped prism of glass and found that the beam spread out into a band made up of red, orange, yellow, green, blue, and violet light, each color fading gently into the next (figure 2.6). (The phenomenon itself, of course, has always been familiar in the form of the rainbow, the result of sunlight passing through water droplets, which act like tiny prisms.)

Figure 2.6. Newton's experiment splitting the spectrum of white light.

WHITE LIGHT

PRISM

RED
ORANGE
YELLOW
GREEN
BLUE
VIOLET

What Newton showed was that sunlight, or *white light,* is a mixture of many specific radiations (now recognized as wave forms of varying wavelengths) which impress the eye as so many different colors. A prism separates the colors because, on passing from air into glass, and from glass into air, light is bent, or *refracted,* and each wavelength undergoes a different amount of refraction —the shorter the wavelength, the greater the refraction. The short wavelengths of violet light are refracted most; the long wavelengths of red, least.

This phenomenon explains, among other things, an important flaw in the very earliest telescopes, which was that objects viewed through them were surrounded by obscuring rings of color, which were spectra caused by the dispersion of light as it passed through the lenses.

Newton despaired of correcting this effect as long as lenses or any sort were used. He therefore designed and built a *reflecting telescope* in which a parabolic mirror, rather than a lens, was used to magnify an image. Light of all wavelengths was reflected alike, so that no spectra were formed on reflection, and rings of color (*chromatic aberration*) were not to be found.

In 1757, the English optician John Dollond prepared lenses of two different kinds of glass, one kind canceling out the spectrum-forming tendency of the other. In this way, *achromatic* ("no color") lenses could be built. Using such lenses, *refracting telescopes* became popular again. The largest such telescope, with a 40-inch lens, is at Yerkes Observatory and was built in 1897. No larger refracting telescopes have been built since or are likely to be built, for still larger lenses would absorb so much light as to cancel their superior magnifying powers. The giant telescopes of today are, in consequence, all of the reflecting variety, since the reflecting surface of a mirror absorbs little light.

THE SPECTROSCOPE

In 1814, a German optician, Joseph von Fraunhofer, went beyond Newton. He passed a beam of sunlight through a narrow slit before allowing it to be refracted by a prism. The spectrum that resulted was actually a series of images of the slit in light of every possible wavelength. There were so many slit images that they melted together to form the spectrum. Fraunhofer's prisms were so excellently made and produced such sharp slit images that it was possible to see that some of the slit images were missing. If particular wavelengths of light were missing in sunlight, no slit image would be formed at that wavelength, and the sun's spectrum would be crossed by dark lines.

Fraunhofer mapped the location of the dark lines he detected, and recorded over 700. They have been known as *Fraunhofer lines* ever since. In 1842, the lines of the solar spectrum were first photographed by the French physicist Alexandre Edmond Becquerel. Such photography greatly facilitated spectral studies; and, with the use of modern instruments, more than 30,000 dark lines have been detected in the solar spectrum, and their wavelengths measured.

In the 1850s, a number of scientists toyed with the notion that the lines were characteristic of the various elements present in the sun. The dark lines would represent absorption of light, at the wavelengths in question, by certain elements; bright lines would represent characteristic emissions of light by elements. About 1859, the German chemists Robert Wilhelm Bunsen and Gustav Robert Kirchhoff worked out a system for identifying elements in this way. They heated various substances to incandescence, spread out their glow into spectra, measured the location of the lines (in this case, bright lines of emission, against a dark background) on a background scale, and matched up each line with a particular element. Their *spectroscope* was quickly applied to discovering new elements by means of new spectral lines not identifiable with known elements. Within a couple of years, Bunsen and Kirchhoff discovered cesium and rubidium in this manner.

The spectroscope was also applied to the light of the sun and the stars and soon turned up an amazing quantity of new information chemical and otherwise. In 1862, the Swedish astronomer Anders Jonas Angstrom identified hydrogen in the sun by the presence of spectral lines characteristic of that element.

Hydrogen can also be detected in the stars, although, by and large, the spectra of the stars vary among themselves because of differences in their chemical constitution (and other properties, too). In fact, stars can be classified according to the general nature of their spectral line pattern. Such a classification was first worked out by the Italian astronomer Pietro Angelo Secchi in 1867, on the basis of 4,000 spectra. By the 1890s, the American astronomer Edward Charles Pickering was studying stellar spectra by the tens of thousands, and the spectral classification could be made finer with the painstaking assistance of Annie J. Cannon and Antonia C. Maury.

Originally, the classification was by capital letters in alphabetical order, but as more was learned about the stars, it became necessary to alter that order to put the spectral classes into a logical arrangement. If the letters are arranged in order of stars of decreasing temperature, we have O, B, A, F, G, K, M, R, N, and S. Each classification can be further subdivided by numbers from 1 to 10. The sun is a star of intermediate temperature with a spectral class of G-0, while Alpha Centauri is G-2. The somewhat hotter Procyon is F-5, while the considerably hotter Sirius is A-0.

Just as the spectroscope could locate new elements on earth, so it could locate them in the heavens. In 1868, the French astronomer Pierre Jules César Janssen was observing a total eclipse of the sun in India and reported sighting a spectral line he could not identify with any produced by any known element. The English astronomer Sir Norman Lockyer, sure that the line represented a new element, named it *helium*, from the Greek word for "sun." Not until nearly thirty years later was helium found on the earth.

The spectroscope eventually became a tool for measuring the radial velocity of stars, as we saw earlier in this chapter, and for exploring many other matters —the magnetic characteristics of a star, its temperature, whether the star is single or double, and so on.

Moreover, the spectral lines were a veritable encyclopedia of information about atomic structure, which, however, could not properly be utilized until after the 1890s, when the subatomic particles within the atom were first discovered. For instance, in 1885, the German physicist Johann Jakob Balmer showed that hydrogen produces a whole series of lines that are regularly spaced according to a rather simple formula. This was used, a generation later, to deduce an important picture of the structure of the hydrogen atom (see chapter 8).

Lockyer himself showed that the spectral lines produced by a given element alter at high temperatures. This indicated some change in the atoms. Again, this was not appreciated until it was later found that an atom consists of smaller particles, some of which are driven off at high temperatures, altering the atomic structure and the nature of the lines the atom produced. (Such altered lines were sometimes mistaken for indications of new elements, but—alas—helium remained the only new element ever discovered in the heavens.)

PHOTOGRAPHY

When, in 1830, the French artist, Louis Jacques Mandé Daguerre produced the first *daguerreotypes* and thus introduced photography, this, too, soon became an invaluable instrument for astronomy. Through the 1840s, various American astronomers photographed the moon; and one picture, by the American astronomer George Phillips Bond, was a sensation at the Great Exhibition of 1851 in London. They also photographed the sun. In 1860, Secchi made the first photograph of a total eclipse of the sun. By 1870, photographs of such eclipses had proved that the corona and prominences are part of the sun and not of the moon.

Meanwhile, beginning in the 1850s, astronomers were also making pictures of the distant stars. By 1887, the Scottish astronomer David Gill was making stellar photography routine. Photography was well on its way to becoming more important than the human eye in observing the universe.

The technique of photography with telescopes steadily improved. A major stumbling block was the fact that a large telescope can cover only a very small field. If an attempt is made to enlarge the field, distortion creeps in at the edges. In 1930, the Russian-German optician Bernard Schmidt designed a method for introducing a correcting lens that would prevent such distortion. With such a lens, a wide swatch of sky can be photographed at one swoop and studied for interesting objects that can then be studied intensely by an ordinary telescope. Since such telescopes are almost invariably used for photographic work, they are called *Schmidt cameras*.

The largest Schmidt cameras now in use are a 53-inch instrument, first put to use in 1960 in Tautenberg, East Germany, and a 48-inch instrument used in conjunction with the 200-inch Hale telescope on Mount Palomar. The third largest is a 39-inch instrument put into use at an observatory in Soviet Armenia in 1961.

About 1800, William Herschel (the astronomer who first guessed the shape of our galaxy) performed a very simple but interesting experiment. In a beam of sunlight transmitted through a prism, he held a thermometer beyond the red end of the spectrum. The mercury climbed! Plainly some form of invisible radiation existed at wavelengths below the visible spectrum. The radiation Herschel had discovered became known as *infrared*—below the red; and, as we now know, fully 60 percent of the sun's radiation is in the infrared.

In 1801, the German physicist Johann Wilhelm Ritter was exploring the other end of the spectrum. He found that silver nitrate, which breaks down to metallic silver and darkens when it is exposed to blue or violet light, would break down even more rapidly if it were placed beyond the point in the spectrum where violet fades out. Thus, Ritter discovered the "light" now called *ultraviolet* ("beyond the violet"). Between them, Herschel and Ritter had widened the time-honored spectrum and crossed into new realms of radiation.

These new realms bear promise of yielding much information. The ultraviolet portion of the solar spectrum, invisible to the eye, shows up in nice detail by way of photography. In fact, if a quartz prism is used (quartz transmits ultraviolet light, whereas ordinary glass absorbs most of it), quite a complicated ultraviolet spectrum can be recorded, as was first demonstrated in 1852 by the British physicist George Gabriel Stokes. Unfortunately, the atmosphere transmits only the *near ultraviolet*—that part with wavelength almost as long as violet light. The *far ultraviolet*, with its particularly short wavelengths, is absorbed in the upper atmosphere.

RADIO ASTRONOMY

In 1860, the Scottish physicist James Clerk Maxwell worked out a theory that predicted a whole family of radiation associated with electric and magnetic phenomena *(electromagnetic radiation)*—a family of which ordinary light was only one small portion. The first definite evidence bearing out his prediction came a quarter of a century later, seven years after Maxwell's premature death through cancer. In 1887, the German physicist Heinrich Rudolf Hertz, generating an oscillating current from the spark of an induction coil, produced and detected radiation of extremely long wavelengths—much longer than those of ordinary infrared. These came to be called *radio waves*.

The wavelengths of visible light can be measured in *micrometers* (millionths of a meter). They range from 0.39 micrometers (extreme violet) to 0.78 micrometers (extreme red). Next come the *near infrared* (0.78 to 3 micrometers), the *middle infrared* (3 to 30 micrometers), and then the *far infrared* (30 to 1,000 micrometers). It is here that radio waves begin: the so-called *microwaves* run from 1,000 to 160,000 micrometers and long-wave radio goes as high as many billions of micrometers.

Radiation can be characterized not only by wavelength but also by *frequency*, the number of waves of radiation produced in each second. This value is so high for visible light and the infrared that it is not commonly used in these cases. For the radio waves, however, frequency reaches down into lower figures and comes into its own. One thousand waves per second is a *kilocycle*, while 1 million waves per second is a *megacycle*. The microwave region runs from 300,000 megacycles down to 1,000 megacycles. The much longer radio waves used in ordinary radio stations are down in the kilocycle range.

Within a decade after Hertz's discovery, the other end of the spectrum opened up similarly. In 1895, the German physicist Wilhelm Konrad Roentgen accidentally discovered a mysterious radiation which he called *X rays*. Their wavelengths turned out to be shorter than ultraviolet. Later, *gamma rays*, associated with radioactivity, were shown by Rutherford to have wavelengths even smaller than those of X rays.

The short-wave half of the spectrum is now divided roughly as follows: the wavelengths from 0.39 down to 0.17 micrometers belong to the near ultraviolet; from 0.17 down to 0.01 micrometers, to the far ultra-violet, from 0.01 to 0.00001 micrometers, to X rays; and gamma rays range from this down to less than one billionth of a micrometer.

Newton's original spectrum was thus expanded enormously. If we consider each doubling of wavelength as equivalent to 1 octave (as is the case in sound), the electromagnetic spectrum over the full range studied amounts to almost 60 octaves. Visible light occupies just 1 octave near the center of the spectrum.

With a wider spectrum, of course, we can get a fuller view of the stars. We know, for instance, that sunshine is rich in ultraviolet and in infrared. Our atmosphere cuts off most of these radiations; but in 1931, quite by accident, a radio window to the universe was discovered.

Karl Jansky, a young radio engineer at the Bell Telephone Laboratories, was studying the static that always accompanies radio reception. He came across a very faint, very steady noise which could not be coming from any of the usual sources. He finally decided that the static was caused by radio waves from outer space.

At first, the radio signals from space seemed strongest in the direction of the sun; but day by day, the direction of strongest reception slowly drifted away from the sun and made a circuit of the sky. By 1933, Jansky decided the radio waves were coming from the Milky Way and, in particular, from the direction of Sagittarius, toward the center of the galaxy.

Thus was born *radio astronomy*. Astronomers did not take to it immediately, for it had serious drawbacks. It gave no neat pictures—only wiggles on a chart which were not easy to interpret. More important, radio waves are much too long to resolve a source as small as a star. The radio signals from space had wavelengths hundreds of thousands, and even millions, of times the wavelength of light, and no ordinary radio receiver could give anything more than a general idea of the direction they were coming from. A *radio telescope* would have to have a receiving "dish" a million times as wide as the mirror of an optical telescope to produce as sharp a picture of the sky. For a radio dish to be the equivalent of the 200-inch telescope it would have to be 3,150 miles across and have twice the area of the United States—manifestly impossible.

These difficulties obscured the importance of the new discovery, but a young radio ham named Grote Reber carried on, for no reason other than personal curiosity. Through 1937 he spent time and money building in his backyard a small radio telescope with a parabolic dish about 30 feet in diameter to receive and concentrate the radio waves. Beginning in 1938, he found a number of sources of radio waves other than the one in Sagittarius—one in the constellation Cygnus, for instance, and another in Cassiopeia. (Such sources of radiation were at first called *radio stars,* whether the sources were actually stars or not, but are now usually called *radio sources.*)

During the Second World War, while British scientists were developing radar, they discovered that the sun was interfering by sending out signals in the microwave region. This aroused their interest in radio astronomy, and after the war the British pursued their tuning-in on the sun. In 1950, they found that many of the sun's radio signals were associated with sunspots. (Jansky had conducted his experiments during a period of minimal sunspot activity, which is why he detected the galactic radiation rather than that of the sun.)

What is more, since radar technology made use of the same wavelengths as radio astronomy did, by the end of the Second World War, astronomers had available a large array of instruments adapted to the manipulation of microwaves which did not exist before the war. These were rapidly improved, and interest in radio astronomy leap-frogged.

The British pioneered in building large antennae to sharpen reception and pinpoint radio stars. Their 250-foot dish at Jodrell Bank in England, built under the supervision of Sir Bernard Lovell, was the first really large radio telescope.

Ways to sharpen reception were found. It was not necessary to build impossibly huge radio telescopes to get high resolution. Instead, one might build a sizable radio telescope in one place and another one a long distance away. If both dishes are timed by superaccurate *atomic clocks* and are made to move in unison by clever computerization, the two together can give results similar to those produced by a single large dish of the combined width, over the distance of separation. Such combinations of dishes are said to be *long baseline* and even *very long baseline* radio telescopes. Australian astronomers, with a large, relatively empty land at their disposal, pioneered this advance; and, by now, cooperating dishes in California and Australia have produced a baseline of 6,600 miles.

Hence, radio telescopes are *not* fuzz producers far behind the sharp-eyed optical telescopes. Radio telescopes can actually make out *more* detail than optical telescopes can. To be sure, such very long baseline radio telescopes have gone about as far as they can on the earth's surface, but astronomers are dreaming of radio telescopes in space cooperating with one another and with dishes on the earth to make still longer baselines.

Nevertheless, long before radio telescopes advanced to present levels, they were making important discoveries. In 1947 the Australian astronomer John C. Bolton narrowed down the third strongest radio source in the sky, which proved to be none other than the Crab Nebula. Of the radio sources detected here and there in the sky, this was the first to be pinned down to an actual visible object. It seemed unlikely that a star was giving rise to such intense radiation, since other stars did not. The source was much more likely to be the cloud of expanding gas in the nebula.

This discovery strengthened other evidence that cosmic radio signals arise primarily from turbulent gas. The turbulent gas of the outer atmosphere of the sun gives rise to radio waves, so that what is called the *radio sun* is much larger than the visible sun. Then, too, Jupiter, Saturn, and Venus, each with a turbulent atmosphere, have been found to be radio emitters.

Jansky, who started it all, was largely unappreciated in his lifetime and died in 1950 at the age of 44, just as radio astronomy was hitting its stride. He received posthumous recognition in that the strength of radio emission is now measured in *janskies*.

LOOKING BEYOND OUR GALAXY

Radio astronomy probed far out into space. Within our galaxy, there is a strong radio source (the strongest outside the solar system), which is called Cass because it is located in Cassiopeia. Walter Baade and Rudolph Minkowski at Palomar trained the 200-inch telescope on the spot where this source was pinpointed by British radio telescopes, and found streaks of turbulent gas. It is possible that these may be remnants of the supernova of 1572, which Tycho observed in Cassiopeia.

A still more distant discovery was made in 1951. The second strongest radio source lies in the constellation Cygnus. Reber first reported it in 1944. As radio telescopes later narrowed down its location, it began to appear that this radio source was outside our galaxy—the first to be pinpointed beyond the Milky Way. Then, in 1951, Baade, studying the indicated portion of the sky with the 200-inch telescope, found an odd galaxy in the center of the field. It had a double center and seemed to be distorted. Baade at once suspected that this odd, distorted, double-centered galaxy was not one galaxy but two, joined broadside-to like a pair of clashing cymbals. Baade thought they were two colliding galaxies—a possibility he had already discussed with other astronomers. The evidence seemed to support the view; and for a while, colliding galaxies were accepted as fact. Since most galaxies exist in rather compact clusters in which they move like bees in a swarm, there seemed nothing unlikely about such collisions.

The radio source in Cygnus was adjusted to be about 260 million light-years away, yet the radio signals were stronger than those of the Crab Nebula in our own stellar neighborhood. This was the first indication that radio telescopes would be able to penetrate greater distances than optical telescopes could. Even the 250-foot Jodrell Bank radio telescope, tiny by present standards, could outrange the 200-inch optical telescope.

And yet as the number of radio sources found among the distant galaxies increased and passed the hundred mark, astronomers grew uneasy. Surely they could not all be brought about by colliding galaxies. That would be overdoing a good thing.

In fact, the whole notion of galactic collisions in the sky grew shaky. The Soviet astrophysicist Victor Amazaspovich Ambartsumian advanced theoretical reasons in 1955 for supposing that radio galaxies were exploding rather than colliding.

This possibility has been greatly strengthened by the discovery, in 1963, that the galaxy M-82, in the constellation of Ursa Major (a strong radio source about 10 million light-years away), is such an *exploding galaxy*.

Investigation of M-82 with the 200-inch Hale telescope, making use of the light of a particular wavelength, showed great jets of matter up to 1,000 light-years long emerging from the galactic center. From the amount of matter exploding outward, the distance it had traveled, and its rate of travel, it seems likely that the explosion took place about 1,500,000 years ago.

It now seems that galactic cores are generally active; that turbulent and very violent events take place there, so that the universe generally is a more exciting place than we dreamed of before the coming of radio astronomy. The apparent utter serenity of the sky as seen by the unaided eye is only the product of our limited vision (which sees only the stars of our own quiet neighborhood) over a limited time.

At the very center of our own galaxy even, there is a tiny region, only a few light-years across at most, that is an intensely active radio source.

And, incidentally, the fact that exploding galaxies exist, and that active galactic cores are common and may be universal, does not necessarily put the notion of galactic collisions out of court. In any cluster of galaxies, it seems likely that large galaxies grow at the expense of small ones; and often one galaxy is considerably larger than any of the others in the cluster. There are signs that it has achieved its size by colliding with and absorbing smaller galaxies. One large galaxy has been photographed that shows signs of several different cores, all but one of which are not its own but were once parts of independent galaxies. The phrase *cannibal galaxy* has thus come into use.

The New Objects

By the 1960s, it might have been easy for astronomers to suppose that there were few surprises left among the physical objects in the heavens. New theories, new insights, yes; but surely little in the way of startling new varieties of stars, galaxies, or anything else could remain after three centuries of observation with steadily more sophisticated instruments.

Any astronomers of this opinion were due for enormous shocks—the first coming as a result of the investigation of certain radio sources that looked unusual but not surprising.

QUASARS

The radio sources first studied in deep space seemed to exist in connection with extended bodies of turbulent gas: the Crab Nebula, distant galaxies, and so on. A few radio sources, however, seemed unusually small. As radio telescopes grew more refined, and as the view of the radio sources was sharpened, it began to seem possible that radio waves were being emitted by individual stars.

Among these compact radio sources were several known as 3C48, 3C147, 3C196, 3C273, and 3C286. The *3C* is short for "Third Cambridge Catalog of Radio Stars," a listing compiled by the English astronomer Martin Ryle and his co-workers; while the remaining numbers denote the placing of the source on that list.

In 1960, the areas containing these compact radio sources were combed by the American astronomer, Allen Sandage with the 200-inch telescope; and in each case, a star did indeed seem to be the source. The first star to be detected was that associated with 3C48. In the case of 3C273, the brightest of the objects, the precise position was obtained by Cyril Hazard, in Australia, who recorded the moment of radio blackout as the moon passed before it.

The stars involved had been recorded on previous photographic sweeps of the sky and had always been taken to be nothing more than faint members of our own galaxy. Painstaking photographing, spurred by their unusual radio emission, now showed, however, that that was not all there was to it. Faint nebulosities proved to be associated with some of the objects, and 3C273 showed signs of a tiny jet of matter emerging from it. In fact, there were two radio sources in connection with 3C273: one from the star and one from the jet. Another point of interest that arose after close inspection was that these stars were unusually rich in ultraviolet light.

It would seem, then, that the compact radio sources, although they looked like stars, might not be ordinary stars after all. They eventually came to be called *quasi-stellar radio sources* (*quasi-stellar* means "star-resembling"). As the term became more important to astronomers, it became too inconvenient a mouthful and, in 1964, was shortened by the Chinese-American physicist Hong Yee Chiu to *quasar,* an uneuphonious word that is now firmly embedded in astronomic terminology.

Clearly, the quasars were interesting enough to warrant investigation with the full battery of astronomic techniques, including spectroscopy. Such astronomers as Allen Sandage, Jesse L. Greenstein, and Maarten Schmidt labored to obtain the spectra. When they accomplished the task in 1960, they found themselves with strange lines they could not identify. Furthermore, the lines in the spectra of one quasar did not match those in any other.

In 1963, Schmidt returned to the spectrum of 3C273, which, as the brightest of these puzzling objects, showed the clearest spectrum. Six lines were present, of which four were spaced in such a way as to seem to resemble a series of hydrogen lines—except that no such series ought to exist in the place where they were found. What, though, if those lines were located elsewhere but were found where they were because they had been displaced toward the red end of the spectrum? If so, it was a large displacement, one that indicated a recession at the velocity of over 25,000 miles per second. This seemed unbelievable; and yet, if such a displacement existed, the other two lines could also be identified: one represented oxygen minus two electrons; the other magnesium minus two electrons.

Schmidt and Greenstein turned to the other quasar spectra and found that the lines there could also be identified, provided huge red shifts were assumed.

Such enormous red shifts could be brought about by the general expansion of the universe; but if the red shift was equated with distance in accordance with Hubble's law, it turned out that the quasars could not be ordinary stars of our own galaxy at all. They had to be among the most distant objects known —billions of light-years away.

By the end of the 1960s, a concentrated search had uncovered 150 quasars. The spectra of about 110 of them were studied. Every single one of these showed a large red shift—larger indeed, than that of 3C273. The distance of a couple of them is estimated to be about 9 billion light-years.

If the quasars are indeed as far away as the red shift makes them seem, astronomers are faced with some puzzling and difficult points. For one thing, these quasars must be extraordinarily luminous to appear as bright as they do at such a distance; they must be anywhere from 30 to 100 times as luminous as an entire ordinary galaxy.

Yet if this is so, and if the quasars have the form and appearance of a galaxy, they ought to contain up to 100 times as many stars as an ordinary galaxy and be up to 5 or 6 times as large in each dimension. Even at their enormous distances they ought to show up as distinct oval blotches of light in large telescopes. Yet they do not. They remain starlike points in even the largest telescope and thus, despite their unusual luminosity, may be far smaller in size than ordinary galaxies.

The smallness in size was accentuated by another phenomenon; for as early as 1963, the quasars were found to be variable in the energy they emitted, both in the visible-light region and in the radio-wave region. Increases and decreases of as much as three magnitudes were recorded over the space of a few years.

For radiation to vary so markedly in so short a time, a body must be small. Small variations might result from brightenings and dimmings in restricted regions of a body, but large variations must involve the body as a whole. If the body is involved as a whole, some effect must make itself felt across the full width of the body within the time of variation. But no effect can travel faster than light; so that if a quasar varies markedly over a period of a few years, it cannot be more than a light-year or so in diameter. Actually, some calculations indicate quasars may be as little as a light-week (500 billion miles) in diameter.

Bodies that are at once so small and so luminous must be expending energy at a rate so great that the reserves cannot last long (unless there is some energy source as yet undreamed of, which is not impossible). Some calculations indicate that a quasar can only deliver energy at this enormous rate for a million years or so. In that case, the quasars we see only became quasars a short time ago, cosmically speaking; and there must be objects that were once quasars but are quasars no longer.

Sandage, in 1965, announced the discovery of objects that may indeed be aged quasars. They seemed like ordinary bluish stars but possessed huge red shifts as quasars do. They were as distant, as luminous, as small as quasars; but they lacked the radio-wave emission. Sandage called them *blue stellar objects*, which can be abbreviated *BSOs*.

The BSOs seem to be more numerous than quasars: a 1967 estimate places the total number of BSOs within reach of our telescopes at 100,000. There are many more BSOs than quasars because the bodies last much longer in BSO form than in quasar form.

The belief that quasars are far-distant objects is not universal among astronomers. There is the possibility that the enormous red shifts of quasars are not *cosmological:* that is, that they are not a consequence of the general expansion of the universe; that they are perhaps relatively near objects that are hastening away from us for some local reason—having been ejected from a galactic core at tremendous velocities, for instance.

The most ardent proponent of this viewpoint is the American astronomer Halton C. Arp, who has presented cases of quasars that seem to be physically connected with galaxies nearby in the sky. Since the galaxies have a relatively low red shift, the greater red shift of the quasars (which, if connected, must be at the same distance) cannot be cosmological.

Another puzzle has been the discovery in the late 1970s that radio sources inside quasars (which can be separately detected by present-day long baseline radio telescopes) seem to be separating at speeds that are several times the speed of light. To exceed the speed of light is considered impossible in present-day physical theory, but such a *superluminal velocity* would exist only if the quasars are indeed as far away as they are thought to be. If they are actually closer, then the apparent rate of separation would translate into speeds less than that of light.

Nevertheless, the view that quasars are relatively near (which would also mean they were less luminous and produced less energy, thus relieving *that* puzzle) has not won over most astronomers. The general view is that the evidence in favor of cosmological distances is overwhelming, that Arp's evidence of physical connections is insufficiently strong, and that the apparent superluminal velocities are the result of an optical illusion (and several plausible explanations have already been advanced).

But, then, if quasars are indeed as distant as their red shifts make them appear, if they are indeed as small and yet as luminous and energetic as such distances would make necessary, what are they?

The most likely answer dates back to 1943, when the American astronomer Carl Seyfert observed an odd galaxy, with a very bright and very small nucleus. Other galaxies of the sort have since been observed, and the entire group is now referred to as *Seyfert galaxies*. Though only a dozen were known by the end of the 1960s, there is reason to suspect that as many as 1 percent of all galaxies may be of the Seyfert type.

Can it be that Seyfert galaxies are objects intermediate between ordinary galaxies and quasars? Their bright centers show light variations that would make those centers almost as small as quasars. If the centers were further intensified and the rest of the galaxy further dimmed, they would become indistinguishable from a quasar; and one Seyfert galaxy, 3C120, is almost quasarlike in appearance.

The Seyfert galaxies have only moderate red shifts and are not enormously distant. Can it be that the quasars are very distant Seyfert galaxies—so distant that we can see only the luminous and small centers; and so distant that we can only see the largest galaxies which thus give us the impression that quasars are extraordinarily luminous, whereas we should rightly suspect that they are very large Seyfert galaxies that we can see despite their distance?

Indeed, recent photographs have shown signs of haze about quasars, seeming to indicate the dim galaxy that surrounds the small, active, and very luminous center. Presumably, then, the far reaches of the universe beyond a billion light-years are as filled with galaxies as are the nearer regions. Most of those galaxies, however, are far too dim to make out optically, and we see only the bright centers of the most active and largest individuals among them.

NEUTRON STARS

If radio-wave radiation had given rise to that peculiar and puzzling astronomical body, the quasar, research at the other end of the spectrum suggested another body just as peculiar.

In 1958, the American astrophysicist Herbert Friedman discovered that the sun produces a considerable quantity of X rays. These could not be detected from the earth's surface, for the atmosphere absorbs them; but rockets, shooting beyond the atmosphere and carrying appropriate instruments, could detect the radiation with ease.

For a while, the source of solar X rays was a puzzle. The temperature of the sun's surface is only 6,000° C—high enough to vaporize any form of matter but not high enough to produce X rays. The source had to lie in the sun's corona, a tenuous halo of gases stretching outward from the sun in all directions for many millions of miles. Although the corona delivers fully half as much light as the full moon, it is completely masked by the light of the sun itself and is visible only during eclipses, at least under ordinary circumstances. In 1930, the French astronomer Bernard Ferdinand Lyot invented a telescope that, at high altitudes and on clear days, could observe the inner corona even in the absence of an eclipse.

The corona was felt to be the X-ray source because, even before the rocket studies of X rays, it had been suspected of possessing unusually high temperatures. Studies of the spectrum of the corona (during eclipses) had revealed lines that could not be associated with any known element. A new element was suspected and named *coronium*. In 1941, however, it was found that the lines of coronium can be produced by iron atoms that have had many subatomic particles broken away from them. To break off all those particles, however, requires a temperature of something like a million degrees, and such a temperature would certainly be enough to produce X rays.

X-ray radiation increases sharply when a solar flare erupts into the corona. The X-ray intensity at that time implies a temperature as high as 100 million degrees in the corona above the flare. The reason for such enormous temperatures in the thin gas of the corona is still a matter of controversy. (Temperature here has to be distinguished from heat. The temperature is a measure of the kinetic energy of the atoms or particles in the gas; but since the particles are few, the actual heat content per unit of volume is low. The X rays are produced by collisions between the extremely energetic particles.)

X rays come from beyond the solar system, too. In 1963, rocket-borne instruments were launched by Bruno Rossi and other astronomers to see whether solar X rays were reflected from the moon's surface. They detected, instead, two particularly concentrated X-ray sources elsewhere in the sky. The weaker (Tau X-1, because it is in the constellation Taurus) was quickly associated with the Crab Nebula. In 1966, the stronger, in the constellation Scorpio (Sco X-1) was found to be associated with an optical object which seemed the remnant (like the Crab Nebula) of an old nova. Since then, many other X-ray sources have been detected in the sky.

To be giving off energetic X rays with an intensity sufficient to be detected across an interstellar gap required a source of extremely high temperature and large mass. The concentration of X rays emitted by the sun's corona would not do at all.

To be at once massive and have a temperature of a million degrees suggested something even more condensed and extreme than a white dwarf. As long ago as 1934, Zwicky had suggested that the subatomic particles of a white dwarf might, under certain conditions, combine into uncharged particles called *neutrons*. These could then be forced together until actual contact was made. The result would be a sphere no more than 10 miles across which would yet retain the mass of a full-sized star. In 1939, the properties of such a *neutron star* were worked out in some detail by the American physicist J. Robert Oppenheimer. Such an object would attain so high a surface temperature, at least in the initial stages after its formation, as to emit X rays in profusion.

The search by Friedman for actual evidence of the existence of such neutron stars centered on the Crab Nebula, where it was felt that the terrific explosion that had formed it might have left behind, not a condensed white dwarf, but a supercondensed neutron star. In July 1964, the moon passed across the Crab Nebula, and a rocket was sent beyond the atmosphere to record the X-ray emission. If it were coming from a neutron star, then the X-ray emission would be cut off entirely and at once as the moon passed before the tiny object. If the X-ray emission were from the Crab Nebula generally, then it would drop off gradually as the moon eclipsed the nebula bit by bit.

The latter proved to be the case, and the Crab Nebula seemed to be but a larger and much more intense corona.

For a moment, the possibility that neutron stars might actually exist and be detectable dwindled; but in the same year that the Crab Nebula failed its test, a new discovery was made in another direction. The radio waves from certain sources seemed to indicate a very rapid fluctuation in intensity. It was as though there were "radio twinkles" here and there.

Astronomers quickly designed instruments that were capable of catching very short bursts of radio-wave radiation, and that they felt would make it possible to study these fast changes in greater detail. One astronomer making use of such a radio telescope was Anthony Hewish at Cambridge University Observatory. He supervised the construction of 2,048 separate receiving devices spread out in an array that covered an area of nearly 3 acres; and in July 1967, the array was put to work. Within a month, a young British graduate student, Jocelyn Bell, who was at the controls, detected bursts of radio-wave energy from a place midway between Vega and Altair. It was not difficult to detect and would have been found years earlier if astronomers had expected to find such short bursts and had developed the equipment to detect them. The bursts were, as it happened, astonishingly brief, lasting only one-thirtieth of a second. Even more astonishing, the bursts followed one another with remarkable regularity at intervals of 1.33 seconds. The intervals were so regular, in fact, that the period could be worked out to one hundred-millionth of a second: it was 1.33730109 seconds.

Naturally, there was no way of telling, at least at first, what these pulses represented. Hewish could only think of it as a *pulsating star*, each pulsation giving out a burst of energy. This name was shortened almost at once to *pulsar*, and by it the new object came to be known.

One should speak of the new objects in the plural, for once Hewish found the first, he searched for others. By February 1968, when he announced the discovery, he had located four and eventually, as a result, received a share of the 1974 Nobel Prize in physics. Other astronomers avidly began searching, and 400 pulsars are now known. It is possible there may be as many as 100,000 in our galaxy altogether. Some may be as close as 100 light-years or so. (There is no reason to suppose they do not exist in other galaxies; but at that distance they are probably too faint to detect.)

All the pulsars are characterized by extreme regularity of pulsation, but the exact period varies from pulsar to pulsar. One had a period as long as 3.7 seconds. In November 1968, astronomers at Green Bank, West Virginia, detected a pulsar in the Crab Nebula that had a period of only 0.033089 seconds. It was pulsing 30 times a second.

Naturally, the question was, What can produce such short flashes with such fantastic regularity? Some astronomical body must be undergoing some very regular change at intervals rapid enough to produce the pulses. Could it be a planet that circles a star in such a way that once each revolution it moves beyond the star (as seen from the direction of earth) and, as it emerges, emits a powerful flash of radio waves? Or else could a planet be rotating and, each time it does so, would some particular spot on its surface, which leaks radio waves in vast quantity, sweep past our direction?

To do this, however, a planet must revolve about a star or rotate about its axis in a period of seconds or in fractions of a second, and this was unthinkable. For pulses as rapid as those of pulsars, some object must be rotating or revolving at enormous velocities, which would require very small size combined with huge temperatures, or huge gravitational fields, or both.

This instantly brought white dwarfs to mind, but even white dwarfs cannot revolve about each other, or rotate on their axes, or pulsate, with a period short enough to account for pulsars. White dwarfs are still too large, and their gravitational fields too weak.

Thomas Gold at once suggested that a neutron star was involved. He pointed out that a neutron star is small enough and dense enough to be able to rotate about its axis in 4 seconds or less. What's more, it had already been theorized that a neutron star would have an enormously intense magnetic field, with magnetic poles that need not be at the pole of rotation. Electrons would be held so tightly by the neutron star's gravity that they could emerge only at the magnetic poles. As they were thrown off, they would lose energy, in the form of radio waves. Hence, there would be a steady sheaf of radio waves emerging from two opposite points on the neutron star's surface.

If, as the neutron star rotates, one or both of those sheafs of radio waves sweeps past our direction, then we will detect a short burst of radio-wave energy once or twice each revolution. If this is so, we would detect only pulsars that happen to rotate in such a way as to sweep at least one of the magnetic poles in our direction. Some astronomers estimate that only 1 neutron star out of 100 would do so. If there are indeed as many as 100,000 neutron stars in the galaxy, then only 1000 might be detectable from earth.

Gold went on to point out that if his theory were correct, the neutron star would be leaking energy at the magnetic poles and its rate of rotation would be slowing down. Thus, the shorter the period of a pulsar, the younger it is and the more rapidly it would be losing energy and slowing down.

The most rapid pulsar at that time known was in the Crab Nebula. It might well be the youngest, since the supernova explosion that would have left the neutron star behind took place less than 1,000 years ago.

The period of the Crab Nebula pulsar was studied carefully, and it was indeed found to be slowing, just as Gold had predicted. The period was increasing by 36.48 billionths of a second each day. The same phenomenon was discovered in other pulsars as well; and as the 1970s opened, the neutron-star hypothesis was widely accepted.

Sometimes a pulsar will suddenly speed up its period very slightly, then resume the slowing trend. Some astronomers suspect this may be the result of a *starquake*, a shifting of mass distribution within the neutron star. Or it might be the result of some sizable body plunging into the neutron star and adding its own momentum to the star's.

There was no reason the electrons emerging from the neutron star should lose energy only as microwaves. This phenomenon should produce waves all along the spectrum. It should produce visible light, too.

Keen attention was focused on the sections of the Crab Nebula where visible remnants of the old explosion might exist. Sure enough, in January 1969, it was noted that the light of a dim star within the Nebula did flash on and off in precise time with the microwave pulses. It would have been detected earlier if astronomers had had the slightest idea that they ought to search for such rapid alternations of light and darkness. The Crab Nebula pulsar was the first optical pulsar discovered—the first visible neutron star.

The Crab Nebula pulsar released X rays, too. About 5 percent of all the X rays from the Crab Nebula emerged from that tiny flickering light. The connection between X rays and neutron stars, which seemed extinguished in 1964, thus came triumphantly back to life.

It might have seemed that no further surprises were to be expected from neutron stars; but in 1982, astronomers at the 300-meter Arecibo radio telescope in Puerto Rico located a pulsar that was pulsing at 642 times a second, twenty times faster than the Crab Nebula pulsar. It is probably smaller than most pulsars—not more than 3 miles in diameter; and with a mass of perhaps two or three times that of our sun, its gravitational field must be enormously intense. Even so, so rapid a rotation must come close to tearing it apart. Another puzzle is that its rate of rotation is not slowing nearly as fast as it ought considering the vast energies being expended.

A second such *fast pulsar* has been detected, and astronomers are busily speculating about the reasons for its existence.

BLACK HOLES

Nor is even the neutron star the limit. When Oppenheimer worked out the properties of the neutron star in 1939, he predicted also that it was possible for a star that was massive enough (more than 3.2 times the mass of our sun) to collapse altogether to a point or *singularity*. When such collapse proceeded past the neutron-star stage, the gravitational field would become so intense that no matter and, in fact, not even light could escape from it. Since anything caught in its unimaginably intense gravitational field would fall into it without hope of return, it could be pictured as an infinitely deep "hole" in space. Since not even light could escape, it was a *black hole*—a term first used by the American physicist John Archibald Wheeler in the 1960s.

Only about one star in a thousand is massive enough to have any chance of forming a black hole on collapse; and, of such stars, most may lose enough mass in the course of a supernova explosion to avoid that fate. Even so, there may be tens of millions of such stars in existence right now; and in the course of the galaxy's existence, there may well have been billions. Even if only one out of a thousand of these massive stars actually form a black hole on collapse, there should still be a million of them here and there in the galaxy. If so, where are they?

The trouble is that black holes are enormously difficult to detect. They cannot be seen in the ordinary way since they cannot give off light or any form of radiation. And although their gravitational field is vast in their immediate vicinity, at stellar distances the intensity of the field is no greater than for ordinary stars.

In some cases, however, a black hole can exist under specialized conditions that make detection possible. Suppose a black hole is part of a binary-star system; that it and a companion revolve about a mutual center of gravity, and that the companion is a normal star.

If the two are close enough to each other, matter from the normal star may little by little drift toward the black hole and take up an orbit about it. Such matter in orbit about a black hole is called an *accretion disk*. Little by little the matter in the accretion disk would spiral into the black hole and, in so doing, would (by a well-known process) give off X rays.

It is necessary, then, to search for an X-ray source in the sky where no star is visible, but a source that seems to orbit another nearby star that *is* visible.

In 1965, a particularly intense X-ray source was detected in the constellation Cygnus and was named Cygnus X-1. It is thought to be about 10,000 light-years from us. It was just another X-ray source until an X-ray-detecting satellite was launched from the coast of Kenya in 1970 and, from space, detected 161 new X-ray sources. In 1971, the satellite detected irregular changes in the intensity of X rays from Cygnus X-1. Such irregular changes would be expected of a black hole as matter entered from an accretion disk in spurts.

Cygnus X-1 was at once investigated with great care and was found to exist in the immediate neighborhood of a large, hot, blue star about 30 times as massive as our sun. The astronomer C. T. Bolt, at the University of Toronto, showed that this star and Cygnus X-1 were revolving about each other. From the nature of the orbit, Cygnus X-1 had to be 5 to 8 times as massive as our sun. If Cygnus X-1 were a normal star, it would be seen. Since it was not seen, it had to be a very small object. Since it was too massive to be a white dwarf or even a neutron star, it had to be a black hole. Astronomers are not yet completely certain of this assumption, but most are satisfied with the evidence and believe Cygnus X-1 to be the first black hole to be discovered.

Black holes, it would seem, might most likely be formed in places where stars were most thickly strewn and where huge masses of material might most likely accumulate in one place. Because high intensities of radiation are associated with the central regions of such star accumulations as globular clusters and galactic cores, astronomers are coming more and more to the belief that there are black holes at the centers of such clusters and galaxies.

Indeed, a compact and energetic microwave source has been detected at the center of our own galaxy. Could that represent a black hole? Some astronomers speculate that it does, and that our galactic black hole has the mass of 100 million stars, or 1/1,000 that of the entire galaxy. It would have a diameter 500 times that of the sun (or equal to that of a huge red-giant star) and would be large enough to disrupt stars through tidal effects, or to gulp them down whole before they break up, if the approach were fast enough.

Actually, it now appears that it is possible for matter to escape from a black hole, although not in the ordinary way. The English physicist Stephen Hawking, in 1970, showed that the energy content of a black hole might occasionally produce a pair of subatomic particles, one of which might escape. In effect, this would mean that a black hole would *evaporate*. Star-sized black holes evaporate in this fashion in so slow a manner that inconceivable times would have to elapse (trillions of trillions of times the total lifetime of the universe so far) before they would evaporate totally.

The evaporation rate would increase, however, as the mass became smaller. A *mini-black hole*, no more massive than a planet or an asteroid (and such tiny objects could exist if they are sufficiently dense: that is, squeezed into a small enough volume) would evaporate rapidly enough to give off appreciable amounts of X rays. Furthermore, as it evaporated and grew less massive, the rate of evaporation and the rate of X-ray production would steadily increase. Finally, when the mini-black hole was small enough, it would explode and give off a pulse of gamma rays of characteristic nature.

But what could compress small amounts of matter of the fearfully high densities required for mini-black hole formation? Massive stars can be compressed by their own gravitational fields, but that will not work for a planet-sized object, and the latter would require greater densities for black-hole formation than the former would.

In 1971, Hawking suggested that mini-black holes were formed at the time of the big bang when conditions were far more extreme than they have been at any other time. Some of those mini-black holes may have been of such a size that only now, after 15 billion years of existence, have they evaporated to the point of explosion, and astronomers might detect gamma-ray bursts that would serve as evidence for their existence.

The theory is attractive, but so far no such evidence has been reported.

"EMPTY" SPACE

But if there are objects in the universe that surprise us, there are also surprises in the vast not-so-empty spaces between the stars. The non-emptiness of "empty" space has proven to be a matter of difficulty for astronomers in observations relatively close to home.

In a sense, the galaxy hardest for us to see is our own. For one thing, we are imprisoned within it, while the other galaxies can be viewed as a whole from outside. It is like the difference between trying to view a city from the roof of a low building and seeing it from an airplane. Furthermore, we are far out from the center and, to make matters worse, lie in a spiral arm clogged with dust. In other words, we are on a low roof on the outskirts of the city on a foggy day.

The space between stars, generally speaking, is not a perfect vacuum under the best of conditions. There is a thin gas spread generally through interstellar space within galaxies. Spectral absorption lines due to such *interstellar gas* were first detected in 1904 by the German astronomer Johannes Franz Hartmann. In the outskirts of a galaxy, the concentration of gas and dust becomes much thicker. We can see such dark fogs of dust rimming the nearer galaxies.

We can actually "see" the dust clouds, in a negative way, within our own galaxy as dark areas in the Milky Way. Examples are the dark Horsehead Nebula, outlined starkly against the surrounding brilliance of millions of stars, and the even more dramatically named Coalsack in the Southern Cross, a region of scattered dust particles 30 light-years in diameter and about 400 light-years away from us.

Although the gas and dust clouds hide the spiral arms of the galaxy from direct vision, they do not hide the structure of the arms from the spectroscope. Hydrogen atoms in the clouds are ionized (broken up into electrically charged subatomic particles) by the energetic radiation from the bright Population I stars in the arms. Beginning in 1951, streaks of ionized hydrogen were found by the American astronomer William Wilson Morgan, marking out the lines of the blue giants—that is, the spiral arms. Their spectra were similar to the spectra shown by the spiral arms of the Andromeda galaxy.

The nearest such streak of ionized hydrogen includes the blue giants in the constellation of Orion, and this streak is therefore called the Orion Arm. Our solar system is in that arm. Two other arms were located in the same way. One lies farther out from the galactic center than our own and includes giant stars in the constellation Perseus (the Perseus Arm). The other lies closer to the galactic center and contains bright clouds in the constellation Sagittarius (the Sagittarius Arm). Each arm seems to be about 10,000 light-years long.

Then radio came along as a still more powerful tool. Not only could it pierce through the obscuring clouds, but it made the clouds themselves tell their story —through their own voice. This came about as a result of the work of the Dutch astronomer Hendrik Christoffel Van de Hulst. In 1944, the Netherlands was ground under the heavy boot of the Nazi army, and astronomic observation was nearly impossible. Van de Hulst, confining himself to pen and paper work, studied the characteristics of ordinary hydrogen atoms, of which most of the interstellar gas is composed.

He suggested that, every once in a while, such atoms, on colliding, might change their energy state and, in so doing, emit a weak radiation in the radio part of the spectrum. A particular hydrogen atom might do so only once in 11 million years; but among the vast numbers present in intergalactic space, enough would be radiating each moment to produce a continuously detectable emission.

Van de Hulst calculated that the wavelength of the radiation should be 21 centimeters. Sure enough, with the development of new radio techniques after the war, this "song of hydrogen" was detected in 1951 by Edward Mills Purcell and Harold Irving Ewen at Harvard University.

By tuning in on the 21-centimeter radiation of collections of hydrogen, astronomers were able to trace out the spiral arms and follow them for long distances—in most cases, nearly all the way around the galaxy. More arms were found, and maps of the concentration of hydrogen show half a dozen or more streaks.

What is more, the song of hydrogen told something about its movements. Like all waves, this radiation is subject to the Doppler-Fizeau effect. It allows astronomers to measure the velocity of the moving hydrogen clouds and, thereby, to explore, among other things, the rotation of our galaxy. This new technique confirmed that the galaxy rotates in a period (at our distance from the center) of 200 million years.

In science, each new discovery unlocks doors leading to new mysteries. And the greatest progress comes from the unexpected—the discovery that overthrows previous notions. An interesting example at the moment is a puzzling phenomenon brought to light by radio study of a concentration of hydrogen at the center of our galaxy. The hydrogen seems to be expanding yet is confined to the equatorial plane of the galaxy. The expansion itself.is surprising, because there is no theory to account for it. And if the hydrogen is expanding, why has it not all dissipated away during the long lifetime of the galaxy? Is it a sign perhaps that, some 10 million years ago, as Oort suspects, its center exploded, as that of M-82 did much more recently? Then, too, the plane of hydrogen is not perfectly flat. It bends downward on one end of the galaxy and upward on the other. Why? No good explanation has yet been offered.

Hydrogen is not, or should not, be unique as far as radio waves are concerned. Every different atom, or combination of atoms, is capable of emitting characteristic radio-wave radiation or of absorbing characteristic radio-wave radiation from a general background. Naturally, then, astronomers sought to find the telltale fingerprints of atoms other than the supremely common hydrogen.

Almost all the hydrogen that occurs in nature is of a particularly simple variety called *hydrogen-1*. There is a more complex form, which is *deuterium* or *hydrogen-2*. The radio-wave radiation from various spots in the sky were combed for the wavelengths that theory predicted. In 1966, it was detected, and the indications are that the quantity of hydrogen-2 in the universe is about 5 percent that of hydrogen-1.

Next to the varieties of hydrogen, as common components of the universe, are helium and oxygen. An oxygen atom can combine with a hydrogen atom to form a *hydroxyl group*. This combination would not be stable on earth, for the hydroxyl group is very active and would combine with almost any other atom or molecule it encountered. It would, notably, combine with a second hydrogen atom to form a molecule of water. In interstellar space, however, where the atoms are spread so thin that collisions are few and far between, a hydroxyl group, once formed, would persist undisturbed for long periods of time—as was pointed out in 1953 by the Soviet astronomer I. S. Shklovskii.

Such a hydroxyl group would, calculations showed, emit or absorb four particular wavelengths of radio waves. In October 1963, two of them were detected by a team of radio engineers at Lincoln Laboratory of M.I.T.

Since the hydroxyl group is some 17 times as massive as the hydrogen atom alone, it is more sluggish and moves at only one-fourth the velocity of the hydrogen atom at any given temperature. In general, movement blurs the wavelengths so that the hydroxyl wavelengths are sharper than those of hydrogen. Its shifts are easier to determine, and it is easier to tell whether a gas cloud, containing hydroxyl, is approaching or receding.

Astronomers were pleased, but not entirely astonished, at finding evidence of a two-atom combination in the vast reaches between the stars. Automatically, they began to search for other combinations, but not with a great deal of hope. Atoms are spread so thin in interstellar space that there seemed to be only a remote chance of more than two atoms coming together long enough to form a combination. The chance that atoms less common than oxygen (such as those of carbon and nitrogen, which are next most common of those able to form combinations) would be involved seemed out of the question.

But then, beginning in 1968, came the real surprises. In November of that year, they discovered the telltale radio-wave fingerprints of water molecules (H_2O). Those molecules were made up of 2 hydrogen atoms and 1 oxygen atom—3 atoms altogether. In the same month, even more astonishingly, ammonia molecules (NH_3) were detected. These were composed of 4-atom combinations: 3 atoms of hydrogen and 1 of nitrogen.

In 1969, another 4-atom combination, including a carbon atom, was detected. This was formaldehyde (H_2CO).

In 1970, a number of new discoveries were made, including the presence of a 5-atom molecule, cyanoacetylene, which contained a chain of 3 carbon atoms (HCCCN) and methyl alcohol, with a molecule of 6 atoms (CH_3OH).

In 1971, the 7-atom combination of methylacetylene (CH_3CCH) was detected; and by 1982, a 13-atom combination was detected. This was cyanodecapenta-yne, which consists of a chain of 11 carbon atoms in a row, with a hydrogen atom at one end and a nitrogen atom at the other ($HC_{11}N$).

Astronomers found themselves with a totally new, and unexpected, subdivision of the science before them: *astrochemistry*.

How those atoms come together to form complicated molecules, and how such molecules manage to remain in being despite the flood of hard radiation from the stars, which ordinarily might be expected to smash them apart, astronomers cannot say. Presumably these molecules are formed under conditions that are not quite as empty as we assumed interstellar space to be— perhaps in regions where dust clouds are thickening toward star formation.

If so, still more complicated molecules may be detected, and their presence may revolutionize our views on the development of life on planets, as we shall see in later chapters.

Chapter 3

The Solar System

Birth of the Solar System

However glorious and vast the unimaginable depths of the universe, we cannot remain lost in its glories forever. We must return to the small family of worlds within which we live. We must return to our sun—a single star among the hundreds of billions that make up our galaxy—and to the worlds that circle it, of which Earth is one.

By the time of Newton, it had become possible to speculate intelligently about the creation of Earth and the solar system as a separate problem from the creation of the universe as a whole. The picture of the solar system showed it to be a structure with certain unifying characteristics (figure 3.1).

1. All the major planets circle the sun in approximately the plane of the sun's equator. In other words, if you were to prepare a three-dimensional model of the sun and its planets, you would find it could be made to fit into a very shallow cake pan.

2. All the major planets circle the sun in the same direction—counterclockwise if you were to look down on the solar system from the direction of the North Star.

3. Each major planet (with some exceptions) rotates around its axis in the same counterclockwise sense as its revolution around the sun, and the sun itself also rotates counterclockwise.

Figure 3.1. The solar system, drawn schematically, with an indication of the hierarchy of planets according to relative size.

4. The planets are spaced at smoothly increasing distances from the sun and have nearly circular orbits.

5. All the satellites, with some exceptions, revolve about their respective planets in nearly circular orbits in the plane of the planetary equator and in a counterclockwise direction.

The general regularity of this picture naturally suggested that some single process had created the whole system.

What, then, is the process that produced the solar system? All the theories so far proposed fall into two classes: catastrophic and evolutionary. The catastrophic view is that the sun was created in single blessedness and gained a family, at some comparatively late stage in its history, as the result of some violent event. The evolutionary ideas hold that the whole system, sun and planets alike, came into being in an orderly way at the very start.

In the eighteenth century, when scientists were still under the spell of the Biblical stories of such great events as the Flood, it was fashionable to assume that the history of the earth was full of violent catastrophes. Why not one supercatastrophe to start the whole thing going? One popular theory was the proposal of the French naturalist Georges Louis Leclerc de Buffon, in 1745, that the solar system had been created out of the débris resulting from a collision between the sun and a comet.

Buffon, of course, implied a collision between the sun and another body of comparable mass. He called the other body a *comet* for lack of another name. We now know comets to be tiny bodies surrounded by insubstantial wisps of gas and dust, but Buffon's principle would remain if we called the colliding body by some other name; and in later times, astronomers returned to his notion.

To some, though, it seemed more natural, and less fortuitous, to imagine a long-drawn-out and noncatastrophic process as occasioning the birth of the solar system. This would somehow fit the majestic picture Newton had drawn of natural law governing the motions of the worlds of the universe.

Newton himself had suggested that the solar system might have been formed from a thin cloud of gas and dust that slowly condensed under gravitational attraction. As the particles came together, the gravitational field would become more intense, the condensation would be hastened, and finally the whole mass would collapse into a dense body (the sun), made incandescent by the energy of the contraction.

In essence, this is the basis of the most popular theories of the origin of the solar system today. But a great many thorny problems had to be solved to answer specific questions. How, for instance, could a highly dispersed gas be brought together by the extremely weak force of gravitation? In recent years, astronomers have proposed that the initiating force might be a supernova explosion. Imagine that a vast cloud of dust and gas which has already existed, relatively unchanged, for billions of years, happens to have moved into the neighborhood of a star that has just exploded as a supernova. The shock wave of that explosion, the vast gust of dust and gas that forces its way through the nearly quiescent cloud I have mentioned compresses that cloud, thus intensifying its gravitational field and initiating the condensation that results in the formation of a star.

If this is the way the sun was created, what about the planets? Where did they come from? The first attempts at an answer were put forward by Immanuel Kant in 1755 and independently by the French astronomer and mathematician Pierre Simon de Laplace in 1796. Laplace's picture was the more detailed.

As Laplace described it, the vast, contracting cloud of matter was rotating to start with. As it contracted, the speed of its rotation increased, just as a skater spins faster when he pulls in his arms. (This effect is due to the *conservation of angular momentum:* since angular momentum is equal to the speed of motion times the distance from the center of rotation, when the distance from the center decreases the speed of motion increases in compensation.) And as the rotating cloud speeded up, according to Laplace, it began to throw off a ring of material from its rapidly rotating equator, thus removing some of the angular momentum. As a result, the remaining cloud slowed down; but, as it contracted further, it again reached a speed at which it threw off another ring of matter. So the coalescing sun left behind a series of rings —doughnut-shaped clouds of matter. These rings, Laplace suggested, slowly condensed to form the planets; and along the way, they themselves threw off small rings that formed their satellites.

Because, by this view, the solar system began as a cloud, or nebula, and because Laplace, as an example, pointed to the Andromeda Nebula (not then known to be a vast galaxy, but thought to be a spinning cloud of dust and gas), this suggestion became known as the *nebular hypothesis.*

Laplace's nebular hypothesis seemed to fit the main features of the solar system very well—and even some of its details. For instance, the rings of Saturn might be satellite rings that had failed to coagulate. (Put all together, they would indeed form a satellite of respectable size.) Similarly, the asteroids, circling around the sun in a belt between Mars and Jupiter, might be products of sections of a ring that had not united to form a planet. And when Helmholtz and Kelvin worked up theories attributing the sun's energy to its slow contraction, that, too, seemed to fit right in with Laplace's picture.

The nebular hypothesis held the field through most of the nineteenth century. But apparently fatal flaws began to appear well before its end. In 1859, James Clerk Maxwell, analyzing Saturn's rings mathematically, showed that a ring of gaseous matter thrown off by any body could only condense to a collection of small particles like the rings of Saturn; it would never form a solid body, because gravitational forces would pull the ring apart before such a condensation materialized.

The problem of angular momentum also arose. It turned out that the planets, making up only a little more than 0.1 percent of the mass of the whole solar system, carry 98 percent of its total angular momentum! Jupiter alone possesses 60 percent of all the angular momentum of the solar system. The sun, then, retains only a tiny fraction of the angular momentum of the original cloud. How did almost all of the angular momentum get shoved into the small rings split off the nebula? The problem is all the more puzzling since, in the case of Jupiter and Saturn which have satellite systems that seem like miniature solar systems and have, presumably, been formed in the same way, the central planetary body retains most of the angular momentum.

By 1900, the nebular hypothesis was so dead that the idea of any evolutionary process at all seemed discredited. The stage was set for the revival of a catastrophic theory. In 1905, two American scientists, Thomas Chrowder Chamberlin and Forest Ray Moulton, using a better term than *comet*, explained the planets as the result of a near collision between our sun and another star. The encounter pulled gaseous matter out of both suns, and the clouds of material left in the vicinity of our sun afterward condensed into small *planetesimals*, and these into planets. This is the *planetesimal hypothesis*. As for the problem of angular momentum, the British scientists James Hopwood Jeans and Harold Jeffreys proposed, in 1918, a *tidal hypothesis*, suggesting that the passing sun's gravitational attraction had given the dragged-out masses of gas a kind of sidewise yank (put "English" on them, so to speak) and thus imparted angular momentum to them. If such a catastrophic theory were true, then planetary systems would have to be extremely scarce. Stars are so widely spaced that stellar collisions are 10,000 times less common than are supernovae, which are themselves not common. It is estimated that, in the lifetime of the galaxy, there has been time for only ten encounters of the type that would produce solar systems by this theory.

However, these initial attempts at designing catastrophes failed when put to the test of mathematical analysis. Russell showed that, in any such near collision, the planets would have to end up thousands of times as far from the sun as they actually are. Furthermore, attempts to patch up the theory by imagining a variety of actual collisions, rather than near misses, had little success. During the 1930s, Lyttleton speculated about the possibility of a three-star collision, and later Hoyle had suggested that the sun had had a companion that had "gone" supernova and left planets as a legacy. In 1939, however, the American astronomer Lyman Spitzer showed that any material ejected from the sun under any circumstances would be so hot that it would not condense into planetesimals but would merely expand into a thin gas. That seemed to end all thought of catastrophe (although, in 1965, a British astronomer, M. M. Woolfson, suggested that the sun may have drawn its planetary material from a very diffuse, cool star, so that extreme temperatures need not be involved).

And so, after the planetesimal theory had come to a dead end, astronomers returned to the evolutionary idea and took another look at Laplace's nebular hypothesis.

By that time, their view of the universe had expanded enormously. They now had to account for the formation of galaxies, which called for much bigger clouds of gas and dust than Laplace had envisaged as the parent of the solar system. And it now appeared that such vast collections of matter would experience turbulence and would break up into eddies, each of which could condense into a separate system.

In 1944, the German astronomer Carl F. von Weizsäcker made a thorough analysis of this idea. He calculated that the largest eddies would contain enough matter to form galaxies. During the turbulent contraction of such an eddy, subeddies would develop. Each subeddy would be large enough to give birth to a solar system (with one or more suns). On the outskirts of the solar eddy itself, subsubeddies might give rise to planets. Thus, at junctions where subsubeddies met, moving against each other like meshing gears, forming dust particles would collide and coalesce, first planetesimals and then planets (figure 3.2).

The Weizsäcker theory, in itself, did not solve the matter of the angular momentum of the planets any more than had the much simpler Laplacian version. The Swedish astrophysicist Hannes Alfven took into account the magnetic field of the sun. As the young sun whirled rapidly, its magnetic field

---

Figure 3.2. Carl F. von Weizsäcker's model of the origin of the solar system. His theory holds that the great cloud from which it was formed broke up into eddies and subeddies which then coalesced into the sun, the planets, and their satellites.

---

acted as a brake, slowing it up, and the angular momentum was passed on to the planets. Hoyle elaborated on this notion so that the Weizsäcker theory, modified to include magnetic as well as gravitational forces, seems the best one yet to account for the origin of the solar system.

The Sun

The sun is clearly the source of light, of heat, and of life itself on Earth, and even prehistoric humanity must have deified it. The Pharaoh, Ikhnaton, who came to the Egyptian throne in 1379 B.C., and was the first monotheist we know of, considered the sun to be the one god. In medieval times, the sun was the symbol of perfection and, though not considered to be itself a god, was certainly taken as representing the perfection of the Almighty.

The ancient Greeks were the first to get a notion of its actual distance, and Aristarchus' observations showed that it must be several million miles away at the least and thus, to judge by its apparent size, that it must be larger than the Earth. Mere size, however, was not impressive in itself, since it was easy to suppose that the sun was merely a vast ball of insubstantial light.

Not till Newton's time did it became obvious that the sun has to be not only larger but much more massive than the Earth, and that the Earth orbits around the sun precisely because the former is bound by the latter's intense gravitational field. We now know that the sun is about 93,000,000 miles from Earth, and that it is 865,000 miles in diameter, or 110 times the diameter of Earth. Its mass is 330,000 times that of Earth and, indeed, is 745 times that of all the planetary material put together. In other words, the sun contains about 99.86 percent of all the matter in the solar system and is overwhelmingly its chief member.

Yet we must not allow sheer size to overimpress us. It is certainly not a perfect body, if by perfection we mean (as the medieval scholars did) that it is uniformly bright and spotless.

Toward the end of 1610, Galileo used his telescope to observe the sun during the sunset haze and saw dark spots on the sun's disk every day. By observing the steady progression of the spots across the surface of the sun and their foreshortening as they approached the edge, he decided that they were part of the solar surface and that the sun was rotating on its axis in a little over twenty-five earth-days.

Naturally, Galileo's findings met with considerable opposition; for by the older view, they seemed blasphemous. A German astronomer, Christoph Scheiner, who also observed the spots, suggested they were not part of the sun but were small bodies that orbited about the sun and showed up darkly against its glowing disk. Galileo won that debate, however.

In 1774, a Scottish astronomer, Alexander Wilson, noted that a large sunspot near the edge of the sun, when it was seen sideways, looked concave, as though it were a crater on the sun. This point was taken up in 1795 by Herschel, who suggested that the sun was a dark, cold body with a flaming layer of gases all about it. The sunspots, by this view, were holes through which the cold body below could be seen. Herschel speculated that the cold body might even be inhabited by living beings. (Note how even brilliant scientists can come up with daring suggestions that seem reasonable in the light of the knowledge of the time, but nevertheless turn out to be ludicrously wrong as further evidence on the subject accumulates.)

Actually, sunspots are not really black. They are areas of the solar surface that are cooler than the rest so that they look dark in comparison. If, however, Mercury or Venus moves between us and the sun, each shows up on the solar disk as a small, *really* black circle; and if that circle moves near a sunspot, one can then see that the spot is not truly black.

Still, even totally wrong notions can be useful, for Herschel's idea served to increase interest in the sunspots.

The real breakthrough, however, came with a German pharmacist, Heinrich Samuel Schwabe, whose hobby was astronomy. Since he worked all day, he could not sit up all night looking at the stars. He cast about for a daytime task and decided to observe the solar disk and look for planets near the sun that might demonstrate their existence by crossing in front of it.

In 1825, he started observing the sun, and could not help noting the sunspots. After a while, he forgot about the planets and began sketching the sunspots, which changed in position and shape from day to day. He spent no less than seventeen years observing the sun on every day that was not completely cloudy.

By 1843, he was able to announce that the sunspots did not appear utterly at random: there was a cycle. Year after year, there were more and more sunspots till a peak was reached. Then the number declined until there were almost none; whereupon a new cycle started. We now know that the cycle is somewhat irregular but averages out to about eleven years. Schwabe's announcement was ignored (he was only a pharmacist, after all) until the well-known scientist Alexander von Humboldt mentioned the cycle in 1851 in his book *Kosmos*, a large overview of science.

At this time, the Scottish-German astronomer Johann von Lamont was measuring the intensity of Earth's magnetic field and found that it was rising and falling in regular fashion. In 1852, a British physicist, Edward Sabine, pointed out that this cycle kept time with the sunspot cycle.

It thus appeared that sunspots affect Earth, and they began to be studied with intense interest. Each year came to be given a *Zurich sunspot number* according to a formula first worked out in 1849 by a Swiss astronomer, Rudolf Wolf, who worked in Zurich. (He was the first to point out that the incidence of auroras also rose and fell in time to the sunspot cycle.)

The sunspots seem to be connected with the sun's magnetic field and to appear at the point of emergence of magnetic lines of force. In 1908, three centuries after the discovery of sunspots, G. E. Hale detected a strong magnetic field associated with sunspots. Why the magnetic field of the sun should behave as it does, emerge from the surface at odd times and places, increase and decrease in intensity in a somewhat irregular cycle still remains among the solar puzzles that have so far defied solution.

In 1893, the English astronomer Edward Walter Maunder was checking through early reports in order to set up data for the sunspot cycle in the first century after Galileo's discovery. He was astonished to find that there were virtually no reports on sunspots between the years 1645 and 1715. Important astronomers, such as Cassini, looked for them and commented on their failure to see any. Maunder published his findings in 1894, and again in 1922, but no attention was paid to his work. The sunspot cycle was so well established that it seemed unbelievable that there could have been a seven-decade period in which hardly any appeared.

In the 1970s, the American astronomer John A. Eddy came across this report and, checking into it, discovered that there actually seemed to have been what came to be called a *Maunder minimum*. He not only repeated Maunder's researches but investigated reports of naked-eye sightings of particularly large sunspots from many regions, including the Far East—data that had been unavailable to Maunder. Such records go back to the fifth century B.C. and generally yield five to ten sightings per century. There are gaps, and one of those gaps spans the Maunder minimum.

Eddy checked reports on auroras, too. These rise and fall in frequency and intensity with the sunspot cycle. It turned out there were many reports after 1715, and quite a few before 1645, but just about none in between.

Again, when the sun is magnetically active and there are many sunspots, the corona is full of streamers of light and is very beautiful. In the absence of sunspots, the corona seems a rather featureless haze. The corona can be seen during solar eclipses; and while few astronomers traveled to view such eclipses in the seventeenth century, such reports as existed during the Maunder minimum were invariably of coronas of the kind associated with few or no sunspots.

Finally, at the time of sunspot maxima, there is a chain of events that succeeds in producing carbon-14 (a variety of carbon that I shall mention in the next chapter) in smaller quantities than usual. It is possible to analyze tree rings for carbon-14 content and to judge the existence of sunspot maxima and minima by fall and rise of carbon-14 content, respectively. Such analysis also produced evidence for the existence of the Maunder minimum and, indeed, numerous Maunder minima in earlier centuries.

Eddy reported that there seem to have been some twelve periods over the last five thousand years in which there were Maunder minima enduring from fifty to a couple of hundred years each. There was one such between 1400 and 1510, for instance.

Since sunspot cycles have an effect on Earth, we might ask what effect Maunder minima have. It may be that they are associated with cold periods. The winters were so cold in Europe in the first decade of the 1700s that it was called the *little ice age*. It was also cold during the 1400–1510 minimum, when the Norse colony in Greenland died out because the weather simply got too bad for survival.

The Moon

When, in 1543, Copernicus placed the sun at the center of the solar system, only the moon was left to owe allegiance to Earth which, for so long previously, had been assumed to be the center.

The moon circles Earth (relative to the stars) in 27.32 days. It turns on its own axis in precisely that same period. This equality between its period of revolution and rotation results in its perpetually presenting the same face to Earth. This equality of revolution and rotation is not a coincidence. It is the result of Earth's tidal effect on its moon, as I shall explain later.

The moon's revolution with respect to the stars is the *sidereal month*. However, as the moon revolves about Earth, Earth revolves about the sun. By the time the moon has made one revolution about Earth, the sun has moved somewhat in its sky because of Earth's motion (which has dragged the moon with it). The moon must continue its revolution for about 2½ days before it catches up with the sun and is back in the same spot relative to the sun it was in before. The moon's revolution about Earth with respect to the sun is the *synodic month*, which is 29.53 days long.

The synodic month was more important to humanity than the sidereal, for as the moon revolves about Earth, the face we see experiences a steadily changing angle of sunlight, and that angle depends on its revolution with respect to the sun. It undergoes a succession of *phases*. At the beginning of a month, the moon is located just east of the sun and appears as a very thin crescent visible just after sunset.

From night to night it moves farther from the sun, and the crescent thickens. Eventually, the lighted portion of the moon is a semicircle, and then it moves beyond that. When the moon has moved so that it is in that portion of the sky directly opposite to that of the sun, the sunlight shines upon the moon over Earth's shoulder (so to speak) and the entire visible face of the moon is lit up: that full circle of light is the *full moon*.

Next the shade encroaches from the side of the moon where the crescent first appeared. Night after night, the moon's lighted portion shrinks, until it is a half-moon again, with the light on the side opposite to where it was on the earlier half-moon. Finally, the moon ends up just west of the sun and appears in the sky just before dawn as a crescent curving in the opposite direction from that which it had formed at first. The moon then moves past the sun and shows up as a crescent just after sunset, and the whole set of changes starts over.

The entire cycle of phase change lasts 29½ days, the length of the synodic month, and formed the basis of humanity's earliest calendars.

Human beings first assumed that the moon was really waxing and waning, growing and fading as the phases changed. It was even assumed that, each time a crescent appeared in the western sky after sunset, it was literally a *new moon*, and it is still called that today.

The ancient Greek astronomers realized, however, that the moon must be a globe, that the changes in phase arose from the fact that it shone only by reflecting sunlight, and that the changing position of the moon in the sky with respect to the sun accounted for the phases exactly. This was a most important fact. The Greek philosophers, notably Aristotle, tried to differentiate Earth from the heavenly bodies by demonstrating that the properties of Earth were altogether different from those the heavenly bodies held in common. Thus, Earth was dark and gave off no light, while the heavenly bodies all gave off light. Aristotle thought the heavenly bodies were made of a substance he called *aether* (from a Greek word for "glowing" or "blazing"), which was fundamentally different from the materials that made up Earth. And yet the cycle of the phases of the moon showed that the moon, like Earth, gave off no light of its own and glowed only because it reflected sunlight. Thus, the moon at least was Earthlike in this respect.

What's more, occasionally the sun and the moon were so precisely on opposite sides of Earth that the sun's light was blocked by Earth and could not reach the moon. The moon (always at full moon) passed into Earth's shadow and was eclipsed.

In primitive times, it was thought the moon was being swallowed by some malign force and would disappear altogether and forever. It was a frightening phenomenon; and it was an early victory of science to be able to predict an eclipse and to show that it was a natural phenomenon with an easily understood explanation. (It is thought by some that Stonehenge was, among other things, a primitive Stone Age observatory which could be used to predict the coming of lunar eclipses by the shifting of positions of the sun and the moon relative to the regularly placed stones of the structure.)

In fact, when the moon is a crescent, it is sometimes possible to see its remainder dimly outlined in ruddy light. It was Galileo who suggested that Earth, like the moon, must reflect sunlight and shine, and that the portion of the moon unlit by the sun was dimly lit by Earthlight. This would be visible only when so little of the sunlit portion could be seen that its light would not wash out the much dimmer Earthlight. Not only, then, was the Moon nonluminous like Earth, but Earth reflected sunlight and would show phases like the moon (if viewed from the moon).

Another supposed fundamental difference between Earth and the heavenly bodies was that Earth was flawed, imperfect, and forever changing while the heavenly bodies were perfect and unchanging.

Only the sun and the moon appear to the unaided eye to be anything more than dots of light. Of the two, the sun appears to be a perfect circle of perfect light. The moon, however—even discounting the phases—is not perfect. When the full moon shines, and the moon seems a perfect circle of light, it is nevertheless clearly not perfect. There are smudges upon its softly glowing surface, which detract from the notion of perfection. Primitive man made pictures out of the smudges, each different culture coming up with a different picture. Human self-love is such that people frequently saw the smudges as forming the picture of a human being, and we still speak of the "man in the moon."

It was Galileo who, in 1609, looked through a telescope at the sky for the first time and turned it on the moon to see mountains, craters, and flat areas (which he took to be seas or, in Latin, *maria*). This was the final indication that the moon was not a "perfect" heavenly body, fundamentally different from Earth, but was an Earthlike world.

This realization did not in itself totally demolish the older view, however. The Greeks had noted that there were several objects in the sky that steadily shifted position against the stars generally, and that, of them all, the moon shifted position most rapidly. They assumed that it did so because it was closer to Earth than any other heavenly body was (and in this the Greeks were right). It might be argued that the moon, because of its closeness to Earth, was somewhat polluted by Earth's imperfections, that it suffered from proximity. It was not till Galileo discovered spots on the sun that the notion of heavenly perfection really shivered.

MEASURING THE MOON

But if the moon was the closest body to Earth, how close was it? Of the ancient Greek astronomers who tried to determine that distance, Hipparchus worked out essentially the right answer. Its average distance from Earth is now known to be 238,900 miles, or about 9.6 times Earth's circumference.

If the moon's orbit were circular, that would be its distance at all times. The moon's orbit, however, is somewhat elliptical, and Earth is not at the center of the ellipse but at one of the foci, which are off-center. The moon approaches Earth slightly in one-half of its orbit and recedes from it in the other half. At its closest point (*perigee*), the moon is but 221,500 miles from Earth, and at its farthest point (*apogee*), 252,700 miles.

The moon is, as the Greeks surmised, by far the closest to Earth of all the heavenly bodies. Even if we forget the stars and consider only the solar system, the moon is, relatively speaking, in our backyard.

The moon's diameter (judging from its distance and its apparent size) is 2,160 miles. Earth's globe is 3.65 times as broad, and the sun's is 412 times as broad. It just happens that the sun's distance from Earth is about 390 times that of the moon on the average, so that differences in distance and diameter nearly cancel out, and the two bodies, so different in real size, *appear* almost equally large in the sky. It is for this reason that, when the moon gets in front of the sun, the smaller, nearer body can so nearly fit over the larger, farther one, making the total eclipse of the sun the wonderful spectacle it is. It is an astonishing coincidence from which we benefit.

GOING TO THE MOON

The comparative nearness of the moon and its prominent appearance in the sky has long acted as a spur to the human imagination. Was there some way of reaching it? (One might equally wonder about reaching the sun, but the sun's obviously intense heat served to cool one's desire to do so. The moon was clearly a much more benign target as well as a much closer one.)

In early times, reaching the moon would not seem an insuperable task, since it was assumed that the atmosphere extended up to the heavenly bodies, so that anything that lifted you up in the air might well carry you up to the moon in extreme cases.

Thus, in the second century A.D., the Syrian writer Lucian of Samosata wrote the first story of space travel that we know of. In it, a ship is caught in a waterspout which lifts it high into the air, high enough to reach the moon.

Again, in 1638, there appeared *Man in the Moone* by an English clergyman, Francis Godwin (who died before its publication). Godwin has his hero carried to the moon in a chariot pulled by large geese who migrate to the moon annually.

In 1643, however, the nature of air pressure came to be understood, and it was rapidly seen that Earth's atmosphere could not extend more than a comparatively few miles above its surface. Most of the space between Earth and the moon was vacuum into which waterspouts could not penetrate and across which geese could not fly. The problem of reaching the moon was suddenly much more formidable, yet still not insuperable.

In 1650, there appeared (again posthumously) *Voyage to the Moon* by the French writer and duelist Cyrano de Bergerac. In his tale, Cyrano lists seven ways by which it might be possible to reach the moon. Six of them were quite wrong for one reason or another, but the seventh method was through the use of rockets. Rockets were indeed the one method then known (or now, for that matter) whereby the vacuum could be crossed.

It was not till 1687, however, that the rocket principle was understood. In that year, Newton published his great book *Principia Mathematica* in which, among other things, he listed his three laws of motion. The third law is popularly known as the law of action and reaction: when a force is applied in one direction, there is an equal and opposite force in the other. Thus, if a rocket ejects a mass of matter in one direction, the rest of the rocket moves in the other, and will do so in a vacuum as well as in air. In fact, it will do so with greater ease in a vacuum where there is no air resistance to motion. (The general feeling that a rocket must need "something to push against" is wrong.)

ROCKETRY

Nor were rockets a matter of theory only. They were in existence centuries before Cyrano wrote and Newton theorized.

The Chinese, as long ago as the thirteenth century, invented and used small rockets for psychological warfare—to frighten the enemy. Modern Western civilization adapted rockets to a bloodier purpose. In 1801, a British artillery expert, William Congreve, having learned about rockets in the Orient, where Indian troops used them against the British in the 1780s, devised a number of deadly missiles. Some were used against the United States in the War of 1812, notably at the bombardment of Fort McHenry in 1814, which inspired Francis Scott Key to write the "Star-Spangled Banner," singing of "the rockets' red glare." Rocket weapons faded out in the face of improvements in range, accuracy, and power of conventional artillery. However, the Second World War saw the development of the American bazooka and the Soviet "Katusha," both of which are essentially rocket-propelled packets of explosives. Jet planes, on a much larger scale, also make use of the rocket principle of action and reaction.

Around the beginning of the twentieth century, two men independently conceived a new and finer use of rockets—exploring the upper atmosphere and space. They were a Russian, Konstantin Eduardovich Tsiolkovsky, and an American, Robert Hutchings Goddard. (It is odd indeed, in view of later developments, that a Russian and an American were the first heralds of the age of rocketry, though an imaginative German inventor, Hermann Ganswindt, also advanced even more ambitious, though less systematic and scientific, speculations at this time.)

The Russian was the first in print; he published his speculations and calculations in 1903 to 1913, whereas Goddard did not publish until 1919. But Goddard was the first to put speculation into practice. On 16 March 1926, from a snow-covered farm in Auburn, Massachusetts, he fired a rocket 200 feet into the air. The remarkable thing about his rocket was that it was powered by a liquid fuel, instead of gunpowder. Then, too, whereas ordinary rockets, bazookas, jet planes, and so on make use of the oxygen in the surrounding air, Goddard's rocket, designed to work in outer space, had to carry its own oxidizer in the form of liquid oxygen (lox, as it is now called in missile-man slang).

Jules Verne, in his nineteenth-century science fiction, had visualized a cannon as a launching device for a trip to the moon, but a cannon expends all its force at once and at the start, when the atmosphere is thickest and offers the greatest resistance. The total acceleration required, moreover, is attained at the very start and is great enough to crush any human beings inside the spaceship into a bloody mash of flesh and bone.

Goddard's rockets moved upward slowly at first, gaining speed and expending final thrust high in the thin atmosphere, where resistance is low. The gradual attainment of speed means that acceleration is kept at bearable levels, an important point for manned vessels.

Unfortunately Goddard's accomplishment got almost no recognition, except from his outraged neighbors, who managed to have him ordered to take his experiments elsewhere. Goddard went off to shoot his rockets in greater privacy; and, between 1930 and 1935, his vehicles attained speeds of as much as 550 miles an hour and heights of a mile and a half. He developed systems for steering a rocket in flight and gyroscopes to keep a rocket headed in the proper direction. Goddard also patented the idea of multistage rockets. Because each successive stage sheds part of the original weight and starts at a high velocity imparted by the preceding stage, a rocket divided into a series of stages can attain much higher speeds and greater heights than could a rocket with the same quantity of fuel all crammed into a single stage.

During the Second World War, the United States Navy halfheartedly supported further experiments by Goddard. Meanwhile, the German government threw a major effort into rocket research, using as its corps of workers a group of youngsters who had been inspired primarily by Hermann Oberth, a Rumanian mathematician who, in 1923, had written on rockets and spacecraft independently of Tsiolkovsky and Goddard. German research began in 1935 and culminated in the development of the V-2. Under the guidance of the rocket expert Wernher von Braun (who, after the Second World War, placed his talents at the disposal of the United States), the first true rocket missile was shot off in 1942. The V-2 came into combat use in 1944, too late to win the war for the Nazis, although they fired 4,300 of them altogether, of which 1,230 hit London. Von Braun's missiles killed 2,511 Englishmen and seriously wounded 5,869 others.

On 10 August 1945, almost on the very day of the war's end, Goddard died —just in time to see his spark blaze into flame at last. The United States and the Soviet Union, stimulated by the successes of the V-2, plunged into rocket research, each carrying off as many German experts in rocketry as could be lured to its side.

At first, the United States used captured V-2's to explore the upper atmosphere; but by 1952, the stock of these rockets was used up. By then, larger and more advanced rocket-boosters were being built in both the United States and the Soviet Union, and progress continued.

EXPLORING THE MOON

A new era began when, on 4 October 1957 (within a month of the hundredth anniversary of Tsiolkovsky's birth), the Soviet Union put the first man-made satellite (Sputnik I) in orbit. Sputnik I traveled around Earth in an elliptical orbit—156 miles above the surface (or 4,100 miles from Earth's center) at perigee and 560 miles away at apogee. An elliptical orbit is something like the course of a roller coaster. In going from apogee to perigee, the satellite slides downhill, so to speak, and loses gravitational potential. Thus, velocity increases, so that at perigee the satellite starts uphill again at top speed, as a roller coaster does. The satellite loses velocity as it climbs (as does the roller coaster) and is moving at its slowest speed at apogee, before it turns downhill again.

Sputnik I at perigee passed through wispy bits of the upper atmosphere; and the air resistance, though slight, was sufficient to slow the satellite a bit on each trip. On each successive revolution, it failed to attain its previous apogee height. Slowly, it spiraled inward. Eventually it lost so much energy that it yielded to Earth's pull sufficiently to dive into the denser atmosphere, there to be burned up by friction with the air.

The rate at which a satellite's orbit decays in this way depends partly on the mass of the satellite, partly on its shape, and partly on the density of the air through which it passes. Thus, the density of the atmosphere at that level can be calculated. The satellites have given us the first direct measurements of the density of the upper atmosphere. The density proved to be higher than had been thought; but at the altitude of 150 miles, for instance, it is still only 1 ten-millionth of that at sea level and, at 225 miles, only 1 trillionth.

These wisps of air ought not be dismissed too readily, however. Even at a height of 1,000 miles, where the atmospheric density is only 1 quadrillionth the sea-level figure, that faint breath of air is a billion times as dense as the gases in outer space itself. Earth's envelope of gases spreads far outward.

The Soviet Union did not remain alone in this field but, within four months, was joined by the United States, which, on 30 January 1958, placed in orbit its first satellite, *Explorer 1.*

Once satellites had been placed in orbit about Earth, eyes turned more longingly than ever toward the moon. To be sure, the moon had lost some of its glamour, for though it was a world and not just a light in the sky, it was no longer the world it was thought to be in earlier times.

Prior to Galileo's telescope, it had always been assumed that if the heavenly bodies were worlds, they would surely be filled with living things, even intelligent humanoid living things. The early science-fiction stories about the moon made this assumption, as did later ones, right into the twentieth century.

In 1835, an English writer named Richard Adams Locke wrote a series of articles for the *New York Sun* which purported to describe serious scientific studies of the moon's surface, which discovered many kinds of living things. The descriptions were detailed and were promptly believed by millions of people.

And yet it had not been long after Galileo looked at the moon through his telescope that it began to seem clear that life could not exist on the moon. The moon's surface was never obscured by cloud or mist. The dividing line between light and dark hemispheres was always sharp, so that there was no detectable twilight. The dark "seas" that Galileo thought to be bodies of water were found to be speckled with small craters; they were, at best, relatively smooth bodies of sand. It was soon clear that the moon contained no water and no air—therefore, no life.

Still, it was perhaps too easy to come so quickly to this conclusion. What about the moon's hidden side that human beings never saw? Might there not be scraps of water under the surface, which, if insufficient to support large forms of life, might support the equivalent of bacteria? Or, if there were no life at all, might there not be chemicals in the soil that represented a slow and possibly aborted evolution toward life? And even if there were nothing of that kind, there were still questions to be answered about the moon that had nothing to do with life. Where was it formed? What was its mineralogical structure? How old was it?

It was therefore not long after the launching of Sputnik I that the new technique began to be used to explore the moon. The first successful *moon probe*—that is, the first satellite to pass near the moon—was sent up by the Soviet Union on 2 January 1959. It was *Lunik I,* the first man-made object to take up an orbit about the sun. Within two months, the United States had duplicated the feat.

On 12 September 1959, the Soviets sent up *Lunik II* and aimed it to hit the moon. For the first time in history, a man-made object rested on the surface of another world. Then, a month later, the Soviet satellite *Lunik III* slipped beyond the moon and pointed a television camera at the side we never see from Earth. Forty minutes of pictures of the other side were sent back from a distance of 40,000 miles above the lunar surface. They were fuzzy and of poor quality but showed something interesting. The other side of the moon had scarcely any maria of the type that are so prominent a feature of our side. Why this asymmetry should exist is not entirely clear. Presumably the maria were formed comparatively late in the moon's history, when one side already faced Earth forever and the large meteors that formed the seas were slanted toward the near face of the moon by Earth's gravity.

But lunar exploration was only beginning. In 1964, the United States launched a moon probe, *Ranger 7,* which was designed to strike the moon's surface, taking photographs as it approached. On 31 July 1964, it completed its mission successfully, taking 4,316 pictures of an area now named Mare Cognitum ("known sea"). In early 1965, *Ranger 8* and *Ranger 9* had even greater success, if that were possible. These moon probes revealed the moon's surface to be hard (or crunchy, at worst) and not covered by the thick layer of dust some astronomers had suspected might exist. The probes showed even those areas that seemed most flat, when seen through a telescope, to be covered by craters too small to be seen from the Earth.

The Soviet probe *Luna IX* succeeded in making a *soft landing* (one not involving the destruction of the object making the landing) on the moon on 3 February 1966 and sent back photographs from ground levels. On 3 April 1966, the Soviets placed *Luna X* in a three-hour orbit about the moon; it measured radioactivity from the lunar surface, and the pattern indicated the rocks of the lunar surface were similar to the basalt that underlies Earth's oceans.

American rocketmen followed this lead with even more elaborate rocketry. The first American soft landing on the moon was that of *Surveyor 1* on 1 June 1966. By September 1967, *Surveyor 5* was handling and analyzing lunar soil under radio control from Earth. It did indeed prove to be basaltlike and to contain iron particles that were probably meteoric in origin.

On 10 August 1966, the first of the American Lunar Orbiter probes were sent circling around the moon. The Lunar Orbiters took detailed photographs of every part of the moon, so that its surface features everywhere (including the part forever hidden from Earth's surface) came to be known in fine detail. In addition, startling photographs were taken of Earth as seen from the neighborhood of the moon.

The lunar craters, by the way, have been named for astronomers and other great men of the past. Since most of the names were given by the Italian astronomer Giovanni Battista Riccioli about 1650, it is the older astronomers —Copernicus, Tycho, and Kepler—as well as the Greek astronomers Aristotle, Archimedes, and Ptolemy, who are honored by the larger craters.

The other side, first revealed by *Lunik III*, offered a new chance. The Russians, as was their right, pre-empted some of the more noticeable features. They named craters not only after Tsiolkovsky, the great prophet of space travel, but also after Lomonosov and Popov, two Russian chemists of the late eighteenth century. They have awarded craters to Western personalities, too, including Maxwell, Hertz, Edison, Pasteur, and the Curies, all of whom are mentioned in this book. One very fitting name placed on the other side of the moon is that of the French pioneer-writer of science fiction, Jules Verne.

In 1970, the other side of the moon was sufficiently well known to make it possible to name its features systematically. Under the leadership of the American astronomer Donald Howard Menzel, an international body assigned hundreds of names, honoring great men of the past who had contributed to the advance of science in one way or another. Very prominent craters were allotted to such Russians as Mendeleev (who first developed the periodic table that I will discuss in chapter 6) and Gagarin, who was the first man to be placed in orbit about Earth and who had since died in an airplane accident. Other prominent features were used to memorialize the Dutch astronomer Hertzsprung, the French mathematician Galois, the Italian physicist Fermi, the American mathematician Wiener, and the British physicist Cockcroft. In one restricted area, we can find Nernst, Roentgen, Lorentz, Moseley, Einstein, Bohr, and Dalton, all of great importance in the development of the atomic theory and subatomic structure.

Reflecting Menzel's interest in science writing and science fiction is his just decision to allot a few craters to those who helped rouse the enthusiasm of an entire generation for space flight when orthodox science dismissed it as a chimera. For that reason, there is a crater honoring Hugo Gernsback, who published the first magazines in the United States devoted entirely to science fiction; and another to Willy Ley, who, of all writers, most indefatigably and accurately portrayed the victories and potentialities of rocketry.

ASTRONAUTS AND THE MOON

And yet unmanned exploration of the moon, however dramatic and successful, is not enough. Could not human beings accompany the rockets? Indeed, it took only three and a half years after the launching of Sputnik I for the first step in this direction to be taken.

On 12 April 1961, the Soviet cosmonaut Yuri Alexeyevich Gagarin was launched into orbit and returned safely. Three months later, on 6 August, another Soviet cosmonaut, Gherman Stepanovich Titov, flew seventeen orbits before landing, spending 24 hours in free flight. On 20 February 1962, the United States put its first man in orbit when the astronaut John Herschel

Glenn circled Earth 3 times. Since then dozens of men have left Earth and, in some cases, remained in space for months. A Soviet woman cosmonaut, Valentina V. Tereshkova, was launched on 16 June 1963 and remained in free flight for 71 hours, making 17 orbits altogether. In 1983, the astronaut Sally Ride became the first American woman to be placed in orbit.

Rockets have left Earth carrying two and three men at a time. The first such launching was that of the Soviet cosmonauts Vladimir M. Komarov, Konstantin P. Feokstistov, and Boris G. Yegorov, on 12 October 1964. The Americans launched Virgil I. Grissom and John W. Young in the first multimanned U.S. rocket on 23 March 1965.

The first man to leave his rocket ship in space was the Soviet cosmonaut Aleksei A. Leonov, who did so on 18 March 1965. This *space walk* was duplicated by the American astronaut Edward H. White on 3 June 1965.

Although most of the space "firsts" through 1965 had been made by the Soviets, the Americans thereafter went into the lead. Manned vehicles maneuvered in space, rendezvoused with each other, docked, and began to move farther and farther out.

The space program, however, did not continue without tragedy. In January 1967, three American astronauts—Grissom, White, and Roger Chaffee—died on the ground in a fire that broke out in their space capsule during routine tests. Then, on 23 April 1967, Komarov died when his parachute fouled during re-entry. He was the first man to die in the course of a space flight.

The American plans to reach the moon by means of three-man vessels (the Apollo program) were delayed by the tragedy while the space capsules were redesigned for greater safety, but the plans were not abandoned. The first manned Apollo vehicle, *Apollo 7*, was launched on 11 October 1968, with its three-man crew under the command of Walter M. Schirra. *Apollo 8*, launched on 21 December 1968, under the command of Frank Borman, approached the moon, circling it at close quarters. *Apollo 10*, launched on 18 May 1969, also approached the moon, detached the lunar module, and sent it down to within nine miles of the lunar surface.

Finally, on 16 July 1969, *Apollo 11* was launched under the command of Neil A. Armstrong. On 20 July, Armstrong was the first human being to stand on the soil of another world.

Since then six other Apollo vehicles have been launched. Five of them—*12, 14, 15, 16, and 17*—completed their missions with outstanding success. *Apollo 13* had trouble in space and was forced to return without landing on the moon, but *did* return safely without loss of life.

The Soviet space program has not yet included manned flights to the moon. However, on 12 September 1970, an unmanned vessel was fired to the moon. It soft-landed safely, gathered up specimens of soil and rock, then safely brought these back to Earth. Still later, an automatic Soviet vehicle landed on the moon and moved about under remote control for months, sending back data.

The most dramatic result obtained from studies on the moon rocks brought back by the landings on the moon, manned and unmanned, is that the moon seems to be totally dead. Its surface seems to have been exposed to great heat, for it is covered with glassy bits, which seem to imply the surface rock has been melted. No trace of any water has been found, nor any indication that water may exist under the surface or even did in the past. There is no life, and not even any sign of chemicals that may be related to life.

There have been no moon landings of any kind since December 1971; and none are, at the moment, planned. There is no question, however, that human technology is capable of placing human beings and their machines on the lunar surface at any time that seems desirable, and the space program continues in other ways.

Venus and Mercury

Of the planets that circle the sun, two—Venus and Mercury—are closer to it than Earth is. Whereas Earth's average distance from the sun is 92,-900,000 miles, the figure for Venus is 67,200,000 miles, and that for Mercury, 36,000,000 miles.

The result is that we never see Venus or Mercury very far from the sun. Venus can never be more than 47 degrees, from the sun as seen from Earth, and Mercury can never be more than 28 degrees from the sun. When east of the sun, Venus or Mercury shows up in the evening in the western sky after sunset and sets soon after, becoming then the *evening star*.

When Venus or Mercury is on the other side of its orbit and is west of the sun, it shows up before dawn, rising in the east not long before sunrise and then disappearing in the solar blaze when the sun rises not long after—and becoming then the *morning star*.

At first, it seemed natural to believe that the two evening stars and two morning stars were four different bodies. Gradually it was borne in on observers that when one of the evening stars was in the sky, the corresponding morning star never was; and vice versa. It began to seem that there were two planets, each of which shuttled from side to side of the sun, serving as evening star and morning star alternately. The first Greek to express this idea was Pythagoras in the sixth century B.C.—and he may have learned it from the Babylonians.

Of the two planets, Venus is by far the easier to observe. In the first place, it is closer to Earth. When Earth and Venus are on the same side of the sun, the two can be separated by a distance of as little as 25 million miles. Venus is then just about 100 times as far from us as the moon is. No sizable body (except the moon) approaches us more closely than Venus does. Mercury's average distance from Earth, when both are on the same side of the sun, is 57 million miles.

Not only is Venus closer to Earth (at least, when both planets are on the same side of the sun), but it is the larger body and catches more light. Venus has a diameter of 7,526 miles, while Mercury's diameter is only 3,014 miles. Finally, Venus has clouds and reflects a far larger fraction of the sunlight that falls upon it than Mercury does. Mercury has no atmosphere and (like the moon) has only bare rock to reflect light.

The result is that Venus, at its brightest, has a magnitude of -4.22. It is then 12.6 times as bright as Sirius, the brightest star, and is indeed the brightest object in the sky except for the sun and the moon. Venus is so bright that, on a dark, moonless night, it can cast a detectable shadow. At its brightest, Mercury has a magnitude of only -1.2, which makes it nearly as bright as Sirius but, still, only one-seventeenth as bright as Venus at its brightest.

Mercury's closeness to the sun means that it is visible only near the horizon and at times when the sky is still bright with twilight or dawn. Hence, despite its brightness, the planet is hard to observe. It is frequently said that Copernicus himself never observed Mercury.

The fact that Venus and Mercury are always found close to the sun, and oscillate from side to side of that body, would naturally make some people suppose that the two planets circle the sun rather than Earth. This notion was first suggested by the Greek astronomer Heracleides about 350 B.C. but was not accepted, until Copernicus raised the idea again, not only for Mercury and Venus but for all the planets, nineteen centuries later.

If Copernicus were correct, and if Venus were an opaque body shining by the reflected light of the sun (as the moon did), then, as observed from Earth, Venus ought to show phases like the moon. On 11 December 1610, Galileo, observing Venus through his telescope, saw that its sphere was only partly lit. He observed it from time to time and found that it did show phases like the moon. That was just about the last nail in the coffin of the older geocentric picture of the planetary system, which could not explain the phases of Venus as those were actually observed. Mercury, too, was eventually observed to show phases.

MEASURING THE PLANETS

Both planets were difficult to observe telescopically. Mercury was so close to the sun, so small and so distant, that very little could be made out in the way of markings on its surface. The Italian astronomer Giovanni Virginio Schiaparelli studied those markings carefully from time to time, however, and, on the basis of the way they changed with time, announced in 1889 that Mercury rotated on its axis in 88 days.

This statement seemed to make sense, for Mercury revolved about the sun in 88 days, too. It was close enough to the sun to be gravitationally locked by it, as the moon by Earth, so that Mercury's period of rotation and revolution should be the same.

Venus, though larger and nearer, was even more difficult to observe because it was perpetually obscured by a thick and unbroken cloud layer and presented a featureless white expanse to all viewers. No one knew anything about its rotation period, although some thought that Venus, too, might be gravitationally locked to the sun, with a rotation period equal to its period of revolution of 224.7 days.

What changed the situation was the development of techniques for handling radar, for emitting beams of microwaves, which could be reflected from objects, and then detecting those reflected beams. During the Second World War, radar could be used to detect airplanes, but beams of microwaves could be bounced off heavenly bodies as well.

In 1946, for instance, a Hungarian scientist, Zoltan Lajos Bay, bounced a microwave beam off the moon and received the echoes.

The moon, however, was a comparatively easy target. In 1961, three different American groups, one British group, and one Soviet group all succeeded in sending microwave beams to Venus and back. Those beams traveled at the speed of light, which was then precisely known. From the time taken by the beam to reach Venus and return, it was possible to calculate the distance of Venus at that time with greater accuracy than had hitherto been possible. From that determination, all the other solar-system distances could be recalculated, since the relative configuration of the planets was well known.

In addition, all objects that are not actually at absolute zero (and no object is) continually emit beams of microwaves. From the wavelength spread of the beam, it is possible to calculate the temperature of the emitting body.

In 1962, microwaves were detected being emitted by the night side of Mercury, the portion of the visible sphere that was not in sunlight. If Mercury's period of rotation was really 88 days, one face of the planet would be forever facing the sun and would be very hot, while the opposite face would be forever away from the sun and would be very cold. From the nature of the emitted microwaves, though, the night side had a temperature considerably higher than one would expect, and thus must at some time or other get sunlight.

When a beam of microwaves is bounced off a rotating body, the beam undergoes certain changes in reflection because of the motion of the surface; and the nature of the changes allows one to calculate the speed of the moving surface. In 1965, two American electrical engineers, Rolf Buchanan Dyce and Gordon H. Pettengill, working with microwave beam reflection, discovered that Mercury's surface was turning faster than expected: Mercury was rotating on its axis in 59 days, so that every bit of its surface was in sunlight at one time or another.

The exact figure for the rotation proved to be 58.65 days—just two-thirds of the revolution period of 88 days. This, too, indicates a gravitational lock, though one less extreme than when rotation and revolution are equal.

THE VENUS PROBES

Venus offered even more startling surprises. Because it was nearly the same size as Earth (with a diameter of 7,526 miles, compared with Earth's 7,927 miles), it was often viewed as Earth's "twin sister." Venus was closer to the sun but had a shielding layer of clouds that might keep it from becoming too hot. It was assumed the clouds were composed of water droplets, and that Venus itself therefore had an ocean, perhaps an even more extensive one than Earth did, and might therefore be rich in sea life. Many science-fiction stories were written (including some by me) concerning such a water-rich, life-rich planet.

In 1956 came the first shock. A team of American astronomers, headed by Cornell H. Mayer, studied the microwaves emitted by Venus's dark side and came to the conclusion that that side had to be at a temperature far above the boiling point of water. Venus had to be very hot and, therefore, very high in radiation.

This conclusion was almost incredible. Something more impressive than a feeble beam of microwaves seemed to be required. Once rockets could be sent successfully to the neighborhood of the moon, it seemed logical to try for similar probes to various planets.

On 27 August 1962, the first successful Venus probe, *Mariner 2*, was launched by the United States. It bore instruments capable of detecting and analyzing microwaves being emitted by Venus, and forwarding the results across tens of millions of miles of vacuum to Earth.

On 14 December 1962, *Mariner 2* passed within 22,000 miles of Venus's cloud layer, and there could be no further doubt. Venus was hellishly hot all over its surface, near the poles as well as at the equator, and on the night side as well as on the day side. The surface temperature is something like 475 degrees C—more than hot enough to melt tin and lead and to boil mercury.

That was not all for 1962. Microwaves can penetrate clouds. Microwaves that were beamed at Venus went right through the clouds to Venus's solid surface and bounced off it. These waves could "see" the surface as human beings, dependent on light-waves, cannot. In 1962, from the distortion of the reflected beam, Roland L. Carpenter and Richard M. Goldstein found that Venus was rotating in a period of something like 250 earth-days. Later analysis by the American physicist Irwin Ira Shapiro showed it to be 243.09 days. This slow rotation was not the result of a gravitational lock on the sun, for the period of revolution was 224.7. Venus rotates on its axis *more slowly* than it revolves about the sun.

What is more, Venus rotates on its axis in the "wrong direction." Whereas the general direction of spin, when viewed (in imagination) from a point high above Earth's north pole, is counterclockwise, Venus rotates on its axis in a clockwise direction. There is no good explanation so far for this *retrograde* rotation.

Every time Venus is at its closest to us, it has spun on its axis, the wrong way, exactly five times since its previous approach and thus always has the same face in the direction of Earth at closest approach. Apparently, Venus is in gravitational lock with Earth, but the latter would seem far too small to influence Venus across the distance between the two.

After *Mariner 2*, other Venus probes were launched by both the United

States and the Soviet Union. Those of the Soviet Union were so designed as to penetrate Venus's atmosphere and then parachute to a soft landing. Conditions were so extreme that none of the Soviet's *Venera probes* lasted long after entry, but they did gain certain information about the atmosphere.

In the first place, the atmosphere was surprisingly dense, about 90 times as dense as that of Earth, and consisted chiefly of carbon dioxide (a gas present in Earth's atmosphere only in a very small amount). Venus's atmosphere is 96.6 percent carbon dioxide and 3.2 percent nitrogen. (Still, with Venus's atmosphere as dense as it is, the total quantity of nitrogen in it is about three times that in Earth's.)

On 20 May 1978, the United States launched *Pioneer Venus* which arrived at Venus on 4 December 1978 and went into orbit about it. *Pioneer Venus* passed very nearly over Venus's poles. Several probes left *Pioneer Venus* and entered Venus's atmosphere, confirming and extending Soviet data.

The main cloud layer on Venus is about 2 miles thick and is about 30 miles above the surface. The cloud layer consists of water containing a quantity of sulfur; and above the main cloud layer is a mist of corrosive sulfuric acid.

Below the cloud layer is a haze down to a height of 20 miles above the surface; and below that, Venus's atmosphere seems completely clear. The lower atmosphere seems stable, without storms or weather changes—just incredibly steady heat everywhere. There are only gentle winds; but considering the density of the air, even a gentle wind would have the force of an earthly hurricane. All in all, one can scarcely think of a more unpleasant world than Earth's "twin sister."

Of the sunlight striking Venus, almost all is either reflected or absorbed by the clouds, but 3 percent penetrates to the clear lower reaches, and perhaps 2.5 percent reaches the ground. Allowing for the fact that Venus is closer to the sun and gets brighter sunlight to begin with, Venus's surface receives about one-sixth the light that Earth's does, despite the former's thick and permanent cloud layer. Venus may be dim compared with Earth; but, if we could somehow survive there, we could see perfectly well on Venus's surface.

Indeed, after landing, one of the Soviet probes was able to take photographs of Venus's surface. These showed a scattering of rocks, which had sharp edges, indicating that not much erosion had taken place.

Microwaves striking Venus's surface and reflecting back can be used to "see" the surface, just as light-waves can, if the reflected beams can be detected and analyzed by instruments as light-waves are by eye or photograph. Microwaves, which are much longer than light-waves, "see" more fuzzily but are better than nothing. Pioneer Venus was able to map Venus's surface by microwaves.

Most of Venus's surface seems to be the kind we associate with continents, rather than with sea bottoms. Whereas Earth has a vast sea bottom (water-filled) making up seven-tenths of the planetary surface, Venus has a huge supercontinent that covers about five-sixths of the total surface, with small regions of lowland (no water) making up the remaining sixth.

The supercontinent that covers Venus seems to be level, with some indications of craters, but not many. The thick atmosphere may have eroded them away. There are, however, raised portions on the supercontinent, two of them being of huge size.

In what on Earth would be the arctic region, on Venus is a large plateau, which is named Ishtar Terra and is about as large in area as the United States. On the eastern portion of Ishtar Terra is the mountain range Maxwell Montes, with some peaks reaching a height of 7.3 miles above the general level outside the plateau. Such peaks are distinctly higher than any of Earth's mountain peaks.

In the equatorial region of Venus, there is another and even larger plateau called Aphrodite Terra. Its highest peaks are not quite as high as those on Ishtar Terra.

It is hard to tell whether any of the mountains of Venus are actually volcanoes. Two may be—at least extinct ones; and of them, Rhea Mons, spreads out across an area as large as New Mexico.

THE MERCURY PROBES

Mercury's surface does not present the problems Venus's does. There is no atmosphere on Mercury, no cloud layer. It is only necessary to send out a Mercury probe.

On 3 November 1973, *Mariner 10* was launched. It passed close by Venus on 5 February 1974, from which neighborhood it sent back useful data, and then moved on toward Mercury.

On 29 March 1974, *Mariner 10* passed within 435 miles of Mercury's surface. It then moved into orbit about the sun in such a way as to make one revolution in 176 days, or just twice Mercury's year. That brought it back to Mercury in the same spot as before, because for each of *Mariner 10*'s circuits of the sun, Mercury completed two. On 21 September 1974, *Mariner 10* passed Mercury a second time; and on 16 March 1975, it passed a third time, coming within 203 miles of Mercury's surface. By then, *Mariner 10* had consumed the gas that kept it in a stable position, and was thereafter useless for further study of the planet.

In the three passes, *Mariner 10* photographed about three-eighths of the surface of Mercury and showed a landscape that looked much like the surface of the moon. There were craters everywhere, with the largest about 125 miles in diameter. Mercury has very few "seas," however. The largest region that is relatively crater-free is about 870 miles across. It is called Caloris ("heat") because it is almost directly under the sun when Mercury is at its closest approach (*perihelion*) to that body.

Mercury also possesses long cliffs, 100 miles or more long and about 1.5 miles high.

Mars

Mars is the fourth planet from the sun, the one just beyond Earth. Its average distance from the sun is 141,600,000 miles. When Earth and Mars are on the same side of the sun, the two planets can approach within 50,-000,000 miles of each other on the average. Because Mars's orbit is rather elliptical, there are times when Mars and Earth are separated by only 30,-000,000 miles. Such close approaches take place every thirty-two years.

Whereas the sun and the moon change their positions more or less steadily moving from west to east, against the background of the stars, the planets have a more complicated motion. Most of the time, they do move west to east, relative to the stars, from night to night. At some points the movement of each planet slows; it comes to a complete halt and then starts moving "backward," from east to west. This *retrograde motion* is never as great as the forward motion, so that, on the whole, each planet moves from west to east and eventually makes a complete circuit of the sky. The retrograde motion is largest and most prominent in the case of Mars.

Why does this happen? The older picture of the planetary system with Earth at the center had great trouble explaining this retrograde motion. The Copernican system, with the sun at the center, explained it easily. Earth, moving in an orbit closer to the sun than that of Mars, has a shorter distance to cover in completing its revolution. When Earth is on the same side of the Sun as Mars is, it overtakes Mars so that Mars *seems* to move backward. Comparison of Earth's orbital motion with that of any other planet can explain all the retrograde appearances—a great factor in forcing the acceptance of the sun-centered planetary system.

Mars is farther from the Sun than Earth is, and gets less intense sunlight. It is a small planet, only 4,220 miles in diameter (a little over half that of Earth), and has a very thin atmosphere so that it does not reflect much of the light it does get. On the other hand, it has one advantage compared with Venus. When Venus is closest to us, it is between us and the sun, and we see only its dark side. Mars, however, when it is closest to us, is beyond us, being farther from the sun, and we see its sunlit side (a kind of "full Mars"), which adds to its brightness. At its very brightest, Mars has a magnitude of -2.8, which makes it, at that time, brighter than any object in the sky except the sun, the moon, and Venus. That brightness is only attained every thirty-two years, however, when Mars is unusually close. When it is in that part of its orbit that places it on the other side of the sun from us, it is quite far away and is only as bright as a reasonably bright star.

From 1580 on, the Danish astronomer Tycho Brahe made careful observations of Mars (without a telescope, it not yet having been invented) in order to study its movements and make more accurate predictions of its future positions. After Tycho died, his assistant, the German astronomer Johann Kepler used those observations to work out the orbit of Mars. He found he had to abandon the notion of circular orbits, which astronomers had held for 2,000 years, and, in 1609, showed that the planets had to move in elliptical orbits. The Keplerian version of the planetary system still holds today and will undoubtedly hold, in essence, permanently.

Another basic contribution of Mars to the plan of the solar system came in 1673 (as I stated earlier) when Cassini determined the parallax of Mars and, for the first time, got an idea of the true distances of the planets.

Thanks to the telescope, Mars became more than a point of light. Christian Huyghens, in 1659, observed a dark, triangular marking which he named Syrtis Major (that is "large bog"). By following this marking, he was able to show that Mars rotated on its axis in about 24½ hours. (The present-day figure is 24.623 hours.)

Mars, being farther from the sun than Earth is, has a longer orbit and travels more slowly under the sun's gravitational pull. It takes it 687 earth-days (1.88 earth-years) to complete a revolution, or 668.61 Mars-days.

Mars is the only planet we know that has a rotation period so close to that of Earth. Not only that, but in 1781, William Herschel showed that the Martian axis was tipped very much in the way that Earth's is. Earth's axis is tipped 23.45 degrees from the vertical, so that the Northern Hemisphere has spring and summer when the North Pole is slanted toward the sun, and fall and winter when the North Pole is slanted away; while the Southern Hemisphere has its seasons reversed, because the South Pole slants away from the sun when the North Pole slants toward it; and vice versa.

The Martian axis is tipped 25.17 degrees from the vertical, as Herschel could tell by observing closely the direction in which the markings on Mars move as the planet turns. Thus, Mars has seasons just as Earth does, except that each season is almost twice as long as on Earth and is, of course, colder.

Another similarity showed up in 1784, when Herschel noted that Mars has ice caps at its north and south pole. On the whole, Mars is more like Earth than any other world we have ever observed in the sky. Unlike the moon and Mercury, Mars has an atmosphere (first observed by Herschel) but not a thick, cloud-laden atmosphere as Venus has.

The similarity of Mars to Earth does not extend to satellites. Earth has a large satellite, the moon, but Mercury and Venus have no satellites at all. Mars, too, seemed to have no satellites at first. At least, more than two and a half centuries of telescopic observation revealed none.

In 1877, though, when Mars was going to make one of its close approaches to Earth, the American astronomer Asaph Hall decided to search the Martian neighborhood for any sign of satellites. Since none had yet been found, he felt any had to be very small and very close to Mars and were probably obscured by the planet's light.

Night after night he observed; and on 11 August 1877, he decided to give up. His wife Angelina Stickney Hall urged him to try one more night—and on that one more night, he did discover two tiny satellites close to Mars. He named them Phobos and Deimos after the name of the sons of Mars in the myths. (The names mean "fear" and "terror," appropriate for the sons of the war god.)

Phobos, the inner of the two satellites, is only 5,810 miles from the center of Mars and is, therefore, only 3,700 miles above the Martian surface. It completes one turn about its small orbit in 7.65 hours—or less than one-third the time it takes Mars to turn on its axis, so that as Phobos speeds along, it continually overtakes the Martian surface. Phobos therefore rises in the west and sets in the east when observed from Mars. Deimos, the farther of the satellites, is 14,600 miles from the center of Mars and completes one revolution about the planet in 30.3 hours.

As the satellites were too small to show up as anything but points of light in even the best telescopes, for a century after their discovery, nothing was known about them but their distance from Mars and times of revolution. From the distance and motion of the satellites, it was easy to calculate the strength of Mars's gravitational field and, therefore, of its mass. Mars turned out to have almost exactly one-tenth the mass of the Earth, and its surface gravity was just three-eighths that of Earth. A person weighing 150 pounds on Earth would weigh 56¼ pounds on Mars.

Nevertheless, Mars is a distinctly larger world than the moon. Mars has 8.7 times the mass of the moon, and the surface gravity on Mars is 2.25 times that on the moon. Roughly speaking, Mars is just about intermediate in these respects between the moon and Earth. (Venus and Mercury, having no satellites, could not have their mass determined so easily. We now know Venus's mass to be four-fifths that of Earth and Mercury's to be one-eighteenth that of Earth. Mercury, with only about half the mass of Mars, is the smallest of the eight major planets.)

Knowing the size and the mass of a world, we can easily calculate its density. Mercury, Venus, and Earth all have densities that are more than five times the density of water: 5.48, 5.25, and 5.52, respectively. These are more than would be expected if the worlds were built up of solid rock, and each planet is therefore thought to have a metallic core. (This subject will be taken up in more detail in the next chapter.)

The moon has a density 3.34 times that of water and may be made up of rocky material through and through. Mars is intermediate. Its density is 3.93 times that of water, and it may have a very small metallic core.

MAPPING MARS

It was natural that astronomers would try to map Mars, to draw the dark and light pattern of the spots and patches on its surface. This could be done well for the moon, but Mars, even at its closest, is 150 times as far from us as the moon is, and has a thin, but obscuring atmosphere, which the moon lacks.

In 1830, however, a German astronomer, Wilhelm Beer, who had been mapping the moon in detail, turned his attention to Mars. He produced the first map of Mars that showed a pattern of dark and light. He assumed the dark areas to be water and the light areas land. The trouble was that other astronomers tried their hand at map making also, and each astronomer came up with a different map.

The most successful of the map makers of Mars, however, was Schiaparelli (who was later, and wrongly, to fix the rotation of Mercury at eighty-eight days). In 1877, during the close approach of Mars that made it possible for Hall to discover its two satellites, Schiaparelli drew a map of Mars that looked altogether different from anything that had been drawn before. This time, though, astronomers agreed. Telescopes had been steadily improving, and now they all saw more or less what Schiaparelli saw, and the new map of Mars lasted for nearly a century. To various regions of Mars, Schiaparelli gave names drawn from the mythology and geography of ancient Greece, Rome, and Egypt.

Schiaparelli, in observing Mars, noted that there were thin dark lines connecting larger dark areas in the way that straits or channels connect two seas. Schiaparelli called these lines *channels*, making use of the Italian word *canali* for the purpose. The word was mistranslated as *canals* in English, and that made all the difference: channels are a natural phenomenon, while canals are man-made.

Schiaparelli's observations at once created a new interest in Mars. The planet had long been thought of as very Earth-like, but it was smaller than Earth, with a weaker gravitational field. Mars might not have been able to hold on to much of an atmosphere or to much of its water, so that it might have been drying out over many millions of years. Any intelligent life that might have evolved on Mars would be fighting to survive desiccation.

It became easy for people to think that not only was there intelligent life on Mars, but that it might display a more advanced technology than our own. The Martians might have built canals to bring water from the icecaps down to their farms in the milder equatorial regions.

Other astronomers began to detect the canals, and the most enthusiastic of these was the American Percival Lowell. A rich man, he opened a private observatory in Arizona in 1894. There in the clean, mile-high desert air, far from city lights, visibility was excellent, and Lowell began to draw maps in much greater detail than Schiaparelli had. Eventually, he plotted over 500 canals and wrote books that popularized the notion of life on Mars.

In 1897, the English science-fiction writer Herbert George Wells published a serialized novel, *War of the Worlds*, in a popular magazine, and that further publicized the notion. Large numbers of people came to take life on Mars for granted; and on 30 October 1938, Orson Welles produced a radio dramatization of *War of the Worlds*, with the Martians landing in New Jersey, so realistically that large numbers of people, imagining the show to be an actual news report, fled in terror.

Nonetheless, many astronomers denied the reality of Lowell's canals. They could not see the canals themselves, and Maunder (who had first described the periods of sunspot lack, or Maunder minima) felt they were optical illusions. In 1913, he set up circles within which he put smudgy irregular spots and then placed schoolchildren at distances from which they could barely see what was inside the circles. He asked them to draw what they saw, and they drew straight lines very much like Lowell's canals.

Furthermore, straightforward observation seemed to lessen the similarity of Mars to Earth. In 1926, two American astronomers, William Weber Coblentz and Carl Otto Lampland, managed to take measurements of the surface temperature of Mars. It was colder than had been thought. During the day, there was some indication that the Martian equator might be fairly mild at perihelion time, when Mars was closest to the sun, but the Martian night seemed everywhere to be as cold as Antarctica at its coldest. The difference between day and night temperatures made it seem that Mars's atmosphere was thinner than had been thought.

In 1947, the Dutch-American astronomer Gerard Peter Kuiper analyzed the infrared portion of the light arriving from Mars and concluded that the Martian atmosphere was chiefly carbon dioxide. He could find no sign of nitrogen, oxygen, or water vapor. The chance of complex life forms anything like those on Earth seemed dim indeed. Nevertheless, a nagging belief in Martian vegetation and even in Martian canals lingered.

THE MARS PROBES

Once rockets began to rise into and beyond Earth's atmosphere, however, hopes for a solution to the century-old problem rose with them.

The first successful Mars probe, *Mariner 4*, was launched on 28 November 1964. On 14 July 1965, *Mariner 4* passed within 6,000 miles of the Martian surface. As it did so, it took a series of 20 photographs, which were turned into radio signals, beamed back to Earth and there converted into photographs again. What those photographs showed were craters—no sign of any canals.

As *Mariner 4* passed behind Mars, its radio signals, before disappearing, passed through the Martian atmosphere, indicating that the Martian atmosphere is thinner than anyone had supposed: it is less than 1/100 as dense as Earth's.

Mariner 6 and *Mariner 7*, more elaborate Mars probes, were launched on 24 February and 27 March 1969, respectively. These passed within 2,000 miles of the Martian surface and sent back 200 photographs altogether. Wide portions of the surface were photographed; and it was shown that while some regions were heavily cratered like the moon, others were relatively featureless, and still others were jumbled and chaotic. Apparently Mars has a complex geological development.

However, there were no signs of canals anywhere, the atmosphere was at least 95 percent carbon dioxide, and the temperature was even lower than the measurements of Coblentz and Lampland had indicated. All hope for intelligent life on Mars—or any kind of complex life—seemed gone.

More remained to be done, however. The next successful Mars probe was *Mariner 9*, which was launched on 30 May 1971. It reached Mars on 13 November 1971 and, instead of passing it, went into orbit about it. It was fortunate that it did so, for midway on its journey to Mars, a planet-wide duststorm rose and, for many months, photographs would have yielded nothing but a haze. In orbit, the probe could outwait the storm; and in December, the Martian atmosphere cleared, and the probe got to work.

It mapped all of Mars as clearly as the moon was mapped; and, after a century, the canal mystery was settled once and for all. There are no canals. Those that were "seen," as Maunder had insisted, were the result of optical illusions. Everything was dry, and the dark areas were merely darker drifts of dust particles, as the American astronomer Carl Sagan had suggested a couple of years earlier.

Half the planet, mostly in its southern hemisphere, was cratered like the moon. The other half seemed to have had its craters obliterated by volcanic action, and some large mountains that were clearly volcanoes (though perhaps long-inactive ones) were located. The largest of these was named Olympus Mons in 1973. It reaches a height of 15 miles above general ground level, and its large central crater is 40 miles across. It is far larger than any volcano on Earth.

There is one crack in the Martian surface that might have given the illusion of being a canal. It is a large canyon, now named Valles Marineris, and is about 1,900 miles long, up to 310 miles wide, and 1¼ miles deep. It is 9 times as long as the Grand Canyon, 14 times as wide, and twice as deep. It may have been the result of volcanic action about 200 million years ago.

There are also markings on Mars that meander across the Martian surface and have tributaries strongly resembling dried river beds. Could it be that Mars is now suffering an ice age with all the water frozen into the icecaps and the subsoil? Was there a time in the reasonably recent past, and would a time come in the reasonably near future, when conditions would ameliorate, water would appear in liquid form, and rivers would flow once more? If so, might very simple forms of life still precariously exist in the Martian soil?

What was needed was a soft landing on Mars. *Viking 1* and *Viking 2* were launched on 20 August and 9 September 1975, respectively. *Viking 1* went into orbit about Mars on 19 June 1976 and sent down a *lander*, which came to rest successfully on the Martian surface on 20 July. Some weeks later, *Viking 2* sent down a lander in a more northerly position.

As they passed through the Martian atmosphere, the landers analyzed it and found that, in addition to carbon dioxide, there was 2.7 percent nitrogen and 1.6 percent argon. There was the merest trace of oxygen.

On the surface, the landers found the maximum daytime temperature to be $-20°$ F. There seemed no chance that the surface temperature ever reached the melting point of ice anywhere on Mars, which meant no liquid water anywhere. It was too cold for life, just as Venus was too hot for life. Or, at least, it was too cold for any but the simplest forms of life. It was so cold that even carbon dioxide froze in the coldest regions, and it would seem the icecaps were at least partly frozen carbon dioxide.

The landers sent back photographs of the Martian surface, and analyzed the soil. It turned out that the Martian soil is richer in iron and poorer in aluminum than earthly soil is. About 80 percent of the Martian soil is an iron-rich clay, and the iron present may be in the form of *limonite*, an iron compound that is responsible for the color of red bricks. Mars's ruddy color, which roused fear in early human beings because of the association with blood, has nothing to do with it: Mars is simply a rusty world.

Most important, the landers were equipped with small chemical laboratories capable of testing the soil to see if it would react in such a way as to make it seem that living cells were present. Three different experiments were performed; and in none were the results clear-cut. It seemed that life might conceivably exist, but real certainty was lacking. What made scientists uncertain was that analysis of the soil showed that there were no detectable quantities of organic compounds—that is, the type of compounds associated with life. Scientists were simply not ready to believe that non-organic life could be present, and the solution to the problem will have to be deferred until such time as more elaborate landers can be placed on the planet, or better yet, when human beings themselves can reach it.

THE MARTIAN SATELLITES

Originally it had not been planned to have the Mars probes make detailed studies of the small Martian satellites; but when *Mariner 9* found itself in orbit about Mars with no pictures to take because of the sandstorm, its cameras were turned on the satellites. The photographs of the satellites showed them to be irregular in outline. (Astronomical objects are usually thought of as spheres, but they are spheres only if they are large enough for their gravitational fields to be strong enough to flatten any major irregularities.) In fact, each satellite looked much like a baking potato in shape and even had craters which had an uncanny resemblance to the "eyes" of potatoes.

The diameter of Phobos, the larger of the two, varied from 12 to 17 miles; and of Deimos, from 6 to 10 miles. They were merely mountains flying about Mars. In each case, the longest diameter points toward Mars at all times, so that each is gravitationally locked by Mars, as the moon is by Earth.

The two largest craters on Phobos are named Hall and Stickney in honor of their discoverer and of his wife, who urged him to try one more night. The two largest craters on Deimos are named Voltaire and Swift: the former, for the French satirist; and the latter, for Jonathan Swift, the English satirist, since both in their fiction had imagined Mars as having two satellites.

Jupiter

Jupiter, the fifth planet from the sun, is the giant of the planetary system. It is 88,700 miles in diameter, 11.2 times that of Earth. Its mass is 318.4 times that of Earth. In fact, it is more than twice as massive as all the other planets put together. Nevertheless, it is still a pygmy compared to the sun, which has a mass 1,040 times that of Jupiter.

On the average, Jupiter is 483 million miles from the sun, or 5.2 times Earth's distance from the sun. Jupiter never gets closer than about 390 million miles to us even when both it and Earth are on the same side of the sun, and the sunlight that Jupiter receives is only one twenty-seventh as bright as that which we receive. Even so, because of its huge size, it shines bright in our sky.

Its magnitude, at its brightest, is −2.5, which is considerably brighter than any star. Venus and Mars at their brightest can outdo Jupiter (Venus by a considerable margin). On the other hand, Venus and Mars are often far dimmer, when moving to the farther portion of their orbits. Jupiter, on the other hand, dims only slightly as it moves away from Earth since its orbit is so distant that it scarcely makes a difference whether it is on our side of the sun or not. Jupiter is often, therefore, the brightest object in the sky except for the sun and the moon (especially since it can be in the sky all night long, while Venus never can) and so is well named for the king of the gods in Graeco-Roman mythology.

JOVIAN SATELLITES

When Galileo constructed his first telescope and turned it on the sky, he did not neglect Jupiter. On 7 January 1610, he studied Jupiter and almost at once noticed three little sparks of light near it—two on one side and one on the other, all in a straight line. Night after night, he returned to Jupiter, and always those three little bodies were there, their positions changing as they oscillated from one side of the planet to the other. On 13 January, he noticed a fourth object.

He came to the conclusion that four small bodies were circling Jupiter, just as the moon circles Earth. These were the first objects in the solar system, invisible to the unaided eye, to be discovered by the telescope. Also, here was visible proof that there are some bodies in the solar system that do not revolve about Earth.

Kepler coined the word *satellite* for these four objects, after a Latin word for people who serve in the entourage of some rich or powerful man. Since then, objects that circle a planet have been called by that name. The moon is Earth's satellite, and *Sputnik I* was an artificial satellite.

These four satellites of Jupiter are lumped together as the *Galilean satellites.* Shortly after Galileo's discovery, they were given individual names by a Dutch astronomer, Simon Marius. From Jupiter outward, they are Io, Europa, Ganymede, and Callisto, each name that of someone associated with Jupiter (Zeus, to the Greeks) in the myths.

Io, the nearest of the Galileans, is 262,000 miles from Jupiter's center, about the distance of the moon from Earth's center. However, Io circles Jupiter in 1.77 days—not the 27.32 days the moon takes to circle Earth. Io moves so much more rapidly because it is in the grip of Jupiter's gravitational attraction, which—owing to Jupiter's greater mass—is far more intense than Earth's. (Indeed, it is from Io's speed that Jupiter's mass can be calculated.)

Europa, Ganymede, and Callisto, respectively, are 417,000, 665,000, and 1,171,000 miles from Jupiter and circle it in 3.55 days, 7.16 days, and 16.69 days. Jupiter and its four Galilean satellites are like a miniature solar system, and their discovery made the Copernican scheme of the planets much more believable.

Once the satellites made it possible to determine the mass of Jupiter, the great surprise was that its mass is so low. It might be 318.4 times that of Earth, but its volume is 1,400 times that of Earth. If Jupiter takes up 1,400 times as much room as Earth, why should it not have 1,400 times as much matter as Earth does and therefore be 1,400 times as massive? The answer is that each part of Jupiter has a smaller mass than an equivalent part of Earth has. Jupiter has a smaller density.

In fact, Jupiter's density is only 1.34 times that of water, or only one-fourth the density of Earth. Clearly, Jupiter must be made up of material less dense than rocks and metal.

The satellites themselves are comparable to our moon. Europa, the smallest of the four, is about 1,940 miles in diameter, or a bit smaller than the moon. Io, which is 2,270 miles across, is just about the size of the moon. Callisto and Ganymede are each larger than the moon. Callisto has a diameter of 3,010 miles; and Ganymede, 3,260 miles.

Ganymede is actually the largest satellite in the solar system and has a mass 2½ times that of the moon. In fact, Ganymede is distinctly larger than the planet Mercury, while Callisto is just about Mercury's size. Mercury, however, is made of denser materials than Ganymede is, so that the larger body of Ganymede has only about three-fifths the mass of Mercury. Io and Europa, the two inner satellites, are about as dense as the moon and must be made up of rocky material. Ganymede and Callisto have densities much like that of Jupiter and must be made up of lighter materials.

It is not surprising that Jupiter has four large satellites and Earth only one, considering how much larger the former is. In fact, if there is any surprise, it should be that Jupiter does not have still more, or Earth still less.

The four Galilean satellites together have 6.2 times the mass of the moon but only 1/4,200 the mass of Jupiter, the planet they circle. The moon, all by itself, has 1/81 the mass of Earth, the planet it circles.

Planets generally have satellites that are tiny in comparison with themselves —as Jupiter has. Of the small planets, Venus and Mercury have no satellites at all (even though Venus is almost the size of Earth), and Mars has two satellites, but very tiny ones. Earth's satellite is so large that the two make up what might almost be considered a double planet. (Until recently, Earth was thought to be unique in this respect—but wrongly so, as we will see later in this chapter.)

For nearly three centuries after Galileo's discovery, no further satellites were discovered for Jupiter; although during that time, fifteen satellites were discovered for other planets.

Finally, in 1892, the American astronomer Edward Emerson Barnard detected a speck of light near Jupiter, so dim that it was almost impossible to see it in the glare of Jupiter's light. It was a fifth satellite of Jupiter and the last satellite to be discovered by eye observation. Since then, satellites have been discovered from photographs taken either from Earth or by a probe.

This fifth satellite was named Amalthea (after a nymph who was supposed to have nursed Zeus as an infant). The name was made official only in the 1970s.

Amalthea is only 112,000 miles from Jupiter's center and circles it in 11.95 hours. It is closer than any of the Galilean satellites—one reason it took so long to be discovered; Jupiter's light is blinding at that distance. For another, its diameter is only about 155 miles, only one-thirteenth that of the smallest Galilean, so that it is very dim.

Jupiter turned out, though, to have many other satellites, even smaller than Amalthea and therefore even dimmer. Most of these are located far from Jupiter, far outside the orbit of any of the Galileans. In the twentieth century, eight of these *outer satellites* were detected: the first in 1904, and the eighth in 1974. In that time, they were denoted only by Roman numerals in the order of their discovery, from Jupiter VI to Jupiter XIII.

The American astronomer Charles Dillon Perrine discovered Jupiter VI in December 1904 and Jupiter VII in January 1905. Jupiter VI is about 60 miles in diameter; and Jupiter VII, about 20 miles in diameter.

Jupiter VIII was discovered in 1908 by the British astronomer P. J. Melotte; while the American astronomer Seth B. Nicholson discovered Jupiter IX in 1914, Jupiter X and Jupiter XI in 1938, and Jupiter XII in 1951. These latter four are each about 15 miles across.

Finally, on 10 September 1974, the American astronomer Charles T. Kowal discovered Jupiter XIII, which is only 10 miles across.

These outer satellites can be divided into two groups. The inner four—VI, VII, X, and XIII— are at average distances from Jupiter in the neighborhood of 7 million miles, so that they are about six times as far from Jupiter as Callisto (the outermost Galilean) is. The outer four are, on the average, about 14 million miles from Jupiter and are thus twice as far away as the inner four.

The Galilean satellites all move about Jupiter in the plane of the planet's equator and in almost exactly circular orbits. This is an expected state of affairs and is brought about by Jupiter's tidal effect (which I shall discuss further in the next chapter) on the satellites. If a satellite's orbit is not in the equatorial plane (that is, it is *inclined*), or if it is not circular (that is, it is *eccentric*), the tidal effect, given time, draws the satellite into the orbital plane and makes the orbit circular.

While tidal effect is proportionate to the affecting object's mass, it weakens rapidly over distance and is inversely proportionate to the size of the object affected. Hence, despite its huge mass, Jupiter exerts only a weak tidal effect on the small outer satellites. Thus, even though four of them are at about the same distance from Jupiter (on the average), and four others are all at about another distance, there is no imminent danger of collisions. With each orbit differently inclined and differently eccentric, none ever approaches any other as all circle the planet.

The outer group of four of these outer satellites have orbits inclined to such a degree that they have been twisted upside down, so to speak. They revolve about Jupiter in retrograde fashion, moving clockwise (as viewed from above Jupiter's north pole) rather than counterclockwise, as do all the other satellites of Jupiter.

It is possible that these small outer satellites are captured asteroids (which I shall discuss later in this chapter) and, as such, their irregular orbits could be due to their having been part of Jupiter's satellite system for relatively short times—only since their capture—with tidal effects having less time to modify their orbits. Besides, it can be shown that it is easier for a planet to capture a satellite if that satellite approaches in such a way as to move about the planet in a retrograde orbit.

The satellite that recedes farthest from Jupiter is Jupiter VIII (now called Pasiphae, for all the outer satellites were given official names—obscure mythological ones—in recent years). Its orbit is so eccentric that at its farthest point, Pasiphae is 20.6 million miles from Jupiter, over 80 times as far as the moon ever gets from Earth. This is the farthest any known satellite gets from the planet it circles.

Jupiter IX (Sinope) has a slightly larger average distance than Pasiphae and therefore takes longer to circle Jupiter. Sinope goes once around Jupiter in 758 days, or almost exactly two years and one month. No other known satellite has so long a period of revolution.

JUPITER'S SHAPE AND SURFACE

What about Jupiter itself? In 1691, Cassini, studying Jupiter in his telescope, noted that it was not a circle of light, but was, rather a definite ellipse. This observation meant, three dimensionally, that Jupiter was not a sphere but an oblate spheroid, rather like a tangerine.

This was astonishing since the sun and the moon (the latter when full) are perfect circles of light and seemed therefore perfect spheres. However, Newton's theories (then quite new) explained the situation perfectly. As we shall see in the next chapter, a *rotating* sphere is likely to be an oblate spheroid. Rotation causes a spinning sphere to bulge in the equatorial regions and flatten at the poles; the faster the rotation, the more extreme the departure from the spherical.

Hence, the diameter from one point on the equator to another point on it at the other side (the *equatorial diameter*) must be longer than the diameter from the north pole to the south pole (the *polar diameter*). Jupiter's equatorial diameter, the usual diameter given in astronomy books, is 88,700 miles, but the polar diameter is only 83,300 miles. The difference between the two is 5,400 miles (about two-thirds the total diameter of Earth); and this difference divided by the equatorial diameter gives a figure known as *oblateness*. The oblateness of Jupiter is 0.062 or, in fractions, about one-sixteenth.

Mercury, Venus, and our moon, which rotate very slowly, have no measurable oblateness. While the sun does rotate at a moderate speed, its enormous gravitational pull keeps it from bulging much, and it, too, has no measurable oblateness. Earth rotates moderately quickly and has a small oblateness of 0.0033. Mars also has a moderate speed of rotation and a smaller gravitational pull to keep it from bulging; and its oblateness is 0.0052.

Jupiter has an oblateness nearly nineteen times that of Earth despite a much greater gravitational pull, so we must expect Jupiter to spin much more quickly on its axis. And so it does. Cassini himself, in 1665, had followed markings on Jupiter's surface as they moved steadily about the globe, and noted the period of rotation to be just under 10 hours. (The present figure is 9.85 hours, or two-fifths of an earth-day.)

Although Jupiter has a much shorter rotational period than Earth has, the former is the far larger of the two. A point on Earth's equator travels 1,040 miles an hour as it makes a complete circuit in 24 hours. A point on Jupiter's equator would have to travel 28,000 miles an hour to complete a circuit of the planet in 9.85 hours.

The spots noted by Cassini (and by other astronomers after him) were always changing and so were not likely to be part of a solid surface. What these astronomers were seeing was more likely to be a cloud layer, as in the case of Venus, and the spots would be various storm systems. There are also colored streaks parallel to Jupiter's equator which might be the result of prevailing winds. For the most part, Jupiter is yellow in color, while the colored streaks vary from orange to brown, with occasional bits of white, blue, or gray.

The most notable marking on Jupiter's surface was first seen by the English scientist Robert Hooke in 1664; and in 1672, Cassini made a drawing of Jupiter which showed this marking as a large round spot. The spot showed up in other drawings in later years; but it was not until 1878 that it was dramatically described by a German astronomer, Ernst Wilhelm Tempel. It seemed quite red to him at the time, and it has ever since been known as the Great Red Spot. The color changes with time and some times is so pale the spot can hardly be noticed with a poor telescope. It is an oval 30,000 miles across from east to west and 8,000 miles from north to south, as seen from Earth.

Some astronomers wondered if the Great Red Spot was a vast tornado. In fact, Jupiter is so large and massive that there was some speculation that it might be much hotter than other planets—hot enough to be nearly red hot. The Great Red Spot might actually be a red-hot region. Nevertheless, although Jupiter must undoubtedly be extremely hot in its interior, its surface is not. In 1926, an American astronomer, Donald Howard Menzel, showed that Jupiter's temperature at the cloud layer we can see is $-135°$ C.

JUPITER'S SUBSTANCE

Because of its low density, Jupiter must be rich in material that is less dense than rocks and metals.

The most common materials in the universe generally are hydrogen and helium. Hydrogen atoms make up about 90 percent of all the atoms there are, and helium atoms make up another 9 percent. This fact may not be surprising when one considers that hydrogen atoms are the simplest in existence, with helium atoms second simplest. Of the atoms that remain, carbon, oxygen, nitrogen, neon, and sulfur make up the bulk. Hydrogen and oxygen atoms combine to form water molecules; hydrogen and carbon atoms combine to form methane molecules; hydrogen and nitrogen atoms combine to form ammonia molecules.

The density of all these substances under ordinary conditions is equal to or less than that of water. Under great pressures, as would prevail in Jupiter's interior, their densities might rise to be greater than that of water. If Jupiter consisted of such substances, they would account for its low density.

In 1932, a German astronomer, Rupert Wildt, studied the light reflected from Jupiter and found that certain wavelengths were absorbed—just those wavelengths that would be absorbed by ammonia and methane. He reasoned that these two substances, at least, are present in Jupiter's atmosphere.

In 1952, Jupiter was going to pass in front of the star Sigma Arietis—an event closely observed by two American astronomers, William Alvin Baum and Arthur Dodd Code. As the star approached Jupiter's globe, its light passed through the thin atmosphere above Jupiter's cloud layer. From the manner in which the light was dimmed, it was possible to show that the atmosphere was principally hydrogen and helium. In 1963, studies by an American astronomer, Hyron Spinrad, showed neon present as well.

All these substances are gases under earthly conditions; and if they make up a major portion of Jupiter's structure, it came to seem fair to call Jupiter a *gas giant*.

The first Jupiter probes were *Pioneer 10* and *Pioneer 11*, which were launched on 2 March 1972 and on 5 April 1973, respectively. *Pioneer 10* passed only 85,000 miles above Jupiter's visible surface on 3 December 1973. *Pioneer 11* passed only 26,000 miles above it one year later, on 2 December 1974—passing over the planet's north pole, which human beings thus saw for the first time.

The next pair of probes, more advanced, were *Voyager 1* and *Voyager 2*, which, respectively, were launched on 20 August and 5 September 1977. They passed by Jupiter in March and July of 1979.

These probes confirmed the earlier deductions about Jupiter's atmosphere. It was largely hydrogen and helium in about a ratio of 10 to 1 (just about the situation in the universe generally). Components not detected from Earth included ethane and acetylene (both combinations of carbon and hydrogen), water, carbon monoxide, phosphine, and germane.

Undoubtedly Jupiter's atmosphere has a very complicated chemistry, and we will not know enough about it until a probe can be sent into it and made to survive long enough to send back information. The Great Red Spot is (as most astronomers had suspected) a gigantic, more-than-Earth-size, and more or less permanent hurricane.

The whole planet seems to be liquid. The temperature rises rapidly with depth, and the pressures serve to turn the hydrogen into a red-hot liquid. At the center, there may be a core of white-hot metallic hydrogen in solid form. (Conditions in Jupiter's deep interior are too extreme to duplicate on Earth so far, and it may be some time before we can make firm estimates about it.)

THE JUPITER PROBES

The Jupiter probes took photographs of the four Galilean satellites at close quarters; and for the first time, human eyes saw them as something more than tiny, featureless disks.

More accurate information was obtained about their actual size and mass. These proved to involve only minor corrections, although Io, the innermost Galilean, was found to be a quarter more massive than had been thought.

Ganymede and Callisto, as one might have guessed from their low densities, are made up of light substances, such as water. At the low temperature one would expect from their distance from the sun (and as small bodies, without the great internal heat of Jupiter or even Earth), these substances are in solid form and are therefore referred to as *ices*. Both satellites are littered with numerous craters.

The satellites could be heated by the tidal influences of Jupiter, which tend to flex the substance of a satellite, creating heat by friction. Tidal influence decreases rapidly as distance increases. Ganymede and Callisto are far enough from Jupiter for tidal heating to be insignificant, and remain icy.

Europa is closer and was too warm at some early stage in its history to gather much in the way of ices; or, if it did, much of them melted, vaporized, and was lost to space in the course of that history. (The gravitational fields of the Galilean satellites are too small to hold an atmosphere in the presence of tidal heating.) It may be the inability of collecting the plentiful ices, or the loss after collecting, that makes Europa and Io distinctly smaller than Ganymede and Callisto.

Europa has retained enough of the ices to have a worldwide ocean (as Venus was once thought to have). At Europa's temperature the ocean is in the form of a worldwide glacier. What's more, this glacier is remarkably smooth (Europa is the smoothest solid world astronomers have yet encountered), although it is crisscrossed with thin, dark markings that make it look remarkably like Lowell's maps of the planet Mars.

The fact that the glacier is smooth and not punctured with craters leads one to suppose that it may be underlaid with liquid water, melted by tidal heating.

Meteoric strikes may (if large enough) break through the icy coating, but water will then well up and freeze, healing the break. Smaller strikes may cause fissures, which come and go; or the fissures may be caused by tidal effects or other factors. On the whole, though, the surface remains smooth.

Io, the innermost Galilean, receives the most tidal heating and is apparently completely dry. Even before the coming of the probes, it seemed puzzling. In 1974, the American astronomer Robert Brown reported Io to be surrounded by a yellow haze of sodium atoms. Indeed, it seemed to travel through a thin haze that filled its entire orbit like a doughnut circling Jupiter. Io had to be the source of the haze, but no one knew how.

The Pioneer probes showed that Io actually has a thin atmosphere about 1/20,000 the density of Earth's, and the Voyager probes solved the problem by taking photographs that showed that Io possesses active volcanoes. They are the only active volcanoes known to exist other than Earth's. Apparently regions of molten rock (heated by Jupiter's tidal action) lie under Io's surface and have, in various places, burst through the crust in sprays of sodium and sulfur, resulting in the atmosphere and the orbital doughnut. Io's surface is caked with sulfur, giving it a yellow to brown color. Io is not rich in craters, since these have been filled with volcano material. Only a few dark markings indicate craters too recent to have been filled in.

Within the orbit of Io is satellite Amalthea, which cannot be seen from Earth as anything but a dot of light. The Voyager probes showed Amalthea to be an irregular body, like the two satellites of Mars, but much larger. Amalthea's diameters vary from 165 to 87 miles.

Three additional satellites were discovered, each closer to Jupiter than Amalthea, and each considerably smaller than Amalthea. They are Jupiter XIV, Jupiter XV, and Jupiter XVI and have diameters estimated to be 15, 50, and 25 miles, respectively. Under present conditions, none of these satellites can possibly be seen from Earth, considering their size and their closeness to Jupiter's blaze.

Jupiter XVI is the closest to Jupiter, at a distance of only 80,000 miles from its center—that is, only 36,000 miles above its cloud surface. It races about Jupiter in 7.07 hours. Jupiter XIV is only slightly farther and completes an orbit in 7.13 hours. Both move about Jupiter faster than it rotates on its axis, and, if they could be observed from Jupiter's cloud layer, would (as in the case of Phobos, seen from Mars) seem to rise in the west and set in the east.

Within the orbit of the innermost satellite, there is débris which shows up as a thin, sparse ring of bits and pieces circling Jupiter. It is too thin and sparse to be seen from Earth in the ordinary fashion.

Saturn

Saturn was the most distant planet known to the ancients, for despite its distance it shines with considerable brightness. At its brightest, it has a magnitude of -0.75 and is then brighter than any star but Sirius. It is also brighter than Mercury and, in any case, easier to observe because, being farther from the sun than we are, it need not remain in its vicinity but can shine in the midnight sky.

Its average distance from the sun is 886.7 million miles, which makes it 1-5/6 times as far from the sun as Jupiter is. It revolves about the sun in 29.458 years, compared with a revolutionary period of 11.862 years for Jupiter. The Saturnian year is therefore 2½ times as long as the Jovian year.

In many respects, Saturn plays second fiddle to Jupiter. In size, for instance, it is the second largest planet after Jupiter. Its equatorial diameter is 74,600 miles, only about five-sixths that of Jupiter. It is this smaller size, together with its great distance, that makes the sunlight bathing it only half as intense as the sunlight on Jupiter, making it much dimmer than Jupiter. On the other hand, Saturn is still large enough to make a respectable showing.

Saturn's mass is 95.1 times that of Earth, making it the second most massive planet after Jupiter. Its mass is only three-tenths that of Jupiter, and yet its volume is six-tenths of Jupiter's.

To have so little a mass in so large a volume, Saturn's density must be very low; and, indeed, it is the least dense object we know in the solar system, having on the whole, a density only 0.7 times that of water. If we could imagine Saturn wrapped in plastic to keep it from dissolving or dispersing, and if we could find an ocean large enough, and if we placed Saturn in the ocean, it would float. Presumably, Saturn is made up of material that is even richer in very light hydrogen, and poorer in everything else, than Jupiter is. Then, too, Saturn's weaker gravity cannot compress the substance composing it as tightly as Jupiter can compress its substance.

Saturn rotates quickly; but, even though it is the somewhat smaller body, it does not rotate as quickly as Jupiter. Saturn rotates on its axis in 10.67 days, so that its day is 8 percent longer than Jupiter's.

Even though Saturn spins more slowly than Jupiter does, Saturn's outer layers are less dense than Jupiter's, and it has a smaller gravitational pull to hold them. As a result, Saturn has the larger equatorial bulge and is the most oblate object in the solar system. Its oblateness is 0.102: it is 1.6 times as oblate as Jupiter and 30 times as oblate as Earth. Although Saturn's equatorial diameter is 74,600, its polar diameter is only 67,100. The difference is 7,500 miles, nearly the total diameter of Earth.

SATURN'S RINGS

In another respect, Saturn turns out to be unique—and uniquely beautiful. When Galileo first looked at Saturn through his primitive telescope, it seemed to him to have an odd shape, as though its globe was flanked by two small globes. He continued to observe, but the two small globes grew progressively harder to see and finally, toward the end of 1612, disappeared altogether.

Other astronomers also reported something peculiar in connection with Saturn, but it was not till 1656 that Christian Huygens interpreted the matter rightly. He reported that Saturn was encircled by a bright, thin ring that nowhere touched it.

Saturn's axis of rotation is tipped as Earth's is; Saturn's axial tipping is 26.73 degrees, compared with Earth's 23.45 degrees. Saturn's ring is in its equatorial plane, so it is tipped with respect to the sun (and to us). When Saturn is at one end of its orbit, we look down from above upon the near side of the ring, while the far side remains hidden. When Saturn is at the other end of its orbit, we look up from below to the near side of the ring, while the far side remains hidden. It takes a little over 14 years for Saturn to go from one side of its orbit to the other. During that time, the ring slowly shifts from far down to far up. Halfway along the road, the ring is exactly halfway between, and we see it edge on. Then, during the other half of the orbit, when Saturn is traveling from the other side back to the starting point, the ring slowly shifts from far up to far down again; and halfway between, we see it edge on again. Twice every Saturnian orbit, or every fourteen years and a bit, the ring is seen edge on. The ring is so thin that, at edge-on times, it simply disappears. Such was the situation when Galileo was observing, at the end of 1612; and, out of chagrin (according to one story), he never looked at Saturn again.

In 1675, Cassini noticed that Saturn's ring was not an unbroken curve of light. There was a dark line all around the ring, dividing it into an outer and an inner section. The outer section was narrower and not as bright as the inner one. There were two rings, it seemed, one inside the other; and ever since, Saturn's rings have been referred to in the plural (figure 4.9). The dark line is now called Cassini's Division.

The German-Russian astronomer Friedrich G. W. von Struve called the outer ring Ring A in 1826, and the inner one Ring B. In 1850, an American astronomer, William Cranch Bond, reported a dim ring still closer to Saturn than Ring B. This dim ring is Ring C, and there is no noticeable division between it and Ring B.

There is nothing like Saturn's rings anywhere in the solar system or, for that matter, anywhere that we can see with any of our instruments. To be sure, we now know that there is a thin ring of matter around Jupiter, and it is possible that any gas-giant planet, like Jupiter or Saturn, may have a ring of débris close to itself. If Jupiter's ring is typical, however, they are poor, puny things; but Saturn's ring system is magnificent. From extreme end to extreme end Saturn's ring system, as seen from Earth, stretches across a distance of 167,600 miles. This is 21 times the width of Earth and, in fact, almost twice the width of Jupiter.

What are Saturn's rings? Cassini thought they were smooth, solid objects like thin quoits. In 1785, though, Laplace (who was later to advance the nebular hypothesis) pointed out that different parts of the rings were at different distances from Saturn's center and would be subject to different degrees of pull from Saturn's gravitational field. Such difference in gravitational pull is the tidal effect I have mentioned earlier, and would tend to pull the ring apart. Laplace thought the rings might be a series of very thin rings set so close together that they would look solid from the distance of Earth.

In 1855, however, Maxwell (who was later to predict the existence of a broad band of electromagnetic radiation) showed that even this suggestion would not suffice. The only way the rings could resist disruption by tidal effect was for them to consist of relatively small particles of countless meteorites distributed about Saturn in such a way as to give the impression of solid rings from the distance of Earth. There has been no doubt since that Maxwell was correct in this hypothesis.

Working on the matter of tidal effects in another way, a French astronomer, Edouard Roche, showed that any solid body approaching another considerably larger body would feel powerful tidal forces that would eventually tear the former into fragments. The distance at which the smaller body would be torn apart is the *Roche limit* and is usually given as 2.44 times the *equatorial radius* (the distance from the center to a point on the equator) of the larger body.

Thus, Saturn's Roche limit is 2.44 times the planet's equatorial radius of 37,300 miles (half the equatorial diameter), or 91,000 miles. The outermost edge of Ring A is 84,800 miles from Saturn's center, so that the entire ring system lies within the Roche limit. (Jupiter's ring lies within its Roche limit, too.)

Apparently, Saturn's rings represent débris that could never coalesce into a satellite (as débris beyond the Roche limit would—and apparently did) or was from a satellite that had ventured too close for some reason and was torn apart. Either way, they remained a collection of small bodies. (Tidal effect diminishes as the body being affected decreases in size; at some point, the fragments are so small that further fragmentation stops, except perhaps through the occasional collision of two small bodies.) According to some estimates, if the material in the rings of Saturn were collected into one body, the result would be a sphere slightly larger than our moon.

THE SATURNIAN SATELLITES

In addition to the rings, Saturn, like Jupiter, has a family of satellites. A Saturnian satellite was discovered for the first time by Huygens in 1656, the same year he discovered the rings. Two centuries later, the satellite received the name Titan, which was the class of deity to which Saturn (Cronos) belonged in the Greek myths. Titan is a large body, almost (though not quite) the size of Ganymede. It is, moreover, less dense than Ganymede, so that the discrepancy in mass is still greater. It is, nevertheless, the second largest known satellite in the solar system, whether diameter or mass is taken as the criterion.

In one respect, Titan is (so far) at the head of the class. Farther from the sun, and therefore colder, than Jupiter's satellites, it is better able to hold the molecules of gas, rendered sluggish by cold, despite its small surface gravity. In 1944, the Dutch-American astronomer, Gerard Peter Kuiper, was able to detect an undeniable atmosphere about Titan and found it to contain methane. The molecules of methane are made up of 1 carbon atom and 4 hydrogen atoms (CH_4), and it is the chief constituent of natural gas on Earth.

At the time of the discovery of Titan, five other satellites were known altogether: the moon, and Jupiter's four Galilean satellites. All were roughly the same size, much more similar in size than the known planets were. Between 1671 and 1684, however, Cassini discovered no fewer than four additional Saturnian satellites, each with a diameter considerably less than that of Europa, the smallest of the Galileans. The diameters ranged from 900 miles for the largest of Cassini's discoveries (now known as Iapetus) to 650 miles for the smallest (Tethys). From then on, it was understood that satellites could be quite small.

By the end of the nineteenth century, nine satellites of Saturn were known. The last of the nine to be discovered was Phoebe, first detected by the American astronomer William Henry Pickering. It is by far the farthest of the satellites and is at an average distance from Saturn of 8 million miles. It revolves about Saturn in 549 days in the retrograde direction. It is also the smallest of the satellites (hence, its late discovery, since smallness implies dimness), with a diameter of about 120 miles.

Between 1979 and 1981, three probes that had previously passed Jupiter— *Pioneer 11, Voyager 1,* and *Voyager 2*—offered closeup looks at Saturn itself, its rings, and the satellites.

Titan was, of course, a prime target, because of its atmosphere. Some radio signals from *Voyager 1* skimmed through Titan's atmosphere on their way to Earth. Some signal energy was absorbed; and from the detail of that absorption, it was calculated that Titan's atmosphere was unexpectedly dense. From the quantity of methane detected from Earth, it had been thought that Titan might have an atmosphere as dense as that of Mars. Not so. It was 150 times as dense as the Martian atmosphere and, indeed, was perhaps 1.5 times as dense as Earth's.

The reason for this surprising figure was that only methane had been detected from Earth, and if it were the only constituent of Titan's atmosphere, the atmosphere would have been thin. However, methane makes up only 2 percent of the Titanian atmosphere, the rest being nitrogen, a gas difficult to detect by its absorption characteristics.

The thick Titanian atmosphere is smog-filled, and no view of the solid surface was possible. That very smog is full of interest, however. Methane is a molecule that can easily *polymerize*—that is, combine with itself to form larger molecules. Thus, scientists are free to speculate on a Titan that may have oceans or sludge made up of fairly complicated carbon-containing molecules. In fact, we can even amuse ourselves with the possibility that Titan is coated with asphalt, with outcroppings of solidified gasoline, and has sparkling lakes of methane and ethane.

The other Saturnian satellites are, as might be expected, cratered. Mimas, the innermost of the nine satellites, has one so large (considering the satellite's size) that the impact that produced it must have nearly shattered the world.

Enceladus, the second of the nine, is comparatively smooth, however, and may have been partially melted through tidal heating. Hyperion is the least spherical and has a diameter that varies from 70 to 120 miles. It is shaped rather like the Martian satellites but is much larger, of course—large enough for one to suppose that it ought to be reasonably spherical as a result of its own gravitational pull. Perhaps it was recently fractured.

Iapetus, from its original discovery in 1671, had its peculiarity, being five times as bright when west of Saturn as when east. Since Iapetus always keeps one face turned toward Saturn, we see one hemisphere when it is on one side of Saturn, and the other hemisphere when it is on the other side. The natural guess was that one hemisphere happens to reflect sunlight five times as efficiently as the other. Photographs by *Voyager 1* confirmed this guess. Iapetus is light and dark, as though one side were icy and the other coated with dark dust. The reason for this difference is not known.

The Saturn probes succeeded in finding eight small satellites that were too small to be detected from Earth, bringing the total number of Saturnian satellites to seventeen. Of these eight new satellites, five are closer to Saturn than Mimas is. The closest of these satellites is only 85,000 miles from Saturn's center (48,000 miles above the Saturnian cloud cover) and revolves about the planet in 14.43 hours.

Two satellites that are just inside Mimas's orbit are unusual in being *co-orbital:* that is, they share the same orbit, chasing each other around Saturn endlessly. It was the first known example of such co-orbital satellites. They are at a distance of 94,000 miles from Saturn's center and revolve about the planet in 16.68 hours. In 1967, a French astronomer, Audouin Dollfus, reported a satellite inside Mimas's orbit and named it Janus. This, which was probably the result of sighting one or another of the intra-Mimas satellites, yielded erroneous orbital data because different ones may have been noted at different times. Janus is no longer included on the Saturnian satellite list.

The three remaining newly discovered satellites also represent unprecedented situations. The long-known satellite Dione, one of Cassini's discoveries, was found to have a tiny co-orbital companion. Whereas Dione has a diameter of 700 miles, the companion (Dione-B) has a diameter of only about 20 miles. Dione-B, in circling Saturn, remains at a point 60 degrees ahead of Dione. As a result, Saturn, Dione, and Dione-B are always at the apices of an equilateral triangle. This is a *Trojan situation,* for reasons I shall explain later.

Such a situation, only possible when the third body is much smaller than the first two, can take place when the small body is 60 degrees ahead or behind the larger body. Ahead, it is in the L-4 position, behind, it is in the L-5 position. Dione-B is in the L-4 position. (*L* stands for the Italian-French astronomer Joseph Louis Lagrange, who, in 1772, worked out the fact that such a configuration was gravitationally stable.)

Then there is Tethys, still another of Cassini's satellites. It has two co-orbital companions: Tethys-B in the L-4 position and Tethys-C in the L-5 position.

Clearly, the Saturnian satellite family is the richest and most complex in the solar system, as far as we now know.

The Saturnian rings are also far more complex than had been thought. From a close view, they consist of hundreds, perhaps even thousands of thin ringlets, looking like the grooves on a phonograph record. In places, dark streaks show up at right angles to the ringlets, like spokes on a wheel. Then, too, a faint outermost ring seems to consist of three intertwined ringlets. None of this can be explained so far, though the general feeling is that a straightforward gravitational explanation must be complicated by electrical effects.

The Outermost Planets

In the days before the telescope, Saturn was the farthest planet known and the one that moved most slowly. It was also the dimmest, but it was still a first-magnitude object. For thousands of years after the recognition that planets existed, there seems to have been no speculation about the possibility that there might be planets too distant, and therefore too dim, to be visible.

URANUS
Even after Galileo had demonstrated that there are myriads of stars too dim to be seen without a telescope, the possibility of dim planets does not seem to have made much of a stir.

And then, on 13 March 1781, William Herschel (not yet famous) was making measurements of star positions when, in the constellation of Gemini, he found himself staring at an object that was not a point of light but, instead, showed a small disk. At first, he assumed it was a distant comet, for comets were the only objects, other than planets, that showed up as disks under telescopic observation. However, comets are hazy, and this object showed sharp edges. Furthermore, it moved against the starry background more slowly than Saturn and, therefore, had to be farther away. It was a distant planet, much farther away than Saturn, and much dimmer. The planet was eventually named Uranus (*Ouranos,* in the Greek form) for the god of the sky and the father of Saturn (Cronos) in the Greek myths.

Uranus is 1,783,000,000 miles from the sun on the average and is thus just about exactly twice as far from the sun as Saturn is. Uranus is, furthermore, smaller than Saturn, with a diameter of 32,200 miles. This is four times the diameter of Earth, and Uranus is still a gas giant like Jupiter and Saturn but is much smaller than those two. Its mass is 14.5 times that of Earth, but only 1/6.6 that of Saturn and 1/22 that of Jupiter.

Because of its distance and its relatively small size, Uranus is much dimmer in appearance than either Jupiter or Saturn. It is not, however, totally invisible to the unaided eye. If one looks in the right place on a dark night, Uranus is visible as a very faint star, even without a telescope.

Might not astronomers have detected it, then, even in ancient times? They undoubtedly did, but a very dim star did not attract attention when planets were assumed to be bright. And even if it were looked at in successive nights, its motion is so small that its change of position might not have been noticed. What's more, early telescopes were not very good and, even when pointed in the right direction, did not show Uranus's small disk clearly.

Still, in 1690, the English astronomer John Flamsteed listed a star in the constellation Taurus and even gave it the name of 34 Tauri. Later astronomers could not locate that star; but once Uranus was discovered and its orbit worked out, a backward calculation showed that it was in the place that Flamsteed had reported 34 Tauri to be. And half a century later, the French astronomer Pierre Charles Lemonnier saw Uranus on thirteen different occasions and recorded it in thirteen different places, imagining he had seen thirteen different stars.

There are conflicting reports on its period of rotation. The usual figure is 10.82 hours; but in 1977, the period was claimed to be 25 hours. We probably will not be certain until probe data is received.

One certainty about Uranus's rotation rests with its axial tipping. The axis is tipped through an angle of 98 degrees, or just a little more than a right angle. Thus, Uranus, as it revolves about the sun once every eighty-four years, seems to be rolling on its side; and each pole is exposed to continuous sunlight for forty-two years and then to continuous night for forty-two years.

At Uranus's distance from the sun, that makes very little difference. If Earth rotated in this fashion, however, the seasons would be so extreme that it is doubtful whether life would ever have developed upon it.

After Herschel discovered Uranus, he kept observing it at intervals and, in 1787, detected two satellites, which were eventually named Titania and Oberon. In 1851, the English astronomer William Lassell discovered two more satellites, closer to the planet, which were named Ariel and Umbriel. Finally, in 1948, Kuiper detected a fifth planet, closer still; it is Miranda.

All the Uranian satellites revolve about Uranus in the plane of its equator, so that not only the planet, but its satellite system, seem to be turned on its side. The satellites move north and south of the planet, rather than east and west as is usual.

The Uranian satellites are all fairly close to Uranus. There are no distant ones (at least, that we can see). The farthest of the five that are known is Oberon, which is 364,000 miles from Uranus's center, only half again as far as the moon is from Earth. Miranda is only 80,800 miles from Uranus's center.

None of the satellites are large ones after the fashion of the Galilean satellites, Titan, or the moon. The largest is Oberon, which is just about 1,000 miles across; while the smallest is Miranda, with a diameter of 150 miles.

For a long time, there seemed to be nothing particularly exciting about the Uranian satellite system; but then, in 1973, a British astronomer, Gordon Tayler, calculated that Uranus would move in front of a ninth-magnitude star, SA0158687. This event excited astronomers, for as Uranus passed in front of the star, there would come a period, just before the star was blanked out, when its light would pass through the upper atmosphere of the planet. Again, just as the star emerged from behind the planet, it would pass through its upper atmosphere. The fate of the starlight as it passed through the atmosphere might well tell astronomers something about the temperature, the pressure, and the composition of Uranus's atmosphere. The *occultation* was scheduled to take place on 10 March 1977. In order to observe it, on that night an American astronomer, James L. Elliot, and several associates were in an airplane carrying them high above the distorting and obscuring effects of the lower atmosphere.

Before Uranus reached the star, the starlight suddenly dimmed for about 7 seconds and then brightened again. As Uranus continued to approach, there were four more brief episodes of dimming for 1 second each. When the star emerged on the other side, there were the same episodes of dimming in reverse order. The only way of explaining this phenomenon was to suppose that there were thin rings of matter about Uranus—rings not ordinarily visible from Earth because they were too thin, too sparsely filled, too dark.

Careful observation of Uranus during occultations of other stars, such as one on 10 April 1978, showed a total of nine rings. The innermost one is 25,200 miles from the center of Uranus, and the outermost one is 30,500 miles from the center. The entire ring system is well within Roche's limit.

It can be calculated that the Uranian rings are so thin, so sparse, and so dark that they are only 1/3,000,000 as bright as Saturn's rings. It is no surprise that the Uranian rings are hard to detect in any fashion but this indirect one.

Later, when Jupiter's ring was detected, it began to seem that rings were not such an unusual phenomenon after all. Perhaps all gas giants have a ring system in addition to numerous satellites. The only thing that makes Saturn unique is not that it has rings, but that those rings are so extensive and bright.

NEPTUNE

Soon after Uranus was discovered, its orbit was worked out. However, as the years passed, it was found that Uranus was not following the orbit as calculated —not quite. In 1821, the French astronomer Alexis Bouvard recalculated Uranus's orbit, taking into account early observations such as that of Flamsteed. Uranus did not quite follow the new orbit either.

The tiny pull on Uranus of the other planets (*perturbations*) slightly affected Uranus's motion, causing it to lag behind, or pull ahead, of its theoretical position by a very small amount. These effects were recalculated carefully, but still Uranus did not behave correctly. The logical conclusion was that, beyond Uranus, there might be an unknown planet exerting a gravitational pull that was not being allowed for.

In 1841, a twenty-two-year-old mathematics student at Cambridge University in England tackled the problem and worked at it in his spare time. His name was John Couch Adams; and by September 1845, he had finished. He had calculated where an unknown planet ought to be located if it were to travel in such a way as to account for the missing factor in Uranus's orbit. However, he could not get English astronomers interested in his project.

Meanwhile, a young French astronomer, Urbain Jean Joseph Leverrier, was also working on the problem quite independently. He completed his work about half a year after Adams and got just about the same answer that Adams did. Leverrier was fortunate enough to get a German astronomer, Johann Gottfried Galle, to check the indicated region of the sky for the presence of an unknown planet. Galle happened to have a new chart of the stars in that portion of the sky. He began his search on the night of 23 September 1846, and he and his assistant, Heinrich Ludwig D'Arrest, had barely been working an hour when they found an eighth-magnitude object that was not on the chart.

It was the planet! And it was nearly at the spot where the calculations had said it would be. It was eventually named after Neptune, the god of the sea, because of its greenish color. The credit for its discovery is nowadays divided equally between Adams and Leverrier.

Neptune travels about the sun in an orbit that places it about 2,800,000,000 miles away, so that it is more than half again as far from the sun as Uranus is (and 30 times as far from the sun as Earth is). It completes one revolution about the sun in 164.8 years.

Neptune is the twin of Uranus (much as Venus is the twin of Earth; at least in dimensions). Neptune's diameter is 30,800 miles, just a bit smaller than that of Uranus, but the former is denser and is 18 percent more massive than Uranus. Neptune is 17.2 times as massive as Earth and is the fourth gas giant circling the sun.

On 10 October 1846, less than three weeks after Neptune was first sighted, a Neptunian satellite was detected and named Triton, after a son of Neptune (Poseidon) in the Greek myths. Triton turned out to be another of the large satellites, with a mass nearly equal to that of Titan. It was the seventh such satellite to be discovered, and the first since the discovery of Titan nearly two centuries before.

Its diameter is about 2,400 miles, making it a bit larger than our moon; and its distance from Neptune's center is 221,000 miles, almost the distance of Earth from its moon. Because of Neptune's greater gravitational pull, Triton completes one revolution in 5.88 days, or in about one-fifth the time our moon takes.

Triton revolves about Neptune in the retrograde direction. It is not the only satellite to revolve in this way. The others, however (Jupiter's four outermost satellites, and Saturn's outermost satellite), are all very small and very distant from the planet they circle. Triton is large and is close to its planet. Why it should follow a retrograde orbit remains a mystery.

For over a century, Triton remained Neptune's only known satellite. Then, in 1949, Kuiper (who had discovered Miranda the year before) detected a small and very dim object in Neptune's neighborhood. It was another satellite and was named Nereid (the sea nymphs of the Greek myths).

Nereid has a diameter of about 150 miles and travels about Neptune in direct fashion. It has, however, the most eccentric orbit of any known satellite. At its closest approach to Neptune, it is 864,000 miles away; but at the other end of its orbit, it is 6,050,000 miles away. It is, in other words, seven times as far away from Neptune at one end of its orbit as at the other end. Its period of revolution is 365.21 days, or 45 minutes less than 1 Earth-year.

Neptune has not yet been visited by a probe, so it is no surprise that we know of no other satellites or of a ring system. We do not even know whether Triton has an atmosphere, although since Titan does, Triton may well have one, too.

PLUTO

Neptune's mass and position accounted for most of the discrepancy in Uranus's orbit. Still, to account for the rest, some astronomers thought that an unknown planet even more distant than Neptune ought to be searched for. The astronomer most assiduous in his calculations and search was Lowell (who had become famous for his views on the Martian canals).

The search was not easy. Any planet beyond Neptune would be so dim that it would be lost in the crowds of equally dim ordinary stars. What's more, such a planet would move so slowly that its change in position would not be easy to detect. By the time Lowell died in 1916, he had not found the planet.

Astronomers at the Lowell Observatory in Arizona, however, continued the search after Lowell's death. In 1929, a young astronomer, Clyde William Tombaugh, took over the search, using a new telescope that could photograph a comparatively large section of the sky very sharply.

He also made use of a *blink comparator*, which could project light through one photographic plate taken on a certain day, and then through another plate of the same star region taken a few days later, and so on in rapid alternation. The plates were adjusted so that the stars on each were focused on the same spot. The true stars would remain perfectly steady as the light flashed through first one plate, then the other. Any dim planet present, however, would have altered position so as to be present here, there, here, there, in rapid alternation. It would blink.

Discovery was not easy even so, for a particular plate would have many tens of thousands of stars on it and would have to be narrowly scanned at every part to see if one of those myriads was blinking.

But at 4 P.M. on 18 February 1930, Tombaugh was studying a region in the constellation of Gemini and found a blink. He followed that object for nearly a month and, on 13 March 1930, announced he had found the new planet. It was named Pluto, after the god of the underworld, since it was so far from the light of the sun. Besides, the first two letters of the name were the initials of Percival Lowell.

Pluto's orbit was worked out and turned out to have numerous surprises. It was not as far from the sun as Lowell and other astronomers had thought it would be. Its average distance from the sun turned out to be only 3,670,000,000 miles, only 30 percent farther than Neptune.

Furthermore, the orbit was more eccentric than that of any other planet. At its farthest point from the sun, Pluto was 4,600,000,000 miles away; but at the opposite end of its orbit, when nearest the sun, it was only 2,700,000,000 miles away.

At perihelion, when Pluto is nearest the sun, it is actually closer to the sun than Neptune is by about 100,000,000 miles. Pluto circles the sun in 247.7 years; but in each of these revolutions, there is a twenty-year period when it is closer than Neptune and is not the farthest planet. As it happens, one of those periods is the last two decades of the twentieth century, so that now, as I write, Pluto is closer than Neptune.

Pluto's orbit does not really cross Neptune's, however, but is strongly tilted compared with the other planets. It is inclined to Earth's orbit by about 17.2 degrees, while Neptune's orbit is inclined only slightly to Earth's. When Pluto and Neptune's orbits cross, therefore, and both are at the same distance from the sun, one is far below the other. As a consequence, the two planets never approach each other at a distance of less than 1,500,000,000 miles.

The most disturbing thing about Pluto, however, was its unexpected dimness, which indicated at once that it is no gas giant. If it were anywhere near the size of Uranus or Neptune, it would have been considerably brighter. The initial estimate was that it might be the size of Earth.

Even this turned out to be an overestimate. In 1950, Kuiper managed to see Pluto as a tiny disk; and when he measured the width of the disk, he felt that it could only be 3,600 miles in diameter, rather less than the diameter of Mars. Some astronomers were reluctant to believe this estimate; but on 28 April 1965, Pluto passed very close to a faint star and did not get in front of it. If Pluto were larger than Kuiper had estimated, it would have obscured the star.

Thus, it was clear that Pluto is too small to influence Uranus's orbit in any perceptible way. If a distant planet accounted for the last bit of discrepancy in the Uranian orbit, Pluto was not it.

In 1955, it was noted that Pluto's brightness varied in a regular way that repeats itself every 6.4 days. It was assumed that Pluto rotates on its orbit in 6.4 days—an unusually long period of rotation. Mercury and Venus have still longer periods but are strongly affected by tidal influences of the nearby sun. What was Pluto's excuse?

Then, on 22 June 1978 came a discovery that seemed to provide it. On that day, the American astronomer James W. Christy, examining photographs of Pluto, noticed a distinct bump on one side. He examined other photographs and finally decided that Pluto has a satellite. It is quite close to Pluto, not more than 12,500 miles away, center to center. At the distance of Pluto, that is a very slight separation to detect; hence, the long delayed discovery. Christy named the satellite Charon, after the ferryman, in the Greek myths, who takes shades of the dead across the River Styx to Pluto's underworld kingdom.

Charon circles Pluto in 6.4 days, which is just the time it takes for Pluto to turn on its axis. This is not a coincidence. It must be that the two bodies, Pluto and Charon, have slowed each other by tidal action until each always faces the same side to the other. They now revolve about a common center of gravity, like the two halves of a dumbbell held together by gravitational pull.

This is the only planet-satellite combination to revolve dumbbell-fashion. Thus in the case of the moon and Earth, the moon always faces one side to Earth, but Earth has not yet been slowed to the point of always facing one side to the moon, because the former is much larger and would take much more slowing. If Earth and the moon were more equal in size, the dumbbell fashion of revolution might have resulted.

From the distance between them and the time of revolution, it is possible to work out the total mass of both bodies: it turns out to be only about one-eighth the mass of the moon. Pluto is far smaller than even the most pessimistic estimates.

From the comparative brightness of the two, Pluto seems to be only 1,850 miles in diameter, almost the size of Europa, the smallest of the seven large satellites. Charon is 750 miles in diameter, about the size of Saturn's satellite Dione.

The two objects are not far apart in size. Pluto is probably only 10 times as massive as Charon; whereas Earth is 81 times as massive as the moon. That size differential accounts for why Pluto and Charon revolve about each other dumbbell-fashion, while Earth and the moon do not. Pluto/Charon is the closest thing in the solar system that we know of to a "double planet." Until 1978, it had been thought that Earth/moon was.

Asteroids

ASTEROIDS BEYOND MARS'S ORBIT

Each planet, with one exception, is somewhere between 1.3 and 2.0 times as far from the sun as the next nearer planet. The one exception is Jupiter, the fifth planet: it is 3.4 times as far from the sun as Mars, the fourth planet, is.

This extraordinary gap puzzled astronomers after the discovery of Uranus (at which time, the possibility of new planets became exciting). Could there be a planet in the gap—a 4½th planet, so to speak, one that had evaded notice all this time? A German astronomer, Heinrich W. M. Olbers, led a group who planned to engage in a systematic search of the skies for such a planet.

While they were making their preparations, an Italian astronomer, Giuseppe Piazzi, who was observing the heavens without any thought of new planets, came across an object that shifted position from day to day. From the speed of its movement, it seemed to lie somewhere between Mars and Jupiter; and from its dimness, it had to be very small. The discovery was made on 1 January 1801, the first day of the new century.

From Piazzi's observations, the German mathematician Johann K. F. Gauss was able to calculate the object's orbit; and, indeed, it was a new planet with an orbit lying between that of Mars and Jupiter, exactly where it ought to have been to make the distribution of the planets even. Piazzi, who had been working in Sicily, named the new planet Ceres, after a Roman goddess of grain who had been particularly associated with the island.

From its dimness and distance, it was calculated that Ceres had to be very small indeed, far smaller than any other planet. The latest figures show it to be about 620 miles in diameter. Ceres probably has a mass only about one-fiftieth that of our moon and is much smaller than the larger satellites.

It did not seem possible that Ceres was all there was in the gap between Mars and Jupiter, so Olbers continued the search despite Piazzi's discovery. By 1807, sure enough, three more planets were discovered in the gap. They were named Pallas, Juno, and Vesta; and each is even smaller than Ceres. Juno, the smallest, may be only 60 miles in diameter.

These new planets are so small that in even the best telescope of the time they did not show a disk. They remained points of light, as the stars did. In fact, for this reason, Herschel suggested they be called asteroids ("starlike"), and the suggestion was adopted.

It was not till 1845 that a German astronomer, Karl L. Hencke, discovered a fifth asteroid, which he named Astraea; but after that, further discoveries were made steadily. By now, over 1,600 asteroids have been detected, every one of them considerably smaller than Ceres, the first to be detected; and undoubtedly thousands more are as yet undetected. Almost all of them are in the gap between Mars and Jupiter, a gap now referred to as the asteroid belt.

Why should the asteroids exist? Quite early, when only four asteroids were known, Olbers suggested that they were the remnants of an exploded planet. Astronomers are, however, dubious about this possibility. They consider it more likely that the planet never formed: whereas in other regions the matter of the original nebula gradually coalesced into planetesimals (equivalent to asteroids) and these into individual planets (with the last ones joining leaving their marks as craters), in the asteroid belt, coalescence never went past the planetesimal stage. The feeling is that the perturbing effect of giant Jupiter, nearby, was responsible.

By 1866, enough asteroids had been discovered to show that they were not spread evenly through the gap. There were regions where asteroidal orbits were absent. There were no asteroids with an average distance from the sun of 230 million miles, or 275 million miles, or 305 million miles, or 340 million miles.

An American astronomer, Daniel Kirkwood, suggested in 1866 that in these orbits, asteroids would circle the sun in a period that was a simple fraction of that of Jupiter. Under such conditions, Jupiter's perturbing effect would be unusually large, and any asteroid circling there would be forced either closer to the sun or farther from it. These *Kirkwood gaps* made it clearer that Jupiter's influence was pervasive and could prevent coalescence.

A still closer connection between Jupiter and the asteroids became clear later. In 1906, a German astronomer, Max Wolf, discovered asteroid 588. It was unusual because it moved at a surprisingly slow speed and therefore had to be surprisingly far from the sun. It was, in fact, the farthest asteroid yet discovered. It was named Achilles after the Greek hero of the Trojan War. (Though asteroids are usually given feminine names, those with unusual orbits are given masculine names.)

Careful observation showed Achilles to be moving in Jupiter's orbit, 60 degrees ahead of Jupiter. Before the year was over, asteroid 617 was discovered in Jupiter's orbit, 60 degrees behind Jupiter, and was named Patroclus, after Achilles' friend in Homer's *Iliad*. Other asteroids were discovered clustering about each of these, and all were named after heroes of the Trojan War. This was the first case of the discovery of actual examples of stability when three bodies are found at the apices of an equilateral triangle. Hence, the situation came to be called *Trojan positions,* and the asteroids *Trojan asteroids.* Achilles and its group occupy the L-4 position, and Patroclus and its group the L-5 position.

The outer satellites of Jupiter, which seem to be captured satellites, may once have been Trojan asteroids.

Saturn's outermost satellite, Phoebe, and Neptune's outer satellite, Nereid, may conceivably also be captured satellites—an indication that at least a scattering of asteroids exist in the regions beyond Jupiter. Perhaps these originally existed in the asteroid belt and, through particular perturbations, were forced outward, where eventually they were captured by particular planets.

In 1920, for instance, Baade discovered asteroid 944, which he called Hidalgo. When its orbit was calculated, this asteroid was found to move outward far beyond Jupiter and to have an orbital period of 13.7 years—three times that of the average asteroid and even longer than Jupiter's.

It has a high orbital eccentricity of 0.66. At perihelion it is only about 190 million miles from the sun, so that it is neatly within the asteroid belt at that time. At aphelion, however, it is 895 million miles from the sun—as far then from the sun as Saturn is. Hidalgo's orbit is so tipped, however, that when it is at aphelion, it is far below Saturn and is in no danger of being captured; but another satellite on such a far-flung orbit might be closer to Saturn and eventually might be captured by it or by another of the outermost planets.

Might not an asteroid be so affected by gravitational perturbation as to take up an orbit far beyond the asteroid belt at all times? In 1977, the American astronomer Charles Kowall detected a very dim speck of light that moved against the starry background, but at only one-third the speed of Jupiter. It had to be far outside Jupiter's orbit.

Kowall followed it for a period of days, worked out an approximate orbit, then started looking for it in older photographic plates. He located it on some thirty plates, one dating back to 1895, and had enough position to plot an accurate orbit.

It is a sizable asteroid, perhaps 120 miles in diameter. When closest to the sun, it is as near to the sun as Saturn is. At the opposite end of its orbit, it is as far from the sun as Uranus is. It seems to shuttle between Saturn and Uranus, although, because its orbit is tipped, it does not approach very close to either.

Kowall gave it the name Chiron, after one of the centaurs (half-man, half-horse) in Greek myth. Its period of revolution is 50.7 years and at the moment is close to its aphelion point. In a couple of decades, it will be at less than half the distance from us, and we may be able to see it more clearly.

EARTH GRAZERS AND APOLLO OBJECTS

If asteroids penetrate beyond Jupiter's orbit, might there not be others that penetrate within Mars's orbit, closer in to the sun?

The first such case was discovered on 13 August 1898 by a German astronomer, Gustav Witt. He detected asteroid 433 and found that its period of revolution to be only 1.76 years—44 days less than that of Mars. Hence, its average distance from the sun has to be less than that of Mars. The new asteroid was named Eros.

Eros, it turned out, has a fairly high orbital eccentricity. At aphelion, it is well within the asteroid belt; but at perihelion, it is only 105 million miles from the sun, not much more than the distance of Earth from the sun. Because its orbit is tipped to that of Earth, it does not approach the latter as closely as it would if both orbits were in the same plane.

Still, if Eros and Earth are at the proper points in their orbits, the distance between them could be only 14 million miles. This is only a little over half the minimum distance of Venus from Earth and means that, if we do not count our own moon, Eros was, at the time of its discovery, our closest known neighbor in space.

It is not a large body. Judging from changes in its brightness, it is brick-shaped, and its average diameter is ten miles across. Still, this is not to be sneezed at. If it were to collide with Earth, it would be an unimaginable catastrophe.

In 1931, Eros approached a point only 16 million miles from Earth; and a vast astronomical project was set up to determine its parallax accurately, so that the distances of the solar system could be determined more accurately than ever. The project succeeded, and the results were not improved upon until radar beams were reflected from Venus.

An asteroid that can approach Earth more closely than Venus can is called (with some exaggeration) an *Earth grazer*. Between 1898 and 1932, only three more Earth grazers were discovered, and each of those approached Earth less closely than Eros did.

The record was broken, however, on 12 March 1932, when a Belgian astronomer, Eugène Delporte, discovered asteroid 1221 and found that though its orbit was similar to that of Eros, it managed to approach within 10 million miles of Earth's orbit. He named the new asteroid Amor (the Latin equivalent of Eros).

On 24 April 1932, just six weeks later, the German astronomer Karl Reinmuth discovered an asteroid he named Apollo, because it was another Earth grazer. It is an astonishing asteroid, for at perihelion, it is only 60 million miles from the sun. It moves not only inside Mars's orbit but inside Earth's as well, and even inside Venus's. However, its eccentricity is so great that at aphelion it is 214,000,000 miles from the sun, farther out than Eros ever goes. Apollo's period of revolution is therefore 18 days longer than that of Eros. On 15 May 1932, Apollo approached within 6,800,000 miles of Earth, less than 30 times the distance of the moon. Apollo is less than a mile across —large enough to make it none too great a "graze." Since then, any object that approaches the sun more closely than Venus does has been called an *Apollo object.*

In February 1936, Delporte, who had detected Amor four years earlier, detected another Earth grazer, which he named Adonis. Just a few days before its detection, Adonis had passed only 1,500,000 miles from Earth, or just a little over 6.3 times the distance of the moon from us. What's more, the new Earth grazer has a perihelion of 41 million miles and at that distance is close to the orbit of Mercury. It was the second Apollo object to be discovered.

In November 1937, Reinmuth (the discoverer of Apollo) discovered a third, naming it Hermes. It had passed within 500,000 miles of Earth, only a little more than twice the distance of the moon. Reinmuth, on what data he had, calculated a rough orbit, from which it appeared that Hermes could pass within 190,000 miles of Earth (less than the distance of the moon) if both Hermes and Earth were at appropriate points in their orbit. However, Hermes has never been detected since.

On 26 June 1949, Baade discovered an even more unusual Apollo object. Its period of revolution was 1.12 years, and its orbital eccentricity was the highest known for any asteroid—0.827. At aphelion, it is safely in the asteroid belt between Mars and Jupiter but, at perihelion, is only 17,700,000 miles from the sun—closer than any planet, even Mercury, ever comes. Baade named this asteroid Icarus, after the young man in Greek myth, who, flying through the air on wings his father Daedalus has devised, approached the sun too closely; the sun melted the wax securing the feathers of the wings to his back, and he fell to his death.

Since 1949, other Apollo objects have been discovered. Some have orbital periods of less than a year, and at least one is closer, at every point in its orbit, to the sun than Earth is. In 1983, one was discovered that approached the sun more closely than Icarus does.

Some astronomers estimate that there are in space about 750 Apollo objects with diameters of half a mile and more. It is estimated that in the course of a million years, four sizable Apollo objects strike Earth; three strike Venus; one strikes either Mercury, Mars, or the moon; and seven have their orbits altered in such a way that they leave the solar system altogether. The number of Apollo objects does not, however, diminish with time; it is also likely that new ones are added from time to time by gravitational perturbations of objects in the asteroid belt.

Comets

Another class of members of the solar system can, on occasion, approach the sun closely. These appear to the eye as softly shining, hazy objects that stretch across the sky, as I mentioned in chapter 2, like fuzzy stars with long tails or streaming hair. Indeed, the ancient Greeks called them *aster kometes* ("hairy stars"), and we still call them *comets* today.

Unlike the stars and the planets, the comets seem not to follow easily predictable paths but to come and go without order and regularity. Since people in pre-scientific days felt that the stars and the planets influenced human beings, the erratic comings and goings of comets seemed to be associated with erratic things in life—with unexpected disaster, for instance.

It was not until 1473 that any European did more than shudder when a comet appeared in the sky. In that year, a German astronomer, Regiomontanus, observed a comet and put down its position against the stars night after night.

In 1532, two astronomers—an Italian named Girolamo Fracastoro and a German named Peter Apian—studied a comet that appeared in that year, and pointed out that its tail always pointed away from the sun.

Then, in 1577, another comet appeared, and Tycho Brahe, observing it, tried to determine its distance by parallax. If it were an atmospheric phenomenon, as Aristotle had thought, it should have a parallax larger than the moon. It did not! Its parallax was too small to be measured. The comet was beyond the moon and had to be an astronomical object.

But why did comets come and go with such irregularity? Once Isaac Newton had worked out the law of universal gravitation in 1687, it seemed clear that comets, like other astronomical objects of the solar system, ought to be held in the gravitational grip of the sun.

In 1682, a comet had appeared, and Edmund Halley, a friend of Newton, had observed its path across the sky. Looking back on earlier records, he thought that the comets of 1456, 1531, and 1607 had followed a similar path. These comets had come at intervals of seventy-five or seventy-six years.

It struck Halley that comets circle the sun just as planets do, but in orbits that are extremely elongated ellipses. They spend most of their time in the enormously distant aphelion portion of their orbit, where they are too distant and too dim to be seen, and then flash through their perihelion portion in a comparatively short time. They are visible only during this short time; and since no one can observe the rest of their orbit, their comings and goings seem capricious.

Halley predicted that the comet of 1682 would return in 1758. He did not live to see it, but it did return and was first sighted on 25 December 1758. It was a little behind time because Jupiter's gravitational pull had slowed it as it passed by that planet. This particular comet has been known as Halley's Comet, or Comet Halley, ever since. It returned again in 1832 and 1910 and is slated to return once more in 1986. Indeed, astronomers, knowing where to look, observed it as a faint, faint object, still far away (but approaching) in early 1983.

Other comets have had their orbits worked out since: these are all short-period comets whose entire orbits are within the planetary system. Thus, Comet Halley at perihelion is only 54,600,000 miles from the sun and is then just inside the orbit of Venus. At aphelion, it is 3,280,000,000 miles from the sun and is beyond the orbit of Neptune.

The comet with the smallest orbit is Comet Encke which revolves about the sun in 3.3 years. At perihelion, it is 31,400,000 miles from the sun, rivaling the approach of Mercury. At aphelion, it is 380,000,000 miles from the sun and is within the farther reaches of the asteroid belt. It is the only comet we know whose orbit is entirely inside the orbit of Jupiter.

Long-period comets, however, have aphelia far beyond the planetary system and return to the inner reaches of the solar system only every million years or so. In 1973, the Czech astronomer Lajos Kohoutek discovered a new comet which, promising to be extraordinarily bright (but was not), created a stir of interest. At its perihelion, it was only 23,400,000 miles from the sun—closer than Mercury was. At aphelion, however (if the orbital calculation is correct), it recedes to about 311,000,000,000 miles or 120 times as far from the sun as Neptune is. Comet Kohoutek should complete one revolution about the sun in 217,000 years. Undoubtedly, there are other comets with orbits mightier still.

In 1950, Oort suggested that, in a region stretching outward from the sun from 4 trillion to 8 trillion miles (up to 25 times as far as Comet Kohoutek at aphelion), there are 100 billion small bodies with diameters that are, for the most part, from 0.5 to 5 miles across. All of them together would have a mass of no more than one-eighth that of Earth.

This material is a kind of *cometary shell* left over from the original cloud of dust and gas that condensed nearly 5 billion years ago to form the solar system. Comets differ from asteroids in that while the latter are rocky in nature, the former are made chiefly of icy materials that are as solid as rock at their ordinary distance from the sun but would easily evaporate if they were near some source of heat. (The American astronomer Fred Lawrence Whipple had first suggested, in 1949, that comets are essentially icy objects with perhaps a rocky core or with gravel distributed throughout. This is popularly called the *dirty snowball theory.*)

Ordinarily comets stay in their far-off home, circling slowly about the distant sun with periods of revolution in the millions of years. Once in a while, however, because of collisions or the gravitational influence of some of the nearer stars, some comets are speeded up in their very slow revolution about the sun and leave the solar system altogether. Others are slowed and move toward the sun, circling it and returning to their original position, then dropping down again. Such comets can be seen when (and if) they enter the inner solar system and pass near Earth.

Because comets originate in a spherical shell, they can come into the inner solar system at any angle and are as likely to move in a retrograde direction as in a direct one. Comet Halley, for instance, moves in retrograde direction.

Once a comet enters the inner solar system, the heat of the sun vaporizes the icy materials that compose it, and dust particles trapped in the ice are liberated. The vapor and dust form a kind of hazy atmosphere about the comet (the *coma*) and make it look like a large, fuzzy object.

Thus, Comet Halley, when it is completely frozen, may be only 1.5 miles in diameter. When it passes by the sun, the haze that forms all about it can be as much as 250,000 miles in diameter, taking up a volume that is over 20 times that of giant Jupiter—but the matter in the haze is so thinly spread out that it is nothing more than a foggy vacuum.

Issuing from the sun are tiny particles, smaller than atoms (the subject of chapter 7), that speed outward in all directions. This *solar wind* strikes the haze surrounding the comet and sweeps it outward in a long tail, which can be more voluminous than the sun itself, but in which matter is even more thinly spread. Naturally, this tail has to point away from the sun at all times, as Fracastoro and Apian noted four and a half centuries ago.

At each pass around the sun, a comet loses some of its material as it vaporizes and streams out in the tail. Eventually, after a couple of hundred passes, the comet simply breaks up altogether into dust and disappears. Or else, it leaves behind a rocky core (as Comet Encke is doing) that eventually will seem no more than an asteroid.

In the long history of the solar system, many millions of comets have either been speeded up and driven out of it, or have been slowed and made to drop toward the inner solar system, where they eventually meet their end. There are still, however, many billions left; there is no danger of running out of comets.

Chapter 4

The Earth

Of Shape and Size

The solar system consists of an enormous sun, four giant planets, five smaller ones, over forty satellites, over a hundred thousand asteroids, over a hundred billion comets perhaps, and yet, as far as we know today, on only one of those bodies is there life—on our own earth. It is to the earth, then, that we must now turn.

THE EARTH AS SPHERE

One of the major inspirations of the ancient Greeks was their decision that the earth has the shape of a sphere. They conceived this idea originally (tradition credits Pythagoras with being the first to suggest it about 525 B.C.) on philosophical grounds—for example, that a sphere is the perfect shape. But the Greeks also verified this idea with observations. Around 350 B.C., Aristotle marshaled conclusive evidence that the earth was not flat but round. His most telling argument was that as one traveled north or south, new stars appeared over the horizon ahead, and visible ones disappeared below the horizon behind. Then, too, ships sailing out to sea vanished hull first in whatever direction they traveled, while the cross-section of the earth's shadow on the moon, during a lunar eclipse, was always a circle, regardless of the position of the moon. Both these latter facts could be true only if the earth were a sphere.

Among scholars at least, the notion of the spherical earth never entirely died out, even during the Dark Ages. The Italian poet Dante Alighieri assumed a spherical earth in that epitome of the medieval view, *The Divine Comedy.*

It was another thing entirely when the question of a *rotating* sphere arose. As long ago as 350 B.C., the Greek philosopher Heraclides of Pontus suggested that it was far easier to suppose that the earth rotates on its axis than that the entire vault of the heavens revolves around the earth. This idea, however, most ancient and medieval scholars refused to accept; and as late as 1632, Galileo was condemned by the Inquisition at Rome and forced to recant his belief in a moving earth.

Nevertheless, the Copernican theory made a stationary earth completely illogical, and slowly its rotation was accepted by everyone. It was only in 1851, however, that this rotation was actually demonstrated experimentally. In that year, the French physicist Jean Bernard Léon Foucault set a huge pendulum swinging from the dome of a Parisian church. According to the conclusions of physicists, such a pendulum ought to maintain its swing in a fixed plane, regardless of the rotation of the earth. At the North Pole, for instance, the pendulum would swing in a fixed plane, while the earth rotated under it, counterclockwise, in twenty-four hours. To a person watching the pendulum (who would be carried with the earth, which would seem motionless to him), the pendulum's plane of swing would seem to be turning clockwise through one full revolution every twenty-four hours. At the South Pole, one's experience would be the same except that the pendulum's plane of swing would turn counterclockwise.

At latitudes below the poles, the plane of the pendulum would still turn (clockwise in the Northern Hemisphere and counterclockwise in the Southern), but in longer and longer periods as one moved farther from the poles. At the Equator, the pendulum's plane of swing would not alter at all.

During Foucault's experiment, the pendulum's plane of swing turned in the proper direction and at just the proper rate. Observers could, so to speak, see with their own eyes the earth turn under the pendulum.

The rotation of the earth brings with it many consequences. The surface moves fastest at the Equator, where it must make a circle of 25,000 miles in twenty-four hours, at a speed of just over 1,000 miles an hour. As one travels north (or south) from the Equator, a spot on the earth's surface need travel more slowly, since it must make a smaller circle in the same twenty-four hours. Near the poles, the circle is small indeed; and, at the poles, the surface is motionless.

The air partakes of the motion of the surface of the earth over which it hovers. If an air mass moves northward from the Equator, its own speed (matching that of the Equator) is faster than that of the surface it travels toward. It overtakes the surface in the west-to-east journey and drifts eastward. This drift is an example of the *Coriolis effect,* named for the French mathematician Gaspard Gustave de Coriolis, who first studied it in 1835.

The effect of such Coriolis effects on air masses is to set them to turning with a clockwise twist in the Northern Hemisphere. In the Southern Hemisphere, the effect is reversed, and a counterclockwise twist is produced. In either case, *cyclonic disturbances* are set up. Massive storms of this type are called *hurricanes* in the North Atlantic and *typhoons* in the North Pacific. Smaller but more intense storms of this sort are *cyclones* or *tornadoes.* Over the sea, such violent twisters set up dramatic *sea spouts.*

However, the most exciting deduction obtained from the earth's rotation was made two centuries before Foucault's experiment, in Isaac Newton's time. At that time, the notion of the earth as a perfect sphere had already held sway for nearly 2,000 years, but then Newton took a careful look at what happens to such a sphere when it rotates. He noted the difference in the rate of motion of the earth's surface at different latitudes and considered what it must mean.

The faster the rotation, the stronger the centrifugal effect—that is, the tendency to push material away from the center of rotation. It follows, therefore, that the centrifugal effect increases steadily from zero at the stationary poles to a maximum at the rapidly whirling equatorial belt. Hence, the earth should be pushed out most around its middle: in other words, it should be an *oblate spheroid,* with an *equatorial bulge* and flattened poles. It must have roughly the shape of a tangerine rather than of a golf ball. Newton even calculated that the polar flattening should be about 1/230 of the total diameter, which is surprisingly close to the truth.

The earth rotates so slowly that the flattening and bulging are too slight to be readily detected. But at least two astronomical observations supported Newton's reasoning, even in his own day. First, Jupiter and Saturn were clearly seen to be markedly flattened at the poles, as I pointed out in the previous chapter.

Second, if the earth really bulges at the Equator, the varying gravitational pull on the bulge by the moon, which most of the time is either north or south of the Equator in its circuit around the earth, should cause the earth's axis of rotation to mark out a double cone, so that each pole points to a steadily changing point in the sky. The points mark out a circle about which the pole makes a complete revolution every 25,750 years. In fact, Hipparchus had noted this shift about 150 B.C. when he compared the position of the stars in his day with those recorded a century and a half earlier. The shift of the earth's axis has the effect of causing the sun to reach the point of equinox about 50 seconds of arc eastward each year (that is, in the direction of morning). Since the equinox thus comes to a preceding (that is, earlier) point each year, Hipparchus named this shift the *precession of the equinoxes,* and it is still known by that name.

Naturally scientists set out in search of more direct proof of the earth's distortion. They resorted to a standard device for solving geometrical problems —trigonometry. On a curved surface, the angles of a triangle add up to more than 180 degrees. The greater the curvature, the greater the excess over 180 degrees. Now if the earth was an oblate spheroid, as Newton had said, the excess should be greater on the more sharply curved surface of the equatorial bulge than on the less curved surface toward the poles. In the 1730s, French scientists made the first test by doing some large-scale surveying at separate sites in the north and the south of France. On the basis of these measurements, the French astronomer Jacques Cassini (son of the astronomer who had pointed out the flattening of Jupiter and Saturn) decided that the earth bulged at the poles, not at the Equator! To use an exaggerated analogy, its shape was more like that of a cucumber than of a tangerine.

But the difference in curvature between the north and the south of France obviously was too small to give conclusive results. Consequently, in 1735 and 1736, a pair of French expeditions went forth to more widely separated regions —one to Peru, near the Equator, and the other to Lapland, approaching the Arctic. By 1744, their surveys had given a clear answer: the earth is distinctly more curved in Peru than in Lapland.

Today the best measurements show that the diameter of the earth is 26.68 miles longer through the Equator than along the axis through the poles (7,926.36 miles against 7,899.78 miles).

The eighteenth-century inquiry into the shape of the earth made the scientific community dissatisfied with the state of the art of measurement. No decent standards for precise measurement existed. This dissatisfaction was partly responsible for the adoption, during the French Revolution half a century later, of the logical and scientifically worked-out *metric system* based on the meter. The metric system now is used by scientists all over the world, to their great satisfaction, and it is the system in general public use virtually everywhere but the United States.

The importance of accurate standards of measure cannot be overestimated. A good percentage of scientific effort is continually being devoted to improvement in such standards. The standard meter and standard kilogram were made of platinum-iridium alloy (virtually immune to chemical change) and were kept in a Paris suburb under conditions of great care—in particular, under constant temperature to prevent expansion or contraction.

New alloys such as Invar (short for "invariable"), composed of nickel and iron in certain proportions, were discovered to be almost unaffected by temperature change. These could be used in forming better standards of length, and the Swiss-born, French physicist Charles Edouard Guillaume, who developed Invar, received the Nobel Prize for physics in 1920 for this discovery.

In 1960, however, the scientific community abandoned material standards of length. The General Conference of Weights and Measures adopted as standard the length of a tiny wave of light produced by a particular variety of the rare gas krypton. Exactly 1,650,763.73 of these waves (far more unchanging than anything man-made could be) equal 1 meter, a length that is now a thousand times as exact as it had been before. In 1984, the meter was tied to the speed of light, as the distance travelled by light in an appropriate fraction of a second.

MEASURING THE GEOID

The smoothed-out, sea-level shape of the earth is called the *geoid*. Of course, the earth's surface is pocked with irregularities—mountains, ravines, and so on. Even before Newton raised the question of the planet's overall shape, scientists had tried to measure the magnitude of these minor deviations from a perfect sphere (as they thought). They resorted to the device of a swinging pendulum. Galileo, in 1581, as a seventeen-year-old boy, had discovered that a pendulum of a given length always completed its swing in just about the same time, whether the swing was short or long; he is supposed to have made the discovery while watching the swinging chandeliers in the cathedral of Pisa during services. There is a lamp in the cathedral still called *Galileo's lamp*, but it was not hung until 1584. (Huygens hooked a pendulum to the gears of a clock and used the constancy of its motion to keep the clock going with even accuracy. In 1656, he devised the first modern clock in this way—the *grandfather clock*—and at once increased tenfold the accuracy of timekeeping.)

The period of the pendulum depends both on its length and on the gravitational force. At sea level, a pendulum with a length of 39.1 inches makes a complete swing in just 1 second, a fact worked out in 1644 by Galileo's pupil, the French mathematician Marin Mersenne. The investigators of the earth's irregularities made use of the fact that the period of a pendulum's swing depends on the strength of gravity at any given point. A pendulum that swings perfect seconds at sea level, for instance, will take slightly longer than 1 second to complete a swing on a mountain top, where gravity is slightly weaker because the mountain top is farther from the center of the earth.

In 1673, a French expedition to the north coast of South America (near the Equator) found that, at that location, the pendulum was slowed even at sea level. Newton later took this finding as evidence for the existence of the equatorial bulge, which would lift the camp farther from the earth's center, and weaken the force of gravity. After the expedition to Peru and Lapland had proved his theory, a member of the Lapland expedition, the French mathematician Alexis Claude Clairault, worked out methods of calculating the oblateness of the earth from pendulum swings. Thus, the geoid, or sea-level shape of the earth, can be determined, and it turns out to vary from the perfect oblate spheroid by less than 300 feet at all points. Nowadays gravitational force is also measured by a *gravimeter*, a weight suspended from a very sensitive spring. The position of the weight against a scale in the background indicates the force with which it is pulled downward, and hence measures variations in gravity with great delicacy.

Gravity at sea level varies by about 0.6 percent, being least at the Equator, of course. The difference is not noticeable in ordinary life, but it can affect sports records. Achievements at the Olympic Games depend to some extent on the latitude (and altitude) of the city in which they are conducted.

A knowledge of the exact shape of the geoid is essential for accurate map making; and as late as the 1950s, only 7 percent of the earth's land surface can really be said to have been accurately mapped. The distance between New York and London, for instance, was not known to better than a mile or so, and the locations of some islands in the Pacific were known only within a possible error of several miles. In these days of air travel and (alas!) potential missile aiming, this margin of error is inconvenient. But truly accurate mapping has now been made possible—oddly enough, not by surveys of the earth's surface but by astronomical measurements of a new kind. The first instrument of these new measurements was the man-made satellite called *Vanguard I*, launched by the United States on 17 March 1958. *Vanguard I* revolved around the earth in a period of 2½ hours and, in the first couple of years of its lifetime, had already made more revolutions than the moon had in all the centuries it has been observed with the telescope. By observations of *Vanguard I*'s position at specific times from specific points of the earth, the distances between these observing points can be calculated precisely. In this way, positions and distances not known to within a matter of miles were, in 1959, determined to within a hundred yards or so. (Another satellite named *Transit I-B*, launched by the United States on 13 April 1960, was the first of a series specifically intended to extend this into a system for the accurate location of position on the earth's surface, which could greatly improve and simplify air and sea navigation.)

Like the moon, *Vanguard I* circles the earth in an ellipse that is not in the earth's equatorial plane; and also like the moon, the perigee (closest approach) of *Vanguard I* shifts because of the attraction of the equatorial bulge. Because *Vanguard I* is far closer to the bulge and far smaller than the moon, it is affected to a greater extent; and because of its many revolutions, the effect of the bulge can be well studied. By 1959, it was certain that the perigee shift of *Vanguard I* was not the same in the Northern Hemisphere as in the Southern, and thus that the bulge was not quite symmetrical with respect to the Equator. The bulge seemed to be 25 feet higher (that is, 25 feet more distant from the earth's center) at spots south of the Equator than at spots north of it. Further calculations showed that the South Pole was 50 feet closer to the center of the earth (counting from sea level) than was the North Pole.

Further information, obtained in 1961, based on the orbits of *Vanguard I* and *Vanguard II* (the latter having been launched on 17 February 1959) indicates that the sea-level Equator is not a perfect circle. The equatorial diameter is 1,400 feet (nearly a quarter of a mile) longer in some places than in others.

Newspaper stories have described the earth as "pear-shaped" and the Equator as "egg-shaped." Actually, these deviations from the perfectly smooth curve are perceptible only to the most refined measurements. No one looking at the earth from space would see anything resembling a pear or an egg, but only what would seem a perfect sphere. Besides, detailed studies of the geoid have shown so many regions of very slight flattening and very slight humping that, if the earth must be described dramatically, it had better be called "lumpy shaped."

Eventually satellites, even by methods as direct as taking detailed photographs of the earth's surface, have made it possible to map the entire world to within an accuracy of a few feet.

Airplanes and ships, which would ordinarily determine their position with reference to stars, could eventually do so by fixing on the signals emitted by *navigation satellites*—regardless of weather, since microwaves penetrate clouds and fogs. Even submarines below the ocean surface can do so. This can be done with such accuracy that an ocean liner can calculate the difference in position between its bridge and its galley.

WEIGHING THE EARTH

Knowledge of the exact size and shape of the earth makes it possible to calculate its volume, about 260 billion cubic miles. Calculating the earth's mass, however, is more complex, but Newton's law of gravitation gives us something to begin with. According to Newton, the *gravitational force* (f) between any two objects in the universe can be expressed as follows:

$$f = \frac{gm_1 m_2}{d^2}$$

where m_1 and m_2 are the masses of the two bodies concerned, and d is the distance between them, center to center. As for g, that represents the *gravitational constant*.

What the value of the constant was, Newton could not say. If we can learn the values of the other factors in the equation, however, we can find g; for by transposing the terms, we get:

$$g = \frac{fd^2}{m_1 m_2}$$

To find the value of g, therefore, all we need to do is to measure the gravitational force between two bodies of known mass at the separation of a known distance. The trouble is that gravitational force is the weakest force we know, and the gravitational attraction between two masses of any ordinary size that we can handle is almost impossible to measure.

Nevertheless, in 1798, the English physicist Henry Cavendish, a wealthy, neurotic genius who lived and died in almost complete seclusion but performed some of the most astute experiments in the history of science, managed to make the measurement. Cavendish attached a ball of known mass to each end of a long rod and suspended this dumbbell-like contraption on a fine thread. Then he placed a larger ball, also of known mass, close to each ball on the rod—on opposite sides, so that gravitational attraction between the fixed large balls and the suspended small balls would cause the horizontally hung dumbbell to turn, thus twisting the thread (figure 4.1). The dumbbell did indeed turn slightly. Cavendish now measured how much force was needed to produce this amount of twist of the thread. This told him the value

Figure 4.1. Henry Cavendish's apparatus for measuring gravity. The two small balls are attracted by the larger ones, causing the thread on which they are suspended to twist. The mirror shows the amount of this slight twist by the deflection of reflected light on the scale.

of f. He also knew m_1 and m_2, the masses of the balls, and d, the distance between the attracted balls. So he was able to compute the value of g. Once he had that, he could calculate the mass of the earth, because the earth's gravitational pull (f) on any given body can be measured. Thus Cavendish "weighed" the earth for the first time.

The measurements have since been greatly refined. In 1928, the American physicist Paul R. Heyl at the United States Bureau of Standards determined the value of g to be 0.00000006673 dyne centimeter squared per gram squared —a number since refined to 0.000000066726. You need not be concerned about those units, but note the smallness of the figure. It is a measure of the weakness of gravitational force. Two 1-pound weights placed 1 foot apart attract each other with a force of only one-half of one billionth of an ounce.

The fact that the earth itself attracts such a weight with the force of 1 pound even at a distance of 3,960 miles from its center emphasizes how massive the earth is. In fact, the mass of the earth turns out to be 6,-585,000,000,000,000,000,000 tons or, in metric units, 5,976,000,000,000,-000,000,000,000 kilograms.

From the mass and volume of the earth, its average density is easily calculated. In metric units, the answer comes out to 5.518 grams per cubic centimeter (5.518 times the density of water). The density of the earth's surface rocks averages only about 2.8 grams per cubic centimeter, so the density of the interior must be much greater. Does it increase smoothly all the way down to the center? The first proof that it does not—that the earth is made up of a series of different layers—came from the study of earthquakes.

Earth's Layers

EARTHQUAKES

There are not many natural disasters that can, in five minutes, kill hundreds of thousands of people. Of these, by far the most common is the earthquake.

The earth suffers a million quakes a year, including at least 100 serious ones and 10 disastrous ones. The most murderous quake is supposed to have taken place in the northern province of Shensi in China in 1556, when 830,000 people were killed. Other quakes nearly as bad have also taken place in the Far East. On 30 December 1703, an earthquake killed 200,000 people in Tokyo, Japan; and on 11 October 1737, one killed 300,000 people in Calcutta, India.

In those days, though, when science was developing in western Europe, little attention was paid to events that took place on the other side of the world. But then came a disaster much closer to home.

On 1 November 1755, a great earthquake, possibly the most violent of modern times, struck the city of Lisbon, demolishing every house in the lower part of the city. Then what is called a *tidal wave* swept in from the ocean. Two more shocks followed, and fires broke out. Sixty thousand people were killed, and the city was left a scene of devastation.

The shock was felt over an area of one and a half million square miles, doing substantial damage in Morocco as well as in Portugal. Because it was All Soul's Day, people were in church, and it is said that all over southern Europe those in the cathedrals saw the chandeliers dance and sway.

The Lisbon disaster made a great impression on the scholars of the day. It was an optimistic time when many thinkers felt that the new science of Galileo and Newton would give human beings the means of making the earth a paradise. This blow showed that there were still giant, unpredictable, and apparently malicious forces beyond human control. The earthquake inspired Voltaire, the great literary figure of the time, to write his famous pessimistic satire *Candide*, with its ironical refrain that all is for the best in this best of all possible worlds.

We are accustomed to thinking of dry land as shaking with the effect of an earthquake, but the earth beneath the ocean floor may be set to quivering, too, with even more devastating effects. The vibration sets up long, gentle swells in the ocean which, on reaching the shallow shelves in the neighborhood of land—particularly when driven into the narrowing confines of a harbor— pile up into towers of water, sometimes 50 to 100 feet high. If the waves hit with no warning, thousands of people are drowned. The popular name of "tidal wave" for such earthquake-generated waves is a misnomer. They may resemble monstrous tides, but they have entirely different causes. Nowadays, they are referred to by the Japanese name *tsunami* ("harbor wave"). Japan's coastline is particularly vulnerable to such waves, so this nomenclature is justified.

After the Lisbon disaster, to which a tsunami had added its share of destruction, scientists began turning their thoughts earnestly to the possible causes of earthquakes. The best theory of the ancient Greeks (aside from the thought that they were caused by the angry writhing of giants imprisoned underground) had been Aristotle's suggestion that they was caused by masses of air, imprisoned underground and trying to escape. Modern scientists, however, suspected that earthquakes might be the effect of earth's internal heat on stresses within the solid rock itself.

The English geologist John Michell (who had studied the forces involved in *torsion*, or twisting, later used by Cavendish to measure the mass of the earth) suggested in 1760 that earthquakes are waves set up by the shifting of masses of rock miles below the surface, and it was he who first suggested that tsunamis are the result of undersea earthquakes.

To study earthquakes properly, an instrument for detecting and measuring these waves had to be developed, and this did not come to pass until one hundred years after the Lisbon quake. In 1855, the Italian physicist Luigi Palmieri devised the first *seismograph* (from Greek words meaning "earthquake writing").

Palmieri's invention consisted of a horizontal tube turned up at each end and partly filled with mercury. Whenever the ground shook, the mercury moved from side to side. It responded to an earthquake, of course, but also to any other vibration, such as that of a cart rumbling along a road nearby.

A much better device, and the ancestor of all those used since, was constructed in 1880 by an English engineer, John Milne. Five years before, he had gone to Tokyo to teach geology and mining and there had ample opportunity to study earthquakes, which are common in Japan. His seismograph was the result.

In its simplest form, Milne's seismograph consists of a massive block suspended by a comparatively weak spring from a support firmly fixed in bedrock. When the earth moves, the suspended block remains still, because of its inertia. However, the spring attached to the bedrock stretches or contracts a little with the earth's motion. This motion is recorded on a slowly rotating drum by means of a pen attached to the stationary block, writing on smoked paper. Actually, two blocks are used: one oriented to record the earthquake waves traveling north and south; the other, east and west. Ordinary vibrations, not originating in bedrock, do not affect the seismograph. Nowadays, the most delicate seismographs, such as the one at Fordham University, use a ray of light in place of a pen, to avoid the frictional drag of the pen on the paper. This ray shines on sensitized paper, making tracings that are developed as a photograph.

Milne was instrumental in setting up stations for the study of earthquakes and related phenomena in various parts of the world, particularly in Japan. By 1900, thirteen seismograph stations were in existence, and today there are over 500, spread over every continent including Antarctica. Within ten years after the establishment of the first of these, the correctness of Michell's suggestion that earthquakes are caused by waves propagated through the body of the Earth was clear.

This new knowledge of earthquakes did not mean that they occurred less frequently, or that they were less deadly when they did occur. The 1970s, in fact, were rich in severe earthquakes.

On 27 July 1976, an earthquake in China destroyed a city south of Peking and killed about 650,000 people. This was the worst disaster of the sort since the one in Shensi four centuries before. There were other bad earthquakes in Guatemala, Mexico, Italy, the Philippines, Rumania, and Turkey.

These earthquakes do not mean that our planet is growing less stable. Modern methods of communication make it certain that we hear of all earthquakes everywhere—often with instant eyewitness scenes, thanks to television—where in earlier times (even a few decades ago) distant catastrophes would have gone unreported and unnoticed. What's more, earthquakes are more likely to be catastrophic now than in earlier times (even a century ago), since there are many more people on Earth, crowded much more intensively into cities, and because man-made structures, vulnerable to earthquakes, are much more numerous and expensive.

All the more reason to work out methods for predicting earthquakes before they occur. Seismologists are seeking for significant changes. The ground might hump up in places. Rocks might pull apart or squeeze together, absorbing water or squeezing it out, so that rises and falls in well water might be significant. There might be changes in the natural magnetism of rocks or in electrical conductivity. Animals, aware of tiny vibrations or alterations in the environment, which human beings are too busy to notice, may begin to react in a nervous manner.

The Chinese, in particular, have taken to collecting all reports of anything unusual, even flaking paint, and report that an earthquake in northeastern China on 4 February 1975 was predicted. People therefore left their homes for the open fields outside the city, and thousands of lives were saved. However, the more serious earthquake of 1976 was *not* predicted.

There is also the point that until predictions are more certain than they are now, warnings may do more harm than good. A false alarm could disrupt life and the economy and do more harm than a mild quake could. Furthermore, after one or two false alarms, a correct prediction might be ignored.

The damage an earthquake can do is not surprising. The largest earthquakes are estimated to release a total energy equal to 100,000 ordinary atomic bombs or, if you prefer, 100 large hydrogen bombs. It is only because earthquakes' energies are dissipated over a large area that they are not even more destructive than they are. They can make the earth vibrate as though it were a gigantic tuning fork. The Chilean earthquake of 1960 caused our planet to vibrate at a frequency of just under once an hour (20 octaves below middle C and quite inaudible).

Earthquake intensity is measured on a scale from 0 up through 9, where each number represents an energy release about 31 times that of the number below. (No quake of intensity greater than 9 has ever been recorded, but the Good Friday quake in Alaska in 1964 recorded an intensity of 8.5.) This is called the *Richter scale* because it was introduced in 1935 by the American seismologist Charles Francis Richter.

One favorable aspect of earthquakes is that not all the earth's surface is equally exposed to their dangers (though this is cold comfort to those who live in regions that are so exposed).

About 80 percent of earthquake energy is released in the areas bordering the vast Pacific Ocean. Another 15 percent is released in an east-west band sweeping across the Mediterranean. These earthquake zones (see figure 4.2) are closely associated with volcanic areas—one reason the effect of internal heat was associated with earthquakes.

VOLCANOES

Volcanoes are a natural phenomenon as frightening as earthquakes and longer-lasting, although in most cases their effects are confined to a smaller area. About 500 volcanoes are known to have been active in historical times, two-thirds of them along the rim of the Pacific.

On rare occasions, when a volcano traps and overheats huge quantities of water, appalling catastrophes can take place. On 26–27 August 1883, the small East Indian volcanic island Krakatoa, situated in the strait between Sumatra and Java, exploded with a roar that has been described as the loudest sound ever formed on earth during historic times. The sound was heard by human ears as far away as 3,000 miles and could be picked up by instruments all over the globe. The sound waves traveled several times completely about the planet. Five cubic miles of rock were fragmented, hurled into the air, and fell over an area of 300,000 square miles. Ashes darkened the sky over hundreds of square miles, leaving in the stratosphere dust that brightened sunsets for years. Tsunamis 100 feet in height killed 36,000 people on the shores of Java and Sumatra, and their waves could be detected easily in all parts of the world.

A similar event, with even greater consequences, may have taken place over 3,000 years before in the Mediterranean Sea. In 1967, American archaeologists discovered the ash-covered remains of a city on the small island of Thera, 80 miles north of Crete. About 1400 B.C., apparently it exploded as Krakatoa did but with still greater force, a possibly louder sound, and even more disastrous consequences. The tsunami that resulted struck the island of Crete, then the home of a long-developed and admirable civilization, a crippling blow from which that civilization never recovered. The Cretan control of the seas vanished, and a period of turmoil and darkness eventually followed; recovery took many centuries. The dramatic disappearance of Thera lived on in the minds of survivors, and its tale passed down the line of generations with embellishments. It may very well have given rise to Plato's tale of Atlantis, which was told about eleven centuries after the death of Thera and of Cretan civilization.

Perhaps the most famous single volcanic eruption in the history of the world was minute compared with Krakatoa or Thera. It was the eruption of Vesuvius

SEISMIC MAP OF MAJOR EARTHQUAKES
AS REGISTERED BY SEISMOGRAPHS, 1899-1923

COMPILED FROM COMMITTEE REPORTS
BRITISH ASSOCIATION AND OTHER SOURCES

BY CHESTER A REEDS 1933
AMERICAN MUSEUM OF NATURAL HISTORY

• SINGLE EPICENTER
✦ MULTIPLE EPICENTER

Figure 4.2. Earthquake epicenters 1963–1977. Courtesy National Geophysical Data Center, National Oceanic and Atmospheric Administration.

in 79 A.D. (up to that time it had been considered a dead volcano), which buried the Roman resort cities of Pompeii and Herculaneum. The famous encyclopedist Gaius Plinius Secundus (better known as Pliny) died in that catastrophe, which was described by his nephew, Pliny the Younger, an eye-witness.

Excavations of the buried cities began in serious fashion after 1763. These offered an unusual opportunity to study relatively complete remains of a city that had existed during the most prosperous period of ancient times.

Another unusual phenomenon is the actual birth of a new volcano. Such an awesome event was witnessed in Mexico on 20 February 1943, when in the village of Parícutin, 200 miles west of Mexico City, a volcano began to appear in what had been a quiet cornfield. In eight months, it had built itself up to an ashy cone 1,500 feet high. The village had to be abandoned, of course.

On the whole, Americans have not been very conscious of volcanic erup-tions, which seem, for the most part, to take place in foreign lands. To be sure, the largest active volcano is on the island of Hawaii, which has been an American possession for over eighty years, and an American state for over thirty. Kilauea has a crater with an area of 4 square miles and is frequently in eruption. The eruptions are never explosive, however; and while the lava overflows periodically, it moves slowly enough to ensure little loss of life, even though there is sometimes destruction of property. It was unusually active in 1983.

The Cascade range, which follows the Pacific coast line (about 100 to 150 miles inland) from northern California to southern British Columbia, has numerous famous peaks, such as Mount Hood and Mount Rainier, which are known to be extinct volcanoes. Because they are extinct, they are given little thought, and yet a volcano can lie dormant for centuries and then come roaring back to life.

This fact was brought home to Americans in connection with Mount Saint Helens in south-central Washington State. Between 1831 and 1854, it had been active, but not many people lived there then, and the details are vague. For a century and a third, it had certainly been absolutely quiet, but then on 18 May 1980, after some preliminary rumbling and quaking, it erupted sud-denly. Twenty people, who had not taken the elementary precaution of leaving the region, were killed, and over one hundred were reported missing. It has been active ever since—not much as volcanic eruptions go, but it was the first such eruption in the forty-eight contiguous states in a long time.

There is more to volcanic eruptions than immediate loss of life. In giant eruptions, vast quantities of dust are thrown high into the atmosphere, and years may pass before the dust settles. After the Krakatoa eruption, there were gorgeous sunsets as the dust scattered the light of the setting sun for a long period. A less benign effect is that the dust can reflect sunlight so that less of the sun's warmth reaches the earth's surface for a time.

Sometimes the delayed effect is relatively local but catastrophic. In 1783, the volcano of Laki in south-central Iceland erupted. Lava eventually covered

220 square miles during a two-year eruption but did little direct damage. Ash and sulfur dioxide, however, spewed out over almost all of Iceland and even reached Scotland. The ash darkened the sky, so that the crops, unable to get sunlight, died. The sulfur dioxide fumes killed three-quarters of the domestic animals on the island. With crops gone and animals dead, 10,000 Icelanders, one-fifth of the whole population of the island, died of starvation and disease.

On 7 April 1815, Mount Tambora, on a small island east of Java, exploded. Thirty-six cubic miles of rock and dust were hurled into the upper atmosphere. For that reason, sunlight was reflected to a greater extent than usual, and temperatures on Earth were lower than usual for a year or so. In New England, for instance, 1816 was unusually cold, and there were freezing spells in every month of that year, even July and August. It was called "the year without a summer."

Sometimes volcanoes kill immediately but not necessarily through lava or even ash. On 8 May 1902, Mount Pelée on the island of Martinique in the West Indies erupted. The explosion produced a thick cloud of red-hot gases and fumes. These gases poured quickly down the side of the mountain and headed straight for Saint Pierre, the chief town on the island. In 3 minutes, 38,000 people in the city were dead by asphyxiation. The only survivor was a criminal in an underground prison who would have been hanged that very day, if everyone else had not died.

FORMATION OF EARTH'S CRUST

Modern research in volcanoes and their role in forming much of the earth's crust began with the French geologist Jean Etienne Guettard in the mid-eighteenth century. For a while, in the late eighteenth century, the single-handed efforts of the German geologist Abraham Gottlob Werner popularized the false notion that most rocks were of sedimentary origin, from an ocean that had once been world-wide (*neptunism*). The weight of the evidence, particularly that presented by Hutton, made it quite certain, however, that most rocks were formed through volcanic action (*plutonism*). Both volcanoes and earthquakes would seem the expression of the earth's internal energy, originating for the most part from radioactivity (see chapter 7).

Once seismographs allowed the detailed study of earthquake waves, it was found that those most easily studied came in two general varieties: *surface waves* and *bodily waves*. The surface waves follow the curve of the earth; the bodily waves go through the interior—and, by virtue of this short cut, usually are the first to arrive at the seismograph. These bodily waves, in turn, are of two types: primary (*P waves*) and secondary (*S waves*) (figure 4.3). The primary waves, like sound waves, travel by alternate compression and expansion of the medium (to visualize them, think of the pushing together and pulling apart of an accordion). Such waves can pass through any medium—solid or fluid. The secondary waves, on the other hand, have the familiar form of snakelike wiggles at right angles to the direction of travel and cannot travel through liquids or gases.

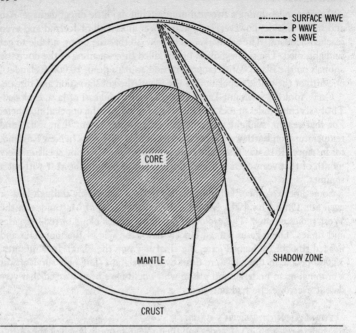

Figure 4.3. Earthquake waves' routes in the earth. Surface waves travel along the crust. The earth's liquid core refracts the P-type bodily waves. S waves cannot travel through the core.

The primary waves move faster than secondary waves and consequently reach a seismograph station sooner. From the time lag of the secondaries, it is possible to estimate the distance of the earthquake. And its location or *epicenter* (the spot on the earth's surface directly above the rock disturbance) can be pinpointed by getting distance bearings at three or more stations: the three radii trace out three circles that will intersect at a single point.

The speed of both the P and the S types of wave is affected by the kind of rock, the temperature, and the pressure, as laboratory studies have shown. Therefore earthquake waves can be used as probes to investigate conditions deep under the earth's surface.

A primary wave near the surface travels at 5 miles per second; 1,000 miles below the surface, judging from the arrival times, its velocity must be nearly 8 miles per second. Similarly, a secondary wave has a velocity of less than 3 miles per second near the surface and of 4 miles per second at a depth of 1,000 miles. Since increase in velocity is a measure of increase in density, we can estimate the density of the rock beneath the surface. At the surface of the earth, as I have mentioned, the average density is 2.8 grams per cubic centimeter; 1,000 miles down, it amounts to 5 grams per cubic centimeter; 1,800 miles down, nearly 6 grams per cubic centimeter.

At the depth of 1,800 miles, there is an abrupt change. Secondary waves are stopped cold. The British geologist Richard Dixon Oldham maintained, in 1906, that therefore the region below is liquid: the waves have reached the boundary of the earth's *liquid core.* And primary waves, on reaching this level, change direction sharply; apparently they are refracted by entering the liquid core.

The boundary of the liquid core is called the *Gutenberg discontinuity,* after the American geologist Beno Gutenberg, who in 1914 defined the boundary and showed that the core extended 2,160 miles from the earth's center. The density of the various deep layers of the earth were worked out in 1936 from earthquake data by the Australian mathematician Keith Edward Bullen. His results were confirmed by the data yielded by the huge Chilean earthquake of 1960. We can therefore say that at the Gutenberg discontinuity, the density of the material jumps from 6 to 9 and, therefore, increases smoothly to 11.5 grams per cubic centimeter at the center.

THE LIQUID CORE

What is the nature of the liquid core? It must be composed of a substance that has a density of from 9 to 11.5 grams per cubic centimeter under the conditions of temperature and pressure in the core. The pressure is estimated to range from 10,000 tons per square inch at the top of the liquid core to 25,000 tons per square inch at the center of the earth. The temperature is less certain. On the basis of the rate at which temperature is known to increase with depth in deep mines and of the rate at which rocks can conduct heat, geologists estimate (rather roughly) that temperatures in the liquid core must be as high as 5,000° C. (The center of the much larger planet Jupiter may be as high as 50,000° C.)

The substance of the core must be some common element—common enough to be able to make up a sphere half the diameter of the earth and one-third its mass. The only heavy element that is at all common in the universe is iron. At the earth's surface, its density is only 7.86 grams per cubic centimeter; but under the enormous pressures of the core, it would have a density in the correct range—9 to 12 grams per cubic centimeter. What is more, under center-of-the-earth conditions it would be liquid.

If more evidence is needed, meteorites supply it. These fall into two broad classes: *stony meteorites,* composed chiefly of silicates, and *iron meteorites,* made up of about 90 percent iron, 9 percent nickel, and 1 percent other elements. Many scientists believe that the meteorites are remnants of shattered asteroids, some of which may have been large enough to separate out into metallic and stony portions. In that case, the metallic portions must have been nickel-iron, and so might be the earth's metallic core. (Indeed, in 1866, long before seismologists had probed the earth's core, the composition of the iron meteorites suggested to the French geologist Gabriel Auguste Daubrée that the core of our planet was made of iron.)

Today most geologists accept the liquid nickel-iron core as one of the facts of life as far as the earth's structure is concerned. One major refinement, however, has been introduced. In 1936, the Danish geologist Inge Lehmann, seeking to explain the puzzling fact that some primary waves show up in a *shadow zone* on the surface from which most such waves are excluded, proposed that a discontinuity within the core about 800 miles from the center introduced another bend in the waves and sent a few careening into the shadow zone. Gutenberg supported this view, and now many geologists differentiate between an *outer core* that is liquid nickel-iron, and an *inner core* that differs from the outer core in some way, perhaps in being solid or slightly different chemically. As a result of the great Chilean earthquakes of 1960, the entire globe was set into slow vibrations at rates matching those predicted by taking the inner core into account. This is strong evidence in favor of its existence.

EARTH'S MANTLE

The portion of the earth surrounding the nickel-iron core is called the *mantle.* It seems to be composed of silicates, but judging from the velocity of earthquake waves passing through them, these silicates are different from the typical rocks of the earth's surface—as was first shown in 1919 by the American physical chemist Leason Heberling Adams. Their properties suggest that they are rocks of the so-called *olivine* type (olive-green in color), which are comparatively rich in magnesium and iron and poor in aluminum.

The mantle does not quite extend to the surface of the earth. A Croatian geologist named Andrija Mohorovicic, while studying the waves produced by a Balkan earthquake in 1909, decided that there was a sharp increase in wave velocity at a point about 20 miles beneath the surface. This *Mohorovicic discontinuity* (known as *Moho* for short) is now accepted to be the boundary of the earth's *crust.*

The nature of the crust and of the upper mantle is best explored by means of the surface waves I mentioned earlier. Like the bodily waves, the surface waves come in two varieties: *Love waves* (named for their discoverer Augustus Edward Hough Love) are horizontal ripples, like the shape of a snake moving over the ground; *Rayleigh waves* (named after the English physicist John William Strutt, Lord Rayleigh) are vertical, like the path of the mythical sea serpent moving through the water.

Analysis of these surface waves (notably by Maurice Ewing of Columbia University) shows that the crust is of varying thickness. It is thinnest under the ocean basins, where the Moho discontinuity in some places is only 8 to 10 miles below sea level. Since the oceans themselves are 5 to 7 miles deep in spots, the solid crust may be as thin as 3 miles under the ocean deeps. Under the continents, on the other hand, the Moho discontinuity lies at an average depth of about 20 miles below sea level (it is about 22 miles under New York City, for instance), and it plunges to a depth of nearly 40 miles beneath mountain ranges. This fact, combined with evidence from gravity measurements, shows that the rock in mountain ranges is less dense than the average.

The general picture of the crust is of a structure composed of two main types of rock—basalt and granite—with the less dense granite riding buoyantly on the basalt, forming continents and, in places where the granite is particularly thick, mountains (just as a large iceberg rises higher out of the water than a small one). Young mountains thrust their granite roots deep into the basalt, but, as the mountains are worn down by erosion, they adjust by floating slowly upward (to maintain the equilibrium of mass called *isostasy*, a name suggested in 1889 by the American geologist Clarence Edward Dutton). In the Appalachians, a very ancient mountain chain, the root is about gone.

The basalt beneath the oceans is covered with one-quarter to one-half mile of sedimentary rock, but little or no granite—the Pacific basin is completely free of granite. The thinness of the crust under the oceans has suggested a dramatic project: Why not drill a hole through the crust down to the Moho discontinuity and tap the mantle to see what it is made of? It would not be an easy task, for it would mean anchoring a ship over an abyssal section of the ocean, lowering drilling gear through miles of water, and then drilling through a greater thickness of rock than anyone has yet drilled. Early enthusiasm for the project evaporated, and the matter now lies in abeyance.

The "floating" of the granite in the basalt inevitably suggests the possibility of *continental drift*. In 1912, the German geologist Alfred Lothar Wegener suggested that the continents were originally a single piece of granite, which he called Pangaea ("all-Earth"). At some early stage of the earth's history, this fractured, and the continents drifted apart. He argued that they were still drifting—Greenland, for instance, moving away from Europe at the rate of a yard a year. What gave him (and others, dating back to Francis Bacon about 1620) the idea was mainly the fact that the eastern coastline of South America seemed to fit like a jigsaw piece into the shape of the western coast of Africa.

For a half-century, Wegener's theory was looked upon with hard disfavor. As late as 1960, when the first edition of this book was published, I felt justified, in view of the state of geophysical opinion at that time, in categorically dismissing it. The most telling argument against it was that the basalt underlying both oceans and continents was simply too stiff to allow the continental granite to drift through it, even in the millions of years allowed for it to do so.

And yet evidence in favor of the supposition that the Atlantic Ocean once did not exist, and that the separate continents once formed a single land mass, grew massively impressive. If the continents were matched, not by their actual shoreline (an accident of the current sea level) but by the central point of the continental slope (the shallow floor of the ocean neighboring the continents which is exposed during ages of low sea level), then the fit is excellent all along the Atlantic, in the north as well as the south. Then, too, rock formations in parts of western Africa match the formations in parts of eastern South America in fine detail. Past wanderings of the magnetic poles look less startling if one considers that the continents, not the poles, wandered.

Nor was there only geographic evidence for Pangaea and its breakup. Biological evidence was even stronger. In 1968, for instance, a 2½-inch fossilized bone from an extinct amphibian was found in Antarctica. Such a creature could not possibly have lived so close to the South Pole, so Antarctica must once have been farther from the pole or, at least, milder in temperature. The amphibian could not have crossed even a narrow stretch of salt water, so Antarctica must have been part of a larger body of land, containing warmer areas. The fossil record, generally (which I shall talk about in chapter 16), is quite in tune with the existence at one time, and the subsequent breakup, of Pangaea.

It is important to emphasize here the basis of geologists' opposition to Wegener. People who pound away at the fringe areas of science frequently justify their dubious theories by insisting that scientists tend to be dogmatic, with their minds closed to new work (true enough in some cases and at some times, though never to the extent the "fringe" theorists claim). They frequently use Wegener and his continental drift as an example, and there they are wrong.

Geologists did not object to the *concept* of Pangaea and its breakup. Indeed, more radical suggestions to account for the manner in which life was spread over the earth were considered hopefully. What they objected to was the specific mechanism Wegener advanced—the notion of large granite blocks drifting through a basalt "ocean." There were serious reasons for objecting to it, and those reasons hold even today. The continents do not drift through the basalt.

Some other mechanism, then, must account for the geographic and biologic indications of continental changes in position—a mechanism that is more plausible and for which there is evidence. I shall discuss the evidence later in the chapter; but about 1960, the American geologist Harry Hammond Hess thought it reasonable, on the basis of new findings, to suggest that molten mantle material might be welling up—along certain fracture-lines running the length of the Atlantic Ocean, for instance—and be forced sideways near the top of the mantle, to cool and harden. The ocean floor is, in this way, pulled apart and stretched. It is not, then, that the continents drift, but that they are pushed apart by a spreading sea floor.

As the story seems now, Pangaea did exist, after all, and was intact as recently as 225 million years ago, when the dinosaurs were coming into prominence. To judge from the evolution and distribution of plants and animals, the breakup must have become pronounced about 200 million years ago. Pangaea then broke into three parts: the northern part (North America, Europe, and Asia) is called Laurasia; the southern part (South America, Africa, and India) is called Gondwana, from an Indian province; Antarctica plus Australia formed the third part.

Some 65 million years ago, with the dinosaurs already extinct and the mammals ruling earth, South America separated from Africa on the west, and India on the east separated and moved up toward southern Asia. Finally,

PERMIAN — 225 MILLION
YEARS AGO

TRIASSIC — 200 MILLION
YEARS AGO

JURASSIC — 135 MILLION
YEARS AGO

CRETACEOUS — 65 MILLION
YEARS AGO

CENOZOIC — PRESENT

Figure 4.4. Geologic eras.

North America split off from Europe, India crunched up into Asia (with the Himalayan Mountains folding up at the junction line), Australia moved away from its connection with Antarctica, and the continental arrangement as we have it at present was seen. (For the continental changes, see figure 4.4.)

THE ORIGIN OF THE MOON

An even more startling suggestion about the changes that may have taken place on the earth over geologic periods dates back to 1879, when the British astronomer George Howard Darwin (a son of Charles Darwin) suggested that the moon was a piece of the earth that had broken loose in early times, leaving the Pacific Ocean as the scar of the separation.

This is an attractive thought, since the moon makes up only a little over

1 percent of the combined earth-moon mass and is small enough for its width to lie within the stretch of the Pacific. If the moon were made up of the outer layers of the earth, it would account for the moon's having no iron core and being much less dense than the earth, and for the Pacific floor's being free of continental granite.

The possibility of an earth-moon breakup seems unlikely on various grounds, however; and virtually no astronomer or geologist now thinks that it can have taken place. Nevertheless, the moon seems certainly to have been closer in the past than it is today.

The moon's gravitational pull produces tides both in the ocean and in the earth's solid crust. As the earth rotates, ocean water is dragged across sections of shallow floor, while layers of rock rub together as they rise and fall. The friction represents a slow conversion into heat of the earth's energy of rotation, so that its rotational period gradually increases. The effect is not great in human terms, for the day lengthens by 1 second in about 62,500 years. As the earth loses rotational energy, the angular momentum must be conserved. What the earth loses, the moon gains. Its speed increases as it revolves about the earth, which means it drifts farther away very slowly.

If one works backward in time toward the far geologic past, we see that the earth's rotation must speed up, the day be significantly shorter, the moon significantly closer, and the whole effect more rapid. Darwin calculated backward to find out when the moon was close enough to earth to form a single body; but even if we don't go that far, we ought to find evidence of a shorter day in the past. For instance, about 570 million years ago—the time of the oldest fossils—the day may have been only a little over 20 hours long, and there may have been 428 days in a year.

Nor is this only theory now. Certain corals lay down bands of calcium carbonate more actively at some seasons than others, so that you can count annual bands just as in tree trunks. It is also suggested that some lay down calcium carbonate more actively by day than by night, so that there are very fine daily bands. In 1963, the American paleontologist John West Wells counted the fine bands in fossil corals and reported there were, on the average, 400 daily bands per annual bands in corals dating back 400 million years and 380 daily bands per annual band in corals dating back only 320 million years.

Of course, the question is, If the moon was much closer to the earth then, and the earth rotated more rapidly, what happened in still earlier periods? If Darwin's theory of an earth-moon separation is not so, what *is* so?

One suggestion is that the moon was captured at some time in the past. Its capture 600 million years ago, for instance, might account for the fact that we find numerous fossils in rocks dating back to about that time, whereas earlier rocks have nothing but uncertain traces of carbon. Perhaps the earlier rocks were washed clean by the vast tides that accompanied the capture of the moon. (There was no land life at the time; if there had been, it would have been destroyed.) If the moon were captured, it would have been closer then than now, and there would be a lunar recession and a lengthening of the day since, but nothing of the sort before.

Another suggestion is that the moon was formed in the neighborhood of the earth, out of the same gathering dust cloud, and has been receding ever since, but never was actually part of the earth.

The study and analysis of the moon rocks brought back to Earth by astronauts in the 1970s might have settled the problem (many people had thought optimistically that it would), but it did not. For instance, the moon's surface is covered with bits of glass, which are not to be found on Earth's surface. The moon's crust is also entirely free of water and is poor in all substances that melt at relatively low temperatures, poorer than Earth is. This is an indication that the moon may at one time have been routinely subjected to high temperatures.

Suppose, then, the moon at the time of its formation had had a highly elliptical orbit with its aphelion at roughly its present distance to the sun and its perihelion in the neighborhood of Mercury's orbit. It might have circled in this way for a few billion years before a combination of positions of itself, Earth, and perhaps Venus resulted in the moon's capture by Earth. The moon would abandon its position as a small planet to become a satellite, but its surface would still show the marks of its earlier Mercurylike perihelion.

On the other hand the glasses could be the result of the local heat produced by the meteoric bombardment that had given birth to the moon's craters. Or, in the very unlikely case of the moon's having fissioned from the earth, they might be the result of the heat produced by that violent event.

All suggestions about the moon's origin seem, in fact, to be equally improbable; and scientists have been heard to mutter that if the evidence for the moon's origin is carefully considered, then the only possible conclusion is that the moon is not really out there—a conclusion, however, that just means they must continue the search for additional evidence. There is an answer, and it will be found.

THE EARTH AS LIQUID

The fact that the earth consists of two chief portions—the silicate mantle and the nickel-iron core (in about the same proportions as the white and yolk of an egg)—has persuaded most geologists that the earth must have been liquid at some time in its early history. It might then have consisted of two mutually insoluble liquids. The silicate liquid, being the lighter, would float to the top and cool by radiating its heat into space. The underlying iron liquid, insulated from direct exposure to space, would give up its heat far more slowly and would thus remain liquid to the present day.

There are at least three ways in which the earth could have become hot enough to melt, even from a completely cold start as a collection of planetesimals. These bodies, on colliding and coalescing, would give up their energy of motion (*kinetic energy*) in the form of heat. Then, as the growing planet was compressed by gravitational force, still more energy would be liberated as heat. Third, the radioactive substances of the earth—uranium, thorium, and potassium—have delivered large quantities of heat over the ages as they have broken down; in the early stages, when there was a great deal more radioactive material than now, radioactivity itself might have supplied enough heat to liquefy the earth.

Not all scientists are willing to accept a liquid stage as an absolute necessity. The American chemist Harold Clayton Urey, in particular, maintained that most of the earth was always solid. He argued that in a largely solid earth an iron core could still be formed by a slow separation of iron; and that even now, iron may be migrating from the mantle into the core at the rate of 50,000 tons a second.

The Ocean

The earth is unusual among the planets of the solar system in possessing a surface temperature that permits water to exist in all three states: liquid, solid, and gas. A number of worlds farther from the sun than Earth are essentially icy—Ganymede and Callisto, for instance. Europa has a worldwide surface glacier and may have liquid water beneath, but all such outer worlds can have only insignificant traces of water vapor above the surface.

The earth is the only body in the solar system, as far as we know, to have oceans—vast collections of liquid water (or any liquid at all, for that matter) exposed to the atmosphere above. Actually, I should say *ocean,* because the Pacific, Atlantic, Indian, Arctic, and Antarctic oceans all comprise one connected body of salt water in which the Europe-Asia-Africa mass, the American continents, and smaller bodies such as Antarctica and Australia can be considered islands.

The statistics of this ocean are impressive. It has a total area of 140 million square miles and covers 71 percent of the earth's surface. Its volume, reckoning the average depth of the oceans as $2\frac{1}{3}$ miles, is about 326 million cubic miles. It contains 97.2 percent of all the H_2O on the earth and is the source of the earth's fresh water supply as well, for 80,000 cubic miles of it are evaporated each year to fall again as rain or snow. As a result of such precipitation, there is some 200,000 cubic miles of fresh water under the continents' surface and about 30,000 cubic miles of fresh water gathered into the open as lakes and rivers.

Viewed in another fashion, the ocean is less impressive. Vast as it is, it makes up only a little over 1/4,000 of the total mass of the earth. If we imagine the earth to be the size of a billiard ball, the ocean would be represented by an unnoticeable film of dampness. If you made your way down to the very deepest part of the ocean, you would only be 1/580 of the distance to the center of the earth—and all the rest of that distance would be first rock and then metal.

And yet that unnoticeable film of dampness means everything to us. The first forms of life originated there; and, from the standpoint of sheer quantity, the oceans still contain most of our planet's life. On land, life is confined to within a few feet of the surface (though birds and airplanes do make temporary sorties from this base); in the oceans, life permanently occupies the whole of a realm as deep as seven miles or more in some places.

And yet, until recent years, human beings have been as ignorant of the ocean depths and particularly of the ocean floor as if the ocean were located on the planet Venus.

THE CURRENTS

The founder of modern oceanography was an American naval officer named Matthew Fontaine Maury. In his early thirties, he was lamed in an accident that, however unfortunate for himself, brought benefits to humanity. Placed in charge of the depot of charts and instruments (undoubtedly intended as a sinecure), he threw himself into the task of charting ocean currents. In particular, he studied the course of the Gulf Stream, which had first been investigated as early as 1769 by the American scholar Benjamin Franklin. Maury gave it a description that has become a classic remark in oceanography: "There is a river in the ocean." It is certainly a much larger river than any on land. It transports a thousand times as much water each second as does the Mississippi. It is 50 miles wide at the start, nearly a half mile deep, and moves at speeds of up to 4 miles an hour. Its warming effect is felt even in the far northern island of Spitzbergen.

Maury also initiated international cooperation in studying the ocean; he was the moving figure behind a historic international conference held in Brussels in 1853. In 1855, he published the first textbook in oceanography, entitled *Physical Geography of the Sea.* The Naval Academy at Annapolis honored his achievements by naming Maury Hall after him.

Since Maury's time, the ocean currents have been thoroughly mapped. They move in large clockwise circles in the oceans of the Northern Hemisphere and in large counterclockwise circles in those of the Southern, thanks to the Coriolis effect. The Gulf Stream is but the western branch of a clockwise circle of current in the North Atlantic. South of Newfoundland, it heads due east across the Atlantic (the *North Atlantic drift*). Part of it is deflected by the European coast around the British Isles and up the Norwegian coast; the rest is deflected southward along the northwest shores of Africa. This last part, passing along the Canary Islands, is the *Canaries current.* The configuration of the African coast combines with the Coriolis effect to send the current westward across the Atlantic (the *north equatorial current*). It reaches the Caribbean, and the circle starts all over.

A larger counterclockwise swirl moves water along the rims of the Pacific Ocean south of the Equator. There, the current, skirting the continents, moves northward from the Antarctic up the western coast of South America, as far as Peru. This portion of the circle is the cold *Peru,* or *Humboldt, current* (named for the German naturalist Alexander von Humboldt, who first described it about 1810).

The configuration of the Peruvian coastline combines with the Coriolis effect to send this current westward across the Pacific just south of the Equator (the *south equatorial current*). Some of this flow finds its way through the waters of the Indonesian archipelago into the Indian Ocean. The rest moves southward past the eastern coast of Australia, and then eastward again.

These swirls of water help to even out the temperature of the ocean somewhat and, indirectly, the continental coasts as well. There are still uneven distributions of temperature, but not as much as there would be without the ocean currents.

Most of the ocean currents move slowly, even more slowly than the Gulf Stream. Even at slow speeds, such large areas of the ocean are involved that enormous volumes of water are moved. Off New York City, the Gulf Stream moves water northeastward past some fixed line at the rate of about 45 million tons per second.

There are water currents in the polar regions as well. The clockwise currents in the Northern Hemisphere and the counterclockwise ones in the Southern both succeed in moving water from west to east on the poleward edge of the circle.

South of the continents of South America, Africa, and Australia, a current circles the continent of Antarctica from west to east across unbroken ocean (the only place on Earth where water can drift from west to east without ever meeting land). This *west-wind drift* in the Antarctic is the largest ocean current on Earth, moving nearly 100 million tons of water eastward past any given line each second.

The west wind drift in the arctic regions is interrupted by land masses, so that there is a *North Pacific drift* and a *North Atlantic drift*. The North Atlantic drift is deflected southward by the western coast of Greenland, and the frigid polar water passes Labrador and Newfoundland, so that that portion is the *Labrador current*. The Labrador current meets the Gulf Stream south of Newfoundland, producing a region of frequent fogs and storms.

The western and eastern sides of the Atlantic Ocean are a study in contrasts. Labrador, on the western side, exposed to the Labrador current, is a desolation with a total population of 25,000. On the eastern side, in precisely the same latitudes, are the British Isles with a population of 55,000,000, thanks to the Gulf Stream.

A current moving directly along the Equator is not subjected to the Coriolis effect and may move in a straight line. Such a thin, straight current was located in the Pacific Ocean, moving due east for several thousand miles along the Equator. It is called the *Cromwell current* after its discoverer, the American oceanographer Townsend Cromwell. A similar current, somewhat slower, was discovered in the Atlantic in 1961 by the American oceanographer Arthur D. Voorhis.

Nor is circulation confined to surface currents only. That the deeps cannot maintain a dead calm is clear from several indirect forms of evidence. For one thing, the life at the top of the sea is continually consuming its mineral nutrients—phosphate and nitrate—and carrying this material down to the depths with itself after death; if there were no circulation to bring it up again, the surface would become depleted of these minerals. For another thing, the oxygen supplied to the oceans by absorption from the air would not percolate down to the depths at a sufficient rate to support life there if there were no conveying circulation. Actually oxygen is found in adequate concentration down to the very floor of the abyss. This can be explained only by supposing that there are regions in the ocean where oxygen-rich surface waters sink.

The engine that drives this vertical circulation is temperature difference. The ocean's surface water is cooled in polar regions and therefore sinks. This continual flow of sinking water spreads out all along the ocean floor, so that even in the tropics the bottom water is very cold—near the freezing point. Eventually the cold water of the depths wells up toward the surface, for it has no other place to go. After rising to the surface, the water warms and drifts off toward the Arctic or the Antarctic, there to sink again. The resulting circulation, it is estimated, would bring about complete mixing of the Atlantic Ocean, if something new were added to part of it, in about 1,000 years. The larger Pacific Ocean would undergo complete mixing in perhaps 2,000 years.

The Antarctic is much more efficient in supplying cold water than the Arctic is. Antarctica has an icecap ten times as large as all the ice in the Arctic, including the Greenland icecap. The water surrounding Antarctica, made frigid by melting ice, spreads northward on the surface till it meets the warm waters carried southward from the tropical regions. The cold water from Antarctica, denser than the warm tropical waters, sinks below it at the line of the *Antarctic convergence,* which in some places extends as far north as 40° S.

The cold Antarctic water spreads through all the ocean bottoms carrying with it oxygen (for oxygen, like all gases, dissolves more easily and in greater quantities in cold water than in warm) and nutrients. Antarctica (the "icebox of the world") thus fertilizes the oceans and controls the weather of the planet.

The continental barriers complicate this general picture. To follow the actual circulation, oceanographers have resorted to oxygen as a tracer. As the polar water, rich in oxygen, sinks and spreads, the oxygen is gradually diminished by organisms that make use of it. So, by sampling the oxygen concentration in deep water at various locations, one can plot the direction of the deep-sea currents.

Such mapping has shown that one major current flows from the Arctic Ocean down the Atlantic under the Gulf Stream and in the opposite direction, another from the Antarctic up the south Atlantic. The Pacific Ocean gets no direct flow from the Arctic to speak of, because the only outlet into it is the narrow and shallow Bering Strait. Hence, it is the end of the line for the deep-sea flow. That the North Pacific is the dead end of the global flow is shown by the fact that its deep waters are poor in oxygen. Large parts of this largest ocean are therefore sparsely populated with life forms and are the equivalent of desert areas on land. The same may be said of nearly land-locked seas like the Mediterranean, where full circulation of oxygen and nutrients is partly choked off.

More direct evidence for this picture of the deep-sea currents was obtained in 1957 during a joint British-American oceanographic expedition. The investigators used a special float, invented by the British oceanographer John Crossley Swallow, which is designed to keep its level at a depth of a mile or more and is equipped with a device for sending out short-wave sound waves. By means of these signals, the float can be tracked as it moves with the deep-sea current. The expedition thus traced the deep-sea current down the Atlantic along its western edge.

THE OCEAN'S RESOURCES

All this information will acquire practical importance when the world's expanding population turns to the ocean for more food. Scientific "farming of the sea" will require knowledge of these fertilizing currents, just as land farming requires knowledge of river courses, ground water, and rainfall. The present harvest of seafood—some 80 million tons in 1980—can, with careful and efficient management, be increased (it is estimated) to something over 200 million tons per year, while leaving sea life enough leeway to maintain itself adequately. (The assumption is, of course, that we do not continue our present course of heedlessly damaging and polluting the ocean, particularly those portions of it—nearest the continental shores—that contain and offer human beings the major portion of sea organisms. So far, we are not only failing to rationalize a more efficient use of the sea for food but are decreasing its ability to yield us the quantity of food we harvest now.)

Food is not the only important resource of the ocean. Sea water contains in solution vast quantities of almost every element. As much as 4 billion tons of uranium, 300 million tons of silver, and 4 million tons of gold are contained in the oceans—but in dilution too great for practical extraction. However, both magnesium and bromine are now obtained from sea water on a commercial scale. Moreover, an important source of iodine is dried seaweed, the living plants having previously concentrated the element out of sea water to an extent that humans cannot yet profitably duplicate.

Much more prosaic material is dredged up from the sea. From the relatively shallow waters bordering the United States, some 20 million tons of oyster shells are obtained each year to serve as a valuable source of limestone. In addition, 50 million cubic yards of sand and gravel are obtained in similar fashion.

Scattered over the deeper portions of the ocean floor are metallic nodules that have precipitated out about some nucleus that may be a pebble or a shark tooth. (It is the oceanic analogue of the formation of a pearl about a sand grain inside an oyster.) These are usually referred to as manganese nodules because they are richest in that metal. It is estimated that there are 31,000 tons of these nodules per square mile of the Pacific floor. Obtaining these in quantity would be difficult indeed, and the manganese content alone would not make it worthwhile under present conditions. However, the nodules contain 1 percent nickel, 0.5 percent copper, and 0.5 percent cobalt. These minor constituents make the nodules far more attractive than they would otherwise be.

And what of the 97 percent of the ocean that is actually water, rather than dissolved material?

Americans use 95,000 cubic feet of water per person per year, for drinking, for washing, for agriculture, for industry. Most nations are less lavish in their use; but for the world generally, the use is 53,000 cubic feet per person per year. All this water, however, must be *fresh* water. Sea water, as is, is useless for any of these purposes.

There is, of course, a great deal of fresh water on Earth in an absolute sense. Less than 3 percent of all the water on Earth is fresh, but that still amounts to about 360 million cubic feet per person. Three-quarters of this is not available for use, to be sure, but is tucked away in the permanent icecaps that cover 10 percent of the planet's land surface.

The liquid fresh water on Earth comes to about 85 million cubic feet per person and is constantly replenished by rainfall that amounts to 4 million cubic feet per person. We might argue that the annual rainfall amounts to 75 times the quantity used by the human race, and that there is therefore plenty of fresh water.

However, most of the rain falls on the ocean or as snow on the ice pack. Some of the rain that falls on land and remains liquid, or becomes liquid when it grows warmer, runs off to sea without being used. A great deal of water in the forests of the Amazon region is virtually not used by human beings at all. And the human population is steadily growing and is also steadily polluting such fresh water supplies as exist.

Fresh water is therefore going to be a scarce commodity before long, and humanity is beginning to turn to the ultimate source, the ocean. It is possible to distill sea water, evaporating and then condensing the water itself, and leaving the dissolved material behind, using, ideally, the heat of the sun for the purpose. Such *desalination* procedures can be used as a fresh-water source and are so used where sunlight is steadily available, or where fuel is cheap, or where needs must. A large ocean liner routinely supplies itself with fresh water by burning its oil in order to distill sea water as well as to run its engines.

There are also suggestions that icebergs be collected in the polar regions and floated to warm, but arid seaports, where what has survived of the ice can be melted down for use.

Undoubtedly, however, the best way of utilizing our fresh-water resources (or any resources) is by wise conservation, the reduction to a minimum of waste and pollution, and the cautious limitation of Earth's human population.

THE OCEAN DEPTHS AND CONTINENTAL CHANGES

What about the direct observation of ocean depths? A lone record from ancient times survives (if it can be trusted). The Greek philosopher Posidonius, about 100 B.C., is supposed to have measured the depth of the

Mediterranean Sea just off the shores of the island of Sardinia and is said to have come up with a depth of about 1.2 miles.

It was not until the eighteenth century, however, that scientists began a systematic study of the depths for the purpose of studying sea life. In the 1770s, a Danish biologist, Otto Frederik Muller, devised a dredge that could be used to bring up specimens of such life from many yards beneath the surface.

One person who used a dredge with particular success was an English biologist, Edward Forbes, Jr. During the 1830s, he dredged up sea life from the North Sea and from other waters around the British Isles. Then, in 1841, he joined a naval ship that was going to the eastern Mediterranean, and there dredged up a starfish from a depth of 450 yards.

Plant life can live only in the uppermost layer of the ocean, since sunlight does not penetrate more than 80 yards or so. Animal life cannot live except (ultimately) upon plant life. It seemed to Forbes, therefore, that animal life could not long remain below the level at which plants were to be found. In fact, he felt that a depth of 450 yards was probably the limit of sea life and that, below it, the ocean was barren and lifeless.

And yet, just as Forbes was deciding this, the British explorer James Clark Ross, who was exploring the shores of Antarctica, dredged up life from as deep as 800 yards, well below Forbes's limit. Antarctica was far away, however; and most biologists continued to accept Forbes's decision.

The sea bottom first became a matter of practical interest to human beings (rather than one of intellectual curiosity to a few scientists) when it was decided to lay a telegraph cable across the Atlantic. In 1850, Maury had worked up a chart of the Atlantic sea bottom for purposes of cable laying. It took fifteen years, punctuated by many breaks and failures, before the Atlantic cable was finally laid—under the incredibly persevering drive of the United States financier Cyrus West Field, who lost a fortune in the process. (More than twenty cables now span the Atlantic.)

But the process, thanks to Maury, marked the beginning of the systematic exploration of the sea bottom. Maury's soundings made it appear that the Atlantic Ocean was shallower in its middle than on either side. The central shallow region, Maury named Telegraph Plateau in honor of the cable.

The British ship *Bulldog* labored to continue and extend Maury's exploration of the sea bottom. It set sail in 1860; and on board was a British physician, George C. Wallich, who used a dredge and brought up thirteen starfish from a depth of 2,500 yards (nearly 1½ miles). Nor were they starfish that had died and sunk to the sea bottom: they were very much alive. Wallich reported this at once and insisted that animal life could exist in the cold darkness of the deep sea, even without plants.

Biologists were still reluctant to believe in this possibility; and a Scottish biologist, Charles W. Thomson, went out dredging in 1868 in a ship called *Lightning.* Dredging through deep waters, he obtained animals of all kinds, and all argument ended. Forbes's idea of a lower limit of sea life ended.

Thomson wanted to determine just how deep the ocean is, and set out on 7 December 1872 in the *Challenger*, remaining at sea for three and a half years for a distance of 78,000 miles altogether. To measure the depth of the oceans the *Challenger* had no better device than the time-honored method of paying out 4 miles of cable with a weight on the end until it reached the bottom. Over 370 soundings were made in this fashion. This procedure, unfortunately, is not only fantastically laborious (for deep sounding) but is also of low accuracy. Ocean-bottom exploration was revolutionized in 1922, however, with the introduction of *echo sounding* by means of sound waves; in order to explain how this works, a digression on sound is in order.

Mechanical vibrations set up longitudinal waves in matter (in air, for instance), and we can detect some of these as sound. We hear different wavelengths as sounds of different pitch. The deepest sound we hear has a wavelength of 22 meters and a frequency of 15 cycles per second. The shrillest sound a normal adult can hear has a wavelength of 2.2 centimeters and a frequency of 15,000 cycles per second. (Children can hear somewhat shriller sounds.)

The absorption of sound by the atmosphere depends on the wavelength. The longer the wavelength, the less sound is absorbed by a given thickness of air. For this reason, foghorn blasts are far in the bass register so that they can penetrate as great a distance as possible. The foghorn of a large liner like the old *Queen Mary* sounds at 27 vibrations per second, about that of the lowest note on the piano. It can be heard at a distance of 10 miles, and instruments can pick up the sound at a distance of 100 to 150 miles.

Sounds also exist deeper in pitch than the deepest we can hear. Some of the sounds set up by earthquakes or volcanoes are in this *infrasonic* range. Such vibrations can encircle the earth, sometimes several times, before being completely absorbed.

The efficiency with which sound is reflected depends on the wavelength in the opposite way. The shorter the wavelength, the more efficient the reflection. Sound waves with frequencies higher than those of the shrillest sounds we hear are even more efficiently reflected. Some animals can hear shriller sounds than we can and make use of this ability. Bats squeak to emit sound waves with *ultrasonic* frequencies as high as 130,000 cycles per second and listen for the reflections. From the direction in which reflections are loudest and from the time lag between squeak and echo, they can judge the location of insects to be caught and twigs to be avoided. They can thus fly with perfect efficiency if they are blinded, but not if they are deafened. (The Italian biologist Lazzaro Spallanzani, who first made this observation in 1793, wondered if bats could see with their ears, and, of course, in a sense, they do.)

Porpoises, as well as guacharos (cave-dwelling birds of Venezuela), also use sounds for *echo-location* purposes. Since they are interested in locating larger objects, they can use the less efficient sound waves in the audible region for the purpose. (The complex sounds emitted by the large-brained porpoises and dolphins may even, it is beginning to be suspected, be used for purposes of general communication—for talking, to put it bluntly. The American biologist John C. Lilly investigated this possibility exhaustively with inconclusive results.)

To make use of the properties of ultrasonic sound waves, humans must first produce them. Small-scale production and use are exemplified by the *dog whistle* (first constructed in 1883). It produces sound in the near ultrasonic range which can be heard by dogs but not by humans.

A route whereby much more could be done was opened by the French chemist Pierre Curie and his brother, Jacques, who in 1880 discovered that pressures on certain crystals produced an electric potential (*piezoelectricity*). The reverse was also true. Applying an electric potential to a crystal of this sort produced a slight constriction as though pressure were being applied (*electrostriction*). When the technique for producing a very rapidly fluctuating potential was developed, crystals could be made to vibrate quickly enough to form ultrasonic waves. This was first done in 1917 by the French physicist Paul Langevin, who immediately applied the excellent reflective powers of this short-wave sound to the detection of submarines—though by the time he was done, the First World War was over. During the Second World War, this method was perfected and became *sonar* ("*so*und *n*avigation *a*nd *r*anging," *ranging* meaning "determining distance").

The determination of the distance of the sea bottom by the reflection of ultrasonic sound waves replaced the sounding line. The time interval from the sending of the signal (a *sharp pulse*) and the return of its echo measures the distance to the bottom. The only thing the operator has to worry about is whether the reading signals a false echo from a school of fish or some other obstruction. (Hence, the instrument is useful to fishing fleets.)

The echo-sounding method not only is swift and convenient but also makes it possible to trace a continuous profile of the bottom over which the vessel moves, so that oceanographers are obtaining a picture of the topography of

---

Figure 4.5. Profile of the Pacific bottom. The great trenches in the sea floor go deeper below sea level than the height of the Himalayas, and the Hawaiian peak stands higher from the bottom than the tallest land mountain.

---

the ocean bottom. More detail could be gathered in five minutes than the *Challenger* could have managed in its entire voyage.

The first ship to use sonar in this way was the German oceanographic vessel *Meteor,* which studied the Atlantic Ocean in 1922. By 1925, it was obvious that the ocean bottom was by no means featureless and flat, and that Maury's Telegraph Plateau was not a gentle rise and fall but was, in fact, a mountain range, longer and more rugged than any mountain range on land. It wound down the length of the Atlantic, and its highest peaks broke through the water surface and appeared as such islands as the Azores, Ascension, and Tristan da Cunha. It was called the Mid-Atlantic Range.

As time went on other dramatic discoveries were made. The island of Hawaii is the top of an underwater mountain 33,000 feet high, measuring from its undersea base—higher than anything in the Himalayas; thus, Hawaii may fairly be called the tallest mountain on the earth. There are also numerous flat-topped cones, called *seamounts* or *guyots.* The latter name honors the Swiss-American geographer Arnold Henry Guyot, who brought scientific geography to the United States when he emigrated to America in 1848. Seamounts were first discovered during the Second World War by the American geologist Harry Hammond Hess, who located 19 in quick succession. At least 10,000 exist, mostly in the Pacific. One of these, discovered in 1964 just south of Wake Island, is over 14,000 feet high.

Moreover, there are the ocean deeps (*trenches*), more than 20,000 feet deep, in which the Grand Canyon would be lost. The trenches, all located alongside island archipelagoes, have a total area amounting to nearly 1 percent of the ocean bottom. This may not seem much, but it is actually equal to one-half the area of the United States, and the trenches contain fifteen times as much water as all the rivers and lakes in the world. The deepest of them are in the Pacific; they are found there alongside the Philippines, the Marianas, the Kuriles, the Solomons, and the Aleutians (figure 4.5). There are other

WAKE   HAWAII   U.S.

great trenches in the Atlantic off the West Indies and the South Sandwich Islands, and there is one in the Indian Ocean off the East Indies.

Besides the trenches, oceanographers have traced on the ocean bottom canyons, sometimes thousands of miles long, which look like river channels. Some of them actually seem to be extensions of rivers on land—notably a canyon extending from the Hudson River into the Atlantic. At least twenty such huge gouges have been located in the Bay of Bengal alone, as a result of oceanographic studies of the Indian Ocean during the 1960s. It is tempting to suppose that these were once river beds on land, when the ocean was lower than now. But some of the undersea channels are so far below the present sea level that it seems altogether unlikely they could ever have been above the ocean. In recent years, various oceanographers—notably William Maurice Ewing and Bruce Charles Heezen—have developed another theory: that the undersea canyons were gouged out by turbulent flows (*turbidity currents*) of soil-laden water in an avalanche down the off-shore continental slopes at speeds of up to 60 miles an hour. One turbidity current, which focused scientific attention on the problem, took place in 1929 after an earthquake off Newfoundland. The current snapped a number of cables, one after the other, and made a great nuisance of itself.

The Mid-Atlantic Range continued to present surprises. Later soundings elsewhere showed that it was not confined to the Atlantic. At its southern end, it curves around Africa and moves up the western Indian Ocean to Arabia. In mid-Indian Ocean, it branches so that the range continues south of Australia and New Zealand and then works northward in a vast circle all around the Pacific Ocean. What began (in men's minds) as the Mid-Atlantic Ridge became the Mid-Oceanic Ridge. And in one rather basic fashion, the Mid-Oceanic Ridge is not like the mountain ranges on the continent: the continental highlands are of folded sedimentary rocks, while the vast oceanic ridge is of basalt squeezed up from the hot lower depths.

After the Second World War, the details of the ocean floor were probed with new energy by Ewing and Heezen. Detailed soundings in 1953 showed, rather to their astonishment, that a deep canyon ran the length of the Ridge and right along its center. This was eventually found to exist in all portions of the Mid-Oceanic Ridge, so that sometimes it is called the Great Global Rift. There are places where the Rift comes quite close to land: it runs up the Red Sea between Africa and Arabia, and it skims the borders of the Pacific through the Gulf of California and up the coast of the state of California.

At first it seemed that the Rift might be continuous, a 40,000-mile crack in the earth's crust. Closer examination, however, showed that it consists of short, straight sections that are set off from each other as though earthquake shocks had displaced one section from the next. And, indeed, it is along the Rift that the earth's quakes and volcanoes have tended to occur.

The Rift was a weak spot up through which heated molten rock (*magma*) welled slowly from the interior—cooling, piling up to form the Ridge, and spreading out farther still. The spreading can be as rapid as 16 centimeters per year, and the entire Pacific Ocean floor could be covered with a new layer in 100 million years. Indeed, sediment drawn up from the ocean floor is rarely found to be older, which would be remarkable in a planetary life forty-five times as long, were it not for the concept of *sea-floor spreading*.

It appeared at once that the earth's crust was divided into large plates, separated from each other by the Great Global Rift and its offshoots. These were called *tectonic plates*, *tectonic* coming from a Greek word for "carpenter," since the plates seemed to be cleverly joined to make a seemingly unbroken crust. The study of the evolution of the earth's crust in terms of these plates is referred to by those words in reverse as *plate tectonics*.

There are six large tectonic plates and a number of smaller ones, and it quickly became apparent that earthquakes commonly take place along their boundaries. The boundaries of the Pacific plate (which includes most of the Pacific Ocean) include the earthquake zones in the East Indies, in the Japanese islands, in Alaska and California, and so on. The Mediterranean boundary between the Eurasian and African plates is second only to the Pacific rim for its well-remembered earthquakes.

Then, too, the *faults* that had been detected in the earth's crust as deep cracks where the rock on one side could, periodically, slide against the rock on the other to produce earthquakes, were also on the boundaries of the plates and on the offshoots of those boundaries. The most famous of all such faults, the San Andreas, which runs the length of coastal California from San Francisco to Los Angeles, is part of the boundary between the American and the Pacific plates.

And what about Wegener's continental drift? If an individual plate is considered, then objects upon it cannot drift or change position. They are locked in place by the stiffness of the basalt (as those who were opposed to Wegener's notions had pointed out). What's more, neighboring plates were so tightly wedged together that it was difficult to see what could make them move.

The answer came from another consideration. The plate boundaries were places where not only earthquakes were common, but volcanoes, too. Indeed, the shores of the Pacific, as one follows the boundary of the Pacific plate, are so marked by volcanoes, both active and inactive, that the whole has been referred to as the *circle of fire*.

Could it be, then, that magma might well up from the hot layers deep in the earth through the cracks between the tectonic plates, these cracks representing weaknesses in Earth's otherwise solid crust? Specifically, magma might be welling up very slowly through the Mid-Atlantic Rift and solidifying on contact with ocean water to form the Mid-Atlantic Range on either side of the Rift.

We can go farther. Perhaps as the magma welled up and solidified, it pushed the plates apart. If so, it would succeed in pushing Africa and South America apart on the south, and Europe and North America apart on the north, breaking up Pangaea, forming the Atlantic Ocean, and making it ever wider.

Europe and Africa would be pushed apart, too, with the Mediterranean and Red seas forming. Because the sea floor would grow wider as a result, this effect was called *sea-floor spreading* and was first proposed by H. H. Hess and Robert S. Dietz in 1960. The continents were not floating or drifting apart, as Wegener had thought; they were fixed to plates that were being *pushed* apart.

How could sea-floor spreading be demonstrated? Beginning in 1963, the rocks obtained from the ocean floor on either side of the Mid-Atlantic Rift were tested for their magnetic properties. The pattern changed with distance from the Rift, and did so in exact correspondence, but as a mirror image, on either side. There was clear evidence that the rocks were youngest near the Rift and increasingly older as one moved away from it on either side.

In this way, it could be estimated that the Atlantic sea floor was spreading, at the moment, at the rate of just under an inch a year. On this basis, the time when the Atlantic Ocean first began to open could be roughly determined. In this and other ways, the movement of tectonic plates has completely revolutionized the study of geology in these last two decades.

Naturally, if two plates are forced apart, each must (in view of the tightness of the fit of all the plates) be jammed into another on the other side. When two plates come together slowly (at a rate of no more than 2 inches or so per year), the crust buckles and bulges both up and down, forming mountains and their *roots*. Thus, the Himalayan Mountains seem to have been formed when the plate bearing India made slow contact with the plate bearing the rest of Asia.

On the other hand, when two plates come together too rapidly to allow buckling, the surface of one plate may gouge its way under the other, forming a deep trench, a line of islands, and a disposition toward volcanic activity. Such trenches and islands are found in the western Pacific, for instance.

Plates push apart under the influence of sea-floor spreading, as well as come together. The Rift passes right through western Iceland, which is (very slowly) being pushed apart. Another place of division is at the Red Sea, which is rather young and exists only because Africa and Arabia have already pushed apart somewhat. (The opposite shores of the Red Sea fit closely if put together.) This process is continuing, so that the Red Sea is, in a sense, a new ocean in the process of formation. Active upwelling in the Red Sea is indicated by the fact that at the bottom of that body of water there are, as discovered in 1965, sections with a temperature of 56° C and a salt concentration at least five times normal.

Presumably, there has been a long, very slow cycle of magma welling up to push plates apart in some places, and plates coming together, pushing crust downward, and converting it to magma. In the process, the continents come together into a single land mass and then split up, not once, but many times, with mountains forming and being worn down, ocean deeps forming and being filled in, volcanoes forming and becoming extinct. The earth is geologically, as well as biologically, alive.

Geologists can now even follow the course of the most recent breakup of Pangaea, though still only in a rough manner. An early break came in an east-west line. The northern half of Pangaea—including what is now North America, Europe, and Asia—is sometimes called Laurasia, because the oldest part of the North American surface rocks, geologically speaking, are those of the Laurentian Highlands north of the St. Lawrence River.

The southern half—including what is now South America, Africa, India, Australia, and Antarctica—is called Gondwanaland (a name invented in the 1890s by an Austrian geologist, Edward Suess, who derived it from a region in India and based it on a theory of geologic evolution that then seemed reasonable but is now known to be wrong).

About 200 million years ago, North America began to be pushed away from Eurasia; and 150 million years ago, South America began to be pushed away from Africa—the two continents eventually connecting narrowly at Central America. The land masses were pushed northward as they separated until the two halves of Laurasia clasped the Arctic region between them.

About 110 million years ago, the eastern portion of Gondwanaland broke into several fragments: Madagascar, India, Antarctica, and Australia. Madagascar stayed fairly close to Africa, but India moved farther than any other land mass in the time since the most recent Pangaea. It moved 5,500 miles northward to push into southern Asia to form the Himalayan Mountains, the Pamirs, and the Tibetan plateau—the youngest, greatest, and most impressive highland area on Earth.

Antarctica and Australia may have separated only 40 million years ago. Antarctica moved southward to its frozen destiny. Australia is still moving northward today.

LIFE IN THE DEEP

After the Second World War, the deeps of the ocean continued to be explored. An underwater-listening device, the *hydrophone,* has, in recent years, shown that sea creatures click, grunt, snap, moan, and in general make the ocean depths as maddeningly noisy as ever the land is.

A new *Challenger* probed the Marianas Trench in the western Pacific in 1951 and found that it (and not one off the Philippine Islands) was the deepest gash in the earth's crust. The deepest portion is now called the *Challenger Deep.* It is over 36,000 feet deep: if Mount Everest were placed in it, a mile of water would roll over its topmost peak. Yet the *Challenger* brought up, from the floor of the abyss, bacteria which look much like bacteria of the surface but cannot live at a pressure of less than 1,000 atmospheres!

The creatures of the trenches are so adapted to the great pressures of these bottoms that they are unable to rise out of their trench; in effect, they are imprisoned in an island. They have experienced a segregated evolution. Yet they are in many respects related to other organisms closely enough that it seems their evolution in the abyss has not gone on for a very long time. One can visualize some groups of ocean creatures being forced into ever lower depths by the pressure of competition, just as other groups were forced ever higher up the continental shelf until they emerged onto the land. The first group had to become adjusted to higher pressures; the second, to the absence of water. On the whole, the latter adjustment was probably the more difficult, so we should not be amazed that life exists in the abyss.

To be sure, life is not as rich in the depths as nearer the surface. The mass of living matter below $4\frac{1}{2}$ miles is only one-tenth as great per unit volume of ocean as it is estimated to be at 2 miles. Furthermore, there are few, if any, carnivores below $4\frac{1}{2}$ miles, since there is insufficient prey to support them. They are scavengers instead, eating anything organic that they can find. The recentness with which the abyss has been colonized is brought out by the disclosure that no species of creature found there has been developed earlier than 200 million years ago, and most have histories of no more than 50 million years. It is only at the beginning of the age of the dinosaurs that the deep sea, hitherto bare of organisms, was finally invaded by life.

Nevertheless, some of the organisms that invaded the deep survived there, whereas their relatives nearer the surface died out—as was demonstrated, most dramatically, in the late 1930s. On 25 December 1938, a trawler fishing off South Africa brought up an odd fish about 5 feet long. What was odd about it was that its fins were attached to fleshy lobes rather than directly to the body. A South-African zoologist, J. L. B. Smith, who had the chance of examining it, recognized it as a matchless Christmas present. It was a *coelacanth*, a primitive fish that zoologists had thought extinct for 70 million years. Here was a living specimen of an animal that was supposed to have disappeared from the earth before the dinosaurs reached their prime.

The Second World War halted the hunt for more coelacanths; but in 1952, another of a different genus was fished up off Madagascar. By now, numbers have been found. Because it is adapted to fairly deep waters, the coelacanth dies soon after being brought to the surface.

Evolutionists have been particularly interested in studying the coelacanth specimens because it was from this fish that the first amphibians developed; in other words, the coelacanth is a direct descendant of our fishy ancestors.

An even more exciting find was made in the late 1970s. There are *hot spots* in the ocean floor, where the hot magma of the mantle rises unusually near the upper boundary of the crust and heats the water above it.

Beginning in 1977, a deep-sea submarine carried scientists down to investigate the sea floor near hot spots east of the Galapagos Islands and at the mouth of the Gulf of California. In the latter hot spot, they found *chimneys*, through which hot gushes of smoky mud surge upward, filling the surrounding sea water with minerals.

The minerals are rich in sulfur, and the neighborhood of these hot spots is also rich in species of bacteria that obtain their energy from chemical reactions involving sulfur plus heat, instead of from sunlight. Small animals feed on these bacteria, and larger animals feed on the smaller ones.

This was a whole new chain of life forms that did not depend upon the plant cells in the uppermost layers of the sea. Even if sunlight did not exist at all anywhere, this chain can exist provided heat and minerals continue to gush upward from the earth's interior; hence, it can exist only near the hot spots.

Clams, crabs, and various kinds of worms, some quite large, were retrieved and studied from these sea-floor areas. All of these flourished in water that would be poisonous to species not adapted to the chemical peculiarities of the region.

DEEP-SEA DIVING

This is an example of the fact that the ideal way to study the ocean deeps is to send human observers down into them. Water is not a suitable environment for us, of course. Since ancient times, divers have practiced their skills and learned to dive down for 60 feet or so and remain underwater for up to 2 minutes. But the unaided body cannot much improve this performance.

In the 1930s, goggles, rubber foot fins, and *snorkels* (short pipes, one end in the mouth and the other sticking up above the surface of the water, from a German word for "snout") made it possible for swimmers to move underwater for longer periods of time and with more efficiency than otherwise. This was *skin diving*, immediately below the surface, or "skin," of the ocean.

In 1943, the French naval officer Jacques-Yves Cousteau developed a system in which skin divers began carrying cylinders of compressed air, which could be exhaled into canisters of chemicals that absorbed the carbon dioxide and rendered the exhaled air fit to breathe again. These were *aqua-lungs*, and the sport, which became popular after the war, was called *scuba diving*, the word *scuba* being an acronym for "self-contained underwater breathing apparatus."

Experienced scuba divers can attain depths of about 200 feet, but that is still very shallow compared with the total depth of the ocean.

The first practical diving suit was designed in 1830 by Augustus Siebe. A diver in a modern diving suit can go down about 300 feet. A diving suit encloses the human body entirely, but a more elaborate enclosure would amount to an entire vessel suited for undersea travel—a *submarine*.

The first submarine that could actually remain beneath water for a reasonable period of time without drowning the person inside was built as long ago as 1620 by a Dutch inventor, Cornelis Drebbel. No submarine could be practical, however, until it could be driven by something other than a hand-turned propeller. Steam power was not useful because one could not burn fuel in the limited atmosphere of an enclosed submarine. What was needed was a motor run by electricity from a storage battery.

The first such electric submarine was built in 1886. Though the battery had to be periodically recharged, the vessel's cruising distance between recharges was something like 80 miles. By the time the First World War began, the major European powers all had submarines and used them as war vessels. These early submarines, however, were fragile and could not descend far.

In 1934, Charles William Beebe managed to get down to about 3,000 feet in his *bathysphere*, a small, thick-walled craft equipped with oxygen and with chemicals to absorb carbon dioxide.

The bathysphere was an inert object suspended from a surface vessel by a cable (a snapped cable meant the end). What was needed was a maneuverable ship of the abyss. Such a ship, the *bathyscaphe*, was invented in 1947 by the Swiss physicist Auguste Piccard. Built to withstand great pressures, it used a heavy ballast of iron pellets (which are automatically jettisoned in case of emergency) to take it down and a "balloon" containing gasoline (which is lighter than water) to provide buoyancy and stability. In its first test off Dakar, West Africa, in 1948, the bathyscaphe (unmanned) descended 4,500 feet. In the same year, Beebe's co-worker, Otis Barton, plumbed to a depth of 4,500 feet, using a modified bathysphere called a *benthoscope*.

Later, Piccard and his son Jacques built an improved version of the bathyscaphe and named the new vessel *Trieste*, because the then Free City of Trieste had helped finance its construction. In 1953, Piccard plunged 2½ miles into the depths of the Mediterranean.

The *Trieste* was bought by the United States Navy for research. On 14 January 1960, Jacques Piccard and a Navy man, Don Walsh, took it to the bottom of the Marianas Trench, plumbing 7 miles to the deepest part of any abyss. There, at the ultimate ocean depth, where the pressure was 1,100 atmospheres, they found water currents and living creatures. In fact, the first creature seen was a vertebrate, a one-foot-long flounderlike fish, with eyes.

In 1964, the French-owned bathyscaphe *Archimède* made ten trips to the bottom of the Puerto Rico Trench, which, with a depth of 5¼ miles, is the deepest abyss in the Atlantic. There, too, every square foot of the ocean floor had its life form. Oddly enough, the bottom did not descend smoothly into the abyss; rather, it seemed terraced, like a giant, spread-out staircase.

The Icecaps

The extremities of our planet have always fascinated human beings, and one of the most adventurous chapters in the history of science has been the exploration of the polar regions. Those regions are charged with romance, spectacular phenomena, and elements of human destiny—the strange auroras in the sky, the extreme cold, and especially the immense icecaps, or glaciers, which hold the key to our world climate and our way of life.

THE NORTH POLE

The actual push to the poles came rather late in human history. It began during the great age of exploration following the discovery of the Americas by Christopher Columbus. The first Arctic explorers went chiefly to find a sea route around the top of North America. Pursuing this will-o'-the-wisp, the English navigator Henry Hudson (in the employ of Holland) found Hudson Bay and his death in 1610. Six years later, another English navigator, William Baffin, discovered what came to be called Baffin Bay, and penetrated to within 800 miles of the North Pole (figure 4.6). Eventually, in the years 1846 to 1848,

Figure 4.6. Map of the North Pole.

the British explorer John Franklin worked his way over the northern coast of Canada and discovered the Northwest Passage (and a most impractical passage for ships it then was). He died on the voyage.

There followed a half-century of efforts to reach the North Pole, motivated in large part by sheer adventure and the desire to be the first to get there. In 1873, the Austrian explorers Julius Payer and Carl Weyprecht reached within

600 miles of the Pole and named a group of islands they found Franz Josef Land, after the Austrian emperor. In 1896, the Norwegian explorer Fridtjof Nansen drifted on the Arctic ice to within 300 miles of the Pole. At length, on 6 April 1909, the American explorer Robert Edwin Peary arrived at the Pole itself.

By now, the North Pole has lost much of its mystery. It has been explored on the ice, from the air, and under water. Richard Evelyn Byrd and Floyd Bennett were the first to fly over it, in 1926; and submarines have traversed its waters.

Meanwhile, the largest northern icecap, which is centered in Greenland, has drawn a number of scientific expeditions. Wegener died in the course of one such expedition in November 1930. The Greenland glacier has been found to cover about 640,000 of that island's 840,000 square miles, and its ice is known to reach a thickness of a mile in some places.

As the ice accumulates, it is pushed down to the sea, where the edges break off, or *calve*, to form icebergs. Some 16,000 icebergs are thus formed in the Northern Hemisphere each year, 90 percent of them breaking off the Greenland icecap. The icebergs work slowly southward, particularly down the west Atlantic. About 400 icebergs per year pass Newfoundland and threaten shipping lanes; between 1870 and 1890, fourteen ships were sunk and forty damaged by collision with icebergs.

The climax came in 1912, when, on its maiden voyage, the luxury liner *Titanic* collided with an iceberg and sank. An international watch over the positions of these inanimate monsters has been maintained ever since. During the years since this Ice Patrol has come into existence, not one ship has been sunk by an iceberg.

THE SOUTH POLE—ANTARCTICA

Far larger than Greenland is the South Pole's great continental glacier. The Antarctic icecap covers seven times the area of the Greenland glacier and has an average thickness of 1½ miles, with 3-mile depths in spots. This is due to the great size of the Antarctic continent—some 5 million square miles, though how much is land and how much is ice-covered sea is still uncertain (figure 4.7). Some explorers believe that western Antarctica, at least, is a group of large islands bound together by ice; but at the moment, the continent theory seems to have the upper hand.

The famous English explorer James Cook (better known as Captain Cook) was the first European to cross the Antarctic Circle. In 1773, he circumnavigated the Antarctic regions. (It was perhaps this voyage that inspired Samuel Taylor Coleridge's *The Rime of the Ancient Mariner*, published in 1798, which described a voyage from the Atlantic to the Pacific by way of the icy regions of Antarctica.)

In 1819, the British explorer Williams Smith discovered the South Shetland Islands, just 50 miles off the coast of Antarctica; in 1821, a Russian expedition, under Fabian Gottlieb Bellingshausen, sighted a small island (Peter I Island)

Figure 4.7. The major continental glaciers are today largely restricted to Greenland and Antarctica. At the height of the last ice age, the glaciers extended over most of northern and western Europe and south of the Great Lakes on the North American continent.

within the Antarctic Circle; and, in the same year, the Englishman George Powell and the American Nathaniel B. Palmer first laid eyes on a peninsula of the Antarctic continent itself—now called Antarctic Peninsula.

In the following decades, explorers inched toward the South Pole. By 1840, the American naval officer Charles Wilkes announced that the land strikes added up to a continental mass; and, subsequently, he was proved right. The Englishman James Weddell penetrated an ocean inlet east of Palmer Peninsula (now called Weddell Sea) to within 900 miles of the Pole. Another British explorer, James Clark Ross, discovered the other major ocean inlet into Antarctica (now called the Ross Sea) and got within 710 miles of the Pole. Between 1902 and 1904, a third Briton, Robert Falcon Scott, traveled over the Ross ice shelf (a section of ice-covered ocean as large as the state of Texas) to within 500 miles of the Pole. And, in 1909, still another Englishman, Ernest Shackleton, crossed the ice to within about 100 miles of it.

On 16 December 1911, the goal was finally reached by the Norwegian explorer Roald Amundsen. Scott, making a second dash of his own, got to the South Pole just three weeks later, only to find Amundsen's flag already planted there. Scott and his men perished on the ice on their way back.

In the late 1920s, the airplane helped to make good the conquest of Antarctica. The Australian explorer George Hubert Wilkins flew over 1,200 miles of its coastline, and Richard Evelyn Byrd, in 1929, flew over the South Pole. By that time the first base, Little America I, had been established in the Antarctic.

THE INTERNATIONAL GEOPHYSICAL YEAR

The North and South polar regions became focal points of the greatest international project in science of modern times. This had its origin in 1882–83, when a number of nations joined in the International Polar Year of exploration and scientific investigation of phenomena such as the aurorae and the earth's magnetism. The project was so successful that, in 1932–33, it was repeated with a second International Polar Year. In 1950, the United States geophysicist Lloyd Berkner (who had been a member of the first Byrd antarctic expedition) proposed a third such year. The proposal was enthusiastically adopted by the International Council of Scientific Unions. This time scientists were prepared with powerful new research instruments and bristling with new questions—about cosmic rays, the upper atmosphere, the ocean depths, even the possibility of the exploration of space. An ambitious International Geophysical Year (IGY) was arranged, and the time selected was 1 July 1957 to 31 December 1958 (a period of maximum sunspot activity). The enterprise enlisted heart-warming international cooperation; even the cold-war antagonists, the Soviet Union and the United States, managed to bury the hatchet for the sake of science.

Although the most spectacular achievement of the IGY, from the standpoint of public interest, was the successful launching of man-made satellites by the Soviet Union and the United States, science reaped many other fruits that were no less important. Outstanding among these was a vast international exploration of Antarctica. The United States alone set up seven stations, probing the depth of the ice and bringing up from miles down samples of the air trapped in it (which must date back millions of years) and of bacterial remnants. Some bacteria, frozen 100 feet below the ice surface and perhaps a century old, were revived and grew normally. In January 1958, the Soviet group established a base at the Pole of Inaccessibility—the spot in Antarctica farthest inland—and there, 600 miles from the South Pole, recorded new lows in temperature. In August 1960—the Antarctic midwinter—a temperature of −127° F, cold enough to freeze carbon dioxide, was recorded. In the following decade, dozens of year-round stations were operating in Antarctica.

In the most dramatic Antarctic feat, a British exploring team headed by Vivian Ernest Fuchs and Edmund Percival Hillary crossed the continent by land for the first time in history (with, to be sure, special vehicles and all the resources of modern science at their disposal). (Hillary had, in 1953, also been the first, along with the Sherpa mountaineer Tenzing Norgay, to climb Mount Everest, the highest mountain on earth.)

The success of the IGY and the warmth generated by this demonstration of cooperation in the midst of the cold war led to an agreement in 1959 among twelve nations to bar all military activities (including nuclear explosions and the dumping of radioactive wastes) from the Antarctic. Thus, Antarctica will be reserved for scientific activities.

GLACIERS

The earth's load of ice, amounting to nearly 9 million cubic miles, covers about 10 percent of its land area. About 86 percent of the ice is piled up in the Antarctic continental glacier and 10 percent in the Greenland glacier. The remaining 4 percent makes up the small glaciers in Iceland, Alaska, the Himalayas, the Alps, and a few other locations.

The Alpine glaciers have been under study for a long time. In the 1820s, two Swiss geologists, Ignatz Venetz and Johann von Charpentier, noticed that rocks characteristic of the central Alps were scattered over the plains to the north. How had they got there? The geologists speculated that the mountain glaciers had once covered a much larger area and had left boulders and piles of débris behind when they retreated.

A Swiss zoologist, Jean Louis Rodolphe Agassiz, looked into this notion. He drove lines of stakes into the glaciers and waited to see whether they moved. By 1840, he had proved beyond doubt that glaciers flow like very slow rivers at a rate of about 225 feet per year. Meanwhile, he had traveled over Europe and found marks of glaciers in France and England. He found boulders foreign to their surroundings in other areas and scoured marks on rock that could only have been made by the grinding of glaciers, carrying pebbles encrusted along their bottoms.

Agassiz went to the United States in 1846 and became a Harvard professor. He found signs of glaciation in New England and the Midwest. By 1850, it seemed obvious that at some time a large part of the Northern Hemisphere must have been under a large continental glacier. The deposits left by the glacier have been studied in detail since Agassiz's time, and these studies have shown that the glacier advanced and retreated a number of times in the last million years, which make up the *Pleistocene epoch*.

The term *Pleistocene glaciation* is now usually used by geologists for something that is popularly known as the *ice ages*. There were, after all, ice ages before the Pleistocene. There was one about 250 million years ago, and another about 600 million years ago, and still another, perhaps, in between, about 400 million years ago. Little is known of these earlier ice ages, since the great time lapse has wiped out much of the geological evidence. On the whole, then, ice ages are uncommon and take up only a few tenths of 1 percent of Earth's total history.

In regard to the Pleistocene glaciation, it would seem that the Antarctic ice sheet, though now the largest by far, was little involved with the progress of this most recent ice age. The Antarctic ice sheet can expand only into the sea and break up there. The floating ice may become more copious and be more effective in cooling the ocean generally, but the land areas of the Southern Hemisphere are too far from Antarctica to be affected to the point of growing ice sheets of their own (except for some glaciation in the southernmost Andes Mountains).

Quite otherwise is the case in the Northern Hemisphere, where great stretches of land crowd close about the pole. It is there that the expansion of the ice sheets is most dramatic; and the Pleistocene glaciation is discussed almost exclusively in connection with the Northern Hemisphere. In addition to the single Arctic ice sheet (Greenland) that now exists, there were three more ice sheets, with an area of 1 million square miles each: Canada, Scandinavia, and Siberia.

Perhaps because Greenland was the seedland of the northern glaciation, nearby Canada was far more glaciated than more distant Scandinavia or still more distant Siberia. The Canadian ice sheet, growing from the northeast, left much of Alaska and the Pacific slope unglaciated but extended southward until the rim of the ice stretched over much of the northern United States. At its maximum southern extension, the boundary of the ice stretched from Seattle, Washington, to Bismark, North Dakota, then veered southeastward, following very much along the line of the modern Missouri River, past Omaha and St. Louis, then eastward past Cincinnati, Philadelphia, and New York. The southern boundary seems to have been right along the full length of what is now Long Island.

All in all, when the ice sheets were at their farthest extent, they covered over 17 million square miles of land in both polar regions or some 30 percent of Earth's present land surface. This is three times as much land as is covered by ice today.

Careful examination of the layers of sediment in the soil of areas where the ice sheets existed show that they advanced and retreated four times. Each of the four glacial periods endured from 50,000 to 100,000 years. Between them were three *interglacial periods* which were mild, even warm, and were also long.

The fourth, and most recent, glaciation reached its maximum extent about 18,000 years ago, when it stood at what is now the Ohio River. There followed a slow retreat. An idea of the slowness can be obtained when one understands that the retreat progressed at but 250 feet a year over some stretches of time. At others, there was even a partial, and temporary, renewed advance.

About 10,000 years ago, when civilization was already beginning in the Middle East, the glaciers began their final retreat. By 8,000 years ago, the Great Lakes were clear; and by 5,000 years ago (at about which time, writing had been invented in the Middle East), the ice had retreated to about where it is today.

The coming and going of glaciers leaves its mark, not only on the climate of the rest of the earth but on the very shape of the continents. For instance, if the now-shrinking glaciers of Greenland and Antarctica were to melt completely, the ocean level would rise nearly 200 feet. It would drown the coastal areas of all the continents, including many of the world's largest cities, with the water level reaching the twentieth story of Manhattan's skyscrapers. On the other hand, Alaska, Canada, Siberia, Greenland, and even Antarctica would become more habitable.

The reverse situation takes place at the height of an ice age. So much water is tied up in the form of land-based icecaps (up to three or four times the present amount) that the sea-level mark is as much as 440 feet lower than it now is. When this is so, the continental shelves are exposed.

The continental shelves are relatively shallow portions of the ocean adjoining the continents. The sea floor slopes more or less gradually to a depth of about 130 meters. After this, the slope is much steeper, and considerably greater depths are achieved rapidly. The continental shelves are, structurally, part of the continents they adjoin: it is the edge of the shelf that is the true boundary of the continent. At the present moment, there is enough water in the ocean basins to flood the borders of the continent.

Nor is the continental shelf small in area. It is much broader in some places than others; there is considerable shelf area off the east coast of the United States, but little off the west coast (which is at the edge of a crustal plate). On the whole, though, the continental shelf is some 50 miles wide on the average and makes up a total area of 10 million square miles. In other words, a potential continental area rather greater than the Soviet Union in size is drowned under the ocean waters.

It is this area that is exposed during periods of maximum glaciation and was indeed exposed in the last great ice ages. Fossils of land animals (such as the teeth of elephants) have been dredged up from the continental shelves, miles from land and under yards of water. What's more, with the northern continental sections ice-covered, rain was more common than now, farther south, so that the Sahara Desert was then grassland. The drying of the Sahara as the icecaps receded took place not long before the beginning of historic times.

There is thus a pendulum of habitability. As the sea level drops, large continental areas become deserts of ice, but the continental shelves become habitable, as do present-day deserts. As the sea level rises, there is further flooding of the lowlands, but the polar regions become habitable, and again deserts retreat.

You can see, then, that the periods of glaciation were not necessarily times of desolation and catastrophe. All the ice in all the ice sheets at the time of the maximum extent of glaciation makes up only about 0.35 percent of the total water in the ocean. Hence, the ocean is scarcely affected by the oscillations in ice. To be sure, the shallow areas are greatly decreased in area, and those areas are rich in life. On the other hand, the tropic ocean waters are anywhere from 2 to 5 degrees cooler than they are now, which means more oxygen in solution and more life.

Then, too, the advance and retreat of the ice is exceedingly slow, and animal life in general can adapt, migrating slowly north and south. There is even time for evolutionary adaptation to take place, so that during the ice ages, the woolly mammoth flourished.

Finally, the oscillations are not as wild as they might seem, for the ice never entirely melts. The Antarctica icecap has been in existence, relatively unchanged, for some 20 million years and limits the fluctuation in sea level and in temperature.

And yet I do not mean to say that the future gives us no cause for worry. There is no reason to think that a fifth glaciation may not eventually come —with its own problems. In the previous glaciation, the few human beings were hunters who could easily drift southward and northward on the tracks of the game they hunted. In the next glaciation, human beings will undoubtedly be (as they are today) great in numbers and relatively fixed to the ground by virtue of their cities and other structures. Furthermore, it is possible that various facets of human technology may hasten the advance or retreat of the glaciers.

CAUSES OF ICE AGES

The major question regarding the ice ages involves their cause. What makes the ice advance and retreat, and why have the glaciations been relatively brief, the present one having occupied only 1 million of the last 100 million years?

It takes only a small change in temperature to bring on or to terminate an ice age—just enough fall in temperature to accumulate a little more snow in the winter than melts in the summer, or enough rise to melt a little more snow in the summer than falls in the winter. It is estimated that a drop in the earth's average annual temperature of only 3.5° C is sufficient to make glaciers grow, whereas a rise of the same amount would melt Antarctica and Greenland to bare rock in a matter of centuries.

A small drop in temperature sufficient to increase the ice cover slightly over a few years serves to make the process continue. Ice reflects light more efficiently than bare rock or soil does; ice reflects 90 percent of the light that falls on it, while bare soil reflects less than 10 percent. A slight increase in ice cover reflects more sunlight and absorbs less, so that the average temperature of the earth would drop a little farther, and the growth of the ice cover would accelerate.

Similarly, if the earth's temperature went up slightly—just enough to force a small retreat in the ice—less sunlight would be reflected and more absorbed, accelerating the retreat.

What, then, is the process that triggers the action either way?

One possibility is that the earth's orbit is not entirely fixed and does not repeat itself exactly over the years. For instance, the time of perihelion is not fixed. Right now, perihelion, the time when the sun is closest to Earth, comes shortly after the winter solstice. However, the position of the perihelion shifts steadily and makes a complete circuit of the orbit in 21,310 years. Then, too, the direction of the axis changes and marks out a circle in the sky (the precession of the equinoxes) in 25,780 days. Then, too, the actual amount of the tilt changes very slightly, growing a tiny bit more, then a tiny bit less, and in a slow oscillation.

All these changes have a small effect on Earth's average temperature—not great, but enough at certain times to pull the trigger for either the advance of the glaciers or their retreat.

In 1920, a Yugoslavian physicist, Milutin Milankovich, suggested a cycle of this sort that was 40,000 years in length, with a "Great Spring," a "Great Summer," a "Great Fall," and a "Great Winter," each 10,000 years long. The earth would, by this theory, be particularly susceptible to glaciation in the time of the "Great Winter" and would actually undergo it when other factors were favorable as well. Once glaciated, the earth would undergo deglaciation most likely in the "Great Summer" if other factors were favorable.

Milankovich's suggestion did not meet with much favor when it was advanced; but in 1976, the problem was tackled by J. D. Hays and John Imbrie of the United States and by N. J. Shackleton of Great Britain. They worked on long cores of sediment dredged up from two different places in the Indian Ocean—relatively shallow places far from land, so that no contaminating material would be brought down from nearby coastal areas or shallower sea bottom.

These cores were made up of material laid down steadily over a period of 450,000 years. The farther down the core one observed, the farther back the year. It was possible to study the skeletons of tiny one-celled animals, which come in different species that flourish at different temperatures. From the nature of the skeleton, the temperature could be determined.

Then, too, oxygen atoms chiefly come in two different varieties, and the ratio of these varieties vary with the temperature. By measuring the ratio at different places in the core, one could determine the ocean temperature at different times.

Both methods of measuring temperature agreed, and both seemed to indicate something much like the Milankovich cycle. It may be, then, that the earth has a glaciated Great Winter at long intervals, just as it has a snow-covered winter every year.

But then why should the Milankovich cycle have worked during the course of the Pleistocene but not for a couple of hundred million years before that when there was no glaciation at all?

In 1953, Maurice Ewing and William L. Donn suggested the reason might lie in the peculiar geography of the Northern Hemisphere. The Arctic region is almost entirely oceanic, but it is a landlocked ocean with large continental masses hemming it in on all sides.

Imagine the Arctic Ocean a trifle warmer than it is today, with little or no sea ice upon it and offering an unbroken stretch of liquid surface. The Arctic Ocean would then serve as a source of water vapor, which, cooling in the upper atmosphere, would fall as snow. The snow that fell back into the ocean would melt, but the snow that fell on the surrounding continental masses would accumulate, and trigger the glaciation: the temperature would drop, and the Arctic Ocean would freeze over.

Ice does not liberate as much water vapor as does liquid water at the same temperature. Once the Arctic Ocean freezes over, then, there is less water vapor in the air and less snowfall. The glaciers start retreating, and if they then trigger deglaciation, the retreat is accelerated.

It may be, then, that the Milankovich cycle sets off periods of glaciation only when there is a landlocked ocean at one or both poles. There may be some hundreds of millions of years when no such landlocked ocean exists and there is no glaciation; then the shifting of the tectonic plates creates such a situation, and there begins a million years or more during which the glaciers advance and retreat regularly. This interesting suggestion is not as yet totally accepted.

There are, to be sure, less regular changes in Earth's temperature and more erratic producers of cooling and warming trends. The American chemist Jacob Bigeleisen, working with H. C. Urey, measured the ratio of the two varieties of oxygen atom in the ancient fossils of sea animals in order to measure the temperature of the water in which the animals lived. By 1950, Urey and his group had developed the technique to so fine a point that, by analyzing the shell layers of a millions-of-years-old fossil (an extinct form of squid), they could determine that the creature had been born during a summer, lived four years, and died in the spring.

This "thermometer" has established that 100 million years ago the average world-wide ocean temperature was about 70° F. It cooled slowly to 61° F 10 million years later and then rose to 70° F again after another 10 million years. Since then, the ocean temperature has declined steadily. Whatever triggered this decline may also be a factor in the extinction of the dinosaurs (which were probably adapted to mild and equable climates) and put a premium on the warm-blooded birds and mammals, which can maintain a constant internal temperature.

Figure 4.8. The record of the ocean temperatures during the last 100 million years.

Cesare Emiliani, using the Urey technique, studied the shells of foraminifera brought up in cores from the ocean floor. He found that the overall ocean temperature was about 50° F 30 million years ago and 43° F 20 million years ago and is now 35° F (figure 4.8).

What caused these long-term changes in temperature? One possible explanation is the so-called *greenhouse effect* of carbon dioxide. Carbon dioxide absorbs infrared radiation rather strongly. Thus, when there are appreciable amounts of it in the atmosphere, it tends to block the escape of heat at night from the sun-warmed earth. The result is that heat accumulates. On the other hand, when the carbon dioxide content of the atmosphere falls, the earth steadily cools.

If the current concentration of carbon dioxide in the air should double (from 0.03 percent of the air to 0.06 percent), that small change would suffice to raise the earth's overall temperature by 3 degrees and would bring about the complete and quick melting of the continental glaciers. If the carbon dioxide dropped to half the present amount, the temperature would drop sufficiently to bring the glaciers down to the area of New York City again.

Volcanoes discharge large amounts of carbon dioxide into the air; the weathering of rocks absorbs carbon dioxide (thus forming limestone). Here, then, is a possible pair of mechanisms for long-term climatic changes. A period of greater-than-normal volcanic action might release a large amount of carbon dioxide into the air and initiate a warming of the earth. Contrariwise, an era of mountain building, exposing large areas of new and unweathered rock to the air, could lower the carbon-dioxide concentration in the atmosphere. The latter process may have happened at the close of the *Mesozoic* (the age of reptiles) some 80 million years ago, when the long decline in the earth's temperature began.

Whatever the cause of the ice ages may have been, it seems now that human beings themselves may be changing our future climate. The American physicist Gilbert N. Plass has suggested that we may be seeing the last of the ice ages, because the furnaces of civilization are loading the atmosphere with carbon dioxide. A hundred million chimneys are ceaselessly pouring carbon dioxide into the air; the total amount is about 6 billions tons a year—200 times the quantity coming from volcanoes. Plass pointed out that, since 1900, the carbon-dioxide content of our atmosphere has increased about 10 percent and may increase as much again by the year 2000. This addition to the earth's "greenhouse" shield against the escape of heat, he calculated, should raise the average temperature by about $1.1°$ C per century. During the first half of the twentieth century, the average temperature has indeed risen at this rate, according to the available records (mostly in North America and Europe). If the warming continues at the same rate, the continental glaciers may disappear in a century or two.

Investigations during the IGY seemed to show that the glaciers are indeed receding almost everywhere. One of the large glaciers in the Himalayas was reported in 1959 to have receded 700 feet since 1935. Others had retreated 1,000 or even 2,000 feet. Fish adapted to frigid waters are migrating northward, and warm-climate trees are advancing in the same direction. The sea level is rising slightly each year, as would be expected if the glaciers are melting. The sea level is already so high that, at times of violent storms at high tide, the ocean is not far from threatening to flood the New York subway system.

And yet there seems to be a slight downturn in temperature since the early 1940s, so that half the temperature increase between 1880 and 1940 has been wiped out. This change may be due to increasing dust and smog in the air since 1940: particles that cut off sunlight and, in a sense, shade the earth. It would seem that two different types of human atmospheric pollution are currently canceling each other's effect, at least in this respect and at least temporarily.

Chapter 5

The Atmosphere

The Shells of Air

Aristotle supposed the world to be made up of four shells, constituting the four elements of matter: earth (the solid ball), water (the ocean), air (the atmosphere), and fire (an invisible outer shell that occasionally became visible in the flashes of lightning). The universe beyond these shells, he said, was composed of an unearthly, perfect fifth element that he called *ether* (from a Latin derivative, the name became *quintessence,* which means "fifth element").

There was no room in this scheme for emptiness: where earth ended, water began; where both ended, air began; where air ended, fire began; and where fire ended, ether began and continued to the end of the universe. "Nature," said the ancients, "abhors a vacuum" (Latin for "emptiness").

MEASURING AIR

The suction pump, an early invention to lift water out of wells, seemed admirably to illustrate this abhorrence of a vacuum (figure 5.1). A piston is fitted tightly within a cylinder. When the pump handle is pushed down, the piston is pulled upward, leaving a vacuum in the lower part of the cylinder. But since nature abhors a vacuum, the surrounding water opens a one-way valve at the bottom of the cylinder and rushes into the vacuum. Repeated pumping lifts the water higher and higher in the cylinder, until it pours out of the pump spout.

Figure 5.1. Principle of the water pump. When the handle raises the piston, a partial vacuum is created in the cylinder, and water rises into it through a one-way valve. After repeated pumping, the water level is high enough for the water to flow out of the spout.

According to Aristotelian theory, it should have been possible in this way to raise water to any height. But miners who had to pump water out of the bottoms of mines found that, no matter how hard and long they pumped, they could never lift the water higher than 33 feet above its natural level.

Galileo grew interested in this puzzle toward the end of his long and inquisitive life. He could come to no conclusion except that apparently nature abhorred a vacuum only up to certain limits. He wondered whether the limit would be lower if he used a liquid denser than water, but he died before he could try this experiment.

Galileo's students Evangelista Torricelli and Vincenzo Viviani did perform it in 1644. Selecting mercury (which is 13½ times as dense as water), they filled a yard-long glass tube with mercury, stoppered the open end, upended the tube in a dish of mercury, and removed the stopper. The mercury began to run out of the tube into the dish; but, when its level had dropped to 30 inches above the level in the dish, it stopped pouring out of the tube and held at that level.

Thus was constructed the first *barometer*. Modern mercury barometers are not essentially different. It did not take long to discover that the height of the mercury column was not always the same. The English scientist Robert Hooke pointed out, in the 1660s, that the height of the mercury column decreased before a storm, thus pointing the way to the beginning of scientific weather forecasting or *meteorology*.

What was holding the mercury up? Viviani suggested that it was the weight of the atmosphere, pressing down on the liquid in the dish. This was a revolutionary thought, for the Aristotelian notion had been that air had no weight, being drawn only to its proper sphere above the earth. Now it became plain that a 33-foot column of water, or a 30-inch column of mercury, measured the weight of the atmosphere—that is, the weight of a column of air of the same cross section from sea level up to as far as the air went.

The experiment also showed that nature does not necessarily abhor a vacuum under all circumstances. The space left in the closed end of the tube after the mercury fell was a vacuum, containing nothing but a very small quantity of mercury vapor. This *Torricellian vacuum* was the first artificially produced vacuum.

The vacuum was pressed into the service of science almost at once. In 1650, the German scholar Athanasius Kircher demonstrated that sound could not be transmitted through a vacuum, thus upholding an Aristotelian theory (for once). In the next decade, Robert Boyle showed that very light objects will fall as rapidly as heavy ones in a vacuum, thus upholding Galileo's theories of motion against the views of Aristotle.

If air has a finite weight, it must have some finite height. The weight of the atmosphere turned out to be 14.7 pounds per square inch; on this basis, the atmosphere was just about 5 miles high—if it was evenly dense all the way up. But, in 1662, Boyle showed that it could not be, because pressure increased air's density. He stood up a tube shaped like the letter J and poured some mercury into the mouth of the tube, on the tall side of the J. The mercury trapped a little air in the closed end on the short side. As he poured in more mercury, the air pocket shrank. At the same time, its pressure increased, Boyle discovered, for it shrank less as the mercury grew weightier. By actual measurement, Boyle showed that reducing the volume of gas to one-half doubled its pressure; in other words, the volume varied in inverse ratio to the pressure (figure 5.2). This historic discovery, known as *Boyle's law,* was the first step

---

Figure 5.2. Diagram of Boyle's experiment. When the left arm of the tube is stoppered and more mercury is poured into the right arm, the trapped air is compressed. Boyle showed that the volume of the trapped air varies inversely with the pressure, thus demonstrating Boyle's law.

---

in the long series of discoveries about matter that eventually led to the atomic theory.

Since air contracts under pressure, it must be densest at sea level and steadily become thinner as the weight of the overlying air declines toward the top of the atmosphere. This notion was first demonstrated in 1648 by the French mathematician Blaise Pascal, who sent his brother-in-law Florin Perier nearly a mile up a mountainside and had him carry a barometer to note how the mercury level dropped as altitude increased.

Theoretical calculations showed that, if the temperature were the same all the way up, the air pressure would decrease tenfold with every 12 miles of rise in altitude. In other words, at 12 miles the column of mercury the air could support would have dropped from 30 inches to 3 inches; at 24 miles it would be .3 of an inch; at 36 miles, .03 of an inch; and so on. At 108 miles, the air pressure would amount to only 0.000000003 of an inch of mercury. This may not sound like much, but over the whole earth, the weight of the air above 108 miles would still total 6 million tons.

Actually all these figures are only approximations, because the air temperature changes with height. Nevertheless, they do clarify the picture, and we can see that the atmosphere has no definite boundary; it simply fades off gradually into the near emptiness of space. Meteor trails have been detected as high as 100 miles where the air pressure is only 1 millionth what it is on the earth's surface, and the air density only 1 billionth. Yet that is enough to heat these tiny bits of matter to incandescence through air resistance. And the aurora borealis (northern lights), formed of glowing wisps of gas bombarded by particles from outer space, has been located as high as 500 to 600 miles above sea level.

AIR TRAVEL

From earliest times, there seems to have been a haunting desire on the part of human beings to travel through the air. The wind can, and does, carry light objects—leaves, feathers, seeds—through the air. More impressive are the gliding animals, such as flying squirrels, flying phalangers, even flying fish, and —to a far greater extent—the true fliers, such as insects, bats, and birds.

The yearning of human beings to follow suit leaves its mark in myth and legend. Gods and demons can routinely travel through air (angels and fairies are always pictured with wings); and there is Icarus, after whom an asteroid was named (see chapter 3); and the flying horse, Pegasus; and even flying carpets in Oriental legend.

The first artificial device that could at least glide at considerable heights for a considerable time was the *kite*, in which paper, or some similar material is stretched over a flimsy wooden framework, equipped with a tail for stability and a long cord by which it can be held. A kite is supposed to have been invented by the Greek philosopher, Archytas in the fourth century B.C.

Kites were used for thousands of years, chiefly for amusement, though practical uses were also possible. A kite can hold a lantern aloft as a signal over a wide area. It can carry a light cord across a river or a ravine; then the cord can be used to pull heavier cords across until a bridge is built.

The first attempt to use kites for scientific purposes came in 1749, when a Scottish astronomer, Alexander Wilson, attached thermometers to kites, hoping to measure temperatures at a height. Much more significant was the kite flying of Benjamin Franklin in 1752, to which I shall return in chapter 9.

Kites (or kindred gliding artifacts) did not become large enough and strong enough to carry human beings for another century and a half, but the problem was solved in another fashion in Franklin's lifetime.

In 1782, two French brothers, Joseph Michel and Jacques Étienne Montgolfier, lit a fire under a large bag with an opening underneath and thus filled the bag with hot air. The bag rose slowly; the Montgolfiers had successfully launched the first balloon. Within a few months, balloons were being made with hydrogen, a gas only 1/14 as dense as air, so that each pound of hydrogen could carry aloft a payload of 13 pounds. Now gondolas went up carrying animals and, soon, men.

Within a year of the launching of the first balloon, an American named John Jeffries made a balloon flight over London with a barometer and other instruments, plus an arrangement to collect air at various heights. By 1804, the French scientist Joseph Louis Gay-Lussac had ascended nearly 4½ miles and brought down samples of the rarefied air. Such adventures were made a little safer by the French balloonist Jean Pierre Blanchard, who, in 1785, at the very onset of the *balloon age,* invented the parachute.

This was nearly the limit for humans in an open gondola; three men rose to 6 miles in 1875, but only one, Gaston Tissandier, survived the lack of oxygen. He was able to describe the symptoms of air deficiency, and that was the birth of *aviation medicine.* Unmanned balloons carrying instruments were designed and put into action in 1892, and these could be sent higher and bring back information on temperature and pressure from hitherto unexplored regions.

In the first few miles of altitude rise, the temperature dropped, as was expected. At 7 miles or so, it was −55° C. But then came a surprise. Above this level, the temperature did not decrease; in fact, it even rose slightly.

The French meteorologist Leon Phillippe Teisserenc de Bort suggested, in 1902, that the atmosphere might have two layers: a turbulent lower layer containing clouds, winds, storms, and all the familiar weather changes (in 1908, he called this layer the *troposphere,* from the Greek for "sphere of change"); and a quiet upper layer containing sublayers of lighter gases, helium, and hydrogen (he named this the *stratosphere,* meaning "sphere of layers"). Teisserenc de Bort called the level at which the temperature ceased to decline the *tropopause* ("end of change"), or the boundary between the troposphere and the stratosphere. The tropopause has since been found to vary from an altitude of about 10 miles above sea level at the Equator to only 5 miles above at the poles.

During the Second World War, high-flying United States bombers discovered a dramatic phenomenon just below the tropopause—the *jet stream*, consisting of very strong, steady, west-to-east winds blowing at speeds up to 500 miles per hour. Actually there are two jet streams; one in the Northern Hemisphere at the general latitude of the United States, the Mediterranean, and north China; and one in the Southern at the latitude of New Zealand and Argentina. The streams meander, often debouching into eddies far north or south of their usual course. Airplanes now take advantage of the opportunity to ride on these swift winds. But far more important is the discovery that the jet streams have a powerful influence on the movement of air masses at lower levels. This knowledge at once helped to advance the art of weather forecasting.

As human beings cannot survive in the thin, cold atmosphere of great heights, it was necessary to develop a sealed cabin, within which the pressures and temperatures of earth's surface air can be maintained. Thus, in 1931, the Piccard brothers (Auguste and Jean Felix), the first of whom later invented the bathyscaphe, rose to 11 miles in a balloon carrying a sealed gondola. Then new balloons of plastic material, lighter and less porous than silk, made it possible to go higher and remain up longer. In 1938, a balloon named *Explorer II* went to 13 miles; and by the 1980s, manned balloons have gone as high as 23½ miles and unmanned balloons to more than 32 miles.

These higher flights showed that the zone of nearly constant temperature does not extend indefinitely upward. The stratosphere comes to an end at a height of about 20 miles, and above it the temperature starts to rise!

This *upper atmosphere*, above the stratosphere, containing only 2 percent of the earth's total air mass, was penetrated in the 1940s, for further progress, by a new type of vehicle altogether—the rocket (see chapter 3).

The most direct way to read instruments that have recorded conditions high in the air is to bring them down and look at them. Instruments carried aloft by kites can easily be brought down, but balloons are less easily managed in this respect, and rockets may not come down at all. Of course, an instrument packet can be ejected from a rocket and may come down independently, but even it is hard to rely on. In fact, rockets alone would have accomplished little in the exploration of the atmosphere had it not been for a companion invention—*telemetering*. Telemetering was first applied to atmospheric research, in a balloon, in 1925 by a Russian scientist named Pyotr A. Molchanoff.

Essentially, this technique of "measuring at a distance" entails translating the conditions to be measured (for example, temperature) into electrical impulses that are transmitted back to earth by radio. The observations take the form of changes in intensity or spacing of the pulses. For instance, a temperature change affects the electrical resistance of a wire and so changes the nature of the pulse; a change in air pressure similarly is translated into a certain kind of pulse by the fact that air cools the wire, the extent of the cooling depending on the pressure; radiation sets off pulses in a detector; and so on. Nowadays, telemetering has become so elaborate that the rockets seem to do everything but talk, and their intricate messages have to be interpreted by rapid computers.

Rockets and telemetering, then, showed that above the stratosphere, the temperature rises to a maximum of some −10° C at a height of 30 miles and then drops again to a low of −90° C at a height of 50 miles. This region of rise and fall in temperature is called the *mesosphere*, a word coined in 1950 by the British geophysicist Sydney Chapman.

Beyond the mesosphere, what is left of the thin air amounts to only a few thousandths of 1 percent of the total mass of the atmosphere. But this scattering of air atoms steadily increases in temperature to an estimated 1,000° C at 300 miles and probably to still higher levels above that height. It is therefore called the *thermosphere* ("sphere of heat")—an odd echo of Aristotle's original sphere of fire. Of course, temperature here does not signify heat in the usual sense: it is merely a measure of the speed of the particles.

Above 300 miles we come to the *exosphere* (a term first used by Lyman Spitzer in 1949), which may extend as high as 1,000 miles and gradually merges into interplanetary space.

Increasing knowledge of the atmosphere may enable us to do something about the weather some day and not merely talk about it. Already, a small start has been made. In the early 1940s, the American chemists Vincent Joseph Schaefer and Irving Langmuir noted that very low temperatures could produce nuclei about which raindrops would form. In 1946, an airplane dropped powdered carbon dioxide into a cloud bank in order to form first nuclei and then raindrops (*cloud seeding*). Half an hour later, it was raining. Bernard Vonnegut later improved the technique when he discovered that powdered silver iodide generated on the ground and directed upward worked even better. Rainmakers, of a new scientific variety, are now used to end droughts—or to attempt to end them, for clouds must first be present before they can be seeded. In 1961, Soviet astronomers were partially successful in using cloud seeding to clear a patch of sky through which an eclipse might be glimpsed.

Other attempts at *weather modification* have included the seeding of hurricanes in an attempt to abort them or at least to moderate their force (seeding clouds in order to abort crop-damaging hailstorms; dissipating fogs, and so on). Results in all cases have been hopeful at best, but never a clear-cut success. Furthermore, any attempt at deliberate modification of weather is bound to help some but hurt others (a farmer might want rain, while an amusement-park owner does not), and lawsuits are an obvious side effect of weather-modification programs. What the future holds in this direction is, therefore, uncertain.

Nor are rockets for exploration only (although those are the only uses mentioned in chapter 3). They can, and already have, been turned to the everyday service of humanity. In fact, even some forms of exploration can be of immediate practical use. If a satellite is rocketed into orbit, it need not look only away from our planet; it can turn its instruments upon Earth itself. In this way, satellites have made it possible, for the first time, to see our planet (or at least a good part of it at any one time) as a unit and to study the air circulation as a whole.

On 1 April 1960, the United States launched the first *weather-eye* satellite, *Tiros I* (*Tiros* standing for "*T*elevision *I*nfrared *O*bservation *S*atellite"). Then *Tiros II* was launched in November and, for ten weeks, sent down over 20,000 pictures of vast stretches of the earth's surface and its cloud cover, including pictures of a cyclone in New Zealand and a patch of clouds in Oklahoma that was apparently spawning tornadoes. *Tiros III,* launched in July 1961, photographed eighteen tropical storms, and, in September, showed hurricane Esther developing in the Caribbean two days before it was located by more orthodox methods. The more sensitive *Nimbus I* satellite, launched on 28 August 1964, could send back cloud photographs taken at night. Eventually hundreds of automatic picture transmission stations were in operation in scores of nations, so that weather forecasting without satellite data has now become unthinkable. Every newspaper can run a cloud-pattern photograph of the United States daily, and weather forecasting, while still not mathematically certain, is not the crude guessing game it was only a quarter-century ago.

Most fascinating and useful is the manner in which meteorologists can now locate and track hurricanes. These severe storms have become far more damaging than in the past, since beachfronts have become much more built up and populous since the Second World War, and were there not a clear knowledge of the position and movements of these storms, there is no question but that loss of life and property would be many times what it is now. (In respect to the usefulness and value of the space program, satellite-tracking of hurricanes alone pays back far more than the program costs.)

Other earthbound uses of satellites have been developed. As early as 1945, the British science-fiction writer Arthur C. Clarke had pointed out that satellites could be used as relays by which radio messages could span continents and oceans, and that as few as three strategically placed satellites could afford world coverage. What then seemed a wild dream began to come true fifteen years later. On 12 August 1960, the United States launched *Echo I,* a thin polyester balloon coated with aluminum, which was inflated in space to a diameter of 100 feet in order to serve as a passive reflector of radio waves. A leader in this successful project was John Robinson Pierce of Bell Telephone Laboratories, who had himself written science-fiction stories under a pseudonym.

On 10 July 1962, *Telstar I* was launched by the United States. It did more than reflect; it received the waves, amplified them, and sent them onward. By use of Telstar, television programs spanned the oceans for the first time (though that did not in itself improve their quality, of course). On 26 July 1963, *Syncom II,* a satellite that orbited at a distance of 22,300 miles above the earth's surface, was put in orbit. Its orbital period was just 24 hours, so that it hovered indefinitely over the Atlantic Ocean, turning in synchronization with the earth. *Syncom III,* placed over the Indian Ocean in similar synchronous fashion, relayed the Olympic Games from Japan to the United States in October 1964.

A still more sophisticated communications satellite, *Early Bird*, was launched 6 April 1965; it made available 240 voice circuits and one television channel. (In that year, the Soviet Union began to send up communications satellites as well.) By the 1970s, television, radio, and radiotelephony had become essentially global, thanks to satellite relays. Technologically, Earth has become "one world," and those political forces that work against that inescapable fact are increasingly archaic, anachronistic, and deadly dangerous.

The fact that satellites can be used to map Earth's surface and study its clouds is obvious. Not quite so obvious but just as true is the fact that satellites can study snow cover, glacier movements, and geological details on a large scale. From geological details, likely regions where oil may exist can be marked off. Crops on the large scale can be studied, as forests can; and regions of abnormality and disease can be pinpointed. Forest fires can be spotted, and irrigation needs located. The ocean can be studied, as can water currents and fish movements. Such *earth resources satellites* are the immediate answer to those critics who question the money spent on space in the face of great problems "right here at home." It is often from space that such problems can best be studied and methods of solution demonstrated.

Finally, there are in orbit numerous *spy satellites* designed to be able to detect military movements, military concentrations and stores, and so on. There are not lacking people who plan to make space another arena for war or to develop *killer satellites* designed to strike down enemy satellites, or to place advanced weapons in space which can strike more quickly than Earth-based weapons. This is the demonic side of space exploration, even though it only marginally increases the speed with which a full-scale thermonuclear war can destroy civilization.

The stated purpose of "keeping the peace" by discouraging the other side from making war is proclaimed by both superpowers, the United States and the Soviet Union. The acronym for this theory of peace by "mutual assured destruction," with each side knowing that starting a war will bring about its own destruction as well as that of the other side, is MAD—and mad it is, for increasing the quantity and the deadliness of armaments has never hitherto prevented war.

The Gases in Air

THE LOWER ATMOSPHERE

Up to modern times, air was considered a simple, homogeneous substance. In the early seventeenth century, the Flemish chemist Jan Baptista van Helmont began to suspect that there were chemically different gases. He studied the vapor given off by fermenting fruit juice (*carbon dioxide*) and recognized it as a new substance. Van Helmont was, in fact, the first to use the term *gas*—a word he is supposed to have coined, about 1620, from *chaos,* the Greek word for the original substance out of which the universe was made. In 1756, the Scottish chemist Joseph Black studied carbon dioxide thoroughly and definitely established it as a gas other than air. He even showed that small quantities of it existed in the air. Ten years later, Henry Cavendish studied a flammable gas not found in the atmosphere. It was eventually named *hydrogen.* The multiplicity of gases was thus clearly demonstrated.

The first to realize that air was a mixture of gases was the French chemist Antoine-Laurent Lavoisier. In experiments conducted in the 1770s, he heated mercury in a closed vessel and found that the mercury combined with part of the air, forming a red powder (*mercuric oxide*), but four-fifths of the air remained a gas. No amount of heating would consume any of this remaining gas. A candle would not burn in it, nor could mice live in it.

Lavoisier decided that air was made up of two gases. The one-fifth that combined with mercury in his experiment was the portion of the air that supports life and combustion: this he called *oxygen.* The remainder he called *azote,* from Greek words meaning "no life." Later it became known as *nitrogen,* because the substance was present in sodium nitrate, commonly called *niter.* Both gases had been discovered in the previous decade. Nitrogen had been discovered in 1772 by the Scottish physician Daniel Rutherford; and oxygen, in 1774 by the English Unitarian minister Joseph Priestley.

This alone is sufficient to demonstrate that Earth's atmosphere is unique in the solar system. Aside from Earth, seven worlds in the solar system are known to have an appreciable atmosphere. Jupiter, Saturn, Uranus, and Neptune (the first two, certainly; the latter two, probably) have hydrogen atmospheres, with helium as a minor constituent. Mars and Venus have carbon dioxide atmospheres, with nitrogen as a minor constituent. Titan has a nitrogen atmosphere with methane as a minor constituent. Earth alone has an atmosphere nearly evenly split between two gases, and Earth alone has oxygen as a major constituent. Oxygen is an active gas and, from ordinary chemical considerations, it would be expected that it would combine with other elements and would disappear from the atmosphere in its free form. This is something we will return to later in the chapter; but for now, let us continue dealing with the further details of the chemical composition of air.

By the mid-nineteenth century, the French chemist Henri Victor Regnault had analyzed air samples from all over the world and discovered the composition of the air to be the same everywhere. The oxygen content was 20.9 percent, and it was assumed that all the rest (except for a trace of carbon dioxide) was nitrogen.

Nitrogen is a comparatively inert gas; that is, it does not readily combine with other substances. It can, however, be forced into combination—for instance, heating it with magnesium metal forms the solid *magnesium nitride.* Some years after Lavoisier's discovery, Henry Cavendish tried to exhaust the nitrogen by combining it with oxygen under the influence of an electric spark. He failed. No matter what he did, he could not get rid of a small bubble of remaining gas, amounting to less than 1 percent of the original quantity. Cavendish thought this might be an unknown gas, even more inert than nitrogen. But not all chemists are Cavendishes, and the puzzle was not followed up, so the nature of this residue of air was not discovered for another century.

In 1882, the British physicist Robert John Strutt, Lord Rayleigh, compared the density of nitrogen obtained from air with the density of nitrogen obtained from certain chemicals and found, to his surprise, that the air nitrogen was definitely denser. Could it be that nitrogen obtained from air was not pure but contained small quantities of another, heavier gas? A Scottish chemist, Sir William Ramsay, helped Lord Rayleigh look further into the matter. By this time, they had the aid of spectroscopy. When they heated the small residue of gas left after exhaustion of nitrogen from air and examined its spectrum, they found a new set of bright lines—lines that belonged to no known element. To their newly discovered, very inert element they gave the name *argon* (from a Greek word meaning "inert").

Argon accounted for nearly all of the approximately 1 percent of unknown gas in air—but there were still several *trace constituents* in the atmosphere, each constituting only a few parts per million. During the 1890s Ramsay went on to discover four more inert gases: *neon* ("new"), *krypton* ("hidden"), *xenon* ("stranger"), and *helium,* which had been discovered more than thirty years before in the sun. In recent decades, the infrared spectroscope has turned up three others: *nitrous oxide* ("laughing gas"), whose origin is unknown; *methane,* a product of the decay of organic matter; and *carbon monoxide.* Methane is released by bogs, and some 45 million tons of the same gas, it has been calculated, are added to the atmosphere each year by the venting of intestinal gases by cattle and other large animals. The carbon monoxide is probably man-made, resulting from the incomplete combustion of wood, coal, gasoline, and so on.

THE STRATOSPHERE

I have so far been discussing the composition of the lowest reaches of the atmosphere. What about the stratosphere? Teisserenc de Bort believed that helium and hydrogen might exist in some quantity up there, floating on the heavier gases underneath. He was mistaken. In the middle 1930s, Russian balloonists brought down samples of air from the upper stratosphere, and it proved to be made up of oxygen and nitrogen in the same 1-to-4 mixture as the air of the troposphere.

But there were reasons to believe some unusual gases existed still higher in the upper atmosphere, and one of the reasons was the phenomenon called the *airglow.* This is the very feeble general illumination of all parts of the night sky, even in the absence of the moon. The total light of the airglow is considerably greater than that of the stars, but is so diffuse that it is not noticeable except to the delicate light-gathering instruments of the astronomer.

The source of the light had been a mystery for many years. In 1928, the astronomer V. M. Slipher succeeded in detecting in the airglow some mysterious spectral lines that had been found in nebulae in 1864 by William Huggins and were thought to represent an unfamiliar element, named *nebulium*. In 1927, through experiments in the laboratory, the American astronomer Ira Sprague Bowen showed that the lines came from *atomic oxygen:* that is, oxygen existing as single atoms and not combined in the normal form of the two-atom molecule. Similarly, other strange spectral lines from the aurora turned out to represent atomic nitrogen. Both atomic oxygen and atomic nitrogen in the upper atmosphere are produced by energetic radiation from the sun, which breaks down the molecules into single atoms—a possibility first suggested in 1931 by Sydney Chapman. Fortunately the high-energy radiation is, in this way, absorbed or weakened before it reaches the lower atmosphere.

The airglow, Chapman maintained, comes from the recombination at night of the atoms that are split apart by solar energy during the day. In recombining, the atoms give up some of the energy they absorbed in splitting, so that the airglow is a kind of delayed and very feeble return of sunlight in a new and specialized form. Experiments in 1956—both in the laboratory and, through rockets, in the upper atmosphere, under the direction of Murray Zelikoff—supplied direct evidence of this theory. Spectroscopes carried by the rockets recorded the green lines of atomic oxygen most strongly at a height of 60 miles. A smaller proportion of the nitrogen was in the atomic form, because nitrogen molecules hold together more strongly than do oxygen molecules; nevertheless, the red light of atomic nitrogen was strong at a height of 95 miles.

Slipher had also found lines in the airglow that were suspiciously like well-known lines emitted by sodium. The presence of sodium seemed so unlikely that the matter was dropped in embarrassment. What would sodium, of all things, be doing in the upper atmosphere? It is not a gas, after all, but a very reactive metal that does not occur alone anywhere on the earth. It is always combined with other elements, most commonly in *sodium chloride* (table salt). But, in 1938, French scientists established that the lines were indeed identical with the sodium lines. Unlikely or not, sodium had to be in the upper atmosphere. Again, rocket experiments clinched the matter: their spectroscopes recorded the yellow light of sodium unmistakably, and most strongly at a height of 55 miles. Where the sodium comes from is still a mystery—perhaps from ocean salt spray or from vaporized meteors. Still more puzzling is the fact that *lithium*—a rare relative of sodium—was also found, in 1958, to be contributing to the airglow.

In the course of their experiments, Zelikoff's team produced an artificial airglow. They fired a rocket that at miles released a cloud of nitric oxide gas. This accelerated the recombination of oxygen atoms in the upper atmosphere.

Observers on the ground easily sighted the bright glow that resulted. A similar experiment with sodium vapor also was successful: it created a clearly visible, yellow glow. When Soviet scientists sent Lunik III in the direction of the moon in October 1959, they arranged for it to expel a cloud of sodium vapor as a visible signal that it had gone into orbit.

At lower levels in the atmosphere, atomic oxygen disappears, but the solar radiation is still energetic enough to bring about the formation of the three-atom variety of oxygen called *ozone*. The ozone concentration is greatest at a height of 15 miles. Even there, in what is called the *ozonosphere* (first discovered in 1913 by the French physicist Charles Fabry), it makes up only 1 part in 4 million of the air, but that is enough to absorb ultraviolet light sufficiently to protect life on the earth.

Ozone is formed by the combination of atomic oxygen (a single atom) with ordinary oxygen molecules (two atoms). Ozone does not accumulate to large amounts, for it is unstable. The three-atom molecule easily breaks down to the much more stable two-atom form by the action of sunlight, by the nitrous oxide that occurs naturally in tiny amounts in the atmosphere, and by other chemicals. The balance between formation and breakdown leaves, in the ozonosphere at all times, the small concentration referred to; and its shield against the sun's ultraviolet (which would break down many of the delicate molecules essential to living tissue) has protected life since oxygen first entered Earth's atmosphere in quantity.

The ozonosphere is not far above the tropopause and varies in height in the same way, being lowest at the poles and highest at the Equator. The ozonosphere is richest in ozone at the poles and poorest at the Equator where the breakdown effect of sunlight is highest.

It would be dangerous if human technology were to produce anything that would accelerate ozone breakdown in the upper atmosphere and weaken the ozonosphere shield. The weakening of the shield would increase the ultraviolet incidence at Earth's surface, which would, in turn, increase the incidence of skin cancer—especially among fair-skinned people. Some have estimated that a 5-percent reduction in the ozone shield could result in 500,000 additional cases of skin cancer each year over the world in general. Ultraviolet light, if increased in concentration, might also affect the microscopic life (*plankton*) in the sea surface with possible fearful consequences, since plankton forms the base of the food chain in the sea and, to a certain extent, on land as well.

There is indeed some danger that human technology will affect the ozonosphere. Increasingly, jet planes are flying through the stratosphere, and rockets are making their way through the entire atmosphere and into space. The chemicals poured into the upper atmosphere by the exhausts of these vehicles might conceivably accelerate ozone breakdown. The possibility was used as an argument against the development of supersonic planes in the early 1970s.

In 1974, spray cans were unexpectedly found to be a possible danger. These cans use imprisoned Freon (a gas that will be mentioned again, in chapter 11) as a source of pressure that serves to drive out the contents of the can (hairspray, deodorants, air-fresheners, or whatever) in a fine spray. Freon itself is, chemically, as harmless as one can imagine a gas to be—colorless, odorless, inert, and unreactive, without any effect on human beings. About 1,700,-000,000 pounds of it were being released into the atmosphere from spray cans and other devices each year at the time its possible danger was pointed out.

The gas, reacting with nothing, spreads slowly through the atmosphere and finally reaches the ozonosphere where it might serve to accelerate the break-down of ozone. This possibility was raised on the basis of laboratory tests. Whether it would actually do this under the conditions of the upper atmosphere is somewhat uncertain, but the possibility represents too great a danger to dismiss in cavalier fashion. The use of spray cans with Freon has vastly decreased since the controversy began.

However, Freon is used to a much greater extent in air-conditioning and in refrigeration, where it has not been easily given up or even replaced. Thus, the ozonosphere remains at hazard, for, once formed, Freon is bound sooner or later to be discharged into the atmosphere.

THE IONOSPHERE

Ozone is not the only atmospheric constituent that is far more prominent at great heights than in the neighborhood of the surface. Further rocket experiments showed that Teisserenc de Bort's speculations concerning layers of helium and hydrogen were not wrong but merely misplaced. From 200 to 600 miles upward, where the atmosphere has thinned out to near-vacuum, there is a layer of helium, now called the *heliosphere*. The existence of this layer was first deduced in 1961 by the Belgian physicist Marcel Nicolet from the frictional drag on the *Echo I* satellite. This deduction was confirmed by actual analysis of the thin-gas surroundings by *Explorer XVII*, launched on 2 April 1963.

Above the heliosphere is an even thinner layer of hydrogen, the *proto-nosphere*, which may extend upward some 40,000 miles before quite fading off into the general density of interplanetary space.

High temperatures and energetic radiation can do more than force atoms apart or into new combinations. They can chip electrons away from atoms and so *ionize* the atoms. What remains of the atom is called an *ion* and differs from an ordinary atom in carrying an electric charge. The word *ion*, first coined in the 1830s by the English scholar William Whewell, comes from a Greek word meaning "traveler." Its origin lies in the fact that when an electric current passes through a solution containing ions, the positively charged ions travel in one direction, and the negatively charged ions in the other.

A young Swedish student of chemistry named Svante August Arrhenius was the first to suggest, in 1884, that ions are charged atoms, as the only means of explaining the behavior of certain solutions that conducted an electric current. His notions, advanced in the thesis he presented for his degree of doctor of philosophy in that year, were so revolutionary that his examiners could scarcely bring themselves to pass him. The charged particles within the atom had not yet been discovered, and the concept of an electrically charged atom seemed ridiculous. Arrhenius got his degree, but with only a minimum passing grade.

When the electron was discovered in the late 1890s (see chapter 6), Arrhenius's theory suddenly made startling sense. He was awarded the Nobel Prize in chemistry in 1903 for the same thesis that nineteen years earlier had nearly lost him his doctoral degree. (This sounds like an improbable movie scenario, I admit, but the history of science contains many episodes that make Hollywood seem unimaginative.)

The discovery of ions in the atmosphere did not emerge until after Guglielmo Marconi started his experiments with wireless. When, on 12 December 1901, he sent signals from Cornwall to Newfoundland, across 2,100 miles of the Atlantic Ocean, scientists were startled. Radio waves travel only in a straight line. How had they managed to go around the curvature of the earth and get to Newfoundland?

A British physicist, Oliver Heaviside, and an American electrical engineer, Arthur Edwin Kennelly, soon suggested that the radio signals might have been reflected back from the sky by a layer of charged particles high in the atmosphere. The *Kennelly-Heaviside layer*, as it has been called ever since, was finally located in 1922. The British physicist Edward Victor Appleton discovered it by paying attention to a curious fading phenomenon in radio transmission. He decided that the fading was the result of interference between two versions of the same signal: one coming directly from the transmitter to his receiver; the other, by a roundabout route via reflection from the upper atmosphere. The delayed wave was out of phase with the first, so the two waves partly canceled each other; hence, the fading.

It was a simple matter then to find the height of the reflecting layer. All Appleton had to do was send signals at such a wavelength that the direct signal completely canceled the reflected one: that is, the two signals arrived at directly opposite phases. From the wavelength of the signal used and the known velocity of radio waves, he could calculate the difference in the distances the two trains of waves had traveled. In this way, he determined, in 1924, that the Kennelly-Heaviside layer was some 65 miles up.

The fading of radio signals generally occurred at night. In 1926, Appleton found that, shortly before dawn, radio waves were not reflected back by the Kennelly-Heaviside layer but were reflected from still higher layers (now sometimes called the *Appleton layers*), which begin at a height of 140 miles (figure 5.3).

For all these discoveries Appleton received the Nobel Prize in physics in 1947. He had defined the important region of the atmosphere called the *ionosphere*, a word introduced in 1930 by the Scottish physicist Robert Alexander Watson-Watt. It includes the later-named mesosphere and thermosphere and is now divided into layers. From the stratopause up to 65 miles or so is the *D region*. Above that is the Kennelly-Heaviside layer, called the *D layer*. Above the D layer, to a height of 140 miles, is the *E region*—an

Figure 5.3. Profile of the atmosphere. The jagged lines indicate the reflection of radio signals from the Kennelly-Heaviside and Appleton layers of the ionosphere. Air density decreases with height and is expressed in percentages of barometric pressure at sea level.

intermediate area relatively poor in ions. This is followed by the Appleton layers: the F_1 *layer* at 140 miles and the F_2 *layer* at 200 miles. The F_1 layer is the richest in ions, the F_2 layer being significantly strong only in the daytime. Above these layers is the *F region*.

These layers reflect and absorb only the long radio waves used in ordinary radio broadcasts. The shorter waves, such as those used in television, pass through, for the most part. Hence, television broadcasting is limited in range

—a limitation that can be remedied by satellite relay stations in the sky, which allow live television to span oceans and continents. The radio waves from space (for example, from radio stars) also pass through the ionosphere, fortunately; if they did not, there would be no radio astronomy possible from Earth's surface.

The ionosphere is strongest at the end of the day, after the day-long effect of the sun's radiation, and weakens by dawn because many ions and electrons have recombined. Storms on the sun, intensifying the streams of particles and high-energy radiation sent to the earth, cause the ionized layers to strengthen and thicken. The regions above the ionosphere also flare up into auroral displays. During these electric storms long-distance transmission of radio waves on the earth is disrupted and sometimes blacked out altogether.

It has turned out that the ionosphere is only one of the belts of radiation surrounding the earth. Outside the atmosphere, in what used to be considered "empty" space, satellites in 1958 disclosed a startling surprise. To understand it, we must make an excursion into the subject of magnetism.

Magnets

Magnets got their name from the ancient Greek town of Magnesia, near which the first *lodestones* were discovered. The lodestone is an iron oxide with natural magnetic properties. Tradition has it that Thales of Miletus, about 550 B.C., was the first philosopher to describe it.

MAGNETISM AND ELECTRICITY

Magnets became something more than a curiosity when it was discovered that a steel needle stroked by a lodestone was magnetized and that, if the needle was allowed to pivot freely in a horizontal plane, it would end up lying approximately along a north-south line. Such a needle was, of course, of tremendous use to mariners; in fact, it became indispensable to ocean navigation, though the Polynesians did manage to cross the Pacific from island to island without a compass.

It is not known who first put such a magnetized needle on a pivot and enclosed it in a box to make a compass. The Chinese are supposed to have done it first and passed it on to the Arabs, who, in turn, passed it on to the Europeans. This is all very doubtful and may be only legend. At any rate, in the twelfth century the compass came into use in Europe and was described in detail in 1269 by a French scholar best known by his Latinized name of Peter Peregrinus. Peregrinus named the end of the magnet that pointed north the *north pole* and the other the *south pole*.

Naturally, people speculated about why a magnetized needle should point north. Because magnets were known to attract other magnets, some thought there was a gigantic lodestone mountain in the far north toward which the needle strained. (Such a mountain is used to great effect in the tale of Sinbad the Sailor, in *The Arabian Nights.*) Others were even more romantic and gave magnets a "soul" and a kind of life.

The scientific study of magnets began with William Gilbert, the court physician of Queen Elizabeth I. It was Gilbert who discovered that the earth itself is a giant magnet. He mounted a magnetized needle so that it could pivot freely in a vertical direction (a *dip needle*), and its north pole then dipped toward the ground (*magnetic dip*). Using a spherical lodestone as a model of the earth, he found that the needle behaved in the same way when it was placed over the northern hemisphere of his sphere. Gilbert published these findings in 1600 in a classic book entitled *De Magnete.*

For a long time, scientists speculated that the earth might have a gigantic iron magnet as its core. Although the earth was indeed found to have an iron core, it is now certain that this core cannot be a magnet, because iron, when heated, loses its strong magnetic properties (*ferromagnetism,* the prefix coming from the Latin word for "iron") at 760° C, and the temperature of the earth's core must be at least 1000° C.

The temperature at which a substance loses its magnetism is called the *Curie temperature,* since it was first discovered by Pierre Curie in 1895. Cobalt and nickel, which resemble iron closely in many respects, are also ferromagnetic. The Curie temperature for nickel is 356° C; for cobalt, it is 1075° C. At low temperatures, certain other metals are ferromagnetic: below −188° C, dysprosium is ferromagnetic, for instance.

In general, magnetism is a property of the atom itself; but in most materials, the tiny atomic magnets are oriented in random directions, so that most of the effect is canceled out. Even so, weak magnetic properties are often evidenced, and the result is *paramagnetism.* The strength of magnetism is expressed in terms of *permeability.* The permeability of a vacuum is 1.00 and that of paramagnetic substances is between 1.00 and 1.01.

Ferromagnetic substances have much higher permeabilities. Nickel has a permeability of 40; cobalt, of 55; and iron, in the thousands. In such substances, the existence of *domains* was postulated in 1907 by the French physicist Pierre Weiss. These are tiny areas, about 0.001 to 0.1 centimeters in diameter (which have actually been detected), within which the atomic magnets are so lined up as to reinforce one another, producing strong, overall fields. In ordinary nonmagnetized iron, the domains themselves are randomly oriented and cancel one another's effect. When the domains are brought into line by the action of another magnet, the iron is magnetized. The reorientation of domains during magnetism actually produces clicking and hissing noises that can be detected by suitable amplification, termed the *Barkhausen effect* after its discoverer, the German physicist Heinrich Barkhausen.

In *antiferromagnetic substances,* such as manganese, the domains also line up, but in alternate directions, so that most of the magnetism is canceled. Above a particular temperature, substances lose antiferromagnetism and become paramagnetic.

If the earth's iron core is not itself a permanent magnet because it is above the Curie temperature, then there must be some other way of explaining the earth's ability to affect a compass needle. What that way might be grew out of the work of the English scientist Michael Faraday, who discovered the connection between magnetism and electricity.

In the 1820s, Faraday started with an experiment that had been first described by Peter Peregrinus (and which still amuses young students of physics). The experiment consists in sprinkling fine iron filings on a piece of paper above a magnet and gently tapping the paper. The shaken filings tend to line up along arcs from the north to the south poles of the magnet. Faraday decided that these marked actual *magnetic lines of force*, forming a magnetic *field*.

Faraday, who had been attracted to the subject of magnetism by the Danish physicist Hans Christian Oersted's observation in 1820 that an electric current flowing in a wire deflected a nearby compass needle, came to the conclusion that the current must set up magnetic lines of force around the wire.

He was all the more convinced since the French physicist André Marie Ampère had gone to study current-carrying wires immediately after Oersted's discovery. Ampère showed that two parallel wires with the current flowing in the same direction attracted each other; with currents flowing in opposite directions, they repelled each other. This was very like the fashion in which two magnetic north poles (or two magnetic south poles) repelled each other while a magnetic north pole attracted a magnetic south pole. Better still, Ampère showed that a cylindrical coil of wire with an electric current flowing through it behaved like a bar magnet. In memory of his work, the unit of intensity of electric current was officially named the *ampere* in 1881.

But if all this were so, thought Faraday (who had one of the most efficient intuitions in the history of science), and if electricity can set up a magnetic field so like the real thing that current-carrying wires can act like magnets, should not the reverse be true? Ought not a magnet produce a current of electricity that would be just like the current produced by chemical batteries?

In 1831, Faraday performed the experiment that was to change human history. He wound a coil of wire around one segment of an iron ring and a second coil of wire around another segment of the ring. Then he connected the first coil to a battery. His reasoning was that if he sent a current through the first coil, it would create magnetic lines of force that would be concentrated in the iron ring, and that this induced magnetism, in turn, would produce a current in the second coil. To detect that current, he connected the second coil to a *galvanometer*—an instrument for measuring electrical currents, which had been devised by the German physicist Johann Salomo Christoph Schweigger in 1820.

The experiment did not work as Faraday had expected. The flow of current in the first coil generated nothing in the second coil. But Faraday noticed that, at the moment when he turned on the current, the galvanometer needle kicked over briefly, and it did the same thing, but in the opposite direction, when he turned the current off. He guessed at once that it was the movement of magnetic lines of force across a wire, not the magnetism itself, that set up the current. When a current began to flow in the first coil, it initiated a magnetic field that, as it spread, cut across the second coil, setting up a momentary electric current there. Conversely, when the current from the battery was cut off, the collapsing lines of magnetic force again cut across the wire of the second coil, causing a momentary surge of electricity in the direction opposite that of the first flow.

Thus, Faraday discovered the principle of electrical induction and created the first *transformer*. He proceeded to demonstrate the phenomenon more plainly by using a permanent magnet and moving it in and out of a coil of wire; although no source of electricity was involved, a current flowed in the coil whenever the magnet's lines of force cut across the wire (figure 5.4).

Figure 5.4. A Faraday experiment on the induction of electricity. When the magnet is moved in or out of the coil of wire, the cutting of its lines of force by the wire produces an electrical current in the coil.

Faraday's discoveries not only led directly to the creation of the dynamo for generating electricity but also laid the foundation for James Clerk Maxwell's *electromagnetic* theory, which linked light and other forms of radiation (such as radio) in a single family of *electromagnetic radiations*.

EARTH'S MAGNETIC FIELD
Now the close connection between magnetism and electricity points to a possible explanation of the earth's magnetism. The compass needle has traced out its magnetic lines of force, which run from the *north magnetic pole,* located off northern Canada, to the *south magnetic pole,* located at the rim of Antarctica, each being about 15 degrees of latitude from the geographic poles. (The earth's magnetic field has been detected at great heights by rockets carrying *magnetometers.*) The new suggestion is that the earth's magnetism may originate in the flow of electric currents deep in its interior.

The physicist Walter Maurice Elsasser has proposed that the rotation of the earth sets up slow eddies in the molten iron core, circling west to east. These eddies have the effect of producing an electric current, likewise circling west to east. Just as Faraday's coil of wire produced magnetic lines of force within

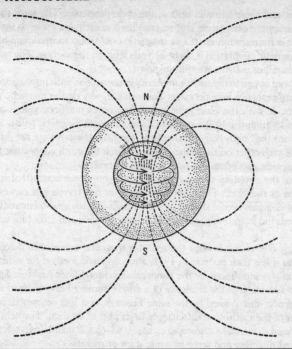

Figure 5.5. Elsasser's theory of the generation of the earth's magnetic field. Movements of material in the molten nickel-iron core set up electric currents which, in turn, generate magnetic lines of force. The dotted lines show the earth's magnetic field.

the coil, so the circling electric current does in the earth's core. It therefore creates the equivalent of an internal magnet extending north and south. This magnet, in turn, accounts for the earth's general magnetic field, oriented roughly along the axis of rotation, so that the magnetic poles are near the north and south geographic poles (figure 5.5).

The sun also has a general magnetic field, which is two or three times as intense as that of the earth, and local fields, apparently associated with the sunspots, which are thousands of times as intense. Studies of these fields (made possible by the fact that intense magnetism affects the wavelength of the light emitted) suggest that there are circular flows of electric charge within the sun.

There are, in fact, many puzzling features concerning sunspots, which may be answered once the causes of magnetic fields on an astronomic scale are worked out. In the course of a sunspot cycle, the spots appear only at certain latitudes, and these latitudes shift as the cycle progresses. The spots show a certain magnetic orientation that reverses itself in each new cycle, so that the total cycle from maximum at one magnetic orientation to maximum at the same magnetic orientation is about 21 years long, on the average. The reasons for this sunspot activity are still unknown.

We need not go to the sun for mysteries in connection with magnetic fields. There are problems here on earth. For instance, why do the magnetic poles not coincide with the geographic poles? The north magnetic pole is about 1,000 miles from the North Pole. Similarly, the south magnetic pole is about 1,000 from the South Pole. Furthermore, the magnetic poles are not directly opposite each other on the globe. A line through the earth connecting them (the *magnetic axis*) does not pass through its center.

Again, the deviation of the compass needle from *true north* (that is, the direction of the North Pole) varies irregularly as one travels east or west. In fact, the compass needle shifted on Columbus's first voyage—a circumstance Columbus hid from his crew lest they become terrified and force him to turn back.

This is one of the reasons the use of a magnetic compass to determine direction is less than perfect. In 1911, a nonmagnetic method for indicating direction was introduced by the American inventor Elmer Ambrose Sperry. It takes advantage of the tendency of a rapidly turning heavy-rimmed wheel (a *gyroscope*, first studied by the same Foucault who had demonstrated the rotation of the earth) to resist changes in its plan of rotation. This tendency can be used to serve as a *gyroscopic compass*, which will maintain its reference to a fixed direction and serve to guide ships or rockets.

But if the magnetic compass is less than perfect, it has been useful enough to serve human beings for centuries. The deviation of the magnetic needle from the true north can be allowed for. A century after Columbus, in 1581, the Englishman Robert Norman prepared the first map indicating the actual direction marked out by a compass needle (*magnetic declination*) in various parts of the world. Lines connecting those points on the planet that show equal declinations (*isogonic lines*) run crookedly from north magnetic pole to south magnetic pole.

Unfortunately, such maps must be periodically changed, for even at one spot the magnetic declination changes with time. For instance, the declination at London shifted 32 degrees of arc in two centuries; it was 8 degrees east of north in 1600 and steadily swung around counterclockwise until it was 24 degrees west of north in 1800. Since then, it has shifted back and, in 1950, was only 8 degrees west of north.

Magnetic dip also changes slowly with time for any given spot on Earth, and the map showing lines of equal dip (*isoclinic lines*) must also be constantly revised. Moreover, the intensity of Earth's magnetic field increases with latitude and is three times as strong near the magnetic poles as in the equatorial regions. This intensity also changes constantly, so that maps showing *isodynamic lines* must also be periodically revised.

Like everything else about the magnetic field, the overall intensity of the field changes. For some time now, the intensity has been diminishing. The field has lost 15 percent of its total strength since 1670; if this loss continues, the intensity will reach zero by about the year 4000. What then? Will it continue decreasing, in the sense that it will reverse with the north magnetic pole in Antarctica and the south magnetic pole in the Arctic? In other words, does Earth's magnetic field periodically diminish, reverse, intensify, diminish, reverse, and so on?

One way of telling whether it indeed can is to study volcanic rocks. When lava cools, the crystals form in alignment with the magnetic field. As long ago as 1906, the French physicist Bernard Brunhes noted that some rocks were magnetized in the direction *opposite* to Earth's present magnetic field. This finding was largely ignored at the time, since it did not seem to make sense; but there is no denying it now. The telltale rocks inform us that not only has Earth's magnetic field reversed, it has done so many times: nine times in the last 4 million years, at irregular intervals.

The most spectacular finding in this respect is on the ocean floor. If melted rock is indeed pushing up through the Global Rift and spreading out, then as one moves east or west from the Rift, one comes across rock that has solidified a progressively longer time ago. By studying the magnetic alignment, one can indeed find reversals occurring in strips, progressively farther from the Rift, at intervals of anywhere from 50,000 to 20 million years with the pattern on one side of the Rift being the mirror-image of that on the other. The only rational explanation so far is to suppose that there *is* sea-floor spreading, and there *are* magnetic-field reversals.

The fact of the reversals is easier to ascertain, however, than the reasons for it.

In addition to long-term drifts of the magnetic field, there are small changes during the course of the day. These suggest some connection with the sun. Furthermore, there are *disturbed days* when the compass needle jumps about with unusual liveliness. The earth is then said to be experiencing a *magnetic storm*. Magnetic storms are identical with electric storms and are usually accompanied by an increase in the intensity of auroral displays, an observation reported as long ago as 1759 by the English physicist John Canton.

The *aurora borealis* (a term introduced in 1621 by the French philosopher Pierre Gassendi, and Latin for "northern dawn") is a beautiful display of moving, colored streamers or folds of light, giving an effect of unearthly splendor. Its counterpart in the Antarctic is called the *aurora australis* ("southern dawn"). In 1741, the Swedish astronomer Anders Celsius noted its connection with Earth's magnetic field. The auroral streamers seem to follow the earth's magnetic lines of force and to concentrate, and become visible, at those points where the lines crowd most closely together—that is, at the magnetic poles. During magnetic storms, the northern aurora can be seen as far south as Boston and New York.

Why the aurora should exist was not hard to understand. Once the ionosphere was discovered, it was understood that something (presumably solar radiation of one sort or another) was energizing the atoms in the upper atmosphere and converting them into electrically charged ions. At night, the ions would lose their charge and their energy, the latter making itself visible in the form of auroral light. It was a kind of specialized air glow, which followed the magnetic lines of force and concentrated near the magnetic poles because that would be expected of electrically charged ions. (The airglow itself involves uncharged atoms and therefore ignores the magnetic field.)

THE SOLAR WIND

But what about the disturbed days and the magnetic storms? Again the finger of suspicion points to the sun.

Sunspot activity seems to generate magnetic storms. How such a disturbance 93 million miles away can affect the earth is not easy to see, yet it must be so, since such storms are particularly common when sunspot activity is high.

The beginning of an answer came in 1859, when an English astronomer, Richard Christopher Carrington, observed a starlike point of light burst out of the sun's surface, last 5 minutes, and subside. This is the first recorded observation of a *solar flare*. Carrington speculated that a large meteor had fallen into the sun, and assumed it to be an extremely unusual phenomenon.

In 1889, however, George E. Hale invented the *spectroheliograph* which allowed the sun to be photographed in the light of a particular spectral region. This picked up solar flares easily and showed that they are common and are associated with sunspot regions. Clearly, solar flares are eruptions of unusual energy that somehow involve the same phenomena that produce the sunspots (hence, the cause of flares is as yet unknown). When the solar flare is near the center of the solar disk, it faces Earth, and anything shot out of it moves in the direction of the Earth. Such central flares are sure to be followed by magnetic storms on Earth after a few days, when particles fired out by the sun reach Earth's upper atmosphere. As long ago as 1896, such a suggestion had been made by the Norwegian physicist Olaf Kristian Birkeland.

As a matter of fact, there was plenty of evidence that, wherever the particles might come from, the earth was bathed in an aura of them extending pretty far out in space. Radio waves generated by lightning had been found to travel along the earth's magnetic lines of force at great heights. (These waves, called *whistlers* because they were picked up by receivers as odd whistling noises, had been discovered accidentally by the German physicist Heinrich Barkhausen during the First World War.) The radio waves could not follow the lines of force unless charged particles were present.

Yet it did not seem that these charged particles emerged from the sun only in bursts. In 1931, when Sydney Chapman was studying the sun's corona, he was increasingly impressed by its extent. What we can see during a total solar eclipse is only its innermost portion. The measurable concentrations of charged particles in the neighborhood of the earth were, he felt, part of the corona. Hence, in a sense, the earth is revolving about the sun within that luminary's extremely attenuated outer atmosphere. Chapman drew the picture of the corona expanding outward into space and being continually renewed at the sun's surface. There would be charged particles continuously streaming out of the sun in all directions, disturbing Earth's magnetic field as it passed.

This suggestion became virtually inescapable in the 1950s, thanks to the work of the German astrophysicist Ludwig Franz Biermann. For half a century, it had been thought that the tails of comets, which always point generally away from the sun and increase in length as the comet approaches the sun, were formed by the pressure of light from the sun. Such light-pressure does exist, but Biermann showed that it is not nearly enough to produce cometary tails. Something stronger and with more of a push was required; this something could scarcely be anything but charged particles. The American physicist Eugene Norman Parker argued further in favor of a steady outflow of particles, with additional bursts at the time of solar flares and, in 1958, named the effect the *solar wind*. The existence of this solar wind was finally demonstrated by the Soviet satellites *Lunik I* and *Lunik II*, which streaked outward to the neighborhood of the moon in 1959 and 1960, and by the American planetary probe *Mariner II*, which in 1962 passed near Venus.

The solar wind is no local phenomenon. There is reason to think it remains dense enough to be detectable at least as far out as the orbit of Saturn. Near the earth the velocity of solar-wind particles varies from 220 to 500 miles per second, and it takes particles three and a half days to travel from the sun to the earth. The solar wind causes a loss to the sun of a million tons of matter per second—a loss that, however huge in human terms, is utterly insignificant on the solar scale. The density of the solar wind is about a quintillionth that of our atmosphere; and in the entire lifetime of the sun, less than 1/100 of 1 percent of its mass has been lost to the solar wind.

The solar wind may well affect our everyday life. Beyond its effect on the magnetic field, the charged particles in the upper atmosphere may ultimately have an effect on the details of Earth's weather. If so, the ebb and flow of the solar wind is still another weapon in the armory of the weather forecast.

THE MAGNETOSPHERE

An unforeseen effect of the solar wind was unexpectedly worked out as a result of satellite launchings. One of the prime jobs given to the artificial satellites was to measure the radiation in the upper atmosphere and nearby space, especially the intensity of the *cosmic rays* (charged particles of particularly high energy). How intense was this radiation up beyond the atmospheric shield? The satellites carried *Geiger counters* (first devised by the German physicist Hans Geiger in 1907 and vastly improved in 1928), which measure particle radiation in the following way: The counter has a box containing gas under a voltage not quite strong enough to send a current through the gas. When a high-energy particle of radiation penetrates into the box, it converts an atom of the gas into an ion. This ion, hurtled forward by the energy of the blow, smashes neighboring atoms to form more ions, which in turn smash their neighbors to form still more. The resulting shower of ions can carry an electric current; and for a fraction of a second, a current pulses through the counter. The pulse is telemetered back to earth. Thus the instrument counts the particles, or flux of radiation, at the location where it happens to be.

When the first successful American satellite, *Explorer I*, went into orbit on 31 January 1958, its counter detected about the expected concentrations of particles at heights up to several hundred miles. But at higher altitudes (and *Explorer I* went as high as 1,575 miles), the count fell off; in fact, at times it dropped to zero! This might have been dismissed as due to some peculiar accident to the counter, but *Explorer III*, launched on 26 March 1958, and reaching an apogee of 2,100 miles, had the same experience. So did the Soviet *Sputnik III*, launched on 15 May 1958.

James A. Van Allen of the State University of Iowa, who was in charge of the radiation program, and his aides came up with a possible explanation. The count fell virtually to zero, they decided, not because there was little or no radiation, but because there was too much. The instrument could not keep up with the particles entering it, and blanked out in consequence. (This would be analogous to the blinding of our eyes by a flash of too-bright light.)

When *Explorer IV* went up on 26 July 1958, it carried special counters designed to handle heavy loads. One of them, for instance, was shielded with a thin layer of lead (analogous to dark sunglasses) which would keep out most of the radiation. And this time the counters did tell another story. They showed that the "too-much-radiation" theory was correct. *Explorer IV*, reaching a height of 1,368 miles, sent down counts that allowing for the shielding, disclosed a radiation intensity far higher than scientists had imagined.

It became apparent that the Explorer satellites had only penetrated the lower regions of this intense field of radiation. In the fall of 1958 the two satellites shot by the United States in the direction of the moon (so-called *moon probes*)—*Pioneer I*, which went out 70,000 miles, and *Pioneer III*, which reached 65,000 miles—showed two main bands of radiation encircling the earth. They were named the *Van Allen radiation belts*, but were later named the *magnetosphere* in line with the names given other sections of space in the neighborhood of the earth (figure 5.6).

It was at first assumed that the magnetosphere was symmetrically placed about the earth, rather like a huge doughnut, and that the magnetic lines of force were themselves symmetrically arranged. This notion was upset when satellite data brought back other news. In 1963, in particular, the satellites *Explorer XIV* and *Imp-I* were sent into highly elliptical orbits designed to carry them beyond the magnetosphere if possible.

It turned out that the magnetosphere has a sharp boundary, the *magnetopause*, which is driven back upon the earth on the side toward the sun by the solar wind, but which loops back around the earth and extends an enormous distance on the night side. The magnetopause is some 40,000 miles from the earth in the direction of the sun, but the teardrop tail on the other

Figure 5.6. The magnetosphere, or Van Allen radiation belts, as traced by satellites. They appear to be made up of charged particles trapped in the earth's magnetic field.

side may extend outward for a million miles or more. In 1966, the Soviet satellite *Luna X,* which circled the moon, detected a feeble magnetic field surrounding that world which may actually have been the tail of earth's magnetosphere sweeping past.

The entrapment of charged particles along the magnetic lines of force had been predicted in 1957 by an American-born Greek amateur scientist, Nicholas Christofilos, who made his living as a salesman for an American elevator firm. He had sent his calculations to scientists engaged in such research, but no one had paid much attention to them. (In science, as in other fields, professionals tend to disregard amateurs.) It was only when the professionals independently came up with the same results that Christofilos achieved recognition and was welcomed into the University of California. His idea about particle entrapment is now called the *Christofilos effect.*

In August and September 1958 to test whether the effect really occurs in space, the United States fired three rockets carrying nuclear bombs 300 miles up and there exploded the bombs—an experiment that was named Project Argus. The flood of charged particles resulting from the nuclear explosions spread out along the lines of force and were indeed trapped there. The resulting band persisted for a considerable time; *Explorer IV* detected it during several hundred of its trips around the earth. The cloud of particles also gave rise to feeble auroral displays and disrupted radar for a while.

This was the prelude to other experiments that affected or even altered Earth's near-space environment, and some of them met with opposition and vast indignation from sections of the scientific community. A nuclear bomb exploded in space on 9 July 1962 introduced marked changes in the magnetosphere, changes that showed signs of persisting for a prolonged interval, as some disapproving scientists (such as Fred Hoyle) had predicted. The Soviet Union carried out similar high-altitude tests in 1962. Such tampering with the natural state of affairs may interfere with our understanding of the magnetosphere, and it is unlikely that this experiment will be soon repeated.

Then, too, attempts were made to spread a layer of thin copper needles into orbit about the earth to test their ability to reflect radio signals, in order to establish an unfailing method for long-distance communication. (The ionosphere is disrupted by magnetic storms every once in a while and then radio communication may fail at a crucial moment.) Despite the objection of radio astronomers who feared interference with the radio signals from space, the project (Project West Ford, after Westford, Massachusetts, where the preliminary work was done) was carried through on 9 May 1963. A satellite containing 400 million copper needles, each three-quarters of an inch long and finer than a human hair—50 pounds' worth altogether—was put into orbit. The needles were ejected and then slowly spread into a world-circling band that was found to reflect radio waves just as had been expected. This band remained in orbit for three years. A much thicker band would be required for useful purposes, however, and it is doubtful whether the objections of the radio astronomers can be overcome for that.

PLANETARY MAGNETOSPHERES

Naturally, scientists were curious to find out whether there were radiation belts about heavenly bodies other than the earth. If Elsasser's theory is correct, a planetary body must fulfill two requirements in order to have a sizable magnetosphere: it must have a liquid, electrically conducting core, in which swirls can be set up; and it must have a fairly rapid period of rotation to set up those swirls. The moon, for instance, is of low density and is small enough not to be very hot at its center, and thus almost certainly contains no liquid metal core. Even if it does, the moon rotates far too slowly to set it swirling. The moon, therefore, should have no magnetic field of any consequence on both scores. Nevertheless, no matter how clear-cut such deduction may be, it always helps to have a direct measurement, and rocket probes can easily be outfitted to make such measurements.

Indeed, the first lunar probes, the Soviet-launched *Lunik I* (2 January 1959) and *Lunik II* (September 1959), found no signs of radiation belts about the moon, and this finding has been confirmed in every approach to the moon since.

Venus is a more interesting case. It is almost as massive and almost as dense as Earth and must certainly have a liquid metallic core much as Earth does. However, Venus rotates very slowly, even more slowly than the moon. The Venus probe, *Mariner 2,* in 1962, and all the Venus probes since have agreed that Venus has virtually no magnetic field. The magnetic field it does have (possibly resulting from conducting effects in the ionosphere of its dense atmosphere) is certainly less than 1/20,000 as intense as Earth's.

Mercury is also dense and must have a metallic core; but, like Venus, it rotates very slowly. *Mariner 10,* which skimmed Mercury in 1973 and 1974, detected a weak magnetic field, somewhat stronger than that of Venus, and with no atmosphere to account for it. Weak as it is, Mercury's magnetic field is too strong to be caused by its slow rotation. Perhaps because of Mercury's size (considerably smaller than that of either Venus or Earth), its metallic core is cool enough to be ferromagnetic and possesses some slight property as a permanent magnet. However, we cannot tell yet whether it does.

Mars rotates reasonably rapidly but is smaller and less dense than Earth. It probably does not have a liquid metallic core of any size, but even a small one may produce some effect, and Mars seems to have a small magnetic field, stronger than Venus's though much weaker than Earth's.

Jupiter is another thing altogether. Its giant mass and its rapid rotation would make it an obvious candidate for a magnetic field if there were certain knowledge of the conducting characteristics of its core. Back in 1955, however, when such knowledge did not exist and no probes had yet been constructed, two American astronomers, Bernard Burke and Kenneth Franklin, detected radio waves from Jupiter that were nonthermal: that is, they did not arise merely from temperature effects. They had to arise from some other cause, perhaps high-energy particles trapped in a magnetic field. In 1959, Frank Donald Drake did so interpret the radio waves from Jupiter.

The first Jupiter probes, *Pioneer 10* and *Pioneer 11,* gave ample confirmation of theory. They had no trouble detecting a magnetic field (for compared with Earth's, it was a giant) even more intense than was to be expected from the huge planet. The magnetosphere of Jupiter is some 1,200 times as large as Earth's. If it were visible to the eye, it would fill an area of the sky (as seen from Earth) that was several times larger than the full moon appears to us. Jupiter's magnetosphere is 19,000 times as intense as Earth's; and if manned space vessels ever reach the planet, it would form a deadly barrier to a close approach, embracing moreover the Galilean satellites.

Saturn also has an intense magnetic field, one that is intermediate in size between that of Jupiter and Earth. We cannot yet tell by direct observation, but it seems reasonable to suppose that Uranus and Neptune also have magnetic fields that may be stronger than Earth's. In all the gas giants, the nature of the liquid, conducting core would be either liquid metal or liquid metallic hydrogen—the latter almost certainly in the case of Jupiter and Saturn.

Meteors and Meteorites

Even the Greeks knew that *shooting stars* were not really stars, because no matter how many fell, the celestial population of stars remained the same. Aristotle reasoned that a shooting star, being a temporary phenomenon, had to be something within the atmosphere (and this time he was right). These objects were therefore called *meteors*, meaning "things in the air." Meteors that actually reach the earth's surface are called *meteorites*.

The ancients even witnessed falls of meteorites to the earth and found some to be lumps of iron. Hipparchus of Nicaea is said to have reported such a fall. The Kaaba, the sacred black stone in Mecca, is supposed to be a meteorite and to have gained its sanctity through its heavenly origin. The *Iliad* mentions a lump of rough iron being awarded as one of the prizes in the funeral games for Patroclus; this must have been meteoric in origin, because the time was the Bronze Age, before the metallurgy of iron ore had been developed. In fact, meteoric iron was probably in use as early as 3000 B.C.

During the eighteenth century, with the Age of Reason in full sway, science made a backward step in this respect. The scorners of superstition laughed at stories of "stones from the sky." Farmers who came to the Académie Française with samples of meteorites were politely, but impatiently, shown the door. When, in 1807, two Connecticut scholars (one of them the young chemist Benjamin Silliman) reported having witnessed a fall, President Thomas Jefferson said that he would sooner believe that two Yankee professors would lie than that stones would fall from heaven.

Jefferson was actually out of date, for reports of meteorite falls in France had finally stirred the physicist Jean Baptiste Biot, in 1803, to investigate such sightings. His investigation, soberly and thoroughly done, went a long way to convincing the scientific world that stones did indeed fall from heaven.

Then, on 13 November 1833, the United States was treated to a meteor shower of the type called Leonids because they seem to radiate from a point in the constellation Leo. For some hours it turned the sky into a Roman-candle display more brilliant than any ever seen before or since. No meteorites reached the ground, as far as is known, but the spectacle stimulated the study of meteors, and astronomers turned to it for the first time in all seriousness.

The very next year, the Swedish chemist Jöns Jakob Berzelius began a program for the chemical analyses of meteorites. Eventually such analyses gave astronomers valuable information on the general age of the solar system and even on the overall chemical makeup of the universe.

METEORS

By noting the times of year when meteors came thickest, and the positions in the sky from which they seemed to come, the meteor watchers were able to work out orbits of various clouds of meteors. In this way, they learned that a meteor shower occurs when the earth's orbit intersects the orbit of a meteor cloud.

Meteor clouds have elongated orbits as comets do, and it makes sense to consider them as the débris of disintegrated comets. Comets can disintegrate to leave dust and gravel behind according to the Whipple picture of comet structure, and some comets have been actually seen to disintegrate.

When such comet dust enters the atmosphere, they can make a brave display, as they did in 1833. A shooting star as bright as Venus comes into the atmosphere as a speck weighing only 1 gram (1/28 of an ounce). Some visible meteors are only 1/10,000 as massive as that!

The total number of meteors hitting the earth's atmosphere can be computed, and turns out to be incredibly large. Each day there are more than 20,000 weighing at least 1 gram, nearly 200 million others large enough to make a glow visible to the naked eye, and many billions more of still smaller sizes.

We know about these very small *micrometeors* because the air has been found to contain dust particles with unusual shapes and a high nickel content, quite unlike ordinary terrestrial dust. Another evidence of the presence of micrometeors in vast quantities is the faint glow in the heavens called *zodiacal light* (first discovered about 1700 by G. D. Cassini)—so called because it is most noticeable in the neighborhood of the plane of the earth's orbit, where the constellations of the zodiac occur. The zodiacal light is very dim and cannot be seen even on a moonless night unless conditions are favorable. It is brightest near the horizon where the sun has set or is about to rise; and on the opposite side of the sky, there is a secondary brightening called the Gegenschein (German for "opposite light"). The zodiacal light differs from the airglow: its spectrum has no lines of atomic oxygen or atomic sodium, but is just that of reflected sunlight and nothing more. The reflecting agent presumably is dust concentrated in space in the plane of the planets' orbits —in short, micrometeors. Their number and size can be estimated from the intensity of the zodiacal light.

Micrometeors have now been counted with new precision by means of such satellites as *Explorer XVI*, launched in December 1962, and *Pegasus I*, launched 16 February 1965. To detect them, some of the satellites are covered with patches of a sensitive material that signals each meteoric hit through a change in electrical resistance. Others record the hits by means of a sensitive microphone behind the skin, picking up the "pings." The satellite counts have indicated that 3,000 tons of meteoric matter enter our atmosphere each day, five-sixths of it consisting of micrometeors too small to be detected as shooting stars. These micrometeors may form a thin dust cloud about the earth, one that stretches out, in decreasing density, for 100,000 miles or so before fading out to the usual density of material in interplanetary space.

The Venus probe *Mariner 2* showed the dust concentration in space generally to be only 1/10,000 the concentration near Earth—which seems to be the center of a dustball. Fred Whipple suggests that the moon may be the source of the cloud, the dust being flung up from the moon's surface by the meteorite beating it has had to withstand. Venus, which has no moon, also has no dustball.

The geophysicist Hans Petterson, who has been particularly interested in this meteoric dust, took some samples of air in 1957 on a mountaintop in Hawaii, which is as far from industrial dust-producing areas as one can get on the earth. His findings led him to that about 5 million tons of meteoric dust fall on the earth each year. (A similar measurement by James M. Rosen in 1964, making use of instruments borne aloft by balloons, set the figure at 4 million tons, though still others find reason to place the figure at merely 100,000 tons per year.) Hans Petterson tried to get a line on this fall in the past by analyzing cores brought up from the ocean bottom for high-nickel dust. He found that, on the whole, there was more in the upper sediments than in the older ones below; thus—though the evidence is still scanty—the rate of meteoric bombardment may have increased in recent ages. This meteoric dust may possibly be of direct importance to all of us, for, according to a theory advanced by the Australian physicist Edward George Bowen in 1953, this dust serves as nuclei for raindrops. If so, then the earth's rainfall pattern reflects the rise and fall of the intensity with which micrometeorites bombard us.

METEORITES

Occasionally pieces of matter that are larger than tiny bits of gravel, even substantially large, penetrate Earth's atmosphere. They may be large enough to survive the heat of air resistance as they race through the atmosphere at anywhere from 8 to 45 miles per second, and to reach the ground. These, as I have said, are meteorites. Such meteorites are thought to be small asteroids —specifically, earth grazers that have grazed too closely and come to grief.

Most of the meteorites found on the ground (about 1,700 are known altogether, of which 35 weigh over a ton each) have been iron, and it seemed that iron meteorites must far outnumber the stony type. This theory proved to be wrong, however. A lump of iron lying half-buried in a stony field is very noticeable, whereas a stone among other stones is not; a stony meteorite, once investigated, however, shows characteristic differences from earthly stones.

When astronomers made counts of meteorites found that were actually seen to fall, they discovered that the stony meteorites outnumbered iron ones 9 to 1. (For a time, most stony meteorites were discovered in Kansas, which may seem odd until one realizes that, in the stoneless, sedimentary soil of Kansas, a stone is as noticeable as a lump of iron would be elsewhere.)

These two types of meteorites are thought to arise in the following manner: Asteroids, in the youth of the solar system, may have been larger, on the average, than they now are. Once formed, and prevented from further consolidation by the perturbations of Jupiter, they underwent collisions among themselves and breakups. Before that happened, however, the asteroids may have grown hot enough, on forming, to allow a certain separation of components, with iron sinking to the center and stone forced into the outer layer. Then, when such asteroids were fragments, there were both stony and metallic débris, making for meteorites of each type on Earth now.

There is a third type of meteorite—*carbonaceous chondrites*—that is quite rare. These will be discussed, more appropriately, in chapter 13.

Meteorites seldom do damage. Although about 500 substantial meteorites strike the earth annually (with only some 20 recovered, unfortunately), the earth's surface is large, and only small areas are thickly populated. No human being has ever been killed by a meteorite so far as is known, although a woman in Alabama reported being bruised by a glancing blow on 30 November 1955. In 1982, a meteorite flashed through a home in Wethersfield, Connecticut, without hurting the occupants. Oddly enough, Wethersfield had been struck eleven years earlier without harm.

Yet meteorites have a devastating potentiality. In 1908, for instance, a strike in central Siberia gouged out craters up to 150 feet in diameter and knocked down trees for 20 miles around. Fortunately, the meteorite fell in a wilderness and, while it destroyed a herd of deer, did not kill a single human being. Had it fallen from the same part of the sky five hours later in the earth's rotation, it might have hit St. Petersburg (Leningrad), then the capital of Russia. If it had, the city would have been wiped out as thoroughly as by a hydrogen bomb. One estimate is that the total weight of the meteorite was 40,000 tons.

This *Tunguska event* (so-called from the locality of the strike) has presented mysteries. The inaccessibility of the locality, and the confusion of war and revolution that took place soon after, made it impossible to investigate the area for many years. Once investigated, it offered no trace of meteoric material. In recent years, a Soviet science-fiction writer invented radioactivity at the site as part of a story—an invention that was taken as a sober finding by many people who had a natural affection for the sensational. As a result, many wild theories evolved—from a strike by a mini-black hole to an extraterrestrial nuclear explosion. The most likely rational explanation is that the incoming meteor was icy in nature, and probably a very small comet, or a piece of a larger one (possibly Comet Encke). It exploded in air before striking and did immense damage without producing any meteoric matter of stone or metal.

The largest strike since then, near Vladivostok (again in Siberia), was in 1947.

There are signs of even heavier strikes in prehistoric times. In Coconino County in Arizona, there is a round crater about four-fifths of a mile across and 600 feet deep, surrounded by a lip of earth 100 to 150 feet high. It looks like a miniature crater of the moon. It was long assumed to be an extinct volcano, but a mining engineer named Daniel Moreau Barringer insisted it was the result of a meteoric collision, and the hole now bears the name Barringer Crater. The crater is surrounded by lumps of meteoric iron—thousands (perhaps millions) of tons of it altogether. Although only a small portion has been recovered so far, more meteoric iron has already been extracted from it and its surroundings than in all the rest of the world. The meteoric origin of the crater was also borne out by the discovery there, in 1960, of forms of silica that could have been produced only by the momentary enormous pressures and temperatures accompanying meteoric impact.

Barringer Crater, formed in the desert an estimated 25,000 years ago by an iron meteorite about 150 feet across, has been preserved fairly well. In most parts of the world, similar craters would have been obliterated by water and plant overgrowth. Observations from airplanes, for instance, have sighted previously unnoticed circular formations, partly water-filled and partly overgrown, which are almost certainly meteoric. Several have been discovered in Canada, including Brent Crater in central Ontario and Chubb Crater in northern Quebec, each of which is 2 miles or more in diameter; and Ashanti Crater in Ghana, which is 6 miles in diameter. These are perhaps more than a million years old. Some seventy such *fossil craters* are known, with diameters of up to 85 miles or so.

The craters of the moon range from tiny holes to giants 150 miles or more across. The moon—lacking air, water, or life—is a nearly perfect museum for craters since they are subject to no wear except from the very slow action of temperature change resulting from the two-week alternation of lunar day and lunar night. Perhaps the earth would be pockmarked like the moon were it not for the healing action of wind, water, and growing things.

It had been felt, at first, that the craters of the moon were volcanic in origin, but they do not really resemble earthly volcanic craters in structure. By the 1890s, the view that the craters had originated from meteoric strikes came into prominence and has gradually become accepted.

The large "seas" or maria, which are vast, roughly circular stretches that are relatively craterfree, would, in this view, result from the impact of particularly large meteors. This view was bolstered in 1968 when satellites placed in orbit about the moon showed unexpected deviations in their circumlunar flights. The nature of these deviations forced the conclusion that parts of the lunar surface are denser than average and produce a slight increase in gravitational attraction, to which the satellite flying over them responded. These denser-than-average areas, which seemed to coincide with the maria, received the name *mascons* (short for "mass concentration"). The most obvious deduction was that the sizable iron meteors that formed the seas are still buried beneath them and are considerably denser than the rocky material that generally makes up the moon's crust. At least a dozen mascons were detected within a year of their initial discovery.

The view of the moon as a "dead world" where no volcanic action is possible is, on the other hand, overdrawn. On 3 November 1958, the Russian astronomer N. A. Kozyrev observed a reddish spot in the crater Alphonsus. (William Herschel had reported seeing reddish spots on the moon as early as 1780.) Kozyrev's spectroscopic studies seemed to make it clear that gas and dust had been emitted. Since then, other red spots have been momentarily seen, and it seems certain that volcanic activity does occasionally take place on the moon. During the total lunar eclipse in December 1964, it was found that as many as 300 craters were hotter than the surrounding landscape—although, of course, they were not hot enough to glow.

Airless worlds generally, such as Mercury and the satellites of Mars, Jupiter, and Saturn, are thickly spread with craters which commemorate the bombardment that took place 4 billion and more years ago when the worlds were formed by accretion of planetesimals. Nothing has occurred since to remove those markings.

Venus is poor in craters, perhaps because of the erosive effects of its thick atmosphere. One hemisphere of Mars is poor in craters, perhaps because volcanic action has built up a fresh crust. Io has virtually no craters because of the lava built up by its active volcanoes. Europa has no craters because meteoric impacts break through the encircling glacier into the liquid beneath, whereupon the liquid exposed quickly refreezes and "heals" the break.

Meteorites, as the only pieces of extraterrestrial matter we can examine, are exciting not only to astronomers, geologists, chemists, and metallurgists but also to cosmologists, who are concerned with the origins of the universe and the solar system. Among the meteorites are puzzling glassy objects found in several places on earth. The first were found in 1787 in what is now western Czechoslovakia. Australian examples were detected in 1864. They received the name *tektites,* from a Greek word for "molten," because they appear to have melted in their passage through the atmosphere.

In 1936, the American astronomer Harvey Harlow Ninninger suggested that tektites are remnants of splashed material forced away from the moon's surface by the impact of large meteors and caught by Earth's gravitational field. A particularly widespread strewing of tektites is to be found in Australia and southeast Asia (with many dredged up from the floor of the Indian Ocean). These seem to be the youngest of the tektites, only 700,000 years old. Conceivably, these could have been produced by the great meteoric impact that formed the crater Tycho (the youngest of the spectacular lunar craters) on the moon. The fact that this strike seems to have coincided with the most recent reversal of Earth's magnetic field has caused some speculation that the strikingly irregular series of such reversals may mark other such earth-moon catastrophes.

Another unusual classification of meteorites are those that may be found in Antarctica. For one thing, any meteorite, whether stony or metallic, if lying on the vast Antarctica icecap is inevitably noticeable. In fact, any solid object anywhere on that continent, if not ice and not of human origin, is bound to be a meteorite. And once it lands, it remains untouched (at least over the last 20 million years) unless it is buried in snow or stumbled over by an emperor penguin.

Not many human beings are present in Antarctica at any time, and not much of the continent has been peered at closely, so that up to 1969, only four meteorites were found—all by accident. In 1969, a group of Japanese geologists came across nine closely spaced meteorites. These roused the interest of scientists generally, and ever more meteorites were found. By 1983, more than 5,000 meteoric fragments had been found on the frozen continent, more by far than in all the rest of the world. (Antarctica is not especially chosen out for strikes, but meteorites are much more easily spotted there.)

Some of the Antarctic meteorites are strange indeed. In January 1982, a greenish-tan meteoritic fragment was discovered and, upon analysis, proved to have a composition remarkably like some of the moon rocks brought back by the astronauts. There is no easy way of demonstrating how a piece of lunar material could have been blasted into space and come to Earth, but certainly that is a possibility.

Then, too, some meteoric fragments in Antarctica have, when heated, given off gases, which proved to have a composition much like the Martian atmosphere. What's more, these meteorites seemed to be only 1,300,000,000 years old rather than 4,500,000,000 years old as ordinary meteorites are. About 1,300,000,000 years ago, the Martian volcanoes may have been violently active. It may be that some meteorites are pieces of Martian lava somehow blown to Earth.

The ages of meteorites (computed by methods that will be described in chapter 7) are important tools, by the way, in the determination of the age of the earth and of the solar system generally.

Air: Keeping It and Getting It

Perhaps before we wonder how the earth got its atmosphere, we should consider how it has managed to hang on to it through all the eons of whirling and wheeling through space. The answer to the latter question involves something called *escape velocity*.

ESCAPE VELOCITY

If an object is thrown upward from the earth, the pull of gravity gradually slows it until it comes to a momentary halt and then falls back. If the force of gravity were the same all the way up, the height reached by the object would be proportional to its initial upward velocity: that is, it would reach four times as high when launched at a speed of 2 miles an hour as it would when it started at 1 mile an hour (energy increases as the square of the velocity).

But the force of gravity does not remain constant: it weakens slowly with height. (To be exact, it weakens as the square of the distance from the earth's center.) Let us say we shoot an object upward with a velocity of 1 mile per second: it will reach a height of 80 miles before turning and falling (if we ignore air resistance). If we were to fire the same object upward at 2 miles per second, it would climb higher than four times that distance. At the height of 80 miles, the pull of the earth's gravity is appreciably lower than at ground level, so that the object's further flight would be subject to a smaller gravitational drag. In fact, the projectile would rise to 350 miles, not 320.

Given an initial upward velocity of 6.5 miles per second, an object will climb 25,800 miles. At that point the force of gravity is not more than 1/40 as strong as it is on the earth's surface. If we added just 1/10 of a mile per second to the object's initial speed (that is, launched it at 6.6 miles per second), it would go up to 34,300 miles.

It can be calculated that an object fired up at an initial speed of 6.98 miles per second will never fall back to the earth. Although the earth's gravity will gradually slow the object's velocity, its effect will steadily decline, so that it will never bring the object to a halt (zero velocity) with respect to the earth. (So much for the cliché that "everything that goes up must come down.")

The speed of 6.98 miles per second, then, is the earth's escape velocity. The velocity of escape from any astronomical body can be calculated from its mass and size. From the moon, it is only 1.5 miles per second; from Mars, 3.2 miles per second; from Saturn, 23 miles per second; from Jupiter, the most massive planet in the solar system, it is 38 miles per second.

Now all this has a direct bearing on the earth's retention of its atmosphere. The atoms and molecules of the air are constantly flying about like tiny missiles. Their individual velocities vary a great deal, and the only way they can be described is statistically: for example, giving the fraction of the molecules moving faster than a particular velocity, or giving the average velocity under given conditions. The formula for doing this was first worked out in 1860 by James Clerk Maxwell and the Austrian physicist Ludwig Boltzmann, and it is called the *Maxwell-Boltzmann law.*

The mean velocity of oxygen molecules in air at room temperature turns out to be 0.3 mile per second. The hydrogen molecule, being only 1/16 as heavy, moves on the average four times as fast, or 1.2 miles per second, because, according to the Maxwell-Boltzmann law, the velocity of a particular particle at a particular temperature is inversely proportional to the square root of its molecular weight.

It is important to remember that these are only average velocities. Half the molecules go faster than the average; a certain percentage go more than twice as fast as the average; a smaller percentage more than three times as fast; and so on. In fact, a tiny percentage of the oxygen and hydrogen molecules in the atmosphere go faster than 6.98 miles per second, the escape velocity.

In the lower atmosphere, these speedsters cannot actually escape, because collisions with their slower neighbors slow them down. But in the upper atmosphere, their chances are much better. First of all, the unimpeded radiation of the sun up there excites a large proportion of them to enormous energy and great speeds. In the second place, the probability of collisions is greatly reduced in the thinner air. Whereas a molecule at the earth's surface travels only 4 millionths of an inch (on the average) before colliding with a neighbor, at a height of 65 miles its average free path before colliding is 4 inches; and at 140 miles, it is 1,100 yards. There the average number of collisions encountered by an atom or molecule is only 1 per second, against 5 billion per second at sea level. Thus, a fast particle at a height of 100 miles or more stands a good chance of escaping from the earth. If it happens to be moving upward, it is moving into regions of ever less density and experiences an ever smaller chance of collision, so that it may in the end depart into interplanetary space, never to return.

In other words, the earth's atmosphere leaks. But the leakage applies mainly to the lightest molecules. Oxygen and nitrogen are heavy enough so that only a tiny fraction of them achieves the escape velocity, and not much oxygen or nitrogen has been lost from the earth since their original formation. On the other hand, hydrogen and helium are easily raised to escape velocity. Consequently it is not surprising that no hydrogen or helium to speak of remains in the atmosphere of the earth today.

The more massive planets, such as Jupiter and Saturn, can hold even hydrogen and helium, so they may have large and deep atmospheres composed mostly of these elements (which, after all, are the most common substances in the universe). The hydrogen present in vast quantities would react with other elements present, so that carbon, nitrogen, and oxygen would be present only in the form of hydrogen-containing compounds: methane (CH_4), ammonia (NH_3), and water (H_2O), respectively. The ammonia and methane in Jupiter's atmosphere, although present as relatively small-concentration impurities, were first discovered (in 1931, by the German-American astronomer Rupert Wildt) because these compounds produce noticeable absorption bands in the spectra, whereas hydrogen and helium do not. The presence of hydrogen and helium were detected by indirect methods in 1952. And, of course, the Jupiter probes, from 1973 on, confirmed these findings and gave us further details.

Working in the other direction, a small planet like Mars is less able to hold even the comparatively heavy molecules and has an atmosphere only 1 hundredth as dense as our own. The moon, with a smaller escape velocity, cannot hold any atmosphere to speak of and is airless.

Temperature is just as important a factor as gravity. The Maxwell-Boltzmann equation says that the average speed of particles is proportional to the square root of the absolute temperature. If the earth were at the temperature of the sun's surface, all the atoms and molecules in its atmosphere would be speeded up four to five times, and the earth could no more hold on to its oxygen and nitrogen than it could to hydrogen or helium.

Thus, Mercury has 2.2 times the surface gravity of the moon and should do a better job at holding an atmosphere. Mercury is, however, considerably hotter than the moon and so ends up just as airless as the moon.

Mars has a surface gravity only slightly greater than that of Mercury but is considerably colder than Mercury and even than Earth or the moon. That Mars manages to have a thin atmosphere is more because of its low temperature than of its moderately high surface gravity. The satellites of Jupiter are still colder than Mars, but also have a surface gravity in the range of the moon and so do not hold an atmosphere. Titan, the large satellite of Saturn, is so cold, however, that it can hold a thick nitrogen atmosphere. Perhaps Triton, the large satellite of Neptune, may do so also.

THE ORIGINAL ATMOSPHERE

The earth's possession of an atmosphere is a strong point against the theory that it and the other planets of the solar system originated from some catastrophic accident, such as near-collision between another sun and ours. It argues, rather, in favor of the dust-cloud and planetesimal theory. As the dust and gas of the cloud condensed into planetesimals and these in turn collected to form a planetary body, gas might have been trapped within a spongy mass, like air in a snowbank. The subsequent gravity contraction of the mass might then have squeezed out the gases toward the surface. Whether a particular gas would be held in the earth would depend in part on its chemical reactivity. Helium and neon, though they must have been among the most common gases in the original cloud, are so inert chemically that they form no compounds and would have escaped as gases in short order. Therefore the concentrations of helium and neon on the earth are insignificant fractions of their concentrations in the universe generally. It has been calculated, for instance, that the earth has retained only 1 out of every 50 billion neon atoms present in the original cloud of gas, and our atmosphere has even fewer, if any, of the original helium atoms. I say "if any" because, while there is a little helium in the atmosphere today, all of it may come from the breakdown of radioactive elements and leakage of helium trapped in cavities underground.

On the other hand, hydrogen, though lighter than helium or neon, has been captured with greater efficiency because it has combined with other substances, notably with oxygen to form water. It is estimated that the earth still has 1 out of every 5 million hydrogen atoms that were in the original cloud.

Nitrogen and oxygen illustrate the chemical aspect even more neatly. Although the nitrogen molecule and the oxygen molecule are about equal in mass, the earth has held on to 1 out of 6 of the original atoms of highly reactive oxygen but on to only 1 out of every 800,000 of inert nitrogen.

When we speak of gases of the atmosphere, we have to include water vapor, and here we get into the interesting question of how the oceans originated. In the early stages of the earth's history, even if our planet was then only moderately hot, all the water must have been in the form of vapor. Some geologists believe that the water was then concentrated in the atmosphere as a dense cloud of vapor, and that, after the earth cooled, it fell in torrents to form the ocean. On the other hand, some geologists maintain that our oceans have been built up mainly by water seeping up from the earth's interior. Volcanoes show that there still is a great deal of water in the crust, for the gas they discharge is mostly water vapor. If that is so, the oceans may still be growing, albeit slowly.

But was the earth's atmosphere always what it is today, at least since its formation in the first place? It seems unlikely. For one thing, molecular oxygen, which makes up one-fifth of the volume of the atmosphere, is so active a substance that its presence in free form is extremely unlikely, unless it were continuously being produced. Furthermore, no other planet has an atmosphere anything like our own, so that one is strongly tempted to conclude that

Earth's atmosphere is the result of unique events (as, for instance, the presence of life on this planet, but not on the others).

Harold Urey has presented detailed arguments in favor of the idea that the original atmosphere was composed of ammonia and methane. Hydrogen, helium, carbon, nitrogen, and oxygen are the predominant elements in the universe, with hydrogen far and away the most common. In the presence of such a preponderance of hydrogen, carbon would be likely to combine with hydrogen to form methane (CH_4), nitrogen with hydrogen to form ammonia (NH_3), and oxygen with hydrogen to form water (H_2O). Helium and excess hydrogen would, of course, escape; the water would form the oceans; the methane and ammonia, as comparatively heavy gases, would be held by the earth's gravity and so constitute the major portion of the atmosphere.

If all the planets with sufficient gravity to hold an atmosphere at all began with atmospheres of this type, they would nevertheless not all keep it. Ultraviolet radiation from the sun would introduce changes. These changes would be minimal for the outer planets, which, in the first place, received comparatively little radiation from the distant sun and, in the second place, had vast atmospheres capable of absorbing considerable radiation without being perceptibly changed. The outer planets, therefore, would keep the hydrogen/helium/ammonia/methane atmospheres to the present day.

Not so the five inner worlds of Mars, Earth, our moon, Venus, and Mercury. Of these, the moon and Mercury are too small, too hot, or both to retain any perceptible atmosphere. This leaves Mars, Earth, and Venus, with thin atmospheres of chiefly ammonia, methane, and water to begin with. What would happen?

Ultraviolet radiation striking water molecules in the upper primordial atmosphere of the earth would break them apart to hydrogen and oxygen (*photodissociation*). The hydrogen would escape, leaving oxygen behind. Being reactive, however, the molecules would react with almost any other molecule in the neighborhood. They would react with methane (CH_4) to form carbon dioxide (CO_2) and water (H_2O). They would react with ammonia (NH_3) to form free nitrogen (N_2) and water. Very slowly, but steadily, the atmosphere would be converted from methane and ammonia to nitrogen and carbon dioxide. The nitrogen would tend to react slowly with the minerals of the crust to form nitrates, leaving carbon dioxide as the major portion of the atmosphere.

Will water continue to photodissociate, however? Will hydrogen continue to escape into space, and will oxygen continue to collect in the atmosphere? And if oxygen does collect and finds nothing to react with (it cannot react further with carbon dioxide), then will it not add a proportion of molecular oxygen to the carbon dioxide present (thus accounting for earth's atmospheric oxygen)? The answer is a resounding No.

Once carbon dioxide becomes the major component of the atmosphere, ultraviolet radiation does not bring about further changes through dissociation of the water molecule. When the oxygen begins to collect in free form, a thin ozone layer is formed in the upper atmosphere. This absorbs the ultraviolet, blocking it from the lower atmosphere and preventing further photodissociation. A carbon-dioxide atmosphere is stable.

But carbon dioxide introduces the greenhouse effect (see chapter 4). If the carbon-dioxide atmosphere is thin and is relatively far from the sun, and there is very little water in any case, the effect is small, as is the case with Mars, for instance.

Suppose, though, that a planet's atmosphere is more like that of Earth, and it is as close to the sun (or closer). The greenhouse effect will then be enormous: temperatures will rise, vaporizing the oceans to an ever greater extent. The water vapor will add to the greenhouse effect, accelerating the change, forcing ever more carbon dioxide into the air as well as through temperature effects on the crust. In the end, the planet will be enormously hot, will have all its water in the atmosphere in the form of a vapor that will forever hide its surface under eternal clouds, and will have a thick atmosphere of carbon dioxide.

This was precisely the case of Venus, which had to endure a *runaway greenhouse effect*. The little bit of additional heat it received through its being closer to the sun than Earth is served as a trigger and began the process.

Earth did not move in the direction of either Mars or Venus. The nitrogen content of its atmosphere did not soak into the crust, leaving a thin, cold carbon-dioxide wind as on Mars. Nor did the greenhouse effect turn it into a choking desert world of great heat as on Venus. Something else happened, and that something was the development of life, perhaps even while the atmosphere was still in its ammonia/methane stage.

Life-induced reactions in earth's oceans broke down nitrogen compounds to liberate molecular nitrogen and thus kept that gas in the atmosphere in large quantities. Furthermore, cells developed the capacity to break down the water molecules to hydrogen and oxygen by using the energy of *visible* light, which is not blocked by ozone. The hydrogen was combined with carbon dioxide to form the complicated molecules that made up the cell, while the oxygen was liberated into the atmosphere. In this way, thanks to life, earth's atmosphere altered from nitrogen-and-carbon-dioxide to nitrogen-and-oxygen. The greenhouse effect was reduced to very little; the earth remained cool, capable of retaining its unique possession of an ocean of liquid water and an atmosphere containing large quantities of free oxygen.

In fact, our oxygenated atmosphere may be a characteristic only of the last 10 percent of earth's existence; and even as recently as 600 million years ago, our atmosphere may have had only one-tenth as much oxygen as it has now.

But we do have it now, and we may be thankful for the life that made the free atmospheric oxygen possible, and for the life that such oxygen in turn makes possible.

Chapter 6

The Elements

The Periodic Table

So far I have dealt with the sizable bodies of the universe—the stars and galaxies, the solar system, and Earth and its atmosphere. Now let us consider the nature of the substances that compose them all.

EARLY THEORIES

The early Greek philosophers, whose approach to most problems was theoretical and speculative, decided that the earth was made of a very few *elements*, or basic substances. Empedocles of Akragas, about 430 B.C., set the number at four—earth, air, water, and fire. Aristotle, a century later, supposed the heavens to consist of a fifth element, *aether*. The successors of the Greeks in the study of matter, the medieval alchemists, got mired in magic and quackery, but they came to shrewder and more reasonable conclusions than the Greeks because they at least handled the materials they speculated about.

Seeking to explain the various properties of substances, the alchemists attached these properties to certain controlling elements that they added to the list. They identified mercury as the element that imparted metallic properties to substances, and sulfur as the element that imparted the property of flammability. One of the last and best of the alchemists, the sixteenth-century Swiss physician Theophrastus Bombastus von Hohenheim, better known as Paracelsus, added salt as the element that imparted resistance to heat.

The alchemists reasoned that one substance could be changed into another by merely adding and subtracting elements in the proper proportions. A metal such as lead, for instance, might be changed into gold by adding the right amount of mercury to the lead. The search for the precise technique of converting *base metal* to gold went on for centuries. In the process, the alchemists discovered substances vastly more important than gold—such as the mineral acids and phosphorus.

The mineral acids—nitric acid, hydrochloric acid, and, particularly, sulfuric acid (first prepared about 1300)—introduced a virtual revolution in alchemical experiments. These substances were much stronger acids than the strongest previously known (the acetic acid of vinegar); and with them, substances could be decomposed without the use of high temperatures and long waits. Even today, the mineral acids, particularly sulfuric acid, are of vital use in industry. It is said that the extent of the industrialization of a nation can be judged by its annual consumption of sulfuric acid.

Nevertheless, few alchemists allowed themselves to be diverted by these important side issues from what they considered the main quest. Unscrupulous members of the craft indulged in outright fakery, producing gold by sleight-of-hand, to win what we would call today "research grants" from rich patrons. This brought the profession into such disrepute that the very word *alchemist* had to be abandoned. By the seventeenth century, *alchemist* had become *chemist*, and *alchemy* had graduated to a science called *chemistry*.

In the bright birth of science, one of the first of the new chemists was Robert Boyle, the author of Boyle's law of gases (see chapter 5). In his *The Sceptical Chymist*, published in 1661, Boyle first laid down the specific modern criterion of an element: a basic substance that can be combined with other elements to form *compounds*, and that, conversely, cannot be broken down to any simpler substance after it is isolated from a compound.

Boyle retained a medieval view about what the actual elements were, however. For instance, he believed that gold was not an element and could be formed in some way from other metals. So, in fact, did his contemporary Isaac Newton, who devoted a great deal of time to alchemy. (Indeed, Emperor Francis Joseph of Austria-Hungary subsidized experiments for making gold as late as 1867.)

In the century after Boyle, practical chemical work began to make clear which substances could be broken down into simpler substances and which could not. Henry Cavendish showed that hydrogen would combine with oxygen to form water, so water could not be an element. Later Lavoisier resolved the supposed element air into oxygen and nitrogen. It became plain that none of the Greek elements was an element by Boyle's criterion.

As for the elements of the alchemists, mercury and sulfur did indeed turn out to be elements "according to Boyle." But so did iron, tin, lead, copper, silver, gold, and such nonmetals as phosphorus, carbon, and arsenic. And Paracelsus's "element" salt eventually was broken down into two simpler substances.

Of course, the definition of elements depended on the chemistry of the time. As long as a substance could not be broken down by the chemical techniques of the day, it could still be considered an element. For instance, Lavoisier's list of thirty-three elements included such items as lime and magnesia. But fourteen years after Lavoisier's death on the guillotine in the French Revolution, the English chemist Humphry Davy, using an electric current to split the substances, divided lime into oxygen and a new element he called *calcium*, and similarly split magnesia into oxygen and another new element he named *magnesium*.

On the other hand, Davy was able to show that a green gas that the Swedish chemist Carl Wilhelm Scheele had made from hydrochloric acid was not a compound of hydrochloric acid and oxygen, as had been thought, but a true element, and he named it *chlorine* (from the Greek word for "green").

ATOMIC THEORY

At the beginning of the nineteenth century, there developed a radically new way of looking at elements which harked back to some of the Greeks, who had, after all, contributed what has turned out to be perhaps the most important single concept in the understanding of matter.

The Greeks argued about whether matter was continuous or discrete: that is, whether it could be divided and subdivided indefinitely into ever finer dust or would be found in the end to consist of indivisible particles. Leucippus of Miletus and his pupil Democritus of Abdera insisted, about 450 B.C., that the latter was the case. Democritus, in fact, gave the particles a name: he called them *atoms* (meaning "nondivisible"). He even suggested that different substances were composed of different atoms or combinations of atoms, and that one substance could be converted into another by rearranging the atoms. Considering that all this was only an intelligent guess, one is thunderstruck by the correctness of his intuition. Although the idea may seem obvious today, it was so far from obvious at the time that Plato and Aristotle rejected it out of hand.

It survived, however, in the teachings of Epicurus of Samos, who wrote about 300 B.C., and in the philosophic school (Epicureanism) to which he gave rise. An important Epicurean was the Roman philosopher Lucretius, who, about 60 B.C., embodied atomic notions in a long poem *On the Nature of Things.* One battered copy of Lucretius's poem survived through the Middle Ages, and the poem was one of the earliest works to be printed once that technique had been invented.

The notion of atoms thus never entirely passed out of the consciousness of Western scholarship. Prominent among the atomists in the dawn of modern science were the Italian philosopher Giordano Bruno and the French philosopher Pierre Gassendi. Bruno had many unorthodox scientific views, such as a belief in an infinite universe with the stars distant suns about which planets revolved, and expressed himself boldly. He was burned as a heretic in 1600 —the outstanding martyr to science of the Scientific Revolution. The Russians have named a crater on the other side of the moon in his honor.

Gassendi's views impressed Boyle, whose own experiments showing that gases could easily be compressed and expanded seemed to show that these gases must be composed of widely spaced particles. Both Boyle and Newton were therefore among the convinced atomists of the seventeenth century.

In 1799, the French chemist Joseph Louis Proust showed that copper carbonate contained definite proportions by weight of copper, carbon, and oxygen, however it might be prepared. The proportions were in the ratio of small whole numbers: 5 to 4 to 1. He went on to show a similar situation for a number of other compounds.

That situation could best be explained by assuming that compounds are formed by the union of small numbers of bits of each element that could combine only as intact objects. The English chemist John Dalton pointed this out in 1803 and, in 1808, published a book in which all the new chemical information gathered in the past century and a half was shown to make sense if all matter were supposed to be composed of indivisible atoms. (Dalton kept the old Greek word as a tribute to the ancient thinkers.) It did not take long for this *atomic theory* to persuade most chemists.

According to Dalton, each element possesses a particular kind of atom, and any quantity of the element is made up of identical atoms of this kind. What distinguishes one element from another is the nature of its atoms. And the basic physical difference between atoms is in their weight. Thus sulfur atoms are heavier than oxygen atoms, which in turn are heavier than nitrogen atoms; they, in turn, heavier than carbon atoms; and these, in turn, heavier than hydrogen atoms.

The Italian chemist Amedeo Avogadro applied the atomic theory to gases in such a way as to show that it makes sense to suppose that equal volumes of gas (of whatever nature) are made up of equal numbers of particles. This is *Avogadro's hypothesis*. These particles were at first assumed to be atoms but eventually were shown to be composed, in most cases, of small groups of atoms called *molecules*. If a molecule contains atoms of different kinds (like the water molecule, which consists of an oxygen atom and two hydrogen atoms), it is a molecule of a *chemical compound*.

Naturally it became important to measure the relative weights of different atoms—to find the *atomic weights* of the elements, so to speak. The tiny atoms themselves were hopelessly beyond the reach of nineteenth-century weighing techniques. But by weighing the quantity of each element separated from a compound, and making deductions from an element's chemical behavior, it was possible to work out the relative weights of the atoms. The first to go about this systematically was the Swedish chemist Jöns Jacob Berzelius. In 1828, he published a list of atomic weights based on two standards—one giving the atomic weight of oxygen the arbitrary value of 100, the other taking the atomic weight of hydrogen as equal to 1.

Berzelius's system did not catch on at once; but in 1860, at the first International Chemical Congress in Karlsruhe, Germany, the Italian chemist

Stanislao Cannizzaro presented new methods for determining atomic weights, making use of Avogadro's hypothesis, which had hitherto been neglected. Cannizzaro described his views so forcefully that the world of chemistry was won over.

The weight of oxygen rather than hydrogen was adopted as the standard at that time, because oxygen can more easily be brought into combination with various elements (and combination with other elements was the key step in the usual method of determining atomic weights). Oxygen's atomic weight was arbitrarily taken by the Belgian chemist Jean Servais Stas, in 1850, as exactly 16, so that the atomic weight of hydrogen, the lightest known element, would be just about 1—1.0080, to be exact.

Ever since Cannizzaro's time, chemists have sought to work out atomic weights with ever greater accuracy. This reached a climax, as far as purely chemical methods were concerned, in the work of the American chemist Theodore William Richards, who, in 1904 and thereafter, determined the atomic weights with an accuracy previously unapproached. For this he received the Nobel Prize in chemistry in 1914. On the basis of later discoveries about the physical constitution of atoms, Richards's figures have since been corrected to still more refined values.

Throughout the nineteenth century, although much work was done on atoms and molecules, and scientists generally were convinced of their reality, there existed no direct evidence that they were anything more than convenient abstractions. Some prominent scientists, such as the German chemist Wilhelm Ostwald, refused to accept them in any other way. To him, they were useful but not "real."

The reality of molecules was made clear by *Brownian motion*. This was first observed in 1827 by the Scottish botanist Robert Brown, who noted that pollen grains suspended in water jiggled erratically. At first it was thought that the jiggling was because of the life in the pollen grains, but equally small particles of completely inanimate dyes also showed the motion.

In 1863, it was first suggested that the movement was due to unequal bombardment of the particles by surrounding water molecules. For large objects, a slight inequality in the number of molecules striking from left and from right would not matter. For microscopic objects, bombarded by perhaps only a few hundred molecules per second, a few in excess—this side or that —can induce a perceptible jiggle. The random movement of the tiny particles is almost visible proof of the *graininess* of water, and of matter generally.

Einstein worked out a theoretical analysis of this view of Brownian motion and showed how one could work out the size of the water molecules from the extent of the little jiggling movements of the dye particles. In 1908, the French physicist Jean Perrin studied the manner in which particles settle downward through water under the influence of gravity. The settling is opposed by molecular collisions from below, so that a Brownian movement is opposing gravitational pull. Perrin used this finding to calculate the size of the water molecules by means of the equation Einstein had worked out, and even

Ostwald had to give in. For his investigations Perrin received the Nobel Prize for physics in 1926.

So atoms have steadily been translated from semimystical abstractions into almost tangible objects. Indeed, today we can say that we have at last "seen" the atom. This is accomplished with the *field ion microscope*, invented in 1955 by Erwin W. Mueller of Pennsylvania State University. His device strips positively charged ions off an extremely fine needle tip and shoots them to a fluorescent screen in such a way as to produce a 5 million-fold magnified image of the needle tip. This image actually makes the individual atoms composing the tip visible as bright little dots. The technique was improved to the point where images of single atoms could be obtained. The American physicist Albert Victor Crewe reported the detection of individual atoms of uranium and thorium by means of a scanning electron-microscope in 1970.

MENDELEEV'S PERIODIC TABLE

As the list of elements grew in the nineteenth century, chemists began to feel as if they were becoming entangled in a thickening jungle. Every element had different properties, and they could see no underlying order in the list. Since the essence of science is to try to find order in apparent disorder, scientists hunted for some sort of pattern in the properties of the elements.

In 1862, after Cannizzaro had established atomic weight as one of the important working tools of chemistry, a French geologist, Alexandre Emile Béguyer de Chancourtois, found that he could arrange the elements in the order of increasing atomic weight in a tabular form, so that elements with similar properties fell in the same vertical column. Two years later, a British chemist, John Alexander Reina Newlands, independently arrived at the same arrangement. But both scientists were ignored or ridiculed. Neither could get his suggestions properly published at the time. Many years later, after the importance of the periodic table had become universally recognized, their papers were published at last. Newlands even got a medal.

It was the Russian chemist Dmitri Ivanovich Mendeleev who got the credit for finally bringing order into the jungle of the elements. In 1869, he and the German chemist Julius Lothar Meyer proposed tables of the elements, making essentially the same point that de Chancourtois and Newlands had already made. But Mendeleev received the recognition because he had the courage and confidence to push the idea farther than the others.

In the first place, Mendeleev's *periodic table* (so called because it showed the periodic recurrence of similar chemical properties) was more complicated than that of Newlands and nearer what we now believe to be correct (see table 6.1). Second, where the properties of an element placed it out of order according to its atomic weight, Mendeleev boldly switched the order, on the ground that the properties are more important than the atomic weight. He was eventually proved correct, as we shall see later in this chapter. For instance, tellurium, with an atomic weight of 127.61, should, on the weight basis, come after iodine, whose atomic weight is 126.91. But in the columnar

1 Hydrogen (H) 1.008								
3 Lithium (Li) 6.939	**4** Beryllium (Be) 9.012							
11 Sodium (Na) 22.990	**12** Magnesium (Mg) 24.312							
19 Potassium (K) 39.102	**20** Calcium (Ca) 40.08	**21** Scandium (Sc) 44.956	**22** Titanium (Ti) 47.90	**23** Vanadium (V) 50.942	**24** Chromium (Cr) 51.996	**25** Manganese (Mn) 54.938	**26** Iron (Fe) 55.847	**27** Cobalt (Co) 58.933
37 Rubidium (Rb) 85.47	**38** Strontium (Sr) 87.62	**39** Yttrium (Y) 88.905	**40** Zirconium (Zr) 91.22	**41** Niobium (Nb) 92.906	**42** Molybdenum (Mo) 95.94	**43**° Technetium (Tc) 98.91	**44** Ruthenium (Ru) 101.07	**45** Rhodium (Rh) 102.905
55 Cesium (Cs) 132.905	**56** Barium (Ba) 137.34	57 Lanthanum (La) 138.91	58 Cerium (Ce) 140.12	59 Praseodymium (Pr) 140.907	60 Neodymium (Nd) 144.24	61° Promethium (Pm) 145	62 Samarium (Sm) 150.35	63 Europium (Eu) 151.96
			72 Hafnium (Hf) 178.49	**73** Tantalum (Ta) 180.948	**74** Tungsten (W) 183.85	**75** Rhenium (Re) 186.2	**76** Osmium (Os) 190.2	**77** Iridium (Ir) 192.2
87° Francium (Fr) 223	**88**° Radium (Ra) 226.05	89° Actinium (Ac) 227	90° Thorium (Th) 232.038	91° Protactinium (Pa) 231	92° Uranium (U) 238.03	93° Neptunium (Np) 237	94° Plutonium (Pu) 242	95° Americium (Am) 243
			104° Rutherfordium (Rf) 259	**105**° Hahnium (Ha) 260				

Table 6.1. The periodic table of the elements. The shaded areas of the table represent the two rare-earth series: the lanthanides and the actinides, named after their respective first members. The number in the lower right-hand corner of each box indicates the atomic weight of the element. An asterisk marks elements that are radioactive. Each element's atomic number appears at top center of its box.

								2 Helium (He) 4.003
			5 Boron (B) 10.811	6 Carbon (C) 12.011	7 Nitrogen (N) 14.007	8 Oxygen (O) 15.999	9 Fluorine (F) 18.998	10 Neon (Ne) 20.183
			13 Aluminum (Al) 26.982	14 Silicon (Si) 28.086	15 Phosphorus (P) 30.974	16 Sulfur (S) 32.064	17 Chlorine (Cl) 35.453	18 Argon (A) 39.948
28 Nickel (Ni) 58.71	29 Copper (Cu) 63.54	30 Zinc (Zn) 65.37	31 Gallium (Ga) 69.72	32 Germanium (Ge) 72.59	33 Arsenic (As) 74.922	34 Selenium (Se) 78.96	35 Bromine (Br) 79.909	36 Krypton (Kr) 83.80
46 Palladium (Pd) 106.4	47 Silver (Ag) 107.870	48 Cadmium (Cd) 112.40	49 Indium (In) 114.82	50 Tin (Sn) 118.69	51 Antimony (Sb) 121.75	52 Tellurium (Te) 127.60	53 Iodine (I) 126.904	54 Xenon (Xe) 131.30
64 Gadolinium (Gd) 157.25	65 Terbium (Tb) 158.924	66 Dysprosium (Dy) 162.50	67 Holmium (Ho) 164.930	68 Erbium (Er) 167.26	69 Thulium (Tm) 168.934	70 Ytterbium (Yb) 173.04	71 Lutetium (Lu) 174.97	
78 Platinum (Pt) 195.09	79 Gold (Au) 196.967	80 Mercury (Hg) 200.59	81 Thallium (Tl) 204.37	82 Lead (Pb) 207.19	83 Bismuth (Bi) 208.98	84* Polonium (Po) 210	85* Astatine (At) 210	86* Radon (Rn) 222
96* Curium (Cm) 244	97* Berkelium (Bk) 245	98* Californium (Cf) 246	99* Einsteinium (Es) 253	100* Fermium (Fm) 255	101* Mendelevium (Md) 256	102* Nobelium (No) 255	103* Lawrencium (Lw) 257	

table, putting tellurium ahead of iodine places it under selenium, which it closely resembles, and similarly puts iodine under its cousin bromine.

Finally, and most important, where Mendeleev could find no other way to make his arrangement work, he did not hesitate to leave holes in the table and to announce, with what seemed infinite gall, that elements must be discovered that belonged in those holes. He went farther. For three of the holes, he described the element that would fit each, utilizing as his guide the properties of the elements above and below the hole in the table. And here Mendeleev had a stroke of luck. Each of his three predicted elements was found in his own lifetime, so that he witnessed the triumph of his system. In 1875, the French chemist Lecoq de Boisbaudran discovered the first of these missing elements and named it *gallium* (after the Latin name for France). In 1879, the Swedish chemist Lars Fredrik Nilson found the second and named it *scandium* (after Scandinavia). And in 1886, the German chemist Clemens Alexander Winkler isolated the third and named it *germanium* (after Germany, of course). All three elements had almost precisely the properties predicted by Mendeleev.

ATOMIC NUMBERS

With the discovery of X rays by Roentgen, a new era opened in the history of the periodic table. In 1911, the British physicist Charles Glover Barkla discovered that when X rays are scattered by a metal, the scattered rays have a sharply defined penetrating power, depending on the metal; in other words, each element produces its own *characteristic X rays*. For this discovery Barkla was awarded the Nobel Prize in physics for 1917.

There was some question whether X rays were streams of tiny particles or consisted of wavelike radiations after the manner of light. One way of checking was to see whether X rays could be *diffracted* (that is, forced to change direction) by a *diffraction grating* consisting of a series of fine scratches. However, for proper diffraction, the distance between the scratches must be roughly equal to the size of the waves in the radiation. The most finely spaced scratches that could be prepared sufficed for ordinary light, but the penetrating power of X rays made it likely that, if X rays were wavelike, the waves would have to be much smaller than those of light. Therefore, no ordinary diffraction gratings would suffice to diffract X rays.

However, it occurred to the German physicist Max Theodore Felix von Laue that crystals are a natural diffraction grating far finer than any artificial one. A *crystal* is a solid with a neat geometric shape, with its plane faces meeting at characteristic angles, and with a characteristic symmetry. This visible regularity is the result of an orderly array of atoms making up its structure. There were reasons for thinking that the space between one layer of atoms and the next was about the size of an X-ray wavelength. If so, crystals would diffract X rays.

Laue experimented and found that X rays passing through a crystal were indeed diffracted and formed a pattern on a photographic plate that showed them to have the properties of waves. Within the same year, the English physicist William Lawrence Bragg and his equally distinguished father, William Henry Bragg, developed an accurate method of calculating the wavelength of a particular type of X ray from its diffraction pattern. Conversely, X-ray diffraction patterns were eventually used to determine the exact orientation of the atom layers that do the diffracting. In this way, X rays opened the door to a new understanding of the atomic structure of crystals. For their work on X rays, Laue received the Nobel Prize for physics in 1914, while the Braggs shared the Nobel Prize for physics in 1915.

Then, in 1914, the young English physicist Henry Gwyn-Jeffreys Moseley determined the wavelengths of the characteristic X rays produced by various metals, and made the important discovery that the wavelength decreased in a regular manner as one went up the periodic table.

This pinned the elements into definite position in the table. If two elements, supposedly adjacent in the table, yielded X rays that differ in wavelength by twice the expected amount, then there must be a gap between them belonging to an unknown element. If they differ by three times the expected amount, there must be two missing elements. If, on the other hand, the two elements' characteristic X rays differ by only the expected amount, one can be certain that there is no missing element between the two.

It was now possible to give the elements definite numbers. Until then there had always been the possibility that some new discovery might break into the sequence and throw any adopted numbering system out of kilter. Now there could no longer be unsuspected gaps.

Chemists proceeded to number the elements from 1 (hydrogen) to 92 (uranium). These *atomic numbers* were found to be significant in connection with the internal structures of the atoms (see chapter 7) and to be more fundamental than the atomic weight. For instance, the X-ray data proved that Mendeleev had been right in placing tellurium (atomic number 52) before iodine (53), in spite of tellurium's higher atomic weight.

Moseley's new system proved its worth almost at once. The French chemist Georges Urbain, after discovering *lutetium* (named after the old Latin name of Paris), had later announced that he had discovered another element which he called *celtium.* According to Moseley's system, lutetium was element 71 and celtium should be 72. But when Moseley analyzed celtium's characteristic X rays, it turned out to be lutetium all over again. Element 72 was not actually discovered until 1923, when the Danish physicist Dirk Coster and the Hungarian chemist Georg von Hevesy detected it in a Copenhagen laboratory and named it *hafnium* (from the Latinized name of Copenhagen).

Moseley was not present for this verification of the accuracy of his method; he had been killed at Gallipoli in 1915 at the age of twenty-eight—certainly one of the most valuable lives lost in the First World War. Moseley probably lost a Nobel Prize through his early death. The Swedish physicist Karl Manne George Siegbahn extended Moseley's work, discovering new series of X rays and accurately determining X-ray spectra for the various elements. He was awarded the Nobel Prize for physics in 1924.

In 1925, Walter Noddack, Ida Tacke, and Otto Berg of Germany filled another hole in the periodic table. After a three-year search through ores containing elements related to the one they were hunting for, they turned up element 75 and named it *rhenium,* in honor of the Rhine River. This left only four holes: elements 43, 61, 85, and 87.

It was to take two decades to track those four down. Although chemists did not realize it at the time, they had found the last of the stable elements. The missing ones were unstable species so rare on the earth today that all but one of them would have to be created in the laboratory to be identified. And thereby hangs a tale.

Radioactive Elements

IDENTIFYING THE ELEMENTS

After the discovery of X rays in 1895, many scientists were impelled to investigate these new and dramatically penetrating radiations. One of them was the French physicist Antoine-Henri Becquerel. His father, Alexandre Edmond (the physicist who had first photographed the solar spectrum), had been particularly interested in *fluorescence,* which is visible radiation given off by substances after exposure to the ultraviolet rays in sunlight.

The elder Becquerel had, in particular, studied a fluorescent substance called *potassium uranyl sulfate* (a compound made up of molecules each containing an atom of uranium). Henri wondered whether the fluorescent radiations of the potassium uranyl sulfate contained X rays. The way to find out was to expose the sulfate to sunlight (whose ultraviolet light would excite the fluorescence), while the compound lay on a photographic plate wrapped in black paper. Since the sunlight could not penetrate the black paper, it would not itself affect the plate, but, if the fluorescence it excited contained X rays, they *would* penetrate the paper and darken the plate. Becquerel tried the experiment in 1896, and it worked. Apparently there were X rays in the fluorescence. Becquerel even got the supposed X rays to pass through thin sheets of aluminum and copper, and thus seemed to clinch the matter, for no radiation except X rays was known to do this.

But then, by a great stroke of good fortune, although Becquerel undoubtedly did not view it as such at the time, a siege of cloudy weather intervened. Waiting for the return of sunlight, Becquerel put away his photographic plates, with pinches of sulfate lying on them, in a drawer. After several days, he grew impatient and decided to develop his plates anyway, with the thought that even without direct sunlight some trace of X rays might have been produced. When he saw the developed pictures, Becquerel experienced one of those moments of deep astonishment and delight that are the dream of all scientists. The photographic plate was deeply darkened by strong radiation!

Something other than fluorescence or sunlight was responsible for it. Becquerel decided (and experiments quickly proved) that this "something" was the uranium in the potassium uranyl sulfate.

This discovery further electrified scientists, already greatly excited by the recent discovery of the X rays. One of the scientists who at once set out to investigate the strange radiation from uranium was a young Polish-born chemist named Marie Sklodowska, who just the year before had married Pierre Curie, the discoverer of the Curie temperature (see chapter 5).

Pierre Curie, in collaboration with his brother Jacques, had discovered that certain crystals, when put under pressure, develop a positive electric charge on one side and a negative charge on the other. This phenomenon is called *piezoelectricity* (from a Greek word meaning "to press"). Marie Curie decided to measure the radiation given off from uranium by means of piezoelectricity. She set up an arrangement whereby this radiation would ionize the air between two electrodes, a current would then flow, and the strength of this small current would be measured by the amount of pressure that had to be placed on a crystal to produce a balancing countercurrent. This method worked so well that Pierre Curie dropped his own work at once and, for the rest of his life, joined Marie as an eager second.

It was Marie Curie who suggested the term *radioactivity* to describe the ability of uranium to give off radiations, and who went on to demonstrate the phenomenon in a second radioactive substance—thorium. In fast succession, enormously important discoveries were made by other scientists as well. The penetrating radiations from radioactive substances proved to be even more penetrating and more energetic than X rays; they are now called *gamma rays*. Radioactive elements were found to give off other types of radiation also, which led to discoveries about the internal structure of the atom, but this is a story for another chapter (see chapter 7). What has the greatest bearing on this discussion of the elements is the discovery that the radioactive elements, in giving off the radiation, changed to other elements—a modern version of *transmutation*.

Marie Curie was the first to come on the implications of this phenomenon, and she did so accidentally. In testing pitchblende for its uranium content, to see if samples of the ore had enough uranium to be worth the refining effort, she and her husband found, to their surprise, that some of the pieces had more radioactivity than they ought to have even if they had been made of pure uranium. The implication was, of course, that there had to be other radioactive elements in the pitchblende. These unknown elements could only be present in small quantities, because ordinary chemical analysis did not detect them, so they must be very radioactive indeed.

In great excitement, the Curies obtained tons of pitchblende, set up shop in a small shack, and—under primitive conditions and with only their unbeatable enthusiasm to drive them on—they proceeded to struggle through the heavy, black ore for the trace quantities of new elements. By July of 1898, they had isolated a trace of black powder 400 times as intensely radioactive as the same quantity of uranium.

This contained a new element with chemical properties like those of tellurium, and it therefore probably belonged beneath it in the periodic table. (It was later given the atomic number 84.) The Curies named it *polonium*, after Marie's native land.

But polonium accounted for only part of the radioactivity. More work followed; and, by December of 1898, the Curies had a preparation that was even more intensely radioactive than polonium. It contained still another element, which had properties like those of barium (and was eventually placed beneath barium and was found to have the atomic number 88). The Curies called it *radium*, because of its intense radioactivity.

They worked on for four more years to collect enough pure radium so that they could see it. Then Marie Curie presented a summary of her work as her Ph.D. dissertation in 1903. It was probably the greatest doctoral dissertation in scientific history. It earned her not one but two Nobel Prizes. Marie and her husband, along with Becquerel, received the Nobel Prize for physics in 1903 for their studies of radioactivity; and, in 1911, Marie alone (her husband having died in a traffic accident in 1906) was awarded the Nobel Prize for chemistry for the discovery of polonium and radium.

Polonium and radium are far more unstable than uranium or thorium, which is another way of saying that they are far more radioactive. More of their atoms break down each second. Their lifetimes are so short that practically all the polonium and radium in the universe should have disappeared within a matter of a million years or so. Why do we still find them in the billions-of-years-old earth? The answer is that radium and polonium are continually being formed in the course of the breakdown of uranium and thorium to lead. Wherever uranium and thorium are found, small traces of polonium and radium are likewise to be found. They are intermediate products on the way to lead as the end product.

Three other unstable elements on the path from uranium and thorium to lead were discovered by means of the careful analysis of pitchblende or by researches into radioactive substances. In 1899, André Louis Debierne, on the advice of the Curies, searched pitchblende for other elements and came up with one he called *actinium* (from the Greek word for "ray"), which eventually received the atomic number 89. The following year, the German physicist Friedrich Ernst Dorn demonstrated that radium, when it broke down, formed a gaseous element. A radioactive gas was something new! Eventually the element was named *radon* (from radium and from argon, its chemical cousin) and given the atomic number 86. Finally, in 1917, two different groups—Otto Hahn and Lise Meitner in Germany, and Frederick Soddy and John Arnold Cranston in England—isolated from pitchblende element 91, named *protactinium*.

FINDING THE MISSING ELEMENTS

By 1925, then, the score stood at eighty-eight identified elements—eighty-one stable and seven unstable. The search for the missing four—numbers 43, 61, 85, 87—became avid indeed.

Since all the known elements from number 84 to 92 were radioactive, it was confidently expected that 85 and 87 would be radioactive as well. On the other hand, 43 and 61 were surrounded by stable elements, and there seemed no reason to suspect that they were not themselves stable; consequently, they should be found in nature.

Element 43, lying just above rhenium in the periodic table, was expected to have similar properties and to be found in the same ores. In fact, the team of Noddack, Tacke, and Berg, who had discovered rhenium, felt certain that they had also detected X rays of a wavelength that went along with element 43. So they announced its discovery, too, and named it *masurium*, after a region in East Prussia. However, their identification was not confirmed: and in science, a discovery is not a discovery unless and until it has been confirmed by at least one independent researcher.

In 1926, two University of Illinois chemists announced that they had found element 61 in ores containing its neighboring elements (60 and 62), and they named their discovery *illinium*. The same year, a pair of Italian chemists at the University of Florence thought that they had isolated the same element and named it *florentium*. But other chemists could not confirm the work of either group.

A few years later, an Alabama Polytechnic Institute physicist, using a new analytical method of his own devising, reported that he had found small traces of element 87 and of element 85; he called them *virginium* and *alabamine*, after his native and adopted states, respectively. But these discoveries could not be confirmed either.

Events were to show that the "discoveries" of elements 43, 61, 85, and 87 had been mistaken.

The first of the four to be identified beyond doubt was element 43. The American physicist Ernest Orlando Lawrence, who was to receive the Nobel Prize in physics for his invention of the cyclotron (see chapter 7), made the element in his accelerator by bombarding *molybdenum* (element 42) with high-speed particles. His bombarded material developed radioactivity, and Lawrence sent it for analysis to the Italian chemist Emilio Gino Segrè, who was interested in the element-43 problem. Segrè and his colleague Carlo Perrier, after separating the radioactive part from the molybdenum, found that it resembled rhenium in its properties but was not rhenium. They decided that it could only be element 43, and that element 43, unlike its neighbors in the periodic table, was radioactive. Because it is not being produced as a breakdown product of a higher element, virtually none of it is left in the earth's crust, and so Noddack and company were undoubtedly mistaken in thinking they had found it. Segrè and Perrier eventually were given the privilege of naming element 43; they called it *technetium*, from a Greek word meaning "artificial," because it was the first laboratory-made element. By 1960, enough technetium had been accumulated to determine its melting point—close to 2200° C. (Segrè was later to receive a Nobel Prize for quite another discovery, having to do with another laboratory-made bit of matter—see chapter 7.)

In 1939, element 87 was finally discovered in nature. The French chemist Marguerite Perey isolated it from among the breakdown products of uranium. Element 87 was present in extremely small amounts, and only improvements in technique enabled it to be found where earlier it had been missed. She later named the new element *francium*, after her native land.

Element 85, like technetium, was produced in the cyclotron, by bombardment of *bismuth* (element 83). In 1940, Segrè, Dale Raymond Corson, and Kenneth Ross MacKenzie isolated element 85 at the University of California, Segrè having by then emigrated from Italy to the United States. The Second World War interrupted their work on the element, but after the war they returned to it and, in 1947, proposed the name *astatine* for the element, from a Greek word meaning "unstable." (By that time, tiny traces of astatine had, like francium, been found in nature among the breakdown products of uranium.)

Meanwhile, the fourth and final missing element, number 61, had been discovered among the products of the fission of uranium, a process that is explained in chapter 10. (Technetium, too, turned up among these products.) Three chemists at the Oak Ridge National Laboratory—J. A. Marinsky, L. E. Glendenin, and Charles DuBois Coryell—isolated element 61 in 1945. They named it *promethium*, after the Greek demigod Prometheus, who had stolen fire for mankind from the sun. Element 61, after all, had been stolen from sunlike fires of the atomic furnace.

So the list of elements, from 1 to 92, was at last complete. And yet, in a sense, the strangest part of the adventure had only begun. For scientists had broken through the bounds of the periodic table; uranium was not the end.

TRANSURANIUM ELEMENTS

A search for elements beyond uranium—*transuranium elements*—had actually begun as early as 1934. Enrico Fermi in Italy had found that when he bombarded an element with a newly discovered subatomic particle called the *neutron* (see chapter 7), this often transformed the element into the one of the next higher atomic number. Could uranium be built up to element 93— a totally synthetic element that, as far as was then known, did not exist in nature? Fermi's group proceeded to attack uranium with neutrons and got a product that they thought was indeed element 93. They called it *uranium X*.

In 1938, Fermi received the Nobel Prize in physics for his studies in neutron bombardment. At the time, the real nature of his discovery, or its consequences for humanity, was not even suspected. Like that other Italian, Columbus, he had found, not what he was looking for, but something far more important of which he was not aware.

Suffice it to say here that, after a series of chases up a number of false trails, it was finally discovered that what Fermi had done was, not to create a new element, but to split the uranium atom into two nearly equal parts. When physicists turned in 1940 to studies of this process, element 93 cropped up as an almost casual result of their experiments. In the mélange of elements that came out of the bombardment of uranium by neutrons, there was one that at first defied identification. Then it dawned on Edwin McMillan of the University of California that perhaps the neutrons released by fission had converted some of the uranium atoms to a higher element, as Fermi had hoped would happen. McMillan and Philip Abelson, a physical chemist, were able to prove that the unidentified element was in fact element 93. The proof of its existence lay in the nature of its radioactivity, as was to be the case in all subsequent discoveries.

McMillan suspected that another transuranium element might be mixed with element 93. The chemist Glenn Theodore Seaborg, together with his co-workers Arthur Charles Wahl and Joseph William Kennedy, soon showed that this was indeed fact, and that the element was number 94.

Since uranium, the supposed end of the periodic table, had been named, at the time of its discovery, for the then newly discovered planet, Uranus, elements 93 and 94 were now named for Neptune and Pluto, planets discovered after Uranus: *neptunium* and *plutonium*, respectively. It turned out that these elements exist in nature, for small traces of neptunium and plutonium were later found in uranium ores. So uranium was not the heaviest natural element after all.

Seaborg and a group at the University of California, in which Albert Ghiorso was prominent, went on to build more transuranium elements, one after the other. In 1944, by bombarding plutonium with subatomic particles, they created elements 95 and 96, named, respectively, *americium* (after America) and *curium* (after the Curies). When they had manufactured a sufficient quantity of americium and curium to work with, they bombarded those elements and successfully produced number 97 in 1949 and number 98 in 1950. These they named *berkelium* and *californium*, after Berkeley and California. In 1951, Seaborg and McMillan shared the Nobel Prize in chemistry for this train of achievements.

The next elements were discovered in more catastrophic fashion. Elements 99 and 100 emerged in the first hydrogen-bomb explosion, detonated in the Pacific in November 1952. Although their existence was detected in the explosion débris, the elements were not confirmed and named until after the University of California group made small quantities of both in the laboratory in 1955. The names given them were *einsteinium* and *fermium*, for Albert Einstein and Enrico Fermi, both of whom had died some months before. Then the group bombarded a small quantity of einsteinium and formed element 101, which they called *mendelevium*, after Mendeleev.

The next step came through a collaboration between California and the Nobel Institute in Sweden. The institute carried out a particularly complicated type of bombardment that produced a small quantity of element 102. It was named *nobelium,* in honor of the institute. The element has been formed by methods other than those described by the first group of workers, so that there was a delay before nobelium was officially accepted as the name of the element.

In 1961, a few atoms of element 103 were detected at the University of California, and it was given the name *lawrencium,* after E. O. Lawrence, who had recently died. In 1964, a group of Soviet scientists under Georgii Nikolaevich Flerov reported the formation of element 104, and, in 1967, the formation of element 105. In both cases, the methods used to form the elements could not be confirmed, and American teams under Albert Ghiorso formed them in other ways. The Soviet group named 104 *kurchatovium,* after Igor Vasilievich Kurchatov, who had led the Soviet team who developed their atomic bomb, and had died in 1960. The American group named 104 *rutherfordium* and 105 *hahnium,* after Ernest Rutherford and Otto Hahn, both of whom made key discoveries in subatomic structure. Elements as high as 109 have been reported.

SUPER-HEAVY ELEMENTS

Each step in this climb up the transuranium scale was harder than the one before. At each successive stage, the element became harder to accumulate and more unstable. When mendelevium was reached, identification had to be made on the basis of seventeen atoms, no more. Fortunately, radiation-detecting techniques were marvelously refined by 1955. The Berkeley scientists actually hooked up their instruments to a firebell, so that every time a mendelevium atom was formed, the characteristic radiation it emitted on breaking down announced the event by a loud and triumphant ring of the bell. (The fire department soon put a stop to this.)

The higher elements were detected under even more rarefied conditions. A single atom of a desired element can be detected by noting the breakdown products in detail.

Is there any point in trying to go ever farther, aside from the thrill of breaking a record and getting your name in the record book as the discoverer of an element? (Lavoisier, the greatest of all chemists, never managed such a discovery, and his failure bothered him greatly.)

One important possible discovery remains to be made. The increase in instability as one goes up the scale of atomic numbers is not uniform. The most complex of the stable atoms is bismuth (83). After it, the six elements from 84 to 89 inclusive are so unstable that any amount present at the time of the formation of the earth would be all gone by now. And then, rather surprisingly, there follow thorium (90) and uranium (92) which are almost stable. Of the original thorium and uranium existing on Earth at the time of its formation, 80 percent of the former and 50 percent of the latter still exist today. Physicists have worked out theories of atomic structure to account for this (as I shall explain in the next chapter); and if those theories are correct, then elements 110 and 114 ought to be more stable than would be expected from their high atomic numbers. There is, therefore, considerable interest in getting to these elements, as a way of testing the theories.

In 1976, there was a report that certain *halos* (circular black markings in mica) might indicate the presence of these *super-heavy elements.* The halos arise from the radiation given off by small bits of thorium and uranium, but there are a few extra-large halos that must arise from more energetically radioactive atoms that are yet sufficiently stable to have persisted down to modern times. These might be the super-heavies. Unfortunately the deductions were not supported by scientists generally, and the suggestion was dropped. Scientists are still looking.

Electrons

When Mendeleev and his contemporaries found that they could arrange the elements in a periodic table composed of families of substances showing similar properties, they had no notion why the elements fell into such groups or why the properties were related. Eventually a clear and rather simple answer emerged, but it came only after a long series of discoveries that at first seemed to have nothing to do with chemistry.

It all began with studies of electricity. Faraday performed every experiment with electricity he could think of, and one of the things he tried to do was to send an electric discharge through a vacuum. He was not able to get a vacuum good enough for the purpose. But, by 1854, a German glass blower named Heinrich Geissler had invented an adequate vacuum pump and produced a glass tube enclosing metal electrodes in an unprecedentedly good vacuum. When experimenters succeeded in producing electric discharges in the *Geissler tube,* they noticed that a green glow appeared on the tube wall opposite the negative electrode. The German physicist Eugen Goldstein suggested, in 1876, that this green glow was caused by the impact on the glass of some sort of radiation originating at the negative electrode, which Faraday had named the *cathode.* Goldstein called the radiation *cathode rays.*

Were the cathode rays a form of electromagnetic radiation? Goldstein thought so, but the English physicist William Crookes and some others said no: they were a stream of particles of some kind. Crookes designed improved versions of the Geissler tube (called *Crookes tubes*), and with these he was able to show that the rays were deflected by a magnet. Thus, they were probably made up of electrically charged particles.

In 1897, the physicist Joseph John Thomson settled the question beyond doubt by demonstrating that the cathode rays could also be deflected by electric charges. What, then, were these cathode "particles"? The only negatively charged particles known at the time were the negative ions of atoms. Experiments showed that the cathode-ray particles could not possibly be such ions, for they were so strongly deflected by an electromagnetic field that they must have an unthinkably high electric charge or else must be extremely light particles with less than 1/1,000 the mass of a hydrogen atom. The latter interpretation turned out to fit the evidence best. Physicists had already guessed that the electric current was carried by particles, and so these cathode-ray particles were accepted as the ultimate particles of electricity. They were called *electrons*—a name that had been suggested in 1891 by the Irish physicist George Johnstone Stoney. The electron was finally determined to have 1/1,837 the mass of a hydrogen atom. (For establishing its existence, Thomson was awarded the Nobel Prize in physics in 1906.)

The discovery of the electron at once suggested that it might be a subparticle of the atom—in other words, that atoms were not the ultimate, indivisible units of matter that Democritus and John Dalton had pictured them to be.

This was a hard pill to swallow, but the lines of evidence converged inexorably. One of the most convincing items was Thomson's showing that negatively charged particles that came out of a metal plate when it was struck by ultraviolet radiation (the *photoelectric effect*) were identical with the electrons of the cathode rays. The photoelectric electrons must have been knocked out of the atoms of the metal.

THE PERIODICITY OF THE PERIODIC TABLE

Since electrons could easily be removed from atoms (by other means as well as by the photoelectric effect), it was natural to conclude that they were located in the outer regions of the atom. If so, there must be a positively charged region within the atom balancing the electrons' negative charges, because the atom as a whole was normally neutral. It was at this point that investigators began to close in on the solution of the mystery of the periodic table.

To remove an electron from an atom takes a little energy. Conversely, when an electron falls into the vacated place in the atom, it must *give up* an equal amount of energy. (Nature is usually symmetrical, especially when it comes to considerations of energy.) This energy is released in the form of electromagnetic radiation. Now, since the energy of radiation is measured in terms of wavelength, the wavelength of the radiation emitted by an electron falling into a particular atom will indicate the force with which the electron is held by that atom. The energy of radiation increases with shortening wavelength: the greater the energy, the shorter the wavelength.

We arrive, then, at Moseley's discovery that metals (that is, the heavier elements) produced X rays, each at a characteristic wavelength, which decreased in regular fashion as one went up the periodic table. Each successive element, it seemed, held its electrons more strongly than the one before—another way of saying that each had a successively stronger positive charge in its internal region.

Assuming that each unit of positive charge corresponded to the negative charge on an electron, it followed that the atom of each successive element must have one more electron than the one before. The simplest way of picturing the periodic table, then, was to suppose that the first element, hydrogen, had 1 unit of positive charge and 1 electron; the second element, helium, 2 positive charges and 2 electrons; the third, lithium, 3 positive charges and 3 electrons; and so on all the way up to uranium, with 92 positive charges and 92 electrons. So the atomic numbers of the elements turned out to represent the number of electrons in their intact atoms.

One more major clue and the atomic scientists had the answer to the periodicity of the periodic table. It developed that the electronic radiation of a given element is not necessarily restricted to a single wavelength; it might emit radiations at two, three, four, or even more different wavelengths. These sets of radiations were named the *K-series*, the *L-series*, the *M-series*, and so on. The investigators concluded that the electrons are arrayed in *shells* around the positively charged core of the atom. The electrons of the innermost shell are most strongly held, and their removal takes the most energy. An electron falling into this shell would emit the most energetic radiation, that is, of the shortest wavelengths, or the K-series. The electrons of the next innermost shell are responsible for the L-series of radiations; the next shell produces the M-series; and so on. Consequently, the shells were called the *K-shell*, the *L-shell*, the *M-shell*, and so on.

By 1925, the Austrian physicist Wolfgang Pauli advanced his *exclusion principle*, which explained just how electrons are distributed within each shell, since no two electrons can possess, according to this principle, exactly the same values of *quantum numbers*. For this work, Pauli received the Nobel Prize for physics in 1945.

THE NOBLE, OR INERT, GASES

In 1916, the American chemist Gilbert Newton Lewis worked out the kinships of properties and the chemical behavior of some of the simpler elements on the basis of their shell structure. There was ample evidence, to begin with, that the innermost shell was limited to two electrons by Pauli's exclusion principle. Hydrogen has only one electron; therefore the shell is unfilled. The atom's tendency is to fill this K-shell, and it can do so in a number of ways. For instance, two hydrogen atoms can pool their single electrons and, by sharing the two electrons, mutually fill their K-shells. Hence, hydrogen gas almost always exists in the form of a pair of atoms—the hydrogen molecule. To separate the two atoms and free them as *atomic hydrogen* takes a good deal of energy. Irving Langmuir of the General Electric Company, who independently worked out a similar scheme involving electrons and chemical behavior, presented a practical demonstration of the strong tendency of the hydrogen atom to keep its electron shell filled. He made an *atomic hydrogen torch* by blowing hydrogen gas through an electric arc, which split the molecules' atoms apart; when the atoms recombined after passing the arc, they liberated the energy they had absorbed in splitting apart, and thus yielded temperatures up to 3400° C!

In helium, element 2, the K-shell is filled with 2 electrons; helium atoms therefore are stable and do not combine with other atoms. When we come to lithium, element 3, we find that 2 of its electrons fill the K-shell, and the third starts the L-shell. The succeeding elements add electrons to this shell one by one: beryllium has 2 electrons in the L-shell, boron has 3, carbon 4, nitrogen 5, oxygen 6, fluorine 7, and neon 8. Eight is the limit for the L-shell, as Pauli had shown; and therefore neon corresponds to helium in having its outermost electron shell filled. And sure enough, it, too, is an inert gas with properties like helium's.

Every atom with an unfilled outer shell has a tendency to enter into combination with other atoms in such a manner as to leave it with a filled outer shell. For instance, the lithium atom readily surrenders its one L-shell electron so that its outer shell is the filled K, while fluorine tends to seize an electron to add to its seven and complete the L-shell. Therefore lithium and fluorine have an affinity for each other; when they combine, lithium donates its L-electron to fluorine to fill the latter's L-shell. Since the atoms' interior positive charges do not change, lithium, with one electron subtracted, now carries a net positive charge, while fluorine, with one extra electron, carries a net negative charge. The mutual attraction of the opposite charges holds the two ions together. The compound is called lithium fluoride (see figure 6.1).

L-shell electrons can be shared as well as transferred. For instance, each of two fluorine atoms can share one of its electrons with the other, so that each atom has a total of eight in its L-shell, counting the two shared electrons. Similarly, two oxygen atoms will pool a total of four electrons to complete their L-shells; and two nitrogen atoms will share a total of six. Thus fluorine, oxygen, and nitrogen all form two-atom molecules.

The carbon atom, with only four electrons in its L-shell, will share each of them with a different hydrogen atom, thereby filling the K-shells of the four hydrogen atoms and in turn filling its own L-shell by sharing *their* electrons. This stable arrangement is the methane molecule, CH_4.

In the same way, a nitrogen atom will share electrons with three hydrogen atoms to form ammonia; an oxygen atom will share electrons with two hydrogen atoms to form water; a carbon atom will share electrons with two oxygen atoms to form carbon dioxide; and so on. Almost all the components formed by the elements in the first part of the periodic table can be accounted for on the basis of this tendency to complete the outermost shell by giving up electrons, accepting electrons, or sharing electrons.

The element after neon—sodium—has 11 electrons, and the eleventh must start a third shell. Then follow magnesium, with 2 electrons in the M-shell, aluminum with 3, silicon with 4, phosphorus with 5, sulfur with 6, chlorine with 7, and argon with 8.

Now each element in this group corresponds to one in the preceding series. Argon, with 8 electrons in the M-shell, is like neon (with 8 electrons in the

FLUORINE   LITHIUM

FLUORINE   FLUORINE

Figure 6.1. Transfer and sharing of electrons. Lithium transfers the electron
in its outer shell to fluorine in the combination of lithium fluoride; each atom
then has a full outer shell. In the fluorine molecule, two electrons are shared,
filling both atoms' outer shells.

L-shell) and is an inert gas. Chlorine, having 7 electrons in its outer shell,
resembles fluorine closely in chemical properties. Likewise, silicon resembles
carbon; sodium resembles lithium; and so on.

So it goes right through the periodic table. Since the chemical behavior of
every element depends on the configuration of electrons in its outermost
shell, all those with, say, one electron in the outer shell will react in much
the same way chemically. Thus, all the elements in the first column of the
periodic table—lithium, sodium, potassium, rubidium, cesium, and even the
radioactive element francium—are remarkably alike in their chemical properties. Lithium has 1 electron in the L-shell, sodium 1 in the M-shell, potassium 1 in the N-shell, rubidium 1 in the O-shell, cesium 1 in the P-shell, and
francium 1 in the Q-shell. Again, all the elements with 7 electrons in their
respective outer shells—fluorine, chlorine, bromine, iodine, and astatine—
resemble one another. The same is true of the last column in the table—the

closed-shell group that includes helium, neon, argon, krypton, xenon, and radon.

The Lewis-Langmuir concept works so well that it still serves in its original form to account for the more simple and straightforward varieties of behavior among the elements. However, not all the behavior is quite as simple and straightforward as might be thought.

For instance, each of the inert gases—helium, neon, argon, krypton, xenon, and radon—has eight electrons in the outermost shell (except for helium, which has two electrons in its only shell), and this is the most stable possible situation. Atoms of these elements have a minimum tendency to lose or gain electrons and therefore a minimum tendency to engage in chemical reactions. The gases would be *inert,* as their name proclaims.

However, a "minimum tendency" is not really the same as "no tendency," but most chemists forgot this truth and acted as though it was ultimately impossible for the inert gases to form compounds. This was not true of all of them. As long ago as 1932, the American chemist Linus Pauling considered the ease with which electrons could be removed from different elements, and noted that all elements without exception, even the inert gases, can be deprived of electrons. This deprivation, however, requires more energy in the case of the inert gases than in that of other elements near them in the periodic table.

The amount of energy required to remove electrons among the elements in any particular family decreases with increasing atomic weight, and the heaviest inert gases, xenon and radon, do not have unusually high requirements. It is no more difficult to remove an electron from a xenon atom, for instance, than from an oxygen atom.

Pauling therefore predicted that the heavier inert gases might form chemical compounds with elements that are particularly prone to accept electrons. The element most eager to accept electrons is fluorine, and that seemed to be the natural target.

Now radon, the heaviest inert gas, is radioactive and is unavailable in any but trace quantities. Xenon, however, the next heaviest, is stable and occurs in small quantities in the atmosphere. The best chance, therefore, would be to attempt to form a compound between xenon and fluorine. However, for thirty years nothing was done in this respect, chiefly because xenon was expensive and fluorine very hard to handle, and chemists felt they had better things to do than chase this particular will-o'-the-wisp.

In 1962, however, the British-Canadian chemist Neil Bartlett—working with a new compound, platinum hexafluoride (PtF_6)—found that it was remarkably avid for electrons, almost as much as was fluorine itself. This compound would take electrons away from oxygen, an element that is normally avid to gain electrons rather than lose them. If PtF_6 could take electrons from oxygen, it ought to be able to take them from xenon, too. The experiment was tried, and xenon fluoroplatinate ($XePtF_6$), the first compound of an inert gas, was reported.

Other chemists at once sprang into the fray, and a number of xenon compounds with fluorine, with oxygen, or with both were formed, the most stable being xenon difluoride (XeF_2). A compound of krypton and fluorine, krypton tetrafluoride (KrF_4), has also been formed, as well as a radon fluoride. Compounds with oxygen were also formed. There were, for instance, xenon oxytetrafluoride ($XeOF_4$), xenic acid (H_2XeO_4), and sodium perxenate (Na_4XeO_6). Most interesting, perhaps, was xenon trioxide (Xe_2O_3), which explodes easily and is dangerous. The smaller inert gases—argon, neon, and helium—are more resistant to sharing their electrons than the larger ones, and remain inert for all chemists can do even yet.

Chemists quickly recovered from the initial shock of finding that the inert gases can form compounds: such compounds fit into the general picture after all. Consequently, there is now a reluctance to speak of the gases as *inert gases*. The alternate name of *noble gases* is preferred, and one speaks of *noble gas compounds* and *noble gas chemistry*. (I think this is a change for the worse. After all, the gases are still inert, even if not completely so. The concept *noble*, in this context, implies "standoffish" or "disinclined to mix with the common herd," and is just as inappropriate as *inert* and, moreover, does not suit a democratic society.)

THE RARE-EARTH ELEMENTS

In addition to the fact that the Lewis-Langmuir scheme was applied too rigidly to the inert gases, it can scarcely be applied at all to many of the elements with atomic numbers higher than 20. In particular, refinements had to be added to deal with a very puzzling aspect of the periodic table having to do with the so-called *rare earths*—elements 57 to 71, inclusive.

To go back a bit, the early chemists considered any substance that was insoluble in water and unchanged by heat to be an *earth* (a hangover of the Greek view of "earth" as an element). Such substances included what we would today call calcium oxide, magnesium oxide, silicon dioxide, ferric oxide, aluminum oxide, and so on—compounds that actually constitute about 90 percent of the earth's crust. Calcium oxide and magnesium oxide are slightly soluble and, in solution, display *alkaline* properties (that is, opposite to those of acids), and so were called the *alkaline earths;* when Humphry Davy isolated the metals calcium and magnesium from these earths, they were named *alkaline earth metals*. The same name was eventually applied to all the elements that fall into the column of the periodic table containing magnesium and calcium: that is, to beryllium, strontium, barium, and radium.

The puzzle to which I have referred began in 1794, when a Finnish chemist, Johan Gadolin, examined an odd rock that had been found near the Swedish hamlet Ytterby and decided that it was a new "earth." Gadolin gave this "rare earth" the name *yttria*, after Ytterby. Later the German chemist Martin Heinrich Klaproth found that yttria could be divided into two "earths," for one of which he kept the name yttria, while he named the other *ceria* (after the newly discovered planetoid Ceres). But the Swedish chemist Carl Gustav

Mosander subsequently broke these down further into a series of different earths. All eventually proved to be oxides of new elements named the *rare-earth metals*. By 1907, fourteen such elements had been identified. In order of increasing atomic weight they are:

lanthanum (from a Greek word meaning "hidden")
cerium (from Ceres)
praseodymium (from the Greek for "green twin," after a green line in its spectrum)
neodymium ("new twin")
samarium (from "samarskite," the mineral in which it was found)
europium (from Europe)
gadolinium (in honor of Johan Gadolin)
terbium (from Ytterby)
dysprosium (from a Greek word meaning "hard to get at")
holmium (from Stockholm)
erbium (from Ytterby)
thulium (from Thule, an old name for Scandinavia)
ytterbium (from Ytterby)
lutetium (from Lutetia, an old name for Paris).

On the basis of their X-ray properties, these elements were assigned the atomic numbers from 57 (lanthanum) to 71 (lutetium). As I related earlier, there was a gap at 61 until the missing element, promethium, emerged from the fission of uranium. It made the fifteenth in the list.

Now the trouble with the rare-earth elements is that they apparently cannot be made to fit into the periodic table. It is fortunate that only four of them were definitely known when Mendeleev proposed the table; if they had all been on hand, the table might have been altogether too confusing to be accepted. There are times, even in science, when ignorance is bliss.

The first of the rare-earth metals, lanthanum, matches up all right with yttrium, number 39, the element above it in the table (figure 6.2). (Yttrium, though found in the same ores as the rare earths and similar to them in properties, is not a rare-earth metal. It is, however, named after Ytterby. Four elements honor that hamlet—which is overdoing it.) The confusion begins with the rare earth after lanthanum—namely, cerium—which ought to resemble the element following yttrium—that is, zirconium. But it does nothing of the sort; instead, it resembles yttrium again. And the same is true of all fifteen of the rare-earth elements: they strongly resemble yttrium and one another (in fact, they are so alike chemically that at first they could not be separated except by the most tedious procedures), but they are not related to any other elements preceding them in the table. We have to skip the whole rare-earth group and go on to hafnium, element 72, to find the element related to zirconium, the one after yttrium.

Baffled by this state of affairs, chemists could do no better than to group all the rare-earth elements into one box beneath yttrium and list them individually in a kind of footnote to the table.

Figure 6.2. The electron shells of lanthanum. Note that the fourth subshell of the N-shell has been skipped and is empty.

THE TRANSITIONAL ELEMENTS

The answer to the puzzle finally came as a result of details added to the Lewis-Langmuir picture of the electron-shell structure of the elements.

In 1921, C. R. Bury suggested that the shells were not necessarily limited to 8 electrons apiece. Eight always sufficed to satisfy the outer shell. But a shell might have a greater capacity when it was not on the outside. As one shell built on another, the inner shells might absorb more electrons, and each succeeding shell might hold more than the one before. Thus the K-shell's total capacity would be 2 electrons, the L-shell's 8, the M-shell's 18, the N-shell's 32, and so on—the step-ups going according to a pattern of successive squares multiplied by two (that is, 2×1, 2×4, 2×9, 2×16, etc.).

This view was backed up by a detailed study of the spectra of the elements. The Danish physicist Niels Henrik David Bohr showed that each electron shell was made up of subshells at slightly different energy levels. In each succeeding shell, the spread of the subshells was greater, so that soon the shells overlapped. As a result, the outermost subshell of an interior shell (say, the M-shell) might actually be farther from the center, so to speak, than the innermost subshell of the next shell beyond it (say, the N-shell). This being so, the N-shell's inner subshell might fill with electrons while the M-shell's outer subshell was still empty.

An example will make this clearer. The M-shell, according to the theory, is divided into three subshells, whose capacities are 2, 6, and 10 electrons, respectively, making a total of 18. Now argon, with 8 electrons in its M-shell, has filled only two inner subshells. And, in fact, the M-shell's third, or outermost, subshell will not get the next electron in the element-building process, because it lies beyond the innermost subshell of the N-shell: that is, in potassium, the element after argon, the nineteenth electron goes, not into the outermost subshell of M, but into the innermost subshell of N. Potassium, with 1 electron in its N-shell, resembles sodium, which has 1 electron in its M-shell. Calcium, the next element (20), has 2 electrons in the N-shell and resembles magnesium, which has 2 in the M-shell. But now the innermost subshell of the N-shell, having room for only 2 electrons, is full. The next electrons to be added can start filling the outermost subshell of the M-shell, which so far has not been touched. Scandium (21) begins the process, and zinc (30) completes it. In zinc, the outermost subshell of the M-shell has at last acquired its complement of 10 electrons. The 30 electrons of zinc are distributed as follows: 2 in the K-shell, 8 in the L-shell, 18 in the M-shell, and 2 in the N-shell. At this point, electrons can resume the filling of the N-shell. The next electron gives the N-shell 3 electrons and forms gallium (31), which resembles aluminum, with 3 in the M-shell.

The point is that elements 21 to 30, formed on the road to filling a subshell that had been skipped temporarily, are *transitional* elements. Note that calcium resembles magnesium, and gallium resembles aluminum. Now magnesium and aluminum are adjacent members of the periodic table (numbers 12 and 13). But calcium (20) and gallium (31) are not. Between them lie the transitional elements, and these introduce a complication in the periodic table.

The N-shell is larger than the M-shell and is divided into four subshells instead of three: they can hold 2, 6, 10, and 14 electrons, respectively. Krypton, element 36, fills the two innermost subshells of the N-shell, but here the innermost subshell of the overlapping O-shell intervenes, and, before electrons can go on to N's two outer subshells, they must fill that one. The element after krypton, rubidium (37), has its thirty-seventh electron in the O-shell. Strontium (38) completes the filling of the two-electron O-subshell. Thereupon a new series of transitional elements proceeds to fill the skipped third subshell of the N-shell. With cadmium (48) this is completed; now N's fourth and outermost subshell is skipped, while electrons fill O's second innermost subshell, ending with xenon (54).

But even now N's fourth subshell must bide its turn; for by this stage, the overlapping has become so extreme that even the P-shell interposes a subshell that must be filled before N's last. After xenon come cesium (55) and barium (56), with 1 and 2 electrons, respectively, in the P-shell. It is still not N's turn: the fifty-seventh electron, surprisingly, goes into the third subshell of the O-shell, creating the element lanthanum (figure 6.3). Then, and only then, an electron at long last enters the outermost subshell of the N-shell. One by one the rare-earth elements add electrons to the N-shell until element 71, lutetium, finally fills it. Lutetium's electrons are arranged thus: 2 in the K-shell,

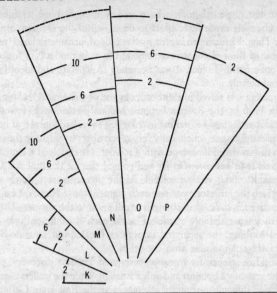

Figure 6.3. Schematic representation of the overlapping of electron shells and subshells in lanthanum. The outermost subshell of the N-shell has yet to be filled.

8 in the L-shell, 18 in the M-shell, 32 in the N-shell, 9 in the O-shell (two subshells full plus 1 electron in the next subshell), and 2 in the P-shell (innermost subshell full).

Now at last we begin to see why the rare-earth elements, and some other groups of transitional elements, are so alike. The decisive thing that differentiates elements, as far as their chemical properties are concerned, is the configuration of electrons in their outermost shell. For instance, carbon, with four electrons in its outermost shell, and nitrogen, with five, are completely different in their properties. On the other hand, in sequences where electrons are busy filling inner subshells while the outermost shell remains unchanged, the properties vary less. Thus iron, cobalt, and nickel (elements 26, 27, and 28), all of which have the same outer-shell electronic configuration—an N-subshell filled with two electrons—are a good deal alike in chemical behavior. Their internal electronic differences (in an M-subshell) are largely masked by their surface electronic similarity. And this goes double for the rare-earth elements. Their differences (in the N-shell) are buried under, not one, but two outer electronic configurations (in the O-shell and the P-shell), which in all these elements are identical. Small wonder that the elements are chemically as alike as peas in a pod.

Because the rare-earth metals have so few uses, and are so difficult to separate, chemists made little effort to do so—until the uranium atom was fissioned. Then it became an urgent matter indeed, because radioactive varieties of some of these elements were among the main products of fission; and in the atomic bomb project, it was necessary to separate and identify them quickly and cleanly.

The problem was solved in short order by use of a chemical technique first devised in 1906 by the Russian botanist Mikhail Semenovich Tswett, who named it *chromatography* ("writing in color"). Tswett had found that he could separate plant pigments, chemically very much alike, by washing them down a column of powdered limestone with a solvent. He dissolved his mixture of plant pigments in petroleum ether and poured this on the limestone. Then he proceeded to pour in clear solvent. As the pigments were slowly washed down through the limestone powder, each pigment moved down at a different rate, because each differed in strength of adhesion to the powder. The result was that they separated into a series of bands, each of a different color. With continued washing, the separated substances trickled out separately at the bottom of the column, one after the other.

The world of science for many years ignored Tswett's discovery, possibly because he was only a botanist and only a Russian, while the leaders of research on separating difficult-to-separate substances at the time were German biochemists. But, in 1931, a German biochemist, Richard Willstätter, rediscovered the process, whereupon it came into general use. (Willstätter had received the 1915 Nobel Prize in chemistry for his excellent work on plant pigments. Tswett, so far as I know, has gone unhonored.)

Chromatography through columns of powder was found to work on almost all sorts of mixtures—colorless as well as colored. Aluminum oxide and starch proved to be better than limestone for separating ordinary molecules. Where ions are separated, the process is called *ion exchange;* and compounds known as *zeolites* were the first efficient agents applied for this purpose. Calcium and magnesium ions could be removed from *hard water,* for instance, by pouring the water through a zeolite column. The calcium and magnesium ions adhere to the zeolite and are replaced in solution by the sodium ions originally present on the zeolite, so *soft water* drips out of the bottom of the column. The sodium ions of zeolite have to be replenished from time to time by pouring in a concentrated solution of salt (sodium chloride). In 1935, a refinement came with the development of *ion-exchange resins.* These synthetic substances can be designed for the job to be done. For instance, certain resins will substitute hydrogen ions for positive ions, while others substitute hydroxyl ions for negative ions; a combination of both types will remove most of the salts from sea water. Kits containing such resins were part of the survival equipment on life rafts during the Second World War.

It was the American chemist Frank Harold Spedding who adapted ion-exchange chromatography to the separation of the rare earths. He found that these elements came out of an ion-exchange column in the reverse order of their atomic number, so that they were not only quickly separated but also identified. In fact, the discovery of promethium, the missing element 61, was confirmed in this way from the tiny quantities found among the fission products.

Thanks to chromatography, purified rare-earth elements can now be prepared by the pound or even by the ton. It turns out that the rare earths are not particularly rare: the rarest of them (excepting promethium) are more common than gold or silver, and the most abundant—lanthanum, cerium, and neodymium—are more plentiful than lead. Together the rare-earth metals make up a larger percentage of the earth's crust than copper and tin combined. So scientists have pretty well dropped the term *rare earths* and now call this series of elements the *lanthanides,* after its lead-off member. To be sure, the individual lanthanides have not been much used in the past, but the ease of separation now has multiplied their uses, and by the 1970s, 25,000,000 pounds a year were being used. Mischmetal, a mixture that consists chiefly of cerium, lanthanum, and neodymium, makes up three-fourths the weight of cigarette-lighter flints. A mixture of the oxides is used in polishing glass, and different oxides are added to glass to produce certain desirable properties. Certain mixtures of europium and yttrium oxides are used as red-sensitive phosphors in color television, and so on.

THE ACTINIDES

Nor are the rewards that arise from the better understanding of the lanthanides confined to their practical uses. The new knowledge also provided a key to the chemistry of the elements at the end of the periodic table, including the synthetic ones.

The series of heavy elements in question begins with actinium, number 89. In the table it falls under lanthanum. Actinium has two electrons in the Q-shell, just as lanthanum has two electrons in the P-shell. Actinium's eighty-ninth and last electron entered the P-shell, just as lanthanum's fifty-seventh and last entered the O-shell. Now the question is: Do the elements after actinium continue to add electrons to the P-shell and remain ordinary transition elements? Or do they, perchance, follow the pattern of the elements after lanthanum, where the electrons dive down to fill the skipped subshell below? If the latter is true, then actinium may start a new series of rare-earth metals, which would be called *actinides* after the first member.

The natural elements in this series of actinides are actinium, thorium, protactinium, and uranium. They were not much studied before 1940. What little was known about their chemistry suggested that they were ordinary transition elements. But when the man-made elements neptunium and plutonium were added to the list and studied intensively, these two showed a strong chemical resemblance to uranium. Glenn Seaborg was therefore prompted to propose that the heavy elements were in fact following the lanthanide pattern and filling the buried unfilled fourth subshell of the O-shell. With lawrencium that subshell is filled, and the fifteen actinides exist, in perfect analogy to the fifteen lanthanides. One important confirmation is that ion-exchange chromatography separates the actinides in just the same way it separates lanthanides.

Elements 104 (rutherfordium) and 105 (hahnium) are *transactinides* and, chemists are quite sure, come underneath hafnium and tantalum, the two elements that follow the lanthanides.

Gases

From the dawn of chemistry, it was recognized that many substances could exist in the form of gas, liquid, or solid, depending on the temperature. Water is the most common example: sufficiently cooled, it becomes solid ice; and sufficiently heated, it becomes gaseous steam. Van Helmont, who first used the word *gas*, differentiated between substances that are gases at ordinary temperatures, such as carbon dioxide, and those that, like steam, are gases only at elevated temperatures. He called the latter *vapors*, and we still speak of *water vapor* rather than *water gas*.

The study of gases, or vapors, continued to fascinate chemists, partly because they lent themselves to quantitative studies. The rules governing their behavior were simpler and more easily worked out than those governing the behavior of liquids and solids.

LIQUEFACTION
In 1787, the French physicist Jacques Alexandre César Charles discovered that, when a gas is cooled, each degree of cooling causes its volume to contract by about 1/273 of its volume at 0° C; and, conversely, each degree of warming causes it to expand by the same 1/273. The expansion with warmth raised no logical difficulties, but, if shrinkage with cold were to continue according to *Charles's law* (as it is called to this day), at −273° C, a gas should have shrunk to nothing! This paradox did not particularly bother chemists, for they were sure that Charles's law would not hold all the way down, since the gases would condense to liquids as the temperature dropped, and liquids do not contract as drastically as gases do with falling temperature. Still, chemists did not, at first, have any way of getting to very low temperatures to see what actually happens.

The development of the atomic theory, picturing gases as collections of molecules, presented the situation in new terms. The volume was now seen to depend on the velocity of the molecules. The higher the temperature, the faster they move, the more "elbow room" they require, and the greater the volume. Conversely, the lower the temperature, the more slowly they move, the less room they require, and the smaller the volume. In the 1860s, the British physicist William Thomson, who had just been raised to the peerage

	F	C	K
TEMPERATURE OF SUN SURFACE	10,832°	6,000°	6,273°
BOILING	212°	100°	373°
	200°	90°	363°
	180°	80°	353°
	160°	70°	343°
	140°	60°	333°
	120°	50°	323°
	100°	40°	313°
	80°	30°	303°
ROOM TEMPERATURE	68°	20°	293°
	60°	10°	283°
FREEZING	40° 32°	0°	273°
LIQUEFACTION OF OXYGEN (AT ORDINARY AIR PRESSURE)	−297°	−183°	90°
	−378°	−228°	45°
ABSOLUTE ZERO	−459°	−273°	0°

Figure 6.4. A comparison of the Fahrenheit, Celsius (or centigrade), and Kelvin thermometric scales.

as Lord Kelvin, suggested that it was the molecules' average energy content that declined by 1/273 for every degree of cooling. Whereas volume could not be expected to disappear completely, energy could. Thomson maintained that, at −273° C, the energy of molecules would sink to zero. Therefore −273° C must represent the lowest possible temperature. So this temperature (now put at −273.16° C according to refined modern measurements) would be *absolute zero*, or, as it is often stated, *zero Kelvin*. On this absolute scale, the melting point of ice is 273° K. (See figure 6.4 for the Fahrenheit, Celsius, and Kelvin scales.)

This view made it even more certain that gases would all liquefy as absolute zero approached. With ever less energy available, the gas molecules would require so little elbow room that they would collapse upon each other and be in contact. In other words, they would become liquids, for the properties of liquids can be explained by supposing that they consist of molecules in contact, but that the molecules still contain enough energy to slip and slide freely over, under, and past each other. For that reason, liquids can pour and can easily change their shape to suit a particular container.

As energy continues to decrease with drop in temperature, the molecules eventually possess too little to make their way past each other but come to occupy some fixed position about which they can vibrate but from which they cannot move bodily. In other words, the liquid has frozen to a solid. It seemed clear then to Kelvin that, as one approached absolute zero, all gases would not only liquefy but freeze.

Naturally, among chemists there was a desire to demonstrate the accuracy of Kelvin's suggestion by lowering the temperature to the point where all the gases would first liquefy, then freeze, on the way to actually attaining absolute zero. (There is something about any distant horizon that calls for conquest.)

Scientists had been exploring extremes of coldness even before Kelvin had defined the ultimate goal. Michael Faraday had found that, even at ordinary temperatures, some gases could be liquefied under pressure. He used a strong glass tube bent into boomerang shape. In the closed bottom, he placed a substance that would yield the gas he was after. He then sealed the open end. The end with the solid material he placed into hot water, thus liberating the gas in increasingly greater quantity; and since the gas was confined within the tube, it developed increasingly greater pressure. The other end of the tube Faraday kept in a beaker filled with crushed ice. At that end the gas would be subjected to both high pressure and low temperature and would liquefy. In 1823, Faraday liquefied the gas chlorine in this manner. Chlorine's normal liquefaction point is $-34.5°$ C ($238.7°$ K).

In 1835, a French chemist, C. S. A. Thilorier, used the Faraday method to form liquid carbon dioxide under pressure, using metal cylinders, which would bear greater pressures than glass tubes. He prepared liquid carbon dioxide in considerable quantity and then allowed it to escape from the tube through a narrow nozzle.

Naturally, under these conditions, the liquid carbon dioxide, exposed to normal temperatures would evaporate quickly. When a liquid evaporates, its molecules are pulling away from those by which it is surrounded and become single entities moving freely about. The molecules of a liquid have a force of attraction among themselves, and to pull free against that attraction requires energy. If the evaporation is rapid, there is no time for sufficient energy (in the form of heat) to enter the system, and the only remaining source of energy to feed the evaporation is the liquid itself. When a liquid evaporates quickly, therefore, the temperature of the residue of the liquid drops.

(This phenomenon is experienced by us, for the human body always per-spires gently, and the evaporation of the thin layer of water on our skin withdraws heat from the skin and keeps us cool. The warmer it is, the more we must perspire; and if the air is humid so that evaporation cannot take place, the perspiration collects on our body and we become uncomfortable indeed. Exercise, by multiplying the heat-producing reactions within our body, also increases perspiration, and we are then also uncomfortable under humid conditions.)

When Thilorier (to get back to him) allowed liquid carbon dioxide to evaporate, the temperature of the liquid dropped as evaporation proceeded, until the carbon dioxide froze. For the first time, solid carbon dioxide was formed.

Liquid carbon dioxide is stable only under pressure. Solid carbon dioxide exposed to ordinary pressures will *sublime*—that is, evaporate directly to gas without melting. The sublimation point of solid carbon dioxide is $-78.5°$ C $(194.7°$ K).

Solid carbon dioxide has the appearance of cloudy ice (though it is much colder); and since it does not form a liquid, it is called *dry ice.* Some 400,000 tons of it are produced each year, and most of it is used in preserving food through refrigeration.

Cooling by evaporation revolutionized human life. Prior to the nineteenth century, ice, when obtainable, could be used for preserving food. Ice might be stored away in the winter and preserved, under insulation, through the summer; or it might be brought down from the mountains. At best, it was a tedious and difficult process, and most people had to make do with summer heat (or year-round heat, for that matter).

As early as 1755, the Scottish chemist, William Cullen, had produced ice by forming a vacuum over quantities of water, enforcing rapid evaporation which cooled the water to the freezing point. This could not compete with natural ice, however. Nor could the process be used indirectly simply to cool food because ice would form and clog the pipes.

Nowadays, an appropriate gas is liquefied by a compressor and is allowed to come to room temperature. It is then circulated in coiled pipes around a chamber in which food is contained. As it evaporates, it withdraws heat from the chamber. The gas that emerges is again liquefied by a compressor, allowed to cool, and recirculated. The process is continuous, and heat is pumped out of the enclosed chamber into the outside atmosphere. The result is a *refrigerator,* replacing the older *icebox.*

In 1834, an American inventor, Jacob Perkins, patented (in Great Britain) the use of ether as a refrigerant. Other gases such as ammonia and sulfur dioxide also came into use. All these refrigerants had the disadvantage of being poisonous or flammable. In 1930, however, the American chemist Thomas Midgley discovered dichlorodifluoromethane (CF_2Cl_2), better known under the trade-name of Freon. This is nontoxic (as Midgley demonstrated by filling his lungs with it in public) and nonflammable and suits the purpose perfectly. With Freon, home refrigeration became widespread and commonplace.

(Although Freon and other *fluorocarbons* have always proven totally harmless to human beings, doubts did arise, in the 1970s, about their effect on the ozonosphere, as described in the previous chapter.)

Refrigeration applied, in moderation, to large volumes is *air conditioning,* so called because the air is also conditioned—that is, filtered and dehumidified. The first practical air-conditioning unit was designed in 1902 by the American inventor Willis Haviland Carrier; since the Second World War air conditioning has become nearly universal in major American cities.

To get back to Thilorier once again, he added solid carbon dioxide to a liquid called *diethyl ether* (best known today as an anesthetic; see chapter 11). Diethyl ether is low-boiling and evaporates quickly. Between it and the low temperature of the solid carbon dioxide, which was subliming, a temperature of −110° C (163.2° K) was attained.

In 1845, Faraday returned to the task of liquefying gases under the combined effect of low temperature and high pressure, making use of solid carbon dioxide and diethyl ether as his cooling mixture. Despite this mixture and his use of higher pressures than before, there were six gases he could not liquefy. They were hydrogen, oxygen, nitrogen, carbon monoxide, nitric oxide, and methane; and he named these *permanent gases*. To the list, we might add five more gases that Faraday did not know about. One of them was fluorine, and the other four are the noble gases: helium, neon, argon and krypton.

In 1869, however, the Irish physicist Thomas Andrews deduced from his experiments that every gas has a *critical temperature* above which it cannot be liquefied even under pressure. This assumption was later put on a firm theoretical basis by the Dutch physicist, Johannes Diderik Van der Waals, who, as a result, earned the 1910 Nobel Prize for physics.

To liquefy any gas one had to be certain, therefore, that one was working at a temperature below the critical value, or it was labor thrown out. Efforts were made to reach still lower temperatures to conquer the stubborn gases. A *cascade* method—lowering temperatures by steps—turned the trick. First, liquefied sulfur dioxide, cooling through evaporation, was used to liquefy carbon dioxide; then the liquid carbon dioxide was used to liquefy a more resistant gas; and so on. In 1877, the Swiss physicist Raoul Pictet finally managed to liquefy oxygen, at a temperature of −140° C (133° K) and under a pressure of 500 atmospheres (7,500 pounds per square inch). The French physicist Louis Paul Cailletet, at about the same time, liquefied not only oxygen but also nitrogen and carbon monoxide. Naturally these liquids made it possible to go on at once to still lower temperatures. The liquefaction point of oxygen at ordinary air pressure was eventually found to be −183° C (90° K); that of carbon monoxide, −190° C (83° K); and that of nitrogen, −195° C (78° K).

In 1895, the English chemical engineer William Hampson and the German physicist Karl von Linde independently devised a way of liquefying air on a large scale. The air was first compressed and cooled to ordinary temperatures. It was then allowed to expand and, in the process, to become quite cold. This cold air was used to bathe a container of compressed air until *it* was quite cold. The compressed air was *then* allowed to expand so that it became much colder. This process was repeated, air getting colder and colder, until it liquefied.

Liquid air, in quantity and cheap, was easily separated into liquid oxygen and liquid nitrogen. The oxygen could be used in blowtorches and for medicinal purposes; the nitrogen, under conditions where its inertness was useful.

Thus, incandescent lightbulbs filled with nitrogen allowed the filaments to remain at white-hot temperature for longer periods before slow metal evaporation broke them, than if those same filaments were burning in evacuated bulbs. Liquid air could also be used as a source for minor components such as argon and the other noble gases.

Hydrogen resisted all efforts at liquefaction until 1900. The Scottish chemist James Dewar then accomplished the feat by bringing a new stratagem into play. Lord Kelvin (William Thomson) and the English physicist James Prescott Joule had shown that, even in the gaseous state, a gas can be cooled simply by letting it expand and preventing heat from leaking into the gas from outside, provided the temperature is low enough to begin with. Dewar therefore cooled compressed hydrogen to a temperature of $-200°$ C in a vessel surrounded by liquid nitrogen, let this superfrigid hydrogen expand and cool further, and repeated the cycle again and again by conducting the ever-cooling hydrogen back through pipes. The compressed hydrogen, subjected to this *Joule-Thomson effect*, finally became liquid at a temperature of about $-240°$ C ($33°$ K). At still lower temperatures, Dewar managed to obtain solid hydrogen.

To preserve his superfrigid liquids, he devised special silver-coated glass flasks. These were double-walled with a vacuum between. Heat could be lost (or gained) through a vacuum only by the comparatively slow process of radiation, and the silver coating reflected the incoming (or, for that matter, outgoing) radiation. Such *Dewar flasks* are the direct ancestor of the household Thermos bottle.

ROCKET FUEL

With the coming of rocketry, liquefied gases suddenly rose to new heights of glamour. Rockets require an extremely rapid chemical reaction, yielding large quantities of energy. The most convenient type of fuel is a combination of a liquid combustible, such as alcohol or kerosene, and liquid oxygen. Oxygen, or some alternate oxidizing agent, must be carried by the rocket in any case, because it runs out of any natural supply of oxygen when it leaves the atmosphere. And the oxygen must be in liquid form, since liquids are denser than gases and more oxygen can be squeezed into the fuel tanks in liquid form than in gaseous. Consequently, liquid oxygen has come into high demand in rocketry.

The efficiency of a mixture of fuel and oxidizer is measured by a quantity known as the *specific impulse.* This represents the number of pounds of thrust produced by the combustion of 1 pound of the fuel-oxidizer mixture in 1 second. For a mixture of kerosene and oxygen, the specific impulse is equal to 242. Since the payload a rocket can carry depends on the specific impulse, there has been an avid search for more efficient combinations. The best liquid fuel, from this point of view, is liquid hydrogen. Combined with liquid oxygen, it can yield a specific impulse equal to 350 or so. If liquid ozone or liquid fluorine could be used in place of oxygen, the specific impulse could be raised to something like 370.

Certain light metals, such as lithium, boron, magnesium, aluminum, and, particularly, beryllium, deliver more energy on combining with oxygen than even hydrogen does. Some of these are rare, however, and all involve technical difficulties in the burning—difficulties arising from smokiness, oxide deposits, and so on.

There are also solid fuels that serve as their own oxidizers (like gunpowder, which was the first rocket propellant, but much more efficient). Such fuels are called *monopropellants*, since they need no separate supply of oxidizer and make up the one propellant requried. Fuels that also require oxidizers are *bipropellants* (two-propellants). Monopropellants would be easy to store and handle and would burn in a rapid but controlled fashion. The principal difficulty is probably that of developing a monopropellant with a specific impulse approaching those of the bipropellants.

Another possibility is atomic hydrogen, which Langmuir put to use in his blowtorch. It had been calculated that a rocket engine operating on the recombination of hydrogen atoms into molecules could develop a specific impulse of more than 1,300. The main problem is how to store the atomic hydrogen. So far the best hope seems to be to cool the free atoms very quickly and very drastically immediately after they are formed. Researches at the National Bureau of Standards seem to show that free hydrogen atoms are best preserved if trapped in a solid material at extremely low temperatures—say, frozen oxygen or argon. If we could arrange to push a button, so to speak, to let the frozen gases start warming up and evaporating, the hydrogen atoms would be freed and allowed to recombine. If such a solid could hold even as much as 10 percent of its weight in free hydrogen atoms, the result would be a better fuel than any we now possess. But, of course, the temperature would have to be very low indeed—considerably below that of liquefied hydrogen. These solids would have to be kept at about $-272°$ C, or just 1 degree above absolute zero.

In another direction altogether lies the possibility of driving ions backward (rather than the exhaust gases of burned fuel). The individual ions, of tiny mass, would produce tiny impulses, but could be continued over long periods. A ship placed in orbit by the high but short-lived force of chemical fuel could then, in the virtually frictionless medium of space, slowly accelerate under the long-lived lash of ions to nearly light's velocity. The material best suited to such an ionic drive is cesium, the substance that can most easily be made to lose electrons and form cesium ion. An electric field can then be made to accelerate the cesium ion and shoot it out the rocket opening.

SUPERCONDUCTORS AND SUPERFLUIDS

But to return to the world of low temperature. Even the liquefaction and solidification of hydrogen did not represent the final victory. By the time hydrogen yielded, the inert gases had been discovered; of these the lightest, helium, remained a stubborn holdout against liquefaction at the lowest temperatures attainable. Then, in 1908, the Dutch physicist Heike Kammerlingh

Onnes finally subdued helium. He carried the Dewar system one step further. Using liquid hydrogen, he cooled helium gas under pressure to about $-255°$ C ($18°$ K) and then let the gas expand to cool itself further. By this method he liquefied the gas. Thereafter, by letting the liquid helium evaporate, he got down to the temperature at which helium could be liquefied under normal atmospheric pressure ($4.2°$ K), a temperature at which *all* other substances are solid, and even to temperatures as low as $0.7°$ K. For his low-temperature work, Onnes received the Nobel Prize in physics in 1913. (Nowadays the liquefaction of helium is a simpler matter. In 1947, the American chemist Samuel Cornette Collins invented the *cryostat*, which, by alternate compressions and expansions, can produce as much as 2 gallons of liquid helium an hour.)

Onnes, however, did more than reach new depths of temperature. He was the first to show that unique properties of matter existed at those depths. One of these properties is the strange phenomenon called *superconductivity*. In 1911, Onnes was testing the electrical resistance of mercury at low temperatures. It was expected that resistance to an electric current would steadily decrease as the removal of heat reduced the normal vibration of the atoms in the metal. But at $4.12°$ K the mercury's electrical resistance suddenly disappeared altogether! An electric current coursed through it without any loss of strength. It was soon found that other metals also could be made superconductive. Lead, for instance, became superconductive at $7.22°$ K. An electric current of several hundred amperes set up in a lead ring, kept at that temperature by liquid helium, went on circling through the ring for two and a half years with absolutely no detectable decrease in quantity.

As temperatures were pushed lower and lower, more metals were added to the list of superconductive materials. Tin became superconductive at $3.73°$ K; aluminum, at $1.20°$ K; uranium, at $0.8°$ K; titanium, at $0.53°$ K; hafnium, at $0.35°$ K. (Some 1,400 different elements and alloys are now known to display superconductivity.) But iron, nickel, copper, gold, sodium, and potassium must have still lower transition points—if they can be made superconductive at all—because they have not been reduced to this state at the lowest temperatures reached. The highest transition point found for a metallic element is that of technetium, which becomes superconductive at temperatures under $11.2°$ K.

A low-boiling liquid can easily maintain substances immersed in it at the temperature of its boiling point. To attain lower temperatures, the aid of a still-lower-boiling liquid must be called upon. Liquid hydrogen boils at $20.4°$ K, and it would be most useful to find a superconducting substance with a transition temperature at least this high. Only then can superconductivity be studied in systems cooled by liquid hydrogen. Failing that, only the one lower-boiling liquid, liquid helium—much rarer, more expensive, and harder to handle—must be used. A few alloys, particularly those involving the metal niobium, have transition temperatures higher than those of any pure metal. Finally, in 1968, an alloy of niobium, aluminum, and germanium was found that remained superconductive at $21°$ K. Superconductivity at liquid-hydrogen temperatures became feasible—but just barely.

A useful application of superconductivity suggests itself at once in connection with magnetism. A current of electricity through a coil of wire around an iron core can produce a strong magnetic field: the greater the current, the stronger the field. Unfortunately, the greater the current, the greater the heat produced under ordinary circumstances; and thus, there is a limit to what can be done. In superconductive wires, however, electricity flows without producing heat; and, it would seem, more and more electric current could be squeezed into the wires to produce unprecedentedly strong *electromagnets* at only a fraction of the power that must be expended under ordinary conditions. There is, however, a catch.

Along with superconductivity goes another property involving magnetism. At the moment that a substance becomes superconductive, it also becomes perfectly *diamagnetic:* that is, it excludes the lines of force of a magnetic field. This phenomenon was discovered by the German physicist Walther Meissner in 1933 and is therefore called the *Meissner effect.* By making the magnetic field strong enough, however, one can destroy the substance's superconductivity and the hope for supermagnetism, even at temperatures well below its transition point. It is as if, once enough lines of force have been concentrated in the surroundings, some at last manage to penetrate the substance; and then, gone is the superconductivity as well.

Attempts have been made to find superconductive substances that will tolerate high magnetic fields. There is, for instance, a tin-niobium alloy with the high transition temperature of 18° K. It can support a magnetic field of some 250,000 gauss, which is high indeed. This fact was discovered in 1954, but it was only in 1960 that techniques were developed for forming wires of this ordinarily brittle alloy. A compound of vanadium and gallium can do even better, and superconductive electromagnets reaching field intensities of 500,-000 gauss have been constructed.

Another startling phenomenon at low temperatures was discovered in helium itself. It is called *superfluidity.*

Helium is the only known substance that cannot be frozen solid, even at absolute zero. There is a small irreducible energy content, even at absolute zero, which cannot possibly be removed (so that the energy content is "zero" in a practical sense) but is enough to keep the extremely "nonsticky" atoms of helium free of each other and, therefore, liquid. Actually, the German physicist Hermann Walther Nernst showed, in 1905, that it is not the energy of a substance that becomes zero at absolute zero, but a closely related property: *entropy.* For this work he received the 1920 Nobel Prize in chemistry. I do not mean, however, that solid helium does not exist under any conditions: in 1926, it was produced at temperatures below 1° K, by a pressure of about 25 atmospheres.

In 1935, Willem Hendrik Keesom, who had managed the solidification of helium, and his sister, A. P. Keesom, working at the Onnes laboratory in Leyden, found that liquid helium at a temperature below 2.2° K conducts heat almost perfectly. It conducts heat so quickly—at the speed of sound, in fact

—that all parts of the helium are always at the same temperature. It will not boil—as any ordinary liquid will by reason of localized hot spots forming bubbles of vapor—because there are no localized hot spots in the liquid helium (if you can speak of hot spots in connection with a liquid below 2° K). When it evaporates, the top of the liquid simply slips off quietly—peeling off, so to speak, in sheets.

The Russian physicist Peter Leonidovich Kapitza went on to investigate this property and found that the reason helium conducts heat so well was that it flows with remarkable ease, carrying the heat from one part of itself to another almost instantaneously, at least 200 times as rapidly as copper, the next best heat conductor. It flows even more easily than a gas, having a viscosity only 1/1,000 that of gaseous hydrogen, and leaks through apertures so tiny that they stop a gas. Furthermore, the superfluid liquid forms a film on glass and flows along it as quickly as it pours through a hole. If an open container of the liquid is placed in a larger container filled to a lower level, the fluid will creep up the side of the glass and over the rim into the outer container, until the levels in both are equalized.

Helium is the only substance that exhibits this phenomenon of superfluidity. In fact, the superfluid behaves so differently from the way helium itself does above 2.2° K that it has been given a separate name, *helium II,* to distinguish it from liquid helium above that temperature, called *helium I.*

Since only helium permits investigation of temperatures close to absolute zero, it has become a very important element in both pure and applied science. The atmospheric supply is negligible, and the most important sources are natural gas wells into which helium, formed from uranium and thorium breakdown in the earth's crust, sometimes seeps. The gas produced by the richest known well (in New Mexico) is 7.5 percent helium.

CRYOGENICS

Spurred by the odd phenomena discovered in the neighborhood of absolute zero, physicists have naturally made every effort to get down as close to absolute zero as possible and expand their knowledge of what is now known as *cryogenics.* The evaporation of liquid helium can, under special conditions, produce temperatures as low as 0.5° K. (Temperatures at such a level, by the way, are measured by special methods involving electricity—for example, by the size of the current generated in a thermocouple, by the resistance of a wire made of some nonsuperconductive metal, by changes in magnetic properties, or even by the speed of sound in helium. The measurement of extremely low temperatures is scarcely easier than their attainment.) Temperatures substantially lower than 0.5° have been reached by a technique first suggested in 1925 by the Dutch physicist Peter Joseph Wilhelm Debye. A *paramagnetic substance* (that is, a substance that concentrates lines of magnetic force) is placed almost in contact with liquid helium, separated from it by helium gas, and the temperature of the whole system is reduced to about 1° K. The system is then placed within a magnetic field. The molecules of the paramagnetic substance line up parallel to the field's lines of force and, in doing so, give off heat. This heat is removed by further slight evaporation of the surrounding helium. Now the magnetic field is removed. The paramagnetic molecules immediately fall into a random orientation. In going from an ordered to a random orientation, the molecules must absorb heat, the only source of which is the liquid helium. The temperature of the liquid helium therefore drops.

This process can be repeated over and over, each time lowering the temperature of the liquid helium—a technique perfected by the American chemist William Francis Giauque, who received the Nobel Prize for chemistry in 1949 in consequence. In this way, a temperature of 0.00002° K was reached in 1957.

In 1962, the German-British physicist Heinz London and his co-workers, suggested the possibility of using a new device to attain still lower temperatures. Helium occurs in two varieties, *helium 4* and *helium 3*. Ordinarily they mix perfectly; but at temperatures below about 0.8° K, they separate, with helium 3 in a top layer. Some of the helium 3 is in the bottom layer with the helium 4, and it is possible to cause helium 3 to shift back and forth across the boundary, lowering the temperature each time in a fashion analogous to the shift between liquid and vapor in the case of an ordinary refrigerant such as Freon. Cooling devices making use of this principle were first constructed in the Soviet Union in 1965.

The Russian physicist Isaak Yakovievich Pomeranchuk suggested, in 1950, a method of deep cooling using other properties of helium 3; while as long ago as 1934, the Hungarian-British physicist Nicholas Kurti suggested the use of magnetic properties similar to those taken advantage of by Giauque, but involving the atomic nucleus—the innermost structure of the atom—rather than entire atoms and molecules.

As a result of the use of these new techniques, temperatures as low as 0.000001° K have been attained. And as long as physicists find themselves within a millionth of a degree of absolute zero, might they not just get rid of what little entropy is left and finally reach the mark itself?

No! Absolute zero is unattainable—as Nernst demonstrated in his Nobel-Prize-winning treatment of the subject (sometimes referred to as *the third law of thermodynamics*). In any lowering of temperature, only part of the entropy can be removed. In general, removing half of the entropy of a system is equally difficult regardless of what the total is. Thus it is just as hard to go from 300° K (about room temperature) to 150° K (colder than any temperature Antarctica attains) as to go from 20° K to 10° K. It is then just as hard to go from 10° K to 5° K and from 5° K to 2.5° K, and so on. Having attained a millionth of a degree above absolute zero, the task of going from that to half-a-millionth of a degree is as hard as going from 300° K to 150° K, and if that is attained, it is an equally difficult task to go from half-a-millionth to a quarter-of-a-millionth, and so on forever. Absolute zero lies at an infinite distance no matter how closely it seems to be approached.

The final stages of the quest for absolute zero has, by the way, resulted in the close study of helium 3, an extremely rare substance. Helium is itself not at all common on Earth; and when it is isolated, only 13 atoms out of every 10,000,000 are helium 3, the remainder being helium 4.

Helium 3 is a somewhat simpler atom than helium 4 and has only three-fourths of the mass of the more common variety. The liquefaction point of helium 3 is 3.2° K, a full degree below that of helium 4. What's more, it was at first thought that whereas helium 4 becomes superfluid at temperatures below 2.2° K, helium 3 (a less symmetrical molecule, even though simpler) shows no sign of superfluidity at all. It was only necessary to keep trying. In 1972, it was discovered that helium 3 changes to a superfluid helium II liquid form at temperatures below 0.0025° K.

HIGH PRESSURES

One of the new scientific horizons opened up by the work on liquefaction of gases was the development of an interest in producing high pressures. It seemed that putting various kinds of matter (not only gases) under great pressure might bring out fundamental information about the nature of matter and also about the interior of the earth. At a depth of 7 miles, for instance, the pressure is 1,000 atmospheres; at 400 miles, 200,000 atmospheres; at 2,000 miles, 1,400,000 atmospheres; and at the center of the earth, 4,000 miles down, it reaches 3,500,000 atmospheres. (Of course, Earth is a rather small planet. The central pressures within Saturn are estimated to be over 50,-000,000 atmospheres; within the even larger Jupiter, 100,000,000.)

The best that nineteenth-century laboratories could do was about 3,000 atmospheres, attained by Emile Hilaire Amagat in the 1880s. But, in 1905, the American physicist Percy Williams Bridgman began to devise new methods that soon reached pressures of 20,000 atmospheres and burst the tiny metal chambers he used for his experiments. He went to stronger materials and eventually succeeded in producing pressures of half a million atmospheres. For his work on high pressure he received the Nobel Prize in physics in 1946.

Under his ultrahigh pressures, Bridgman was able to force the atoms and molecules of a substance into more compact arrangements, which were some-times retained after the pressure was released. For instance, he converted ordinary yellow phosphorus, a nonconductor of electricity, into a black, con-ducting form of phosphorus. He brought about startling changes even in water. Ordinary ice is less dense than liquid water. Using high pressure, Bridgman produced a series of ices (ice-II, ice-III, and so on) that not only were denser than the liquid but were ice at temperatures well above the normal freezing point of water. Ice-VII is a solid at temperatures higher than the boiling point of water.

The word *diamond* brings up the most glamorous of all the high-pressure feats. Diamond, of course, is crystallized carbon, as is also graphite. When an element appears in two different forms, these forms are *allotropes*. Diamond and graphite are the most dramatic example of the phenomenon. Ozone and ordinary oxygen are another example. Yellow phosphorus and black phosphorus, mentioned in the previous paragraph (there is red phosphorus, too), are still another example.

Allotropes can seem entirely different in appearance and properties, and there is no more startling example of an allotrope than graphite and diamond —except, possibly, coal and diamond (anthracite coal is, chemically speaking, a sloppy version of graphite).

That diamond is but graphite (or coal) with a different organization of atoms seems, at first sight, completely unbelievable, but the chemical nature of diamond was first proved as long ago as 1772 by Lavoisier and some fellow French chemists. They pooled their funds to buy a diamond and proceeded to heat it to a temperature high enough to burn it up. The gas that resulted was found to be carbon dioxide. Later the British chemist Smithson Tennant showed that the amount of carbon dioxide measured could be produced only if diamond was pure carbon, as graphite is; and in 1799, the French chemist Guyton de Morveau clinched the case by converting a diamond into a lump of graphite.

That was an unprofitable maneuver, but now why could not matters be reversed? Diamond is 55 percent denser than graphite. Why not put graphite under pressure and force the atoms composing it into the tighter packing characteristic of diamond?

Many efforts were made; and, like the alchemists, a number of experimenters reported successes. The most famous was the claim of the French chemist Ferdinand Frédéric Henri Moissan. In 1893, he dissolved graphite in molten cast iron and reported that he found small diamonds in the mass after it cooled. Most of the objects were black, impure, and tiny, but one was colorless and almost a millimeter long. These results were widely accepted; and, for a long time, Moissan was considered to have manufactured synthetic diamonds. However, his results were never successfully repeated.

The search for synthetic diamonds was not without its side victories, however. In 1891, the American inventor Edward Goodrich Acheson, while heating graphite under conditions he thought might form diamond, stumbled upon silicon carbide, to which he gave the trade name Carborundum. This proved harder than any substance then known but diamond; and ever since, it has been a much-used *abrasive*—that is, a substance used for grinding and polishing.

The efficiency of an abrasive depends on its hardness. An abrasive can polish or grind substances less hard than itself, and diamond, as the hardest substance, is the most useful in this respect. The hardness of various substances is commonly measured on the *Mohs scale*, introduced by the German mineralogist Friedrich Mohs in 1818. This assigns minerals numbers from 1, for talc, to 10, for diamond. A mineral of a particular number is able to scratch all minerals with lower numbers. On the Mohs scale, Carborundum is given the number 9. The divisions are not equal, however. On an absolute scale, the difference in hardness between 10 (diamond) and 9 (Carborundum) is four times greater than the difference between 9 (Carborundum) and 1 (talc).

The reason for all this is not hard to see. In graphite, the carbon atoms are arranged in layers. In each individual layer, the carbon atoms are arranged in tessellated hexagons, like the tiles on a bathroom floor. Each carbon atom is bonded to three others in equal fashion; and since carbon is a small atom, the neighbors are close together and strongly held. The tessellation is hard to pull apart but is very thin and easily broken. A tessellation is a comparatively large distance from the next tessellation above and below so that the bonds between layers are weak, and one layer can easily be made to slide upon the next. For that reason, graphite is not only not particularly hard but can actually be used as a lubricant.

In diamond, however, carbon atoms are arranged with absolute three-dimensional symmetry. Each carbon atom is bonded to *four* others at equal distances, each of the four being at the apices of a tetrahedron of which the carbon atom under consideration forms the center. This is a very compact arrangement, so that diamond is substantially denser than graphite. Nor will it pull apart in any direction except under overwhelming force. Other atoms will take up the *diamond configuration;* but, of these, the carbon atom is the smallest and holds together tightest. Thus diamond is harder than any other substance under the conditions of Earth's surface.

In silicon carbide, half the carbon atoms are replaced by silicon atoms. As the silicon atoms are considerably larger than the carbon atoms, they do not hug their neighbors as close, and their bonds are weaker. Thus, silicon carbide is not as hard as diamond (though it is hard enough for many purposes).

Under the surface conditions on Earth, the graphite arrangement of carbon atoms is more stable than the diamond arrangement. Hence, there is a tendency for diamond to turn spontaneously into graphite. You are, however, in no danger of waking up some morning to find your splendid diamond ring has become worthless overnight. The carbon atoms, even in their unstable arrangement, hold together so tight that it would take many millions of years for the change to take place.

This difference in stability makes it all the harder to change graphite to diamond. It was not till the 1930s that chemists finally worked out the pressure requirements for converting graphite to diamond. It turned out that the conversion called for a pressure of at least 10,000 atmospheres, and even then it would be impracticably slow. Raising the temperature would speed the conversion but would also raise the pressure requirements. At 1500° C, a pressure of at least 30,000 atmospheres would be necessary. All this proved that Moissan and his contemporaries, under the conditions they used, could no more have produced diamonds than the alchemists could have produced gold. (There is some evidence that Moissan was actually a victim of one of his assistants, who, tiring of the tedious experiments, decided to end them by planting a real diamond in the cast-iron mixture.)

Aided by Bridgman's pioneering work in attaining the necessary high temperatures and pressures, scientists at the General Electric Company finally accomplished the feat in 1955. Pressures of 100,000 atmospheres or more were produced, along with temperatures of up to 2500° C. In addition, a small quantity of metal, such as chromium, was used to form a liquid film across the graphite. It was on this film that the graphite turned to diamond. In 1962, a pressure of 200,000 atmospheres and a temperature of 5000° C could be attained. Graphite was then turned to diamond directly, without the use of a catalyst.

Synthetic diamonds are too small and impure to be used as gems, but they are now produced commercially as abrasives and cutting tools and, indeed, are a major source of such products. By the end of the decade, an occasional small diamond of gem quality could be produced.

A newer product made by the same sort of treatment can supplement the use of diamond. A compound of boron and nitrogen (*boron nitride*) is very similar in properties to graphite (except that boron nitride is white instead of black). Subjected to the high temperatures and pressures that convert graphite to diamond, the boron nitride undergoes a similar conversion. From a crystal arrangement like that of graphite, the atoms of boron nitride are converted to one like that of diamond. In its new form it is called *borazon*. Borazon is about four times as hard as Carborundum. In addition it has the great advantage of being more resistant to heat. At a temperature of 900° C, diamond burns up but borazon comes through unchanged.

Boron has one electron fewer than carbon; nitrogen, one electron more. The two in combination, alternately, set up a situation closely resembling the carbon-carbon arrangement, but there is a tiny departure from the perfect symmetry of diamond. Boron nitride is therefore not quite as hard as diamond.

Bridgman's work on high pressure is not the last word, of course. As the 1980s began, Peter M. Bell of the Carnegie Institution made use of a device that squeezes materials between two diamonds, and has managed to reach pressures of 1,500,000 atmospheres, over two-fifths that at the Earth's center. He believes it is possible for the instrument to go to 17,000,000 atmospheres before the diamonds themselves fail.

At the California Institute of Technology, shock waves are used to produce momentary pressures that are higher still—up to 75,000,000 atmospheres perhaps.

Metals

Most of the elements in the periodic table are metals. As a matter of fact, only about 20 of the 102 elements can be considered definitely nonmetallic. Yet the use of metals came relatively late in the history of the human species. One reason is that, with rare exceptions, the metallic elements are combined in nature with other elements and are not easy to recognize or extract. Primitive people at first used only materials that could be manipulated by simple treatments such as carving, chipping, hacking, and grinding; and thus their materials were restricted to bones, stones, and wood.

Primitive people may have been introduced to metals through discoveries of meteorites, or of small nuggets of gold, or of metallic copper in the ashes of fires built on rocks containing a copper ore. In any case, people who were curious enough (and lucky enough) to find these strange new substances and look into ways of handling them would discover many advantages in them. Metal differs from rock in that it has an attractive luster when polished. It can be beaten into sheets and drawn into wire. It can be melted and poured into a mold to solidify. It is much more beautiful and adaptable than rock and ideal for ornaments. Metals probably were fashioned into ornaments long before they were put to any other use.

Because they were rare, attractive, and did not alter with time, these metals were valued and bartered until they became a recognized medium of exchange. Originally, pieces of metal (gold, silver, or copper) had to be weighed separately in trading transactions, but, by 700 B.C., standardized weights of metal stamped in some official government fashion were issued in the Asia Minor kingdom of Lydia and the Aegean island of Aegina. Coins are still with us today.

What really brought metals into their own was the discovery that some of them would take a sharper cutting edge than stone could, and would maintain that edge under conditions that would ruin a stone ax. Moreover, metal was tough. A blow that would splinter a wooden club or shatter a stone ax would only slightly deform a metal object of similar size. These advantages more than compensated for the fact that metal is heavier than stone and was harder to obtain.

The first metal obtained in reasonable quantity was copper, which was in use by 4000 B.C. Copper itself is too soft to make useful weapons or armor (though it will make pretty ornaments), but it was often found alloyed with a little arsenic or antimony, which resulted in a substance harder than the pure metal. Then samples of copper ore must have been found that contained tin. The copper-tin alloy (*bronze*) was hard enough for purposes of weaponry. Men soon learned to add the tin deliberately. The Bronze Age replaced the Stone Age in Egypt and western Asia about 3000 B.C. and in southeastern Europe by 2000 B.C. Homer's *Iliad* and *Odyssey* commemorate that period of culture.

Iron was known as early as bronze; but for a long time, meteorites were its only source. It remained no more than a precious metal, limited to occasional use, until methods were discovered for smelting iron ore and thus obtaining iron in unlimited quantities. The difficulty lay in working with fires hot enough and methods suitable enough to add carbon to the iron and harden it into the form we now call *steel.* Iron smelting began somewhere in Asia Minor about 1400 B.C. and developed and spread slowly.

An iron-weaponed army could rout a bronze-armed one, for iron swords would cut through bronze. The Hittites of Asia Minor were the first to use iron weapons to any extent, and they had a period of power in western Asia Then the Assyrians succeeded the Hittites. By 800 B.C., they had a completely ironized army which was to dominate western Asia and Egypt for two and a half centuries. At about the same time, the Dorians brought the Iron Age to Europe by invading Greece and defeating the Achaeans, who committed the error of clinging to the Bronze Age.

IRON AND STEEL

Iron is obtained essentially by heating iron ore (usually a ferric oxide) with carbon. The carbon atoms carry off the oxygen of the ferric oxide, leaving behind a lump of pure iron. In ancient times, the temperatures used did not melt the iron, and the product was a tough metal that could be worked into the desired shape by hammering—that is, *wrought iron.* Iron metallurgy on a larger scale came into being in the Middle Ages. Special furnaces were used, and higher temperatures that melted the iron. The molten iron could be poured into molds to form castings, so it was called *cast iron.* This was much less expensive than wrought iron and much harder, too, but it was brittle and could not be hammered. Increasing demand for iron of either form helped to deforest England, for instance, consuming its wood in the iron-smelting furnaces. But then, in 1780, the English ironworker Abraham Darby showed that *coke* (carbonized coal) would work as well as, or better than, *charcoal* (carbonized wood). The pressures on the forests eased in this direction, and the more-than-century-long domination of coal as an energy source began.

It was not until late in the eighteenth century that chemists, thanks to the French physicist René Antoine Ferchault de Réaumur, finally realized that it was the carbon content that dictates the toughness and hardness of iron. To maximize those properties, the carbon content ought to be between 0.2 percent and 1.5 percent; the steel that then results is harder and tougher and generally stronger than either cast iron or wrought iron. But until the mid-nineteenth century, high-quality steel could be made only by the complicated procedure of carefully adding the appropriate quantity of carbon to wrought iron (itself comparatively expensive). Steel remained therefore a luxury metal, used only where no substitute could be found—as in swords and springs.

The Age of Steel was ushered in by a British engineer named Henry Bessemer. Originally interested primarily in cannon and projectiles, Bessemer invented a system of rifling intended to enable cannon to shoot farther and more accurately. Napoleon III of France was interested and offered to finance further experiments. But a French artillerist killed the idea by pointing out that the propulsive explosion Bessemer had in mind would shatter the cast-iron cannons used in those days. Bessemer, chagrined, turned to the problem of creating stronger iron. He knew nothing of metallurgy, so he could approach the problem with a fresh mind. Cast iron was brittle because of its carbon content. Therefore the problem was to reduce the carbon.

Why not burn the carbon away by melting the iron and sending a blast of air through it? This seemed at first a ridiculous idea. Would not the air blast cool the molten metal and cause it to solidify? Bessemer tried it anyway and found that quite the reverse was true. As the air burned the carbon, the combustion gave off heat and the temperature of the iron rose rather than fell. The carbon burned off nicely. By proper controls, steel could be produced in quantity and comparatively cheaply.

In 1856, Bessemer announced his *blast furnace*. Ironmakers adopted the method with enthusiasm, then dropped it in anger when they found that inferior steel was being formed. Bessemer discovered that the iron ore used by the industry contained phosphorus (which had been absent from his own ore samples). Although Bessemer explained to the ironmakers that phosphorus had betrayed them, they refused to be twice-bitten. Bessemer therefore had to borrow money and set up his own steel works in Sheffield. Importing phosphorus-free iron ore from Sweden, he speedily produced steel at a price that undersold the other ironmakers.

In 1875, the British metallurgist Sidney Gilchrist Thomas discovered that by lining the interior of the furnace with limestone and magnesia, he could easily remove the phosphorus from the molten iron. After this, almost any iron ore could be used in the manufacture of steel. Meanwhile, in 1868, the German-British inventor Karl Wilhelm Siemens developed the *open-hearth method*, in which pig iron was heated with iron ore; this process also could take care of the phosphorus content.

The Age of Steel then got under way. The name is no mere phrase. Without steel, skyscrapers, suspension bridges, great ships, railroads, and many other modern constructions would be almost unthinkable; and, despite the rise of other metals, steel still remains the preferred metal in a host of everyday uses, from automobile bodies to knives.

(It is a mistake, of course, to think that any single advance can bring about a major change in the way of life of humanity. Such change is always the result of a whole complex of interrelated advances. For instance, all the steel in the world could not make skyscrapers practical without the existence of that too-often-taken-for-granted device, the elevator. In 1861, the American inventor Elisha Graves Otis patented a hydraulic elevator; and in 1889, the company he founded installed the first electrically run elevators in a New York commercial building.)

With steel cheap and commonplace, it became possible to experiment with the addition of other metals (*alloy steel*) to see whether it could be still further improved. The British metallurgist Robert Abbott Hadfield pioneered in this direction. In 1882, he found that adding manganese to steel to the extent of 13 percent produced a harder alloy, which could be used in machinery for particularly brutal jobs, such as rock crushing. In 1900, a steel alloy containing tungsten and chromium was found to retain its hardness well at high temperatures, even red heat; this alloy proved a boon for high-speed tools. Today, for particular jobs, there are innumerable other alloy steels, employing such metals as molybdenum, nickel, cobalt, and vanadium.

The great difficulty with steel is its vulnerability to corrosion—a process that returns iron to the crude state of the ore whence it came. One way of combating this is to shield the metal by painting it or by plating it with a metal less likely to corrode—such as nickel, chromium, cadmium, or tin. A more effective method is to form an alloy that does not corrode. In 1913, the British metallurgist Harry Brearley discovered such an alloy by accident. He was looking for steel alloys that would be particularly suitable for gun barrels. Among the samples he discarded as unsuitable was a nickel-chromium alloy. Months later, he happened to notice that these particular pieces in his scrap heap were as bright as ever, although the rest were rusted. That was the birth of *stainless steel.* It is too soft and too expensive for use in large-scale construction, but serves admirably in cutlery and small appliances where nonrusting is more important than hardness.

Since something like a billion dollars a year is spent over the world in the not too successful effort to keep iron and steel from corroding, the search for a general rust inhibitor goes on unabated. One interesting recent discovery is that *pertechnetates* (compounds containing technetium) protect iron against rusting. Of course, this rare, laboratory-made element may never be common enough to be used on any substantial scale, but it offers an invaluable research tool. Its radioactivity allows chemists to follow its fate and to observe what happens to it on the iron surface.

One of iron's most useful properties is its strong ferromagnetism. Iron itself is an example of a *soft magnet.* It is easily magnetized under the influence of an electric or magnetic field—that is, its magnetic domains (see chapter 5) are easily lined up. It is also easily demagnetized when the field is removed, and the domains fall into random orientation again. This ready loss of magnetism can be useful, as in electromagnets, where the iron core is magnetized easily with the current on, but *should* be as easily demagnetized when the current goes off.

Since the Second World War, a new class of soft magnets has been developed. These are the *ferrites,* an example being nickel ferrite ($NiFe_2O_4$) and manganese ferrite ($MnFe_2O_4$), which are used in computers as elements that must gain or lose magnetism with the utmost ease and rapidity.

Hard magnets, with domains that are difficult to orient or that, once oriented, to disorient, will, once magnetized, retain the property over long periods. Various steel alloys are the commonest examples, though particularly strong, hard magnets have been found among alloys that contain little or no iron. The best known example is *alnico,* discovered in 1931, one variety of which is made of aluminum, nickel, and cobalt (the name of the alloy being derived from the first two letters of each of the substances), plus a bit of copper.

In the 1950s, techniques were developed to use powdered iron as a magnet, the particles being so small as to consist of individual domains. These could be oriented in molten plastic, which would then be allowed to solidify, holding the domains fixed in their orientation. Such *plastic magnets* are very easy to shape and mold but can be made adequately strong as well.

NEW METALS

We have seen in recent decades the emergence of enormously useful new metals—ones that were almost useless and even unknown up to a century or so ago and in some cases up to our own generation. The most striking example is aluminum. Aluminum is the most common of all metals—60 percent more common than iron. But it is also exceedingly difficult to extract from its ores In 1825, Hans Christian Oersted (who had discovered the connection between electricity and magnetism) separated a little aluminum in impure form. Thereafter, many chemists tried unsuccessfully to purify the metal, until in 1854 the French chemist Henri Etienne Sainte-Claire Deville finally devised a method of obtaining pure aluminum in reasonable quantities. Aluminum is so active chemically that he had to use metallic sodium (even more active) to break aluminum's grip on its neighboring atoms. For a while aluminum sold for a hundred dollars a pound, making it practically a precious metal. Napoleon III indulged himself in aluminum cutlery and had an aluminum rattle fashioned for his infant son; and in the United States, as a mark of the nation's great esteem for George Washington, the Washington Monument was capped with a slab of solid aluminum in 1885.

In 1886, Charles Martin Hall, a young student of chemistry at Oberlin College, was so impressed by his professor's statement that anyone who could discover a cheap method of making aluminum would make a fortune, that he decided to try his hand at it. In a home laboratory in his woodshed, Hall set out to apply Humphry Davy's early discovery that an electric current sent through a molten metal can separate the metal ions by depositing them on the cathode plate. Looking for a material that could dissolve aluminum, he stumbled across *cryolite*, a mineral found in reasonable quantity only in Greenland. (Nowadays synthetic cryolite is available.) Hall dissolved aluminum oxide in cryolite, melted the mixture, and passed an electric current through it. Sure enough, pure aluminum collected on the cathode. Hall rushed to his professor with his first few ingots of the metal. (To this day, they are treasured by the Aluminum Company of America.)

As it happened, a young French chemist named Paul Louis Toussaint Héroult, who was just Hall's age (twenty-two), discovered the same process in the same year. (To complete the coincidence, Hall and Héroult both died in 1914.)

Although the *Hall-Héroult process* made aluminum an inexpensive metal, it was never to be as cheap as steel, because useful aluminum ore is less common than useful iron ore, and because electricity (the key to aluminum) is more expensive than coal (the key to steel). Nevertheless, aluminum has two great advantages over steel. First, it is light—only one-third the weight of steel. Second, in aluminum, corrosion merely takes the form of a thin, transparent film over its surface, which protects deeper layers from corrosion without affecting the metal's appearance.

Pure aluminum is rather soft, but alloying can modify that. In 1906, the

German metallurgist Alfred Wilm made a tough alloy by adding a bit of copper and a smaller bit of magnesium to the aluminum. He sold his patent rights to the Durener Metal Works in Germany, and they gave the alloy the name Duralumin.

Engineers quickly realized the value of a light but strong metal for aircraft. After the Germans introduced Duralumin in zeppelins during the First World War, and the British learned its composition by analyzing the alloy in a crashed zeppelin, use of this new metal spread over the world. Because Duralumin was not quite as corrosion-resistant as aluminum itself, metallurgists covered it with thin sheets of pure aluminum, forming the product called Alclad.

Today there are aluminum alloys that, weight for weight, are stronger than some steels. Aluminum has tended to replace steel wherever lightness and corrosion resistance are more important than brute strength. It has become, as everyone knows, almost a universal metal, used in airplanes, rockets, railway trains, automobiles, doors, screens, house siding, paint, kitchen utensils, foil wrapping, and so on.

And now we have *magnesium*, a metal even lighter than aluminum. Its main use is in airplanes, as you might expect; as early as 1910, Germany was making use of magnesium-zinc alloys for that purpose. After the First World War, magnesium-aluminum alloys came into increasing use.

Only about one-fourth as abundant as aluminum and more active chemically, magnesium is harder to obtain from ores. But fortunately there is a rich source in the ocean. Magnesium, unlike aluminum or iron, is present in sea water in quantity. The ocean carries dissolved matter to the amount of 3.5 percent of its mass. Of this dissolved material, 3.7 percent is magnesium ion. The ocean as a whole, therefore, contains about 2 quadrillion (2,000,000,-000,000,000) tons of magnesium, or all we could use for the indefinite future.

The problem was to get it out. The method chosen was to pump sea water into large tanks and add calcium oxide (also obtained from the sea, from oyster shells). The calcium oxide reacts with the water and the magnesium ion to form magnesium hydroxide, which is insoluble and therefore precipitates out of solution. The magnesium hydroxide is converted to magnesium chloride by treatment with hydrochloric acid, and the magnesium metal is then separated from the chlorine by means of an electric current.

In January 1941, the Dow Chemical Company produced the first ingots of magnesium from sea water, and the stage was laid for a tenfold increase in magnesium production during the war years.

As a matter of fact, any element that can be extracted profitably from sea water may be considered in virtually limitless supply since, after use, it eventually returns to the sea. It has been estimated that if 100 million tons of magnesium were extracted from sea water each year for a million years, the magnesium content of the ocean would drop from its present figure of 0.13 to 0.12 percent.

If steel was the "wonder metal" of the mid-nineteenth century, aluminum of the early twentieth century, and magnesium of the mid-twentieth century, what will the next new wonder metal be? The possibilities are limited. There are only seven really common metals in the earth's crust. Besides iron, aluminum, and magnesium, they are sodium, potassium, calcium, and titanium. Sodium, potassium, and calcium are far too active chemically to be used as construction metals. (For instance, they react violently with water.) That leaves titanium, which is about one-eighth as abundant as iron.

Titanium has an extraordinary combination of good qualities. It is only a little more than half as heavy as steel; it is stronger, weight for weight, than aluminum or steel; it is resistant to corrosion and able to withstand high temperatures. For all these reasons, titanium is now being used in aircraft, ships, and guided missiles wherever these properties can be put to good use.

Why was the value of titanium so slow to be discovered? The reason is much the same as for aluminum and magnesium: titanium reacts too readily with other substances and, in its impure forms—combined with oxygen or nitrogen —is an unprepossessing metal, brittle and seemingly useless. Its strength and other fine qualities emerge only when it is isolated in really pure form (in a vacuum or under an inert gas). The effort of metallurgists has succeeded to the point where a pound of titanium that would have cost $3,000 in 1947 cost $2 in 1969.

The search need not, however, be for new wonder metals. The older metals (and some nonmetals, too) can be made far more "wonderful" than they are now.

In Oliver Wendell Holmes's poem "The Deacon's Masterpiece," the story is told of a "one-hoss shay" which was carefully made in such a way as to have no weakest point. In the end, the one-horse buggy went all at once—decomposing into a powder. But it had lasted a hundred years.

The atomic structure of crystalline solids, both metal and nonmetal, is rather like the "one-hoss shay" situation. A metal's crystals are riddled with submicroscopic clefts and scratches. Under pressure, a fracture will start at one of these weak points and spread through the crystal. If, like the deacon's wonderful "one-hoss shay," a crystal could be built with no weak points, it would have great strength.

Such no-weak-point crystals do form as tiny fibers called *whiskers* on the surface of crystals. Tensile strengths of carbon whiskers have been found to run as high as 1,400 tons per square inch—or, from 15 to 70 times the tensile strength of steel. If methods could be designed for manufacturing defect-free metal in quantity, we would find ourselves with materials of astonishing strength. In 1968, for instance, Soviet scientists produced a tiny defect-free crystal of tungsten that would sustain a load of 1,635 tons per square inch, as compared with 213 tons per square inch for the best steel. And even if defect-free substances were not available in bulk, the addition of defect-free fibers to ordinary metals would reinforce and strengthen them.

Then, too, as late as 1968, an interesting new method was found for combining metals. The two methods of historic interest were *alloying,* where two or more metals are melted together and form a more-or-less-homogeneous mixture, and *plating,* where one metal is bound firmly to another (a thin layer of expensive metal is usually bound to the surface of a bulky volume of cheaper metal, so that the surface is, for instance, as beautiful and corrosion-resistant as gold but the whole nearly as cheap as copper).

The American metallurgist Newell C. Cook and his associates were attempting to plate a silicon layer on a platinum surface, using molten alkali fluoride as the liquid in which the platinum was immersed. The expected plating did not occur. What happened, apparently, was that the molten fluoride removed the very thin film of bound oxygen ordinarily present on even the most resistant metals, and presented the platinum surface "naked" to the silicon atoms. Instead of binding themselves to the surface on the other side of the oxygen atoms, they worked their way *into* the surface. The result was that a thin outer layer of the platinum became an alloy.

Cook followed this new direction and found that many substances can be combined in this way to form a "plating" of alloy on pure metal (or on another alloy). Cook called the process *metalliding* and quickly showed its usefulness. Thus, copper to which 2 percent to 4 percent of beryllium is added in the form of an ordinary alloy, becomes extraordinarily strong. The same result can be achieved if copper is *beryllided* at the cost of much less of the relatively rare beryllium. Again, steel metallided with boron (*boriding*) is hardened. The addition of silicon, cobalt, and titanium, also produces useful properties.

Wonder metals, in other words, if not found in nature can be created by human ingenuity.

Chapter 7

The Particles

The Nuclear Atom

As I pointed out in the preceding chapter, it was known by 1900 that the atom was not a simple, indivisible particle but contained at least one subatomic particle—the electron, identified by J. J. Thomson. Thomson suggested that electrons were stuck like raisins in the positively charged main body of the atom.

IDENTIFYING THE PARTICLES

But very shortly it developed that there were other particles within the atom. When Becquerel discovered radioactivity, he identified some of the radiation emitted by radioactive substances as consisting of electrons, but other emissions were discovered as well. The Curies in France and Ernest Rutherford in England found one that was less penetrating than the electron stream. Rutherford called this radiation *alpha rays* and gave the electron emission the name *beta rays*. The flying electrons making up the latter radiation are, individually, *beta particles*. The alpha rays were also found to be made up of particles and these were called *alpha particles*. *Alpha* and *beta* are the first two letters of the Greek alphabet.

Meanwhile the French chemist Paul Ulrich Villard discovered a third form of radioactive emission, which was named *gamma rays* after the third letter of the Greek alphabet. The gamma rays were quickly identified as radiation resembling X rays, but with shorter wavelengths.

Rutherford learned by experiment that a magnetic field deflected alpha particles much less than it did beta particles. Furthermore, they were deflected in the opposite direction; hence, the alpha particle had a positive charge, as opposed to the electron's negative one. From the amount of deflection, it could be calculated that the alpha particle must have at least twice the mass of the hydrogen ion, which possessed the smallest known positive charge. The amount of deflection would be affected both by the particle's mass and by its charge. If the alpha particle's positive charge was equal to that of the hydrogen ion, its mass would be two times that of the hydrogen ion; if its charge was double that, it would be four times as massive as the hydrogen ion; and so on (figure 7.1).

Figure 7.1. Deflection of particles by a magnetic field.

Rutherford settled the matter in 1909 by isolating alpha particles. He put some radioactive material in a thin-walled glass tube surrounded by a thick-walled glass tube, with a vacuum between. The alpha particles could penetrate the thin inner wall but not the thick outer one. They bounced back from the outer wall, so to speak, and, in so doing, lost energy and therefore were no longer able to penetrate the thin walls either. Thus they were trapped between. Now Rutherford excited the alpha particles by means of an electric discharge so that they glowed. They then showed the spectral lines of helium. (It has become evident that alpha particles produced by radioactive substances in the soil are the source of the helium in natural-gas wells.) If the alpha particle is helium, its mass must be four times that of hydrogen. Hence, its positive charge amounts to two units, taking the hydrogen ion's charge as the unit.

Rutherford later identified another positive particle in the atom. This one had actually been detected, but not recognized, many years before. In 1886, the German physicist Eugen Goldstein, using a cathode-ray tube with a perforated cathode, had discovered a new radiation that streamed through the holes of the cathode in the direction opposite to the cathode rays themselves. He called it *Kanalstrahlen* ("channel rays"). In 1902, this radiation served as the first occasion when the Doppler-Fizeau effect (see chapter 2) was detected in any earthly source of light. The German physicist Johannes Stark placed a spectroscope in such a fashion that the rays raced toward it and demonstrated the violet shift. For this research, he was awarded the Nobel Prize for physics in 1919.

Since channel rays move in a direction opposite to the negatively charged cathode rays, Thomson suggested that this radiation be called *positive rays*. It turned out that the particles of the *positive rays* could easily pass through matter. They were therefore judged to be much smaller in volume than ordinary ions or atoms. The amount of their deflection by a magnetic field indicated that the smallest of these particles had the same charge and mass as a hydrogen ion, assuming that this ion carried the smallest possible unit of positive charge. The positive-ray particle was therefore deduced to be the fundamental positive particle—the opposite number of the electron. Rutherford named it *proton* (from the Greek word for "first").

The proton and the electron do indeed carry equal, though opposite, electric charges, although the proton is 1,836 times as massive as the electron. It seemed likely, then, that an atom was composed of protons and electrons, mutually balancing their charges. It also appeared that the protons were in the interior of the atom, for whereas electrons can easily be peeled off, protons cannot. But now the big question was: what sort of structure do these particles of the atom form?

THE ATOMIC NUCLEUS

Rutherford himself came upon the beginning of the answer. Between 1906 and 1908, he kept firing alpha particles at a thin foil of metal (such as gold or platinum) to probe its atoms. Most of the projectiles passed right through undeflected (as bullets might pass through the leaves of a tree). But not all did: Rutherford found that, on the photographic plate that served as his target behind the metal, there was an unexpected scattering of hits around the central spot, and some particles bounced back! It was as if some of the bullets had not passed through leaves alone but had ricocheted off something more substantial.

Rutherford decided that they had hit some sort of dense core, which occupied only a very small part of the volume of the atom. Most of an atom's volume, it seemed, must be occupied by electrons. As alpha particles charged through the foil of metal, they usually encountered only electrons, and they brushed aside this froth of light particles without being deflected. But once in a while an alpha particle might happen to hit an atom's denser core, and then it was deflected. That this happened only very occasionally showed that the atomic cores must be very small indeed, because a projectile passing through the metal foil must encounter many thousands of atoms.

It was logical to suppose that the hard core was made up of protons. Rutherford pictured the protons of an atom as crowded into a tiny *atomic nucleus* at the center. (It has since been demonstrated that this nucleus has a diameter of little more than 1/100,000 that of the whole atom.)

This, then, is the basic model of the atom: a positively charged nucleus taking up very little room, but containing almost all the mass of the atom, surrounded by a froth of electrons taking up nearly all the volume of the atom, but containing practically none of its mass. For his extraordinary pioneering work on the ultimate nature of matter, Rutherford received the Nobel Prize in chemistry in 1908.

It now became possible to describe specific atoms and their behavior in more definite terms. For instance, the hydrogen atom possesses but a single electron. If this is removed, the proton that remains immediately attaches itself to some neighboring molecule. But when the bare hydrogen nucleus does not find an electron to share in this fashion, it acts as a proton—that is to say, a subatomic particle—and in that form it can penetrate matter and react with other nuclei if it has enough energy.

Helium, with two electrons, does not give one up so easily. As I mentioned in the preceding chapter, its two electrons form a closed shell, and the atom is therefore inert. If helium is stripped of both electrons, however, it becomes an alpha particle—that is, a subatomic particle carrying two units of positive charge.

The third element, lithium, has three electrons in its atom. Stripped of one or two, it is an ion. If all three of its electrons are removed, it, too, becomes a bare nucleus, carrying a three-unit positive charge.

The number of units of positive charge in the nucleus of an atom has to be exactly equal to the number of electrons it normally contains, for the atom as a whole is ordinarily neutral. And, in fact, the atomic numbers of the elements are based on their units of positive, rather than negative, charge, because the number of an atom's electrons may easily be made to vary in ion formation, whereas the number of its protons can be altered only with great difficulty.

This scheme of the construction of atoms had hardly been worked out when a new conundrum arose. The number of units of positive charge on a nucleus did not balance at all with the nucleus's mass, except in the case of the hydrogen atom. The helium nucleus, for instance, had a positive charge of two but was known to have four times the mass of the hydrogen nucleus. And the situation got worse and worse as one went down the table of elements, until one reached uranium with a mass equal to 238 protons but a charge equal only to 92.

How could a nucleus containing four protons (as the helium nucleus was supposed to) have only two units of positive charge? The first, and simplest, guess was that two units of its charge were neutralized by the presence in the nucleus of negatively charged particles of negligible weight. Naturally the electron sprang to mind. The puzzle might be straightened out if one assumed the helium nucleus to consist of four protons and two neutralizing electrons, leaving a net positive charge of two—and so on all the way to uranium, whose nucleus would have 238 protons and 146 electrons, netting 92 units of positive charge. The whole idea was given encouragement by the fact that radioactive nuclei were actually known to emit electrons—that is, beta particles.

This view of matter prevailed for more than a decade, until a better answer came in a roundabout way from other investigations. But, in the meantime, some serious objections to the hypothesis arose. For one thing, if the nucleus was built essentially of protons, with the light electrons contributing practically nothing to the mass, how was it that the relative masses of the various nuclei did not come to whole numbers? According to the measured atomic weights, the nucleus of the chlorine atom, for instance, had a mass of 35½ times that of the hydrogen nucleus. Did it, then, contain 35½ protons? No scientist (then or now) could accept the idea of half a proton.

Actually, this particular question has an answer that was discovered even before the main issue was solved. It makes an interesting story in itself.

Isotopes

UNIFORM BUILDING BLOCKS

As early as 1816, an English physician named William Prout had suggested that all atoms were built up from the hydrogen atom. As time went on and the atomic weights were worked out, Prout's theory fell by the wayside, because it developed that many elements had fractional weights (taking oxygen as the standard at 16). Chlorine has an atomic weight of 35.453. Other examples are antimony, 121.75; barium, 137.34; boron, 10.811; cadmium, 112.40.

Around the turn of the century there came a series of puzzling observations that was to lead to the explanation. The Englishman William Crookes (he of the Crookes tube) separated from uranium a small quantity of a substance that proved much more radioactive than uranium itself. He suggested that uranium was not radioactive at all—only this impurity, which he called *uranium X*. Henri Becquerel, on the other hand, discovered that the purified, feebly radioactive uranium somehow increased in radioactivity with time. After it was left standing for a while, the active uranium X could be extracted from it again and again. In other words, uranium was converted by its own radioactivity to the still more active uranium X.

Then Rutherford similarly separated a strongly radioactive *thorium X* from thorium and found that thorium, too, went on producing more thorium X. It was already known that the most famous radioactive element of all, radium, broke down to the radioactive gas radon. So Rutherford and his assistant, the chemist Frederick Soddy, concluded that radioactive atoms, in the process of emitting their particles, generally transformed themselves into other varieties of radioactive atoms.

Chemists began searching for such transformations and came up with an assortment of new substances, giving them such names as *radium A, radium B, mesothorium I, mesothorium II,* and *actinium C.* All of them were grouped into three series, depending on their atomic ancestry. One series arose from the breakdown of uranium; another, from that of thorium; and a third, from that of actinium (later it turned out that actinium itself had a predecessor, named *protactinium*). Altogether, some forty members of these series were identified, each distinguished by its own peculiar pattern of radiation. But the end product of all three series was the same: each chain of substances eventually broke down to the same stable element—lead.

Now obviously these forty substances could not all be separate elements; between uranium (92) and lead (82) there were only ten places in the periodic table, and all but two of these belonged to known elements. The chemists found, in fact, that though the substances differed in radioactivity, some of them were identical with one another in chemical properties. For instance, as early as 1907, the American chemists Herbert Newby McCoy and William Horace Ross showed that *radiothorium,* one of the disintegration products of thorium, showed precisely the same chemical behavior as thorium. *Radium D* behaved chemically exactly like lead; in fact, it was often called *radiolead.* All this suggested that the substances in question were actually varieties of the same element: radiothorium, a form of thorium; radiolead, a member of a family of leads; and so on.

In 1913, Soddy gave clear expression of this idea and developed it further. He showed that when an atom emits an alpha particle, it changes into an element two places lower in the list of elements; when it emits a beta particle, it changes into an element one place higher. On this basis, radiothorium would indeed fall in thorium's place in the table, and so would the substances called *uranium X_1* and *uranium Y:* all three would be varieties of element 90. Likewise, radium D, radium B, thorium B, and actinium B would all share lead's place as varieties of element 82.

To the members of a family of substances sharing the same position in the periodic table, Soddy gave the name *isotope* (from Greek words meaning "same position"). He received the Nobel Prize in chemistry in 1921.

The proton-electron model of the nucleus (which, nevertheless, eventually proved to be wrong), fitted in beautifully with Soddy's isotope theory. Removal of an alpha particle from a nucleus would reduce the positive charge of that nucleus by two—exactly what was needed to move it two places down in the periodic table. On the other hand, the ejection of an electron (beta particle) from a nucleus would leave an additional proton unneutralized and thus increase the nucleus's positive charge by one unit. The effect was to raise the atomic number by one, so the element would move to the next higher position in the periodic table.

How is it that when thorium breaks down to radiothorium, after going through not one but three disintegrations, the product is still thorium? Well, in the process the thorium atom loses an alpha particle, then a beta particle, then a second beta particle. If we accept the proton building-block idea, the thorium atom has lost four electrons (two supposedly contained in the alpha particle) and four protons. (The actual situation differs from this picture but in a way that does not affect the result.) The thorium nucleus started with 232 protons and 142 electrons (supposedly). Having lost four protons and four electrons, it is reduced to 228 protons and 138 electrons. In either case, the number of unbalanced protons—232 − 142, or 228 − 138—is 90. This still leaves the atomic number 90, the same as before, therefore. So radiothorium, like thorium, has ninety planetary electrons circling around the nucleus. Since the chemical properties of an atom are controlled by the number of its planetary electrons, thorium and radiothorium behave the same chemically, regardless of their difference in atomic weight (232 against 228).

The isotopes of an element are identified by their atomic weight, or *mass number*. Thus, ordinary thorium is called *thorium 232*, while radiothorium is *thorium 228*. Similarly, the radioactive isotopes of lead are known as *lead 210* (radium D), *lead 214* (radium B), *lead 212* (thorium B), and *lead 211* (actinium B).

The notion of isotopes was found to apply to stable elements as well as to radioactive ones. For instance, it turned out that the three radioactive series I have mentioned ended in three different forms of lead. The uranium series ended in lead 206; the thorium series, in lead 208; and the actinium series, in lead 207. Each of these was an "ordinary," stable isotope of lead, but the three leads differed in atomic weight.

Proof of the existence of stable isotopes came from a device invented by an assistant of J. J. Thomson named Francis William Aston. It was an arrangement that separated isotopes very sensitively by virtue of the difference in deflection of their ions by a magnetic field; Aston called it a *mass spectrograph*. In 1919, using an early version of this instrument, Thomson showed that neon was made up of two varieties of atom: one with a mass number of 20, the other with a mass number of 22. Neon 20 was the common isotope; neon 22 came with it in the ratio of 1 atom in 10. (Later a third isotope, neon 21, was discovered, amounting to only 1 atom in 400 in the neon of the atmosphere.)

Now the reason for the fractional atomic weights of the elements at least became clear. Neon's atomic weight of 20.183 represented the composite mass of the three different isotopes making up the element as it was found in nature. Each individual atom had an integral mass number, but the average mass number—the atomic weight—was fractional.

Aston proceeded to show that several common stable elements were indeed mixtures of isotopes. He found that chlorine, with a fractional atomic weight of 35.453, was made up of chlorine 35 and chlorine 37, in the *abundance ratio* of 3 to 1. Aston was awarded the Nobel Prize in chemistry in 1922.

In his address accepting the prize, Aston clearly forecast the possibility of making use of the energy bound in the atomic nucleus, foreseeing both nuclear power plants and nuclear bombs (see chapter 10). In 1935, the Canadian-American physicist Arthur Jeffrey Dempster used Aston's instrument to take a long step in that direction. He showed that, although 993 of every 1,000 uranium atoms were uranium 238, the remaining seven were uranium 235. This was a discovery fraught with a significance soon to be realized.

Thus, after a century of false trails, Prout's idea was finally vindicated. The elements *are* built of uniform building blocks—if not of hydrogen atoms, at least of units with hydrogen's mass. The reason the elements do not bear this out in their weights is that they are mixtures of isotopes containing different numbers of building blocks. In fact, even oxygen, whose atomic weight of 16 was used as the standard for measuring the relative weights of the elements, is not a completely pure case. For every 10,000 atoms of common oxygen 16, there are twenty atoms of an isotope with a weight equal to 18 units and four with the mass number 17.

Actually there are a few elements consisting of a *single isotope*. (This is a misnomer: to speak of an element as having only one isotope is like saying a woman has given birth to a "single twin.") The elements of this kind include beryllium, all of whose atoms have the mass number 9; fluorine, made up solely of fluorine 19; aluminum, solely aluminum 27; and a number of others. A nucleus with a particular structure is now called a *nuclide*, following the suggestion made in 1947 by the American chemist Truman Paul Kohman. One can properly say that an element such as aluminum is made up of a single nuclide.

TRACKING PARTICLES

Ever since Rutherford identified the first nuclear particle (the alpha particle), physicists have busied themselves poking around in the nucleus, trying either to change one atom into another or to break it up to see what it is made of. At first they had only the alpha particle to work with. Rutherford made excellent use of it.

One of the fruitful experiments Rutherford and his assistants carried out involved firing alpha particles at a screen coated with zinc sulfide. Each hit produced a tiny scintillation (an effect first discovered by Crookes in 1903), so that the arrival of single particles could be witnessed and counted with the naked eye. Pursuing this technique, the experimenters put up a metal disk that would block the alpha particles from reaching the zinc sulfide screen so that the scintillations stopped. When hydrogen was introduced into the apparatus, scintillations appeared on the screen despite the blocking metal disk. Moreover, these new scintillations differed in appearance from those produced by alpha particles. Since the metal disk stopped alpha particles, some other radiation must be penetrating it to reach the screen. The radiation, it was decided, must consist of fast protons. In other words, the alpha particles would now and then make a square hit on the nucleus of a hydrogen atom (which consists of a proton, remember) and send it careening forward, as one billiard ball might send another forward on striking it. The struck protons, being relatively light, would shoot forward at great velocity and so could penetrate the metal disk and strike the zinc sulfide screen.

This detection of single particles by scintillation is an example of a *scintillation counter*. To make such counts, Rutherford and his assistants first had to sit in the dark for 15 minutes in order to sensitize their eyes and then make their painstaking counts. Modern scintillation counters do not depend on the human eye and mind. Instead, the scintillations are converted to electric pulses that are then counted electronically. The final result need merely be read off from appropriate dials. The counting may be made more practical where scintillations are numerous, by using electric circuits that allow only one in two or in four (or even more) scintillations to be recorded. Such *scalers* (which scale down the counting, so to speak) were first devised by the English physicist Charles Eryl Wynn-Williams in 1931. Since the Second World War, organic substances have substituted for zinc sulfide and have proved preferable.

In Rutherford's original scintillation experiments, there came an unexpected development. When his experiment was performed with nitrogen instead of hydrogen as the target for the alpha-particle bombardment, the zinc sulfide screen still showed scintillations exactly like those produced by protons. Rutherford could only conclude that the bombardment had knocked protons out of the nitrogen nucleus.

To try to find out just what had happened, Rutherford turned to the *Wilson cloud chamber*, a device invented in 1895 by the Scottish physicist Charles Thomson Rees Wilson. A glass container fitted with a piston is filled with moisture-saturated air. When the piston is pulled outward, the air abruptly expands and therefore cools. At the reduced temperature, it is supersaturated with the moisture. Under such conditions, any charged particle will cause the water vapor to condense on it. If a particle dashes through the chamber, ionizing atoms in it, a foggy line of droplets will therefore mark its wake.

The nature of this track can tell a great deal about the particle. The light beta particle leaves a faint, wavering path; the particle is knocked about even in passing near electrons. The much more massive alpha particle makes a straight, thick track. If it strikes a nucleus and rebounds, the path has a sharp bend in it. If it picks up two electrons and becomes a neutral helium atom, its track ends. Aside from the size and character of its track, there are other ways of identifying a particle in the cloud chamber. Its response to an applied magnetic field tells whether it is positively or negatively charged, and the amount of curve indicates its mass and energy. By now physicists are so familiar with photographs of all sorts of tracks that they can read them off as if they were primer print. For the development of his cloud chamber, Wilson shared the Nobel Prize in physics in 1927.

The cloud chamber has been modified in several ways since its invention, and "cousin" instruments have been devised. The original cloud chamber was not usable after expansion until the chamber had been reset. In 1939, Alexander Langsdorf, in the United States, devised a *diffusion cloud chamber,* in which warm alcohol vapor diffused into a cooler region in such a way that there was always a supersaturated region, and tracks could be observed continuously.

Then came the *bubble chamber,* a device similar in principle. In it, superheated liquids under pressure are used rather than supersaturated gas. The path of the charged particle is marked by a line of vapor bubbles in the liquid rather than by liquid droplets in vapor. The inventor, the American physicist Donald Arthur Glaser, is supposed to have gotten the idea by studying a glass of beer in 1953. If so, it was a most fortunate glass of beer for the world of physics and for him, for Glaser received the Nobel Prize for physics in 1960 for the invention of the bubble chamber.

The first bubble chamber was only a few inches in diameter. Within the decade, bubble chambers 6 feet long were being used. Bubble chambers, like diffusion cloud chambers, are constantly set for action. In addition, since many more atoms are present in a given volume of liquid than of gas, more ions are produced in a bubble chamber, which is thus particularly well adapted to the study of fast and short-lived particles. Within a decade of its invention, bubble chambers were producing hundreds of thousands of photographs per week. Ultra-short-lived particles were discovered in the 1960s that would have gone undetected without the bubble chamber.

Liquid hydrogen is an excellent liquid with which to fill bubble chambers, because the single-proton hydrogen nucleus is so simple as to introduce a minimum of added complication. In 1973, a bubble chamber was built at Wheaton, Illinois, that was 15 feet in diameter and contained 7,300 gallons of liquid hydrogen. Some bubble chambers contain liquid helium.

Although the bubble chamber is more sensitive to short-lived particles than the cloud chamber, it has its shortcomings. Unlike the cloud chamber, the bubble chamber cannot be triggered by desired events. It must record everything wholesale, and uncounted numbers of tracks must be searched through for those of significance. The search was on, then, for some method of detecting tracks that combined the selectivity of the cloud chamber with the sensitivity of the bubble chamber.

This need was met eventually by the *spark chamber,* in which incoming particles ionize gas and set off electric currents through neon gas that is crossed by many metal plates. The currents show up as a visible line of sparks, marking the passage of the particles, and the device can be adjusted to react only to those particles under study. The first practical spark chamber was constructed in 1959 by the Japanese physicists Saburo Fukui and Shotaro Miyamoto. In 1963, Soviet physicists improved it further, heightening its sensitivity and flexibility. Short streamers of light are produced that, seen on end, make a virtually continuous line (rather than the separate sparks of the spark chamber). The modified device is therefore a *streamer chamber.* It can detect events that take place within the chamber, and particles that streak off in any direction, where the original spark chamber fell short in both respects.

TRANSMUTATION OF ELEMENTS

But, leaving modern sophistication in studying the flight of subatomic particles, we must turn back half a century to see what happened when Rutherford bombarded nitrogen nuclei with alpha particles within one of the original Wilson cloud chambers. The alpha particle would leave a track that would end suddenly in a fork—plainly, a collision with a nitrogen nucleus. One branch of the fork would be comparatively thin, representing a proton shooting off. The other branch, a short, heavy track, represented what was left of the nitrogen nucleus, rebounding from the collision. But there was no sign of the alpha particle itself. It seemed that it must have been absorbed by the nitrogen nucleus, and this supposition was later verified by the British physicist Patrick Maynard Stuart Blackett, who is supposed to have taken more than 20,000 photographs in the process of collecting eight such collisions (surely an example of superhuman patience, faith, and persistence). For this and other work in the field of nuclear physics, Blackett received the Nobel Prize in physics in 1948.

The fate of the nitrogen nucleus could now be deduced. When it absorbed the alpha particle, its mass number of 14 and positive charge of 7 were raised to 18 and 9, respectively. But since the combination immediately lost a proton, the mass number dropped to 17 and the positive charge to 8. Now the element with a positive charge of 8 is oxygen, and the mass number 17 belongs to the isotope oxygen 17. In other words, Rutherford had, in 1919, transmuted nitrogen into oxygen. This was the first man-made transmutation in history. The dream of the alchemists had been fulfilled, though in a manner they could not possibly have foreseen or duplicated with their primitive techniques.

As projectiles, alpha particles from radioactive sources had limits: they were not nearly energetic enough to break into nuclei of the heavier elements, whose high positive charges exercise a strong repulsion against positively charged particles. But the nuclear fortress had been breached, and more energetic attacks were to come.

New Particles

The matter of attacks on the nucleus brings us back to the question of the makeup of the nucleus. The proton-electron theory of nuclear structure, although it explained isotopes perfectly, fell afoul of certain other facts. Subatomic particles generally have a property visualized as *spin*, something like astronomical objects rotating on their axis. The units in which such spin is measured are so taken that both protons and electrons turn out to have spins of either $+\frac{1}{2}$ or $-\frac{1}{2}$. Hence, an even number of electrons or protons (or both), if all confined within a nucleus, should lend that nucleus a spin of 0 or of some whole number—$+1$, -1, $+2$, -2, and so on. If an odd number of electrons or protons (or both) make up a nucleus, the total spin should be a *half-number*, such as $+\frac{1}{2}$, $-\frac{1}{2}$, $+1\frac{1}{2}$, $-1\frac{1}{2}$, $+2\frac{1}{2}$, $-2\frac{1}{2}$, and so on. If you try adding up an even number of positive or negative halves (or a mixture), and then do the same with an odd number, you will see this is, and must be, so.

Now as it happens, the nitrogen nucleus has an electric charge of $+7$ and a mass of 14. By the proton-electron theory, its nucleus must contain 14 protons to account for the mass, and 7 electrons to neutralize half of the charge and leave $+7$. The total number of particles in such a nucleus is 21, and the overall spin of the nitrogen nucleus should be a half-number—but it is not. It is a whole number.

This sort of discrepancy turned up in other nuclei as well, and it seemed that the proton-electron theory just would not do. As long as those were the only subatomic particles known, however, physicists were helpless at finding a substitute theory.

THE NEUTRON

In 1930, however, two German physicists, Walter Bothe and Herbert Becker, reported that they had released from the nucleus a mysterious new radiation of unusual penetrating power. They had produced it by bombarding beryllium atoms with alpha particles. The year before, Bothe had devised methods for using two or more counters in conjunction—*coincidence counters*. These could be used to identify nuclear events taking place in a millionth of a second. For this and other work, he shared in the Nobel Prize for physics in 1954.

Two years later the Bothe-Becker discovery was followed by the French physicists Frédéric and Irène Joliot-Curie. (Irène was the daughter of Pierre and Marie Curie, and Joliot had added her name to his on marrying her.) They used the new-found radiation from beryllium to bombard paraffin, a waxy substance composed of hydrogen and carbon. The radiation knocked protons out of the paraffin.

The English physicist James Chadwick quickly suggested that the radiation consisted of particles. To determine their size, he bombarded boron atoms with them; and from the increase in mass of the new nucleus, he calculated that the particle added to the boron had a mass about equal to the proton. Yet the particle itself could not be detected in a Wilson cloud chamber. Chadwick decided that the explanation must be that the particle had no electric charge (an uncharged particle produces no ionization and therefore condenses no water droplets).

So Chadwick concluded that a completely new particle had turned up—a particle with just about the same mass as a proton but without any charge, or, in other words, electrically neutral. The possibility of such a particle had already been suggested, and a name had even been proposed—*neutron*. Chadwick accepted that name. For his discovery of the neutron, he was awarded the Nobel Prize in physics in 1935.

The new particle at once solved certain doubts that theoretical physicists had had about the proton-electron model of the nucleus. The German theoretical physicist Werner Heisenberg announced that the concept of a nucleus consisting of protons and neutrons, rather than of protons and electrons, gave a much more satisfactory picture. Thus, the nitrogen nucleus could be visualized as made up of seven protons and seven neutrons. The mass number would then be 14, and the total charge (atomic number) would be +7. What's more, the total number of particles in the nucleus would be fourteen—an even number—rather than twenty-one (an odd number) as in the older theory. Since the neutron, like the proton, has a spin of either $+\frac{1}{2}$ or $-\frac{1}{2}$, an even number of neutrons and protons would give the nitrogen nucleus a spin equal to a whole number, and fits the observed facts. All the nuclei that had spins that could not be explained by the proton-electron theory, turned out to have spins that *could* be explained by the proton-neutron theory. The proton-neutron theory was accepted at once and has remained accepted ever since. There are no electrons within the nucleus after all.

Furthermore, the new model fitted the facts of the periodic table of elements just as neatly as the old one had. The helium nucleus, for instance, would consist of two protons and two neutrons, which explained its mass of 4 and nuclear charge of 2 units. And the concept accounted for isotopes in very simple fashion. For example, the chlorine-35 nucleus would have seventeen protons and eighteen neutrons; the chlorine-37 nucleus, seventeen protons and twenty neutrons. They would both, therefore, have the same nuclear charge, and the extra weight of the heavier isotope would lie in its two extra neutrons. Likewise, the three isotopes of oxygen would differ only in their numbers of neutrons: oxygen 16 would have eight protons and eight neutrons; oxygen 17, eight protons and nine neutrons; oxygen 18, eight protons and ten neutrons (figure 7.2).

Figure 7.2. Nuclear makeup of oxygen 16, oxygen 17, and oxygen 18. They contain eight protons each and, in addition, eight, nine, and ten neutrons, respectively.

 PROTON ◯ NEUTRON

In short, every element could be defined simply by the number of protons in its nucleus, which is equivalent to the atomic number. All the elements except hydrogen, however, also had neutrons in the nucleus, and the mass number of a nuclide was the sum of its protons and neutrons. Thus, the neutron joined the proton as a basic building block of matter. For convenience, both are now lumped together under the general term *nucleons*, a term first used in 1941 by the Danish physicist Christian Moller. From this came *nucleonics*, suggested in 1944 by the American engineer Zay Jeffries to represent the study of nuclear science and technology.

This new understanding of nuclear structure has resulted in additional classifications of nuclides. Nuclides with equal numbers of protons are, as I have just explained, isotopes. Similarly, nuclides with equal numbers of neutrons (as, for instance, hydrogen 2 and helium 3, each containing one neutron in the nucleus) are *isotones*. Nuclides with equal total number of nucleons, and therefore of equal mass numbers—such as calcium 40 and argon 40—are *isobars*.

The proton-electron theory of nuclear structure left unexplained, just at first, the fact that radioactive nuclei could emit beta particles (electrons). Where did the electrons come from if there were none in the nucleus? That problem was cleared up, however, as I shall shortly explain.

THE POSITRON

In a very important respect the discovery of the neutron disappointed physicists. They had been able to think of the universe as being built of just two fundamental particles—the proton and the electron. Now a third had to be added. To scientists, every retreat from simplicity is regrettable.

The worst of it was that, as things turned out, this was only the beginning. Simplicity's backward step quickly became a headlong rout. There were more particles to come.

For many years, physicists had been studying the mysterious *cosmic rays* from space, first discovered in 1911 by the Austrian physicist Victor Francis Hess on balloon flights high in the atmosphere.

The presence of such radiation was detected by an instrument so simple as to hearten those who sometimes feel that modern science can progress only by use of unbelievably complex devices. The instrument was an *electroscope*, consisting of two pieces of thin gold foil attached to a metal rod within a metal housing fitted with windows. (The ancestor of this device was constructed as long ago as 1706 by the English physicist Francis Hauksbee.)

If the metal rod is charged with static electricity, the pieces of gold foil separate. Ideally, they would remain separated forever, but ions in the surrounding atmosphere slowly conduct away the charge so that the leaves gradually collapse toward each other. Energetic radiation—such as X rays, gamma rays, or streams of charged particles—produces the ions necessary for such charge leakage. Even if the electroscope is well shielded, there is still a slow leakage, indicating the presence of a very penetrating radiation not directly related to radioactivity. It was this penetrating radiation, which increased in intensity, the higher Hess rose in the atmosphere. Hess shared the Nobel Prize for physics in 1936 for this discovery.

The American physicist Robert Andrews Millikan, who collected a great deal of information on this radiation (and gave it the name *cosmic rays*), decided that it must be a form of electromagnetic radiation. Its penetrating power was such that some of it could even pass through several feet of lead. To Millikan this suggested that the radiation was like the penetrating gamma rays, but with an even shorter wavelength.

Others, notably the American physicist Arthur Holly Compton, contended that the cosmic rays were particles. There was a way to investigate the question. If they were charged particles, they should be deflected by the earth's magnetic field as they approached the earth from outer space. Compton studied the measurements of cosmic radiation at various latitudes and found that it did indeed curve with the magnetic field: it was weakest near the magnetic equator and strongest near the poles, where the magnetic lines of force dipped down to the earth.

The *primary* cosmic particles, as they enter our atmosphere, carry fantastically high energies. Most of them are protons, but some are nuclei of heavier elements. In general, the heavier the nucleus, the rarer it is among the cosmic particles. Nuclei as complex as those making up iron atoms were detected quickly enough; and in 1968, nuclei as complex as those of uranium were detected. The uranium nuclei make up only 1 particle in 10 million. A few very high-energy electrons are also included.

When the primary particles hit atoms and molecules of the air, they smash these nuclei and produce all sorts of *secondary* particles. It is this secondary radiation (still very energetic) that we detect near the earth, but balloons sent to the upper atmosphere have recorded the primary radiation.

Now it was as a result of cosmic-ray research that the next new particle—after the neutron—was discovered. This discovery had actually been predicted by a theoretical physicist. Paul Adrien Maurice Dirac had reasoned, from a mathematical analysis of the properties of subatomic particles, that each particle should have an *antiparticle.* (Scientists like nature to be not only simple but also symmetrical.) Thus there ought to be an *antielectron,* exactly like the electron except that it had a positive instead of a negative charge, and an *antiproton* with a negative instead of a positive charge.

Dirac's theory did not make much of a splash in the scientific world when he proposed it in 1930. But, sure enough, two years later the antielectron actually turned up. The American physicist Carl David Anderson was working with Millikan on the problem of whether cosmic rays were electromagnetic radiation or particles. By then, most people were ready to accept Compton's evidence that they were charged particles, but Millikan was an extraordinarily hard loser and was not satisfied that the issue was settled. Anderson undertook to find out whether cosmic rays entering a Wilson cloud chamber would be bent by a strong magnetic field. To slow down the rays sufficiently so that the curvature, if any, could be detected, Anderson placed in the chamber a lead barrier about ¼ inch thick. He found that the cosmic radiation crossing the chamber after it came through the lead did make a curved track. But he also found something else. In their passage through the lead, the energetic cosmic rays knocked particles out of the lead atoms. One of these particles made a track just like that of an electron. But it curved in the wrong direction! Same mass but opposite charge. There it was—Dirac's antielectron. Anderson called his discovery the *positron*. It is an example of the secondary radiation produced by cosmic rays; but in 1963, it was found that positrons were included among the primary radiations as well.

Left to itself, the positron is as stable as the electron (why not, since it is identical with the electron except for electric charge?) and could exist indefinitely. It is not, however, left to itself, for it comes into existence in a universe filled with electrons. As it streaks along, it almost immediately (say, within a millionth of a second) finds itself in the neighborhood of one.

For a moment, there may be an electron-positron association—a situation in which the two particles circle each other about a mutual center of force. In 1945, the American physicist Arthur Edward Ruark suggested that this two-particle system be called *positronium*, and in 1951, the Austrian-American physicist Martin Deutsch was able to detect positronium through the characteristic gamma-radiation it gave up.

However, even if a positronium system forms, it remains in existence for only a 10-millionth of a second, at most. The dance ends in the combination of the electron and positron. When the two opposite bits of matter combine, they cancel each other, leaving no matter at all (*mutual annihilation*); only energy, in the form of gamma rays, is left behind. Thus was confirmed Albert Einstein's suggestion that matter could be converted into energy and vice versa. Indeed, Anderson soon succeeded in detecting the reverse phenomenon: gamma rays suddenly disappearing and giving rise to an electron-positron pair. This is called *pair production*. (Anderson, along with Hess, received the Nobel Prize in physics in 1936.)

The Joliot-Curies shortly afterward came across the positron in another connection, and in so doing, made an important discovery. Bombarding aluminum atoms with alpha particles, they found that the procedure produced not only protons but also positrons. This in itself was interesting but not fabulous. When they stopped the bombardment, however, the aluminum kept right on emitting positrons! The emission faded off with time. Apparently they had created a new radioactive substance in the target.

The Joliot-Curies interpreted what had happened in this way: When an aluminum nucleus absorbed an alpha particle, the addition of two protons changed aluminum (atomic number 13) to phosphorus (atomic number 15). Since the alpha particle contained four nucleons altogether, the mass number would go up by four—from aluminum 27 to phosphorus 31. Now if the reaction knocked a proton out of this nucleus, the reduction of its atomic number and mass number by one would change it to another element—namely, silicon 30.

Since an alpha particle is the nucleus of helium, and a proton the nucleus of hydrogen, we can write the following equation of this *nuclear reaction:*

aluminum 27 + helium 4 → silicon 30 + hydrogen 1

Notice that the mass numbers balance: 27 plus 4 equals 30 plus 1. So do the atomic numbers, for aluminum's is 13 and helium's 2, making 15 together, while silicon's atomic number of 14 and hydrogen's 1 also add up to 15. This balancing of both mass numbers and atomic numbers is a general rule of nuclear reactions.

The Joliot-Curies assumed that neutrons as well as protons had been formed in the reaction. If phosphorus 31 emitted a neutron instead of a proton, the atomic number would not change, though the mass number would go down one. In that case the element would remain phosphorus but become phosphorus 30. This equation would read:

aluminum 27 + helium 4 → phosphorus 30 + neutron 1

Since the atomic number of phosphorus is 15 and that of the neutron is 0, again the atomic numbers on both sides of the equation also balance.

Both processes—alpha absorption followed by proton emission, and alpha absorption followed by neutron emission—take place when aluminum is bombarded by alpha particles. But there is one important distinction between the two results. Silicon 30 is a perfectly well-known isotope of silicon, making up a little more than 3 percent of the silicon in nature. But phosphorus 30 does not exist in nature. The only known natural form of phosphorus is phosphorus 31. Phosphorus 30, in short, is a radioactive isotope with a brief lifetime that exists today only when it is produced artificially; in fact, it was the first such isotope made in the laboratory. The Joliot-Curies received the Nobel Prize in chemistry in 1935 for their discovery of artificial radioactivity.

The unstable phosphorus 30 that the Joliot-Curies had produced by bombarding aluminum quickly broke down by emitting positrons. Since the positron, like the electron, has practically no mass, this emission did not change the mass number of the nucleus. However, the loss of one positive charge did reduce its atomic number by one, so that it was converted from phosphorus to silicon.

Where does the positron come from? Are positrons among the components of the nucleus? The answer is no. What happens is that a proton within the nucleus changes to a neutron by shedding its positive charge, which is released in the form of a speeding positron.

Now the emission of beta particles—the puzzle we encountered earlier in the chapter—can be explained. This comes about as the result of a process just the reverse of the decay of a proton into a neutron: that is, a neutron changes into a proton. The proton-to-neutron change releases a positron; and, to maintain the symmetry, the neutron-to-proton change releases an electron (the beta particle). The release of a negative charge is equivalent to the gain of a positive charge and accounts for the formation of a positively charged proton from an uncharged neutron. But how does the uncharged neutron manage to dig up a negative charge and send it flying outward?

Actually, if it were just a negative charge, the neutron could not do so. Two centuries of experience have taught physicists that neither a negative electric charge nor a positive electric charge can be created out of nothing. Neither can either type of charge be destroyed. This is the law of *conservation of electric charge*.

However, a neutron does not create only an electron in the process of producing a beta particle; it creates a proton as well. The uncharged neutron disappears, leaving in its place a positively charged proton and a negatively charged electron. The two new particles, *taken together*, have an over-all electric charge of zero. No *net* charge has been created. Similarly, when a positron and electron meet and engage in mutual annihilation, the charge of the positron and electron, *taken together*, is zero to begin with.

When a proton emits a positron and changes into a neutron, the original particle (the proton) is positively charged, and the final particles (the neutron and positron), taken together, have a positive charge.

It is also possible for a nucleus to absorb an electron. When this happens, a proton within the nucleus changes to a neutron. An electron plus a proton (which, taken together, have a charge of zero) form a neutron, which has a zero charge. The electron captured is from the innermost electron shell of the atom, since the electrons of that shell are closest to the nucleus and most easily gathered in. As the innermost shell is the K-shell (see chapter 6), the process is called *K-capture*. An electron from the L-shell then drops into the vacant spot, and an X ray is emitted. It is by these X rays that K-capture can be detected. This was first accomplished in 1938 by the American physicist Luis Walter Alvarez. Ordinary nuclear reactions involving the nucleus alone are usually not affected by chemical change, which affects electrons only. Since K-capture affects electrons as well as nuclei, the chance of its occurring can be somewhat altered as a result of chemical change.

All of these particle interactions satisfy the law of conservation of electric charge and must also satisfy other conservation laws. Any particle interaction that violates none of the conservation laws will eventually occur, physicists suspect, and an observer with the proper tools and proper patience will detect it. Those events that violate a conservation law are "forbidden" and will not take place. Nevertheless, physicists are occasionally surprised to find that what had seemed a conservation law is not as rigorous or as universal as had been thought—as we shall see.

RADIOACTIVE ELEMENTS

Once the Joliot-Curies had created the first artificial radioactive isotope, physicists proceeded merrily to produce whole tribes of them. In fact, radioactive varieties of every single element in the periodic table have now been formed in the laboratory. In the modern periodic table, each element is really a family, with stable and unstable members, some found in nature, some only in the laboratory.

For instance, hydrogen comes in three varieties. First there is ordinary hydrogen, containing a single proton. In 1932, the chemist Harold Urey succeeded in isolating a second by slowly evaporating a large quantity of water, on the theory that he would be left in the end with a concentration of the heavier form of hydrogen that was suspected to exist. Sure enough, when he examined the last few drops of unevaporated water spectroscopically, he found a faint line in the spectrum in exactly the position predicted for *heavy hydrogen*.

Heavy hydrogen's nucleus is made up of one proton and one neutron. Having a mass number of two, the isotope is hydrogen 2. Urey named the atom *deuterium*, from a Greek word meaning "second," and the nucleus a *deuteron*. A water molecule containing deuterium is called *heavy water*. Because deuterium has twice the mass of ordinary hydrogen, heavy water has higher boiling and freezing points than ordinary water. Whereas ordinary water boils at 100° C, and freezes at 0° C, heavy water boils at 101.42° C and freezes at 3.79° C. Deuterium itself has a boiling point of 23.7° K as compared with 20.4° K for ordinary hydrogen. Deuterium occurs in nature in the ratio of 1 part to 6,000 parts of ordinary hydrogen. For his discovery of deuterium, Urey received the Nobel Prize in chemistry in 1934.

The deuteron turned out to be a valuable particle for bombarding nuclei. In 1934, the Australian physicist Marcus Lawrence Elwin Oliphant and the Austrian chemist Paul Harteck, attacking deuterium itself with deuterons, produced a third form of hydrogen, made up of one proton and two neutrons. The reaction went:

hydrogen 2 + hydrogen 2 → hydrogen 3 + hydrogen 1

The new "superheavy" hydrogen was named *tritium*, from the Greek word for "third," and its nucleus is a *triton*. Its boiling point is 25.0° K, and its melting point 20.5° K. Pure tritium oxide (*superheavy water*) has been prepared, and its melting point is 4.5° C. Tritium is radioactive and breaks down comparatively rapidly. It exists in nature, being formed as one of the products of the bombardment of the atmosphere by cosmic rays. In breaking down, it emits an electron and changes to helium 3, a stable but rare isotope of helium, mentioned in the previous chapter (figure 7.3).

Figure 7.3. Nuclei of ordinary hydrogen, deuterium, and tritium.

Of the helium in the atmosphere, only about 1 atom out of 800,000 is helium 3, all originating, no doubt, from the breakdown of hydrogen 3 (tritium) which is itself formed from the nuclear reactions taking place when cosmic-ray particles strike atoms in the atmosphere. The tritium that remains at any one time is even rarer. It is estimated that only 3½ pounds exist all told in the atmosphere and oceans. The helium-3 content of helium obtained in natural gas wells, where cosmic rays have had less opportunity to form tritium, is even smaller in percentage.

These two isotopes, helium 3 and helium 4, are not the only heliums. Physicists have created two radioactive forms: helium 5, one of the most unstable nuclei known; and helium 6, also very unstable.

And so it goes. By now the list of known isotopes has grown to about 1,400 altogether. Over 1,100 of these are radioactive, and many of them have been created by new forms of atomic artillery far more potent than the alpha particles from radioactive sources which were the only projectiles at the disposal of Rutherford and the Joliot-Curies.

The sort of experiment performed by the Joliot-Curies in the early 1930s seemed a matter of the scientific ivory tower at the time, but it has come to have a highly practical application. Suppose a set of atoms of one kind, or of many, are bombarded with neutrons. A certain percentage of each kind of atom will absorb a neutron, and a radioactive atom will generally result. This radioactive element will decay, giving off subatomic radiation in the form of particles or gamma rays.

Every different type of atom will absorb neutrons to form a different type of radioactive atom, giving off different and characteristic radiation. The radiation can be detected with great delicacy. From its type and from the rate at which its production declines, the radioactive atom giving it off can be identified and, therefore, so can the original atom before it absorbed a neutron. Substances can be analyzed in this fashion (*neutron-activation analysis*) with unprecedented precision: amounts as small as a trillionth of a gram of a particular nuclide are detectable.

Neutron-activation analysis can be used to determine delicate differences in the impurities contained in samples of particular pigments from different centuries and, in this way, can determine the authenticity of a supposedly old painting, using only the barest fragment of its pigment. Other delicate decisions of this sort can be made: even hair from Napoleon's century-and-a-half-old corpse was studied and found to contain quantities of arsenic—though whether murderous, medicinal, or fortuitous is hard to say.

PARTICLE ACCELERATORS

Dirac had predicted not only an antielectron (the positron) but also an antiproton. To produce an antiproton, however, would take vastly more energy. The energy needed was proportional to the mass of the particle. Since the proton was 1,836 times as massive as the electron, the formation of an antiproton called for at least 1,836 times as much energy as the formation of a positron. The feat had to wait for the development of a device for accelerating subatomic particles to sufficiently high energies.

At the time of Dirac's prediction, the first steps in this direction had just been taken. In 1928, the English physicists John Douglas Cockcroft and Ernest Thomas Sinton Walton, working in Rutherford's laboratory, developed a *voltage multiplier*, a device for building up electric potential, which could drive the charged proton up to an energy of nearly 400,000 electron volts. (One *electron volt* is equal to the energy developed by an electron accelerated across an electric field with a potential of 1 volt.) With protons accelerated in this machine they were able to break up the lithium nucleus and, for this work, were awarded the Nobel Prize for physics in 1951.

Meanwhile the American physicist Robert Jemison Van de Graaff was creating another type of accelerating machine. Essentially, it operated by separating electrons from protons and depositing them at opposite ends of the apparatus by means of a moving belt. In this way the *Van de Graaff electrostatic generator* developed a very high electric potential between the opposite ends; Van de Graaff got it up to 8 million volts. Electrostatic generators can easily accelerate protons to a speed amounting to 24 million electron volts (physicists now invariably abbreviate million electron volts to *Mev*).

The dramatic pictures of the Van de Graaff electrostatic generator producing huge sparks caught the popular imagination and introduced the public to the *atom smasher*. It was popularly viewed as a device to produce "man-made lightning," although, of course, it was much more than that. (A generator designed to produce artificial lightning and nothing more had actually been built in 1922 by the German-American electrical engineer Charles Proteus Steinmetz.)

The energy that can be reached in such a machine is restricted by practical limits on the attainable potential. However, another scheme for accelerating particles shortly made its appearance. Suppose that, instead of firing particles with one big shot, you accelerated them with a series of small pushes. If each successive push was timed just right, it would increase the speed each time, just as pushes on a child's swing will send it higher and higher if they are applied "in phase" with the swing's oscillations.

This idea gave birth, in 1931, to the *linear accelerator* (figure 7.4). The particles are driven down a tube divided into sections. The driving force is an alternating electric field, so managed that as the particles enter each successive section, they get another push. Since the particles speed up as they go along, each section must be longer than the one before, so that the particles will take the same time to get through it and will be in phase with the timing of the pushes.

It is not easy to keep the timing just right, and anyway there is a limit to the length of a tube it is practical to make, so the linear accelerator did not catch on in the 1930s. One of the things that pushed it into the background was that Ernest Orlando Lawrence of the University of California conceived a better idea.

Figure 7.4. Principle of the linear accelerator. A high-frequency alternating charge alternately pushes and pulls the charged particles in the successive drive tubes, accelerating them in one direction.

Instead of driving the particles down a straight tube, why not whirl them around in a circular path? A magnet could bend them in such a path. Each time they completed a half-circle, they would be given a kick by the alternating field; and in this setup, the timing would not be so difficult to control. As the particles speeded up, their path would be bent less sharply by the magnet, so they would move in ever wider circles and perhaps take the same time for each round trip. At the end of their spiraling flight, the particles would emerge from the circular chamber (actually divided into semicircular halves, called *dees*) and strike their target.

Lawrence's compact new device was named the *cyclotron* (figure 7.5). His first model, less than 1 foot in diameter, could accelerate protons to energies of nearly 1.25 Mev. By 1939 the University of California had a cyclotron, with magnets 5 feet across, capable of raising particles to some 20 Mev, twice the speed of the most energetic alpha particles emitted by radioactive sources. In that year Lawrence received the Nobel Prize in physics for his invention.

The cyclotron itself had to stop at about 20 Mev, because at that energy the particles were traveling so fast that the mass increase with velocity—an effect predicted by Einstein's theory of relativity—became appreciable. This increase in mass caused the particles to start lagging and falling out of phase with the electrical kicks. But there was a cure for this, and it was worked out in 1945 independently by the Soviet physicist Vladimir Iosifovich Veksler and the California physicist Edwin Mattison McMillan. The cure was simply to synchronize the alternations of the electric field with the increase in mass of the particles. This modification of the cyclotron was called the *synchrocyclotron.* By 1946 the University of California had built one that accelerated particles to energies of 200 to 400 Mev. Later larger synchrocyclotrons in the United States and in the Soviet Union raised the energies to 700 to 800 Mev.

Meanwhile the acceleration of electrons had been getting separate attention. To be useful in smashing atoms, the light electrons had to be raised to much higher speeds than protons (just as a ping-pong ball has to be moving much faster than a golf ball to do as much damage). The cyclotron would not work for electrons, because at the high velocities needed to make the electrons effective, their increase in mass was too great. In 1940 the American physicist Donald William Kerst designed an electron-accelerating device which balanced the increasing mass with an electric field of increasing strength. The electrons were kept in the same circular path instead of spiraling outward. This instrument was named the *betatron,* after beta particles. Betatrons now generate electron velocities up to 340 Mev.

They have been joined by another instrument of slightly different design called the *electron synchrotron.* The first of these was built in England in 1946

Figure 7.5. Principle of the cyclotron, shown in top view (*above*) and side view (*below*). Particles injected from the source are given a kick in each dee by the alternating charge and are bent in their spiral path by a magnet.

A.C.

+

−

DEE   DEE   TARGET

PARTICLE PATH

NORTH POLE

SOUTH POLE

by F. K. Goward and D. E. Barnes. These raise electron energies to the 1,000 Mev mark, but cannot go higher because electrons moving in a circle radiate energy at increasing rates as velocity is increased. This radiation produced by an accelerating particle is called *Bremsstrahlung*, a German word meaning "braking radiation."

Taking a leaf from the betatron and electron synchrotron, physicists working with protons began about 1947 to build *proton synchrotrons*, which likewise kept their particles in a single circular path. This helped save on weight. Where particles move in outwardly spiraling paths, a magnet must extend the entire width of the spiral to keep the magnetic force uniform throughout. With the path held in a circle, the magnet need be only large enough to cover a narrow area.

Because the more massive proton does not lose energy with motion in a circular path as rapidly as does the electron, physicists set out to surpass the 1,000-Mev mark with a proton synchroton. This value of 1,000 Mev is equal to 1 billion electron volts—abbreviated to *Bev*. (In Great Britain a billion is a million million, so Bev does not mean the same thing as in the United States; for 1,000 Mev the British use the shorthand *Gev*, the G from *giga*, Greek for "giant.")

In 1952, the Brookhaven National Laboratory on Long Island completed a proton synchroton that reached 2 to 3 Bev. They called it the *cosmotron*, because it had arrived at the main energy range of particles in the cosmic rays. Two years later, the University of California brought in its *Bevatron*, capable of producing particles of between 5 and 6 Bev. Then, in 1957, the Soviet Union announced that its *phasotron* had got to 10 Bev.

But by now these machines seem puny in comparison with accelerators of a newer type, called the *strong-focusing synchrotron*. The limitation on the bevatron type is that particles in the stream fly off into the walls of the channel in which they travel. The new type counteracts this tendency by means of alternating magnetic fields of different shape which keep focusing the particles in a narrow stream. The idea was first suggested by Christofilos, whose "amateur" abilities outshone the professionals here as well as in the case of the Christofilos effect. This, incidentally, further decreased the size of the magnet required for the energy levels attained. Where particle energy was increased fiftyfold, the weight of the magnet involved was less than doubled.

In November 1959, the European Committee for Nuclear Research (CERN), a cooperative agency of twelve nations, completed in Geneva a strong-focusing synchrotron which reached 24 Bev and produced large pulses of particles (containing 10 billion protons) every 3 seconds. This synchrotron is nearly three city blocks in diameter, and one round trip through it is two-fifths of a mile. In the 3-second period during which the pulse builds up, the protons travel half a million times around that track. The instrument has a magnet weighing 3,500 tons and costs 30 million dollars.

The advance continued. Higher and higher energies were sought in order to produce more and more unusual particle interactions, forming more and more massive particles, and learning more and more about the ultimate structure of matter. For instance, instead of accelerating a stream of particles and having them collide with some fixed target, why not set up two streams of particles, circling in opposite directions in *storage rings*, where the speed is simply maintained for some period of time. At appropriate times, the two streams are so directed that they will collide with each other head on. The effective energy of collision is four times that of either colliding with a fixed target. At Fermilab (Fermi National Accelerator Laboratory) near Chicago, an accelerator working on this principle went into operation in 1982 and should reach 1,000 Bev. It is called the *Tevatron*, the *T* standing for "trillion," of course. Other accelerators are being planned that may eventually reach as high as 20,000 Bev.

The linear accelerator, or *linac*, has also undergone a revival. Improvements in technique have removed the difficulties that plagued the early models. For extremely high energies, a linear accelerator has some advantages over the cyclic type. Since electrons do not lose energy when traveling in a straight line, a linac can accelerate electrons more powerfully and focus beams on targets more sharply. Stanford University has built a linear accelerator 2 miles long which can reach energies of perhaps 45 Bev.

With merely the Bevatron, man at last came within reach of creating the antiproton. The California physicists set out deliberately to produce and detect it. In 1955, Owen Chamberlain and Emilio G. Segrè, after bombarding copper with protons of 6.2 Bev hour after hour, definitely caught the antiproton—in fact, sixty of them. It was far from easy to identify them. For every antiproton produced, 40,000 particles of other types came into existence. But by an elaborate system of detectors, so designed and arranged that only an antiproton could touch all the bases, they recognized the particle beyond question. For their achievement, Chamberlain and Segrè received the Nobel Prize in physics in 1959.

The antiproton is as evanescent as the positron—at least in our universe. Within a tiny fraction of a second after it is created, the particle is snatched up by some normal, positively charged nucleus. There the antiproton and one of the protons of the nucleus annihilate each other, turning into energy and minor particles. In 1965, enough energy was concentrated to reverse the process and produce a proton-antiproton pair.

Once in a while, a proton and an antiproton have only a near collision instead of a direct one. When that happens, they mutually neutralize their respective charges. The proton is converted to a neutron, which is fair enough. But the antiproton becomes an *antineutron*! What can an *antineutron* be? The positron is the opposite of the electron by virtue of its opposite charge, and the antiproton is likewise "anti" by virtue of its charge. But what gives the uncharged antineutron the quality of oppositeness?

PARTICLE SPIN

Here we must bring up the matter of particle spin again, a property first suggested, by the way, in 1925, by the Dutch physicists George Eugene Uhlenbeck and Samuel Abraham Goudsmit. In spinning, the particle generates a tiny magnetic field; such fields have been measured and thoroughly explored, notably by the German physicist Otto Stern and the American physicist Isidor Isaac Rabi who received the Nobel Prizes in physics in 1943 and 1944, respectively, for their work on this phenomenon.

Those particles—like the proton, the neutron, and the electron—which have spins that can be measured in half-numbers can be dealt with according to a system of rules worked out independently, in 1926, by Fermi and Dirac. These are therefore called *Fermi-Dirac statistics*. Particles that obey these are *fermions*, so that the proton, the electron, and the neutron are all fermions.

There also exist particles whose spin can be expressed as whole numbers. They can be dealt with by another set of rules devised by Einstein and by the Indian physicist Satyendranath Bose. Particles that follow the *Bose-Einstein* statistics are *bosons*. The alpha particle, for instance, is a boson.

These classes of particles have different properties. For instance, the Pauli exclusion principle (see chapter 5) applies not only to electrons but to all fermions. It does not, however, apply to bosons.

It is easy to understand how a charged particle sets up a magnetic field, but not so easy to see why the uncharged neutron should. Yet it unquestionably does. The most direct evidence is that when a neutron beam strikes magnetized iron, it behaves differently from the way it does when the iron is not magnetized. The neutron's magnetism arises from the strong probability that (as we shall see) the particle is made up of other particles that *do* carry electric charge. These cancel each other out over the neutron as a whole but somehow manage to set up a magnetic field when the particle spins.

In any case, the spin of the neutron gives us the answer to the question of what the antineutron is. It is simply a neutron with its spin direction reversed; its south magnetic pole, say, is up instead of down. Actually the proton and antiproton and the electron and positron show exactly the same pole-reversed phenomenon.

Antiparticles can undoubtedly combine to form *antimatter*, as ordinary particles form ordinary matter (figure 7.6). The first actual example of antimatter was produced at Brookhaven in 1965. There the bombardment of a

Figure 7.6. An atom of hydrogen and an atom of its antimatter counterpart, consisting of an antiproton and a positron.

beryllium target with 7 Bev protons produced combinations of antiprotons and antineutrons, something that was an *antideuteron*. *Antihelium-3* has since been produced; and undoubtedly, if enough pains are taken, still more complicated antinuclei can be formed. The principle is clear, however, and no physicist doubts it. Antimatter can exist.

But does it exist in actuality? Are there masses of antimatter in the universe? If there were, they would not betray themselves from a distance. Their gravitational effects and the light they produce would be exactly like that of ordinary matter. If, however, they encountered ordinary matter, the massive annihilation reactions that result ought to be most noticeable. It ought to be, perhaps, but it is not. Astronomers have not spied any energy bursts anywhere in the sky that can be identified unequivocally as the result of matter-antimatter annihilation. Can it be, then, that the universe is almost entirely matter, with little or no antimatter? If so, why? Since matter and antimatter are equivalent in all respects but that of electromagnetic oppositeness, any force that would create one would have to create the other, and the universe should be made of equal quantities of each.

This is a dilemma. Theory tells us there should be antimatter out there; and observation refuses to back it up. Can we be sure that observation is failing us? What about the cores of active galaxies and, even more so, quasars? Might those energetic phenomena be the result of matter-antimatter annihilation? Probably not! Even such annihilation does not seem enough, and astronomers prefer to accept the notion of gravitational collapse and black hole phenomena as the only known mechanism that would produce the required energy.

COSMIC RAYS

What about cosmic rays, then? Most of the cosmic-ray particles have energies between 1 and 10 Bev. This might be accounted for by matter-antimatter interaction, but a few cosmic particles run much higher: 20 Bev, 30 Bev, 40 Bev (see figure 7.7). Physicists at the Massachusetts Institute of Technology have even detected some with the colossal energy of 20 billion Bev. Numbers such as this are more than the mind can grasp, but we may get some idea of what that energy means when we calculate that the amount of energy represented by 20 billion Bev would be enough to enable a single submicroscopic particle to raise a 4-pound weight 2 inches.

Ever since cosmic rays were discovered, people have wondered where they came from and how they arise. The simplest concept is that somewhere in the galaxy—perhaps in our sun, perhaps farther away—there are nuclear reactions going on which shoot forth particles with the huge energies we find them possessing. Indeed, bursts of mild cosmic rays occur every other year or so (as was first discovered in 1942) in connection with flares from the sun. What, then, of such sources as supernovae, pulsars, and quasars? But there is no known nuclear reaction that could produce anything like 20 billion Bev. The mutual annihilation of the heaviest nuclei of matter and antimatter would liberate speeding particles with energies of, at most, 250 Bev.

Figure 7.7. Smashing of a silver atom by a 30,000-Bev cosmic ray. The collision of the cosmic particle with the silver nucleus produced ninety-five nuclear fragments, whose tracks form the star.

The alternative is to suppose, as Fermi did, that some force in space accelerates the cosmic particles. They may come originally with moderate energies from explosions such as supernovae and gradually be accelerated as they travel through space. The most popular theory at present is that they are accelerated by cosmic magnetic fields, acting like gigantic synchrotrons. Magnetic fields do exist in space, and our galaxy as a whole is thought to possess one, although this can at best be but 1/20,000 as intense as the magnetic field associated with the earth.

Traveling through this field, the cosmic particles would be slowly ac-

THE PARTICLES   313

celerated in a curved path. As they gained energy, their paths would swing out
wider and wider until the most energetic ones would whip right out of the
galaxy. Although most of the particles would never reach this escape trajectory,
because they would lose energy by collisions with other particles or with large
bodies, some would. Indeed, the most energetic cosmic particles that reach
us may be passing through our galaxy after having been hurled out of other
galaxies in this fashion.

THE STRUCTURE OF THE NUCLEUS

Now that so much has been learned about the general makeup and nature
of the nucleus, there is great curiosity as to its structure, particularly the fine
structure inside. First of all, what is its shape? Because it is so small and so
tightly packed with neutrons and protons, physicists naturally assume that it
is spherical. The fine details of the spectra of atoms suggest that many nuclei
have a spherical distribution of charge. Some do not: they behave as if they
have two pairs of magnetic poles, and these nuclei are said to have *quadrupole
moments*. But their deviation from the spherical is not large. The most
extreme case is that of the nuclei of the lanthanides, in which the charge
distribution seems to make up a prolate spheroid (football-shaped, in other
words). Even here the long axis is not more than 20 percent greater than the
short axis.

As for the internal structure of the nucleus, the simplest model pictures it
as a tightly packed collection of particles much like a drop of liquid, where
the particles (molecules) are packed closely with little space between, where
the density is virtually even throughout, and where there is a sharp surface
boundary.

This *liquid-drop model* was first worked out in detail in 1936 by Niels Bohr.
It suggests a possible explanation of the absorption and emission of particles
by some nuclei. When a particle enters the nucleus, one can suppose, it
distributes its energy of motion among all the closely packed particles, so that
no one particle receives enough energy immediately to break away. After
perhaps a quadrillionth of a second, when there has been time for billions of
random collisions, some particle accumulates sufficient energy to fly out of the
nucleus.

The model could also account for the emission of alpha particles by the
heavy nuclei. These large nuclei may quiver as liquid drops do if the particles
making them up move about and exchange energy. All nuclei would so quiver,
but the larger nuclei would be less stable and more likely to break up. For that
reason, portions of the nucleus in the form of the two-proton, two-neutron
alpha particle (a very stable combination) may break off spontaneously from
the surface of the nucleus. The nucleus becomes smaller as a result, less liable
to break up through quivering, and is finally stable.

The quivering may result in another kind of instability, too. When a large
drop of liquid suspended in another liquid is set wobbling by currents in the
surrounding fluid, it tends to break up into smaller spheres, often into roughly

equal halves. It was eventually discovered in 1939 (a discovery I will describe quite fully in chapter 10) that some large nuclei could indeed be made to break down in this fashion by bombardment with neutrons. This is called *nuclear fission.*

In fact, such nuclear fission ought to take place sometimes without the introduction of a disturbing particle from outside. The internal quivering should, every once in a while, cause the nucleus to split in two. In 1940, the Soviet physicists G. N. Flerov and K. A. Petrjak actually detected such *spontaneous fission* in uranium atoms. Uranium exhibits instability mainly by emitting alpha particles, but in a pound of uranium there are four spontaneous fissions per second while about 8 million nuclei are emitting alpha particles.

Spontaneous fission also takes place in protactinium, in thorium, and, more frequently, in the transuranium elements. As nuclei get larger and larger, the probability of spontaneous fission increases. In the heaviest elements of all it becomes the most important method of breakdown, far outweighing alpha-particle emission.

Another popular model of the nucleus likens it to the atom as a whole, picturing the nucleons within the nucleus, like the electrons around the nucleus, as occupying shells and subshells, each affecting the others only slightly. This is called the *shell model.*

By analogy with the situation in the atom's electronic shells, one may suppose that the nuclei with filled outer nucleonic shells should be more stable than those whose outer shells are not filled. The simplest theory would indicate that nuclei with 2, 8, 20, 40, 70, or 112 protons or neutrons, would be particularly stable. This, however, does not fit observation. The German-American physicist Maria Goeppert Mayer took account of the spin of protons and neutrons and showed how this would affect the situation. It turned out that nuclei containing 2, 8, 20, 50, 82, or 126 protons or neutrons would then be particularly stable—as fitted the observations. Nuclei with 28 or 40 protons or neutrons would be fairly stable. All others would be less stable, if stable at all. These *shell numbers* are sometimes called *magic numbers* (with 28 or 40 occasionally referred to as *semi-magic numbers.*)

Among the magic-number nuclei are helium 4 (2 protons and 2 neutrons), oxygen 16 (8 protons and 8 neutrons), and calcium 40 (20 protons and 20 neutrons), all especially stable and more abundant in the universe than other nuclei of similar size.

As for the higher magic numbers, tin has ten stable isotopes, each with 50 protons, and lead has four, each with 82 protons. There are five stable isotopes (each of a different element) with 50 neutrons each, and seven stable isotopes with 82 neutrons each. In general, the detailed predictions of the nuclear-shell theory work best near the magic numbers. Midway between (as in the case of the lanthanides and actinides), the fit is poor. But just in the midway regions, nuclei are farthest removed from the spherical (and shell theory assumes spherical shape) and are most markedly ellipsoidal. The 1963 Nobel Prize for physics was awarded to Goeppert Mayer and to two others: Wigner, and the German physicist Johannes Hans Daniel Jensen, who also contributed to the theory.

In general, as nuclei grow more complex, they become rarer in the universe, or less stable, or both. The most complex stable isotopes are lead 208 and bismuth 209, each with the magic number of 126 neutrons, and lead, with the magic number of 82 protons in addition. Beyond that, all nuclides are unstable and, in general, grow more unstable as the size of the nucleus increases. A consideration of magic numbers, however, explains the fact that thorium and uranium possess isotopes that are much more nearly stable than other nuclides of similar size. The theory also predicts that some isotopes of elements 110 and 114 (as I mentioned earlier) might be considerably less unstable than other nuclides of that size. For this last, we must wait and see.

Leptons

The electron and the positron are notable for their small masses—only 1/1,836 that of the proton, the neutron, the antiproton, or the antineutron—and hence are lumped together as *leptons* (from the Greek *leptos,* meaning "thin").

Although the electron was first discovered nearly a century ago, no particle has yet been discovered that is less massive than the electron (or positron) and yet carries an electric charge. Nor is any such discovery expected. It may be that the electric charge, whatever it is (we know what it does and how to measure its properties, but we do not know what it *is*), has associated with itself a minimum mass, and that that is what shows up in the electron. In fact, there may be nothing to the electron *but* the charge; and when the electron behaves as a particle, the electric charge on that particle seems to have no extension but occupies a mere point.

To be sure, some particles have no mass associated with them at all (actually, no *rest-mass,* which I shall explain in the next chapter), but these have no electric charge. For instance, waves of light and other forms of electromagnetic radiation can behave as particles (see the next chapter). This particle manifestation of what we ordinarily think of as a wave is called *photon* from the Greek word for "light."

The photon has a mass of 0, and an electric charge of 0, but it has a spin of 1, so that it is a boson. How can one tell what the spin is? Photons take part in nuclear reactions, being absorbed in some cases, given off in others. In such nuclear reactions, the total spin of the particles involved before and after the reaction must remain unchanged (*conservation of spin*). The only way for this to happen in nuclear reactions involving photons is to suppose that the photon has a spin of 1. The photon is not considered a lepton, that term being reserved for fermions.

There are theoretical reasons for supposing that, when masses undergo acceleration (as when they move in elliptical orbits about another mass or undergo gravitational collapse), they give off energy in the form of gravitational waves. These waves, too, can possess a particle aspect, and such a gravitational particle is called a *graviton*.

The gravitational force is much, much weaker than the electromagnetic force. A proton and an electron attract each other gravitationally with only about $1/10^{39}$ as much force as they attract each other electromagnetically. The graviton must be correspondingly less energetic than the photon and must therefore be unimaginably difficult to detect.

Nevertheless, the American physicist Joseph Weber began the formidable task of trying to detect the graviton in 1957. Eventually he made use of a pair of aluminum cylinders 153 centimeters long and 66 centimeters wide, suspended by a wire in a vacuum chamber. The gravitons (which would be detected in wave form) would displace those cylinders slightly, and a measuring system for detecting a displacement of a hundred-trillionth of a centimeter is used. The feeble waves of the gravitons, coming from deep in space, ought to wash over the entire planet, and cylinders separated by great distances ought to be affected simultaneously. In 1969, Weber announced he had detected the effects of gravitational waves. This produced enormous excitement, for it lent support to a particularly important theory (Einstein's theory of general relativity). Unfortunately, not all scientific tales have happy endings. Other scientists could not duplicate Weber's results no matter how they tried, and the general feeling is that gravitons are still undetected. Nevertheless, physicists are confident enough of the theory to be sure they exist. They are particles with a mass of 0, a charge of 0, and a spin of 2 and are also bosons. The gravitons, too, are not listed among the leptons.

Photons and gravitons do not have antiparticles; or, rather, each is its own antiparticle. One way of visualizing this is to imagine a paper folded lengthwise, then opened, so that there is a crease running down its center. If you put a little circle to the left of the crease, and another an equal distance to the right, they would represent an electron and a positron. The photon and the graviton would be right on the crease.

NEUTRINOS AND ANTINEUTRINOS

So far, then, it would seem there are two leptons: the electron and the positron. Physicists would have been content with that; there seemed to be no overwhelming need for any more—except that there *was* such a need. There were complications that had to do with the emission of beta particles by radioactive nuclei.

The particle emitted by a radioactive nucleus generally carries a considerable amount of energy. Where does the energy come from? It is created by conversion into energy of a little of the nucleus's mass; in other words, the nucleus always loses a little mass in the act of expelling the particle. Now physicists had long been troubled by the fact that often the beta particle emitted in a nucleus's decay did not carry enough energy to account for the amount of mass lost by the nucleus. In fact, the electrons were not all equally deficient. They emerged with a wide spectrum of energies, the maximum (attained by very few electrons) being almost right, but all the others falling short to a smaller or greater degree. Nor was this a necessary concomitant of subatomic particle-emission. Alpha particles emitted by a particular nuclide possessed equal energies in expected quantities. What, then, was wrong with beta-particle emission? What had happened to the missing energy?

Lise Meitner, in 1922, was the first to ask this question with suitable urgency; and by 1930, Niels Bohr, for one, was ready to abandon the great principle of conservation of energy, at least as far as it applied to subatomic particles. In 1931, however, Wolfgang Pauli, in order to save conservation of energy (see chapter 8), suggested a solution to the riddle of the missing energy. His solution was very simple: another particle carrying the missing energy comes out of the nucleus along with the beta particle. This mysterious second particle has rather strange properties: it has no charge and no mass; all it has, as it speeds along at the velocity of light, is a certain amount of energy. This particle looked, in fact, like a fictional item created just to balance the energy books.

And yet, no sooner had it been proposed than physicists were sure that the particle existed. When the neutron was discovered and found to break down into a proton, releasing an electron which, as in beta decay, also carried a deficiency of energy, they were still surer. Enrico Fermi in Italy gave the putative particle a name—*neutrino*, Italian for "little neutral one."

The neutron furnished physicists with another piece of evidence for the existence of the neutrino. As I have mentioned, almost every particle has a spin. The amount of spin is expressed in multiples of one-half, plus or minus, depending on the direction of the spin. Now the proton, the neutron, and the electron have each a spin of $\frac{1}{2}$. If, then, the neutron, with spin of $\frac{1}{2}$, gives rise to a proton and an electron, each with spin of $\frac{1}{2}$, what happens to the law of conservation of spin? There is something wrong here. The proton and the electron may total their spins to 1 (if both spin in the same direction) or to 0 (if their spins are opposite); but any way you slice it, their spins cannot add up to $\frac{1}{2}$. Again, however, the neutrino comes to the rescue. Let the spin of the neutron be $+ \frac{1}{2}$. Let the proton's spin be $+ \frac{1}{2}$ and the electron's $- \frac{1}{2}$, for a net of 0. Now give the neutrino the spin $+ \frac{1}{2}$, so that it, too, is a fermion (and therefore a lepton)—and the books are neatly balanced.

$$+ \tfrac{1}{2}(n) = + \tfrac{1}{2}(p) - \tfrac{1}{2}(e) + \tfrac{1}{2}(\text{neutrino}).$$

There is still more balancing to do. A single particle (the neutron) has formed two particles (the proton and the electron) and, if we include the neutrino, actually three particles. It seems more reasonable to suppose that the neutron is converted into two particles and an antiparticle, or a net of one particle. In other words, what we really need to balance is not a neutrino but an antineutrino.

The neutrino itself would arise from the conversion of a proton into a neutron. There the products would be a neutron (particle), a positron (antiparticle), and a neutrino (particle). This, too, balances the books.

In other words, the existence of neutrinos and antineutrinos would save not one, but three, important conservation laws: the conservation of energy, the conservation of spin, and the conservation of particle/antiparticles. It is important to save these laws for they seem to hold in all sorts of nuclear reactions that do not involve electrons or positrons, and it would be very useful if they hold in reactions that did involve those particles, too.

The most important proton-to-neutron conversions are those involved in the nuclear reactions that go on in the sun and other stars. Stars therefore emit fast floods of neutrinos, and it is estimated that perhaps 6 percent to 8 percent of their energy is carried off in this way. This, however, is only true for such stars as our sun. In 1961, the American physicist Hong Yee Chiu suggested that, as the central temperatures of a star rise, additional neutrino-producing reactions become important. As a star progresses in its evolutionary course toward a hotter core (see chapter 2), an ever larger proportion of its energy is carried off by neutrinos.

There is crucial importance in this notion. The ordinary method of transmitting energy, by photons, is slow. Photons interact with matter, and they make their way out from the sun's core to its surface only after uncounted myriads of absorptions and re-emissions. Consequently, although the sun's central temperature is 15,000,000° C, its surface is only 6,000° C. The substance of the sun is a good heat insulator.

Neutrinos, however, virtually do not interact with matter. It has been calculated that the average neutrino could pass through 100 light-years of solid lead with only a 50 percent chance of being absorbed. Hence, any neutrinos formed in the sun's core leave at once and at the speed of light, reaching the sun's surface, without interference, in less than 3 seconds and speeding off. (Any neutrinos that move in our direction pass through without affecting us in any way either by day or by night; for at night, when the bulk of the earth is between ourselves and the sun, the neutrinos can pass through the earth and ourselves as easily as through ourselves alone.)

By the time a central temperature of 6,000,000,000° K is reached, Chiu calculates, most of a star's energy is being pumped into neutrinos. The neutrinos leave at once, carrying the energy with them, and the sun's center cools drastically. It is this, perhaps, which leads to the catastrophic contraction that then makes itself evident as a supernova.

TRACKING DOWN THE NEUTRINO

Antineutrinos are produced in any neutron-to-proton conversion, but these do not go on (as far as is known) on the vast scale that leads to such floods of neutrinos from every star. The most important sources of antineutrinos are from natural radioactivity and uranium fission (which I shall discuss in more detail in chapter 10).

Naturally physicists could not rest content until they had actually tracked down the neutrino; scientists are never happy to accept phenomena or laws of nature entirely on faith. But how to detect an entity as nebulous as the neutrino—an object with no mass, no charge, and practically no propensity to interact with ordinary matter?

Still, there was some slight hope. Although the probability of a neutrino reacting with any particle is exceedingly small, it is not/quite zero. To be unaffected in passing through one hundred light-years of lead is just a measure of the average; but some neutrinos will react with a particle before they go that far, and a few—an almost unimaginably small proportion of the total number—will be stopped within the equivalent of 1/10 inch of lead.

In 1953, a group of physicists, led by Clyde Lorrain Cowan and Frederick Reines of the Los Alamos Scientific Laboratory, set out to try the next-to-impossible. They erected their apparatus for detecting neutrinos next to a large fission reactor of the Atomic Energy Commission/on the Savannah River in Georgia. The reactor would furnish streams of neutrons, which, hopefully, would release floods of antineutrinos. To catch them, the experimenter used large tanks of water. The plan was to let the antineutrinos bombard the protons (hydrogen nuclei) in the water and detect the results of the capture of an antineutrino by a proton.

What would happen? When a neutron breaks down, it yields a proton, an electron, and an antineutrino. Now a proton's absorption of an antineutrino should produce essentially the reverse. That is to say, the proton should be converted to a neutron, emitting a positron in the process. So there were two things to be looked for: (1) the creation of neutrons, and (2) the creation of positrons. The neutrons could be detected by dissolving a cadmium compound in the water, for when cadmium absorbs neutrons, it emits gamma rays of a certain characteristic energy. And the positrons could be identified by their annihilating interaction with electrons, which would yield certain other gamma rays. If the experimenters' instruments detected gamma rays of exactly these two telltale energies and separated by the proper time interval, they could be certain that they had caught antineutrinos.

The experimenters arranged their ingenious detection devices, waited patiently; and, in 1956, exactly a quarter-century after Pauli's invention of the particle, they finally trapped the antineutrino. The newspapers and even some learned journals called it simply the *neutrino*.

To get the real neutrino, we need some source that is rich in neutrinos. The obvious one is the sun. What system can be used to detect the neutrino as opposed to the antineutrino? One possibility (following a suggestion of the Italian physicist Bruno Pontecorvo) begins with chlorine 37, which makes up about one-fourth of all chlorine atoms. Its nucleus contains 17 protons and 20 neutrons. If one of those neutrons absorbs a neutrino, it becomes a proton (and emits an electron). The nucleus will then have 18 protons and 19 neutrons and will be argon 37.

To form a sizable target of chlorine neutrons, one might use liquid chlorine, but it is a very corrosive and toxic substance, and keeping it liquid presents a problem in refrigeration. Instead, chlorine-containing organic compounds can be used; one called *tetrachloroethylene* is a good one for the purpose.

The American physicist Raymond R. Davis made use of such a neutrino trap in 1956 to show that there really was a difference between the neutrino and the antineutrino. Assuming the two particles were different, the trap would detect only neutrinos and not antineutrinos. When it was set up near a fission reactor in 1956 under conditions where it would certainly detect antineutrinos (if antineutrinos were identical to neutrinos), it did *not* detect them.

The next step was to try to detect neutrinos from the sun. A huge tank containing 100,000 gallons of tetrachloroethylene was used for the purpose. It was set up in a deep mine in South Dakota. There was enough earth above it to absorb any particles, except neutrinos, emerging from the sun. (Consequently, we have the odd situation that in order to study the sun, we must burrow deep, deep into the bowels of the earth.) The tank was then exposed to the solar neutrinos for several months to allow enough argon 37 to accumulate to be detectable. The tank was then flushed with helium for twenty-two hours, and the tiny quantity of argon 37 in the helium gas determined. By 1968, solar neutrinos were detected, but in not more than one-third the amounts expected from current theories about what is going on inside the sun. This finding proved very disturbing and I will get back to it later in the chapter.

NUCLEAR INTERACTION

Our list of subatomic particles now stands at ten: four massive particles (or *baryons*, from a Greek word for "heavy")—the proton, the neutron, the antiproton, and the antineutron; four leptons—the electron, the positron, the neutrino, and the antineutrino; and two bosons—the photon and the graviton. And yet these are not enough, as physicists decided by the following considerations.

Ordinary attractions between isolated protons and electrons, or repulsions between two protons or two electrons, can easily be explained as the result of *electromagnetic interactions.* The manner in which two atoms hold together, or two molecules, can also be explained by electromagnetic interactions—the attraction of positively charged nuclei for the outer electrons.

As long as the atomic nucleus was thought to be made up of protons and electrons, it seemed reasonable to assume that the electromagnetic interaction —the overall attraction between protons and electrons—would suffice to explain how nuclei held together as well. Once, however, the proton-neutron theory of nuclear structure came to be accepted in 1930, there was an appalled realization that there was no explanation for what holds the nucleus in being.

If protons were the only charged particles present, then the electromagnetic interaction should be represented by very strong repulsions between the protons that were pushed tightly together into the tiny nucleus. Any atomic nucleus should explode with shattering force the instant it was formed (if it ever could be formed in the first place).

Clearly, some other type of interaction must be involved, something much stronger than the electromagnetic interaction and capable of overpowering it. In 1930, the only other interaction known was the *gravitational interaction*, which is so much weaker than the electromagnetic interaction that it can actually be neglected in the consideration of subatomic events, and nobody misses it. No, there must be some *nuclear interaction*, one hitherto unknown but very strong.

The superior strength of the nuclear interaction can be demonstrated by the following consideration. The two electrons of a helium atom can be removed from the nucleus by the application of 54 electron volts of energy. That quantity of energy suffices to handle a strong manifestation of electromagnetic interaction.

On the other hand, the proton and neutron making up a deuteron, one of the most weakly bound of all nuclei, require 2 million electron volts for disruption. Making allowance for the fact that particles within the nucleus are much closer to one another than atoms within a molecule, it is still fair to conclude that the nuclear interaction is about 130 times as strong as the electromagnetic interaction.

But what is the nature of this nuclear interaction? The first fruitful lead came in 1932 when Werner Heisenberg suggested that the protons were held together by *exchange forces*. He pictured the protons and neutrons in the nucleus as continually interchanging identity, so that any given particle is first a proton, then a neutron, then a proton, and so on. This process might keep the nucleus stable in the same way that one holds a hot potato by tossing it quickly from hand to hand. Before a proton could "realize" (so to speak) that it was a proton and try to flee its neighbor protons, it had become a neutron and could stay where it was. Naturally it could get away with this only if the changes took place exceedingly quickly, say within a trillionth of a trillionth of a second.

Another way of looking at this interaction is to imagine two particles, exchanging a third. Each time particle A emits the exchange particle, it moves backward to conserve momentum. Each time particle B accepts the exchange particle, it is pushed backward for the same reason. As the exchange particle bounces back and forth, particles A and B move farther and farther apart so that they seem to experience a repulsion. If, on the other hand, the exchange particle moves around boomerang-fashion, from the rear of particle A to the rear of particle B, then the two particles would be pushed closer together and seem to experience an attraction.

It would seem by Heisenberg's theory that all forces of attraction and repulsion would be the result of exchange particles. In the case of electromagnetic attraction and repulsion, the exchange particle is the photon; and in the case of gravitational attraction (there is apparently no repulsion in the gravitational interaction), the exchange particle is the graviton.

Both the photon and the graviton are without mass, and it is apparently for this reason that electromagnetism and gravitation are forces that decrease only as the square of the distance and can therefore be felt across enormous gaps.

The gravitational interaction and the electromagnetic interaction are *long-distance interactions* and, as far as we know to this day, the only ones of this type that exist.

The nuclear interaction—assuming it existed—could not be one of this type. It had to be very strong within the nucleus if the nucleus were to remain in existence; but it was virtually indetectable outside the nucleus, or it would have been discovered long before. Therefore, the strength of the nuclear interaction dropped very rapidly with distance. With each doubling of distance, it might drop to less than 1/100 of what it was—rather than to merely ¼, as was the case with the electromagnetic and gravitational interactions. For that reason, no massless exchange particle would do.

THE MUON

In 1935, the Japanese physicist Hideki Yukawa mathematically analyzed the problem. An exchange particle possessing mass would produce a short-range force-field. The mass would be in inverse ratio to the range: the greater the mass, the shorter the range. It turned out that the mass of the appropriate particle lay somewhere between that of the proton and the electron; Yukawa estimated it to be between 200 and 300 times the mass of an electron.

Barely a year later, this very kind of particle was discovered. At the California Institute of Technology, Carl Anderson (the discoverer of the positron), investigating the tracks left by secondary cosmic rays, came across a short track that was more curved than a proton's and less curved than an electron's. In other words, the particle had an intermediate mass. Soon more such tracks were detected, and the particles were named *mesotrons*, or *mesons* for short.

Eventually other particles in this intermediate mass range were discovered, and this first one was distinguished as the *mu meson*, or the *muon*. ("Mu" is one of the letters of the Greek alphabet; almost all of which have now been used in naming subatomic particles.) As in the case of the particles mentioned earlier, the muon comes in two varieties, negative and positive.

The negative muon, 206.77 times as massive as the electron (and therefore about one-ninth as massive as a proton) is the particle; the positive muon is the antiparticle. The negative muon and positive muon correspond to the electron and positron, respectively. Indeed, by 1960, it had become evident that the negative muon was identical with the electron in almost every way except mass. It was a *heavy electron*. Similarly, the positive muon was a *heavy positron*.

Positive and negative muons will undergo mutual annihilation and may briefly circle about a mutual center of force before doing so—just as is true of positive and negative electrons. A variation of this situation was discovered in 1960 by the American physicist Vernon Willard Hughes. He detected a system in which the electron circled a positive muon, a system he called *muonium*. (A positron circling a negative muon would be *antimuonium*.)

The muonium atom (if it may be called that) is quite analogous to hydrogen 1, in which an electron circles a positive proton, and the two are similar in many of their properties. Although muons and electrons seem to be identical except for mass, that mass difference is enough to keep the electron and the positive muon from being true opposites, so that one will not annihilate the other. Muonium, therefore, does not have the kind of instability that positronium has. Muonium endures longer and would endure forever (if undisturbed from without) were it not for the fact that the muon itself does not endure since, as I shall shortly point out, it is very unstable.

Another similarity between muons and electrons is this: just as heavy particles may produce electrons plus antineutrinos (as when a neutron is converted to a proton) or positrons plus neutrinos (as when a proton is converted to a neutron), so heavy particles can interact to form negative muons plus antineutrinos or positive muons plus neutrinos. For years, physicists took it for granted that the neutrinos that accompany electrons and positrons and those that accompany negative and positive muons were identical. In 1962, however, it was found that the neutrinos do not cross over, so to speak; the electron's neutrino is not involved in any interaction that would form a muon, and the muon's neutrino is not involved in any interaction that would form an electron or positron.

In short, physicists found themselves with two pairs of chargeless, massless particles, the electron's antineutrino and the positron's neutrino plus the negative muon's antineutrino and the positive muon's neutrino. What the difference between the two neutrinos and between the two antineutrinos might be is more than anyone can tell at the moment, but they are different.

The muons differ from the electron and positron in another respect, that of stability. The electron or positron, left to itself, will remain unchanged indefinitely. The muon is unstable, however, and breaks down after an average lifetime of a couple of millionths of a second. The negative muon breaks down to an electron (plus an antineutrino of the electron variety and a neutrino of the muon variety); while the positive muon does the same in reverse, producing a positron, an electron-neutrino, and a muon-antineutrino.

When a muon decays, then, it forms an electron (or positron) with less than 1/200 of its mass, and a couple of neutrinos with no mass at all. What happens to the remaining 99.5 percent of the mass? Clearly, it turns to energy which may be emitted as photons or expended in the formation of other particles.

In reverse, if enough energy is concentrated on a tiny volume of space, then instead of forming an electron-positron pair, a more bloated pair may form; a pair just like the electron-positron pair except for the energy-bloat which makes its appearance as mass. The adherence of the extra mass to the basic electron or positron is not very strong, so the muon is unstable and quickly sheds that mass and becomes an electron or positron.

THE TAUON

Naturally, if still more energy is concentrated on a tiny volume, a still more massive electron will form. In California, Martin L. Perl made use of an accelerator that smashed high-energy electrons into high-energy positrons head on; and, in 1974, evidence was detected of such a superheavy electron. This he called a *tau electron* (*tau* being another letter of the Greek alphabet), and it is frequently called a *tauon* for short.

As might be expected, the tauon is about 17 times as massive as a muon and, therefore, about 3,500 times as massive as an electron. In fact, the tauon is twice as massive as a proton or a neutron. Despite its mass, the tauon is a lepton for, except for its mass and instability, it has all the properties of an electron. With all its mass, it might be expected to be far more unstable than the muon, and it is. The tauon lasts for only about a trillionth of a second before breaking down to a muon (and then to an electron).

There is, of course, a negative tauon and a positive tauon, and physicists take it for granted that associated with these is a third kind of neutrino and antineutrino, even though these have not yet actually been detected.

THE NEUTRINO'S MASS

There are now twelve leptons known, then: the negative and positive electron (the latter being the positron), the negative and positive muon, the negative and positive tauon, the electron neutrino and antineutrino, the muon neutrino and antineutrino, and the tauon neutrino and antineutrino. Clearly, these are divided into three levels (or, as physicists now say, *flavors*). There is the electron and associated neutrino and their antiparticles; the muon and associated neutrino and their antiparticles; and the tauon and associated neutrino and their antiparticles.

Having these three flavors, there is no reason why there should not be others. It may be that if the amount of energy that could be used could be increased indefinitely, more and more flavors of leptons could be formed, each flavor more massive and more unstable than the one before. Although there may be no theoretical limit to the number of flavors, there would, of course, be a practical limit. Eventually, it might simply take all the energy in the universe to form a lepton of a particularly high level, and there would be no going beyond; and eventually, such a particle would be so unstable that its existence would be meaningless in any sense.

If we confine ourselves to the three flavors now known, the mystery of the neutrinos is compounded. How can there be three massless, chargeless fermion pairs, each distinctly different as far as particle interactions go and yet with no distinguishing property as far as we can tell?

Perhaps, there is a distinguishing property, but we have not looked for it properly. For instance, all three flavors of neutrino are supposed to have zero mass and therefore to be moving, always, at the speed of light. Suppose, though, that each flavor of neutrino has a very tiny mass, different from that of the other two. In that case, their properties would naturally be slightly different one from the other. For instance, they would each travel at very slightly less than the speed of light, and the amount by which that speed would fall short would be slightly different for each.

There are theoretical reasons for arguing, in this case, that any neutrino, as it travels, shifts its identity, being an electron-neutrino at one time, a muon-neutrino at another, and a tauon-neutrino at still another. These shifts represent *neutrino oscillations*—first suggested as a possibility in 1963 by a group of Japanese physicists.

In the late 1970s, Frederick Reines, one of the original detectors of the neutrino, along with Henry W. Sobel and Elaine Pasierb of the University of California, set out to test the matter. They used about 600 pounds of very pure heavy water and bombarded it with neutrinos arising from fissioning uranium. This process should produce *only* electron-neutrinos.

The neutrinos can bring about either of two events. A neutrino can strike the proton-neutron combination of the heavy hydrogen nucleus in the heavy water, splitting them apart and continuing to move on. This is a *neutral-current reaction*, and any of the neutrino flavors can do it. Second, the neutrino, on striking the proton-neutron combination, can induce a change of the proton into a neutron, producing an electron; in this case, the neutrino ceases to exist. This is a *charged-current reaction*, and *only* electron-neutrinos can bring it about.

One can calculate how many of each type of event should take place if the neutrinos did not oscillate and remained only electron-neutrinos, and how many if the neutrinos did oscillate and some had changed over. In 1980, Reines announced that his experiment seemed to demonstrate the existence of neutrino oscillation. (I say "seemed" because the experiment was very nearly at the limit of the detectable, and because other experimenters checking the matter have reported that they have *not* detected signs of such oscillation.)

The matter remains in doubt, but experiments by physicists in Moscow, involving a point that has nothing to do with oscillations, seem to show that the electron-neutrino may have a mass of possibly as much as 40 electron volts. This would give it a mass 1/13,000 that of an electron, so it is no wonder the particle has passed for massless.

If Reines is correct, then, and there is neutrino oscillation, it would explain the shortage of neutrinos from the sun, which I mentioned earlier in the chapter and which is so puzzling to scientists. The device used by Davis to detect solar neutrinos would detect electron-neutrinos only. If the neutrinos emitted from the sun oscillate so that they arrive at Earth in a mixture of the three flavors in perhaps equal quantities, it is no wonder we detect only one-third of the neutrinos we expect.

Then, too, if neutrinos have a small amount of mass, even only 1/13,000 of an electron, there are so many neutrinos in space that all together it is possible to calculate that they far outmatch all the protons and neutrons. More than 99 percent of the mass of the universe would be neutrinos, and they could easily represent the "missing mass" I spoke of in chapter 2. In fact, there would be enough neutrino mass in the universe to close it and to ensure that eventually the expansion would stop and the universe would begin to contract again.

That is, *if* Reines is correct. We do not know yet.

Hadrons and Quarks

Since the muon is a kind of heavy electron, it cannot very well be the nuclear cement Yukawa was looking for. Electrons are not found within the nucleus, and therefore neither should the muon be. This was discovered to be true on a purely experimental basis, long before the near identity of muon and electron was suspected; muons simply showed no tendency to interact with nuclei. For a while, Yukawa's theory seemed to be tottering.

PIONS AND MESONS

In 1947, however, the British physicist Cecil Frank Powell discovered another type of meson in cosmic-ray photographs. It was a little more massive than the muon and proved to possess about 273 times the mass of an electron. The new meson was named a *pi meson* or a *pion*.

The pion was found to react strongly with nuclei and to be just the particle predicted by Yukawa. (Yukawa was awarded the Nobel Prize in physics in 1949, and Powell received it in 1950.) Indeed, there was a positive pion that acted as the exchange force between protons and neutrons, and there was a corresponding antiparticle, the negative pion, which performed a similar service for antiprotons and antineutrons. Both are even shorter-lived than muons; after an average lifetime of about 1/40 microsecond, they break up into muons plus neutrinos of the muon variety. (And, of course, the muon breaks down further to electrons and additional neutrinos.) There is also a neutral pion, which is its own antiparticle. (There is, in other words, only one variety of that particle.) It is extremely unstable, breaking down in less than a quintillionth of a second to form a pair of gamma rays.

Despite the fact that a pion "belongs" within the nucleus, it will fleetingly circle a nucleus before interacting with it, sometimes, to form a *pionic atom* as was detected in 1952. Indeed, any pair of negative and positive particles or particle systems can be made to circle each other; and in the 1960s, physicists studied a number of evanescent "exotic atoms" in order to gain some notion about the details of particle structure.

The pions were the first to be discovered of a whole class of particles, which are lumped together as *mesons*. These do *not* include the muon, although that was the first known particle to be given the name. Mesons interact strongly with protons and neutrons (figure 7.8), while muons do not and have thus lost the right to be included in the group.

Figure 7.8. Meson collision with a nucleus. A high-energy meson from secondary cosmic radiation struck a nucleus and produced a star made up of mesons and alpha particles (*lower left*); the energetic meson then traveled along the wavering path to the upper right, where it was finally stopped by collision with another nucleus.

As an example of particles other than the pion that are members of the group, there are the *K-mesons*, or *kayons*. These were first detected in 1952 by two Polish physicists, Marian Danysz and Jerzy Pniewski. These are about 970 times as massive as an electron and, therefore, about half the mass of a proton or neutron. The kayon comes in two varieties, a positive kayon and a neutral kayon, and each has an antiparticle associated with it. They are unstable, of course, breaking down to pions in about a microsecond.

BARYONS

Above the meson are the baryons (a term I mentioned earlier), which include the proton and the neutron. Until the 1950s, the proton and the neutron were the only specimens known. Beginning in 1954, however, a series of still more massive particles (sometimes called *hyperons*) were discovered. It is the baryon particles that have particularly proliferated in recent years, in fact, and the proton and neutron are but the lightest of a large variety.

There is a *law of conservation of baryon number,* physicists have discovered, for in all particle breakdowns, the net number of baryons (that is, baryons minus antibaryons) remains the same. The breakdown is always from a more massive to a less massive particle and thus explains why the proton is stable and is the *only* baryon to be stable. It happens to be the lightest baryon. If it broke down, it would have to cease being a baryon and thus would break the law of conservation of baryon number. For the same reason, an antiproton is stable, because it is the lightest antibaryon. Of course, a proton and an antiproton can engage in mutual annihilation since, taken together, they make up one baryon plus one antibaryon for a net baryon number of zero.

(There is also a *law of conservation of lepton number,* which explains why the electron and positron are the *only* leptons to be stable. They are the least massive leptons and cannot break down into anything simpler without violating that conservation law. In fact, electrons and positrons have a second reason for not breaking down. They are the least massive particles that can possess an electric charge. If they break down to something simpler, they lose the electric charge—a loss forbidden by the *law of conservation of electric charge.* That is, indeed, a stronger conservation law than the conservation of baryon number, as we shall see, so that electrons and positrons are, in a way, more stable than protons and antiprotons—or, at least, they *may* be more stable.)

The first baryons beyond the proton and neutron to be discovered were given Greek names. There was the *lambda particle,* the *sigma particle,* and the *xi particle.* The first came in one variety, a neutral particle; the second in three varieties, positive, negative, and neutral; the third in two varieties, negative and neutral. Every one of these had an associated antiparticle, making a dozen particles altogether. All were exceedingly unstable; none could live for more than a hundredth of a microsecond or so; and some, such as the neutral sigma particle, broke down after a hundred trillionth of a microsecond.

The lambda particle, which is neutral, can replace a neutron in a nucleus to form a *hypernucleus*—an entity that endures less than a billionth of a second. The first to be discovered was a hypertritium nucleus made up of a proton, a neutron, and a lambda particle. This was located among the products of cosmic radiation by Danysz and Pniewski in 1952. In 1963, Danysz reported hypernuclei containing two lambda particles. What's more, negative hyperons can be made to replace electrons in atomic structure, as was first reported in 1968. Such massive electron-replacements circle the nucleus at such close quarters as to spend their time actually within the nuclear outer regions.

But all these are the comparatively stable particles; they live long enough to be directly detected and to be easily awarded a lifetime and personality of their own. In the 1960s, the first of a whole series of particles was detected by Alvarez (who received the Nobel Prize in physics in 1968 as a result). These were so short-lived that their existence could only be deduced from the necessity of accounting for their breakdown products. Their half-lives are something of the order of a few trillionths of a trillionth of a second, and one might wonder whether they are really individual particles or merely a combination of two or more particles, pausing to nod at each other before flashing by.

These ultra-short-lived entities are called *resonance particles;* and, as physicists came to have at their disposal ever greater energies, they continued to produce ever more particles until 150 and more were known. These were all among the mesons and the baryons, and these two groups were lumped together as *hadrons* (from a Greek word for *bulky*). The leptons remained at a modest three flavors, each flavor containing particle, antiparticle, neutrino, and antineutrino.

Physicists became as unhappy with the multiplicity of hadrons as chemists had been with the multiplicity of elements a century earlier. The feeling grew that the hadrons had to be made up of simpler particles. Unlike the leptons, the hadrons were not points but had definite diameters—not very large ones, to be sure, only around a 10-trillionth of an inch, but that is not a point.

In the 1950s, the American physicist Robert Hofstadter investigated nuclei with extremely energetic electrons. The electrons did not interact with the nuclei but bounced off; and from the bouncing, Hofstadter came to conclusions about hadron structure that eventually proved to be inadequate but were a good start. As a result, he shared in the Nobel Prize in physics in 1961.

THE QUARK THEORY

One thing that seemed needed was a sort of periodic table for subatomic particles—something that would group them into families consisting of a basic member or members with other particles that are excited states of that basic member or members (table 7.1).

Something of the sort was proposed in 1961 by the American physicist Murray Gell-Mann and the Israeli physicist Yuval Ne'emen, who were working independently. Groups of particles were put together in a beautifully symmetric pattern that depended on their various properties—a pattern that Gell-Mann called the *eightfold way* but that is formally referred to as SU # 3. In particular, one such grouping needed one more particle for completion. That particle, if it was to fit into the group, had to have a particular mass and a particular set of other properties. The combination was not a likely one for a particle; yet, in 1964, a particle (the *omega-minus*) was detected with just the predicted set of properties; and in succeeding years, it was detected dozens of times. In 1971 its antiparticle, the *antiomega-minus,* was detected.

Even if baryons are divided into groups and a subatomic periodic table is set up, there would still be enough different particles to give physicists the urge to find something still simpler and more fundamental. In 1964, Gel-Mann—having endeavored to work out the simplest way of accounting for all the baryons with a minimum number of more fundamental *sub-baryonic particles*—came up with the notion of *quarks.* He got this name because he found that

TABLE 7.1

The Longer-lived Subatomic Particles

Family Name	Particle Name	Symbol	Mass	Spin	Electric Charge	Antiparticle	Number of Distinct Particles	Average Lifetime (seconds)	Typical Mode of Decay
	Photon	γ (gamma ray)	0	1	neutral	same particle	1	infinite	—
	Graviton	—	0	2	neutral	same particle	1	infinite	—
Electrons	Electron's neutrino	ν_e	0	½	neutral	$\bar{\nu}_e$	2	infinite	—
	Electron	e^-	1	½	negative	e^+ (positron)	2	infinite	—
Muons	Muon's neutrino	ν_μ	0(?)	½	neutral	$\bar{\nu}_\mu$	2	infinite	—
	Muon	μ^-	206.77	½	negative	μ^+	2	2.212×10^{-6}	$\mu^- \to e^- + \bar{\nu}_e + \nu_\mu$
Mesons	Pion	π^+	273.2	0	positive	π^- } same as the particles	3	2.55×10^{-8}	$\pi^+ \to \mu^+ + \nu_\mu$
		π^-	273.2	0	negative	π^+		2.55×10^{-8}	$\pi^- \to \mu^- + \bar{\nu}_\mu$
		π^0	264.2	0	neutral	π^0		1.9×10^{-16}	$\pi^0 \to \gamma + \gamma$
	Kaon	K^+	966.5	0	positive	$\overline{K^+}$ (negative)	4	1.22×10^{-8}	$K^+ \to \pi^+ + \pi^0$
		K^0	974	0	neutral	$\dfrac{\overline{K^0}}{K^0}$		1.00×10^{-10} and^a 6×10^{-8}	$K^0 \to \pi^+ + \pi^-$
Baryons	Nucleon	p (proton)	1836.12	½	positive	$\bar{p}$ (negative)	4	infinite	
		n (neutron)	1838.65	½	neutral	$\bar{n}$		1013	$n \to p + e^- + \bar{\nu}$
	Lambda	Λ^0	2182.8	½	neutral	$\overline{\Lambda^0}$	2	2.51×10^{-10}	$\Lambda^0 \to p + \pi^-$
	Sigma	Σ^+	2327.7	½	positive	$\overline{\Sigma^+}$ (negative)	6	8.1×10^{-11}	$\Sigma^+ \to n + \pi^+$
		Σ^-	2340.5	½	negative	$\overline{\Sigma^-}$ (positive)		1.6×10^{-10}	$\Sigma^- \to n + \pi^-$
		Σ^0	2332	½	neutral	$\overline{\Sigma^0}$		about 10^{-20}	$\Sigma^0 \to \Lambda^0 + \gamma$
	Xi	Ξ^-	2580	½	negative	$\overline{\Xi^-}$ (positive)	4	1.3×10^{-10}	$\Xi^- \to \Lambda^0 + \pi^-$
		Ξ^0	2570	½	neutral	$\overline{\Xi^0}$		about 10^{-10}	$\Xi^0 \to \Lambda^0 + \pi^0$
							33		

^{a}The K^0 meson has two different lifetimes, all other particles have only one.

Source: *The World of Elementary Particles* by Kenneth W. Ford, © Copyright 1963, Xerox Corporation.

only three quarks in combination were necessary to make up a baryon, and that different combinations of the three quarks were needed to make up all the known baryons. This reminded him of a line from *Finnegan's Wake* by James Joyce: "Three quarks for Musther Mark."

In order to account for the known properties of baryons, the three different quarks had to have specific properties of their own. The most astonishing property was a fractional electric charge. All known particles had either no electric charge, an electric charge exactly equal to that of the electron (or positron), or an electric charge equal to some exact multiple of the electron (or positron). The known charges, in other words, were 0, $+1$, -1, $+2$, -2, and so on. To suggest fractional charges was so odd, that Gell-Mann's notion met with strong initial resistance. It was only the fact that he managed to explain so much that got him a respectful hearing, then a strong following, then a Nobel Prize in physics in 1969.

Gell-Mann started with two quarks, for instance, which are now called *up-quark* and *down-quark*. *Up* and *down* have no real significance but are only a whimsical way of picturing them. (Scientists, particularly young ones, are not to be viewed as soulless and unemotional mental machines. They tend to be as joke-filled, and sometimes as silly, as the average novelist and truck driver.) It might be better to call these *u-quark* and *d-quark*.

The u-quark has a charge of $+\frac{2}{3}$ and the d-quark has one of $-\frac{1}{3}$. There would also be an *anti-u-quark* with a charge of $-\frac{2}{3}$ and an *anti-d-quark* with a charge of $+\frac{1}{3}$.

Two u-quarks and one d-quark would have a charge of $+\frac{2}{3}$, $+\frac{2}{3}$, and $-\frac{1}{3}$—a total of $+1$—and, in combination, would form a proton. On the other hand, two d-quarks and one u-quark would have a charge of $-\frac{1}{3}$, $-\frac{1}{3}$, and $+\frac{2}{3}$—a total of 0—and, in combination, would form a neutron.

Three quarks would always come together in such a way that the total charge would be an integer. Thus, two anti-u-quarks and one anti-d-quark would have a total charge of -1 and would form an antiproton, while two anti-d-quarks and one anti-u-quark would have a total charge of 0 and would form an antineutron.

What's more, the quarks would stick together so firmly, thanks to nuclear interaction, that scientists have been totally unable so far to break protons and neutrons apart into separate quarks. In fact, there are suggestions that the attraction between quarks increases with distance so that there is no conceivable way of breaking up a proton or neutron into its constituent quarks. If so, fractional electric charges may exist, but they can never be detected, which makes Gell-Mann's iconoclastic notion a little easier to take.

These two quarks are insufficient to account for all the baryons, however, or for all the mesons (which are made up of combinations of *two* quarks). Gell-Mann, for instance, originally suggested a third quark, which is now called the *s-quark*. The *s* can be said to stand for "sideways" (to match *up* and *down*) but is more often said to stand for "strangeness" because it had to be used to account for the structure of certain so-called *strange particles*—

strange, because they existed for longer times before breaking down than would be expected.

Eventually, though, physicists investigating the quark hypothesis decided that quarks would have to exist in pairs. If there was an s-quark, there would have to be a companion quark, which they called a *c-quark*. (The *c* stands not for "companion" but for "charm.") In 1974, an American physicist, Burton Richter, and another, Samuel Chao Chung Ting, working independently, with intense energies, isolated particles that had properties requiring the c-quark. (These were particles with "charm.") The two shared the Nobel Prize for physics in 1976, as a result.

The pairs of quarks are flavors; and, in a way, they match the lepton flavors. Each flavor of quark has four members—for instance, the u-quark, the d-quark, the anti-u-quark, and the anti-d-quark—just as each flavor of leptons has four members—for instance, the electron, the neutrino, the antielectron and the antineutrino. In each case, there are three flavors known: electron, muon, and tauon among the leptons; u- and d-quarks, s- and c-quarks, and, finally, t- and b-quarks. The *t-quark* and the *b-quark* stand for "top" and "bottom" in the usual formulation; but among the whimsical, they stand for "truth" and "beauty." The quarks, like the leptons, seem to be particles of point-size and to be fundamental and structureless (but, we can not be sure, for we have been fooled in this respect already, first by the atom, and then by the proton). And it may be that in both cases, there may be an indefinite number of flavors, if we had more and more energy to expend in order to detect them.

One enormous difference between leptons and quarks is that leptons have integral charges, or none at all, and do not combine; whereas quarks have fractional charges and apparently exist only in combination.

The quarks combine according to certain rules. Each different flavor of quark comes in three varieties of property—a property that leptons do not possess. This property is called (metaphorically only) *color*, and the three varieties are called *red, blue,* and *green.*

When quarks get together three at a time to form a baryon, one quark must be red, one blue, and one green, the combination being without color, or *white.* (This is the reason for red, blue, and green; for in the world about us, as on the television screen, that combination will give white.) When quarks get together two at a time to form a meson, one will be a particular color, and the other that particular anticolor, so that the combination is again white. (Leptons have no color, being white to begin with.)

The study of quark combinations in such a way that color is never detected in the final product, just as fractional electric charges are not, is referred to as *quantum chromodynamics, chromo* coming from the Greek word for "color." (This term harks back to a successful modern theory of electromagnetic interactions which is called *quantum electrodynamics.*)

When quarks combine, they do so by means of an exchange particle which, in constantly shifting back and forth, serves to hold them together. This exchange particle is called a *gluon*, for obvious reasons. Gluons have color themselves, which adds complications, and can even stick together to form a product called *glueballs*.

Even though hadrons cannot be pulled apart to form isolated quarks (two in the case of mesons, three in the case of baryons), there are more indirect ways of demonstrating quark existence. Quarks might be formed from scratch if enough energy can be concentrated in a small volume, as by smashing together very energetic streams of electrons and positrons (as sufficed to form the tauon).

The quarks produced in this way would instantly combine into hadrons and antihadrons which would stream off in opposite directions. If there was *enough* energy, there would be three streams forming a three-leaf clover—hadrons, antihadrons, and gluons. The two-leaf clover has been formed; and, in 1979, there were announcements of experiments in which a rudimentary third leaf was just beginning to form. This is considered a strong confirmation of the quark theory.

Fields

Every particle possessing mass is the source of a gravitational field that stretches outward in all directions indefinitely, the intensity of the field decreasing in proportion to the square of the distance from the source.

The intensity of the field is incredibly small where individual particles are concerned, so small that to all intents and purposes the field can be ignored where particle interactions are studied. There is, however, only one kind of mass, and the gravitational interaction between two particles seems always to be an attraction.

What is more, where a system consists of many particles, the gravitational field, from a point outside the system, seems to be the sum of all the individual fields of all the particles. An object such as the sun or the earth behaves as though it has a field of the intensity one would expect if it consisted of a particle containing all the mass of the body located at the center of gravity of the body. (This is precisely true only if the body is perfectly spherical and of uniform density, or of varying density where the variations extend outward from the center in exact spherical symmetry; and all this is almost true for objects like the sun or the earth.)

The result is that the sun and, to a lesser extent, the earth have gravitational fields of enormous intensity, and the two can interact, attract each other, and remain firmly together even though separated by a distance of 93 million miles. Systems of galaxies can hold together though spread over distances of millions of light-years; and if the universe ever starts contracting again, it will do so because of the pull of gravity over the distance of billions of light-years.

Every particle possessing electric charge is the source of an electromagnetic field that stretches outward in all directions indefinitely, the intensity of the field decreasing in proportion to the square of the distance from the source. Every particle possessing both mass and electric charge (and there is no electric charge without mass) is the source of both fields.

ELECTROMAGNETIC INTERACTION

The electromagnetic field is many trillions of trillions of trillions of times as intense as the gravitational field in the case of any given single particle. However, there are two kinds of electric charge, positive and negative, and the electromagnetic field exhibits both attraction and repulsion. Where the two kinds of charge are present in equal numbers, the charges tend to neutralize each other and no electromagnetic field is present outside the system. Thus, normal intact atoms are made up of equal numbers of positive and negative charges and are therefore electrically neutral.

Where one charge or the other is present in excess, an electromagnetic field is present, but the mutual attraction of opposite charges makes it certain that any excess present in either direction is microscopically small so that electromagnetic fields where present cannot compare in intensity with the gravitational fields of bodies of the size of a large asteroid or beyond. Thus, Isaac Newton, who dealt with gravitational interactions *alone*, was able to build a satisfactory explanation of the motions of the bodies of the solar system, one that could be extended to include the motions of stars and galaxies.

Electromagnetic interactions cannot be ignored altogether and play a role in the formation of the solar system, in the transfer of angular momentum from the sun to the planets, and probably in some of the puzzling manifestations of the rings of small particles that circle Saturn, but these are comparatively small refinements.

Every hadron (mesons and baryons and their constituent quarks) is the source of a field that stretches outward in all directions indefinitely, the intensity of the field decreasing so rapidly with distance that it cannot make itself usefully felt at distances greater than the diameter of an atomic nucleus. Such a field, while overpoweringly important within a nucleus, or whenever two speeding particles skim by each other at nuclear distances, can be ignored at greater distances. Such a field plays no role in the general movements of astronomical bodies but is important in consideration of events in the cores of stars, for instance.

Leptons are also the source of a field that can only be felt at nuclear distances. Indeed the range of this field is even shorter than that of the hadron field. They are both nuclear fields, but they are very different, not only in the type of particle they are associated with, but in their intensities. The hadron field is, particle for particle, 137 times as strong as the electromagnetic field. The lepton field is only about a hundred-billionth as strong as the electromagnetic field. The hadron field is therefore usually spoken of as the *strong interaction*, and the lepton field as the *weak interaction*. (Remember that the weak interaction, although weak in comparison with the strong and the electromagnetic interaction, is still about 10,000 trillion trillion times as strong as the gravitational interaction.)

These four interactions, as far as we now know, account for all particle behavior and, by way of it, for all measurable behavior of any kind. There is no indication as yet that any fifth interaction exists or can exist. (Of course, to say that these interactions account for all measurable behavior does not mean, by a long, long shot, that we can as yet understand all measurable behavior. The fact that you may know that a complex mathematical equation has a solution does not mean that you yourself can necessarily find the solution.)

The weak interaction was first dealt with mathematically in 1934 by Fermi; but for decades afterward, it remained the least known of the four interactions. For instance, all four interactions ought to have exchange particles through which the interactions are mediated. There is the photon for the electromagnetic interaction, the graviton for the gravitational interaction, the pion for the strong interaction at the proton-neutron level, and the gluon for the strong interaction at the quark level. Some such particle, called the *W-particle* (*W* for "weak," of course), ought to exist for the weak interaction; but, for over half a century, that W-particle remained elusive.

THE CONSERVATION LAWS

Then, too, there is the question of the conservation laws that set up the rules by which one can judge which particle interactions are possible and which are not; and, therefore, more generally, what can happen in the universe and what cannot. Without the conservation laws, events in the universe would be anarchic and totally incomprehensible.

Nuclear physicsts deal with about a dozen conservation laws. Some are the familiar conservation laws of nineteenth-century physics: the conservation of energy, the conservation of momentum, the conservation of angular momentum, and the conservation of electric charge. Then there are conservation laws that are less familiar: the conservation of strangeness, the conservation of baryon number, the conservation of isotopic spin, and so on.

The strong interactions seem to obey all these conservation laws; and in the early 1950s, physicists took it for granted that the laws were universal and irrevocable. But they were not. In the case of weak interactions, some of the conservation laws are not obeyed.

The particular conservation law that was first shattered was the *conservation of parity*. Parity is a strictly mathematical property that cannot be described in concrete terms; suffice it to say that the property refers to a mathematical function that has to do with the wave characteristics of a particle and its position in space. Parity has two possible values—*odd* and *even*. The key point is that parity has been considered a basic property that, like energy or momentum, is subject to the law of conservation: in any reaction or change, parity must be conserved. That is to say, when particles interact to form new particles, the parity on both sides of the equation (so it was thought) must balance, just as mass numbers must, or atomic numbers, or angular momentum.

Let me illustrate. If an odd-parity particle and an even-parity particle interact to form two other articles, one of the new particles must be odd parity and the other even parity. If two odd-parity particles form two new particles, both of the new ones must be odd or both even. Conversely, if an even-parity particle breaks down to form two particles, both must be even parity or both must be odd parity. If it forms three particles, either all three have even parity or one has even parity and the other two have odd parity. (You may be able to see this more clearly if you consider the odd and even numbers, which follow similar rules. For instance, an even number can only be the sum of two even numbers or of two odd numbers, but never the sum of an even number and an odd one.)

The beginning of the trouble came when it was found that K-mesons sometimes broke down to two pi mesons (which, since the pi meson has odd parity, added up to even parity) and sometimes gave rise to three pi mesons (adding up to odd parity). Physicists concluded that there were two types of K-meson, one of even parity and one of odd parity; they named the two *theta meson* and *tau meson*, respectively.

Now in every respect except the parity result, the two mesons were identical: the same mass, the same charge, the same stability, the same everything. It was hard to believe that there could be two particles with exactly the same properties. Was it possible that the two were actually the same and that there was something wrong with the idea of the conservation of parity? In 1956, two young Chinese physicists working in the United States, Tsung Dao Lee and Chen Ning Yang, made precisely that suggestion. They proposed that, although the conservation of parity held in strong interactions, it might break down in weak interactions, such as are involved in the decay of K-mesons.

As they worked out this possibility mathematically, it seemed to them that if the conservation of parity broke down, the particles involved in weak interactions should show *handedness*, something first pointed out in 1927 by the Hungarian physicist Eugene Wigner. Let me explain.

Your right hand and left hand are opposites. One can be considered the mirror image of the other: in a mirror the right hand looks like a left hand. If all hands were symmetrical in every respect, the mirror image would be no different from the direct image, and there would be no such distinction as "right" and "left" hand in principle (figure 7.9). Very well, then, let us apply this principle to a group of particles emitting electrons. If electrons come out in equal numbers in all directions, the particle in question has no handedness. But if most of them tend to go in a preferred direction—say up rather than down—then the particle is not symmetrical. It shows a handedness: if we look at the emissions in a mirror, the preferred direction will be reversed.

The thing to do, therefore, was to observe a collection of particles that emit electrons in a weak interaction (say, some particle that decays by beta emission) and see if the electrons came out in a preferred direction. Lee and Yang

Figure 7.9. Mirror-image asymmetry and symmetry illustrated by hands.

asked an experimental physicist at Columbia University, Chien-Shiung Wu, to perform the experiment.

She set up the necessary conditions. All the electron-emitting atoms had to be lined up in the same direction if a uniform direction of emission was to be detected; this was done by means of a magnetic field, and the material was kept at a temperature near absolute zero.

Within forty-eight hours, the experiment yielded the answer. The electrons were indeed emitted asymmetrically. The conservation of parity did break down in weak interactions. The *theta meson* and the *tau meson* were one and the same particle, breaking down with odd parity in some cases, with even parity in others. Other experimenters soon confirmed the overthrow of parity; and for their bold conjecture, the theoretical physicists Lee and Yang received the Nobel Prize in physics in 1957.

If symmetry breaks down with respect to weak interactions, perhaps it will break down elsewhere. The universe as a whole may be left-handed (or right-handed) after all. Alternatively, there may be two universes, one left-handed, the other right-handed: one composed of matter; the other, of antimatter.

Physicists are now viewing the conservation laws in general with a new cynicism. Any one of them might, like conservation of parity, apply under some conditions and not under others.

Parity, after its fall, was combined with *charge conjugation,* another mathematical property assigned to subatomic particles, which governed its status as a particle or antiparticle; and the two together were spoken of as *CP con-*

servation, a deeper and more general conservation law than either the conservation of parity (*P*) or the conservation of charge conjugation (*C*) alone. (This sort of thing is not unprecedented. As we shall see in the next chapter, the law of conservation of mass gave way to the deeper and more general *conservation of mass-energy.*)

However, CP conservation proved inadequate, too. In 1964, two American physicists, Val Logsden Fitch and James Watson Cronin, showed that CP conservation was, on rare occasions, also violated in weak interactions. The question of the direction of time (*T*) was therefore added, and people now speak of *CPT symmetry.* For their work, Fitch and Cronin shared the 1980 Nobel Prize in physics.

A UNIFIED FIELD THEORY

Why should there be four different fields, four different ways in which particles might interact? There might be any number, of course, but the urge for simplicity is deeply ingrained in the scientific view. If there must be four (or any number), ought it not to be that all should be different aspects of a single field, a single interaction? If so, the best way of demonstrating this would be to find some mathematical relationship that would express them all, and that would then illuminate some aspects of their properties that would otherwise remain dark. For instance, over a hundred years ago, Maxwell worked out a set of mathematical equations that fit the workings of both electricity and magnetism and showed they were both aspects of a single phenomenon, which we now called the *electromagnetic field.* Might we now not go further?

Einstein began working on a *unified field theory* at a time when only the electromagnetic and gravitational fields were known. He spent decades on the task and failed; and while he was working, the two short-range fields were discovered, and the task was made all the harder.

In the late 1960s, however, the American physicist Steven Weinberg and the Pakistani-British physicist Abdus Salam, working independently, devised a mathematical treatment that covered both the electromagnetic field and the weak field, the two together being called the *electroweak field.* This treatment was then elaborated by the American physicist Sheldon Lee Glashow, who had been a high-school classmate of Weinberg. The theory made it necessary that both electromagnetic interactions and weak interactions should display *neutral currents,* certain particle interactions in which electric charge is not exchanged. Certain of these, not known previously, were found to exist exactly as predicted when searched for—a powerful piece of evidence in favor of the new theory. Weinberg, Salam, and Glashow all shared the 1979 Nobel Prize in physics.

The electroweak theory gave details as to what the missing exchange particles of the weak interaction (particles that had been sought in vain for half a century) ought to be. There ought to be not just a W-particle but three particles—a W^+, a W^-, and something labeled a Z^0, or in other words, a positive, a negative, and a neutral particle. What's more, some of the properties could be specified, if the electroweak theory was correct. They should be about 80 times as massive as the proton, for instance—a property that accounted for their being so elusive. It took enormous energies to bring them into existence and make them detectable. These huge masses, moreover, made the weak interaction *very* short-range, which made it unlikely that two particles should approach each other closely enough for the interaction to take place, which accounted for the weak interaction being so much weaker than the strong one.

By 1983, however, physicists had, at their disposal, energies sufficiently high for the task, and all three particles were finally detected—and with the predicted mass, too. That nailed the electroweak theory into place.

Meanwhile, the same mathematical scheme that seemed to cover both the electromagnetic field and the weak field seemed, to many physicists, to suffice (with some added complications) for the strong field as well. Several ways of doing so have been advanced. If the electroweak theory is a unified theory, one that would include the strong field as well would be a *grand unified theory,* usually abbreviated *GUTs* (because there is more than one).

If the strong field is to be brought under the GUTs umbrella, it would seem that there must be ultramassive exchange particles required beyond the gluons, no less than twelve of them. Because they are more massive than the W's and Z's, they will be harder to detect, and there is no hope for them right now. They will also be far shorter in range than anything that has yet been considered. The range of action of these ultramassive exchange particles of the strong field is less than 1 quadrillionth the diameter of the atomic nucleus.

Now if these ultramassive exchange particles exist, it is possible that one might pass from one quark to another within a proton. Such a passage might destroy one of the quarks, converting it to a lepton. With one of the quarks gone, the proton would become a meson, which would eventually decay to a positron.

However, in order for the exchange to take place, the quarks (which are point particles) must pass close enough to each other to be within the range of action of these ultramassive exchange particles. So incredibly tiny is the range that, even within the close confines of the proton, so close an approach is not likely.

In fact, it has been calculated that the necessary approach would happen so rarely that a proton would be destroyed only after 10^{31} years of existence, on the average. That many years is 600 million trillion times the total existence of the universe up to this point.

Of course, this is an *average* life span. Some protons would live much longer than that; and some much shorter. Indeed, if enough protons could be placed under study, a number of such proton-decays would take place every second. For instance, there might be about 3 billion proton-decays in Earth's oceans every second. (That sounds like a lot but it is a totally insignificant quantity, of course, compared with the total number of protons in the ocean.)

Physicists are anxious to detect such decays and differentiate them clearly from other similar events that might be taking place in far greater numbers.

If the decay could be detected, it would be a powerful piece of evidence in favor of the GUTs; but, as in the case of gravitational waves, the detection required is at the very limit of the possible, and it may take considerable time to settle the matter either way.

The theories involved in these new unifications can be used to work out the details of the big bang with which the universe started. It would seem that at the very start, when the universe had existed for less than a millionth of a trillionth of a trillionth of a trillionth of a second and was far tinier than a proton and had a temperature in the trillions of trillions of trillions of degrees, there was only one field and only one kind of particle interaction. As the universe expanded, and the temperature dropped, the different fields "froze out."

Thus we could imagine the earth, if extremely hot, to be nothing but a gaseous sphere in which all the different kinds of atoms would be evenly mixed so that every portion of the gas would have the same properties as every other. As the gas cooled, however, different substances would separate out first as liquids, then as solids; and eventually there would be a sphere of many different substances existing separately.

So far, though, the gravitational interaction proves intransigent. There seems no way of including it under the umbrella of the kind of mathematics worked out by Weinberg and the rest. The unification that defeated Einstein has so far defeated all his successors as well.

Even so, the GUTs has produced something extremely interesting, indeed. Physicists have wondered how the big bang could produce a universe so lumpy as to have galaxies and stars. Why did not everything simply spread out into a vast haze of gas and dust in all directions? Then, too, why is the universe of such a density that we cannot be quite certain whether it is open or closed? It might have been distinctly open (negatively curved) or closed (positively curved). Instead, it is nearly flat.

An American physicist, Alan Guth, in the 1970s, used GUTs to argue that, when the big bang took place, there was an initial period of exceedingly rapid expansion or inflation. In such an *inflationary universe*, the temperature dropped so rapidly that there was no time for the different fields to separate out or for different particles to form. It is only later in the game, when the universe had become quite large, that the differentiation took place. Hence the flatness of the universe, and so, too, its lumpiness. The fact that GUTs, a theory developed from particles alone, should happen to explain two puzzles that involve the birth of the universe is strong evidence in favor of GUTs being correct.

To be sure, the inflationary universe does not remove all problems, and different physicists have attempted to patch it in different ways to make a better match between predictions and reality—but it is early days yet, and there is considerable hope that some version of GUTs and inflation will work. Perhaps it will, when someone finally works out a way of including the gravitational interaction into the theory, and unification is at last complete.

Chapter 8

The Waves

Light

Until now, I have been dealing with material objects almost entirely—from the very large, such as galaxies, to the very small, such as electrons. Yet there are important immaterial objects, and of these the longest known and the most richly appreciated is light. According to the Bible, the first words of God were, "Let there be light," and the sun and the moon were created primarily to serve as sources of light: "And let them be for lights in the firmament of the heaven to give light upon the earth."

The scholars of ancient and medieval times were completely in the dark about the nature of light. They speculated that it consisted of particles emitted by the glowing object or perhaps by the eye itself. The only facts about it that they were able to establish were that light travels in a straight path, that it is reflected from a mirror at an angle equal to that at which the beam strikes the mirror, and that a light beam is bent (*refracted*) when it passes from air into glass, water, or some other transparent substance.

THE NATURE OF LIGHT

When light enters glass, or some other transparent substance, obliquely—that is, at an angle to the vertical—it is always refracted into a path that forms a smaller angle to the vertical. The exact relationship between the original angle and the refracted angle was first worked out in 1621 by the Dutch physicist Willebrord Snell. He did not publish his finding, and the French philosopher René Descartes discovered the law independently in 1637.

The first important experiments on the nature of light were conducted by Isaac Newton in 1666, as I have already mentioned in chapter 2. He let a beam of sunlight, entering a dark room through a chink in a blind, fall obliquely on one face of a triangular glass prism. The beam was refracted when it entered the glass and then refracted still farther in the same direction when it emerged from a second face of the prism. (The two refractions in the same direction arose because the two sides of the prism met at an angle instead of being parallel, as would have been the case in an ordinary sheet of glass.) Newton caught the emerging beam on a white screen to see the effect of the reinforced refraction. He found that, instead of forming a spot of white light, the beam was spread out in a band of colors—red, orange, yellow, green, blue, and violet, in that order.

Newton deduced that ordinary white light is a mixture of different kinds of light which, separately, affect our eyes so as to produce the sensation of different colors. This band of colors, though it looks real enough, is immaterial, as immaterial as a ghost; and, indeed, Newton's name for it—*spectrum*—comes from a Latin word meaning "ghost."

Newton decided that light consisted of tiny particles (*corpuscles*) traveling at enormous speed. These would explain why light travels in straight lines and casts sharp shadows. It is reflected by a mirror because the particles bounce off the surface, and it is bent on entering a refracting medium (such as water or glass) because the particles travel faster in such a medium than in air.

Still, there were awkward questions. Why should the particles of green light, say, be refracted more than those of yellow light? Why can two beams of light cross without affecting each other—that is, without the particles colliding?

In 1678, The Dutch physicist Christiaan Huygens (a versatile scientist who had built the first pendulum clock and done important work in astronomy) suggested an opposing theory, namely, that light consists of tiny waves. If it is made up of waves, there is no difficulty about explaining the different amount of refraction of different kinds of light through a refracting medium, provided it is assumed that light travels more slowly through the refracting medium than through air. The amount of refraction would vary with the length of the waves: the shorter the wavelength, the greater the refraction. Hence violet light (the most refracted) would have a shorter wavelength than blue light, blue shorter than green, and so on. It is this difference in wavelength, Huygens thought, that distinguishes the colors to the eye. And, of course, if light consists of waves, two beams could cross without trouble. (After all, sound waves and water waves cross without losing their identity.)

But Huygens's wave theory was not very satisfactory either. It did not explain why light rays travel in straight lines and cast sharp shadows, nor why light waves cannot go around obstacles, as water waves and sound waves can. Furthermore, if light consists of waves, how can it travel through a vacuum as it certainly seemed to do in coming to us through space from the sun and stars? What medium was it waving?

For about a century, the two theories contended with each other. Newton's *corpuscular theory* was by far the more popular, partly because it seemed on the whole more logical, and partly because it had the support of Newton's great name. But, in 1801, an English physician and physicist, Thomas Young, performed an experiment that swung opinion the other way. He projected a narrow beam of light through two closely spaced holes toward a screen behind. If light consisted of particles, presumably the two beams emerging through the holes would simply produce a brighter region on the screen where they overlapped and less bright regions where they did not. But this was not what Young found. The screen showed a series of bands of light, each separated from the next by a dark band. It seemed that in these dark intervals, the light of the two beams together added up to darkness!

The wave theory would easily explain this effect. The bright band represented the reinforcement of waves of one beam by waves of the other; in other words, the two sets of waves were *in phase,* both peaks together and strengthening each other. The dark bands, on the other hand, represented places where the waves were *out of phase,* the trough of one canceling the peak of the other. Instead of reinforcing each other, the waves at these places interfered with each other, leaving the net light energy there zero.

From the width of the bands and the distance between the two holes through which the beams issued, it was possible to calculate the length of light-waves—say, of red light or violet or colors between. The wavelengths turned out to be very small indeed. The wavelength of red light, for example, came to about 0.000075 centimeters or 0.000030 inch. (Eventually the wavelengths of light were expressed in a convenient unit suggested by Angstrom. The unit, called the *angstrom*—abbreviated *A*—is 100 millionth of a centimeter. Thus, the wavelength of red light at one end of the spectrum is about 7,500 angstrom units; the wavelength of violet light at the other end is about 3,900 angstrom units; and the color wavelengths of the visible spectrum lie between these numbers.)

The shortness of the wavelengths is very important. The reason light-waves travel in straight lines and cast sharp shadows is that they are incomparably smaller than ordinary objects; waves can curve around an obstruction only when that obstruction is not much larger than the wavelength. Even bacteria, for instance, are vastly wider than a wavelength of light, so light can define them sharply under a microscope. Only objects somewhere near a wavelength of light in size (for example, viruses and other submicroscopic particles) are small enough for light-waves to pass around them.

It was the French physicist Augustin Jean Fresnel who showed (in 1818) that, if an interfering object is small enough, a light-wave will indeed travel around it. In that case, the light produces what is called a *diffraction* pattern. For instance, the very fine parallel lines of a *diffraction grating* act as a series of tiny obstacles that reinforce one another. Since the amount of diffraction depends on the wavelength, a spectrum is produced. From the amount by which any color or portion of the spectrum is diffracted, and from the known separation of the scratches on the glass, the wavelength can again be calculated.

Fraunhofer pioneered in the use of such diffraction gratings, an advance generally forgotten in the light of his more famous discovery of spectral lines. The American physicist Henry Augustus Rowland invented concave gratings and developed techniques for ruling them with as many as 20,000 lines to the inch. It was his work that made it possible for the prism to be supplanted in spectroscopy.

Between such experimental findings and the fact that Fresnel systematically worked out the mathematics of wave motion, the wave theory of light seemed established and the corpuscular theory smashed—apparently for good.

Not only were light waves accepted as existing, their length was measured with increasing precision. By 1827, the French physicist Jacques Babinet was suggesting that the wavelength of light—an unalterable physical quantity—be used as the standard for measurement of length, instead of the various arbitrary standards that were then used. This suggestion did not become practicable, however, until the 1880s, when the German-American physicist Albert Abraham Michelson invented an instrument called the *interferometer*, which could measure the wavelengths of light with unprecedented accuracy. In 1893, Michelson measured the wavelength of the red line in the cadmium spectrum and found it to be $1/1,553,164$ meter long.

A measure of uncertainty still existed when it was discovered that elements consist of different isotopes, each contributing a line of slightly different wavelength. As the twentieth century progressed, however, the spectral lines of individal isotopes were measured. In the 1930s, the lines of krypton 86 were measured. This isotope, being that of gas, could be dealt with at low temperatures where atomic motion is slowed, with less consequent thickening to the line.

In 1960, the krypton-86 line was adopted by the General Conference of Weights and Measures as the fundamental standard of length. The meter has been redefined as equal to $1,650,763.73$ wavelengths of this spectral line. This standard has increased the precision of measurement of length a thousandfold. The old standard meter bar could be measured, at best, to within one part in a million, whereas the light wave can be measured to within one part in a billion.

THE SPEED OF LIGHT

Light obviously travels at tremendous speeds. If you put out a light, it gets dark everywhere at once, as nearly as can be made out. Sound does not travel as fast. If you watch a man in the distance chopping wood, you do not hear the stroke until some moments after the ax has struck. Sound has clearly taken a certain amount of time to travel to the ear. In fact, its speed of travel is easy to measure: 1,090 feet per second, or about 750 miles per hour, in the air at sea level.

Galileo was the first to try to measure the speed of light. Standing on one hill while an assistant stood on another, he would uncover a lantern; as soon as the assistant saw the flash, he would signal by uncovering a light of his own.

Galileo did this at greater and greater distances, assuming that the time it took the assistant to make his response would remain uniform, and therefore that any increase in the interval between his uncovering his own lantern and seeing the responding flash would represent the time taken by the light to cover the extra distance. The idea was sound, but light travels much too fast for Galileo to have detected any difference by this crude method.

In 1676, the Danish astronomer Olaus Roemer did succeed in timing the speed of light—on an astronomical distance scale. Studying Jupiter's eclipses of its four large satellites, Roemer noticed that the interval between successive eclipses became longer when the earth was moving away from Jupiter, and shorter when it was moving toward Jupiter in its orbit. Presumably the difference in eclipse times reflected the difference in distance between the earth and Jupiter: that is, it would be a measure of the distance in the time that light takes to travel between Jupiter and the earth. From a rough estimate of the size of the earth's orbit, and from the maximum discrepancy in the eclipse timing, which Roemer took to represent the time it takes light to cross the full width of the earth's orbit, he calculated the speed of light. His estimate came to 132,000 miles per second, remarkably close to the actual speed for what might be considered a first try, and high enough to evoke the disbelief of his contemporaries.

Roemer's results were, however, confirmed a half-century later from a completely different direction. In 1728, the British astronomer James Bradley found that stars seem to shift position because of the earth's motion—not through parallax, but because the velocity of the earth's motion about the sun is a measurable (though small) fraction of the speed of light. The analogy usually used is that of a man under an umbrella striding through a rainstorm. Even though the drops are falling vertically, the man must tip the umbrella forward, for he is stepping into the drops. The faster he walks, the farther he must tip the umbrella. Similarly, the earth moves into the light rays falling from the stars, and the astronomer must tip the telescope a bit, and in different directions, as the earth changes its direction of motion. From the amount of tip (the *aberration of light*), Bradley could estimate the value of the speed of light at 176,000 miles a second—a higher, and more accurate, value than Roemer's, though still about 5.5 percent too low.

Eventually, scientists obtained still more accurate measurements by applying refinements of Galileo's original idea. In 1849, the French physicist Armand Hippolyte Louis Fizeau set up an arrangement whereby a light was flashed to a mirror 5 miles away and reflected back to the observer. The elapsed time for the 10-mile round trip of the flash was not much more than 1/20,000 of a second, but Fizeau was able to measure it by placing a rapidly rotating toothed wheel in the path of the light beam. When the wheel turned at a certain speed, the flash going out between the two teeth would hit the next tooth when it came back from the mirror, and so Fizeau, behind the wheel, would not see it. When the wheel was speeded up, the returning flash would not be blocked but would come through the next gap between teeth (figure

Figure 8.1. Fizeau's arrangement for measuring the speed of light. Light reflected by the semimirror near the source passes through a gap in the rapidly spinning toothed wheel to a distant mirror (*right*) and is reflected back to the next tooth or the next gap.

8.1). Thus, by controlling and measuring the speed of the turning wheel, Fizeau was able to calculate the elapsed time, and therefore the speed of travel, of the flash of light. He found it to be 196,000 miles a second, which was 5.2 percent too high.

A year later, Jean Foucault (who was soon to perform his pendulum experiment; see chapter 4) refined the measurement by using a rotating mirror instead of a toothed wheel. Now the elapsed time was measured by a slight shift in the angle of reflection by the rapidly turning mirror (figure 8.2). Foucault's best measurement, in 1862, was 185,000 miles per second for the speed of light in air—only 0.7 percent too low. In addition, Foucault used his method to determine the speed of light through various liquids. He found the

Figure 8.2. Foucault's method. The amount of rotation of the mirror, instead of Fizeau's toothed wheel, gave the speed of the light's travel.

speed to be markedly less than the speed of light in air. This finding fitted Huygen's wave theory, too.

Still greater precision in the measurement of light's velocity came with the work of Michelson, who—over a period of more than forty years, starting in 1879—applied the Fizeau-Foucault approach with ever greater refinement. He eventually sent light through a vacuum rather than through air (even air slows it up slightly), using evacuated steel pipes up to a mile long for the purpose. He measured the speed of light in a vacuum to be 186,271 miles per second—only 0.006 percent too low. He was also to show that all wavelengths of light travel at the same speed in a vacuum.

In 1972, a research team under Kenneth M. Evenson made still more precise measurements and found the speed of light to be 186,282.3959 miles per second. Once the speed of light was known with such amazing precision, it became possible to use light, or at least forms of it, to measure distance. (It was practical to do so even when the speed was known less precisely.)

RADAR

Imagine a short pulse of light moving outward, striking some obstacle, being reflected backward, and being received at the point where it has issued forth an instant before. What is needed is a wave form of low enough frequency to penetrate fog, mist, and cloud, but of high enough frequency to be reflected efficiently. The ideal range was found to be in the microwave region, with wavelengths of from 0.2 to 40 inches. From the time lapse between emission of the pulse and return of the echo, the distance of the reflecting object can be estimated.

A number of physicists worked on devices making use of this principle, but the Scottish physicist Robert Alexander Watson-Watt was the first to make it thoroughly practicable. By 1935, he had made it possible to follow an airplane by the microwave reflections it sent back. The system was called *radio detection and ranging*, the word *range* meaning "to determine the distance of." The phrase was abbreviated to *ra.d. a. r.*, or *radar*. (A word, such as *radar*, that is constructed out of the initials of a phrase is called an *acronym*. Acronyms have become common in the modern world, particularly in science and technology.)

The world first became conscious of radar when it was learned that, by using that device, the British had been able to detect oncoming Nazi planes during the Battle of Britain, despite night and fog. To radar therefore belongs at least part of the credit of the British victory.

Since the Second World War, radar has had numerous peacetime uses. It has been used to detect rainstorms and has helped weather forecasters in this respect. It has turned up mysterious reflections called *angels*, which turned out to be, not heavenly messengers, but flocks of birds, so that now radar is used in the study of bird migrations.

And, as I described in chapter 3, it was radar reflections from Venus and Mercury that gave astronomers new knowledge concerning the rotations of those planets and, with regard to Venus, information about the nature of the surface.

LIGHT-WAVES THROUGH SPACE

Through all the mounting evidence of the wave nature of light, a nagging question continued to bother physicists. How is light transmitted through a vacuum? Other kinds of wave—sound, for instance—require a material medium. We derive the sensation of sound by the vibration, back and forth, of the atoms or molecules of the medium through which it travels. (From our observation platform here on Earth, we can never hear an explosion, however loud, on the moon or anywhere else in space because sound waves cannot travel across empty space.) Yet here were light-waves traveling through a vacuum more easily than through matter, and reaching us from galaxies billions of light-years away, although there was nothing there to wave.

Classical scientists were always uncomfortable about the notion of "action at a distance." Newton, for instance, worried about how the force of gravity could operate through space. As a possible explanation, he revived the Greeks' idea of an ether filling the heavens and speculated that perhaps the force of gravity might somehow be conducted by the ether. He avoided the light-problem by supposing light to consist of speeding particles, but that idea fell through when light was eventually found to be a wave phenomenon.

Trying to account for the travel of light-waves through space, physicists decided that light, too, must be conducted by the supposed ether. They began to speak of the *luminiferous* ("light-carrying") *ether*. But this idea at once ran into a serious difficulty. Light-waves are *transverse* waves: that is, they undulate at right angles to the direction of travel, like the ripples on the surface of water, in contrast to the *longitudinal* motion of sound waves, which vibrate back and forth in the direction of travel. Now physical theory said that only a *solid* medium could convey transverse waves. (Transverse water waves travel on the water surface—a special case—but cannot penetrate the body of the liquid.) Therefore the ether had to be solid, not gaseous or liquid—and an extremely rigid solid, too. To transmit waves at the tremendous speed of light, it had to be far more rigid than steel. What is more, this rigid ether had to permeate ordinary matter—not merely the vacuum of space but gases, water, glass, and all the other transparent substances through which light can travel. To cap it all, this solid, super-rigid material had to be so frictionless, so yielding, that it did not interfere in the slightest with the motion of the smallest planetoid or the flicker of an eyelid!

Yet, despite the difficulties introduced by the notion of the ether, it seemed useful. Faraday, who had no mathematical background at all but had marvelous insight, worked out the concept of *lines of force* (lines along which a magnetic field has equal strength) and, visualizing these as elastic distortions of the ether, thus used it to explain magnetic phenomena, too.

In the 1860s, Clerk Maxwell, a great admirer of Faraday, set about supplying the mathematical analysis to account for the lines of force. In doing so, he evolved a set of four simple equations that among them described almost all phenomena involving electricity and magnetism. These equations, advanced in 1864, not only described the interrelationship of the phenomena of electricity and magnetism, but showed the two cannot be separated. Where an electric field exists, there has to be a magnetic field, too, at right angles; and vice versa. There is, in fact, only a single electromagnetic field. (This was the original unified field theory which inspired all the work that followed in the next century.)

In considering the implications of his equations, Maxwell found that a changing electric field has to induce a changing magnetic field, which in turn has to induce a changing electric field, and so on; the two leapfrog, so to speak, and the field progresses outward in all directions. The result is a radiation possessing the properties of a wave-form. In short, Maxwell predicted the existence of *electromagnetic radiation* with frequencies equal to that in which the electromagnetic field waxes and wanes.

It was even possible for Maxwell to calculate the velocity at which such an electromagnetic wave would have to move. He did this by taking into consideration the ratio of certain corresponding values in the equations describing the force between electric charges and the force between magnetic poles. This ratio turned out to be precisely equal to the velocity of light, and Maxwell could not accept that as a mere coincidence. Light was an electromagnetic radiation, and along with it were other radiations with wavelengths far longer, or far shorter, than that of ordinary light—and all these radiations involved the ether.

THE MAGNETIC MONOPOLES

Maxwell's equations, by the way, introduced a problem that is still with us. They seemed to emphasize a complete symmetry between the phenomena of electricity and magnetism: what was true of one is true of the other. Yet in one fundamental way, the two seemed different—a difference that grew all the more puzzling once subatomic particles were discovered and studied. Particles exist that carry one or the other of the two opposed electric charges—positive or negative—but not both. Thus, the electron carries a negative electric charge only, while the positron carries a positive electric charge only. Analogously, ought not there be particles with a north magnetic pole only, and others with a south magnetic pole only? These *magnetic monopoles*, however, have long been sought in vain. Every object—large or small, galaxy or subatomic particle —that has a magnetic field has both a north pole and a south pole.

In 1931, Dirac, tackling the matter mathematically, came to the decision that if magnetic monopoles exist (if even *one* exists anywhere in the universe), it would be necessary for all electric charges to be exact multiples of some smallest charge—as, in fact, they are. And since all electric charges are exact multiples of some smallest charge, must not magnetic monopoles therefore exist?

In 1974, a Dutch physicist, Gerard 't Hooft, and a Soviet physicist, Alexander Polyakov, independently showed that it could be reasoned from the grand unified theories that indeed magnetic monopoles must exist, and that they must be enormous in mass. Although a magnetic monopole would be even smaller than a proton, it would have to have a mass of anywhere from 10 quadrillion to 10 quintillion times that of the proton. It would have the mass of a bacterium, all squeezed into a tiny subatomic particle.

Such particles could only have been formed at the time of the big bang. Never since has there been a sufficiently high concentration of energy to form them. Such huge particles would be moving at 150 miles a second or so, and the combination of huge mass and tiny size would allow it to slip through matter without leaving any signs to speak of. This property may account for the failure to detect magnetic monopoles hitherto.

If, however, the magnetic monopole managed to pass through a coil of wire, it would send a momentary surge of electric current through that coil (a well-known phenomenon that Faraday first demonstrated, see chapter 5). If the coil were at ordinary temperatures, the surge would come and go so quickly it might be missed. If it were superconductive, the surge would remain for as long as the coil was kept cold enough.

The physicist Blas Cabrera, at Stanford University, set up a superconductive niobium coil, kept it thoroughly isolated from stray magnetic fields, and waited four months. On 14 February 1982 at 1:53 P.M., there came a sudden flow of electricity, in just about exactly the amount one would expect if a magnetic monopole had passed through. Physicists are now trying to set up devices to confirm this finding; and until they do, we cannot be certain that the magnetic monopole has been detected at last.

ABSOLUTE MOTION

But back to the ether which, at the height of its power, met its Waterloo as a result of an experiment undertaken to test another classical question as knotty as action at a distance—namely, the question of *absolute motion*.

By the nineteenth century, it had become perfectly plain that the earth, the sun, the stars, and, in fact, all objects in the universe were in motion. Where, then, could you find a fixed reference point, one that was at *absolute rest*, to determine absolute motion—the foundation on which Newton's laws of motion were based? There was one possibility. Newton had suggested that the fabric of space itself (the ether, presumably) was at rest, so that one could speak of *absolute space*. If the ether was motionless, perhaps one could find the absolute motion of an object by determining its motion in relation to the ether.

In the 1880s, Albert Michelson conceived an ingenious scheme to find just that. If the earth is moving through a motionless ether, he reasoned, then a beam of light sent in the direction of its motion and reflected back should travel a shorter distance than one sent out at right angles and reflected back. To make the test, Michelson invented the "interferometer," a device with a *semimirror* that lets half of a light beam through in the forward direction and

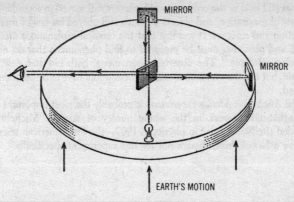

Figure 8.3. Michelson's interferometer. The semimirror (*center*) splits the light beam, reflecting one half and letting the other half go straight ahead. If the two reflecting mirrors (*at right and straight ahead*) are at different distances, the returning beams of light will arrive at the observer out of phase.

reflects the other half at right angles. Both beams are then reflected back by mirrors to an eyepiece at the source. If one beam has traveled a slightly longer distance than the other, they arrive out of phase and form interference bands (figure 8.3). This instrument is an extremely sensitive measurer of differences in length—so sensitive, in fact, that it can measure both the growth of a plant from second to second and the diameter of some stars that seem to be dimensionless points of light in even the largest telescope.

Michelson's plan was to point the interferometer in various directions with respect to the earth's motion and detect the effect of the ether by the amount by which the split beams were out of phase on their return.

In 1887, with the help of the American chemist Edward Williams Morley, Michelson set up a particularly delicate version of the experiment. Stationing the instrument on a stone floating on mercury, so that it could be turned in any direction easily and smoothly, they projected their beam in various directions with respect to the earth's motion. They discovered practically no difference! The interference bands were virtually the same no matter in what direction Michelson and Morley pointed the instrument or how many times they performed the experiment. (It should be said here that more recent experiments along the same line with still more delicate instruments have shown the same negative results.)

The foundations of physics tottered. Either the ether was moving with the earth, which made no sense at all, or there was, perhaps, no such thing as the ether. In either case there was no absolute motion or absolute space. The physics of Newton had had the rug pulled out from under it. Newtonian physics still held in the ordinary world: planets still moved in accordance with his law of gravitation, and objects on earth still obeyed his law of inertia and of action and reaction. It was just that the classical explanations are incomplete, and physicists must be prepared to find phenomena that do not obey the classical "laws." The observed phenomena, both old and new, would remain, but the theories accounting for them would have to be broadened and refined.

The *Michelson-Morley experiment* is probably the most important experiment-that-did-not-work in the whole history of science. Michelson was awarded the Nobel Prize in physics in 1907—the first American scientist to receive a Nobel Prize, though not for this experiment specifically.

Relativity

THE LORENTZ-FITZGERALD EQUATIONS

In 1893, the Irish physicist George Francis FitzGerald came up with a novel explanation to account for the negative results of the Michelson-Morley experiment. He suggested that all matter contracts in the direction of its motion and that the amount of contraction increases with the rate of motion. According to this interpretation, the interferometer is always shortened in the direction of the earth's "true" motion by an amount that exactly compensates for the difference in distance that the light beam has to travel. Moreover, all possible measuring devices, including human sense organs, would be "foreshortened" in just the same way, so that the foreshortening could, in no possible way, be measured, if we move with the object. FitzGerald's explanation almost made it look as if nature conspires to keep us from measuring absolute motion by introducing an effect that just cancels out any differences we might try to use to detect that motion.

This frustrating phenomenon became known as the *FitzGerald contraction*. FitzGerald worked out an equation for it. An object moving at 7 miles per second (about the speed of our fastest present rockets) would contract by only about two parts per billion in the direction of flight. But, at really high speeds, the contraction would be substantial. At 93,000 miles per second (half the speed of light), it would be 15 percent; at 163,000 miles per second (⅞ the speed of light), 50 percent: that is, a 1-foot ruler moving past us at 163,000 miles per second would seem only 6 inches long to us—provided we were not moving along with it and knew a method of measuring its length as it flew by. And at the speed of light, 186,282 miles per second, its length in the direction of motion would be zero. Since presumably there can be no length shorter than zero, it would follow that the speed of light in a vacuum is the greatest possible velocity in the universe.

The Dutch physicist Hendrik Antoon Lorentz soon carried FitzGerald's idea one step further. Thinking about cathode rays, on which Lorentz was working at the time, he reasoned that if the charge of a charged particle were compressed into a smaller volume, the mass of the particle should increase. Therefore a flying particle foreshortened in the direction of its travel by the FitzGerald contraction would have to increase in mass.

Lorentz presented an equation for the mass increase that turned out to be very similar to FitzGerald's equation for shortening. At 93,000 miles per second, an electron's mass would be increased by 15 percent; at 163,000 miles per second, by 100 percent (that is, its mass would be doubled); and at the speed of light, its mass would be infinite. Again it seemed that no speed greater than that of light could be possible, for how could mass be more than infinite?

The FitzGerald length effect and the Lorentz mass effect are so closely connected that the equations are often lumped together as the *Lorentz-FitzGerald equations*.

The change of mass with speed can be measured by a stationary observer far more easily than can the change in length. The ratio of an electron's mass to its charge can be determined from its deflection by a magnetic field. As an electron's velocity increased, the mass would increase, but there was no reason to think that the charge would; therefore, its mass-charge ratio should increase, and its path should become less curved. By 1900, the German physicist Walter Kauffman discovered that this ratio increased with velocity in such a way as to indicate that the electron's mass increases just as predicted by the Lorentz-FitzGerald equations. Later and better measurements showed the agreement to be just about perfect.

In discussing the speed of light as a maximum velocity, we must remember that it is the speed of light in a vacuum (186,282 miles per second) that is important here. In transparent material media, light moves more slowly. Its velocity in such a medium is equal to its velocity in a vacuum divided by the index of refraction of the medium. (The *index of refraction* is a measure of the extent by which a light-beam, entering the material obliquely from a vacuum, is bent.)

In water, with an index of refraction of about 1.3, the speed of light is 186,282 divided by 1.3, or about 143,000 miles per second. In glass (index of refraction about 1.5), the speed of light is 124,000 miles per second; while in diamond (index of refraction, 2.4) the speed of light is a mere 78,000 miles per second.

RADIATION AND PLANCK'S QUANTUM THEORY

It is possible for subatomic particles to travel through a particular transparent medium at a velocity greater than that of light in that medium (though *not* greater than that of light in a vacuum). When particles travel through a medium in this fashion, they throw back a wake of bluish light much as an airplane traveling at supersonic velocities throws back a wake of sound.

The existence of such radiation was observed by the Russian physicist Paul Alekseyevich Cherenkov (his name is also spelled Cerenkov) in 1934; in 1937, the theoretical explanation was offered by the Russian physicists Ilya Mikhailovich Frank and Igor Yevgenevich Tamm. All three shared the Nobel Prize for physics in 1958 as a result.

Particle detectors have been devised to detect the *Cerenkov radiation,* and these *Cerenkov counters* are particularly well adapted to study particularly fast particles, such as those making up the cosmic rays.

While the foundations of physics were still rocking from the Michelson-Morley experiment and the FitzGerald contraction, a second explosion took place. This time the innocent question that started all the trouble had to do with the radiation emitted by matter when it is heated. (Although the radiation in question is usually in the form of light, physicists speak of the problem as *black-body radiation:* that is, they are thinking of an ideal body that absorbs light perfectly—without reflecting any of it away, as a perfectly black body would do—and, in reverse, also radiates perfectly in a wide band of wavelengths.) The Austrian physicist Josef Stefan showed, in 1879, that the total radiation emitted by a body depends only on its temperature (not at all on the nature of its substance), and that, in ideal circumstances, the radiation is proportional to the fourth power of the absolute temperature: that is, doubling the absolute temperature would increase its total radiation $2 \times 2 \times 2 \times 2$, or sixteen-fold (*Stefan's law*). It was also known that, as the temperature rises, the predominant radiation moves toward shorter wavelengths. As a lump of steel is heated, for instance, it starts by radiating chiefly in the invisible infrared, then glows dim red, then bright red, then orange, then yellow-white, and finally, if it could somehow be kept from vaporizing at that point, it would be blue-white.

In 1893, the German physicist Wilhelm Wien worked out a theory that yielded a mathematical expression for the energy distribution of black-body radiation—that is, of the amount of energy radiated at each particular wavelength range. This theory provided a formula that accurately described the distribution of energy at the violet end of the spectrum but not at the red end. (For his work on heat, Wien received the Nobel Prize in physics in 1911.) On the other hand, the English physicists Lord Rayleigh and James Jeans worked up an equation that described the distribution at the red end of the spectrum but failed completely at the violet end. In short, the best theories available could explain one-half of the radiation or the other, but not both at once.

The German physicist Max Karl Ernst Ludwig Planck tackled the problem. He found that, in order to make the equations fit the facts, he had to introduce a completely new notion. He suggested that radiation consists of small units or packets, just as matter is made up of atoms. He called the unit of radiation the *quantum* (after the Latin word for "how much?"). Planck argued that radiation can be absorbed only in whole numbers of quanta. Furthermore, he suggested that the amount of energy in a quantum depends on the wavelength of the radiation. The shorter the wavelength, the more energetic the quantum; or, to put it another way, the energy content of the quantum is inversely proportional to the wavelength.

Now the quantum could be related directly to the frequency of a given radiation—that is, the number of waves emitted in 1 second. Like the quantum's energy content, the frequency is inversely proportional to the radiation's wavelength. The shorter the waves, the more of them can be emitted in 1 second. If both the frequency and the quantum's energy content were inversely proportional to the wavelength, then the two were directly proportional to each other. Planck expressed this relationship by means of his now-famous equation:

$$e = h\nu$$

The symbol e stands for the quantum energy; ν (the Greek letter nu), for the frequency; and h for *Planck's constant,* which gives the proportional relation between quantum energy and frequency.

The value of h is extremely small, and so is the quantum. The units of radiation are so small, in fact, that light looks continuous to us, just as ordinary matter seems continuous. But at the beginning of the twentieth century, the same fate befell radiation as had befallen matter at the beginning of the nineteenth: both now had to be accepted as discontinuous.

Planck's quanta cleared up the connection between temperature and the wavelengths of emitted radiation. A quantum of violet light was twice as energetic as a quantum of red light, and naturally it would take more heat energy to produce violet quanta than red quanta. Equations worked out on the basis of the quantum explained the radiation of a black body very neatly at both ends of the spectrum.

Eventually Planck's quantum theory was to do a great deal more: it was to explain the behavior of atoms, of the electrons in atoms, and of nucleons in the atoms' nuclei. Nowadays, physics before quantum theory is called *classical physics* and since quantum theory, *modern physics.* Planck was awarded the Nobel Prize in physics in 1918.

EINSTEIN'S PARTICLE-WAVE THEORY

Planck's theory made little impression on physicists when it was first announced in 1900. It was too revolutionary to be accepted at once. Planck himself seemed appalled at what he had done. But five years later a young German-born Swiss physicist named Albert Einstein verified the existence of his quanta.

The German physicist Philipp Lenard had found that, when light struck certain metals, it caused the metal surface to emit electrons, as if the force of the light kicked electrons out of the atoms. The phenomenon acquired the name *photoelectric effect,* and for its discovery Lenard received the Nobel Prize for physics in 1905. When physicists began to experiment with it, they found, to their surprise, that increasing the intensity of the light did not give the kicked-out electrons any more energy. But changing the wavelength of light did affect them: blue light, for instance, caused the electrons to fly out at greater speed than yellow light did. A very dim blue light would kick out fewer electrons than a bright yellow light would, but those few *blue-light* electrons would travel with greater speed than any of the *yellow-light* electrons. On the other hand, red light, no matter how bright, failed to knock out any electrons at all from some metals.

None of these phenomena could be explained by the old theories of light. Why should blue light do something red light cannot do?

Einstein found the answer in Planck's quantum theory. To absorb enough energy to leave the metal surface, an electron has to be hit by a quantum of a certain minimum size. In the case of an electron held only weakly by its atom (as in cesium), even a quantum of red light will do. Where atoms hold electrons more strongly, yellow light is required, or blue light, or even ultraviolet. And in any case, the more energetic the quantum, the more speed it gives to the electron it has kicked out.

Here the quantum theory explained a physical phenomenon with perfect simplicity, whereas the prequantum view of light had remained helpless. Other applications of quantum mechanics followed thick and fast. For his explanation of the photoelectric effect (not for his theory of relativity), Einstein was awarded the Nobel Prize in physics in 1921.

In his Special Theory of Relativity, presented in 1905 and evolved in his spare time while he worked as examiner at the Swiss patent office, Einstein proposed a new fundamental view of the universe based on an extension of the quantum theory. He suggested that light travels through space in quantum form (the term *photon* for this unit of light was introduced by Compton in 1928), and thus resurrected the concept of light consisting of particles. But this was a new kind of particle: it has properties of a wave as well as of a particle, and sometimes it shows one set of properties and sometimes the other.

This has been made to seem a paradox, or even a kind of mysticism, as if the true nature of light passes all possible understanding. On the contrary, let me illustrate with an analogy: a man may have many aspects—husband, father, friend, businessman. Depending on circumstances and on his surroundings, he behaves like a husband, a father, a friend, or a businessman. You would not expect him to exhibit his husbandly behavior toward a customer or his businesslike behavior toward his wife, and yet he is neither a paradox nor more than one man.

In the same way, radiation has both corpuscular and wave properties. In some capacities, the corpuscular properties are particularly pronounced; in others, the wave properties. About 1930, Niels Bohr advanced reasons for thinking that any experiment designed to test the wave properties of radiation could not conceivably detect the particle properties, and vice versa. One could deal with one or the other, never with both at the same time. He called this the *principle of complementarity*. This dual set of properties gives a more satisfactory account of radiation than either set of properties alone can.

The discovery of the wave nature of light had led to all the triumphs of nineteenth-century optics, including spectroscopy. But it had also required physicists to imagine the existence of the ether. Now Einstein's particle-wave view kept all the nineteenth-century victories (including Maxwell's equations), but made it unnecessary to assume that the ether exists. Radiation could travel through a vacuum by virtue of its particle attributes, and the ether idea, killed by the Michelson-Morley experiment, could now be buried.

Einstein introduced a second important idea in his special theory of relativity: that the speed of light in a vacuum never varies, regardless of the motion of its source. In Newton's view of the universe, a light beam from a source moving toward an observer should seem to travel more quickly than one from a source moving in any other direction. In Einstein's view, this would *not* seem to happen, and from that assumption he was able to derive the Lorentz-FitzGerald equations. He showed that the increase of mass with velocity, which Lorentz had applied only to charged particles, can be applied to all objects of any sort. Einstein reasoned further that increases in velocity would not only foreshorten length and increase mass but also slow the pace of time: in other words, clocks would slow down along with the shortening of yardsticks.

THE THEORY OF RELATIVITY

The most fundamental aspect of Einstein's theory was its denial of the existence of absolute space and absolute time. This may sound like nonsense: How can the human mind learn anything at all about the universe if it has no point of departure? Einstein answered that all we need to do is to pick a *frame of reference* to which the events of the universe can be related. Any frame of reference (the earth motionless, or the sun motionless, or we ourselves motionless, for that matter) will be equally valid, and we can simply choose the frame that is most convenient. It is more convenient to calculate planetary motions in a frame of reference in which the sun is motionless than in one in which the earth is motionless—but it is no more true.

Thus measurements of space and time are "relative" to some arbitrarily chosen frame of reference—and that is the reason for naming Einstein's idea the *theory of relativity*.

To illustrate. Suppose we on the earth were to observe a strange planet (Planet X), exactly like our own in size and mass, go whizzing past us at 163,000 miles per second relative to ourselves. If we could measure its dimensions as it shot past, we would find it to be foreshortened by 50 percent in the direction of its motion. It would be an ellipsoid rather than a sphere and would, on further measurement, seem to have twice the mass of the earth.

Yet to a man on Planet X, it would seem that he himself and his own planet were motionless. The earth would seem to be moving past *him* at 163,000 miles per second, and it would appear to have an ellipsoidal shape and twice the mass of *his* planet.

One is tempted to ask which planet would *really* be foreshortened and doubled in mass, but the only possible answer depends on the frame of reference. If you find that notion frustrating, consider that a man is small compared with a whale and large compared with a beetle. Is there any point in asking what a man is *really*—large or small?

For all its unusual consequences, relativity explains all the known phenomena of the universe at least as well as prerelativity theories do. But it goes further: it explains easily some phenomena that the Newtonian outlook explained poorly or not at all. Consequently, Einstein has been accepted over Newton, not as a replacement so much as a refinement. The Newtonian view of the universe can still be used as a simplified approximation that works well enough in ordinary life and even in ordinary astronomy, as in placing satellites in orbit. But when it comes to accelerating particles in a synchrotron, for example, we must take account of the Einsteinian increase of mass with velocity to make the machine work.

SPACE-TIME AND THE CLOCK PARADOX

Einstein's view of the universe so mingles space and time that either concept by itself becomes meaningless. The universe is four-dimensional, with time one of the dimensions (but behaving not quite like the ordinary spatial dimensions of length, breadth, and height). The four-dimensional fusion is often referred to as *space-time*. This notion was first developed by one of Einstein's teachers, the Russian-German mathematician Hermann Minkowski, in 1907.

With time as well as space up to odd tricks in relativity, one aspect of relativity that still provokes arguments among physicists is Einstein's notion of the slowing of clocks. A clock in motion, he said, keeps time more slowly than a stationary one. In fact, all phenomena that change with time change more slowly when moving than when at rest, which is the same as saying that time itself is slowed. At ordinary speeds, the effect is negligible; but at 163,000 miles per second, a clock would seem (to an observer watching it fly past) to take two seconds to tick off one second. And at the speed of light, time would stand still.

The time-effect is more disturbing than those involving length and weight. If an object shrinks to half its length and then returns to normal, or if it doubles its weight and then returns to normal, no trace is left behind to indicate the temporary change, and opposing viewpoints need not quarrel.

Time, however, is cumulative. If a clock on Planet X seems to be running at half-time for an hour because of its great speed, and if it is then brought to rest, it will resume its ordinary time-rate, but it will bear the mark of being half an hour slow! Well then, if two ships passed each other, and each considered the other to be moving at 163,000 miles per second and to be moving at half-time, when the two ships came together again, observers on each ship would expect the clock on the other ship to be half an hour slower than their own. But it is not possible for each clock to be slower than the other. What, then, would happen? This problem is called the *clock paradox*.

Actually, it is not a paradox at all. If one ship just flashed by the other and both crews swore the other ship's clock was slow, it would not matter which clock was "really" slow, because the two ships would separate forever. The two clocks would never be brought to the same place at the same time in order to be matched, and the clock paradox would never arise. Indeed, Einstein's Special Theory of Relativity only applies to uniform motion, so it is only the steady separation we are talking about.

Suppose, though, the two ships *did* come together after the flash-past, so that the clocks *could* be compared. In order for that to happen, there must be some new factor. At least one ship must accelerate. Suppose ship B did so —slowing down, traveling in a huge curve to point itself in the direction of A, then speeding up until it catches up with A. Of course, B might choose to consider itself at rest; by its chosen frame of reference, it is A that does all the changing, speeding up backward to come to B. If the two ships were all there were to the universe, then indeed the symmetry would keep the clock paradox in being.

However, A and B are *not* all there is to the universe—and that upsets the symmetry. When B accelerates, it is doing so with reference not only to A but to all the rest of the universe besides. If B chooses to consider itself at rest, it must consider not only A, but all the galaxies without exception, to be accelerating with respect to itself. It is B against the universe, in short. Under these circumstances, it is B's clock that ends up half an hour slow; not A's.

This phenomenon affects notions of space travel. If astronauts leaving Earth speed up to near the speed of light, their rate of time passage would be much slower than ours. They might reach a distant destination and return in what seemed to them weeks, though on the earth many centuries would have passed. If time really slows in motion, one might journey even to a distant star in one's own lifetime. But of course one would have to say good-bye to one's own generation and the world one knew, and return to a world of the future.

GRAVITY AND EINSTEIN'S GENERAL THEORY

In the Special Theory of Relativity, Einstein did not deal with accelerated motion or gravitation. These were treated in his General Theory of Relativity, published in 1915. The General Theory presented a completely altered view of gravitation. It was viewed as a property of space rather than as a force between bodies. As the result of the presence of matter, space becomes curved, and bodies follow the line of least resistance among the curves, so to speak. Strange as Einstein's idea seemed, it was able to explain something that the Newtonian law of gravity had not been able to explain.

The greatest triumph of Newton's law of gravity had come in 1846 with the discovery of Neptune (see chapter 3). After that, nothing seemed capable of shaking Newton's law of gravity. And yet one planetary motion remained unexplained. The planet Mercury's point of nearest approach to the sun, its *perihelion*, changes from one trip to the next; it advances steadily in the course of the planet's revolutions around the sun. Astronomers were able to account for most of this irregularity as due to perturbations of its orbit by the pull of the neighboring planets.

Indeed, there had been some feeling in the early days of work with the theory of gravitation that perturbations arising from the shifting pull of one planet on another might eventually act to break up the delicate mechanism of the solar system. In the earliest decades of the nineteenth century, however, Laplace showed that the solar system was not so delicate. The perturbations are all cyclic, and orbital irregularities never increase to more than a certain amount in any direction. In the long run, the solar system is stable, and astronomers were more certain than ever that all particular irregularities could be worked out by taking perturbations into account.

This assumption, however, did not work for Mercury. After all the perturbations had been allowed for, there was still an unexplained advance of Mercury's perihelion by an amount equal to 43 seconds of arc per century. This motion, discovered by Leverrier in 1845, is not much: in 4,000 years it adds up only to the width of the moon. It was enough, however, to upset astronomers.

Leverrier suggested that this deviation might be caused by a small, undiscovered planet closer to the sun than Mercury. For decades astronomers searched for the supposed planet (called Vulcan), and many were the reports of its discovery. All the reports turned out to be mistaken. Finally it was agreed that Vulcan did not exist.

Then Einstein's General Theory of Relativity supplied the answer. It showed that the perihelion of any revolving body should have a motion beyond that predicted by Newton's law. When this new calculation was applied to Mercury, the planet's shift of perihelion fit it exactly. Planets farther from the sun than Mercury should show a progressively smaller shift of perihelion. In 1960, the perihelion of Venus's orbit had been found to be advancing about 8 seconds of arc per century; this shift fits Einstein's theory almost exactly.

More impressive were two unexpected new phenomena that only Einstein's theory predicted. First, Einstein maintained that an intense gravitational field should slow down the vibrations of atoms. The slowdown would be evidenced by a shift of spectral lines toward the red (the *Einstein shift*). Casting about for a gravitational field strong enough to produce this effect, Eddington suggested the white dwarfs: light leaving such a condensed star against its powerful surface gravity might lose a detectable amount of energy. In 1925, W. S. Adams, who had been the first to demonstrate the enormous density of such stars, studied the spectral lines in the light of white dwarfs and found the necessary red shift.

The verification of Einstein's second prediction was even more dramatic. His theory said a gravitational field would bend light-rays. Einstein calculated that a ray of light just skimming the sun's surface would be bent out of a straight line by 1.75 seconds of arc (figure 8.4). How could that be checked? Well, if stars beyond the sun and just off its edge could be observed during an eclipse of the sun and their positions compared with what they were against the background when the sun did not interfere, any shift resulting from bending of their light should show up. Since Einstein had published his paper

Figure 8.4. The gravitational bending of light waves, postulated by Einstein in the General Theory of Relativity.

on general relativity in 1915, the test had to wait until after the end of the First World War. In 1919, the British Royal Astronomical Society organized an expedition to make the test by witnessing a total eclipse visible from the island of Principe, a small Portuguese-owned island off West Africa. The stars did shift position. Einstein had been verified again.

By this same principle, if one star were directly behind another, the light of the farther star would bend about the nearer in such a way that the farther star would appear larger than it really is. The nearer star would act as a *gravitational lens*. Unfortunately, the apparent size of stars is so minute that an eclipse of a distant star by a much closer one (as seen from Earth) is extremely rare. The discovery of quasars, however, gave astronomers another chance. In the early 1980s, they noted double quasars in which each member has precisely the same property. It is a reasonable supposition that we are seeing only one quasar with its light distorted by a galaxy (or black hole, possibly) that is in the line of sight but invisible to us. The image of the quasar is distorted and made to appear double. (An imperfection in a mirror might have the same effect on our own reflected image.)

TESTING THE GENERAL THEORY

The early victories of Einstein's General Theory were all astronomic in nature. Scientists longed to discover a way to check it in the laboratory under conditions they could vary at will. The key to such a laboratory demonstration arose in 1958, when the German physicist Rudolf Ludwig Mössbauer showed that, under certain conditions, a crystal can be made to produce a beam of gamma rays of sharply defined wavelength. Ordinarily, the atom emitting the gamma ray recoils, and this recoil broadens the band of wavelengths produced. In crystals under certain conditions, a crystal acts as a single atom: the recoil is distributed among all the atoms and sinks to virtually nothing, so that the gamma ray emitted is exceedingly sharp. Such a sharp-wavelength beam can be absorbed with extraordinary efficiency by a crystal similar to the one that produced it. If the gamma rays are of even slightly different wavelength from that which the crystal would naturally produce, it would not be absorbed. This is called the *Mössbauer effect*.

If such a beam of gamma rays is emitted downward so as to fall with gravity, the General Theory of Relativity requires it to gain energy so that its wavelength becomes shorter. In falling just a few hundred feet, it should gain enough energy for the decrease in wavelength of the gamma rays, though very minute, to become sufficiently large that the absorbing crystal will no longer absorb the beam.

Furthermore, if the crystal emitting the gamma ray is moved upward while the emission is proceeding, the wavelength of the gamma ray is increased through the Doppler-Fizeau effect. The velocity at which the crystal is moved upward can be adjusted so as to just neutralize the effect of gravitation on the falling gamma ray, which will then be absorbed by the crystal on which it impinges.

Experiments conducted in 1960 and later made use of the Mössbauer effect to confirm the General Theory with great exactness. They were the most impressive demonstration of its validity that has yet been seen; as a result, Mössbauer was awarded the 1961 Nobel Prize for physics.

Other delicate measurements also tend to support General Relativity: the passage of radar beams past a planet, the behavior of binary pulsars as they revolve about a mutual center of gravity, and so on. All the measurements are borderline, and numerous attempts have been made by physicists to suggest alternate theories. Of all the suggested theories, however, Einstein's is the simplest from the mathematical standpoint. Whenever measurements are made that can possibly distinguish between the theories (and the differences are always minute), it is Einstein's that seems to be supported. After nearly three-quarters of a century, the General Theory of Relativity stands unshaken, although scientists continue (quite properly) to question it. (Mind you, it is the *General* theory that is questioned. The *Special* Theory of Relativity has been verified over and over and over again in so many different ways that no physicist questions it.)

Heat

So far in this chapter I have been neglecting a phenomenon that usually accompanies light in our everyday experience. Almost all luminous objects from a star to a candle give off heat as well as light.

MEASURING TEMPERATURE

Heat was not studied, other than qualitatively, before modern times. It was enough for a person to say, "It is hot," or "It is cold," or "This is warmer than that." To subject temperature to quantitative measure, it was first necessary to find some measurable change that seemed to take place uniformly with change in temperature. One such change was found in the fact that substances expand when warmed and contract when cooled.

Galileo was the first to try to make use of this fact to detect changes in temperature. In 1603, he inverted a glass tube of heated air into a bowl of water. As the air in the tube cooled to room temperature, it contracted and drew water up the tube, and there Galileo had his *thermometer* (from Greek words meaning "heat measure"). When the temperature of the room changed, the water level in the tube changed. If the room warmed, the air in the tube expanded and pushed the water level down; if it grew cooler, the air contracted and the water level moved up. The only trouble was that the basin of water into which the tube had been inserted was open to the air and the air pressure kept changing. That also shoved the water level up and down, independently of temperature, confusing the results. The thermometer was the first important scientific instrument to be made of glass.

By 1654, the Grand Duke of Tuscany, Ferdinand II, had evolved a thermometer that was independent of air pressure. It contained a liquid sealed into a bulb to which a straight tube was attached. The contraction and expansion of the liquid itself was used as the indication of temperature change. Liquids change their volume with temperature much less than gases do; but with a sizable reservoir of liquid and a filled bulb, so that the liquid could expand only up a very narrow tube, the rise and fall within that tube, for even tiny volume changes, could be made considerable.

The English physicist Robert Boyle did much the same thing about the same time, and he was the first to show that the human body had a constant temperature, markedly higher than the usual room temperature. Others demonstrated that certain physical phenomena always take place at some fixed temperature. Before the end of the seventeenth century, such was found to be the case for the melting of ice and the boiling of water.

The first liquids used in thermometry were water and alcohol. Since water froze too soon and alcohol boiled away too easily, the French physicist Guillaume Amontons resorted to mercury. In his device, as in Galileo's, the expansion and contraction of air caused the mercury level to rise or fall.

Then, in 1714, the German physicist Gabriel Daniel Fahrenheit combined the advances of the Grand Duke and of Amontons by enclosing mercury in a tube and using its own expansion and contraction with temperature as the indicator. Furthermore, Fahrenheit put a graded scale on the tube to allow the temperature to be read quantitatively.

There is some argument about exactly how Fahrenheit arrived at the particular scale he used. He set zero, according to one account, at the lowest temperature he could get in his laboratory, attained by mixing salt and melting ice. He then set the freezing point of pure water at 32 and its boiling point at 212. This had two advantages. First, the range of temperature over which water was liquid came to 180, which seemed a natural number to use in connection with *degrees,* as there are 180 degrees in a semicircle. Second, body temperature came near a round 100 degrees; normally it is 98.6° Fahrenheit, to be exact.

So constant is body temperature normally that, if it is more than a degree or so above the average, the body is said to run a fever, and one has a clear feeling of illness. In 1858, the German physician Karl August Wunderlich introduced the procedure of frequent checks on body temperature as an indication of the course of disease. In the next decade, the British physician Thomas Clifford Allbutt invented the *clinical thermometer* which has a constriction in the narrow tube containing the mercury. The mercury thread rises to a maximum when placed in the mouth, but does not fall when the thermometer is removed. The mercury thread simply divides at the constriction, leaving a constant reading in the portion above. In the United States, the Fahrenheit scale is still used. We are familiar with it in everyday affairs such as weather reporting and clinical thermometers.

In 1742, however, the Swedish astronomer Anders Celsius adopted a different scale. In its final form, this set the freezing point of water at 0 and its boiling point at 100. Because of the hundredfold division of the temperature range in which water is liquid, this is called the *centigrade scale,* from Latin words meaning "hundred steps" (see figure 6.5). Most people still speak of measurements on this scale as *degrees centigrade;* but scientists, at an international conference in 1948, renamed the scale after the inventor, following the Fahrenheit precedent. Officially, then, one should speak of the *Celsius scale* and of *degrees Celsius.* The symbol *C* still holds. It was Celsius's scale that won out in the civilized world, and even the United States is attempting to accustom its people to its use. Scientists, in particular, found the Celsius scale convenient.

TWO THEORIES OF HEAT

Temperature measures the intensity of heat but not its quantity. Heat will always flow from a place of higher temperature to a place of lower temperature until the temperatures are equal, just as water will flow from a higher level to a lower one until the levels are equal. Heat behaves so regardless of the relative amounts of heat contained in the bodies involved. Although a bathtub of lukewarm water contains far more heat than a burning match, when the match is placed near the water, heat goes from the match to the water, not vice versa.

Joseph Black, who had done important work on gases (see chapter 5), was the first to make clear the distinction between temperature and heat. In 1760, he announced that various substances were raised in temperature by different amounts when a given amount of heat was poured into them. To raise the temperature of 1 gram of iron by 1 degree Celsius takes three times as much heat as to warm 1 gram of lead by 1 degree. And beryllium requires three times as much heat as iron.

Furthermore, Black showed it was possible to pour heat into a substance without raising its temperature at all. When melting ice is heated, melting is hastened, but the ice does not rise in temperature. Heat will eventually melt all the ice; but during the process, the temperature of the ice itself never goes above 0° C. Likewise, with boiling water at 100° C: as heat is poured into the water, more and more of it boils away as vapor, but the temperature of the liquid does not change while it is boiling.

The development of the steam engine (see chapter 9), which came at about the same time as Black's experiments, intensified the interest of scientists in heat and temperature. They began to speculate about the nature of heat, as earlier they had speculated about the nature of light.

In the case of heat, as of light, there were two theories. One held heat to be a material substance which can be poured or shifted from one substance to another. It was named *caloric,* from the Latin for "heat." According to this view, when wood was burned the caloric in the wood passed into the flame, and from it into a kettle above the flame, and from it into the water in the kettle. As water filled with caloric, it was converted to steam.

As for the other theory, in the late eighteenth century, two famous observations gave rise to the theory of heat as a form of vibration. One was published by the American physicist and adventurer Benjamin Thompson, a Tory who fled the country during the Revolution, was given the title Count Rumford, and then proceeded to knock around Europe. While supervising the boring of cannon in Bavaria in 1798, he noticed that quantities of heat were being produced. He found that enough heat was being generated to bring 18 pounds of water to the boiling point in less than 3 hours. Where was all the caloric coming from? Thompson decided that heat must be a vibration set up and intensified by the mechanical friction of the borer against the cannon.

The next year the chemist Humphry Davy performed an even more significant experiment. Keeping two pieces of ice below the freezing point, he rubbed them together, not by hand but by a mechanical contrivance, so that no caloric could flow into the ice. By friction alone, he melted some of the ice. He, too, concluded that heat must be a vibration and not a material. Actually, this experiment should have been conclusive; but the caloric theory, though obviously wrong, persisted to the middle of the nineteenth century.

HEAT AS ENERGY

Nevertheless, although the nature of heat was misunderstood, scientists learned some important things about it, just as the investigators of light turned up interesting facts about the reflection and refraction of light beams before they knew its nature. The French physicists Jean Baptiste Joseph Fourier, in 1822, and Nicholas Léonard Sadi Carnot, in 1824, studied the flow of heat and made important advances. In fact, Carnot is usually considered the founder of the science of *thermodynamics* (from Greek words meaning "movement of heat"). He placed the working of steam engines on a firm theoretical foundation.

By the 1840s, physicists were concerned with the manner in which the heat that was put into steam could be converted into the mechanical work of moving a piston. Is there a limit to the amount of work that can be obtained from a given amount of heat? And what about the reverse process: How is work converted to heat?

Joule spent thirty-five years converting various kinds of work into heat, doing very carefully what Rumford had earlier done clumsily. He measured the amount of heat produced by an electric current. He heated water and mercury by stirring them with paddle wheels, or by forcing water through narrow tubes. He heated air by compressing it, and so on. In every case, he calculated how much mechanical work had been done on the system and how much heat was obtained as a result. He found that a given amount of work, of any kind, always produces a given amount of heat. Joule had, in other words, determined the *mechanical equivalent of heat.*

Since heat could be converted into work, it must be considered a form of *energy* (from Greek words meaning "containing work"). Electricity, magnetism, light, and motion can all be used to do work, so they, too, are forms of energy. And work itself, being convertible into heat, is a form of energy.

These ideas emphasized something that had been more or less suspected since Newton's time: that energy is *conserved* and can be neither created nor destroyed. Thus, a moving body has *kinetic energy* ("energy of motion"), a term introduced by Lord Kelvin in 1856. Since a body moving upward is slowed by gravity, its kinetic energy slowly disappears. However, as the body loses kinetic energy, it gains energy of position, for, by virtue of its location high above the surface of the earth, it can eventually fall and regain kinetic energy. In 1853, the Scottish physicist William John Macquorn Rankine named this energy of position *potential energy.* It seemed that a body's kinetic energy plus its potential energy (its *mechanical energy*) remain nearly the same during the course of its movement; this constancy was called *conservation of mechanical energy.* However, mechanical energy is not *perfectly* conserved: some is lost to friction, to air resistance, and so on.

What Joule's experiments showed above all was that such conservation could be made exact when heat is taken into account, for, when mechanical energy is lost to friction or air resistance, it appears as heat. Take that heat into account, and one can show, without qualification, that no new energy is created and no old energy destroyed. The first to make this plain was a German physicist, Julius Robert Mayer, in 1842, but his experimental backing was meager, and he lacked strong academic credentials. (Even Joule, who was a brewer by profession and also lacked academic credentials, had difficulty getting his meticulous work published.)

It was not till 1847 that a sufficiently respectable academic figure put this notion into words. In that year, Heinrich von Helmholtz enunciated the *law of conservation of energy:* whenever a certain amount of energy seems to disappear in one place, an equivalent amount must appear in another. This is also called the *first law of thermodynamics.* It remains a foundation block of modern physics, undisturbed by either quantum theory or relativity.

Now, although any form of work can be converted entirely into heat, the reverse is not true. When heat is turned to work, some of it is unusable and is unavoidably wasted. In running a steam engine, the heat of the steam is converted into work only until the temperature of the steam is reduced to the temperature of the environment; after that, although there is much remaining heat in the cold water formed from the steam, no more of it can be converted to work. Even in the temperature range at which work can be extracted, some of the heat does not go into work but is used up in heating the engine and the air around it, in overcoming friction between the piston and the cylinder, and so on.

In any energy conversion—such as, electric energy into light energy, or magnetic energy into energy of motion—some of the energy is wasted. It is not lost—that would be contrary to the first law; but it is converted to heat that is dissipated in the environment.

The capacity of any system to perform work is its *free energy*. The portion of the energy that is unavoidably lost as nonuseful heat is reflected in the measurement of *entropy*—a term first used in 1850 by the German physicist Rudolf Julius Emmanuel Clausius.

Clausius pointed out that, in any process involving a flow of energy, there is always some loss, so that the entropy of the universe is continually increasing. This continual increase of entropy is the *second law of thermodynamics*, sometimes referred to as the "running-down of the universe" or the "heat-death of the universe." Fortunately, the quantity of usable energy (supplied almost entirely by the stars, which are "running down" at a tremendous rate) is so vast that there is enough for all purposes for many billions of years.

HEAT AND MOLECULAR MOTION

A clear understanding of the nature of heat finally came with the understanding of the atomic nature of matter and developed from the realization that the molecules composing a gas are in continual motion, bouncing off one another and off the walls of their container. The first investigator who attempted to explain the properties of gases from this standpoint was the Swiss mathematician Daniel Bernoulli, in 1738, but he was ahead of his times. In the mid-nineteenth century, Maxwell and Boltzmann (see chapter 5) worked out the mathematics adequately and established the *kinetic theory of gases* (*kinetic* comes from a Greek word meaning "motion"). The theory showed heat to be equivalent to the motion of molecules. Thus, the caloric theory of heat received its deathblow. Heat was seen to be a vibrational phenomenon: the movement of molecules in gases and liquids or the jittery to-and-fro trembling of molecules in solids.

When a solid is heated to a point where the to-and-fro trembling is strong enough to break the bonds that hold neighboring molecules together, the solid melts and becomes a liquid. The stronger the bond between neighboring molecules in a solid, the more heat is needed to make it vibrate violently enough to break the bond. Hence, the substance has a higher melting point.

In the liquid state, the molecules can move freely past one another. When the liquid is heated further, the movements of the molecules finally become sufficiently energetic to set them free of the body of the liquid altogether, and then the liquid boils. Again, the boiling point is higher where the intermolecular forces are stronger.

In converting a solid to a liquid, all of the energy of heat goes into breaking the intermolecular bonds. Thus, the heat absorbed by melting ice does not raise the ice's temperature. The same is true of a liquid being boiled.

Now we can distinguish between heat and temperature easily. Heat is the total energy contained in the molecular motions of a given quantity of matter. Temperature represents the average energy of motion per molecule in that matter. Thus, a quart of water at 60° C contains twice as much heat as a pint of water at 60° C (twice as many molecules are vibrating), but the quart and the pint have the same temperature, for the average energy of molecular motion is the same in each case.

There is energy in the very structure of a chemical compound—that is, in the bonding forces that hold an atom or molecule to its neighbor. If these bonds are broken and rearranged into new bonds involving less energy, the excess of energy will make its appearance as heat or light or both. Sometimes the energy is released so quickly as to result in an explosion.

It is possible to calculate the chemical energy contained in any substance and show what the amount of heat released in any reaction must be. For instance, the burning of coal involves breaking the bonds between carbon atoms in the coal and the bonds between the oxygen molecules' atoms, with which the carbon recombines. Now the energy of the bonds in the new compound (carbon dioxide) is less than that of the bonds in the original substances that formed it. This difference, which can be measured, is released as heat and light.

In 1876, the American physicist Josiah Willard Gibbs worked out the theory of *chemical thermodynamics* in such detail that this branch of science was brought from virtual nonexistence to complete maturity at one stroke.

The long paper in which Gibbs described his reasoning was far above the heads of others in America and was published in the *Transactions of the Connecticut Academy of Arts and Sciences* only after considerable hesitation. Even afterward, its close-knit mathematical argument and the retiring nature of Gibbs himself combined to keep the subject under a bushel basket until Ostwald discovered the work in 1883, translated the paper into German, and proclaimed the importance of Gibbs to the world.

As an example of the importance of Gibbs's work, his equations demonstrated the simple, but rigorous, rules governing the equilibrium between different substances existing simultaneously in more than one phase (that is, in both solid form and in solution, in two immiscible liquids and a vapor, and so on). This *phase rule* is the breath of life to metallurgy and to many other branches of chemistry.

Mass to Energy

With the discovery of radioactivity in 1896 (see chapter 6), a totally new question about energy arose at once. The radioactive substances uranium and thorium were giving off particles with astonishing energies. Moreover, Marie

Curie found that radium was continually emitting heat in substantial quantities: an ounce of radium gave off 4,000 calories per hour, and would do so hour
after hour, week after week, decade after decade. The most energetic chemical
reaction known could not produce 1 millionth of the energy liberated by
radium. Was the law of conservation of energy being broken?

And no less surprising was the fact that this production of energy, unlike
chemical reactions, did not depend on temperature: it went on just as well at
the very low temperature of liquid hydrogen as it did at ordinary temperatures!

Quite plainly an altogether new kind of energy, very different from chemical, was involved here. Fortunately physicists did not have to wait long for the
answer. Once again, it was supplied by Einstein, in his Special Theory of
Relativity. Einstein's mathematical treatment of energy showed that mass can
be considered a form of energy—a very concentrated form, for a very small
quantity of mass would be converted into an immense quantity of energy.

Einstein's equation relating mass and energy is now one of the most famous
equations in the world. It is:

$$e = mc^2$$

Here e represents energy (in ergs), m represents mass (in grams) and c
represents the speed of light (in centimeters per second). Other units of
measurement can be used but would not change the nature of the result.

Since light travels at 30 billion centimeters per second, the value of c^2 is
900 billion billion; or, in other words, the conversion of 1 gram of mass energy
will produce 900 billion billion ergs. The *erg* is a small unit of energy not
translatable into any common terms, but we can get an idea of what this
number means when we know that the energy in 1 gram of mass is sufficient
to keep a 1,000-watt electric-light bulb running for 2,850 years. Or, to put it
another way, the complete conversion of 1 gram of mass into energy would
yield as much as the burning of 2,000 tons of gasoline.

Einstein's equation destroyed one of the sacred conservation laws of science.
Lavoisier's law of conservation of mass had stated that matter can be neither
created nor destroyed. Actually, every energy-releasing chemical reaction
changes a small amount of mass into energy: the products, if they could be
weighed with utter precision, would not quite equal the original matter. But
the mass lost in ordinary chemical reactions is so small that no technique
available to the chemists of the nineteenth century could conceivably have
detected it. Physicists, however, were now dealing with a completely different
phenomenon, the nuclear reaction of radioactivity rather than the chemical
reaction of burning coal. Nuclear reactions release so much energy that the
loss of mass is large enough to be measured.

By postulating the interchange of mass and energy, Einstein merged the
laws of conservation of energy and of mass into one law—the *conservation of
mass-energy*. The first law of thermodynamics not only still stood: it was more
unassailable than ever.

The conversion of mass to energy was confirmed experimentally by Aston through his mass spectograph, which could measure the mass of atomic nuclei very precisely by the amount of their deflection by a magnetic field. What Aston did with an improved instrument in 1925 was to show that the various nuclei are not exact multiples of the masses of the neutrons and protons that compose them.

Let us consider the masses of these neutrons and protons for a moment. For a century, the masses of atoms and subatomic particles generally have been measured on the basis of allowing the atomic weight of oxygen to be exactly 16.00000 (see chapter 6). In 1929, however, Giauque had showed that oxygen consists of three isotopes—oxygen 16, oxygen 17, and oxygen 18—and that the atomic weight of oxygen is the weighted average of the mass numbers of these three isotopes.

To be sure, oxygen 16 is by far the most common of the three, making up 99.759 percent of all oxygen atoms. Thus, if oxygen has the over-all atomic weight of 16.00000, the oxygen-16 isotope has a mass number of *almost* 16. (The masses of the small quantities of oxygen 17 and oxygen 18 bring the value up to 16.) Chemists, for a generation after the discovery, did not let this disturb them, but kept the old basis for what came to be called *chemical atomic weights*.

Physicists, however, reacted otherwise. They preferred to set the mass of the oxygen-16 isotope at exactly 16.0000 and determine all other masses on that basis. On this basis, the *physical atomic weights* could be set up. On the oxygen-16 equals 16 standard, the atomic weight of oxygen itself, with its traces of heavier isotopes, is 16.0044. In general the physical atomic weights of all elements would be 0.027 percent higher than their chemical atomic weight counterparts.

In 1961, physicists and chemists reached a compromise and agreed to determine atomic weights on the basis of allowing the carbon-12 isotope to have a mass of 12.0000, thus basing the atomic weights on a characteristic mass number and making them as fundamental as possible. In addition, this base made the atomic weights almost exactly what they were under the old system. Thus, on the carbon-12 equals 12 standard, the atomic weight of oxygen is 15.9994.

Well, then, let us start with a carbon-12 atom, with its mass equal to 12.0000. Its nucleus contains six protons and six neutrons. From mass-spectrographic measurements, it becomes evident that, on the carbon-12 equals 12 standard, the mass of a proton is 1.007825 and that of a neutron is 1.008665. Six protons, then, should have a mass of 6.046950; and six neutrons, 6.051990. Together, the twelve nucleons should have a mass of 12.104940. But the mass of the carbon-12 is 12.00000. What has happened to the missing 0.104940?

This disappearing mass is the *mass defect*. The mass defect divided by the mass number gives the mass defect per nucleon, or the *packing fraction*. The mass has not really disappeared but has been converted into energy, in accordance with Einstein's equation, so that the mass defect is also the *binding energy* of the nucleus. To break a nucleus down into individual protons and neutrons would require the input of an amount of energy equal to the binding energy, since an amount of mass equivalent to that energy would have to be formed.

Aston determined the packing fraction of many nuclei, and he found it to increase rather quickly from hydrogen up to elements in the neighborhood of iron and then to decrease, rather slowly, for the rest of the periodic table. In other words, the binding energy per nucleon is highest in the middle of the periodic table. Thus, conversion of an element at either end of the table into one nearer the middle should release energy.

Take uranium 238 as an example. This nucleus breaks down by a series of decays to lead 206. In the process, it emits eight alpha particles. (It also gives off beta particles, but these are so light they can be ignored.) Now the mass of lead 206 is 205.9745 and that of eight alpha particles totals 32.0208. Altogether these products add up to a mass of 237.9953. But the mass of uranium 238, from which they came, is 238.0506. The difference, or loss of mass, is 0.0553. That loss of mass is just enough to account for the energy released when uranium breaks down.

When uranium breaks down to still smaller atoms, as it does in fission, a great deal more energy is released. And when hydrogen is converted to helium, as it is in stars, there is an even larger fractional loss of mass and a correspondingly richer development of energy.

Physicists began to look upon the mass-energy equivalence as a very reliable bookkeeping. For instance, when the positron was discovered in 1934, its mutual annihilation with an electron produced a pair of gamma rays whose energy was just equal to the mass of the two particles. Furthermore, as Blackett was first to point out, mass could be created out of appropriate amounts of energy. A gamma ray of the proper energy, under certain circumstances, would disappear and give rise to an *electron-positron pair*, created out of pure energy. Larger amounts of energy, supplied by cosmic particles or by particles fired out of proton synchrotons (see chapter 7), would bring about the creation of more massive particles, such as mesons and antiprotons.

It is no wonder that when the bookkeeping did not balance, as in the emission of beta particles of less than the expected energy, physicists invented the neutrino to balance the energy account rather than tamper with Einstein's equation (see chapter 7).

If any further proof of the conversion of mass to energy was needed, nuclear bombs provided the final clincher.

Particles and Waves

In the 1920s, dualism reigned supreme in physics. Planck had shown radiation to be particlelike as well as wavelike. Einstein had shown that mass and energy are two sides of the same coin; and that space and time are inseparable. Physicists began to look for other dualisms.

In 1923, the French physicist Louis Victor de Broglie was able to show that, just as radiation has the characteristics of particles, so the particles of matter, such as electrons, should display the characteristics of waves. The waves associated with these particles, he predicted, would have a wavelength inversely related to the mass times the velocity (that is, the *momentum*) of the particle. The wavelength associated with electrons of moderate speed, de Broglie calculated, ought to be in the X-ray region.

In 1927, even this surprising prediction was borne out. Clinton Joseph Davisson and Lester Halbert Germer of the Bell Telephone Laboratories were bombarding metallic nickel with electrons. As the result of a laboratory accident, which had made it necessary to heat the nickel for a long time, the metal was in the form of large crystals, which were ideal for diffraction purposes because the spacing between atoms in a crystal is comparable to the very short wavelengths of electrons. Sure enough, the electrons passing through those crystals behaved not as particles but as waves. The film behind the nickel showed interference patterns, alternate bands of fogging and clarity, just as it would have shown if X rays rather than electrons had gone through the nickel.

Interference patterns were the very thing that Young had used more than a century earlier to prove the wave nature of light. Now they proved the wave nature of electrons. From the measurements of the interference bands, the wavelength associated with the electron could be calculated, and it turned out to be 1.65 angstrom units, almost exactly what de Broglie had calculated it ought to be.

In the same year, the British physicist George Paget Thomson, working independently and using different methods, also showed that electrons have wave properties.

De Broglie received the Nobel Prize in physics in 1929, and Davisson and Thomson shared the Nobel Prize in physics in 1937.

ELECTRON MICROSCOPY

This entirely unlooked-for discovery of a new kind of dualism was put to use almost at once in microscopy. Ordinary optical microscopes, as I have mentioned, cease to be useful at a certain point because there is a limit to the size of objects that light-waves can define sharply. As objects get smaller, they also get fuzzier, because the light-waves begin to pass around them—something first pointed out by the German physicist Ernst Karl Abbe in 1878. The cure, of course, is to try to find shorter wavelengths with which to resolve the smaller objects. Ordinary-light microscopes can distinguish two dots 1/5,000 millimeter apart, but ultraviolet microscopes can distinguish dots 1/10,000 millimeter apart. X rays would be better still, but there are no lenses for X rays. This problem can be solved, however, by using the waves associated with electrons, which have about the same wavelength as X rays but are easier to manipulate. For one thing, a magnetic field can bend the *electron rays*, because the waves are associated with a charged particle.

Just as the eye can see an expanded image of an object if the light-rays involved are appropriately manipulated by lenses, so a photograph can register an expanded image of an object if electron waves are appropriately manipulated by magnetic fields. And, since the wavelengths associated with electrons are far smaller than those of ordinary light, the resolution obtainable with an *electron microscope* at high magnification is much greater than that available to an ordinary microscope (figure 8.5).

Figure 8.5. Diagram of electron microscope. The magnetic condenser directs the electrons in a parallel beam. The magnetic objective functions like a convex lens, producing an enlarged image, which is then further magnified by a magnetic projector. The image is projected on a fluorescent observation screen or a photographic plate.

ELECTRON SOURCE

ILLUMINATION

MAGNETIC CONDENSER

OBJECT

MAGNETIC OBJECTIVE

FIRST STAGE

INTERMEDIATE IMAGE

MAGNETIC PROJECTOR

SECOND STAGE

OBSERVATION SCREEN
OR PHOTOGRAPHIC
PLATE

A crude electron microscope capable of magnifying 400 times was made in Germany in 1932 by Ernst Ruska and Max Knoll, but the first really usable one was built in 1937 at the University of Toronto by James Hillier and Albert F. Prebus. Their instrument could magnify an object 7,000 times, whereas the best optical microscopes reach their limit with a magnification of about 2,000. By 1939, electron microscopes were commercially available; and eventually Hillier and others developed electron microscopes capable of magnifying up to 2,000,000 times.

Whereas an ordinary electron microscope focuses electrons on the target and has them pass through, another kind has a beam of electrons pass rapidly over the target, scanning it in much the way an electron beam scans the picture tube in a television set. Such a *scanning electron microscope* was suggested as early as 1938 by Knoll, but the first practical device of this sort was built by the British-American physicist Albert Victor Crewe about 1970. The scanning electron microscope is less damaging to the object being viewed, shows the object with a greater three-dimensional effect so that more information is obtained, and can even show the position of individual atoms of the larger varieties.

ELECTRONS AS WAVES

It ought not be surprising should particle-wave dualism work in reverse, so that phenomena ordinarily considered wavelike in nature should have particle characteristics as well. Planck and Einstein had already shown radiation to consist of quanta, which, in a fashion, are particles. In 1923, Compton, the physicist who was to demonstrate the particle nature of cosmic rays (see chapter 7), showed that such quanta possessed some down-to-earth particle qualities. He found that X rays, on being scattered by matter, lose energy and become longer in wavelength. This effect was just what might be expected of a radiation "particle" bouncing off a matter particle: the matter particle is pushed forward, gaining energy; and the X ray veers off, losing energy. This *Compton effect* helped establish the wave-particle dualism.

The matter waves had important consequences for theory, too. For one thing, they cleared up some puzzles about the structure of the atom.

In 1913, Niels Bohr had pictured the hydrogen atom, in the light of the recently propounded quantum theory, as consisting of a central nucleus surrounded by an electron that could circle that nucleus in any one of a number of orbits. These orbits are in fixed positions; if a hydrogen electron drops from an outer orbit to an inner one, it loses energy, emitting that energy in the form of a quantum possessing a fixed wavelength. If the electron were to move from an inner electron to an outer one, it would have to absorb a quantum of energy, but only one of a fixed size and wavelength just enough to move it by the proper amount. Hence, hydrogen can absorb or emit only certain wavelengths of radiation, producing characteristic lines in its spectrum. Bohr's scheme, which was made gradually more complex over the next decade—notably by the German physicist Arnold Johannes Wilhelm Sommerfeld, who introduced elliptical orbits as well—was highly successful in explaining many facts about the spectra of various elements. Bohr was awarded the Nobel Prize in physics in 1922 for his theory. The German physicists James Franck and Gustav Ludwig Hertz (the latter a nephew of Heinrich Hertz), whose studies on collisions between atoms and electrons lent an experimental foundation to Bohr's theories, shared the Nobel Prize in physics in 1925.

Bohr had no explanation of why the orbits were fixed in the positions they held. He simply chose the orbits that would give the correct results, so far as absorption and emission of the actually observed wavelengths of light were concerned.

In 1926, the German physicist Erwin Schrödinger decided to take another look at the atom in the light of the de Broglie theory of the wave nature of particles. Considering the electron as a wave, he decided that the electron does not circle around the nucleus as a planet circles around the sun but constitutes a wave that curves all around the nucleus, so that it is in all parts of its orbit at once, so to speak. It turned out that, on the basis of the wavelength predicted by de Broglie for an electron, a whole number of electron waves would exactly fit the orbits outlined by Bohr. Between the orbits, the waves would not fit in a whole number but would join up *out of phase;* and such orbits could not be stable.

Schrödinger worked out a mathematical description of the atom called *wave mechanics* or *quantum mechanics,* which became a more satisfactory method of looking at the atom than the Bohr system had been. Schrödinger shared the Nobel Prize in 1933 with Dirac, the author of the theory of antiparticles (see chapter 7), who also contributed to the development of this new picture of the atom. The German physicist Max Born, who contributed further to the mathematical development of quantum mechanics, shared in the Nobel Prize in physics in 1954.

THE UNCERTAINTY PRINCIPLE

By this time the electron had become a pretty vague "particle"—a vagueness soon to grow worse. Werner Heisenberg of Germany proceeded to raise a profound question that projected particles, and physics itself, almost into a realm of the unknowable.

Heisenberg had presented his own model of the atom. He had abandoned all attempts to picture the atom as composed either of particles or of waves. He decided that any attempt to draw an analogy between atomic structure and the structure of the world about us is doomed to failure. Instead, he described the energy levels or orbits of electrons purely in terms of numbers, without a trace of picture. Since he used a mathematical device called a *matrix* to manipulate his numbers, his system was called *matrix mechanics.*

Heisenberg received the Nobel Prize in physics in 1932 for his contributions to quantum mechanics, but his matrix system was less popular with physicists than Schrödinger's wave mechanics, since the latter seemed just as useful as Heisenberg's abstractions, and it is difficult for even physicists to force themselves to abandon the attempt to picture what they are talking about.

By 1944, physicists seemed to have done the correct thing, for the Hungarian-American mathematician John von Neumann presented a line of argument that seemed to show that matrix mechanics and wave mechanics are mathematically equivalent: everything demonstrated by one could be equally well demonstrated by the other. Why not, therefore, choose the less abstract version?

After having introduced matrix mechanics (to jump back in time again), Heisenberg went on to consider the problem of describing the position of a particle. How can one determine where a particle is? The obvious answer is: Look at it. Well, let us imagine a microscope that could make an electron visible. We must shine a light or some appropriate kind of radiation on it to see it. But an electron is so small that a single photon of light striking it would move it and change its position. In the very act of measuring its position, we would have changed that position.

This is a phenomenon that occurs in ordinary life. When we measure the air pressure in a tire with a gauge, we let a little air out of the tire and change the pressure slightly in the act of measuring it. Likewise, when we put a thermometer in a bathtub of water to measure the temperature, the thermometer's absorption of heat changes the temperature slightly. A meter measuring electric current takes away a little current for moving the pointer on the dial. And so it goes in every measurement of any kind that we make.

However, in all ordinary measurements, the change in the subject we are measuring is so small that we can ignore it. The situation is quite different when we come to look at the electron. Our measuring device now is at least as large as the thing we are measuring; there is no usable measuring agent smaller than the electron. Consequently our measurement must inevitably have, not a negligible, but a decisive, effect on the object measured. We could stop the electron and so determine its position at a given instant. But, in that case, we could not know its motion or velocity. On the other hand, we might record its velocity, but then we could not fix its position at any given moment.

Heisenberg showed that there is no way of devising a method of pinpointing the position of a subatomic particle unless you are willing to be quite uncertain about its exact motion. And, in reverse, there is no way of pinpointing a particle's exact motion unless you are willing to be quite uncertain about its exact position. To calculate both exactly, at the same instant of time, is impossible.

If Heisenberg was right, then even at absolute zero, there cannot be complete lack of energy. If energy reached zero and particles became completely motionless, then only position need be determined since velocity could be taken as zero. It would be expected, therefore, that some residual *zero-point energy* must remain, even at absolute zero, to keep particles in motion and, so to speak, uncertain. It is this zero-point energy, which cannot be removed, that is sufficient to keep helium liquid even at absolute zero (see chapter 6).

In 1930, Einstein showed that the uncertainty principle, which stated it is impossible to reduce the error in position without increasing the error in momentum, implied that it is also impossible to reduce the error in measurement of energy without increasing the uncertainty of time during which the measurement can take place. He thought he could use this idea as a springboard for the disproof of the uncertainty principle, but Bohr proceeded to show that Einstein's attempted disproof was wrong.

Indeed, Einstein's version of uncertainty proved very useful, since it meant that in subatomic processes, the law of conservation of energy can be violated for very brief periods of time, provided all is brought back to the conservational state by the end of those periods: the greater the deviation from conservation, the briefer the time-interval allowed. (Yukawa used this notion in working out his theory of pions; see chapter 7.) It was even possible to explain certain subatomic phenomena by assuming that particles are produced out of nothing in defiance of energy conservation, but cease to exist before the time allotted for their detection, so that they are only *virtual particles*. The theory of virtual particles was worked out in the late 1940s by three men: the American physicists Julian Schwinger and Richard Phillips Feynman, and the Japanese physicist Sin-itiro Tomonaga. The three were jointly awarded the 1965 Nobel Prize in physics in consequence.

There have even been speculations, since 1976, that the universe began as a tiny, but massive, virtual particle that expanded with extreme quickness and remained in existence. The universe, in this view, formed itself out of Nothing, and we may wonder about there possibly being an infinite number of universes forming (and eventually ending) in an infinite volume of Nothing.

The uncertainty principle has profoundly affected the thinking of physicists and philosophers. It had a direct bearing on the philosophical question of *causality* (that is, the relationship of cause and effect). But its implications for science are not those that are commonly supposed. One often reads that the principle of indeterminacy removes all certainty from nature and shows that science after all does not and never can know what is really going on, that scientific knowledge is at the mercy of the unpredictable whims of a universe in which effect does not necessarily follow cause. Whether this interpretation is valid from the standpoint of philosophy, the principle of uncertainty has in no way shaken the attitude of scientists toward scientific investigation. If, for instance, the behavior of the individual molecules in a gas cannot be predicted with certainty, nevertheless on the average the molecules do obey certain laws, and their behavior can be predicted on a statistical basis, just as insurance companies can calculate reliable mortality tables even though it is impossible to predict when any particular individual will die.

In most scientific observations indeed, the indeterminacy is so small compared with the scale of the measurements involved that it can be neglected for all practical purposes. One can determine simultaneously both the position and the motion of a star, of a planet, of a billiard ball, or even of a grain of sand, with complete satisfactory accuracy.

As for the uncertainty among the subatomic particles themselves this does not hinder but actually helps physicists. It has been used to explain facts about radioactivity and about the absorption of subatomic particles by nuclei, as well as many other subatomic events, more reasonably than would have been possible without the uncertainty principle.

The uncertainty principle means that the universe is more complex than was thought, but not that it is irrational.

Chapter 9

The Machine

Fire and Steam

So far in this book, I have been concerned almost entirely with *pure* science: that is, science as an explanation of the universe about us. Throughout history, however, human beings have been making use of the workings of the universe to increase their own security, comfort, and pleasure. They used those workings, at first, without any proper understanding of them but gradually came to command them through careful observation, common sense, and even hit-and-miss. Such an application of the workings to human uses is *technology*, and it antedates science.

Once science began to grow, however, it became possible to advance technology at ever increasing speed. In modern times, science and technology have grown so intertwined (science advancing technology as it elucidates the laws of nature, and technology advancing science as it produces new instruments and devices for scientists to use) that it is no longer possible to separate them.

EARLY TECHNOLOGY

If we go back to the beginning, consider that though the first law of thermodynamics states that energy cannot be created out of nothing, there is no law against turning one form of energy into another. Our whole civilization has been built upon finding new sources of energy and harnessing it for human use in ever more efficient and sophisticated ways. In fact, the greatest single discovery in human history involved methods for converting the chemical energy of a fuel such as wood into heat and light.

It was perhaps half a million years ago that our hominid ancestors "discovered" fire long before the appearance of *Homo sapiens* (modern man). No doubt they had encountered—and been put to flight by—lightning-ignited brush fires and forest fires before that. But the discovery of fire's virtues did not come until curiosity overcame fear.

There must have come a time when an occasional primitive—perhaps a woman or (most likely) a child—may have been attracted to the quietly burning remnants of such an accidental fire and been amused by playing with it, feeding it sticks, and watching the dancing flames. Undoubtedly, elders would put a stop to this dangerous game until one of them, more imaginative than most, recognized the advantages of taming the flame and turning a childish amusement into adult use. A flame offered light in the darkness and warmth in the cold. It kept predators away. Eventually, people may have found that its heat softened food and made it taste better. (It killed germs and parasites, too, but prehistoric human beings could not know that.)

For hundreds of thousands of years, human beings could only make use of fire by keeping it going constantly. If a flame accidentally went out, it must have been equivalent to an electrical blackout in modern society. A new flame had to be borrowed from some other tribe, or one had to wait for the lightning to do the job. It was only in comparatively recent times that human beings learned how to make a flame at will where no flame had previously existed, and only then was fire truly tamed (figure 9.1). It was *Homo sapiens* who accomplished that task in prehistoric times, but exactly when, exactly where, and exactly how we do not know and may never know.

In the early days of civilization, fire was used not only for light, warmth, protection and cooking but also eventually for the isolation of metals from their ores and for handling the metals thereafter; for baking pottery and brick; and even for making glass.

Other important developments heralded the birth of civilization. About 9000 B.C., human beings began to domesticate plants and animals, beginning the practices of agriculture and herding, and thus increased the food supply and, in animals, found a direct energy source. Oxen, donkeys, camels, and eventually horses (to say nothing of reindeer, yaks, water buffalo, llamas, and elephants in various corners of the world) could bring stronger muscles to bear on necessary tasks while using, as fuel, food too coarse for human beings to eat.

Sometime about 3500 B.C., the wheel was invented (possibly, to begin with, as a potter's wheel for the molding of pottery). Within a few centuries, certainly by 3000 B.C., wheels were placed on sledges, so that loads that had had to be dragged could now be rolled. Wheels were not a direct source of energy, but they made it possible for far less energy to be lost in overcoming friction.

By that time, too, primitive rafts or dugouts were being used to allow the

Figure 9.1. Early firemaking methods.

energy of running water to transport loads. By 2000 B.C. perhaps, sails were used to catch the wind, so that moving air could hasten the transport or even force the ship to move against a slow current. By 1000 B.C., the Phoenicians in their ships were plowing the full length of the Mediterranean Sea.

In 50 B.C. or thereabouts, the Romans began to make use of waterwheels. A quickly running stream could be made to turn a wheel, which could in turn be made to turn other wheels that would do work—grind grain, crush ore, pump water, and so on. Windmills also began to come into use at this time, devices in which moving air rather than moving water turn the wheel. (Quickly running streams are rare, but wind is almost everywhere.) In medieval times, windmills were an important source of energy in western Europe. It was in medieval times, too, that human beings first began to burn the black rock called *coal* in metallurgical furnaces, to employ magnetic energy in the ship's compass (which eventually made possible the great voyages of exploration), and to use chemical energy in warfare.

The first use of chemical energy for destruction (past the simple technique of firing flame-tipped arrows) came about in A.D. 670, when a Syrian alchemist

Callinicus is believed to have invented *Greek fire,* a primitive incendiary bomb composed of sulfur and naphtha, which was credited with saving Constantinople from its first siege by the Moslems in 673. Gunpowder arrived in Europe in the thirteenth century. Roger Bacon described it about 1280, but it had been known in Asia for centuries before that and may have been introduced to Europe by the Mongol invasions beginning in 1240. In any case, artillery powered by gunpowder came into use in Europe in the fourteenth century, and cannons are supposed to have appeared first at the battle of Crécy in 1346.

The most important of all the medieval inventions is the one credited to Johann Gutenberg of Germany. About 1450, he cast the first movable type and thereby introduced printing as a powerful force in human affairs. He also devised printer's ink, in which carbon black was suspended in linseed oil rather than, as hitherto, in water. Together with the replacement of parchment by paper (which had been invented by a Chinese eunuch, Ts'ai Lun—according to tradition—about A.D. 50 and which reached modern Europe, by way of the Arabs, in the thirteenth century), these inventions made possible the large-scale production and distribution of books and other written material. No invention prior to modern times was adopted so rapidly. Within a generation, 40,000 books were in print.

The recorded knowledge of mankind was no longer buried in royal collections of manuscripts but was made accessible in libraries available to all who could read. Pamphlets began to create and give expression to public opinion. (Printing was largely responsible for the success of Martin Luther's revolt against the papacy in 1517, which might otherwise have been nothing more than a private quarrel among monks.) And it was printing that created one of the prime instruments that gave rise to science as we know it. That indispensable instrument is the wide communication of ideas. Science had been a matter of personal communications among a few devotees; now it became a major field of activity, which enlisted more and more workers into an eventually worldwide *scientific community,* elicited the prompt and critical testing of theories, and ceaselessly opened new frontiers.

THE STEAM ENGINE

The great turning point in the harnessing of energy came at the end of the seventeenth century, although there had been a dim foreshadowing in ancient times. The Greek inventor Hero of Alexandria, sometime during the first centuries A.D. (his life cannot be pinned down even to a particular century), built a number of devices that ran on steam power. He used the expanding push of steam to open temple doors, whirl spheres, and so on. The ancient world, then in decline, could not follow up this premature advance.

Then, over fifteen centuries later, a new and vigorously expanding society had a second chance. It arose out of the increasingly acute necessity of pumping water out of mines that were being driven ever deeper. The old hand pump (see chapter 5) made use of a vacuum to lift water; and as the seventeenth century proceeded, human beings came to appreciate, ever more keenly, the great power of a vacuum (or, rather, the power of air pressure called into play by the existence of a vacuum).

In 1650, for instance, the German physicist (and mayor of the city of Magdeburg) Otto von Guericke invented an air pump worked by muscle power. He proceeded to put two flanged metal hemispheres together and to pump the air out from between them through a nozzle that one hemisphere possessed. As the air pressure within dropped lower, the air pressure from without, no longer completely counterbalanced, pushed the hemispheres together more powerfully. At the end, two teams of horses straining in opposite directions could not pull the hemispheres apart; but when air was allowed to re-enter, they fell apart of themselves. This experiment was conducted before important people, including on one occasion the German emperor himself, and it made a big splash.

Now it occurred to several inventors: Why not use steam instead of muscle power to create the vacuum? Suppose one filled a cylinder (or similar vessel) with water and heated the water to a boil. Steam, as it formed, would push out the water. If the vessel was cooled (for example, by means of cold water played on the outside surface), the steam in the vessel would condense to a few drops of water and leave a virtual vacuum. The water that one wanted to raise (as out of a flooded mine) could then rise through a valve into this evacuated vessel.

A French physicist, Denis Papin, saw the power of steam as early as 1679. He developed a *steam digester,* then, in which water was boiled in a vessel with a tight-fitting lid. The accumulating steam created a pressure that raised the boiling point of water and, at this higher temperature, cooked food faster and better. The steam pressure within the digester must have given Papin the notion of making steam do work. He placed a little water at the bottom of a tube and, by heating it, converted it to steam. This expanded forcibly, pushing a piston ahead of it.

The first person to translate this idea into a practical working device, however, was an English military engineer named Thomas Savery. His *steam engine* (the word *engine* originally denoted any ingenious device and comes from the same Greek root as *ingenious*) could be used to pump water out of a mine or a well or to drive a waterwheel, so he called it the "Miner's Friend." But it was dangerous (because the high pressure of the steam might burst the vessels and pipes) and very inefficient (because the heat of the steam was lost each time the container was cooled). Seven years after Savery patented his engine in 1698, an English blacksmith named Thomas Newcomen built an improved engine that operated at low steam pressure; it had a piston in a cylinder and employed air pressure to push down the piston.

Newcomen's engine, too, was not very efficient (it still cooled the chamber after each heating), and the steam engine remained a minor gadget for more than sixty years until a Scottish instrument maker named James Watt found the way to make it effective. Hired by the University of Glasgow to fix a model of a Newcomen engine that was not working properly, Watt fell to thinking

about the device's wasteful use of fuel. Why, after all, should the steam vessel
have to be cooled off each time? Why not keep the steam chamber steam hot
at all times and lead the steam into a separate condensing chamber that could
be kept cold? Watt went on to add a number of other improvements: employing steam pressure to help push the piston, devising a set of mechanical
linkages that kept the piston moving in a straight line, hitching the back-and-
forth motion of the piston to a shaft that turned a wheel, and so on. By 1782,
his steam engine, which got at least three times as much work out of a ton
of coal as Newcomen's, was ready to be put to work as a universal work horse
(figure 9.2).

In the times after Watt, steam-engine efficiency was continually increased,
chiefly through the use of ever hotter steam at ever higher pressure. Carnot's
founding of thermodynamics (see chapter 7) arose mainly out of the realization that the maximum efficiency with which any heat engine can be run is
proportional to the difference in temperature between the hot reservoir
(steam, in the usual case) and the cold.

In the course of the 1700s, various mechanical devices were invented to spin
and weave thread in more wholesale manner. (These replaced the spinning-
wheel, which had come into use in the Middle Ages.) At first this machinery
was powered by animal muscle or a waterwheel; but in 1790 came the crucial
step: it was powered by a steam engine.

Thus, the new textile mills that were being built had neither to be situated
on or near fast-moving streams nor to require animal care. They could be built

Figure 9.2. Watt's steam engine.

anywhere. Great Britain began to undergo a revolutionary change as working people left the land and abandoned home industry to flock into the factories (where working conditions were unbelievably cruel and abominable until society learned, reluctantly, that people ought to be treated no worse than animals).

The same change took place in other countries that adopted the new system of steam-engine power and the Industrial Revolution (a term introduced in 1837 by the French economist Jérôme Adolphe Blanqui).

The steam engine totally revolutionized transportation, too. In 1787, the American inventor John Fitch built a steamboat that worked, but it failed as a financial venture, and Fitch died unknown and unappreciated. Robert Fulton, a more able promoter than Fitch, launched his steamship, the *Clermont,* in 1807 with so much more fanfare and support that he came to be considered the inventor of the steamship, though actually he was no more the builder of the first such ship than Watt was the builder of the first steam engine.

Fulton should perhaps better be remembered for his strenuous attempts to build underwater craft. His submarines were not practical, but they anticipated a number of modern developments. He built one called the *Nautilus,* which probably served as inspiration for Jules Verne's fictional submarine of the same name in *Twenty Thousand Leagues under the Sea,* published in 1870. That, in turn, was the inspiration for the naming of the first nuclear-powered submarine (see chapter 10).

By the 1830s, steamships were crossing the Atlantic and were being driven by the *screw propeller,* a considerable improvement over the side paddle wheels. And by the 1850s, the speedy and beautiful Yankee Clippers had begun to furl their sails and to be replaced by steamers in the merchant fleets and navies of the world.

Later, a British engineer, Charles Algernon Parsons (a son of the Lord Rosse who had discovered the Crab Nebula) thought of a major improvement of the steam engine in connection with ships. Instead of having the steam drive a piston that, in turn, drove a wheel, Parsons thought of eliminating the "middleman" and having a current of steam directed against blades set about the rim of a wheel. The wheel would have to withstand great heat and high speeds; but in 1884, he produced the first practical *steam turbine.*

In 1897, at the Diamond Jubilee of Queen Victoria, the British navy was holding a stately review of its steam-powered warships, when Parsons's turbine-powered ship, *Turbinia,* moved past them, silently, at a speed of 35 knots. Nothing in the British navy could have caught it, and it was the best advertising gimmick one could have imagined. It was not long before both merchant vessels and warships were turbine-powered.

Meanwhile the steam engine had also begun to dominate land transportation. In 1814, the English inventor George Stephenson (owing a good deal to the prior work of an English engineer, Richard Trevithick) built the first practical *steam locomotive.* The in-and-out working of steam-driven pistons could turn metal wheels among steel rails as they could turn paddle wheels in the water. And in 1830, the American manufacturer Peter Cooper built the first steam locomotive in the Western Hemisphere. For the first time in history, land travel became as convenient as sea travel, and overland commerce could compete with seaborne trade. By 1840, the railroad had reached the Mississippi River; and by 1869, the full width of the United States was spanned by rail.

Electricity

In the nature of things, the steam engine is suitable only for large-scale, steady production of power. It cannot efficiently deliver energy in small packages or intermittently at the push of a button: a "little" steam engine, in which the fires are damped down or started up on demand, would be an absurdity. But the same generation that saw the development of steam power also saw the discovery of a means of transforming energy into precisely the form I have mentioned—a ready store of energy that could be delivered anywhere, in small amounts or large, at the push of a button. This form, of course, is electricity.

STATIC ELECTRICITY

The Greek philosopher Thales, about 600 B.C., noted that a fossil resin found on the Baltic shores, which we call amber and they called *elektron,* gained the ability to attract feathers, threads, or bits of fluff when it rubbed with a piece of fur. It was William Gilbert of England, the investigator of magnetism (see chapter 5), who first suggested that this attractive force be called *electricity,* from the Greek word *elektron.* Gilbert found that, in addition to amber, some other materials, such as glass, gained electric properties on being rubbed.

In 1733, the French chemist Charles Francis de Cisternay Du Fay discovered that if two amber rods, or two glass rods, were electrified by rubbing, they repelled each other. However, an electrified glass rod attracted an electrified amber rod. If the two were allowed to touch, both lost their electricity. He felt this showed there were two kinds of electricity, *vitreous* and *resinous.*

The American scholar Benjamin Franklin, who became intensely interested in electricity, suggested that it was a single fluid. When glass was rubbed, electricity flowed into it, making it "positively charged"; on the other hand, when amber was rubbed, electricity flowed out of it, and it therefore became "negatively charged." And when a negative rod made contact with a positive one, the electric fluid would flow from the positive to the negative until a neutral balance was achieved.

This was a remarkably shrewd speculation. If we substitute the word *electrons* for Franklin's *fluid* and reverse the direction of flow (actually electrons flow from the amber to the glass), his guess was essentially correct.

A French inventor named John Théophile Desaguliers suggested, in 1740, that substances through which the electric fluid travels freely (for example, metals) be termed *conductors*, and those through which it does not move freely (for example, glass and amber) be called *insulators*.

Experimenters found that a large electric charge could gradually be accumulated in a conductor if it was insulated from loss of electricity by glass or a layer of air. The most spectacular device of this kind was the *Leyden jar*. It was first devised in 1745 by the German scholar Ewald Georg von Kleist, but it was first put to real use at the University of Leyden in Holland, where it was independently constructed a few months later by the Dutch scholar Peter van Musschenbroek. The Leyden jar is an example of what is today called a *condenser*, or capacitor: that is, two conducting surfaces, separated by a small thickness of insulator, within which one can store a quantity of electric charge.

In the case of the Leyden jar, the charge is built up on tinfoil coating a glass jar, via a brass chain stuck into the jar through a stopper. When you touch the charged jar, you get a startling electric shock. The Leyden jar can also produce a spark. Naturally, the greater the charge on a body, the greater its tendency to escape. The force driving the electrons away from the region of highest excess (the *negative pole*) toward the region of greatest deficiency (the *positive pole*) is the *electromotive force* (EMF), or *electric potential*. If the electric potential becomes high enough, the electrons will even jump an insulating gap between the negative and the positive poles. Thus they will leap across an air gap, producing a bright spark and a crackling noise. The light of the spark is caused by the radiation resulting from the collisions of innumerable electrons with air molecules, and the noise arises from the expansion of the quickly heated air, followed by the clap of cooler air rushing into the partial vacuum momentarily produced.

Naturally one wondered whether lightning and thunder were the same phenomenon, on a vast scale, as the little trick performed by a Leyden jar. A British scholar, William Wall, had made just this suggestion in 1708. This thought was sufficient to prompt Benjamin Franklin's famous experiment in 1752. The kite he flew in a thunderstorm had a pointed wire, to which he attached a silk thread which could conduct electricity down from the thunderclouds. When Franklin put his hand near a metal key tied to the silk thread, the key sparked (figure 9.3). Franklin charged it again from the clouds, then used it to charge a Leyden jar, obtaining the same kind of charged Leyden jar in this fashion as in any other. Thus, Franklin demonstrated that the thunderclouds were charged with electricity, and that thunder and lightning are indeed the effect of a Leyden-jar-in-the-sky in which the clouds form one pole and the earth another.

The luckiest thing about the experiment, from Franklin's personal standpoint, was that he survived. Some others who tried it were killed, because the induced charge on the kite's pointed wire accumulated to the point of producing a fatally intense discharge to the body of the man holding the kite.

METAL TIP

SILK THREAD

KEY

Figure 9.3. Franklin's experiment.

Franklin at once followed up this advance in theory with a practical application. He devised the *lightning rod*, which was simply an iron rod attached to the highest point of a structure and connected to wires leading to the ground. The sharp point bled off electric charges from the clouds above, as Franklin showed by experiment; and, if lightning did strike, the charge was carried safely to the ground.

Lightning damage diminished drastically as the rods rose over structures all over Europe and the American colonies—no small accomplishment. Yet even today, 2 billion lightning flashes strike each year, killing (it is estimated) twenty people a day and hurting eighty more.

Franklin's experiment had two electrifying (please pardon the pun) effects. In the first place, the world at large suddenly became interested in electricity. Second, it put the American colonies on the map, culturally speaking. For the first time an American had actually displayed sufficient ability as a scientist to impress the cultivated Europeans of the Age of Reason. When, a quarter-century later, Franklin represented the infant United States at Versailles and sought assistance, he won respect, not only as the simple envoy of a new republic, but also as a mental giant who had tamed the lightning and brought it humbly to earth. That flying kite contributed more than a little to the cause of American independence.

Following Franklin's work, electrical research advanced by leaps. Quantitative measurements of electrical attraction and repulsion were carried out in 1785 by the French physicist Charles Augustin de Coulomb. He showed that this attraction (or repulsion) between given charges varied inversely as the square of the distance. In this, electrical attraction resembles gravitational attraction. In honor of this finding, the *coulomb* has been adopted as a name for a common unit of quantity of electricity.

DYNAMIC ELECTRICITY

Shortly thereafter, the study of electricity took a new, startling, and fruitful turning. So far I have been discussing *static electricity,* which refers to an electric charge that is placed on an object and then stays there. The discovery of an electric charge that moves, of electric currents or *dynamic electricity,* began with the Italian anatomist Luigi Galvani. In 1791, he accidentally discovered that thigh muscles from dissected frogs would contract if simultaneously touched by two different metals (thus adding the verb *galvanize* to the English language).

The muscles behaved as though they had been stimulated by an electric spark from a Leyden jar, and so Galvani assumed that muscles contain something he called *animal electricity.* Others, however, suspected that the origin of the electric charge might lie in the junction of the two metals rather than in muscle. In 1800, the Italian physicist Alessandro Volta studied combinations of dissimilar metals, connected not by muscle tissue but by simple solutions.

He began by using chains of dissimilar metals connected by bowls half-full of salt water. To avoid too much liquid too easily spilled, he prepared small disks of copper and of zinc, piling them alternately. He also made use of cardboard disks moistened with salt water so that his *voltaic pile* consisted of silver, cardboard, zinc, silver, cardboard, zinc, silver, and so on. From such a setup, electric current could be drawn off continuously.

Any series of similar items indefinitely repeated may be called a battery. Volta's instrument was the first *electric battery* (figure 9.4). It may also be called an *electric cell.* It was to take a century before scientists would understand how chemical reactions involve electron transfers and how to interpret electric currents in terms of shifts and flows of electrons. Meanwhile, however, they made use of the current without understanding all its details.

Humphry Davy used an electric current to pull apart the atoms of tightly bound molecules and was able for the first time, in 1807 and 1808, to prepare such metals as sodium, potassium, magnesium, calcium, strontium, and barium. Faraday (Davy's assistant and protégé) went on to work out the general rules of such molecule-breaking *electrolysis;* and his work, a half century later, was to guide Arrhenius in working out the hypothesis of ionic dissociation (see chapter 5).

The manifold uses of dynamic electricity in the century and a half since Volta's battery seem to have placed static electricity in the shade and to have reduced it to a mere historical curiosity. Not so, for knowledge and ingenuity need never be static. By 1960, the American inventor Chester Carlson had perfected a practical device for copying material by attracting carbon-black to paper through localized electrostatic action. Such copying, involving no solutions or wet media, is called *xerography* (from Greek words meaning "dry writing") and has revolutionized office procedures.

The names of the early workers in electricity have been immortalized in the names of the units used for various types of measurement involving electricity.

COPPER DISC
ZINC DISC
CLOTH SOAKED IN BRINE

+

−

Figure 9.4. Volta's battery. The two different metals in contact give rise to a flow of electrons, which are conducted from one *cell* to the next by the salt-soaked cloth. The familiar *dry battery*, or *flashlight battery*, of today, involving carbon and zinc, was first devised by Bunsen (of spectroscopy fame) in 1841.

I have already mentioned *coulomb* as a unit of quantity of electricity. Another unit is the *faraday:* 96,500 coulombs is equal to 1 faraday. Faraday's name is used a second time: a *farad* is a unit of electrical capacity. Then, too, the unit of electrical intensity (the quantity of electric current passing through a circuit in a given time) is called the *ampere,* after the French physicist Ampère (see chapter 5). One ampere is equal to 1 coulomb per second. The unit of electromotive force (the force that drives the current) is the *volt,* after Volta.

A given EMF did not always succeed in driving the same quantity of electricity through different circuits. It would drive a great deal of current through good conductors, little current through poor conductors, and virtually no current through nonconductors. In 1827, the German mathematician George Simon Ohm studied this *resistance* to electrical flow and showed that it can be precisely related to the amperes of current flowing through a circuit under the push of a known EMF. The resistance can be determined by taking the ration of volts to amperes. This is *Ohm's law,* and the unit of electrical resistance is the *ohm,* 1 ohm being equal to 1 volt divided by 1 ampere.

GENERATING ELECTRICITY

The conversion of chemical energy to electricity, as in Volta's battery and the numerous varieties of its descendants, has always been relatively expensive because the chemicals involved are not common or cheap. For this reason, although electricity could be used in the laboratory with great profit in the early nineteenth century, it could not be applied to large-scale uses in industry.

There have been sporadic attempts to make use of the chemical reactions involved in the burning of ordinary fuels as a source of electricity. Fuels such as hydrogen (or, better still, coal) are much cheaper than metals such as copper and zinc. As long ago as 1839, the English scientist William Grove devised an electric cell running on the combination of hydrogen and oxygen. It was interesting but not practical. In recent years, physicists have been working hard to prepare practical varieties of such *fuel cells*. The theory is all set; only the practical problems must be ironed out, and these are proving most refractory.

When the large-scale use of electricity came into being in the latter half of the nineteenth century, it is not surprising, then, that it did not arrive by way of the electric cell. As early as the 1830s, Faraday had produced electricity by means of the mechanical motion of a conductor across the lines of force of a magnet (figure 9.5; see also chapter 5). In such an *electric generator*, or *dynamo* (from a Greek word for "power"), the kinetic energy of motion could be turned into electricity. Such motion could be kept in being by steam power, which in turn could be generated by burning fuel. Thus, much more indirectly than in a fuel cell, the energy of burning coal or oil (or even wood) could be converted into electricity. By 1844, large, clumsy versions of such generators were being used to power machinery.

What was needed were ever stronger magnets, so that motion across the intensified lines of force could produce larger floods of electricity. These stronger magnets were obtained, in turn, by the use of electric currents. In 1823, the English electrical experimenter William Sturgeon wrapped eighteen

---

Figure 9.5. Faraday's dynamo. The rotating copper disk cuts the magnet's lines of force, inducing a current on the voltmeter.

VOLTMETER

BRASS AXLE

MAGNET

COPPER DISC

turns of bare copper wire about a U-shaped iron bar and produced an *electromagnet*. When the current was on, the magnetic field it produced was concentrated in the iron bar which could then lift twenty times its own weight of iron. With the current off, it was no longer a magnet and would lift nothing.

In 1829, the American physicist Joseph Henry improved this gadget vastly by using insulated wire. Once the wire was insulated, it could be wound in close loops over and over without fear of short circuits. Each loop increased the intensity of the magnetic field and the power of the electromagnet. By 1831, Henry had produced an electromagnet, of no great size, that could lift over a ton of iron.

The electromagnet was clearly the answer to better electrical generators. In 1845, the English physicist Charles Wheatstone made use of such an electromagnet for this purpose. Better understanding of the theory behind lines of force came about with Maxwell's mathematical interpretation of Faraday's work (see chapter 5) in the 1860s; and, in 1872, the German electrical engineer Friedrich von Hefner-Alteneck designed the first really efficient generator. At last electricity could be produced cheaply and in floods, and not only from burning fuel but from falling water.

EARLY APPLICATION OF ELECTRICITY TO TECHNOLOGY

For the work that led to the early application of electricity to technology, the lion's share of the credit must fall to Joseph Henry. Henry's first application of electricity was the invention of *telegraphy*. He devised a system of relays that made it possible to transmit an electric current over miles of wire. The strength of a current declines fairly rapidly as it travels at constant voltage across long stretches of resisting wire; what Henry's relays did was to use the dying signal to activate a small electromagnet that operated a switch that turned on a boost in power from stations placed at appropriate intervals. Thus a message consisting of coded pulses of electricity could be sent for a considerable distance. Henry actually built a telegraph that worked.

Because he was an unworldly man, who believed that knowledge should be shared with the world and therefore did not patent his discoveries, Henry got no credit for this invention. The credit fell to the artist (and eccentric religious bigot) Samuel Finley Breese Morse. With Henry's help, freely given (but later only grudgingly acknowledged), Morse built the first practical telegraph in 1844. Morse's main original contribution to telegraphy was the system of dots and dashes known as the *Morse code*.

Henry's most important development in the field of electricity was the electric motor. He showed that electric current could be used to turn a wheel, just as the turning of a wheel can generate current in the first place. And an electrically driven wheel (or motor) could be used to run machinery. The motor could be carried anywhere; it could be turned on or off at will (without waiting to build up a head of steam); and it could be made as small as one wished (figure 9.6).

The catch was that electricity had to be transported from the generating

Figure 9.6. Henry's motor. The upright bar magnet *D* attracts the wire-wound magnet *B*, pulling the long metal probes *Q* and *R* into the brass thimbles *S* and *T*, which act as terminals for the wet cell *F*. Current flows into the horizontal magnet, producing an electromagnetic field that pulls *A* and *C* together. The whole process is then repeated on the opposite side. Thus the horizontal bars oscillate up and down.

station to the place where the motor was to be used. Some way had to be found to cut down the loss of electrical energy (taking the form of dissipated heat) as it traveled over wires.

One answer was the *transformer*. The experimenters with currents found that electricity suffers far less loss if it is transmitted at a low rate of flow. So the output from the generator was stepped up to a high voltage by means of a transformer that—while multiplying the voltage, say, three times—reduces the current (rate of flow) to one-third. At the receiving station, the voltage can be stepped down again so that the current is correspondingly increased for use in motors.

The transformer works by using the *primary* current to induce a current at high voltage in a *secondary* coil. This induction requires varying the magnetic field through the second coil. Since a steady current will not do this, the current used is a continually changing one that builds up to a maximum and then drops to zero and starts building in the opposite direction—in other words, an *alternating current*.

Alternating current (A.C.) did not win out over direct current (D.C.) without a struggle. Thomas Alva Edison, the greatest name in electricity in the final decades of the nineteenth century, championed direct current and established the first dc generating station in New York in 1882 to supply current for the electric light he had invented. He fought alternating current on the ground that it was more dangerous (pointing out, for instance, that it was used in electric chairs). He was bitterly opposed by Nikola Tesla, an engineer who had worked for Edison and been shabbily treated. Tesla developed a successful system of alternating current in 1888. In 1893, George Westinghouse, also a believer in alternating current, won a crucial victory over Edison by obtaining for his electric company the contract to develop the Niagara Falls power plants on an ac basis. In the following decades, Steinmetz established the theory of alternating currents on a firm mathematical basis. Today alternating current is all but universal in systems of power distribution. (In 1966, to be sure, engineers at General Electric devised a direct-current transformer—long held to be impossible; but it involves liquid-helium temperatures and low efficiency. It is fascinating theoretically, but of no likely commercial use right now.)

Electrical Technology

The steam engine is a *prime mover:* it takes energy already existing in nature (the chemical energy of wood, oil, or coal) and turns it into work. The electric motor is not a prime mover: it converts electricity into work, but the electricity must itself be formed from the energy of burning fuel or falling water. For this reason, electricity is more expensive than steam for heavy jobs. Nevertheless, it can be used for the purpose. At the Berlin Exhibition of 1879, an electric-powered locomotive (using a third rail as its source of current) successfully pulled a train of coaches. Electrified trains are common now, especially for rapid transit within cities, for the added expense is more than made up for by increased cleanliness and smoothness of operation.

THE TELEPHONE

Where electricity really comes into its own, however, is where it performs tasks that steam cannot. There is, for instance, the telephone, patented by the Scottish-born inventor Alexander Graham Bell in 1876. In the telephone mouthpiece, the speaker's sound waves strike a thin steel diaphragm and make it vibrate in accordance with the pattern of the waves. The vibrations of the diaphragm, in turn, set up an analogous pattern in an electric current, which strengthens and weakens in exact mimicry of the sound waves. At the telephone receiver, the fluctuations in the strength of the current actuate an electromagnet that makes a diaphragm vibrate and reproduce the sound waves.

The telephone was crude, at first, and barely worked; but even so, it was the hit of the Centennial Exposition held at Philadelphia in 1876 to celebrate the hundredth anniversary of the Declaration of Independence. The visiting Brazilian emperor, Pedro II, tried it and dropped the instrument in astonishment, saying "It talks!" which made newspaper headlines. Another visitor, Kelvin, was equally impressed, while the great Maxwell was astonished that anything so simple would reproduce the human voice. In 1877, Queen Victoria acquired a telephone, and its success was assured.

Also in 1877, Edison devised an essential improvement. He constructed a mouthpiece containing loose-packed carbon powder. When the diaphragm pressed on the carbon powder, the powder conducted more current; when it moved away, the powder conducted less. In this way, the sound waves of the voice were translated by the mouthpiece into varying pulses of electricity with great fidelity, and the voice one heard in the receiver was reproduced with improved clarity.

Telephone messages could not be carried very far without ruinous investment in thick (therefore low-resistance) copper wire. At the turn of the century, the Yugoslavian-American physicist Michael Idvorsky Pupin developed a method of loading a thin copper wire with inductance coils at intervals. These reinforced the signals and allowed them to be carried across long distances. The Bell Telephone Company bought the device in 1901; and by 1915, long-distance telephony was a fact as the line between New York City and San Francisco was opened.

The telephone operator became an unavoidable and increasing part of life for half a century until her domination (she was almost invariably a woman) began to fade with the beginnings of the dial telephone in 1921. Automation continued to advance until by 1983, hundreds of thousands of telephone employees went out on strike for a couple of weeks, and telephone service continued without interruption. Currently radio beams and communications satellites add to the versatility of the telephone.

RECORDING SOUND

In 1877, a year after the invention of the telephone, Edison patented his *phonograph*. The first records had the grooves scored on tinfoil wrapped around a rotating cylinder. The American inventor Charles Sumner Tainter substituted wax cylinders in 1885, and then Emile Berliner introduced wax-coated disks in 1887. In 1904, Berliner introduced a still more important advance: the flat phonograph record on which the needle vibrates from side to side. Its greater compactness allowed it to replace Edison's cylinder (with a needle vibrating up and down) almost at once.

In 1925, recordings began to be made by means of electricity through the use of a *microphone*, which translated sound into a mimicking electric current via a piezoelectric crystal instead of a metal diaphragm—the crystal allowing a better quality of reproduction of the sound. In the 1930s, the use of radio tubes for amplification was introduced.

In 1948, the Hungarian-American physicist Peter Goldmark developed the *long-playing record*, which turned 33½ times per minute rather than the till-then regulation 78. A single LP record could hold six times the amount of music of the old kind and made it possible to listen to symphonies without the repeated necessity of turning and replacing records.

Electronics made possible *high-fidelity* (*hi-fi*) and *stereophonic* sound, which have had the effect, so far as the sound itself is concerned, of practically removing all mechanical barriers between the orchestra or singer and the listener.

Tape-recording of sound was invented in 1898 by a Danish electrical engineer named Valdemar Poulsen, but had to await certain technical advances to become practical. An electromagnet, responding to an electric current carrying the sound pattern, magnetizes a powder coating on a tape or a wire moving past it, and the playback is accomplished through an electromagnet that picks up this pattern of magnetism and translates it again into a current that will reproduce the sound.

ARTIFICIAL LIGHT BEFORE ELECTRICITY

Of all the tricks performed by electricity, certainly the most popular was its turning night into day. Human beings had fought off the daily crippling darkness-after-sundown with the campfire, the torch, the oil lamp, and the candle; for half a million years or so, the level of artificial light remained dim and flickering.

The nineteenth century introduced some advances in these age-old methods of lighting. Whale oil and then kerosene came to be used in oil lamps, which grew brighter and more efficient. The Austrian chemist Karl Auer, Baron von Welsbach, found that if a fabric cylinder, impregnated with compounds of thorium and cerium were put around a lamp flame, it would glow a brilliant white. Such a *Welsbach mantle,* patented in 1885, greatly increased the brightness of the oil lamp.

Early in the century, gas lighting was introduced by the Scottish inventor William Murdock. He piped coal gas to a jet where it could be allowed to escape and be lit. In 1802, he celebrated a temporary peace with Napoleon by setting up a spectacular display of gas lights; and by 1803, he was routinely lighting his main factory with them. In 1807, some London streets began to use gas lighting, and the custom spread. As the century progressed, large cities grew ever lighter at night, reducing the crime rate and enhancing the security of citizens.

The American chemist Robert Hare found that a hot gas flame played upon a block of calcium oxide (*lime*) produces a brilliant white light. Such *limelight* came to be used to illuminate theater stages to a brighter level than had hitherto been possible. Although this technique has long since been outmoded, people who are in the blaze of publicity are still said to be "in the limelight."

All of these forms of lighting from bonfires to the gas jet involve open flames. Some device must exist to light the fuel—be it wood, coal, oil, or gas —if a flame does not already exist in the vicinity. Prior to the nineteenth century, the least laborious method was to use flint and steel. By striking one against another, a spark could be elicited that might, with luck, light some *tinder* (finely divided inflammable material) which could, in turn, light a candle, and so on.

In the early nineteenth century, chemists began devising methods for coating one end of a piece of wood with chemicals that would burst into flame when the temperature was elevated. Such a piece of wood was a *match*. Friction would raise the temperature, and "striking a match" on a rough surface produced a flame.

The earliest matches smoked horribly, produced a stench, and made use of chemicals that were dangerously poisonous. Matches became really safe to use in 1845, when the Austrian chemist Anton Ritter von Schrötter made use of red phosphorus for the purpose. Eventually *safety matches* were developed in which the red phosphorus is put on a rough strip somewhere on the box or container that holds the matches, while the match itself has the other necessary chemicals in its head. Neither match nor strip can alone burst into flame, but if rubbed on the strip, the match catches fire.

There was also a return to flint and steel, with crucial improvements. In place of the steel is Mischmetal, a mixture of metals (principally cerium) which, on being scraped by a little wheel, yields particularly hot sparks. In place of tinder is easily inflammable *lighter fluid*. The result is the *cigarette lighter*.

ELECTRIC LIGHT

Open flames of one sort or another flicker and are a constant fire hazard. Something totally new was needed, and it had long been noted that electricity could yield light. Leyden jars produced sparks when discharged; electric currents sometimes made wires glow upon passing through them. Both systems have been used for lighting.

In 1805, Humphry Davy forced an electric discharge across the air space between two conductors. By maintaining the current, the discharge was continuous, and he had an *electric arc*. As electricity became cheaper, it became possible to use *arc lamps* for lighting. In the 1870s the streets of Paris and some other cities had such lamps. The light was harsh, flickering, and open, however—and still a fire hazard.

It would be better to have an electric current heat a thin wire, or filament, till it glowed. Naturally, the filament had to be made to glow in the absence of oxygen, or it would not last long before being oxidized. The first attempts to remove oxygen involved the straightforward route of removing air. By 1875, Crookes (in connection with his work on cathode rays; see chapter 7) had devised methods for producing a good enough vacuum for this purpose, and with sufficient speed and economy. Nevertheless, the filaments used remained unsatisfactory, breaking too easily. In 1878, Thomas Edison, fresh from his triumph in creating the phonograph, announced that he would tackle the problem. He was only thirty-one, but such was his reputation as an inventor that his announcement caused the stocks of gas companies to tumble on the New York and London stock exchanges.

After hundreds of experiments and fabulous frustrations, Edison finally found a material that would serve as the filament—a scorched cotton thread. On 21 October 1879, he lit his bulb. It burned for 40 continuous hours. On the following New Year's Eve, Edison put his lamps on triumphant public display by lighting up the main street of Menlo Park, New Jersey, where his laboratory was located. He quickly patented his lamp and began to produce it in quantity.

Yet Edison was not the sole inventor of the incandescent lamp. At least one other inventor had about an equal claim—Joseph Swan of England, who exhibited a carbon-filament lamp at a meeting of the Newcastle-on-Tyne Chemical Society on 18 December 1878, but did not get his lamp into production until 1881.

Edison proceeded to work on the problem of providing houses with a steady and sufficient supply of electricity for his lamps—a task that took as much ingenuity as the invention of the lamp itself. Two major improvements were later made in the lamp. In 1910, William David Coolidge of the General Electric Company adopted heat-resisting metal tungsten as the material for the filament (figure 9.7); and, in 1913, Irving Langmuir introduced the inert

Figure 9.7. Incandescent lamp.

gas nitrogen in the lamp to prevent the evaporation and breaking of the filament that occurs in a vacuum.

Argon (use of which was introduced in 1920) serves the purpose even better than nitrogen, for argon is completely inert. Krypton, another inert gas, is still more efficient, allowing a lamp filament to reach higher temperatures and burn more brightly without loss of life.

For half a century, the clear glass of the light-bulb made the glowing filament within harsh and as difficult to look at as the sun. A chemical engineer, Marvin Pipkin, devised a practical method of etching the glass of the bulb within (on the outside, etching served to collect dust and darken the light). The use of *frosted bulbs* finally produced a soft and pleasant, steady light.

The coming of the electric light had the potential for banishing all open flames from lighting and thus making fires very much a thing of the past. Unfortunately, there are still open flames and probably always will be—in fireplaces, in gas stoves, in gas and oil furnaces. Particularly unfortunate is the fact that hundreds of millions of addicts carry with them open flames in the form of lit cigarettes and frequently used cigarette lighters. The loss of property and of life resulting from cigarette-induced fires (forest fires and brush fires, as well as building fires) is difficult to overestimate.

The glowing filament of the light-bulb (*incandescent* light, since it is induced by sheer heat of the filament as it resists the flow of the electric current) is not the only way of turning electricity into light. For instance, the so-called *neon lights* (introduced by the French chemist Georges Claude in 1910) are tubes in which an electric discharge excites atoms of neon gas to emit a bright, red glow. The *sun lamp* contains mercury vapor which, when excited by a discharge, yields radiation rich in ultraviolet light; this can be used not only to produce a tan but also to kill bacteria or generate fluorescence. And the latter, in turn, leads to *fluorescent lighting*, introduced in its contemporary form in 1939 at the New York World's Fair. Here the ultraviolet light from mercury vapor excites fluorescence in a *phosphor* coating the inside of the tube (figure 9.8). Since this cool light wastes little energy in heat, it consumes less electric power.

PHOSPHOR POWDER COATING — FILAMENT AND MOUNTING

INERT GAS AND MERCURY VAPOR — ELECTRONS

Figure 9.8. Fluorescent lamp. A discharge of electrons from the filament excites the mercury vapor in the tube, producing ultraviolet radiation. The ultraviolet makes the phosphor glow.

A 40-watt fluorescent tube supplies as much light, and far less heat, than a 150-watt incandescent light. Since the Second World War, therefore, there has been a massive swing toward the fluorescent. The first fluorescent tubes made use of beryllium salts as phosphors, which resulted in cases of serious poisoning (*berylliosis*) induced by breathing dusts containing these salts or by introducing the substance through cuts caused by broken tubes. After 1949, other far less dangerous phosphors were used.

The latest promising development is a method that converts electricity directly into light without the prior formation of ultraviolet light. In 1936, the French physicist Georges Destriau discovered that an intense alternating current could make a phosphor, such as zinc sulfide, glow. Electrical engineers are now distributing the phosphor through plastic or glass and are using this phenomenon, called *electroluminescence*, to develop glowing panels. Thus, a luminescent wall or ceiling can light a room, bathing it in a soft, colored glow. The efficiency of electroluminescence is still too low, however, to allow it to compete with other forms of electrical lighting.

PHOTOGRAPHY

Probably no invention involving light has given mankind more enjoyment than photography. This had its earliest beginnings in the observation that light, passing through a pinhole into a small dark chamber (*camera obscura* in Latin), will form a dim, inverted image of the scene outside the chamber. Such a device was constructed about 1550 by an Italian alchemist, Giambattista della Porta. This is the *pinhole camera*.

In a pinhole camera, the amount of light entering is very small. If, however, a lens is substituted for the pinhole, a considerable quantity of light can be brought to a focus, and the image is then much brighter. With that accomplished, it was necessary to find some chemical reaction that will respond to light. A number of men labored in this cause, including, most notably, the Frenchmen Joseph Nicephore Niepce and Louis Jacques Mande Daguerre and the Englishman William Henry Fox Talbot. Niepce tried to make sunlight darken silver chloride in a proper pattern and produced the first primitive photograph in 1822, but an 8-hour exposure was required.

Daguerre went into partnership with Niepce before the latter died, and went on to improve the process. Having had sunlight darken silver salts, he dissolved the unchanged salts in sodium thiosulfate, a process suggested by the scientist John Herschel (the son of William Herschel). By 1839, Daguerre was producing *daguerrotypes,* the first practical photographs, with exposures requiring no more than 20 minutes.

Talbot improved the process still further, producing *negatives* in which the places where light strikes are darkened so that dark remains light while light becomes dark. From such negatives any number of *positives* can be developed, in which the light undergoes another reversal so that light is light and dark, dark, as they should be. In 1844, Talbot published the first book illustrated with photographs.

Photography went on to prove its value in human documentation when, in the 1850s, the British photographed Crimean war scenes and when, in the next decade, the American photographer Matthew Brady, with what we would now consider impossibly primitive equipment, took classic photographs of the American Civil War in action.

For nearly half a century, the *wet plate* had to be used in photography. This consisted of a glass plate, which was smeared with an emulsion of chemical that had to be made up on the spot. The picture had to be taken before the emulsion dried. As long as there was no solution to this limitation, photographs could be taken only by skillful professionals.

In 1878, however, an American inventor, George Eastman, discovered how to mix the emulsion with gelatin, smear it on the plate, and let it dry into a firm gel that would keep for long periods of time. In 1884, he patented photographic *film* in which the gel was smeared first on paper and then, in 1889, on celluloid. In 1888, he invented the Kodak, a camera that would take photographs at the press of a button. The exposed film could then be given away to be developed. Now photography became a popular hobby. As ever more sensitive emulsions came into use, pictures could be taken in a flash of light, and there was no need for a sitter to pose for long periods of time with glazed, unnatural expressions.

One would not suppose that things could be made any simpler, but in 1947, the American inventor Edwin Herbert Land devised a camera with a double roll of film, an ordinary negative film and a positive paper, with sealed containers of chemicals between. The chemicals are released at the proper moment and develop the positive print automatically. A few minutes after you have snapped the camera, you have the completed photograph in your hand.

Throughout the nineteenth century, photographs were black and white, lacking in color. In the early twentieth century, however, a process of color photography was developed by the Luxembourg-born French physicist Gabriel Lippmann, and won him the Nobel Prize for physics in 1908. That proved a false start, however, and practical color photography was not developed until 1936. This second, and successful, try was based on the observation, in 1855, by Maxwell and von Helmholtz that any color in the spectrum can be produced by combining red, green, and blue light. On this principle, the color film is composed of emulsions in three layers—one sensitive to the red, one to the green, and one to the blue components of the image. Three separate but superimposed pictures are formed, each reproducing the intensity of light in its part of the spectrum as a pattern of black-and-white shading. The film is then developed in three successive stages, using red, blue, and green dyes to deposit the appropriate colors on the negative. Each spot in the picture is a specific combination of red, green, and blue, and the brain interprets these combinations to reconstitute the full range of color.

In 1959, Land presented a new theory of color vision. The brain, he maintained, does not require a combination of three colors to create the impression of full color. All it needs is two different wavelengths, or sets of wavelengths, one longer than the other by a certain minimum amount. For instance, one of the sets of wavelengths may be an entire spectrum, or white light. Because the average wavelength of white light is in the yellow-green region, it can serve as the "short" wavelength. Now a picture reproduced through a combination of white light and red light (serving as the long wavelength) comes out in full color. Land has also made pictures in full color with filtered green light and red light and with other appropriate dual combinations.

The invention of motion pictures came from an observation first made by the English physician Peter Mark Roget in 1824. He noted that the eye forms a persistent image, which lasts for an appreciable fraction of a second. After the inauguration of photography, many experimenters, particularly in France, made use of this fact to create the illusion of motion by showing a series of pictures in rapid succession. Everyone is familiar with the parlor gadget consisting of a series of picture cards which, when riffled rapidly, make a figure seem to move and perform acrobatics. If a series of pictures, each slightly different from the one before, is flashed on a screen at intervals of about 1/16 second, the persistence of the successive images in the eye will cause them to blend together and so give the impression of continuous motion.

It was Edison who produced the first *movie*. He photographed a series of pictures on a strip of film and then ran the film through a projector, which showed each in succession with a burst of light. The first motion picture was put on display for public amusement in 1894; and, in 1914, theaters showed the full-length motion picture, *The Birth of a Nation*.

To the silent movies, a *sound track* was added in 1927. The sound track also takes the form of light: the wave pattern of music and the actor's speech is converted, by a microphone, into a varying current of electricity; and this current lights a lamp that is photographed along with the action of the motion picture. When the film, with this track of light at one side, is projected on the screen, the brightening and dimming of the lamp in the pattern of the sound waves is converted back to an electric current by means of a *phototube*, using the photoelectric effect, and the current in turn is reconverted to sound.

Within two years after the first *talking picture*, *The Jazz Singer*, silent movies were a thing of the past, and so, almost, was vaudeville. By the late 1930s, the *talkies* had added color. In addition, the 1950s saw the development of wide-screen techniques and even a short-lived fad for three-dimensional (3D) effects, involving two pictures thrown on the same screen. By wearing polarized spectacles, an observer saw a separate picture with each eye, thus producing a stereoscopic effect.

Internal-Combustion Engines

While *kerosene*, a petroleum fraction, gave way to electricity in the field of artificial illumination, a lighter petroleum fraction, *gasoline*, became indispensable for another technical development that revolutionized modern life as deeply, in its way, as did the introduction of electrical gadgetry. This development was the *internal-combustion engine*, so called because in such an engine, fuel is burned within the cylinder so that the gases formed push the piston directly. Ordinary steam engines are *external-combustion engines*, the fuel being burned outside and the steam being then led, ready-formed, into the cylinder.

THE AUTOMOBILE

This compact device, with small explosions set off within the cylinder, made it possible to apply motive power to small vehicles in ways for which the bulky

GASOLINE-AIR MIXTURE IS
DRAWN INTO CYLINDER

MIXTURE IS
COMPRESSED

INLET VALVE
OPENS

VALVES
CLOSE

PISTON

CAMS

A

B

SPARK PLUG
IGNITES MIXTURE

MIXTURE EXPLODES
FORCING PISTON DOWN

BURNT GASES ARE
PUSHED OUT

BOTH VALVES
CLOSED

EXHAUST VALVE
OPENS

C

D

Figure 9.9. Nikolaus Otto's four-cycle engine, built in 1876.

steam engine was not well suited. To be sure, steam-driven "horseless carriages" were devised as long ago as 1786, when William Murdock, who later introduced gas lighting, built one. A century later, the American inventor Francis Edgar Stanley invented the famous Stanley Steamer, which for a while competed with the early cars equipped with internal combustion machines. The future, however, lay with the latter.

Actually, some internal-combustion engines were built at the beginning of the nineteenth century, before petroleum came into common use. They burned turpentine vapors or hydrogen as fuel. But it was only with gasoline, the one vapor-producing liquid that is both combustible and obtainable in large quantities, that such an engine could become more than a curiosity.

The first practical internal-combustion engine was built in 1860 by a French inventor Etienne Lenoir, who hitched it to a small conveyance which became the first "horseless carriage" with such an engine. In 1876, the German technician Nikolaus August Otto, having heard of the Lenoir engine, built a *four-cycle* engine (figure 9.9). First a piston fitting tightly in a cylinder is pushed outward, so that a mixture of gasoline and air is sucked into the vacated cylinder. Then the piston is pushed in again to compress the vapor. At the point of maximum compression the vapor is ignited and explodes. The explosion drives the piston outward, and it is this powered motion that drives the engine. It turns a wheel which pushes the piston in again to expel the burned residue or *exhaust*—the fourth and final step in the cycle. Now the wheel moves the piston outward to start the cycle over again.

A Scottish engineer named Dugald Clerk almost immediately added an improvement. He hooked up a second cylinder, so that its piston was being driven while the other was in the recovery stage: this device made the power output steadier. Later, the addition of more cylinders (eight is now a common number) increased the smoothness and power of this *reciprocating engine.*

Such an engine was essential if automobiles were to be made practical, but auxiliary inventions were also necessary. The ignition of the gasoline-air mixture at just the right moment presented a problem. All sorts of ingenious devices were used; but by 1923, it became common to depend on electricity. The supply comes from a *storage battery,* which, like any other battery, delivers electricity as the result of a chemical reaction. But it can be recharged by sending an electric current through it in the direction opposite to the discharge; this current reverses the chemical reaction and allows the chemicals to produce more electricity. The reverse current is provided by a small generator driven by the engine.

The most common type of storage battery has plates of lead and lead oxide in alternation, with cells of fairly concentrated sulfuric acid. It was invented by the French physicist Gaston Planté in 1859 and was put into its modern form in 1881 by the American electrical engineer Charles Francis Brush. More rugged and more compact storage batteries have been invented since —for instance, a nickel-iron battery developed by Edison about 1905—but for economy, none can compete with the lead battery.

The electric voltage supplied by the storage battery is stored in the magnetic field of a transformer called an *induction coil,* and the collapse of this field provides the stepped-up voltage that produces the ignition spark across the gap in the familiar spark plugs.

Once an internal-combustion engine starts firing, inertia will keep it moving between power strokes. But outside energy must be supplied to start the engine. At first it was started by hand (for example, the automobile crank), and outboard motors and power lawn mowers are still started by yanking a cord. The automobile crank required a strong hand. When the engine began turning, it was not uncommon for the crank to be yanked out of the hand holding it, then to turn and break the arm. In 1912, the American inventor Charles Franklin Kettering invented a *self-starter* that eventually did away with the crank. The self-starter is powered by the storage battery, which supplies the energy for the first few turns of the engine.

The first practical automobiles were built, independently, in 1885 by the German engineers Gottlieb Daimler and Karl Benz. But what really made the automobile, as a common conveyance, was the invention of *mass production.*

The prime originator of this technique was Eli Whitney, who merits more credit for it than for his more famous invention of the cotton gin. In 1789, Whitney received a contract from the Federal Government to make guns for the army. Up to that time, guns had been manufactured individually, each from its own fitted parts. Whitney conceived the notion of making the parts uniform, so that a given part would fit any gun. This single, simple innovation —manufacturing standard, interchangeable parts for a given type of article— was perhaps as responsible as any other factor for the creation of modern mass-production industry. When power tools came in, they made it possible to stamp out standard parts in practically unlimited numbers.

It was the American engineer Henry Ford who first exploited the concept to the full. He had built his first automobile (a two-cylinder job) in 1892, then had gone to work for the Detroit Automobile Company in 1899 as chief engineer. The company wanted to produce custom-made cars, but Ford had another notion. He resigned in 1902 to produce cars on his own—in quantity. In 1909, he began to turn out the Model T; and by 1913, he began to manufacture it on the Whitney plan—car after car, each just like the one before, and all made with the same parts.

Ford saw that he could speed up production by using human workers as one used machines, performing the same small job over and over with uninterrupted regularity. The American inventor Samuel Colt (who had invented the revolver or "six-shooter") had taken the first steps in this direction in 1847; and the automobile manufacturer Ransom E. Olds had applied the system to the motor car in 1900. Olds lost his financial backing, however, and it fell to Ford to carry this movement to its fruition. Ford set up the *assembly line,* with workers adding parts to the construction as it passed them on moving belts until the finished car rolled off at the end of the line. Two economic advances were achieved by this system: high wages for the workers, and cars that could be sold at amazingly low prices.

By 1913, Ford was manufacturing 1,000 Model T's a day. Before the line was discontinued in 1927, 15 million had been turned out, and the price had dropped to 290 dollars. The passion for yearly change then won out, and Ford was forced to join the parade of variety and superficial novelty that has raised the price of automobiles enormously and lost Americans much of the advantage of mass production.

In 1892, the German mechanical engineer Rudolf Diesel introduced a modification of the internal-combustion engine which was simpler and more economical of fuel. He put the fuel-air mixture under high pressure, so that the heat of compression alone was enough to ignite it. The *diesel engine* made it possible to use higher-boiling fractions of petroleum, which do not knock. Because of the higher compression used, the engine must be more solidly constructed and is therefore considerably heavier than the gasoline engine. Once an adequate fuel-injection system was developed in the 1920s it began to gain favor for trucks, tractors, buses, ships, and locomotives and is now undisputed king of heavy transportation.

Improvements in gasoline itself further enhanced the efficiency of the internal-combustion engine. Gasoline is a complex mixture of molecules made up of carbon and hydrogen atoms (*hydrocarbons*), some of which burn more quickly than others. Too quick a burning rate is undesirable, for then the gasoline-air mixture explodes in too many places at once, producing *engine knock*. A slower rate of burning produces an even expansion of vapor that pushes the piston smoothly and effectively.

The amount of knock produced by a given gasoline is measured as its octane rating, by comparing it with the knock produced by a hydrocarbon called *iso-octane*, which is particularly low in knock production, mixed with *normal heptane*, which is particularly high in knock production. One of the prime functions of gasoline refining is, among many other things, to produce a hydrocarbon mixture with a high octane rating.

Automobile engines have been designed through the years with an increasingly high *compression ratio:* that is, the gasoline-air mixture is compressed to greater densities before ignition. This compression milks the gasoline of more power, but also encourages knock, so that gasoline of continually higher octane rating has had to be developed.

The task has been made easier by the use of chemicals that, when added in small quantities to the gasoline, reduce knock. The most efficient of these *anti-knock compounds* is *tetraethyl lead*, a lead compound whose properties were noted by the American chemist Thomas Midgley, and which was first introduced for the purpose in 1925. Gasoline containing it is *leaded gasoline* or *ethyl gas*. If tetraethyl lead were present alone, the lead oxides formed during gasoline combustion would foul and ruin the engine. For this reason, ethylene bromide is also added. The lead atom of tetraethyl lead combines with the bromide atom of ethylene bromide to form lead bromide, which, at the temperature of the burning gasoline, is vaporized and expelled with the exhaust.

Diesel fuels are tested for ignition delay after compression (too great a delay is undesirable) by comparison with a hydrocarbon called *cetane*, which contains sixteen carbon atoms in its molecule as compared with eight for iso-octane. For diesel fuels, therefore, one speaks of a *cetane number*.

Improvements continued to be made. Low-pressure "balloon" tires arrived in 1923, and tubeless tires in the early 1950s, making blowouts less common. In the 1940s, cars became air-conditioned, and automatic drives came into use so that gear shifting began to drop out of use. Power steering and power brakes arrived in the 1950s. The automobile has become so integral a part of the American way of life that, despite the rising cost of gasoline and the rising danger of air pollution, there seems no way short of absolute catastrophe of putting an end to it.

THE AIRPLANE

Larger versions of the automobile were the bus and the truck, and oil replaced coal on the great ships, but the greatest triumph of the internal-combustion engine came in the air. By the 1890s, humans had achieved the age-old dream—older than Daedalus and Icarus—of flying on wings. Gliding had become an avid sport of the aficionados. The first man-carrying *glider* was built in 1853 by the English inventor George Cayley. The "man" it carried, however, was only a boy. The first important practitioner of this form of endeavor, the German engineer Otto Lilienthal, was killed in 1896 during a glider flight. Meanwhile, a violent urge to take off in powered flight had begun, although gliding as a sport remains popular.

The American physicist and astronomer Samuel Pierpont Langley tried, in 1902 and 1903, to fly a glider powered by an internal-combustion engine, and came within an ace of succeeding. Had his money not given out, he might have got into the air on the next try. As it was, the honor was reserved for the brothers Orville and Wilbur Wright, bicycle manufacturers who had taken up gliders as a hobby.

On 17 December 1903, at Kitty Hawk, North Carolina, the Wright brothers got off the ground in a propeller-driven glider and stayed in the air for 59 seconds, flying 852 feet. It was the first airplane flight in history, and it went almost completely unnoticed by the world at large.

There was considerably more public excitement after the Wrights had achieved flights of 25 miles and more, and when, in 1909, the French engineer Louis Blériot crossed the English Channel in an airplane. The air battles and exploits of the First World War further stimulated the imagination; and the *biplanes* of that day, with their two wings held precariously together by struts and wires, were familiar to a generation of postwar moviegoers. The German engineer Hugo Junkers designed a successful *monoplane* just after the war; and the thick single wing, without struts, took over completely. (In 1939, the Russian-American engineer Igor Ivan Sikorsky built a multiengined plane and designed the first *helicopter*, a plane with upper vanes that made vertical takeoffs and landings and even hovering practical.)

But, through the early 1920s, the airplane remained more or less a curiosity —merely a new and more horrible instrument of war and a plaything of stunt flyers and thrill seekers. Aviation did not come into its own until, in 1927, Charles Augustus Lindbergh flew nonstop from New York to Paris. The world went wild over the feat, and the development of bigger and safer airplanes began.

Two major innovations have been effected in the airplane engine since it was established as a means of transportation. The first was the adoption of the gas-turbine engine (figure 9.10). In this engine, the hot, expanding gases of

Figure 9.10. A turbojet engine. Air is drawn in, compressed, and mixed with fuel, which is ignited in the combustion chamber. The expanding gases power a turbine and produce thrust.

the fuel drive a wheel by their pressure against its blades, instead of driving pistons in cylinders. The engine is simple, cheaper to run, and less vulnerable to trouble, and it needed only the development of alloys that could withstand the high temperatures of the gases to become practicable. Such alloys were devised by 1939. Thereafter, *turboprop planes*, using a turbine engine to drive the propellers, became increasingly popular.

But they were quickly superseded, at least for long flights, by the second major development—the *jet plane*. In principle the driving force here is the same as the one that makes a toy balloon dart forward when its mouth is opened and the air escapes. This is action and reaction: the motion of the expanding, escaping air in one direction results in equal motion, or thrust, in the opposite direction—just as the forward movement of the bullet in a gun barrel makes the gun kick backward in recoil. In the jet engine, the burning of the fuel produces hot, high-pressure gases that drive the plane forward with great force as they stream backward through the exhaust. A rocket is driven by exactly the same means, except that it carries its own supply of oxygen to burn the fuel (figure 9.11).

Patents for *jet propulsion* were taken out by a French engineer, René Lorin, as early as 1913; but at the time, it was a completely impractical scheme for

Figure 9.11. A simple liquid-fueled rocket.

airplanes. Jet propulsion is economical only at speeds of more than 400 miles an hour. In 1939, an Englishman, Frank Whittle, flew a reasonably practical jet plane; and, in January 1944, jet planes were put into war use by Great Britain and the United States against the *buzz-bombs*, Germany's V-1 weapon, a pilotless robot plane carrying explosives in its nose.

After the Second World War, military jets were developed that approached the speed of sound. The speed of sound depends on the natural elasticity of air molecules, their ability to snap back and forth. When the plane approaches that speed, the air molecules cannot get out of the way, so to speak, and are compressed ahead of the plane, which then undergoes a variety of stresses and strains. There was talk of the *sound barrier* as though it were something physical that could not be approached without destruction. However, tests in wind tunnels led the way to more efficient streamlining; and on 14 October 1947, an American X-1 rocket plane, piloted by Charles Elwood Yeager, "broke the sound barrier." For the first time in history, a human being surpassed the speed of sound. The air battles of the Korean War in the early 1950s were fought by jet planes moving at such velocities that comparatively few planes were shot down.

The ratio of the velocity of an object to the velocity of sound (which is 740 miles per hour at 0° C) in the medium through which the object is moving is the *Mach number*, after the Austrian physicist Ernst Mach, who first investigated, theoretically, the consequences of motion at such velocities in the mid-nineteenth century. By the 1960s, airplane velocities surpassed Mach 5—an achievement of the experimental rocket plane X-15, whose rockets pushed it high enough, for short periods of time, to allow its pilots to qualify as astronauts. Military planes travel at lower velocities, and commercial planes at lower velocities still.

A plane traveling at a *supersonic velocity* (over Mach 1) carries its sound waves ahead of it since it travels more quickly than the sound waves alone would. If close enough to the ground to begin with, the cone of compressed sound waves may intersect the ground with a loud *sonic boom*. (The crack of a bullwhip is a miniature sonic boom, since, properly manipulated, the tip of such a whip can be made to travel at supersonic velocities.)

Supersonic commercial flight was initiated in 1970 by the British-French

Concorde, which could, and did, cross the Atlantic in three hours, traveling at twice the speed of sound. An American version of such SST (*supersonic transport*) flight was aborted in 1971, because of worry over excessive noise at airports and of possible environmental damage. Some people pointed out that this was the first time a feasible technological advance had been stopped for being inadvisable, the first time human beings had said, "We can, but we had better not."

On the whole, it may be just as well, for the gains do not seem to justify the expense. The Concorde has been an economic failure, and the Soviet SST program was ruined by the crash of one of their planes in a 1973 exhibition at Paris.

Electronics

THE RADIO

In 1888, Heinrich Hertz conducted the famous experiments that detected radio waves, predicted twenty years earlier by James Clerk Maxwell (see chapter 8). What he did was to set up a high-voltage alternating current that surged into first one, then another of two metal balls separated by a small air gap. Each time the potential reached a peak in one direction or the other, it sent a spark across the gap. Under these circumstances, Maxwell's equations predicted, electromagnetic radiation should be generated. Hertz used a receiver consisting of a simple loop of wire with a small air gap at one point to detect that energy. Just as the current gave rise to radiation in the first coil, so the radiation ought to give rise to a current in the second coil. Sure enough, Hertz was able to detect small sparks jumping across the gap to his detector coil, placed across the room from the radiating coil. Energy was being transmitted across space.

By moving his detector coil to various points in the room, Hertz was able to tell the shape of the waves. Where sparks came through brightly, the waves were at peak or trough. Where sparks did not come through at all, they were midway. Thus he could calculate the wavelength of the radiation. He found that the waves were tremendously longer than those of light.

In the decade following, it occurred to a number of people that the *Hertzian waves* might be used to transmit messages from one place to another, for the waves were long enough to go around obstacles. In 1890, the French physicist Édouard Branly made an improved receiver by replacing the wire loop with a glass tube filled with metal filings to which wires and a battery were attached. The filings would not carry the battery's current unless a high-voltage alternating current was induced in the filings, as Hertzian waves would do. With this receiver he was able to detect Hertzian waves at a distance of 150 yards. Then the English physicist Oliver Joseph Lodge (who later gained a dubious kind of fame as a champion of spiritualism) modified this device and succeeded in detecting signals at a distance of half a mile and in sending messages in Morse code.

The Italian inventor Guglielmo Marconi discovered that he could improve matters by connecting one side of the generator and receiver to the ground and the other to a wire, later called an *antenna* (because it resembled, I suppose, an insect's feeler). By using powerful generators, Marconi was able to send signals over a distance of 9 miles in 1896, across the English Channel in 1898, and across the Atlantic in 1901. Thus was born what the British still call *wireless telegraphy* and the Americans named *radiotelegraphy*, or *radio* for short.

Marconi worked out a system for excluding *static* from other sources and tuning in only on the wavelength generated by the transmitter. For his inventions, Marconi shared the Nobel Prize in physics in 1909 with the German physicist Karl Ferdinand Braun, who also contributed to the development of radio by showing that certain crystals can act to allow current to pass in only one direction. Thus ordinary alternating current could be converted into direct current such as radios needed. The crystals tended to be erratic; but in the 1910s, people were bending over their *crystal sets* to receive signals.

The American physicist Reginald Aubrey Fessenden developed a special generator of high-frequency alternating currents (doing away with the spark-gap device) and devised a system of *modulating* the radio wave so that it carried a pattern mimicking sound waves. What was modulated was the amplitude (or height) of the waves; consequently this was called *amplitude modulation*, now known as *AM radio*. On Christmas Eve 1906, music and speech came out of a radio receiver for the first time.

The early radio enthusiasts had to sit over their sets wearing earphones. Some means of strengthening, or *amplifying*, the signal was needed, and the answer was found in a discovery that Edison had made—his only discovery in "pure" science.

In one of his experiments, looking toward improving the electric lamp, Edison, in 1883, sealed a metal wire into a light bulb near the hot filament. To his surprise, electricity flowed from the hot filament to the metal wire across the air gap between them. Because this phenomenon had no utility for his purposes, Edison, a practical man, merely wrote it up in his notebooks and forgot it. But the *Edison effect* became very important indeed when the electron was discovered and it became clear that current across a gap meant a flow of electrons. The British physicist Owen Willans Richardson showed, in experiments conducted between 1900 and 1903, that electrons "boil" out of metal filaments heated in vacuum. For this work, he eventually received the Nobel Prize for physics in 1928.

In 1904, the English electrical engineer John Ambrose Fleming put the Edison effect to brilliant use. He surrounded the filament in a bulb with a cylindrical piece of metal (called a *plate*). Now this plate could act in either of two ways. If it was positively charged, it would attract the electrons boiling off the heated filament and so would create a circuit that carried electric current. But if the plate was negatively charged, it would repel the electrons and thus prevent the flow of current. Suppose, then, that the plate was hooked up to a source of alternating current. When the current flowed in one direction, the plate would get a positive charge and pass current in the tube; when the alternating current changed direction, the plate would acquire a negative charge and no current would flow in the tube. Thus, the plate would pass current in only one direction; in effect, it would convert alternating to direct current. Because such a tube acts as a valve for the flow of current, the British logically call it a *valve*. In the United States, it is vaguely called a *tube*. Scientists took to calling it a *diode*, because it has two electrodes—the filament and the plate (figure 9.12).

The tube—or *radio tube*, since that is where it was initially used—controls a stream of electrons through a vacuum rather than an electric current through wire. The electrons can be much more delicately controlled than the current can, so that the tubes (and all the devices descended from it) made a whole new range of *electronic devices* that could do things no mere electrical device

Figure 9.12. Principle of the vacuum-tube diode.

Figure 9.13. Principle of the triode.

could. The study and use of tubes and their descendants is referred to as *electronics*.

The tube, in its simplest form, serves as a rectifier and replaced the crystals used up to that time, since the tubes were much more reliable. In 1907, the American inventor Lee De Forest went a step farther. He inserted a third electrode in the tube, making a *triode* out of it (figure 9.13). The third electrode is a perforated plate (*grid*) between the filament and the plate. The grid attracts electrons and speeds up the flow from the filament to the plate (through the holes in the grid). A small increase in the positive charge on the grid will result in a large increase in the flow of electrons from the filament to the plate. Consequently, even the small charge added by weak radio signals will increase the current flow greatly, and this current will mirror all the variations imposed by the radio waves. In other words, the triode acts as an amplifier. Triodes and even more complicated modifications of the tube became essential equipment, not only for radio sets but for all sorts of electronic equipment.

One more step was needed to make radio sets completely popular. During the First World War, the American electrical engineer Edwin Howard Armstrong developed a device for lowering the frequency of a radio wave. This was intended, at the time, for detecting aircraft but, after the war, was put to use in radio receivers. Armstrong's *superheterodyne receiver* made it possible to tune in clearly on an adjusted frequency by the turn of one dial, where previously it had been a complicated task to adjust reception over a wide range of possible frequencies. In 1921, regular radio programs were begun by a station in Pittsburgh. Other stations were set up in rapid succession; and with the control of sound level and station tuning reduced to the turn of a dial, radio sets became hugely popular. By 1927, telephone conversations could be carried on across oceans, with the help of radio; and *wireless telephony* was a fact.

There remained the problem of static. The systems of tuning introduced by Marconi and his successors minimized "noise" from thunderstorms and other electrical sources, but did not eliminate it. Again it was Armstrong who found an answer. In place of amplitude modulation, which was subject to interference from the random amplitude modulations of the noise sources, he substituted *frequency modulation* in 1935: that is, he kept the amplitude of the radio carrier wave constant and superimposed a variation in frequency on it. Where the sound wave was large in amplitude, the carrier wave was made low in frequency, and vice versa. Frequency modulation (FM) virtually eliminated static, and FM radio came into popularity after the Second World War for programs of serious music.

TELEVISION

Television was an inevitable sequel to radio, just as talking movies were to the silents. The technical forerunner of television was the transmission of pictures by wire, which entailed translating a picture into an electric current. A narrow beam of light passed through the picture on a photographic film to a phototube behind. Where the film was comparatively opaque, a weak current was generated in the phototube; where it was clearer, a large current was formed. The beam of light swiftly *scanned* the picture from left to right, line by line, and produced a varying current representing the entire picture. The current was sent over wires and, at the destination, reproduced the picture on film by a reverse process. Such *wirephotos* were transmitted between London and Paris as early as 1907.

Television is the transmission of a "movie" instead of still photographs—either "live" or from a film. The transmission must be extremely fast, which means that the action must be scanned very rapidly. The light-dark pattern of the image is converted into a pattern of electrical impulses by means of a camera using, in place of film, a coating of metal that emits electrons when light strikes it.

A form of television was first demonstrated in 1926 by the Scottish inventor John Logie Baird. However, the first practical television camera was the *iconoscope*, patented in 1938 by the Russian-born American inventor Vladimir Kosma Zworykin. In the iconoscope, the rear of the camera is coated with a large number of tiny cesium-silver droplets. Each emits electrons as the light beam scans across it, in proportion to the brightness of the light. The iconoscope was later replaced by the *image orthicon*—a refinement in which the cesium-silver screen is thin enough so that the emitted electrons can be sent forward to strike a thin glass plate that emits more electrons. This amplification increases the sensitivity of the camera to light, so that strong lighting is not necessary.

The television receiver is a variety of cathode-ray tube. A stream of electrons shot from a filament (*electron-gun*) strikes a screen coated with a fluorescent substance, which glows in proportion to the intensity of the electron stream. Pairs of electrodes controlling the direction of the stream cause it to sweep across the screen from left to right in a series of hundreds of horizontal lines, each slightly below the one before, and the entire "painting" of a picture on the screen in this fashion is completed in 1/30 second. The beam goes on painting successive pictures at the rate of thirty per second. At no instant of time is there more than one dot on the screen (bright or dark, as the case may be); yet, thanks to the persistence of vision, we see not only complete pictures but an uninterrupted sequence of movement and action.

Experimental television was broadcast in the 1920s, but television did not become practical in the commercial sense until 1947. Since then, it has virtually taken over the field of entertainment.

In the mid-1950s, two refinements were added. By the use of three types of fluorescent material on the television screen, designed to react to the beam in red, blue, and green colors, color television was introduced. And *video tape*, a type of recording with certain similarities to the sound track on a movie film, made it possible to reproduce recorded programs or events with better quality than could be obtained from motion-picture film.

THE TRANSISTOR

In the 1980s, in fact, the world was in the *cassette age*. Just as small cassettes can unwind and rewind their tapes to play music with high fidelity—on batteries, if necessary, so that people can walk around or do their housework, with earphones pinned to their heads, hearing sounds no one else can hear—so there are *video cassettes* that can produce films of any type through one's television set or record programs when shown for replay afterward.

The vacuum tube, the heart of all the electronic devices, eventually became a limiting factor. Usually the components of a device are steadily improved in efficiency as time goes on: that is, they are stepped up in power and flexibility and reduced in size and mass (a process sometimes called *miniaturization*). But the vacuum tube became a bottleneck in the road to miniaturization for it had to remain large enough to contain a sizable volume of vacuum or the various components within would leak electricity across a too-small gap.

It had other shortcomings, too. The tube could break or leak and, in either case, would become unusable. (Tubes were always being replaced in early radio and television sets; and, particularly in the latter case, a live-in repairman seemed all but necessary.) Then, too, the tubes would not work until the filaments were sufficiently heated; hence, considerable current was necessary, and there had to be time for the set to "warm up." And then, quite by accident, an unexpected solution turned up. In the 1940s, several scientists at the Bell Telephone Laboratories grew interested in the substances known as *semiconductors.* These substances, such as silicon and germanium, conduct electricity only moderately well, and the problem was to find out why. The Bell Lab investigators discovered that such conductivity as these substances possess was enhanced by traces of impurities mixed with the element in question.

Let us consider a crystal of pure germanium. Each atom has four electrons in its outermost shell; and in the regular array of atoms in the crystal, each of the four electrons pairs up with an electron of a neighboring germanium atom, so that all the electrons are paired in stable bonds. Because this arrangement is similar to that in diamond, germanium, silicon, and other such substances are called *adamantine,* from an old word for "diamond."

If a little bit of arsenic is introduced into this contented adamantine arrangement, the picture grows more complicated. Arsenic has five electrons in its outermost shell. An arsenic atom taking the place of a germanium atom in the crystal will be able to pair four of its five electrons with the neighboring atoms, but the fifth can find no electron to pair with: it is loose. Now if an electric voltage is applied to this crystal, the loose electron will wander in the direction of the positive electrode. It will not move as freely as would electrons in a conducting metal, but the crystal will conduct electricity better than a nonconductor, such as sulfur or glass.

This is not very startling, but now we come to a case that is somewhat more odd. Let us add a bit of boron, instead of arsenic, to the germanium. The boron atom has only three electrons in its outermost shell. These three can pair up with the electrons of three neighboring germanium atoms. But what happens to the electron of the boron atom's fourth germanium neighbor? That electron is paired with a hole! The word *hole* is used advisedly, because this site, where the electron would find a partner in a pure germanium crystal, does in fact behave like a vacancy. If a voltage is applied to the boron-contaminated crystal, the next neighboring electron, attracted toward the positive electrode, will move into the hole. In doing so, it leaves a hole where it was, and the electron next farther away from the positive electrode moves into *that* hole. And so the hole, in effect, travels steadily toward the negative electrode, moving exactly like an electron, but in the opposite direction. In short, it has become a conveyor of electric current.

To work well, the crystal must be almost perfectly pure with just the right amount of the specified impurity (that is, arsenic or boron). The germanium-arsenic semiconductor, with a wandering electron, is said to be *n-type* (*n* for "negative"). The germanium-boron semiconductor, with a wandering hole that acts as if it were positively charged, is *p-type* (*p* for "positive").

Unlike ordinary conductors, the electrical resistance of semiconductors drops as the temperature rises, because higher temperatures weaken the hold of atoms on electrons and allow them to drift more freely. (In metallic conductors, the electrons are already free enough at ordinary temperatures. Raising the temperature introduces more random movement and impedes their flow in response to the electric field.) By determining the resistance of a semiconductor, one can measure temperatures that are too high to be conveniently measured in other fashions. Such temperature-measuring semiconductors are called *thermistors*.

But semiconductors in combination can do much more. Suppose we make a germanium crystal with one-half p-type and the other half n-type. If we connect the n-type side to a negative electrode and the p-type side to a positive electrode, the electrons on the n-type side will move across the crystal toward the positive electrode, while the holes on the p-type side will travel in the opposite direction toward the negative electrode. Thus, a current flows through the crystal. Now let us reverse the situation—that is, connect the n-type side to the positive electrode and the p-type to the negative electrode. This time the electrons of the n-side travel toward the positive electrode— which is to say, away from the p-side—and the holes of the p-side similarly move in the direction away from the n-side. As a result, the border regions at the junction between the n- and p-sides lose their free electrons and holes, thus effecting to a break in the circuit, and no current flows.

In short, we now have a setup that can act as a rectifier. If we hook up alternating current to this dual crystal, the crystal will pass the current in one direction, but not in the other. Therefore alternating current will be converted to direct current. The crystal serves as a diode, just as a vacuum tube (or *valve*) does.

In a way, electronics had come full circle. The tube had replaced the crystal, and now the crystal had replaced the tube—but it was a new kind of crystal, far more delicate and reliable than those that Braun had introduced nearly half a century before.

The new crystal had impressive advantages over the tube. It required no vacuum, so it could be small. It would not break or leak. Since it worked at room temperature, it required very little current and no warm-up time. It was all advantages and no disadvantages, provided only that it could be made cheaply enough and accurately enough.

Since the new crystals were solid all the way through, they opened the way to what came to be called *solid-state electronics*. The new device was named transistor (the suggestion of John Robinson Pierce of the Bell Lab), because it *trans*fers a signal across a re*sistor* (figure 9.14).

In 1948, William Bradford Shockley, Walter Houser Brattain, and John Bardeen at the Bell Lab went on to produce a transistor that could act as an amplifier. This was a germanium crystal with a thin p-type section sandwiched between two n-type ends. It was in effect a triode with the equivalent of a grid between the filament and the plate. With control of the positive charge in the

Figure 9.14. Principle of the junction transistor.

p-type center, holes could be sent across the junctions in such a manner as to control the electron flow. Furthermore, a small variation in the current of the p-type center would cause a large variation in the current across the semiconductor system. The semiconductor triode could thus serve as an amplifier, just as a vacuum tube triode did. Shockley and his co-workers Brattain and Bardeen received the Nobel Prize in physics in 1956.

However well transistors might work in theory, their use in practice required certain concomitant advances in technology—as is invariably true in applied science. Efficiency in transistors depended very strongly on the use of materials of extremely high purity, so that the nature and concentration of deliberately added impurities could be carefully controlled.

Fortunately, William Gardner Pfann introduced the technique of *zone refining* in 1952. A rod of, let us say, germanium, is placed in the hollow of a circular heating element, which softens and begins to melt a section of the rod. The rod is drawn through the hollow so that the molten zone moves along it. The impurities in the rod tend to remain in the molten zone and are therefore literally washed to the ends of the rod. After a few passes of this sort, the main body of the germanium rod is unprecedentedly pure.

By 1953, tiny transistors were being used in hearing aids, making them so small that they could be fitted inside the ear. In short order, the transistor steadily developed so that it could handle higher frequencies, withstand higher temperatures, and be made ever smaller. Eventually it grew so small that individual transistors were not used. Instead, small chips of silicon were etched microscopically to form *integrated circuits* that would do what large numbers of tubes would do. In the 1970s, these chips were small enough to be thought of as *microchips*.

Such tiny solid-state devices that are now universally used offer perhaps the most astonishing revolution of all the scientific revolutions that have taken place in human history. They have made small radios possible; they have made it possible to squeeze enormous abilities into satellites and probes; most of all, they have made possible the development of ever-smaller and ever-cheaper and ever-more versatile computers and, in the 1980s, robots as well. The last two items will be discussed later in chapter 17.

Masers and Lasers

MASERS

Another recent advance of astonishing magnitude begins with investigations involving the ammonia molecule (NH_3). The three hydrogen atoms of the ammonia molecule can be viewed as occupying the three apexes of an equilateral triangle, whereas the single nitrogen atom is some distance above the center of the triangle.

It is possible for the ammonia molecule to vibrate: that is, the nitrogen atom can move through the plane of the triangle to an equivalent position on the other side, then back to the first side, and so on, over and over. The ammonia molecule can, in fact, be made to vibrate back and forth with a natural frequency of 24 billion times a second.

This vibration period is extremely constant, much more so than the period of any artificial vibrating device—much more constant, even, than the movement of astronomical bodies. Such vibrating molecules can be made to control electric currents, which will in turn control time-measuring devices with unprecedented precision—as was first demonstrated in 1949 by the American physicist Harold Lyons. By the mid-1950s such *atomic clocks* were surpassing all ordinary chronometers. Accuracies in time measurement of 1 second in 1,700,000 years have been reached by making use of hydrogen atoms.

The ammonia molecule, in the course of these vibrations, liberates a beam of electromagnetic radiation with a frequency of 24 billion cycles per second. This radiation has a wavelength of 1.25 centimeters and is in the microwave region. Another way of looking at this fact is to imagine the ammonia molecule to be capable of occupying one of two energy levels, with the energy difference equal to that of a photon representing a 1.25-centimeter radiation. If the ammonia molecule drops from the higher energy level to the lower, it emits a photon of this size. If a molecule in the lower energy level absorbs a photon of this size, it rises to the higher energy level.

But what if an ammonia molecule is already in the higher energy level and is exposed to such photons? As early as 1917, Einstein had pointed out that, if a photon of just the right size struck such an upper-level molecule, the molecule would be nudged back down to the lower level and would emit a photon of exactly the size and moving in exactly the direction of the entering photon. There would be two identical photons where only one had existed before. This theory was confirmed experimentally in 1924.

Ammonia exposed to microwave radiation could, therefore, undergo two possible changes: molecules could be pumped up from lower level to higher or be nudged down from higher level to lower. Under ordinary conditions, the former process would predominate, for only a very small percentage of the ammonia molecules would, at any one instant, be at the higher energy level.

Suppose, though, that some method were found to place all or almost all the molecules in the upper energy level. Then the movement from higher level to lower would predominate. Indeed, something quite interesting would happen. The incoming beam of microwave radiation would supply a photon that would nudge one molecule downward. A second photon would be released, and the two would speed on, striking two molecules, so that two more were released. All four would bring about the release of four more, and so on. The initial photon would let loose a whole avalanche of photons, all of exactly the same size and moving in exactly the same direction.

In 1953, the American physicist Charles Hard Townes devised a method for isolating ammonia molecules in the high-energy level and subjected them to stimulation by microwave photons of the correct size. A few photons entered, and a flood of such photons left. The incoming radiation was thus greatly amplified.

The process was described as "*m*icrowave *a*mplification by *s*timulated *e*mission of *r*adiation"; and from the initials of this phrase, the instrument came to be called a *maser.*

Solid masers were soon developed—solids in which electrons could be made to take up one of two energy levels. The first masers, both gaseous and solid, were intermittent: that is, they had to be pumped up to the higher energy level first; then stimulated. After a quick burst of radiation, nothing more could be obtained until the pumping process had been repeated.

To circumvent this drawback, it occurred to the Dutch-American physicist Nicolaas Bloembergen to make use of a three-level system. If the material chosen for the core of the maser can have electrons in any of three energy levels—a lower, a middle, and an upper—then pumping and emission can go on simultaneously. Electrons are pumped up from the lowest energy level to the highest. Once at the highest, proper stimulation will cause them to drop down—first to the middle level, then to the lower. Photons of different size are required for pumping and for stimulated emission, and the two processes will not interfere with each other. Thus, we end with a continuous maser.

As microwave amplifiers, masers can be used as very sensitive detectors in radio astronomy, where exceedingly feeble microwave beams received from outer space will be greatly intensified with great fidelity to the original radiation characteristics. (Reproduction without loss of original characteristics is to reproduce with little "noise." Masers are extraordinarily "noiseless" in this sense of the word.) They have carried their usefulness into outer space, too. A maser was carried on board the Soviet satellite *Cosmos 97*, launched 30 November 1965, and did its work well.

For his work, Townes received the 1964 Nobel Prize for physics, sharing it with two Soviet physicists, Nicolai Gennediyevich Basov and Aleksandr Mikhailovich Prochorov, who had worked independently on maser theory.

LASERS

In principle, the maser technique could be applied to electromagnetic waves of any wavelength, notably to those of visible light. Townes pointed out the possible route of such applications to light wavelengths in 1958. Such a light-producing maser might be called an *optical maser*. Or, this particular process might be called "*l*ight *a*mplification by *s*timulated *e*mission of *r*adiation," with the resultant popular term, *laser* (figure 9.15).

Figure 9.15. Continuous-wave laser with concave mirrors and Brewster angle windows on discharge tube. The tube is filled with a gas whose atoms are raised to high-energy states by electromagnetic excitation. These atoms are then stimulated to emit energy of a certain wavelength by the introduction of a light beam. Acting like a pipe organ, the resonant cavity builds up a train of coherent waves between the end mirrors. The thin beam that escapes is the laser ray. After a drawing in *Science*, 9 October 1964.

The first successful laser was constructed in 1960 by the American physicist Theodore Harold Maiman. He used, for the purpose, a bar of synthetic ruby, this being essentially aluminum oxide with a bit of chromium oxide added. If the ruby bar is exposed to light, the electrons of the chromium atoms are pumped to higher levels and, after a short while, begin to fall back. The first few photons of light emitted (with a wavelength of 694.3 millimicrons) stimulate the production of other such photons, and the bar suddenly emits a beam of deep red light four times as intense as light at the sun's surface. Before 1960 was over, continuous lasers were prepared by an Iranian physicist, Ali Javan, working at Bell Laboratories. He used a gas mixture (neon and helium) as the light source.

The laser made possible light in a completely new form. The light was the most intense that had ever been produced, and the most narrowly mono-chromatic (single wavelength), but it was even more.

Ordinary light, produced in any other fashion—from a wood fire to the sun or to a firefly—consists of relatively short wave packets. They can be pictured as short bits of waves pointing in various directions. Ordinary light is made up of countless numbers of these.

The light produced by a stimulated laser, however, consists of photons of the same size and moving in the same direction. Hence, the wave packets are all of the same frequency; and since they are lined up precisely end to end, so to speak, they melt together. The light appears to be made up of long stretches of waves of even amplitude (height) and frequency (width). This is *coherent light*, because the wave packets seem to stick together. Physicists had learned to prepare coherent radiation for long wavelengths. It had never been done for light, though, until 1960.

The laser was so designed, moreover, that the natural tendency of the photons to move in the same direction was accentuated. The two ends of the ruby tube were accurately machined and silvered so as to serve as plane mirrors. The emitted photons flashed back and forth along the rod, knocking out more photons at each pass, until they had built up sufficient intensity to burst through the end that was more lightly silvered. Those that did come through were precisely those that had been emitted in a direction exactly parallel to the long axis of the rod, for those would move back and forth, striking the mirrored ends over and over. If any photon of proper energy happened to enter the rod in a different direction (even a very slightly different direction) and started a train of stimulated photons in that different direction, these would quickly pass out the sides of the rod after only a few reflections at most.

A beam of laser light is made up of coherent waves so parallel that it can travel through long distances without diverging to uselessness. It could be focused finely enough to heat a pot of coffee a thousand miles away. Laser beams even reached to the moon, in 1962, spreading out to a diameter of only two miles after having crossed nearly a quarter of a million miles of space!

Once the laser was devised, interest in its further development was nothing short of explosive. Within a few years, individual lasers capable of producing coherent light in hundreds of different wavelengths, from the near ultraviolet to the far infrared, were developed. Laser action was obtained from a wide variety of solids, from metallic oxides, fluorides, tungstates, from semiconduc-tors, from liquids, from columns of gas. Each variety had its advantages and disadvantages.

In 1964, the first *chemical laser* was developed by the American physicist Jerome V. V. Kasper. In such a laser, the source of energy is a chemical reaction (in the case of the first, the dissociation of CF_3I by a pulse of light) The advantage of the chemical laser over the ordinary variety is that the energy-yielding chemical reaction can be incorporated with the laser itself, and no outside energy source is needed. This is analogous to a battery-powered device as compared with one that must be plugged into a wall socket. There is an obvious gain in portability to say nothing of the fact that chemical lasers seem to be considerably more efficient than the ordinary variety (12 percent or more, as compared with 2 percent or less).

Organic lasers—those in which a complex organic dye is used as the source of coherent light—were first developed in 1966 by John R. Lankard and Peter Sorokin. The complexity of the molecule makes it possible to produce light by a variety of electronic reactions and therefore in a variety of wavelengths. A single organic laser can be *tuned* to deliver any wavelength within a range, rather than find itself confined to a single wavelength, as is true of the others.

The narrowness of the beam of laser light means that a great deal of energy can be focused into an exceedingly small area; in that area, the temperature reaches extreme levels. The laser can vaporize metal for quick spectral investigation and analysis and can weld, cut, or punch holes of any desired shape through high-melting substances. By shining laser beams into the eye, surgeons have succeeded in welding loosened retinas so rapidly that surrounding tissues have no time to be affected by heat. In similar fashion, lasers have been used to destroy tumors.

To show the vast range of laser applications, Arthur L. Shawlow developed the trivial (but impressive) *laser-eraser,* which in an intensely brief flash evaporates the typewriter ink of the formed letters without so much as scorching the paper beneath; at the other extreme, laser *interferometers* can make unprecedentedly refined measurements. When earth strains intensify, they can be detected by separated lasers, where shifts in the interference fringes of their light will detect tiny earth movements with a delicacy of one part in a trillion. Then, too, the first men on the moon left a reflector system designed to bounce back laser beams to earth. By such a method, the distance to the moon may be determined with greater accuracy than the distance, in general, from point to point on Earth's surface.

One possible application that created excitement from the beginning has been the use of laser beams as carrier beams in communications. The high frequency of coherent light, as compared with that of the coherent radio waves used in radio and television today, holds forth the promise of being able to crowd many thousands of channels into the space that now holds one channel. The prospect arises that every human being on Earth may have his or her own personal wavelength. Naturally, the laser light must be modulated. Varying electric currents produced by sound must be translated into varying laser light (either through changes in its amplitude on its frequency, or perhaps just by turning it on and off), which can in turn be used to produce varying electric current elsewhere. Such systems are being developed.

It may be that since light is much more subject than radio waves to interference by clouds, mist, fog, and dust, it will be necessary to conduct laser light through pipes containing lenses (to reconcentrate the beam at intervals) and mirrors (to reflect it around corners). However, a *carbon-dioxide laser* has been developed that produces continuous laser beams of unprecedented power that are far enough in the infrared to be little affected by the atmosphere. Atmospheric communication may also be possible then.

Much more immediately practical is the possibility of using modulated laser beams in *optical fibers,* supertransparent glass tubes finer than a human hair, to replace insulated copper wires in telephone communications. Glass is tremendously cheaper and more common than copper and can carry far more information by way of laser light. Already, the bulky copper-wired cables in many places are giving way to the far less bulky optical fiber bundles.

A still more fascinating application of laser beams that is very here-and-now involves a new kind of photography. In ordinary photography, a beam of ordinary light reflected from an object falls on a photographic film. What is recorded is the cross-section of the light, which is by no means all the information it can potentially contain.

Suppose instead that a beam of light is split in two. One part strikes an object and is reflected with all the irregularities that this object would impose on it. The second part is reflected from a mirror with no irregularities. The two parts meet at the photographic film, and the interference of the various wavelengths is recorded. In theory, the recording of this interference would include all the data concerning each light beam. The photograph that records this interference pattern seems to be blank when developed; but if light is shone upon the film and passes through and takes on the interference characteristics, it produces an image containing the complete information. The image is as three-dimensional as was the surface from which light was reflected, and an ordinary photograph can be taken of the image from various angles that show the change in perspective.

This notion was first worked out by the Hungarian-British physicist Dennis Gabor in 1947, when he was trying to work out methods for the sharpening of images produced by electron microscopes. He called it *holography,* from a Latin word meaning "the whole writing."

While Gabor's idea was theoretically sound, it could not be implemented, because ordinary light would not do. With wavelengths of all sizes moving in all directions, the interference fringes produced by the two beams of light would be so chaotic as to yield no information at all. It would be like producing a million dim images all superimposed in slightly different positions.

The introduction of laser light changed everything. In 1965, Emmet N. Leith and Juris Upatnieks, at the University of Michigan, were able to produce the first holograms. Since then, the technique has been sharpened to the point where holography in color has become possible, and where the photographed interference fringes can successfully be viewed with ordinary light. *Microholography* promises to add a new dimension (literally) to biological investigations; and where it will end, none can predict.

Chapter 10

The Reactor

Energy

The rapid advances in technology in the twentieth century have been bought at the expense of a stupendous increase in our consumption of the earth's energy resources. As the underdeveloped nations, with their billions of people, join the already industrialized countries in high living, the rate of consumption of fuel will jump even more spectacularly. Where will we find the energy supplies needed to support our civilization?

We have already seen a large part of the earth's timber disappear. Wood was our first fuel. By the beginning of the Christian era, much of Greece, northern Africa, and the Near East had been ruthlessly deforested, partly for fuel, partly to clear the land for animal herding and agriculture. The uncontrolled felling of the forests was a double-barreled disaster. Not only did it destroy the wood supply, but the drastic uncovering of the land meant a more or less permanent destruction of fertility. Most of these ancient regions, which once supported advanced cultures, are sterile and unproductive now, populated by a ground-down and impoverished people.

The Middle Ages saw the gradual deforestation of western Europe, and modern times have seen the much more rapid deforestation of the North American continent. Almost no great stands of virgin timber remain in the world's temperate zones except in Canada and Siberia.

COAL AND OIL: FOSSIL FUELS

Coal and oil have taken wood's place as fuel. Coal was mentioned by the Greek botanist Theophrastus as long ago as 200 B.C., but the first records of actual coal mining in Europe do not date back before the twelfth century. By the seventeenth century, England, deforested and desperately short of wood for its navy, began to shift to the large-scale use of coal for fuel, inspired perhaps by the fact that the Netherlanders had already begun to dig for coal. (They were not the first. Marco Polo, in his famous book about his travels in China in the late 1200s, had described coal burning in that land, which was then the most technologically advanced in the world.)

By 1660, England was producing 2 million tons of coal each year, or more than 80 percent of all the coal that was then being produced in the world.

At first, it was used chiefly as a household fuel; but in 1603, an Englishman, Hugh Platt, discovered that if coal were heated in such a way that oxygen did not get at it, the tarry, pitchy material it contained would be driven off and burned. Left behind was almost pure carbon, and this residue was called *coke*.

At first coke was not of high quality. It was improved with time and eventually could be used in place of charcoal (from wood) to smelt iron ore. Coke burned at a high temperature, and its carbon atoms combined with the oxygen atoms of iron ore, leaving metallic iron behind. In 1709, an Englishman, Abraham Darby, began to use coke on a large scale for iron making. When the steam engine arrived, coal was used to heat and boil the water; and the Industrial Revolution was, in this way, driven forward.

The shift was slower elsewhere. Even in 1800, wood supplied 94 percent of the fuel needs in the young, forest-rich United States. In 1885, however, wood supplied only 50 percent of the fuel needs and, by the 1980s, less than 3 percent. The balance, moreover, has shifted beyond coal to oil and natural gas. In 1900, the energy supplied by coal in the United States was ten times that supplied by oil and gas together. Half a century later, coal supplied only one-third the energy supplied by oil and gas.

In ancient times, the oil used to burn in lamps for illumination was derived from plant and animal sources. Through the long eons of geologic time, however, the oil-rich tiny animals of the shallow seas have sometimes, in dying, escaped being eaten but mingled with the mud and were buried under sedimentary layers. After slow chemical change, the oil was converted to a complex mixture of hydrocarbons and is now properly called *petroleum* (from Latin, meaning "rock oil"). However, such has been its importance to humanity over the last couple of generations that the simple word *oil* has come to mean nothing else. We can be sure that when *oil* hits the headlines it is not referring to olive oil or coconut oil.

Oil is sometimes found on Earth's surface, particularly in the oil-rich Middle East. It was the *pitch* that Noah was instructed to dub on his ark inside and out to make it waterproof. In the same way, when Moses was set afloat as a baby in his "ark of bulrushes," it, too, was daubed with pitch to keep it from sinking. Lighter fractions of the oil (*naphtha*) were sometimes collected and used in lamps, or for flames used in connection with religious rites.

In the 1850s, inflammable liquids were needed for lamps. There was whale oil, and also coal oil (obtained by heating coal in the absence of air). Another source was shale, a soft material that felt something like wax. When heated, it gave off a liquid called *kerosene*. Such shale was found in western Pennsylvania; and in 1859, an American railway conductor, Edwin Laurentine Drake, tried something new.

Drake knew that people dug wells to obtain water, and that sometimes people dug even deeper to get *brine* (very salty water that could be used as a source of salt). Sometimes, an inflammable oily material came up with the brine. There were reports that, in China and Burma two thousand years ago, this oil was burned and the heat used to drive off the water from the brine, leaving the salt behind.

Why not, then, dig for oil? It was used in those days, not only as a fuel in lamps but for medicinal purposes; and Drake felt there would be a good market for anything he might dig up. He drilled a hole 69 feet under the ground at Titusville in western Pennsylvania and, on 28 August 1859, "struck oil." He had drilled the first *oil well*.

For the first half-century, oil's uses were limited; but, with the coming of the internal-combustion engine, oil came to be in great demand. A liquid fraction, lighter than kerosene (that is, more volatile and more easily converted into vapor) was just the thing to burn in the new engines. The fraction was *gasoline*, and the great oil hunt was on and, over the last century, has never ceased.

The Pennsylvania oil fields were quickly consumed, but much larger ones were discovered in Texas in the early twentieth century; then still larger ones in the Middle East in the middle twentieth century.

Oil has many advantages over coal. Human beings do not have to go underground to gouge oil out of the ground; nor do innumerable freight cars have to be loaded with it; nor does it have to be stored in cellars and shoveled into furnaces; nor does it leave ashes to dispose of. Oil is pumped out of the ground, distributed by pipes (or tankers over sea), stored in underground tanks, and fed into furnaces automatically, with flames that can be started and stopped at will, leaving no ash behind. Particularly after the Second World War, the world as a whole shifted vastly from coal to oil. Coal remained a vital material in the manufacture of iron and steel and for various other purposes, but oil became the great fuel resource of the world.

Oil includes some fractions so volatile that they are vapors at ordinary temperature. This is *natural gas* which is now referred to, usually, simply as *gas*, as petroleum has become simply oil. Gas is even more convenient than oil, and its use has been growing even more rapidly than that of the liquid fractions of oil.

And yet these are limited resources. Gas, oil, and coal are *fossil fuels*, relics of plant and animal life eons old, and cannot be replaced once they are used up. With respect to the fossil fuels, human beings are living on their capital at an extravagant rate.

The oil, particularly, is going fast. The world is now burning over 4 million barrels of oil each hour; and despite all efforts at conservation, the rate of consumption will continue to rise in the near future. Although nearly a trillion barrels remain in the earth, this is not more than a thirty-year supply at present levels of use.

Of course, additional oil can be formed by the combination of the more common coal with hydrogen under pressure. This process was first developed by the German chemist Friedrich Bergius in the 1920s, and he shared in the Nobel Prize for chemistry in 1931 as a result. The coal reserve is large indeed, perhaps as large as 7 trillion tons; but not all of it is easy to mine. By the twenty-fifth century or sooner, coal may become an expensive commodity.

We can expect new finds. Perhaps surprises in the way of coal and oil await us in Australia, in the Sahara, even in Antarctica. Moreover, improvements in technology may make it economical to exploit thinner and deeper coal seams, to plunge more and more deeply for oil, and to extract oil from oil shale and from subsea reserves.

No doubt we shall also find ways to use our fuel more efficiently. The process of burning fuel to produce heat to convert water to steam to drive a generator to create electricity wastes a good deal of energy along the way. Most of these losses could be sidestepped if heat could be converted directly into electricity. The possibility of doing this appeared as long ago as 1823, when a German physicist, Thomas Johann Seebeck, observed that, if two different metals are joined in a closed circuit and if the junction of the two elements is heated, a compass needle in the vicinity will be deflected, indicating that the heat is producing an electric current in the circuit (*thermoelectricity*). Seebeck misinterpreted his own work, however, and his discovery was not then followed up.

With the coming of semiconductor techniques, the old *Seebeck effect* underwent a renaissance. Current thermoelectric devices make use of semiconductors. Heating one end of a semiconductor creates an electric potential in the material: in a p-type semiconductor, the cold end becomes negative; in an n-type it becomes positive. Now if these two types of semiconductor are joined in a U-shaped structure, with the n-p junction at the bottom of the U, heating the bottom will cause the upper end of the p branch to gain a negative charge and the upper end of the n branch to acquire a positive charge. As a result, current will flow from one end to the other, and will be generated so long as the temperature difference is maintained (figure 10.1). (In reverse, the use of a current can bring about a temperature drop, so that a thermoelectric device can also be used as a refrigerator.)

The thermoelectric cell, requiring no expensive generator or bulky steam engine, is portable and can be set up in isolated areas as a small-scale supplier of electricity. All it needs as an energy source is a kerosene heater. Such devices are reported to be used routinely in rural areas of the Soviet Union.

Figure 10.1. The thermoelectric cell. Heating the conductor causes electrons to flow toward the cold end of the n-type semi-conductor and from the cold to the warm region of the p-type. If a circuit is formed, current flows in the direction shown by the arrows. Thus heat is converted to electrical energy.

Notwithstanding all possible increases in the efficiency of using fuel and the likelihood of new finds of coal and oil, these sources of energy are definitely limited. The day will come, and not far in the future, when neither coal nor oil can serve as an important large-scale energy source.

The use of fossil fuels will have to be curtailed, in all probability long before the supplies actually run out, for their increasing use has its dangers. Coal is not pure carbon, and oil is not pure hydrocarbon; in each substance, there are minor quantities of nitrogen and sulfur compounds. In the burning of fossil fuels (particularly coal), oxides of nitrogen and sulfur are released into the air. A ton of coal does not release much; but with all the burning that takes place, some 90 million tons of sulfur oxides were being discharged into the atmosphere each year in the course of the 1970s.

Such impurities are a prime source of air pollution and, under the proper meteorological conditions, of *smog* (that is, "smoky fog"), which blankets cities, damages lungs, and can even kill people who already have pulmonary disease.

Such pollution is washed out of the air by the rain, but this is a solution that merely creates a new and possibly worse problem. The nitrogen and sulfur oxides, dissolving in water, turn that water very slightly acid, so that what falls to the ground is *acid rain.*

The rain is not acid enough to bother us directly, but it falls into ponds and lakes and acidifies them—only slightly, but enough to kill much of the fish and other water life, especially if the lakes do not have beds of limestone which might in part neutralize the acid. The acid rain also damages trees. This damage is worst where coal burning is greatest and the rain falls to the east, thanks to prevailing westerly winds. Thus, eastern Canada suffers from acid rain due to coal burning in the American Midwest, while Sweden suffers from the coal burning in western Europe.

The dangers of such pollution can become great indeed if fossil fuels continue to be burned and in increasing volume. Already, international conferences are being held in connection with the problem.

To correct this, oil and coal must be cleaned before being burned—a process that is possible but that will obviously add to the expense of the fuel. However, even if coal that was pure carbon, and oil that was pure hydrocarbon, were burned, the problems would not end. Carbon would burn to carbon dioxide, while hydrocarbon would burn to carbon dioxide and water. These are relatively harmless in themselves (though some carbon monoxide—which is quite poisonous—is bound to be formed as well), and yet the matter cannot be dismissed.

Both carbon dioxide and water vapor are natural constituents of the atmosphere. The quantity of water vapor varies from time to time and place to place, but carbon dioxide is present in constant amounts of about 0.03 percent by weight. Additional water vapor added to the atmosphere by burning fossil fuel finds its way into the ocean eventually and is, in itself, an insignificant addition. Additional carbon dioxide will dissolve, in part, in the ocean and react, in part, with the rocks, but some will remain in the atmosphere.

The quantity of carbon dioxide in the atmosphere has increased by half again its original amount since 1900, thanks to the burning of coal and oil, and is increasing measurably from year to year. The additional carbon dioxide creates no problem where breathing is concerned and may even be considered as beneficial to plant life. It does, however, add somewhat to the greenhouse effect and raises the overall average temperature of the earth by a small amount. Again, it is scarcely enough to be noticeable, but the added temperature tends to raise the vapor pressure of the ocean and to keep more water vapor in the air, on the whole, and that, too, enhances the greenhouse effect.

It is possible, then, that the burning of fossil fuels may trigger a large enough rise in temperature, to begin melting the ice caps with disastrous results to the continental coastlines. It may also result in long-range climatic changes for the worse. There is even a small possibility that it may initiate a runaway greenhouse effect which would push Earth in the direction of Venus, although we need to know a great deal more about atmospheric dynamics and temperature effects before any predictions we make can be more than guesses.

In any case, however, the continued burning of fossil fuels must be treated with considerable caution.

And yet our energy needs will continue and even be far larger than those of today. What can be done?

SOLAR ENERGY

One possibility is to make increasing use of renewable energy sources: to live on the earth's energy income rather than its capital. Wood can be such a resource if forests are grown and harvested as a crop, though wood alone could not come anywhere near meeting all our energy needs. We could also make much more use of wind power and water power, though these again could never be more than subsidiary sources of energy. The same must be said about certain other potential sources of energy in the earth, such as tapping the heat of the interior (as in hot springs) or harnessing the ocean tides.

Far more important, for the long run, is the possibility of directly tapping some of the vast energy pouring on the earth from the sun. This *insolation* produces energy at a rate that is some 50,000 times as great as our current rate of energy consumption. In this respect, one particularly promising device is the *solar battery,* or *photovoltaic cell,* which makes use of solid-state devices to convert sunlight directly into electricity (figure 10.2).

As developed by the Bell Telephone Laboratories in 1954, the photovoltaic cell is a flat sandwich of n-type and p-type semiconductors that is part of an electric circuit. Sunlight striking the plate knocks some electrons out of place —the usual photoelectric effect. The freed electrons move toward the positive pole and holes move toward the negative pole, thus constituting a current. Not much current is produced as compared with an ordinary chemical battery, but the beauty of the solar battery is that it has no liquids, no corrosive chemicals, no moving parts: it just keeps on generating electricity indefinitely merely by lying in the sun.

Figure 10.2. A cell. Sunlight striking the thin wafer frees electrons, thus forming electron-hole pairs. The p-n junction acts as a barrier, or electric field, separating electrons from holes. A potential difference therefore develops across the junction, and current then flows through the wire circuit.

The artificial satellite *Vanguard I,* launched by the United States on 17 March 1958, was the first to be equipped with a photovoltaic cell to power its radio signals; and those signals were continued for years since there was no "off" switch.

The amount of energy falling upon one acre of a generally sunny area of the earth is 9.4 million kilowatt-hours per year. If substantial areas in the earth's desert regions, such as Death Valley and the Sahara, were covered with solar batteries and electricity-storing devices, they could provide the world with its electricity needs for an indefinite time—for as long, in fact, as the human race is likely to endure, if it does not commit suicide.

One catch is, of course, expense. Pure silicon crystals out of which thin slices can be cut for the necessary cells are expensive. To be sure, since 1958, the price has been cut to 1/250th of what it originally was, but solar electricity is still about ten times as expensive as oil-generated electricity.

Of course, photovoltaic cells may get cheaper still and more efficient, but collecting sunlight is not as easy as it sounds. Sunlight is copious but dilute; and as I mentioned, two paragraphs back, vast areas may have to be coated with them, if they are to serve the world. Then, too, it is night for half the time; and even in the daytime, there may be fog, mist, or cloud. Even clear desert air absorbs a sizable fraction of the solar radiation, especially when the sun is low in the sky. Maintenance of large, exposed areas on Earth would be expensive and difficult.

Some scientists suggest that such solar power stations be placed in orbit about the earth under conditions where nearly unbroken sunlight with no atmospheric interference could increase production per unit area as much as sixtyfold, but this is not likely to come to pass in the immediate future

The Nucleus in War

Between the large-scale use of fossil fuels in the present and the large-scale use of solar energy in the future, there is another source of energy, available in large quantities, which made its appearance rather unexpectedly, less than half a century ago, and which has the potentiality of bridging the gap. This is *nuclear energy,* the energy stored in the tiny atomic nucleus.

Nuclear energy is sometimes called *atomic energy,* but that is a misnomer. Strictly speaking, atomic energy is the energy yielded by chemical reactions, such as the burning of coal and oil, because they involve the behavior of the atom as a whole. The energy released by changes in the nucleus is of a totally different kind and vastly greater in magnitude.

THE DISCOVERY OF FISSION
Soon after the discovery of the neutron by Chadwick in 1932, physicists realized that they had a wonderful key for unlocking the atomic nucleus. Since it had no electric charge, the neutron could easily penetrate the charged

nucleus. Physicists immediately began to bombard various nuclei with neutrons to see what nuclear reactions could be brought about; among the most ardent investigators with this new tool was Enrico Fermi of Italy. In the space of a few months, he had prepared new radioactive isotopes of thirty-seven different elements.

Fermi and his associates discovered that they got better results if they slowed down the neutrons by passing them through water or paraffin first. Bouncing off protons in the water or paraffin, the neutrons are slowed just as a billiard ball is when it hits other billiard balls. When a neutron is reduced to *thermal speed* (the normal speed of motion of atoms), it has a greater chance of being absorbed by a nucleus, because it remains in the vicinity of the nucleus longer. Another way of looking at it is to consider that the length of the wave associated with the neutron is longer, for the wavelength is inversely proportional to the momentum of the particle. As the neutron slows down, its wavelength increases. To put it metaphorically, the neutron grows fuzzier and takes up more volume. It therefore hits a nucleus more easily, just as a bowling ball has more chance of hitting a tenpin than a golf ball would have.

The probability that a given species of nucleus will capture a neutron is called its *cross section*. This term, metaphorically, pictures the nucleus as a target of a particular size. It is easier to hit the side of a barn with a baseball than it is to hit a foot-wide board at the same distance. The cross sections of nuclei under neutron bombardment are reckoned in trillion-trillionths of a square centimeter (10^{-24} square centimeter where 1 square centimeter is a little less than one-sixth of a square inch). That unit, in fact, was named a *barn* by the American physicists M. G. Holloway and C. P. Baker in 1942. The name served to hide what was really going on in those hectic wartime days.

When a nucleus absorbs a neutron, its atomic number is unchanged (because the charge of the nucleus remains the same), but its mass number goes up by one unit. Hydrogen 1 becomes hydrogen 2, oxygen 17 becomes oxygen 18, and so on. The energy delivered to the nucleus by the neutron as it enters may *excite* the nucleus—that is, increase its energy content. This surplus energy is then emitted as a gamma ray.

The new nucleus often is unstable. For example, when aluminum 27 takes in a neutron and becomes aluminum 28, one of the neutrons in the new nucleus soon changes to a proton (by emitting an electron). This increase in the positive charge of the nucleus transforms the aluminum (atomic number 13) to silicon (atomic number 14).

Because neutron bombardment is an easy way of converting an element to the next higher one, Fermi decided to bombard uranium to see if he could form an artificial element—number 93. In the products of the bombardment of uranium, he and his co-workers did find signs of new radioactive substances. They thought they had made element 93, and called it *uranium X*. But how could the new element be identified positively? What sort of chemical properties should it have?

Well, element 93, it was thought, should fall under rhenium in the periodic table, so it ought to be chemically similar to rhenium. (Actually, though no one realized it at the time, element 93 belonged in a new rare-earth series, which meant that it would resemble uranium, not rhenium—see chapter 6. Thus, the search for its identification got off on the wrong foot entirely.) If it were like rhenium, perhaps the tiny amount of "element 93" created might be identified by mixing the products of the neutron bombardment with rhenium and then separating out the rhenium by chemical methods. The rhenium would act as a *carrier*, bringing out the chemically similar "element 93" with it. If the rhenium proved to have radioactivity attached to it, this would indicate the presence of element 93.

Otto Hahn and Lise Meitner, the discoverers of protactinium, working together in Berlin, pursued this line of experiment. Element 93 failed to show up with rhenium. Hahn and Meitner then went on to try to find out whether the neutron bombardment had transformed uranium into other elements near it in the periodic table. At this point, in 1938, Germany occupied Austria, and Meitner, who, until then, as an Austrian national, had been safe despite the fact that she was Jewish, was forced to flee from Hitler's Germany to the safety of Stockholm. Hahn continued his work with the German physicist Fritz Strassman.

Several months later, Hahn and Strassman found that barium, when added to the bombarded uranium, carried off some radioactivity. They decided that this radioactivity must belong to radium, the element below barium in the periodic table. The conclusion was, then, that the neutron bombardment of uranium changed some of it to radium.

But this radium turned out to be peculiar stuff. Try as they would, Hahn and Strassman could not separate it from the barium. In France, Irène Joliot-Curie and her co-worker P. Savitch undertook a similar task and also failed.

And then Meitner, the refugee in Scandinavia, boldly cut through the riddle and broadcast a thought that Hahn was voicing in private but hesitating to publish. In a letter published in the British journal *Nature* in January of 1939, she suggested that the "radium" could not be separated from the barium because no radium was there. The supposed radium was actually radioactive barium: it was barium that had been formed in the neutron bombardment of uranium. This radioactive barium decayed by emitting a beta particle and formed lanthanum. (Hahn and Strassman had found that ordinary lanthanum added to the products brought out some radioactivity, which they assigned to actinium; actually it was radioactive lanthanum.)

But how could barium be formed from uranium? Barium was only a middle-weight atom. No known process of radioactive decay could transform a heavy element into one only about half its weight. Meitner made so bold as to suggest that the uranium nucleus had split in two. The absorption of a neutron had caused it to undergo what she termed *fission*. The two elements into which it had split, she said, were barium and element 43, the element above rhenium in the periodic table. A nucleus of barium and one of element 43 (later named *technetium*) would make up a nucleus of uranium. What made it a particularly daring suggestion was that neutron bombardment only supplied 6 million electron-volts, and the main thought of the day concerning nuclear structure made it seem that hundreds of millions would be required.

Meitner's nephew, Otto Robert Frisch, hastened to Denmark to place the new theory before Bohr, even in advance of publication. Bohr had to face the surprising ease with which this would require the nucleus to split, but fortunately he was evolving the liquid-drop theory of nuclear structure, and it seemed to him that this would explain it. (In later years the liquid-drop theory, taking into account the matter of nuclear shells, was to explain even the fine details of nuclear fission and why the nucleus breaks into unequal halves.)

In any case, theory or not, Bohr grasped the implications at once. He was just leaving to attend a conference on theoretical physics in Washington, and there he told physicists what he had heard in Denmark of the fission suggestion. In high excitement, the physicists went back to their laboratories to test the hypothesis; and within a month half a dozen experimental confirmations were announced. The Nobel Prize for chemistry went to Hahn in 1944 as a result.

THE CHAIN REACTION

The fission reaction released an unusual amount of energy, vastly more than did ordinary radioactivity. But it was not solely the additional energy that made fission so portentous a phenomenon. More important was the fact that it released two or three neutrons. Within two months after the Meitner letter, the awesome possibility of a *nuclear chain reaction* had occurred to a number of physicists.

A chain reaction is a common phenomenon in chemistry. The burning of a piece of paper is a chain reaction. A match supplies the heat required to start it; once the burning has begun, this supplies the very agent, heat, needed to maintain and spread the flame. Burning brings about more burning on an ever-expanding scale.

That is similar to a nuclear chain reaction. One neutron fissions a uranium nucleus, thus releasing two neutrons that can produce two fissions that release four neutrons which can produce four fissions, and so on (figure 10.3). The first atom to fission yields 200 Mev of energy; the next step yields 400 Mev, the next 800 Mev, the next 1,600 Mev, and so on. Since the successive stages take place at intervals of about a 50 trillionth of a second, you see that, within a tiny fraction of a second, a staggering amount of energy will be released. (The actual average number of neutrons produced per fission is 2.47, so matters go even more quickly than this simplified calculation indicates.) The fission of 1 ounce of uranium produces as much energy as the burning of 90 tons of coal or of 2,000 gallons of fuel oil. Peacefully used, uranium fission could, in theory, relieve all our immediate worries about vanishing fossil fuels and our mounting consumption of energy.

But the discovery of fission came just before the world was plunged into an

Figure 10.3. Nuclear chain reaction in uranium. The gray circles are uranium
nuclei; the black dots, neutrons; the wavy arrows, gamma rays; and the small
circles, fission fragments.

all-out war. The fissioning of an ounce of uranium, physicists estimated, would
yield as much explosive power as 600 tons of TNT. The thought of the
consequences of a war fought with such weapons was horrible, but the thought
of a world in which Nazi Germany laid its hands on such an explosive before
the Allies did was even more horrible.

The Hungarian-American physicist Leo Szilard, who had been thinking of
nuclear chain reactions for years, foresaw the possible future with complete
clarity. He and two other Hungarian-American physicists, Eugene Wigner
and Edward Teller, prevailed on the gentle and pacific Einstein in the summer
of 1939 to write a letter to President Franklin Delano Roosevelt, pointing out
the potentialities of uranium fission and suggesting that every effort be made
to develop such a weapon before the Nazis managed to do so.

The letter was written on 2 August 1939 and was delivered to the President
on 11 October 1939. Between those dates, the Second World War had
erupted in Europe. Physicists at Columbia University, under the supervision
of Fermi, who had left Italy for America the previous year, worked to produce
sustained fission in a large quantity of uranium.

Eventually the government of the United States itself took action in the
light of Einstein's letter. On 6 December 1941, President Roosevelt (taking
a huge political risk in case of failure) authorized the organization of a giant
project, under the deliberately noncommittal name of Manhattan Engineer
District, for the purpose of devising an atom bomb. The next day, the Japanese
attacked Pearl Harbor, and the United States was at war.

THE FIRST ATOMIC PILE

As was to be expected, practice did not by any means follow easily from theory. It took a bit of doing to arrange a uranium chain reaction. In the first place, you had to have a substantial amount of uranium, refined to sufficient purity so that neutrons would not be wasted in absorption by impurities. Uranium is a rather common element in the earth's crust, averaging about 2 grams per ton of rock, which makes it 400 times as common as gold. But it is well spread out, and there are few places in the world where it occurs in rich ores or even in reasonable concentration. Furthermore, before 1939 uranium had had almost no uses, and no methods for its purification had been worked out. Less than an ounce of uranium metal had been produced in the United States.

The laboratories at Iowa State College, under the leadership of Spedding, went to work on the problem of purification by ion-exchange resins (see chapter 6) and, in 1942, began to produce reasonably pure uranium metal.

That, however, was only a first step. Now the uranium itself had to be broken down to separate out its more fissionable fraction. The isotope uranium 238 (U-238) has an even number of protons (92) and an even number of neutrons (146). Nuclei with even numbers of nucleons are more stable than those with odd numbers. The other isotope in natural uranium—uranium 235 —has an odd number of neutrons (143). Bohr had therefore predicted that it would fission more readily than uranium 238. In 1940, a research team, under the leadership of the American physicist John Ray Dunning, isolated a small quantity of uranium 235 and showed that Bohr's conjecture was true. U-238 fissions only when struck by fast neutrons of more than a certain energy, but U-235 will undergo fission upon absorbing neutrons of any energy, all the way down to simple thermal neutrons.

The trouble was that in purified natural uranium only one atom in 140 is U-235, the rest being U-238. Thus, most of the neutrons released by fission of U-235 would be captured by U-238 atoms without producing fission. Even if the uranium were bombarded with neutrons fast enough to split U-238, the neutrons released by the fissioning U-238 would not be energetic enough to carry on a chain reaction in the remaining atoms of this more common isotope. In other words, the presence of U-238 would cause the chain reaction to damp and die. It would be like trying to burn wet leaves.

There was nothing for it, then, but to try for a large-scale separation of U-235 from U-238, or at least the removal of enough U-238 to effect a substantial enrichment of the U-235 content in the mixture. The physicists attacked this problem by several methods, each of them offering only thin prospects of success. The one that eventually worked best was *gaseous diffusion.* This remained the method of choice, though fearfully expensive, until 1960. A West German scientist then developed a much cheaper technique of U-235 isolation by *centrifugation,* the heavier molecules being thrown outward and the lighter ones, containing U-235, lagging behind. This process makes nuclear bombs cheap enough for minor powers to manufacture, a consummation not entirely to be desired.

The uranium-235 atom is 1.3 percent less massive than the uranium-238 atom. Consequently, if the atoms were in the form of a gas, the U-235 atoms would move about slightly faster than the U-238 atoms and thus might be separated, by reason of their faster diffusion, through a series of filtering barriers. But first uranium had to be converted to a gas. About the only way to get it in this form was to combine it with fluorine and make *uranium hexafluoride*, a volatile liquid composed of one uranium atom and six fluorine atoms. In this compound, a molecule containing U-235 would be less than 1 percent lighter than one containing U-238, a difference that proved sufficient to make the method work.

The uranium hexafluoride vapor was forced through porous barriers under pressure. At each barrier, the molecules containing U-235 got through a bit faster, on the average; and so with every passage through the successive barriers, the advantage in favor of U-235 grew. To obtain sizable amounts of almost pure uranium-235 hexafluoride required thousands of barriers, but well-enriched concentrations of U-235 could be achieved with a much smaller number of barriers.

By 1942, it was reasonably certain that the gaseous diffusion method (and one or two others) could produce *enriched uranium* in quantity; and separation plants (costing a billion dollars each, and consuming as much electricity as all of New York City) were built at the secret city of Oak Ridge, Tennessee, sometimes called Dogpatch by irreverent scientists, after the mythical town in Al Capp's *Li'l Abner*.

Meanwhile, the physicists were calculating the *critical size* that would be needed to maintain a chain reaction in a lump of enriched uranium. If the lump was small, too many neutrons would escape from its surface before being absorbed by U-235 atoms. To minimize this loss by leakage, the volume of the lump had to be large in proportion to its surface. At a certain critical size, enough neutrons would be intercepted by U-235 atoms to keep a chain reaction going.

The physicists also found a way to make efficient use of the available neutrons. *Thermal* (that is, slow) neutrons, as I have mentioned, are more readily absorbed by uranium 235 than are fast ones. The experimenters therefore used a *moderator* to slow the neutrons from the rather high speeds they had on emerging from the fission reaction. Ordinary water would have been an excellent slowing agent, but unfortunately the nuclei of ordinary hydrogen hungrily snap up neutrons. Deuterium (hydrogen 2) fills the bill much better; it has practically no tendency to absorb neutrons. Consequently the fission experimenters became very interested in preparing supplies of heavy water.

Up to 1943, it was prepared by electrolysis for the most part. Ordinary water split into hydrogen and oxygen more readily than did heavy water, so that, if a large supply of water were electrolyzed, the final bit of water was rich in heavy water and could be preserved. After 1943, careful distillation was the favored method. Ordinary water had the lower boiling point, so that the last bit of unboiled water was rich in heavy water.

Heavy water was indeed valuable in the early 1940s. There is a thrilling story of how Joliot-Curie managed to smuggle France's supply of that liquid out of the country ahead of the invading Nazis in 1940. A hundred gallons of it, which had been prepared in Norway, did fall into the hands of the German Nazis. It was destroyed by a British commando raid in 1942.

Still, heavy water had drawbacks: it might boil away when the chain reaction got hot, and it would corrode the uranium. The scientists seeking to create a chain-reacting system in the Manhattan Project decided to use carbon, in the form of very pure graphite, as the moderator.

Another possible moderator, beryllium, had the disadvantage of toxicity. Indeed, the disease, *berylliosis*, was first recognized in the early 1940s in one of the physicists working on the atom bomb.

Now let us imagine a chain reaction. We start things off by sending a triggering stream of neutrons into the assembly of moderator and enriched uranium. A number of uranium-235 atoms undergo fission, releasing neutrons that go on to hit other uranium-235 atoms. They in turn fission and turn loose more neutrons. Some neutrons will be absorbed by atoms other than uranium 235; some will escape from the pile altogether. But if from each fission one neutron, and exactly one, takes effect in producing another fission, then the chain reaction will be self-sustaining. If the *multiplication factor* is more than one, even very slightly more (for example, 1.001), the chain reaction will rapidly build up to an explosion. This is good for bomb purposes but not for experimental purposes. Some device had to be worked out to control the rate of fissions. That could be done by sliding in rods of a substance such as cadmium, which has a high cross section for neutron capture. The chain reaction develops so rapidly that the damping cadmium rods could not be slid in fast enough, were it not for the fortunate fact that the fissioning uranium atoms do not emit all their neutrons instantly. About 1 neutron in 150 is a *delayed neutron* emitted a few minutes after fission, since it emerges, not directly from the fissioning atoms, but from the smaller atoms formed in fission. When the multiplication factor is only slightly above 1, this delay is sufficient to give time for applying the controls.

In 1941, experiments were conducted with uranium-graphite mixtures, and enough information was gathered to lead physicists to decide that, even without enriched uranium, a chain reaction might be set up if only the lump of uranium were made large enough.

Physicists set out to build a uranium chain reactor of critical size at the University of Chicago. By that time some six tons of pure uranium were available; this amount was eked out with uranium oxide. Alternate layers of uranium and graphite were laid down one on the other, fifty-seven layers in all, with holes through them for insertion of the cadmium control rods. The structure was called a *pile*—a noncommittal code name that did not give away its function. (During the First World War, the newly designed armored vehicles on caterpillar treads were referred to as *tanks* for the same purpose of secrecy. The name *tank* stuck, but *atomic pile* fortunately gave way eventually to the more descriptive name *nuclear reactor.*)

The Chicago pile, built under the football stadium, measured 30 feet wide, 32 feet long, and 21½ feet high. It weighed 1,400 tons and contained 52 tons of uranium, as metal and oxide. (Using pure uranium 235, the critical size would have been, it is reported, no more than 9 ounces.) On 2 December 1942, the cadmium control rods were slowly pulled out. At 3:45 P.M. the multiplication factor reached 1: a self-sustaining fission reaction was under way. At that moment humanity (without knowing it) entered the Nuclear Age.

The physicist in charge was Enrico Fermi, and Eugene Wigner presented him with a bottle of Chianti in celebration. Arthur Compton, who was at the site, made a long-distance telephone call to James Bryant Conant at Harvard, announcing the success. "The Italian navigator," he said, "has entered the new world." Conant asked, "How were the natives?" The answer came at once: "Very friendly!"

It is a curious and interesting that the first Italian navigator discovered one new world in 1492, and the second discovered another in 1942.

THE NUCLEAR AGE

Meanwhile another fissionable fuel had turned up. Uranium 238, upon absorbing a thermal neutron, forms uranium 239, which breaks down quickly to neptunium 239, which in turn breaks down almost as quickly to plutonium 239.

As the plutonium-239 nucleus has an odd number of neutrons (145) and is more complex than uranium 235, it should be highly unstable. It seemed a reasonable guess that plutonium 239, like uranium 235, might undergo fission with thermal neutrons. In 1941, this was confirmed experimentally. Still uncertain whether the preparation of uranium 235 would prove practical, the physicists decided to hedge their bets by trying to make plutonium in quantity.

Special reactors were built in 1943 at Oak Ridge and at Hanford, in the State of Washington, for the purpose of manufacturing plutonium. These reactors were a great advance over the first pile in Chicago. For one thing, the new reactors were designed so that the uranium could be removed from the pile periodically. The plutonium produced could be separated from the uranium by chemical methods; and the fission products, some of them strong neutron absorbers, could also be separated out. In addition, the new reactors were water-cooled to prevent overheating. (The Chicago pile could operate only for short periods, because it was cooled merely by air.)

By 1945, enough purified uranium 235 and plutonium 239 were available for the construction of bombs. This portion of the task was undertaken at a third secret city, Los Alamos, New Mexico, under the leadership of the American physicist J. Robert Oppenheimer.

For bomb purposes it was desirable to make the nuclear chain reaction mount as rapidly as possible. This called for making the reaction go with fast neutrons, to shorten the intervals between fissions, so the moderator was omitted. The bomb was also enclosed in a massive casing to hold the uranium together long enough for a large proportion of it to fission.

Since a critical mass of fissionable material will explode spontaneously (sparked by stray neutrons from the air), the bomb fuel was divided into two or more sections. The triggering mechanism was an ordinary explosive which drove these sections together when the bomb was to be detonated. One arrangement was called the "Thin Man"—a tube with two pieces of uranium 235 at its opposite ends. Another, the "Fat Man," had the form of a ball in which a shell composed of fissionable material was *imploded* toward the center, making a dense critical mass held together momentarily by the force of the implosion and by a heavy outer casing called the *tamper.* The tamper also served to reflect back neutrons into the fissioning mass and, therefore, to reduce the critical size.

To test such a device on a minor scale was impossible. The bomb had to be above critical size or nothing. Consequently, the first test was the explosion of a full-scale nuclear-fission bomb, usually called, incorrectly, an *atom bomb* or *A-bomb.* At 5:30 A.M. on 16 July 1945, at Alamogordo, New Mexico, a bomb was exploded with truly horrifying effect; it had the explosive force of 20,000 tons of TNT. I. I. Rabi, on being asked later what he had witnessed, is reported to have said mournfully, "I can't tell you, but don't expect to die a natural death." (It is only fair to add that the gentleman so addressed by Rabi did die a natural death some years later.)

Two more fission bombs were prepared. One, a uranium bomb called "Little Boy," 10 feet long by 2 feet wide and weighing 4½ tons, was dropped on Hiroshima on 6 August 1945; it was set off by radar echo. Days later, the second, a plutonium bomb, 11 feet by 5 feet, weighing 5 tons, and named "Fat Man," was dropped on Nagasaki. Together, the two bombs had the explosive force of 35,000 tons of TNT. With the bombing of Hiroshima, the Nuclear Age, already nearly three years old, broke on the consciousness of the world.

For four years afterward, Americans lived under the delusion that there was a nuclear-bomb "secret" which could be kept from other nations forever if only security measures were made tight enough. Actually, the facts and theories of nuclear fission had been matters of public record since 1939, and the Soviet Union was fully engaged in research on the subject in 1940. If the Second World War had not occupied that nation's lesser resources to a far greater extent than it occupied the greater resources of the uninvaded United States, the U.S.S.R. might have made a nuclear bomb by 1945, as we did. As it was, the Soviet Union exploded its first nuclear bomb on 22 September 1949, to the dismay and unnecessary amazement of most Americans. It had six times the power of the Hiroshima bomb and an explosive effect equal to 210,000 tons of TNT.

On 3 October 1952, Great Britain became the third nuclear power by exploding a test bomb of its own. On 13 February 1960, France joined the "nuclear club" as the fourth member, setting off a plutonium bomb in the Sahara. On 16 October 1964, the People's Republic of China announced the explosion of a nuclear bomb and became the fifth member. In May 1974, India detonated a nuclear bomb, making use of plutonium that had been surreptitiously removed from a reactor (intended for peaceful power production) given it by Canada, and became the sixth member. Since then, a variety of powers, including Israel, South Africa, Argentina, and Iraq, have been reported to be on the edge of possessing nuclear weapons.

Such *nuclear proliferation* has become a source of alarm to many people. It is bad enough to live under the threat of a nuclear war initiated by one of the two superpowers who (presumably) are uncomfortably aware of the consequences and who, for forty years, have refrained. To be at the mercy of small powers, acting in anger over narrow issues, guided by petty rulers of no great mental breadth, would seem intolerable.

THE THERMONUCLEAR REACTION

Meanwhile the fission bomb had been reduced to triviality. Human beings had succeeded in setting off another energetic nuclear reaction which made much more devastating bombs possible.

In the fission of uranium, 0.1 percent of the mass of the uranium atom is converted to energy. But in the fusion of hydrogen atoms to form helium, fully 0.5 percent of their mass is converted to energy, as had first been pointed out in 1915 by the American chemist William Draper Harkins. At temperatures in the millions of degrees, the energy of protons is high enough to allow them to fuse. Thus two protons may unite and, after emitting a positron and a neutrino (a process that converts one of the protons to a neutron), become a deuterium nucleus. A deuterium nucleus may then fuse with a proton to form a tritium nucleus, which can fuse with still another proton to form helium 4. Or deuterium and tritium nuclei will combine in various ways to form helium 4.

Because such nuclear reactions take place only under the stimulus of high temperatures, they are referred to as *thermonuclear reactions*. In the 1930s, the one place where the necessary temperatures were believed to exist was at the center of stars. In 1938, the German-born physicist Hans Albrecht Bethe (who had left Hitler's Germany for the United States in 1935) proposed that fusion reactions were responsible for the energy that the stars radiated. It was the first completely satisfactory explanation of stellar energy since Helmholtz had raised the question nearly a century earlier.

Now the uranium-fission bomb provided the necessary temperatures on the earth. It could serve as a match hot enough to ignite a fusion chain reaction in hydrogen. For a while it looked very doubtful that the reaction could actually be made to work in the form of a bomb. For one thing, the hydrogen fuel, in the form of a mixture of deuterium and tritium, had to be condensed to a dense mass, which meant that it had to be liquefied and kept at a temperature only a few degrees above absolute zero. In other words, what would be exploded would be a massive refrigerator. Furthermore, even assuming a hydrogen bomb could be made, what purpose would it serve? The fission bomb was already devastating enough to knock out cities; a hydrogen bomb would merely pile on destruction and wipe out whole civilian populations.

Nevertheless, despite the unappetizing prospects, the United States and the Soviet Union felt compelled to go on with it. The United States Atomic Energy Commission proceeded to produce some tritium fuel, set up a 65-ton fission-fusion contraption on a coral atoll in the Pacific, and on 1 November 1952, produced the first thermonuclear explosion (a *hydrogen bomb* or *H-bomb*) on our planet. It fulfilled all the ominous predictions: the explosion yielded the equivalent of 10 million tons of TNT (10 *megatons*)—500 times the puny 20-kiloton energy of the Hiroshima bomb. The blast wiped out the atoll.

The Russians were not far behind; on 12 August 1953, they also produced a successful thermonuclear explosion, and it was light enough to be carried in a plane. We did not produce a portable one until early 1954. Where we developed the fusion bomb seven and one-half years after the fission bomb, the Soviets took only five years.

Meanwhile a scheme for generating a thermonuclear chain reaction in a simpler way and packing it into a portable bomb had been conceived. The key to this reaction was the element lithium. When the isotope lithium 6 absorbs a neutron, it splits into nuclei of helium and tritium, giving forth 4.8 Mev of energy in the process. Suppose, then, that a compound of lithium and hydrogen (in the form of the heavy isotope deuterium) is used as the fuel. This compound is a solid, so there is no need for refrigeration to condense the fuel. A fission trigger would provide neutrons to split the lithium. And the heat of the explosion would cause the fusion of the deuterium present in the compound and of the tritium produced by the splitting of lithium. In other words, several energy-yielding reactions would take place: the splitting of lithium, the fusion of deuterium with deuterium, and the fusion of deuterium with tritium.

Now besides releasing tremendous energy, these reactions would also yield a great number of surplus neutrons. It occurred to the bomb builders: Why not use the neutrons to fission a mass of uranium? Even common uranium 238 could be fissioned with fast neutrons (though less readily than U-235). The heavy blast of fast neutrons from the fusion reactions might fission a considerable number of U-238 atoms. Suppose one built a bomb with a U-235 core (the igniting match), a surrounding explosive charge of lithium deuteride, and around all this a blanket of uranium 238 which would also serve as explosive. That would make a really big bomb. The U-238 blanket could be made almost as thick as one wished, because there is no critical size at which uranium 238 will undergo a chain reaction spontaneously. The result is sometimes called a *U-bomb*.

The bomb was built. It was exploded at Bikini in the Marshall Islands on 1 March 1954 and shook the world. The energy yield was around 15 megatons. Even more dramatic was a rain of radioactive particles that fell on twenty-three Japanese fishermen in a fishing boat named *The Lucky Dragon.* The radioactivity destroyed the cargo of fish, made the fishermen ill, eventually killed one, and did not exactly improve the health of the rest of the world.

Since 1954, thermonuclear bombs have become items in the armaments of the United States, the Soviet Union, and Great Britain. In 1967, China became the fourth member of the "thermonuclear club," having made the transition from fission in only three years. The Soviet Union has exploded hydrogen bombs in the 50- to 100-megaton range and the United States is perfectly capable of building such bombs, or even larger ones, at short notice.

In the 1970s, thermonuclear bombs were developed that minimized the blast effect and maximized radiation, particularly neutrons. Hence, less damage would be done to property and more to human beings. Such *neutron bombs* seem desirable to people who worry about property and hold life cheap.

When the first nuclear bombs were used in the last days of the Second World War, they were delivered by airplane. It is now possible to deliver them by *intercontinental ballistic missiles* (ICBMs), which are rocket-powered and are capable of being aimed with great accuracy from any place on Earth to any other place on Earth. Both the United States and the Soviet Union have great stores of such missiles, all capable of being equipped with nuclear warheads.

For that reason, an all-out thermonuclear war between the two superpowers, if engaged in with insane rage on both sides, can put an end to civilization (and, perhaps, to much of Earth's power to support life) in as little as half an hour. If there was ever, in this world, a sobering thought, that is it.

The Nucleus in Peace

The dramatic use of nuclear power in the form of unbelievably destructive bombs has done more to present the scientist in the role of ogre than anything else that has occurred since the beginnings of science. In a way this portrayal has its justifications, for no arguments or rationalizations can change the fact that scientists did indeed construct the nuclear bomb, knowing from the beginning its destructive powers and that it would probably be put to use.

It is only fair to add that they did this under the stress of a great war against ruthless enemies and with an eye to the frightful possibility that a man as maniacal as Adolf Hitler might get such a bomb first. It must also be added that, on the whole, the scientists working on the bomb were deeply disturbed about it, and that many opposed its use, while some even left the field of nuclear physics afterward in what can only be described as remorse.

In 1945, a group of physicists, under the leadership of the Nobel laureate James Franck (now an American citizen), petitioned the secretary of war against the use of the nuclear bomb on Japanese cities and accurately foretold the dangerous nuclear stalemate that would follow its use. Far fewer pangs of conscience were felt by the political and military leaders who made the actual decision to use the bombs, and who, for some peculiar reason, are viewed as patriots by many people who view the scientists as demons.

Furthermore, we cannot and should not subordinate the fact that, in releasing the energy of the atomic nucleus, scientists put at our disposal a power that can be used constructively as well as destructively. It is important to emphasize this in a world and at a time in which the threat of nuclear destruction has put science and scientists on the shamefaced defensive, and in a country like the United States, which has a rather strong Rousseauan tradition against book learning as a corrupter of the simple integrity of human beings in a state of nature.

Even the explosion of an atomic bomb need not be purely destructive. Like the lesser chemical explosives long used in mining and in the construction of dams and highways, nuclear explosives could be vastly helpful in construction projects. All kinds of dreams of this sort have been advanced: excavating harbors, digging canals, breaking up underground rock formations, preparing heat reservoirs for power—even the long-distance propulsion of spaceships. In the 1960s, however, the furor for such far-out hopes died down. The prospects of the danger of radioactive contamination or of unlooked-for expense, or both, served as dampers.

Yet one constructive use of nuclear power that was realized lay in the kind of chain reaction that was born under the football stadium at the University of Chicago. A controlled nuclear reactor can develop huge quantities of heat, which, of course, can be drawn off by a *coolant,* such as water or even molten metal, to produce electricity or heat a building (figure 10.4).

Figure 10.4. A nuclear power plant of the gas-cooled type, shown in a schematic design. The reactor's heat here is transferred to a gas, which may be a vaporized metal circulating through it, and the heat is then used to convert water to steam.

NUCLEAR-POWERED VESSELS

Experimental nuclear reactors that produced electricity were built in Great Britain and the United States within a few years after the war. The United States now has a fleet of well over 100 nuclear-powered submarines, the first of which, the U.S.S. *Nautilus* (having cost 50 million dollars), was launched in January 1954. This vessel, as important for its day as Fulton's *Clermont* was in its, introduced engines with a virtually unlimited source of power that permits submarines to remain underwater for indefinitely long periods, whereas ordinary submarines must surface frequently to recharge their batteries by means of diesel generators that require air for their working. Furthermore, where ordinary submarines travel at a speed of eight knots, a nuclear submarine travels at twenty knots or more.

The first *Nautilus* reactor core lasted for 62,500 miles; included among those miles was a dramatic demonstration. The *Nautilus* made an underwater crossing of the Arctic Ocean in 1958. This trip demonstrated that the ocean depth at the North Pole was 13,410 feet (2½ miles), far deeper than had been thought previously. A second, larger nuclear submarine, the U.S.S. *Triton*, circumnavigated the globe underwater along Magellan's route in eighty-four days, between February and May of 1960.

The Soviet Union also possesses nuclear submarines and, in December 1957, launched the first nuclear-powered surface vessel, the *Lenin*, an ice-breaker. Shortly before, the United States had laid the keel for a nuclear-powered surface vessel; and in July 1959, the U.S.S. *Long Beach* (a cruiser) and the *Savannah* (a merchant ship) were launched. The *Long Beach* is powered by two nuclear reactors.

Less than ten years after the launching of the first nuclear vessels, the United States had four nuclear surface ships operating, being built, or authorized for future building. And yet, except for submarines, enthusiasm for nuclear propulsion also waned. In 1967, the *Savannah* was retired after two years of life. It took 3 million dollars a year to run, and was considered too expensive.

NUCLEAR REACTORS FOR ELECTRIC POWER

But it is not the military alone who must be served. The first nuclear reactor built for the production of electric power for civilian use was put into action in the Soviet Union in June of 1954. It was a small one, with a capacity of not more than 5,000 kilowatts. By October 1956, Great Britain had its Calder Hall plant in operation, with a capacity of more than 50,000 kilowatts. The United States was third in the field. On 26 May 1958, Westinghouse completed a small nuclear reactor for the production of civilian electric power at Shippingport, Pennsylvania, with a capacity of 60,000 kilowatts. Other reactors quickly followed both in the United States and elsewhere.

Within little more than a decade, there were nuclear reactors in a dozen countries, and nearly half the supply of civilian electricity in the United States was being supplied by fissioning nuclei. Even outer space was invaded, for a satellite powered by a small reactor was launched on 3 April 1965. And yet the problem of radioactive contamination is a serious one. When the 1970s opened, public opposition to the continued proliferation of nuclear power plants was becoming louder.

Then, on 28 March 1979, on Three Mile Island in the Susquehanna River near Harrisburg, there was the most serious nuclear accident in American history. Actually, there was no broadcasting of any significant quantity of radioactivity, and no danger to human life, even though there was near panic for a few days. The reactor was, however, put out of action indefinitely, and any cleanup was going to be very long and expensive.

The chief casualty was the nuclear-energy industry. A wave of antinuclear sentiment swept the United States and various other nations, too. The chances of new nuclear reactors being set into operation in the United States have dimmed drastically.

The accident, by bringing home to Americans the terrors of even the possibility of radioactive contamination, seemed also to strengthen public opinion worldwide against the production (let alone the use) of nuclear bombs, and this, to any rational person, would seem to be a good result.

And yet nuclear energy in its peaceful aspect cannot be easily abandoned. The human need for energy is overpowering; and, as I pointed out earlier in the chapter, it may be that we cannot rely on fossil fuels for long or expect a massive replacement by solar energy for some time. Nuclear energy, on the other hand, is here, and there are not lacking many voices who point out that, with the proper safeguards, it is *not* more dangerous than the fossil fuels but less dangerous. (Even in the particular case of radioactive contamination, it should be remembered that coal contains tiny quantities of radioactive impurities, and that coal burning releases more radioactivity into the air than nuclear reactors do—or so it is argued.)

BREEDER REACTORS

In that case, suppose we consider nuclear fission as an energy source. For how long a period could we count on it? Not very long, if we have to depend entirely on the scarce fissionable material uranium 235. But, fortunately, other fissionable fuels can be created with uranium 235 as a starter.

We have seen that plutonium is one of these man-made fuels. Suppose we build a small reactor with enriched uranium fuel and omit the moderator, so that fast neutrons will stream into a surrounding jacket of natural uranium. These neutrons will convert uranium 238 in the jacket into plutonium. If we arrange things so that few neutrons are wasted, from each fission of a uranium-235 atom in the core we may get more than one plutonium atom manufactured in the jacket. In other words, we will breed more fuel than we consume.

The first such *breeder reactor* was built under the guidance of the Canadian-American physicist Walter Henry Zinn at Arco, Idaho, in 1951. It was called

EBR-1 (Experimental Breeder Reactor-1). Besides proving the workability of the breeding principle, it produced electricity. It was retired as obsolescent (so fast is progress in this field) in 1964.

Breeding could multiply the fuel supply from uranium many times, because all of the common isotope of uranium, uranium 238, would become potential fuel.

The element thorium, made up entirely of thorium 232, is another potential fissionable fuel. Upon absorbing fast neutrons, it is changed to the artificial isotope thorium 233, which soon decays to uranium 233. Now uranium 233 is fissionable by slow neutrons and will maintain a self-sustaining chain reaction. Thus, thorium can be added to the fuel supply, and thorium appears to be about five times as abundant as uranium in the earth. In fact, it has been estimated that the top hundred yards of the earth's crust contains an average of 12,000 tons of uranium and thorium per square mile. Naturally, not all of this material is easily available.

All in all, the total amount of power conceivably available from the uranium and thorium supplies of the earth is about twenty times that available from the coal and oil we have left.

And yet the same concerns that cause people to fear ordinary reactors are redoubled where breeder reactors are concerned. Plutonium is much more dangerous than uranium, and there are some who maintain it is the most poisonous material in the world that has the chance of being produced in massive quantities, and that if some of it were to find its way into the environment, that would be a catastrophe that could not be reversed. There is also the fear that plutonium intended for peaceful reactors can be hijacked or purloined and used to build a nuclear bomb (as India did) that could then be used for criminal blackmail.

These fears are perhaps exaggerated, but they are reasonable, and not only accident and theft gives cause for fear. Even if nuclear reactors work without hint of accident, there will remain danger. To see the reason, let us consider radioactivity and the energetic radiation to which it gives rise.

THE DANGERS OF RADIATION

To be sure, life on Earth has always been exposed to natural radioactivity and cosmic rays. However, the production of X rays in the laboratory and the concentration of naturally radioactive substances, such as radium, which ordinarily exist as greatly diluted traces in the earth's crust, vastly compounded the danger. Some early workers with X rays and radium even received lethal doses: both Marie Curie and her daughter Irène Joliot-Curie died of leukemia from their exposures, and there is the famous case of the watchdial painters in the 1920s who died as the result of pointing their radium-tipped brushes with their lips.

The fact that the general incidence of leukemia has increased substantially in recent decades may be due, partly, to the increasing use of X rays for numerous purposes. The incidence of leukemia in doctors, who are likely to

I. The Solar System

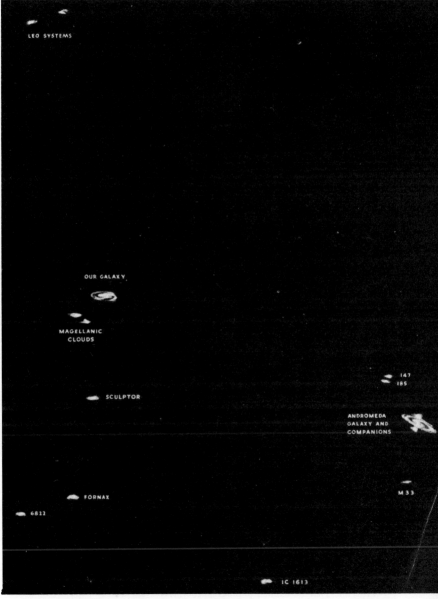

LEO SYSTEMS

OUR GALAXY

MAGELLANIC
CLOUDS

147
185

SCULPTOR

ANDROMEDA
GALAXY AND
COMPANIONS

FORNAX

M 33

6822

IC 1613

Plate I.1   Our region of the universe—a drawing showing the other galaxies in our neighborhood. Courtesy Department of Library Services, American Museum of Natural History.

Plate I.2 Cornell University's radio telescope. The reflector of this radio–radar telescope at Arecibo, Puerto Rico, is 1,000 feet in diameter and is suspended in a natural bowl. Courtesy of Cornell University and Air Force Office of Scientific Research. Courtesy of Arecibo Observatory, National Astronomy and Ionosphere Center (NAIC), Cornell University.

Plate I.3 Halley's Comet, photographed 4 May 1910, with an exposure of 40 minutes. By permission of the Yerkes Observatory, Wisconsin.

Plate I.4 A spiral galaxy in broadside view—the "whirlpool nebula" in Canes Venatici. Courtesy of Palomar Observatory, California.

Plate I.5 A globular cluster in Canes Venatici. Courtesy of Palomar Observatory, California.

Plate I.6 The Crab Nebula, the remains of a supernova, photographed in red light. Courtesy of Palomar Observatory, California.

Plate I.7   The Horsehead Nebula in Orion, south of Zeta Orionis, photographed in red light. Courtesy of Palomar Observatory, California.

Plate I.8   Saturn and its rings: a montage of photographs taken by *Voyager* 1 and *Voyager* 2. Here are pictured all of Saturn's major satellites known before the *Voyager* launches in 1977. The satellites are (*clockwise from upper right*): Titan, Iapetus, Tethys, Mimas, and Rhea. Courtesy of the National Aeronautics and Space Administration.

Plate I.9   Jupiter and its moons in their relative positions: a montage of photographs made by *Voyager* 1 in 1977. The Galilean satellites are Io (*upper left*), Europa (*center*), and Ganymede and Callisto (*lower right*). Courtesy of the National Aeronautics and Space Administration.

Plate I.10 Mars, photographed 19 June from *Viking* 1. Clearly seen are the Tharsis Mountains, three huge volcanoes. Olympus Mons, Mars's largest volcano, is toward the top of the photograph. Courtesy of the National Aeronautics and Space Administration.

Plate I.11   The sun's corona. Courtesy of Mount Wilson Observatory, California.

Plate I.12   A solar flare, stretching 140,000 miles from the sun, photographed in the light of calcium. The white circle represents the earth on a similar scale. Courtesy of Mount Wilson Observatory, California.

Plate I.13   Sun spots photographed with unprecedented sharpness by the U.S. Stratoscope project from a balloon at 80,000 feet. The spots consist of a dark core of relatively cool gases embedded in a strong magnetic field. This group of particularly active spots produced a brilliant aurora and a vigorous magnetic storm on the earth.

Plate I.14   Solar prominences. Courtesy of Mount Wilson Observatory, California.

Plate I.15 Aurora borealis. Courtesy of the National Oceanic and Atmospheric Administration.

Plate I.16 Lunar map. Courtesy of the National Aeronautics and Space Administration.

Plate I.17 This view of the rising earth greeted the Apollo 8 astronauts as they came from behind the moon after orbiting it. On the earth 240,000 statute miles away, the sunset terminator bisects Africa. Courtesy of the National Aeronautics and Space Administration.

II. Earth and Space Travel

Plate II.1 Foucault's famous experiment in Paris in 1851, which showed the rotation of the earth on its axis by means of the swing of a pendulum; the plane of its swings turned clockwise. By permission of the Bettmann Archive.

Plate II.2   The Montgolfier brothers' hot-air balloon, launched at Versailles, 19 September 1783. By permission of the Bettmann Archive.

Plate II.3   Launching of the first U.S. satellite, *Explorer* 1 on 31 January 1958. Courtesy of the United States Army.

Plate II.4   The kneeling figure silhouetted by what seems to be a sparkling halo actually is a mechanic working inside the spacecraft fairing of a McDonnell Douglas Delta rocket. His portable lamp glints on thousands of facets of a triangular "isogrid" pattern milled into the shiny aluminum skin of a fairing to reduce weight while retaining maximum strength. The fairing, 8 feet in diameter and 26 feet long, protects the Delta's payload as it is launched into orbit and from aerodynamic forces and heat during flight through the atmosphere. Courtesy of the McDonnell Douglas Astronautics Company, California.

Plate II.5 Astronaut Edwin E. Aldrin, Jr., lunar module pilot, is photographed walking near the lunar module during *Apollo* 11 extravehicular activity. Courtesy of the National Aeronautics and Space Administration.

Plate 11.6   *Apollo* 11 astronaut Edwin E. Aldrin, Jr., deploys Solar Wind Composition experiment on the moon's surface. Courtesy of the National Aeronautics and Space Administration.

Plate II.7 The Orientale Basin photographed from 1,690 miles above the moon's surface by Lunar Orbiter IV. Courtesy of the National Aeronautics and Space Administration.

Plate II.8   This photograph of the crater Copernicus was taken from 28.4 miles above the surface of the moon by Lunar Orbiter II. Courtesy of the National Aeronautics and Space Administration.

Plate II.9   The *Apollo* 17 space vehicle, 28 August 1972. Courtesy of the National Aeronautics and Space Administration.

Plate II.10 The moon's surface: scientist–astronaut Harrison F. Schmitt stands next to a huge, split lunar boulder, during the *Apollo* 17 expedition. This scene is a composite of three views. Courtesy of the National Aeronautics and Space Administration.

Plate II.11 Model of a future space station. Courtesy of the National Aeronautics and Space Administration.

Plate II.12 Earth, from sunrise to sunset. This sequence was taken by the ATS–III satellite from a point about 22,300 miles above South America, November 1967. The photos show all of that continent and portions of North America, Africa, Europe, and Greenland; clouds cover Antarctica. Courtesy of the National Aeronautics and Space Administration.

Plate II.13 Weather photograph of Earth, showing storms over the Pacific and Caribbean oceans. Courtesy of the United States Department of Commerce.

Plate II.14   Sally Ride, the first woman astronaut, preparing for the STS 7 Space Shuttle launch of 18 June 1983. By permission of United Press International.

Plate II.15   The first free-floating space walk, February 1984: Astronaut Bruce McCandless II, without the use of a tether, at maximum distance from the *Challenger*. Courtesy of the National Aeronautics and Space Administration.

Plate II.16   Earth, as photographed from *Apollo* 17 during the final lunar landing mission. Visible is almost the entire coastline of Africa and the Arabian peninsula. A heavy cloud covers the Antarctic icecap. Courtesy of the National Aeronautics and Space Administration.

III. Aspects of Technology

Plate III.1 Stone tools of early man. The oldest, from the Miocene period, are at the lower left; the most recent, at the lower right. Neg. No. 411257. (Photo: J. Kirschner) Courtesy Department Library Services, American Museum of Natural History.

Plate III.2 Galvani's experiment, which led to the discovery of electric currents. Electricity from his static-electricity machine made the frog's leg twitch; he found that touching the nerve with two different metals also caused the leg to twitch. By permission of the Bettmann Archive.

Plate III.3 A single ice crystal photographed by X-ray diffraction, showing the symmetry and balance of the physical forces holding the structure together. From Franklyn Branley, ed., *Scientist's Choice* (New York: Basic Books, n.d.).

Plate III.4   Electric field around a charged crystal is photographed with the electron microscope by means of a shadow technique. The method uses a fine wire mesh; the distortion of the net, caused by deflection of electrons, shows the shape and strength of the electric field. Courtesy of the National Bureau of Standards.

Plate III.5 Molecular model of titanium oxide in crystalline form, which can serve as a transistor. Removal of one of the oxygen atoms (light balls) will make the material semiconducting. Courtesy of the National Bureau of Standards.

Plate III.6 In a cyclotron, magnets are used to bend a beam of electrically charged particles into a circular path. With the ever-increasing scientific need for beams of higher energies, these accelerators and their magnets have grown in size. Pictured is a super-conducting magnet model developed by Clyde Taylor and co-workers at Lawrence Berkeley Laboratory. Courtesy of Lawrence Berkeley Laboratory, University of California, Berkeley.

Plate III.7 Spinning protons in this schematic drawing are oriented in random directions. The white arrow shows the direction of the spin. Courtesy of the National Bureau of Standards.

Plate III.8 Protons lined up by a steady magnetic field. Those oriented in the opposite-to-normal direction (*arrows pointed downward*) are in the excited spin state. Courtesy of the National Bureau of Standards.

Plate III.9   Tracks of electrons and positrons formed in a bubble chamber by high-energy gamma rays. The circular pattern was made by an electron revolving in a magnetic field. Courtesy of the University of California, Berkeley.

Plate III.10   Fission of a uranium atom. The white streak in the middle of this photographic plate represents the tracks of two atoms flying apart from the central point where the uranium atom split in two. The plate was soaked in a uranium compound and bombarded with neutrons, which produced the fission caught in this picture. The other white dots are randomly developed silver grains. The picture was made in the Eastman Kodak Research Laboratories. By permission of United Press International.

Plate III.11 Radioactivity made visible. On the tray is some tantalum made radioactive in the Brookhaven reactor; the glowing material is shielded here under several feet of water. The radioactive tantalum will be placed in the pipe shown and then transferred to a large lead container for use as a 1,000-curie source of radioactivity for industrial purposes. Courtesy of Brookhaven National Laboratory, New York.

Plate III.12 Drawing of the first chain reactor, built under the Chicago football stadium. Courtesy of Argonne National Laboratory, Illinois.

Plate III.13 The Chicago reactor under construction. This was one of only a few photographs made during the building of the reactor. The rods in the holes are uranium, and the reactor's nineteenth layer, consisting of solid graphite blocks, is being laid on. Courtesy of Argonne National Laboratory, Illinois.

Plate III.14   The mushroom cloud from the atomic bomb dropped by the United States on Hiroshima, Japan, 6 August 1945. This picture was taken by Seizo Yamada, a middle-school student at the time. By permission of the photographer.

Plate III.15   A silo at the nuclear reactor at Three Mile Island, Pennsylvania. Photograph by Sylvia Plachy. Used by permission.

Plate III.16 The life and death of a pinch. This series of pictures shows the brief history of a wisp of plasma in the magnetic field of the Perhapsatron. Each photograph gives two views of the plasma, one from the side and one from below through a mirror. The pinch broke down in a millionths of a second; the number on each picture is in microseconds. Courtesy of Los Alamos Scientific Laboratory, New Mexico.

Plate III.17 The dark streaks are the tracks left by some of the first uranium nuclei ever to be accelerated to near the speed of light. Here you see the last ½ millimeter of three tracks as they came to rest in a special photographic emulsion. The bottom track shows a nucleus splitting into two lighter nuclei. The work was done at the Bevalac, the only accelerator facility in the world that provides ions as heavy as uranium at relativistic energies. The accelerator is located at the University of California's Lawrence Berkeley Laboratory. Courtesy of the Lawrence Berkeley Laboratory, University of California, Berkeley.

Plate III.18 An engineer with the rectifier decks that are part of the high-voltage power supply for a new ion injector system at the Lawrence Berkeley Laboratory's Super-HILAC. The new injector, called Abel, extends the accelerator's capabilities to include high-intensity beams of heavy ions such as uranium. Courtesy of the Lawrence Berkeley Laboratory, University of California, Berkeley.

Plate III.19 The aluminum shell that sits on top of the rectifier decks contains a magnet, and an ion source, where an electrical arc strips away electrons from the atoms. The stripped atoms (called ions) now have a positive charge and can be accelerated by the electrical field in the accelerating columns (visible in its lucite enclosure). The ion beam then receives further acceleration, in a new Wideroe accelerator, by the SuperHILAC before being sent down to the Bevatron. When the SuperHILAC and the Bevatron act in tandem, as they often do, we refer to the combination as the Bevalac. Courtesy of the Lawrence Berkeley Laboratory, University of California, Berkeley.

Plate III.20   The installation of a new vacuum liner which, built in 9-foot sections, was placed in the bore in much the same way as batteries are inserted in a flashlight. The new ultra-high-vacuum upgrade permits the acceleration of uranium ions to energies close to the speed of light. Courtesy of the Lawrence Berkeley Laboratory, University of California, Berkeley.

IV. Aspects of Evolution

Plate IV.1   DNA-protein complex photographed with the electron microscope. The spherical bodies, isolated from the germ cells of a sea animal and magnified 77,500 times, are believed to consist of DNA in combination with protein. By permission of United Press International.

Plate IV.2   Ribonucleoprotein particles (ribosomes) from liver cells in a guinea pig. These particles are the main sites of the synthesis of proteins in the cell. By permission of J.F. Kirsch, Thesis, The Rockefeller University, New York, 1961.

Plate IV.3   Mitochondria, sometimes called "powerhouses of the cell" because they carry out energy-yielding chemical reactions. The mitochondria are the grey crescents around the black bodies, which are lipid droplets used as fuel for energy production. By permission of the Rockefeller Institute, New York (G.E. Palade).

Plate IV.4 Ribosomes, the tiny bodies in the cytoplasm of cells. These were separated from pancreas cells by a centrifuge and magnified about 100,000 times under the electron microscope. They are either free or attached to membrane-bound vesicles, called microsomes. By permission of The Rockefeller Institute, New York (G.E. Palade).

Plate IV.5 Chromosomes damaged by radiation. Some are broken, and one is coiled into a ring. Courtesy of the Brookhaven National Laboratory, New York.

Plate IV.6 Normal chromosomes of *Drosophila*. From Franklyn Branley, ed., *Scientist's Choice* (New York: Basic Books, n.d.). By permission of the publisher.

Plate IV.7   Mutations in fruit flies, shown here in the form of shriveled wings. The mutations were produced by exposure of the male parent to radiation. Courtesy of Brookhaven National Laboratory, New York.

Plate IV.8   In studies on radiation effects, young plants of the Better Times rose were exposed to 5,000 roentgens of gamma rays over 48-hour periods. When the plants flowered twelve months later, a number of mutations were observed: the flower at the bottom is an unstable mutant for the pink sector, and the one at the left is a stable pink mutant. At the right is the flower from an unirradiated control plant of the Better Times rose. Courtesy of Brookhaven National Laboratory, New York.

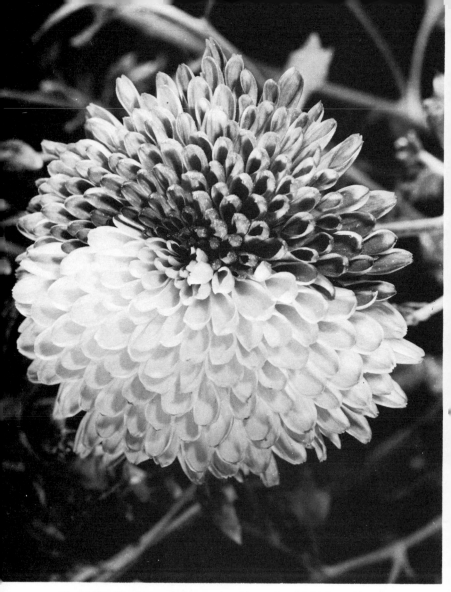

Plate IV.9   The chrysanthemum variety Masterpiece (pink), by accidental spontaneous mutation, produced a bronze "sport", which was named Bronze Masterpiece. Radiation treatment of Masterpiece has duplicated this same process with much higher frequency than occurs in nature. If the variety Bronze Masterpiece is irradiated, it can be caused to revert to the original pink Masterpiece. Courtesy of Brookhaven National Laboratory, New York.

Plate IV.10   Fossil of a bryozoan, a tiny, mosslike water animal, magnified about twenty times. It was brought up from an oil drillhole on Cape Hatteras. By permission of United Press International.

Plate IV.11   Fossil of a foraminifer, also found in a Cape Hatteras drillhole. Chalk and some limestones are composed mainly of the shells of these microscopic, one-celled animals. Notable examples are the White Cliffs of Dover and the stones used in the construction of the pyramids of Egypt. By permission of United Press International.

Plate IV.12   Fossil of a crinoid, or sea lily, a primitive animal of the echinoderm superphylum. This specimen was found in Indiana. Neg. No. 120809 (Photo: Thane Bierwert) Courtesy Department Library Services, American Museum of Natural History.

Plate IV.13 Tyrannosaurus rex, reconstructed from fossilized bones and displayed in the Cretaceous Hall of the American Museum of Natural History in New York City. This big carnivore preyed on dinosaurs with vegetarian diets. Courtesy Department Library Services, American Museum of Natural History.

Plate IV.14   Skeleton of a pterodactyl, an extinct flying reptile. Neg. No. 315134
(Photo: Charles H. Coles and Thane Bierwert) Courtesy Department Library
Services, American Museum of Natural History.

Plate IV.15 Cast of a coelacanth. This ancient fish was found still living in deep water near Madagascar. Courtesy Department Library Services, American Museum of Natural History.

Plate IV.16 An ancient ant delicately preserved in amber. Courtesy Department Library Services, American Museum of Natural History.

Plate IV.17   The evolution of the horse, illustrated by the skull and foot bones. Neg. No. 322448 (Photo: Baltin) Courtesy Department Library Services, American Museum of Natural History.

Plate IV.18 Skull of Pithecanthropus, as reconstructed by Franz Weidenreich. Neg. No. 120979 Courtesy Department Library Services, American Museum of Natural History.

Plate IV.19 Sinanthropus woman, as reconstructed by Franz Weidenreich and Lucile Swan. Neg. No. 322021 Courtesy Department Library Services, American Museum of Natural History.

Plate IV.20   Neanderthal man, according to a restoration by J. H. McGregor. Neg. No. 319951 (Photo: Alex J. Rota) Courtesy Department Library Services, American Museum of Natural History.

Plate IV.21   The original Univac, first of the large electronic computers. By permission of Remington Rand-Univac.

Plate IV.22   The large-scale computer recently ordered by the U.S. Air Force. These computers, with their extensive data communications capacity, will support Air Force combat-mission requirements for handling aircraft parts and inventories and maintenance operations throughout the world. Additionally, the new advanced microprocessor-based systems will perform a wide range of base personnel, financial, civil engineering, and administrative functions. Photo courtesy of Sperry Corporation.

Plate IV.23   A personal computer. This one consists of three basic components—the system unit, the monitor, and the keyboard. Photo courtesy of Sperry Corporation.

Plate IV.24 "Tot", 1982–83. This mobile, programmable, multilingual personal robot contains dual control modes and a sensory system; it acts as sentry and tells time. Designed by Jerome Hamlin; 36″ × 24″ × 12″. Courtesy of ComRo Inc., New York.

be so exposed, is twice that of the general public. In radiologists, who are medical specialists in the use of X rays, the incidence is ten times greater. It is no wonder that attempts are being made to substitute for X rays other techniques, such as those making use of ultrasonic sound. The coming of fission added new force to the danger. Whether in bombs or in power reactors, it unleashes radioactivity on a scale that could make the entire atmosphere, the oceans, and everything we eat, drink, or breathe increasingly dangerous to human life. Fission has introduced a form of pollution that will tax man's ingenuity to control.

When the uranium or plutonium atom splits, its *fission products* take various forms. The fragments may include isotopes of barium, or technetium, or any of a number of other possibilities. All told, some 200 different radioactive fission products have been identified. These are troublesome in nuclear technology, for some strongly absorb neutrons and place a damper on the fission reaction. For this reason, the fuel in a reactor must be removed and purified every once in a while.

In addition, these fission fragments are all dangerous to life in varying degrees, depending on the energy and nature of the radiation. Alpha particles taken into the body, for instance, are more dangerous than beta particles. The rate of decay also is important: a nuclide that breaks down rapidly will bombard the receiver with more radiation per second or per hour than one that breaks down slowly.

The rate of breakdown of a radioactive nuclide is something that can be spoken of only when large numbers of the nuclide are involved. An individual nucleus may break down at any time—the next instant or a billion years hence or any time in between—and there is no way of predicting when it will. Each radioactive species, however, has an average rate of breakdown, so if a large number of atoms is involved, it is possible to predict with great accuracy what proportion of them will break down in any unit of time. For instance, let us say that experiment shows that, in a given sample of an atom we shall call X, the atoms are breaking down at the rate of 1 out of 2 per year. At the end of a year, 500 of every 1,000 original X atoms in the sample would be left as X atoms; at the end of two years, 250; at the end of three years, 125; and so on. The time it takes for half of the original atoms to break down is called that particular atom's *half-life* (an expression introduced by Rutherford in 1904); consequently, the half-life of atom X is one year. Every radioactive nuclide has its own characteristic half-life, which never changes under ordinary conditions. (The only kind of outside influence that can change it is bombardment of the nucleus with a particle or the extremely high temperature in the interior of a star—in other words, a violent event capable of attacking the nucleus per se.)

The half-life of uranium 238 is 4.5 billion years. It is not surprising, therefore, that there is still uranium 238 left on Earth, despite the decay of uranium atoms. A simple calculation will show that it will take a period more than six times as long as the half-life to reduce a particular quantity of a radioactive nuclide to 1 percent of its original quantity. Even 30 billion years from now, there will still be two pounds of uranium left from each ton of it now in the earth's crust.

Although the isotopes of an element are practically identical chemically, they may differ greatly in their nuclear properties. Uranium 235, for instance, breaks down six times as fast as uranium 238; its half-life is only 710 million years. It can be reasoned, therefore, that in eons gone by, uranium was much richer in uranium 235 than it is today. Six billion years ago, for instance, uranium 235 would have made up about 70 percent of natural uranium. Humanity is not, however, just catching the tail end of the uranium 235. Even if we had been delayed another million years in discovering fission, the earth would still have 99.9 percent as much uranium 235 then as it has now.

Clearly any nuclide with a half-life of less than 100 million years would have declined to the vanishing point in the long lifetime of the universe. Hence, we cannot find more than traces of plutonium today. The longest-lived plutonium isotope, plutonium 244, has a half-life of only 70 million years.

The uranium, thorium, and other long-lived radioactive elements thinly spread through the rocks and soil produce small quantities of radiation, which is always present in the air about us. We humans are even slightly radioactive ourselves, for all living tissue contains traces of a comparatively rare, unstable isotope of potassium (potassium 40), which has a half-life of 1.3 billion years. (Potassium 40, as it breaks down, produces some argon 40 and probably accounts for the fact that argon 40 is by far the most common inert-gas nuclide existing on earth. Potassium-argon ratios have been used to test the age of meteorites.)

There is also a radioactive isotope of carbon, carbon 14, which would not ordinarily be expected to occur on Earth since its half-life is only 5,770 years. However, carbon 14 is continually being formed by the impact of cosmic-ray particles on nitrogen atoms of our atmosphere. The result is that there are always traces of carbon 14 present, so that some is constantly being incorporated into the carbon dioxide of the atmosphere. Because it is present in the carbon dioxide, it is incorporated by plants into their tissues and from there spreads to animal life, including ourselves.

The carbon 14 always present in the human body is far smaller in concentration than that of potassium 40; but carbon 14, having the shorter half-life by far, breaks down much more frequently. The total number of carbon-14 breakdowns may be about one-sixth that of potassium-40 breakdowns. However, a certain percentage of carbon 14 is contained in the human genes; and as these break down, profound changes may result in individual cells—changes that would not result in the case of potassium-40 breakdowns.

For this reason, it might well be argued that carbon 14 is the most significant radioactive atom to be found naturally in the human body. This likelihood was pointed out by the Russian-American biochemist Isaac Asimov as early as 1955.

The various naturally occurring radioactive nuclides and energetic radiations (such as cosmic rays and gamma rays) make up the *background radiation*. The constant exposure to natural radiation probably has played a part in evolution by producing mutations and may be partly responsible for the affliction of cancer. But living organisms have lived with it for billions of years. Nuclear radiation has become a serious hazard only in our own time, first as we began to experiment with radium, and then with the coming of fission and nuclear reactors.

By the time the Manhattan Project began, physicists had learned from painful experience the dangers of nuclear radiation. The workers in the project were therefore surrounded with elaborate safety precautions. The "hot" fission products and other radioactive materials were placed behind thick shielding walls and looked at only through lead glass. Instruments were devised to handle the materials by remote control. Each person was required to wear strips of photographic film or other detecting devices to "monitor" his or her accumulated exposure. Extensive animal experiments were carried out to estimate the *maximum permissible exposure*. (Mammals are more sensitive to radiation than are other forms of life; but as mammals, we have average resistance.)

Despite everything, accidents happened, and a few nuclear physicists died of *radiation sickness* from massive doses. Yet there are risks in every occupation, even the safest; the nuclear-energy workers have actually fared better than most, thanks to increasing knowledge of the hazards and care in avoiding them.

But a world full of nuclear power reactors, spawning fission products by the ton and the thousands of tons, will be a different story. How will all that deadly material be disposed of?

A great deal of it is short-lived radioactivity which fades away to harmlessness within a matter of weeks or months; it can be stored for that time and then dumped. Most dangerous are the nuclides with half-lives of one to thirty years. They are short-lived enough to produce intense radiation, yet long-lived enough to be hazardous for generations. A nuclide with a thirty-year half-life will take two centuries to lose 99 percent of its activity.

USING FISSION PRODUCTS

Fission products can be put to good use. As sources of energy, they can power small devices or instruments. The particles emitted by the radioactive isotope are absorbed, and their energy converted to heat, thus in turn producing electricity in thermocouples. Batteries that produce electricity in this fashion are radioisotope power generators, usually referred to as SNAP (Systems for Nuclear Auxiliary Power) or, more dramatically, as *atomic batteries*. They can be as light as 4 pounds, generate up to 60 watts, and last for years. SNAP batteries have been used in satellites—in *Transit 4A* and *Transit 4B*, for instance, which were put in orbit by the United States in 1961 to serve, ultimately, as navigational aids.

The isotope most commonly used in SNAP batteries is strontium 90, which will soon be mentioned in another connection. Isotopes of plutonium and curium are also used in some varieties.

The astronauts who landed on the moon placed such nuclear-powered generators on the surface to power a number of lunar experiments and radio transmission equipment. These continued working faultlessly for years.

Fission products might also have large potential uses in medicine (as in treatment of cancer), in killing bacteria and preserving food, and in many fields of industry, including chemical manufacturing. For instance, the Hercules Powder Company has designed a reactor to use radiation in the production of the antifreeze ethylene glycol.

Yet when all is said and done, no conceivable uses could employ more than a small part of the vast quantities of fission products that power reactors will discharge. This represents an important difficulty in connection with nuclear power generally. It is estimated that every 200,000 kilowatts of nuclear-produced electricity will involve the production of 1½ pounds of fission products per day. What to do with it? Already the United States has stored millions of gallons of radioactive liquid underground; and it is estimated that, by 2000 A.D., as much as half a million gallons of radioactive liquid will require disposal each day! Both the United States and Great Britain have dumped concrete containers of fission products at sea. There have been proposals to drop the radioactive wastes in oceanic abysses, to store them in old salt mines, to incarcerate them in molten glass, and bury the solidified material. But there is always the nervous thought that in one way or another the radioactivity will escape in time and contaminate the soil or the seas. One particularly haunting nightmare is the possibility that a nuclear-powered ship might be wrecked and spill its accumulated fission products into the ocean. The sinking of the American nuclear submarine U.S.S. *Thresher* in the North Atlantic on 10 April 1963, lent new substance to this fear, although in this case such contamination apparently did not take place.

FALLOUT

If radioactive pollution by peaceful nuclear energy is a potential danger, at least it will be kept under control, and probably successfully, by every possible means. But there is a pollution that has already spread over the world and that, indeed, in a nuclear war might be broadcast deliberately. This is the fallout from atomic bombs.

Fallout is produced by all nuclear bombs, even those not fired in anger. Because fallout is carried around the world by the winds and brought to earth by rainfall, it is virtually impossible for any nation to explode a nuclear bomb in the atmosphere without detection. In the event of a nuclear war, fallout in the long run might produce more casualties and do more damage to living things in the world at large than the fire and blast of the bombs themselves would wreak on the countries attacked.

Fallout is divided into three types: *local, tropospheric,* and *stratospheric.* Local fallout results from ground explosions in which radioactive isotopes are adsorbed on particles of soil and settle out quickly within 100 miles of the blast. Air blasts of fission bombs in the kiloton range send fission products into the troposphere. These settle out in about a month, being carried some thousands of miles eastward by the winds in that interval of time.

The huge output of fission products from the thermonuclear superbombs is carried into the stratosphere. Such stratospheric fallout takes a year or more to settle and is distributed over a whole hemisphere, falling eventually on the attacker as well as the attacked.

The intensity of the fallout from the first superbomb, exploded in the Pacific on 1 March 1954, caught scientists by surprise. They had not expected the fallout from a fusion bomb to be so "dirty." Seven thousand square miles were seriously contaminated—an area nearly the size of Massachusetts. But the reason became clear when scientists learned that the fusion core was supplemented with a blanket of uranium 238 that was fissioned by the neutrons. Not only did this multiply the force of the explosion, but it gave rise to a vastly greater cloud of fission products than a simple fission bomb of the Hiroshima type.

The fallout from the bomb tests to date has added only a small amount of radioactivity to the earth's background radiation. But even a small rise above the natural level may increase the incidence of cancer, cause genetic damage, and slightly shorten the average life expectancy. The most conservative estimators of the hazards agree that, by increasing the mutation rate (see chapter 13 for a discussion of mutations), fallout is storing up a certain amount of trouble for future generations.

One of the fission products is particularly dangerous for human life. This is strontium 90 (half-life, twenty-eight years), the isotope so useful in SNAP generators. Strontium 90 falling on the soil and water is taken up by plants and thereafter incorporated into the bodies of those animals (including man) that feed directly or indirectly on the plants. Its peculiar danger lies in the fact that strontium, because of its chemical similarity to calcium, goes to the bones and lodges there for a long time. The minerals in bone have a slow *turnover:* that is, they are not replaced nearly as rapidly as are the substances in the soft tissues. For that reason, strontium 90, once absorbed, may remain in the body for a major part of a person's lifetime (figure 10.5).

Strontium 90 is a brand-new substance in our environment; it did not exist on the earth in any detectable quantity until scientists fissioned the uranium atom. But today, within less than a generation, some strontium 90 has become incorporated in the bones of every human being on earth and, indeed, in all vertebrates. Considerable quantities of it are still floating in the stratosphere, sooner or later to add to the concentration in our bones.

The strontium-90 concentration is measured in *strontium units* (S.U.). One S.U. is 1 micromicrocurie of strontium 90 per gram of calcium in the body. A *curie* is a unit of radiation (named in honor of the Curies, of course) originally meant to be equivalent to that produced by 1 gram of radium in equilibrium with its breakdown product, radon, but is now more generally

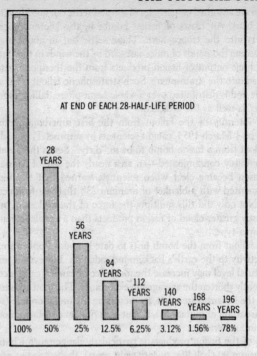

AT END OF EACH 28-HALF-LIFE PERIOD

28 YEARS

56 YEARS

84 YEARS

112 YEARS

140 YEARS

168 YEARS

196 YEARS

100%   50%   25%   12.5%   6.25%   3.12%   1.56%   .78%

Figure 10.5. Decay of strontium 90 over approximately 200 years.

accepted as meaning 37 billion disintegrations per second. A micromicrocurie is 1 trillionth of a curie, or 2.12 disintegrations per minute. A strontium unit would therefore mean 2.12 disintegrations per minute per gram of calcium present in the body.

The concentration of strontium 90 in the human skeleton varies greatly from place to place and among individuals. Some persons have been found to have as much as seventy-five times the average amount. Children average at least four times as high a concentration as adults, because of the higher turnover of material in their growing bones. Estimates of the averages themselves vary, because they are based mainly on estimates of the amounts of strontium 90 found in the diet. (Incidentally, milk is not a particularly hazardous food, from this point of view, because calcium obtained from vegetables has more strontium 90 associated with it. The cow's *filtration system* eliminates some of the strontium it gets in its plant fodder.) The estimates of the average strontium-90 concentration in the bones of people in the United States in 1959, before atmospheric nuclear explosions were banned, ranged from less than 1 strontium until to well over 5 strontium units. (The *maximum permissible* was established by the International Commission on Radiation Protection at 67 S.U.) But the averages mean little, particularly since strontium 90 may collect in *hot spots* in the bones and reach a high enough level there to initiate leukemia or cancer.

The importance of radiation effects has, among other things, resulted in the adoption of a number of types of unit designed to measure these effects. One such, the *roentgen*, named in honor of the discoverer of X rays, is based on the number of ions produced by the X rays or gamma rays being studied. More recently, the *rad* (short for "radiation") has been introduced. It represents the absorption of 100 ergs per gram of any type of radiation.

The nature of the radiation is of importance. A rad of massive particles is much more effective in inducing chemical change in tissues than a rad of light particles; hence, energy in the form of alpha particles is more dangerous than the same energy in the form of electrons.

Chemically, the damage done by radiation is caused chiefly by the breakdown of water molecules (which make up most of the mass of living tissue) into highly active fragments (*free radicals*) that, in turn, react with the complicated molecules in tissue. Damage to bone marrow, interfering with blood-cell production, is a particularly serious manifestation of radiation sickness, which, if far enough advanced, is irreversible and leads to death.

Many eminent scientists firmly believe that the fallout from the bomb tests is an important peril to the human race. The American chemist Linus Pauling has argued that the fallout from a single superbomb may lead to 100,000 deaths from leukemia and other diseases in the world, and he pointed out that radioactive carbon 14, produced by the neutrons from a nuclear explosion, constitutes a serious genetic danger. He was, for this reason, extremely active in pushing for cessation of testing of nuclear bombs; he endorsed all movements designed to lessen the danger of war and to encourage disarmament. On the other hand, some scientists, including the Hungarian-American physicist Edward Teller, minimized the seriousness of the fallout hazard. The sympathy of the world generally lies with Pauling, as might be indicated by the fact that he was awarded the Nobel Peace Prize in 1963.

In the fall of 1958, the United States, the Soviet Union, and Great Britain suspended bomb testing by a gentleman's agreement (which, however, did not prevent France from exploding its first nuclear bomb in the atmosphere in the spring of 1960). For three years, things looked rosy; the concentration of strontium 90 reached a peak and leveled off about 1960 at a point well below what is estimated to be the maximum consistent with safety. Even so, some 25 million curies of strontium 90 and cesium 137 (another dangerous fission product) had been delivered into the atmosphere during the thirteen years of nuclear testing when some 150 bombs of all varieties were exploded. Only two of these were exploded in anger, but the results were dire indeed.

In 1961, without warning, the Soviet Union ended the moratorium and began testing again. Since the U.S.S.R. exploded thermonuclear bombs of unprecedented power, the United States felt forced to begin testing again. World public opinion, sharpened and concentrated by the relief of the moratorium, reacted with great indignation.

On 10 October 1963, therefore, the three chief nuclear powers signed a partial test-ban treaty (*not* a mere gentleman's agreement) in which nuclear-bomb explosions in the atmosphere, in space, or underwater were banned. Only underground explosions were permitted since these do not produce fallout. This has been the most hopeful move in the direction of human survival since the opening of the Nuclear Age.

Controlled Nuclear Fusion

For more than thirty years, nuclear physicists have had in the back of their minds a dream even more attractive than turning fission to constructive uses: it is the dream of harnessing fusion energy. Fusion, after all, is the engine that makes our world go round: the fusion reactions in the sun are the ultimate source of all our forms of energy and of life itself. If somehow we could reproduce and control such reactions on the earth, all our energy problems would be solved. Our fuel supply would be as big as the ocean, for the fuel would be hydrogen.

Oddly enough, this would not be the first use of hydrogen as a fuel. Not long after hydrogen was discovered and its properties studied, it gained a place as a chemical fuel. The American scientist Robert Hare devised an oxyhydrogen torch in 1801, and the hot flame of hydrogen burning in oxygen has served industry ever since.

Liquid hydrogen has also been used as an immensely important fuel in rocketry, and there have even been suggestions about using hydrogen as a particularly clean fuel for the generation of electricity and in powering automobiles and similar vehicles. (In the latter cases, the problem of its ease of explosion in the air remains.) It is, however, as a nuclear-fusion fuel that the future beckons most glitteringly.

Fusion power would be immensely more convenient than fission power. Pound for pound, a fusion reactor would deliver about five to ten times as much power as a fission reactor. A pound of hydrogen, on fusion, could produce 35 million kilowatt-hours of energy. Furthermore, fusion would depend on hydrogen isotopes which could be easily obtained from the ocean in large quantities, whereas fission requires the mining of uranium and thorium —a comparatively much more difficult task. Then, too, while fusion produces such things as neutrons and hydrogen 3, these are not expected to be as dangerous to control as fission products are. Finally, and perhaps most important, a fusion reaction, in the event of any conceivable malfunction, would only collapse and go out, whereas a fission reaction might get out of hand (a *nuclear excursion*), produce a *meltdown* of its uranium (though this has never yet happened), and spread radioactivity dangerously.

If controlled nuclear fusion could be made feasible then, considering the availability of the fuel and the richness of the energy it would produce, it could provide a useful energy supply that could last billions of years—as long as the earth would last. The one dangerous result would then be *thermal pollution*—the general addition of fusion energy to the total heat arriving at the surface of the earth. This could raise the temperature slightly and have results similar to that of a greenhouse effect. This would also be true of solar power obtained from any source other than that of solar radiation reaching Earth in natural fashion. Solar power stations, operating in space, for instance, would add to the natural heat income of Earth's surface. In either case, humanity would have to limit its energy use or devise methods for getting rid of heat from Earth into space at more than the natural rate.

However, all this is of theoretical interest only if controlled nuclear fusion can be brought to the laboratory and then made a practical commercial process. After a generation of work, we have not yet reached that point.

Of the three isotopes of hydrogen, hydrogen 1 is the most common also the one most difficult to force into fusion. It is the particular fuel of the sun, but the sun has it by the trillions of cubic miles, together with an enormous gravity field to hold it together and central temperatures in the many millions of degrees. Only a tiny percentage of the hydrogen within the sun is fusing at any given moment; but given the vast mass present, even a tiny percentage is enough.

Hydrogen 3 is the easiest to bring to fusion, but it exists in such tiny quantities and can be made only at so fearful an expenditure of energy that it is hopeless to think of it, as yet, as a practical fuel all by itself.

That leaves hydrogen 2, which is easier to handle than hydrogen 1 and much more common that hydrogen 3. In all the hydrogen of the world, only one atom out of 6,000 is deuterium, but that is enough. There is 35 trillion tons of deuterium in the ocean, enough to supply man with ample energy for all the foreseeable future.

Yet there are problems. That might seem surprising, since fusion bombs exist. If we can make hydrogen fuse, why can't we make a reactor as well as a bomb? Ah, but to make a fusion bomb, we need to use a fission-bomb igniter and then let it go. To make a fusion reactor, we need a gentler igniter, obviously, and we must then keep the reaction going at a constant, controlled —and nonexplosive—rate.

The first problem is the less difficult. Heavy currents of electricity, high-energy sound waves, laser beams, and so on can all produce temperatures in the millions of degrees very briefly. There is no doubt that the required temperature will be reached.

Maintaining the temperature while keeping the (it is to be hoped) fusing hydrogen in place is something else. Obviously no material container can hold a gas at anything like a temperature of over 100 million degrees. Either the container would vaporize or the gas would cool. The first step toward a solution is to reduce the density of the gas to far below normal pressure, thus cutting down the heat content, though the energy of the particles remains high. The second step is a concept of great ingenuity. A gas at very high temperature has all the electrons stripped off its atoms; it is a *plasma* (a term introduced by Irving Langmuir in the early 1930s) made up of electrons and bare nuclei. Since it consists entirely of charged particles, why not use a strong magnetic field, taking the place of a material container, to hold it? The fact that magnetic fields could restrain charged particles and *pinch* a stream of them together had been known since 1907, when it was named the *pinch effect*. The *magnetic bottle* idea was tried and worked—but only for the briefest instant (figure 10.6). The wisps of plasma pinched in the bottle immediately writhed like a snake, broke up, and died out.

Figure 10.6. Magnetic bottle designed to hold a hot gas of hydrogen nuclei (a plasma). The ring is called a *torus*.

Another approach is to have a magnetic field stronger at the ends of the tube so that plasma is pushed back and kept from leaking. This is also found wanting, though it does not seem by much. If only plasma at 100 million degrees can be held in place for about a second, the fusion reaction would start, and energy would pour out of the system. That energy could be used to make the magnetic field firmer and more powerful and to keep the temperature at the proper level. The fusion reaction would then be self-sustaining, with the very energy it produced serving to keep it going. But to keep the plasma from leaking for just that little second is more than can be done as yet.

Since the plasma leakage takes place with particular ease at the end of the tube, why not remove the ends by giving the tube a doughnut shape? A particularly useful design is a doughnut-shaped tube (torus) twisted into a figure eight. This figure-eight device was first designed in 1951 by Spitzer and is called a *stellarator*. An even more hopeful device was designed by the Soviet physicist Lev Andreevich Artsimovich. It is called Toroidal Kamera Magnetic, which is abbreviated Tokamak.

American physicists are also working with Tokamaks and, in addition, with a device called Scyllac, which is designed to hold denser gas and therefore require a smaller containment period.

For nearly twenty years, physicists have been inching toward fusion power. Progress has been slow, but as yet no signs of a definite dead end have appeared.

Meanwhile, practical applications of fusion research are to be found. Plasma torches emitting jets at temperatures up to 50,000° C in absolute silence can far outdo ordinary chemical torches. And it is suggested that the plasma torch is the ultimate waste-disposal unit. In its flame everything—*everything*—would be broken down to its constituent elements, and all the elements would be available for recycling and for conversion into useful materials again.

PART II

The Biological Sciences

Chapter 11

The Molecule

Organic Matter

The term *molecule* (from a Latin word meaning "small mass") originally meant the ultimate, indivisible unit of a substance; and in a sense, it *is* an ultimate particle, because it cannot be broken down without losing its identity. To be sure, a molecule of sugar or of water can be divided into single atoms or groups, but then it is no longer sugar or water. Even a hydrogen molecule loses its characteristic chemical properties when it is broken down into its two component hydrogen atoms.

Just as the atom has furnished chief excitement in twentieth-century physics, so the molecule has been the subject of equally exciting discoveries in chemistry. Chemists have been able to work out detailed pictures of the structure of even very complex molecules, to identify the roles of specific molecules in living systems, to create elaborate new molecules, and to predict the behavior of a molecule of a given structure with amazing accuracy.

By the mid-twentieth century, the complex molecules that form the key units of living tissue, the proteins and nucleic acids, were being studied with all the techniques made possible by an advanced chemistry and physics. The two sciences *biochemistry* (the study of the chemical reactions going on in living tissue) and *biophysics* (the study of the physical forces and phenomena involved in living processes) merged to form a brand new discipline—*molecular biology*. Through the findings of molecular biology, modern science has in a single generation of effort all but wiped out the borderline between life and nonlife.

Yet less than a century and a half ago, the structure of not even the simplest molecule was understood. About all the chemists of the early nineteenth century could do was to separate all matter into two great categories. They had long been aware (even in the days of the alchemists) that substances fall into two sharply distinct classes with respect to their response to heat. One group—for example, salt, lead, water—remain basically unchanged after being heated. Salt might glow red hot when heated, lead might melt, water might vaporize—but when they are cooled back to the original temperature, they are restored to their original form, none the worse, apparently, for their experience. On the other hand, the second group of substances—for example, sugar, olive oil—are changed permanently by heat. Sugar becomes charred when heated and remains charred after being cooled again; olive oil is vaporized, and the vapor does not condense on cooling. Eventually the scientists noted that the heat-resisting substances generally came from the inanimate world of the air, ocean, and soil, while the combustible substances usually came from the world of life, either from living matter directly or from dead remains. In 1807, Berzelius, who invented the chemical symbols and was to prepare the first adequate list of atomic weights (see chapter 6) named the combustible substances *organic* (because they were derived, directly or indirectly, from the living organisms) and all the rest *inorganic*.

Early chemistry focused mainly on the inorganic substances. It was the study of the behavior of inorganic gases that led to the development of the atomic theory. Once that theory was established, it soon clarified the nature of inorganic molecules. Analysis showed that inorganic molecules generally consist of a small number of different atoms in definite proportions. The water molecule contains two atoms of hydrogen and one of oxygen; the salt molecule contains one atom of sodium and one of chlorine; sulfuric acid contains two atoms of hydrogen, one of sulfur, and four of oxygen; and so on.

When the chemists began to analyze organic substances, the picture seemed quite different. Two substances might have exactly the same composition and yet show distinctly different properties. (For instance, ethyl alcohol is composed of two carbon atoms, one oxygen atom, and six hydrogen atoms; so is dimethyl ether—yet one is a liquid at room temperature, while the other is a gas.) The organic molecules contained many more atoms than the simple inorganic ones, and there seemed to be no rhyme or reason in the way they were combined. Organic compounds simply could not be explained by the straightforward laws of chemistry that applied so beautifully to inorganic substances.

Berzelius decided that the chemistry of life was something apart which obeyed its own set of subtle rules. Only living tissue, he said, could make an organic compound. His point of view is an example of *vitalism*.

Then, in 1828, the German chemist Friedrich Wöhler, a student of Berzelius, produced an organic substance in the laboratory! He was heating a compound called *ammonium cyanate*, which was then generally considered inorganic. Wöhler was thunderstruck to discover that, on being heated, this material turned into a white substance identical in properties with *urea*, a component of urine. According to Berzelius's views, only living tissue could form urea; and yet Wöhler had formed it from inorganic material merely by applying a little heat.

Wöhler repeated the experiment many times before he dared publish his discovery. When he finally did, Berzelius and others at first refused to believe it. But other chemists confirmed the results. Furthermore, they proceeded to synthesize many other organic compounds from inorganic precursors. The first to bring about the production of an organic compound from its elements was the German chemist Adolph Wilhelm Hermann Kolbe, who produced *acetic acid* (the substance that gives vinegar its taste) in this fashion in 1845. It was this work that really killed Berzelius's version of vitalism. More and more it became clear that the same chemical laws applied to inorganic and organic molecules alike. Eventually the distinction between organic and inorganic substances was given a simple definition: all substances containing carbon (with the possible exceptions of a few simple compounds, such as carbon dioxide) are called organic; the rest are inorganic.

CHEMICAL STRUCTURE

To deal with the complex new chemistry, chemists needed a simple shorthand for representing compounds, and fortunately Berzelius had already suggested a convenient, rational system of symbols. The elements were designated by abbreviations of their Latin names. Thus C would stand for carbon, O for oxygen, H for hydrogen, N for nitrogen, S for sulfur, P for phosphorus, and so on. Where two elements began with the same letter, a second letter was used to distinguish them: for example, Ca for calcium, Cl for chlorine, Cd for cadmium, Co for cobalt, Cr for chromium, and so on. In only a comparatively few cases are the Latin or Latinized names (and initials) different from the English, thus: iron (*ferrum*) is Fe; silver (*argentum*), Ag; gold (*aurum*), Au; copper (*cuprum*), Cu; tin (*stannum*), Sn; mercury (*hydragyrum*) Hg; antimony (*stibium*), Sb; sodium (*natrium*), Na; and potassium (*kalium*), K.

With this system it is easy to symbolize the composition of a molecule. Water is written H_2O (thus indicating the molecule to consist of two hydrogen atoms and one oxygen atom); salt, $NaCl$; sulfuric acid, H_2SO_4, and so on. This is the *empirical formula* of a compound; it tells what the compound is made of but says nothing about its structure—that is, the manner in which the atoms of a molecule are connected.

In 1831, Baron Justus von Liebig, a co-worker of Wöhler's, went on to work out the composition of a number of organic chemicals, thus applying *chemical analysis* to the field of organic chemistry. He would carefully burn a small quantity of an organic substance and trap the gases formed (chiefly CO_2 and water vapor, H_2O) with appropriate chemicals. Then he would weigh the chemicals used to trap the combustion products to see how much weight had been added by the trapped products. From that weight he could determine the amount of carbon, hydrogen, and oxygen in the original substance. It was then an easy matter to calculate, from the atomic weights, the numbers of each type of atom in the molecule. In this way, for instance, he established that the molecule of ethyl alcohol had the formula C_2H_6O.

Liebig's method could not measure the nitrogen present in organic compounds; but in 1833, the French chemist Jean Baptiste André Dumas devised a combustion method that did collect the gaseous nitrogen released from substances. He made use of his methods to analyze the gases of the atmosphere with unprecedented accuracy in 1841.

The methods of *organic analysis* were made more and more delicate until veritable prodigies of refinement were reached in the *microanalytical* methods of the Austrian chemist Fritz Pregl. He devised techniques, beginning in 1909, for the accurate analysis of quantities of organic compounds barely visible to the naked eye and received the Nobel Prize for chemistry in 1923 in consequence.

Unfortunately, determining only the empirical formulas of organic compounds was not very helpful in elucidating their chemistry. In contrast to inorganic compounds, which usually consist of two or three atoms or at most a dozen, the organic molecules are often huge. Liebig found that the formula of morphine was $C_{17}H_{19}O_3N$, and of strychnine, $C_{21}H_{22}O_2N_2$.

Chemists were pretty much at a loss to deal with such large molecules or make head or tail of their formulas. Wöhler and Liebig tried to group atoms into smaller collections called *radicals* and to work out theories to show that various compounds were made up of specific radicals in different numbers and combinations. Some of the systems were most ingenious, but none really explained enough. It was particularly difficult to explain why two compounds with the same empirical formula, such as ethyl alcohol and dimethyl ether, should have different properties.

This phenomenon was first dragged into the light of day in the 1820s by Liebig and Wöhler. The former was studying a group of compounds called *fulminates;* the latter, a group called *isocyanates*—and the two turned out to have identical empirical formulas. The elements were present in equal parts, so to speak. Berzelius, the chemical dictator of the day, was told of this finding and was reluctant to believe it until, in 1830, he discovered some examples for himself. He named such compounds, with different properties but with elements present in equal parts, *isomers* (from Greek words meaning "equal parts"). The structure of organic molecules was indeed a puzzle in those days.

The chemists, lost in the jungle of organic chemistry, began to see daylight in the 1850s when they noted that each atom could combine with only a certain number of other atoms. For instance, the hydrogen atom apparently could attach itself to only one atom: it could form hydrogen chloride, HCl, but never HCl_2. Likewise chlorine and sodium could each take only a single partner, so they formed NaCl. An oxygen atom, on the other hand, could take two atoms as partners—for instance, H_2O. Nitrogen could take on three: for example, NH_3 (ammonia). And carbon could combine with as many as four: for example, CCl_4 (carbon tetrachloride).

In short, it looked as if each type of atom had a certain number of hooks by which it could hang on to other atoms. The English chemist Edward Frankland, in 1852, was the first to express this theory clearly, and he called these hooks *valence* bonds, from a Latin word meaning "power," to signify the combining powers of the elements.

The German chemist Friedrich August Kekulé von Stradonitz saw that if carbon were given a valence of 4, and if it were assumed that carbon atoms could use those valences, in part at least, to join up in chains, then a map could be drawn through the organic jungle. His technique was made more visual by the suggestion of a Scottish chemist, Archibald Scott Couper, that these combining forces between atoms (*bonds,* as they are usually called) be pictured in the form of small dashes. In this way, organic molecules could be built up like so many "Tinkertoy" structures.

In 1861, Kekulé published a textbook with many examples of this system, which proved its convenience and value. The *structural formula* became the hallmark of the organic chemist.

For instance, the methane (CH_4), ammonia (NH_3), and water (H_2O) molecules, respectively, can be pictured this way:

$$
\begin{array}{ccc}
\begin{array}{c} H \\ | \\ H-C-H \\ | \\ H \end{array}
&
\begin{array}{c} H \\ | \\ H-N-H \end{array}
&
H-O-H
\end{array}
$$

Organic molecules can be represented as chains of carbon atoms with hydrogen atoms attached along the sides. Thus, butane (C_4H_{10}) has the structure:

$$
\begin{array}{cccc}
H & H & H & H \\
| & | & | & | \\
H-C & -C & -C & -C-H \\
| & | & | & | \\
H & H & H & H
\end{array}
$$

Oxygen or nitrogen might enter the chain in the following manner, picturing the compounds methyl alcohol (CH_4O) and methylamine (CH_5N), respectively:

$$
\begin{array}{cc}
\begin{array}{c} H \\ | \\ H-C-O-H \\ | \\ H \end{array}
&
\begin{array}{c} H \quad H \\ | \quad | \\ H-C-N-H \\ | \\ H \end{array}
\end{array}
$$

An atom possessing more than one hook, such as carbon with its four, need not use each of them for a different atom: it might form a double or triple bond with one of its neighbors, as in ethylene (C_2H_4) or acetylene (C_2H_2):

$$\begin{array}{ccc} \text{H} & \text{H} & \\ | & | & \\ \text{H} - \text{C} = \text{C} - \text{H} & & \text{H} - \text{C} \equiv \text{C} - \text{H} \end{array}$$

Now it became easy to see how two molecules can have the same number of atoms of each element and still differ in properties. The two isomers must differ in the arrangement of those atoms. For instance, the structural formulas of ethyl alcohol and dimethyl ether, respectively, can be written:

$$\begin{array}{cccc} \text{H} & \text{H} & & \text{H} \qquad\quad \text{H} \\ | & | & & | \qquad\qquad | \\ \text{H} - \text{C} - \text{C} - \text{O} - \text{H} & & \text{H} - \text{C} - \text{O} - \text{C} - \text{H} \\ | & | & & | \qquad\qquad | \\ \text{H} & \text{H} & & \text{H} \qquad\quad \text{H} \end{array}$$

The greater the number of atoms in a molecule, the greater the number of possible arrangements and the greater the number of isomers. For instance, heptane, a molecule made up of seven carbon atoms and sixteen hydrogen atoms, can be arranged in nine different ways; in other words, there can be nine different heptanes, each with its own properties. These nine isomers resemble one another fairly closely, but it is only a family resemblance. Chemists have prepared all nine of these isomers but have never found a tenth— good evidence in favor of the Kekulé system.

A compound containing forty carbon atoms and eighty-two hydrogen atoms can exist in some 62.5 trillion arrangements, or isomers. And organic molecules of this size are by no means uncommon.

Only carbon atoms can hook to one another to form long chains. Other atoms do well if they can form a chain as long as half a dozen or so. Hence, inorganic molecules are usually simple and rarely have isomers. The greater complexity of the organic molecule introduces so many possibilities of isomerism that millions of organic compounds are known, new ones are being formed daily, and a virtually limitless number await discovery.

Structural formulas are now universally used as indispensable guides to the nature of organic molecules. As a short cut, chemists often write the formula of a molecule in terms of the groups of atoms, or radicals, that make it up, such as the methyl (CH_3) and methylene (CH_2) radicals. Thus the formula for butane can be written as $CH_3CH_2CH_2CH_3$.

The Details of Structure

In the latter half of the nineteenth century, chemists discovered a particularly subtle kind of isomerism which was to prove very important in the chemistry of life. The discovery emerged from the oddly asymmetrical effect that certain organic compounds had on rays of light passing through them.

OPTICAL ACTIVITY

A cross section of a ray of ordinary light will show that the innumerable waves of which it consists undulate in all planes—up and down, from side to side, and obliquely. Such light is called *unpolarized*. But when light passes through a crystal of the transparent substance called *Iceland spar*, for instance, it is refracted in such a way as to emerge *polarized*. It is as if the array of atoms in the crystal allows only certain planes of undulation to pass through (just as the palings of a fence might allow a person moving sideways to squeeze through but not one coming up broadside to them). There are devices, such as the *Nicol prism*, invented by the Scottish physicist William Nicol in 1829, that let light through in only one plane (figure 11.1). This has now been

Figure 11.1. The polarization of light. The waves of light normally oscillate in all planes (*top*). The Nicol prism (*bottom*) lets through the oscillations in only one plane, reflecting away the others. The transmitted light is plane-polarized.

replaced, for most purposes, by materials such as Polaroid (crystals of a complex of quinine sulfate and iodine, lined up with axes parallel and embedded in nitrocellulose), first produced in 1932 by Edwin Land.

Reflected light is often partly plane-polarized, as was first discovered in 1808 by the French physicist Étienne Louis Malus. (He invented the term *polarization* through the application of a remark of Newton's about the poles of light particles—one occasion where Newton was wrong—but the name remains anyway.) The glare of reflected light from windows of buildings and cars and even from paved highways can therefore be cut to bearable levels by the use of Polaroid sunglasses.

In 1815, the French physicist Jean Baptiste Biot had discovered that when plane-polarized light passes through quartz crystals, the plane of polarization is twisted: that is, the light goes in undulating in one plane and comes out undulating in a different plane. A substance that performs thus is said to display *optical activity.* Some quartz crystals twist the plane clockwise (*dextrorotation*) and some counterclockwise (*levorotation*). Biot found that certain organic compounds, such as camphor and tartaric acid, do the same thing. He thought it likely that some kind of asymmetry in the arrangement of the atoms in the molecules was responsible for the twisting of light. But for several decades, this suggestion remained purely speculative.

In 1844, Louis Pasteur (only twenty-two at the time) took up this interesting question. He studied two substances: tartaric acid and racemic acid. Both had the same chemical composition, but tartaric acid rotated the plane of polarized light, while racemic acid did not. Pasteur suspected that the crystals of salts of tartaric acid would prove to be asymmetric and those of racemic acid would be symmetric. Examining both sets of crystals under the microscope, he found to his surprise that both were asymmetric. But the racemate crystals had two versions of the asymmetry: half of them were the same shape as those of the tartrate, and the other half were mirror images. Half of the racemate crystals were left-handed and half right-handed, so to speak.

Pasteur painstakingly separated the left-handed racemate crystals from the right-handed and then dissolved each kind separately and sent light through each solution. Sure enough, the solution of the crystals possessing the same asymmetry as the tartrate crystals twisted the plane of polarized light just as the tartrate did, and by the same amount. Those crystals *were* tartrate. The other set twisted the plane of polarized light in the opposite direction, with the same amount of rotation. The reason the original racemate had shown no rotation of light, then, was that the two opposing tendencies canceled each other.

Pasteur next reconverted the two separated types of racemate salt to acid again by adding hydrogen ions to the respective solutions. (A *salt*, by the way, is a compound in which some hydrogen ions of the acid molecule are replaced by other positively charged ions, such as those of sodium or potassium.) He found that each of these racemic acids was now optically active—one rotating polarized light in the same direction as tartaric acid did (for it *was* tartaric acid), and the other in the opposite direction.

Other pairs of such mirror-image compounds (*enantiomorphs,* from Greek words meaning "opposite shapes") were found. In 1863, the German chemist Johannes Wislicenus found that lactic acid (the acid of sour milk) forms such a pair. Furthermore, he showed the properties of the two forms to be identical *except* for the action on polarized light. This property has turned out to be generally true of enantiomorphs.

So far, so good, but where did the asymmetry lie? What was there about the two molecules that made them mirror images of each other? Pasteur could not say. And although Biot, who had suggested the existence of molecular asymmetry, lived to be eighty-eight, he did not live long enough to see his intuition vindicated.

It was in 1874, twelve years after Biot's death, that the answer was finally presented. Two young chemists, a twenty-two-year-old Dutchman named

Jacobus Hendricus Van't Hoff and a twenty-seven-year-old Frenchman named Joseph Achille Le Bel, independently advanced a new theory of the carbon valence bonds that explained how mirror-image molecules could be constructed. (Later in his career, Van't Hoff studied the behavior of substances in solution and showed how the laws governing their behavior resembled the laws governing the behavior of gases. For this achievement he was the first man, in 1901, to be awarded the Nobel Prize in chemistry.)

Kekulé had drawn the four bonds of the carbon atom all in the same plane, not necessarily because this was the way they were actually arranged but because it was the convenient way of drawing them on a flat piece of paper. Van't Hoff and Le Bel now suggested a three-dimensional model in which the bonds were directed in two mutually perpendicular planes, two in one plane and two in the other. A good way to picture this model is to imagine the carbon atom as standing on any three of its bonds as legs, in which case the fourth bond points vertically upward. If you suppose the carbon atom to be at the center of a *tetrahedron* (a four-sided geometrical figure with triangular sides), then the four bonds point to the four vertexes of the figure. The model is therefore called the *tetrahedral carbon atom*. (see figure 11.2).

Figure 11.2. The tetrahedral carbon atom.

Now let us attach to these four bonds two hydrogen atoms, a chlorine atom, and a bromine atom. Regardless of which atom we attach to which bond, we will always come out with the same arrangement. Try it and see. With four toothpicks stuck into a marshmallow (the carbon atom) at the proper angles, you can represent the four bonds. Now suppose you stick two black olives (the hydrogen atoms), a green olive (chlorine), and a cherry (bromine) on the ends of the toothpicks in any order. Let us say that when you stand this on three legs with a black olive on the fourth pointing upward, the order on the three standing legs in the clockwise direction is black olive, green olive, cherry. You might now switch the green olive and cherry so that the order runs black olive, cherry, green olive. But all you need to do to see the same order as before is to turn the structure over so that the black olive serving as one of the supporting legs sticks up in the air and the one that was in the air rests on the table. Now the order of the standing legs again is black olive, green olive, cherry.

In other words, when at least two of the four atoms (or groups of atoms) attached to carbon's four bonds are identical, only one structural arrangement is possible. (Obviously this is true when three or all four of the attachments are identical.)

But when all four of the attached atoms (or groups of atoms) are different, the situation changes. Now two different structural arrangements are possible —one the mirror image of the other. For instance, suppose you stick a cherry on the upward leg and a black olive, a green olive, and a cocktail onion on the three standing legs. If you then switch the black olive and green olive so that the clockwise order runs green olive, black olive, onion, there is no way you can turn the structure to make the order come out black olive, green olive, onion, as it was before you made the switch. Thus with four different attachments you can always form two different structures, mirror images of each other. Try it and see.

Van't Hoff and Le Bel thus solved the mystery of the asymmetry of optically active substances. The mirror-image substances that rotated light in opposite directions were substances containing carbon atoms with four different atoms or groups of atoms attached to the bonds. One of the two possible arrangements of these four attachments rotated polarized light to the right; the other rotated it to the left.

More and more evidence beautifully supported Van't Hoff's and Le Bel's tetrahedral model of the carbon atom; and, by 1885, their theory (thanks, in part, to the enthusiastic support of the respected Wislicenus) was universally accepted.

The notion of three-dimensional structure also was applied to atoms other than carbon. The German chemist Viktor Meyer applied it successfully to nitrogen, while the English chemist William Jackson Pope applied it to sulfur, selenium, and tin. The German-Swiss chemist Alfred Werner added other elements and, indeed, beginning in the 1890s, worked out a *coordination theory* in which the structure of complex inorganic substances was explained by careful consideration of the distribution of atoms and atom groupings about some central atom. For this work, Werner was awarded the Nobel Prize in chemistry for 1913.

The two racemic acids that Pasteur had isolated were named "*d*-tartaric acid" (for "dextrorotatory") and "*l*-tartaric acid" (for "levorotatory"), and mirror-image structural formulas were written for them. But which was which? Which was actually the right-handed and which the left-handed compound? There was no way of telling at the time.

To provide chemists with a reference, or standard of comparison, for distinguishing right-handed and left-handed substances, the German chemist Emil Fischer chose a simple compound called *glyceraldehyde,* a relative of the sugars, which were among the most thoroughly studied of the optically active compounds. He arbitrarily assigned left-handedness to one form which he named "L-glyceraldehyde," and right-handedness to its mirror image, named "D-glyceraldehyde." His structural formulas for them were:

$$CHO$$
$$H - \underset{|}{\overset{|}{C}} - OH$$
$$CH_2OH$$

D-glyceraldehyde

$$CHO$$
$$HO - \underset{|}{\overset{|}{C}} - H$$
$$CH_2OH$$

L-glyceraldehyde

Any compound that could be shown by appropriate chemical methods (rather careful ones) to have a structure related to L-glyceraldehyde would be considered in the "L-series" and would have the prefix "L" attached to its name, regardless of whether it was levorotatory or dextrorotatory as far as polarized light was concerned. As it turned out, the levorotatory form of tartaric acid was found to belong to the D-series instead of the L-series. Nowadays, a compound that falls in the D-series structurally but rotates light to the left has its name prefixed by "D(−)"; similarly, we have "D(+)," "L(−)," and "L(+)."

This preoccupation with the minutiae of optical activity has turned out to be more than a matter of idle curiosity. As it happens, almost all the compounds occurring in living organisms contain asymmetric carbon atoms. And in every such case, the organism makes use of only one of the two mirror-image forms of the compound. Furthermore, similar compounds generally fall in the same series. For instance, virtually all the simple sugars found in living tissue belong to the D-series, while virtually all the amino acids (the building blocks of proteins) belong to the L-series.

In 1955, a Dutch chemist named Johannes Martin Bijvoet finally determined what structure tends to rotate polarized light to the left, and vice versa. It turned out that Fischer had, by chance, guessed right in naming the levorotatory and dextrorotatory forms.

THE PARADOX OF THE BENZENE RING

For some years after the secure establishment of the Kekulé system of structural formulas, one compound with a rather simple molecule resisted formulation. That compound was benzene (discovered in 1825 by Faraday). Chemical evidence showed it to consist of six carbon atoms and six hydrogen atoms. What happened to all the extra carbon bonds? (Six carbon atoms linked to one another by single bonds could hold fourteen hydrogen atoms, and they do in the well-known compound called hexane, C_6H_{14}.) Evidently the carbon atoms in benzene were linked by double or triple bonds. Thus, benzene might have a structure such as $CH \equiv C - CH = CH - CH = CH_2$. But the trouble was that the known compounds with that sort of structure had properties quite different from those of benzene. Besides, all the chemical evidence seemed to indicate that the benzene molecule was very symmetrical, and six carbons and six hydrogens could not be arranged in a chain in any reasonably symmetrical fashion.

In 1865, Kekulé himself came up with the answer. He related some years later that the vision of the benzene molecule came to him while he was riding on a bus and sunk in a reverie, half asleep. In his dream, chains of carbon atoms seemed to come alive and dance before his eyes, and then suddenly one coiled on itself like a snake. Kekulé awoke from his reverie with a start and could have cried "Eureka!" He had the solution: the benzene molecule is a ring.

Kekulé suggested that the six carbon atoms of the molecule are arranged as follows:

Here at last was the required symmetry. It explained, among other things, why the substitution of another atom for one of benzene's hydrogen atoms always yielded just one unvarying product. Since all the carbons in the ring are indistinguishable from one another in structural terms, no matter where you make the substitution for a hydrogen atom on the ring you will get the same product. Second, the ring structure showed that there are just three ways in which you can replace two hydrogen atoms on the ring: you can make the substitutions on two adjacent carbon atoms in the ring, on two separated by a single skip, or on two separated by a double skip. Sure enough, it was found that just three doubly substituted benzene isomers can be made.

Kekulé's blueprint of the benzene molecule, however, presented an awkward question. Generally, compounds with double bonds are more reactive, which is to say more unstable, than those with only single bonds. It is as if the extra bond were ready and more than willing to desert the attachment to the carbon atom and form a new attachment. Double-bonded compounds readily add on hydrogen or other atoms and can even be broken down without much difficulty. But the benzene ring is extraordinarily stable—more stable than carbon chains with only single bonds. (In fact, it is so stable and common in organic matter that molecules containing benzene rings make up an entire class of organic compounds, called *aromatic*, all the rest being lumped together as the *aliphatic* compounds.) The benzene molecule resists taking on more hydrogen atoms and is hard to break down.

The nineteenth-century organic chemists could find no explanation for this queer stability of the double bonds in the benzene molecule, and it disturbed them. The point may seem a small one, but the whole Kekulé system of structural formulas was endangered by the recalcitrance of the benzene molecule. The failure to explain this one conspicuous paradox made all the rest uncertain.

The closest approach to a solution prior to the twentieth century was that of the German chemist Johannes Thiele. In 1899, he suggested that when double bonds and single bonds alternate, the nearer ends of a pair of double bonds somehow neutralize each other and cancel each other's reactive nature. Consider, as an example, the compound *butadiene*, which contains, in simplest form, the case of two double bonds separated by a single bond (*conjugated double bonds*). Now if two atoms are added to the compound, they add onto the end carbons, as shown in the following formula. Such a view explained the nonreactivity of benzene, since the three double bonds of the benzene rings, being arranged in a ring, neutralize each other completely.

$$CH_2 = CH - CH = CH_2$$

......atoms add......
here

Some forty years later, a better answer was found by way of the new theory of chemical bonds that pictured atoms as linked by sharing electrons.

The chemical bond, which Kekulé had drawn as a dash between atoms, came to be looked upon as representing a shared pair of electrons (see chapter 6). Each atom that forms a combination with a partner shares one of its electrons with the partner, and the partner reciprocates by donating one of *its* electrons to the bond. Carbon, with four electrons in its outer shell, could form four attachments; hydrogen could donate its one electron to a bond with one other atom; and so on.

Now the question arose: How are the electrons shared? Obviously, two carbon atoms share the pair of electrons between them equally, because each atom has an equal hold on electrons. On the other hand, in a combination such as H_2O, the oxygen atom, which has a stronger hold on electrons than a hydrogen atom, takes possession of the greater share of the pair of electrons it has in common with each hydrogen atom. Hence, the oxygen atom, by virtue of its excessive portion of electrons, has a slight excess of negative charge. By the same token, the hydrogen atom, suffering from an electron deficiency, has a slight excess of positive charge. A molecule containing an oxygen-hydrogen pair, such as water or ethyl alcohol, possesses a small concentration of negative charge in one part of the molecule and a small concentration of positive charge in another. It possesses two poles of charge, so to speak, and is called a *polar molecule*.

This view of molecular structure was first proposed in 1912 by Peter Debye, (who later suggested the magnetic method of attaining very low temperatures; see chapter 6). He used an electric field to measure the amount of separation of poles of electric charge in a molecule. In such a field, polar molecules line themselves up with the negative ends pointing toward the positive pole and the positive ends toward the negative pole, and the ease with which this is done is the measure of the *dipole moment* of the molecule. By the early 1930s, measurements of dipole moments had become routine; and in 1936, for this and other work, Debye was awarded the Nobel Prize in chemistry.

The new picture explained a number of things that earlier views of molecular structure could not explain—for instance, some anomalies of the boiling points of substances. In general, the greater the molecular weight, the higher the boiling point. But this rule is commonly broken. Water, with a molecular weight of only 18, boils at 100° C, whereas propane, with more than twice this molecular weight (44), boils at the much lower temperature of −42° C. Why the difference? The answer is that water is a polar molecule with a high dipole moment, while propane is *nonpolar*—it has no poles of charge. Polar molecules tend to orient themselves with the negative pole of one molecule adjacent to the positive pole of its neighbor. The resulting electrostatic attraction between neighboring molecules makes it harder to tear the molecules apart, and substances have relatively high boiling points. Hence, ethyl alcohol has a much higher boiling point (78° C) than its isomer dimethyl ether, which boils at −24° C, although both substances have the same molecular weight (46). Ethyl alcohol has a large dipole moment, and dimethyl ether only a small one. Water has a dipole moment even larger than that of ethyl alcohol.

When de Broglie and Schrödinger formulated the new view of electrons not as sharply defined particles but as packets of waves (see chapter 8), the idea of the chemical bond underwent a further change. In 1939, the American chemist Linus Pauling presented a quantum-mechanical concept of molecular bonds in a book entitled *The Nature of the Chemical Bond.* His theory finally explained, among other things, the paradox of the stability of the benzene molecule.

Pauling pictured the electrons that form a bond as *resonating* between the atoms they join. He showed that under certain conditions it is necessary to view an electron as occupying any one of a number of positions (with varying probability). The electron, with its wavelike properties, might then best be presented as being spread out into a kind of blur, representing the weighted average of the individual probabilities of position. The more evenly the electron is spread out, the more stable the compound. Such *resonance stabilization* was most likely to occur when the molecule possesses conjugated bonds in one plane and when the existence of symmetry allows a number of alternative positions for the electron (viewed as a particle). The benzene ring is planar and symmetrical, and Pauling showed that the bonds of the ring were not really double and single in alternation, but that the electrons were smeared out, so to speak, into an equal distribution which results in all the bonds being alike and in all being stronger and less reactive than ordinary single bonds.

The resonance structures, though they explain chemical behavior satisfactorily, are difficult to present in simple symbolism on paper. Therefore the old Kekulé structures, although now understood to represent only approximations of the actual electronic situation, are still universally used and will undoubtedly continue to be used through the foreseeable future.

Organic Synthesis

After Kolbe had produced acetic acid, there came in the 1850s a chemist who went systematically and methodically about the business of synthesizing organic substances in the laboratory. He was the Frenchman Pierre Eugène Marcelin Berthelot. He prepared a number of simple organic compounds from still simpler inorganic compounds such as carbon monoxide. Berthelot built his simple organic compounds up through increasing complexity until he finally had ethyl alcohol, among other things. It was *synthetic ethyl alcohol*, to be sure, but absolutely indistinguishable from the "real thing," because it *was* the real thing.

Ethyl alcohol is an organic compound familiar to all and highly valued by most. No doubt the thought that the chemist could make ethyl alcohol from coal, air, and water (coal to supply the carbon, air the oxygen, and water the hydrogen), without the necessity of fruits or grain as a starting point, must have created enticing visions and endowed the chemist with a new kind of reputation as a miracle worker. At any rate, it put organic synthesis on the map.

For chemists, however, Berthelot did something even more significant. He began to form products that did not exist in nature. He took *glycerol*, a compound discovered by Scheele in 1778 and obtained from the breakdown of the fats of living organisms, and combined it with acids not known to occur naturally in fats (although they occur naturally elsewhere). In this way he obtained fatty substances that were not quite like those that occur in organisms.

Thus Berthelot laid the groundwork for a new kind of organic chemistry—the synthesis of molecules that nature cannot supply. This meant the possible formation not only of a kind of *synthetic* which might be a substitute—perhaps an inferior substitute—for some natural compound that is hard or impossible to get in the needed quantity, but are also of *synthetics* which are improvements on anything in nature.

This notion of improving on nature in one fashion or another, rather than merely supplementing it, has grown to colossal proportions since Berthelot showed the way. The first fruits of the new outlook were in the field of dyes.

THE FIRST SYNTHESIS

The beginnings of organic chemistry were in Germany. Wöhler and Liebig were both German, and other men of great ability followed them. Before the middle of the nineteenth century, there were no organic chemists in England even remotely comparable to those in Germany. In fact, English schools had so low an opinion of chemistry that they taught the subject only during the lunch recess, not expecting (or even perhaps desiring) many students to be interested. It is odd, therefore, that the first feat of synthesis with worldwide repercussions was actually carried through in England.

It came about in this way. In 1845, when the Royal College of Science in London finally decided to give a good course in chemistry, it imported a young German to do the teaching. He was August Wilhelm von Hofmann, only twenty-seven at the time, and he was hired at the suggestion of Queen Victoria's husband, the Prince Consort Albert (who was himself of German birth).

Hofmann was interested in a number of things, among them coal tar, which he had worked with on the occasion of his first research project under Liebig. Coal tar is a black, gummy material given off by coal when it is heated strongly in the absence of air. The tar is not an attractive material, but it is a valuable source of organic chemicals. In the 1840s, for instance, it served as a source of large quantities of reasonably pure benzene and of a nitrogen-containing compound called *aniline*, related to benzene, which Hofmann had been the first to obtain from coal tar.

About ten years after he arrived in England, Hofmann came across a seventeen-year-old boy studying chemistry at the college. His name was William Henry Perkin. Hofmann had a keen eye for talent and knew enthusiasm when he saw it. He took on the youngster as an assistant and set him to work on coal-tar compounds. Perkin's enthusiasm was tireless. He set up a laboratory in his home and worked there as well as at school.

Hofmann, who was also interested in medical applications of chemistry, mused aloud one day in 1856 on the possibility of synthesizing quinine, a natural substance used in the treatment of malaria. Now those were the days before structural formulas had come into their own. The only thing known about quinine was its atomic composition, and no one at the time had any idea of just how complicated its structure is. (It was not till 1908 that the structure was correctly deduced.)

Blissfully ignorant of its complexity, Perkin, at the age of eighteen, tackled the problem of synthesizing quinine. He began with allyltoluidine, one of his coal-tar compounds. This molecule seemed to have about half the numbers of the various types of atoms that quinine has in its molecule. If he put two of these molecules together and added some missing oxygen atoms (say, by mixing in some potassium dichromate, known to add oxygen atoms to chemicals with which it is mixed), Perkin thought he might get a molecule of quinine.

Naturally this approach got Perkin nowhere. He ended with a dirty, red-brown goo. Then he tried aniline in place of allyltoluidine and got a blackish goo. This time, though, it seemed to him that he caught a purplish glint in it. He added alcohol to the mess, and the colorless liquid turned a beautiful purple. At once Perkin thought of the possibility of his having discovered something that might be useful as a dye.

Dyes had always been greatly admired, and expensive, substances. There were only a handful of good dyes—dyes that stained fabric permanently and brilliantly and did not fade or wash out. There was dark blue indigo, from the indigo plant and the closely related *woad* for which Britain was famous in early Roman times; there was *Tyrian purple*, from a snail (so called because ancient Tyre grew rich on its manufacture; in the later Roman Empire, the royal children were born in a room with hangings dyed with Tyrian purple, whence the phrase "born to the purple"); and there was reddish *alizarin*, from the madder plant (*alizarin* came from Arabic words meaning "the juice"). To these inheritances from ancient and medieval times, later dyers had added a few tropical dyes and inorganic pigments (today used chiefly in paints).

Hence, Perkin's excitement about the possibility that his purple substance might be a dye. At the suggestion of a friend, he sent a sample to a firm in Scotland which was interested in dyes, and quickly the answer came back that the purple compound had good properties. Could it be supplied cheaply? Perkin proceeded to patent the dye (there was considerable argument about whether an eighteen-year-old could obtain a patent, but eventually he obtained it), to quit school, and to go into business.

His project was not easy. Perkin had to start from scratch, preparing his own starting materials from coal tar with equipment of his own design. Within six months, however, he was producing what he named *aniline purple*—a compound not found in nature and superior to any natural dye in its color range.

French dyers, who took to the new dye more quickly than did the more conservative English, named the color *mauve*, from the mallow (Latin *malva*); and the dye itself came to be known as *mauveine*. Quickly it became the rage (the period being sometimes referred to as the Mauve Decade), and Perkin grew rich. At the age of twenty-three, he was the world authority on dyes.

The dam had broken. A number of organic chemists, inspired by Perkin's astonishing success, went to work synthesizing dyes, and many succeeded. Hofmann himself turned to this new field and, in 1858, synthesized a red-purple dye which was later given the name *magenta* by the French dyers (then arbiters of the world's fashions). The dye was named for the Italian city where the French defeated the Austrians in a battle in 1859.

Hofmann returned to Germany in 1865, carrying his new interest in dyes with him. He discovered a group of violet dyes still known as *Hofmann's violets*. By the mid-twentieth century, no less than 3,500 synthetic dyes were in commercial use.

Chemists also synthesized the natural dyestuffs in the laboratory. Karl Graebe of Germany and Perkin both synthesized alizarin in 1869 (Graebe applying for the patent one day sooner than Perkin); and in 1880, the German chemist Adolf von Baeyer worked out a method of synthesizing indigo. (For his work on dyes, von Baeyer received the Nobel Prize in chemistry in 1905.)

Perkin retired from business in 1874, at the age of thirty-five, and returned to his first love—research. By 1875, he had managed to synthesize *coumarin* (a naturally occurring substance which has the pleasant odor of new-mown hay)—and thus began the synthetic perfume industry.

Perkin alone could not maintain British supremacy against the great development of German organic chemistry; and by the turn of the century, "synthetics" became almost a German monopoly. It was a German chemist, Otto Wallach, who carried on the work on synthetic perfumes that Perkin had started. In 1910, Wallach was awarded the Nobel Prize in chemistry for his investigations. The Croatian chemist Leopold Ruzicka, teaching in Switzerland, first synthesized musk, an important component of perfumes. He shared the Nobel Prize in chemistry in 1938. However, during the First World War, Great Britain and the United States, shut off from the products of the German chemical laboratories, were forced to develop chemical industries of their own.

ALKALOIDS AND PAIN DEADENERS

Achievements in synthetic organic chemistry could not have proceeded at anything better than a stumbling pace if chemists had had to depend upon fortunate accidents such as the one that had been seized upon by Perkin. Fortunately the structural formulas of Kekulé, presented three years after Perkin's discovery, made it possible to prepare blueprints, so to speak, of the organic molecule. No longer did chemists have to prepare quinine by sheer guesswork and hope; they had methods for attempting to scale the structural heights of the molecule step by step, with advance knowledge of where they were headed and what they might expect.

Chemists learned how to alter one group of atoms to another; to open up rings of atoms and to form rings from open chains; to split groups of atoms in two; and to add carbon atoms one by one to a chain. The specific method of doing a particular architectural task within the organic molecule is still often referred to by the name of the chemist who first described the details. For instance, Perkin discovered a method of adding a two-carbon atom group by heating certain substances with chemicals named *acetic anhydride* and *sodium acetate*. This is still called the *Perkin reaction*. Perkin's teacher, Hofmann, discovered that a ring of atoms which included a nitrogen could be treated with a substance called *methyl iodide* in the presence of a silver compound in such a way that the ring was eventually broken and the nitrogen atom removed. This is the *Hofmann degradation*. In 1877, the French chemist Charles Friedel, working with the American chemist James Mason Crafts, discovered a way of attaching a short carbon chain to a benzene ring by the use of heat and aluminum chloride. This is now known as the *Friedel-Crafts reaction*.

In 1900, the French chemist Victor Grignard discovered that magnesium metal, properly used, could bring about a rather large variety of different joinings of carbon chains; he presented the discovery in his doctoral dissertation. For the development of these *Grignard reactions* he shared in the Nobel Prize in chemistry in 1912. The French chemist Paul Sabatier, who shared it with him, had discovered (with Jean Baptiste Senderens) a method of using finely divided nickel to bring about the addition of hydrogen atoms in those places where a carbon chain possessed a double bond. This is the *Sabatier-Senderens reduction*.

In 1928, the German chemists Otto Diels and Kurt Alder discovered a method of adding the two ends of a carbon chain to opposite ends of a double bond in another carbon chain, thus forming a ring of atoms. For the discovery of this *Diels-Alder reaction*, they shared the Nobel Prize for chemistry in 1950.

In other words, by noting the changes in the structural formulas of substances subjected to a variety of chemicals and conditions, organic chemists worked out a slowly growing set of ground rules on how to change one compound into another at will. It was not easy. Every compound and every change had its own peculiarities and difficulties. But the main paths were blazed, and the skilled organic chemist found them clear signs toward progress in what had formerly seemed a jungle.

Knowledge of the manner in which particular groups of atoms behave could also be used to work out the structure of unknown compounds. For instance, when simple alcohols react with metallic sodium and liberate hydrogen, only the hydrogen linked to an oxygen atom is released, not the hydrogens linked to carbon atoms. On the other hand, some organic compounds will take on hydrogen atoms under appropriate conditions while others will not. It turns out that compounds that add hydrogen generally possess double or triple bonds and add the hydrogen at those bonds. From such information a whole new type of chemical analysis of organic compounds arose; the nature of the atom groupings was determined, rather than just the numbers and kinds of various atoms present. The liberation of hydrogen by the addition of sodium signified the presence of an oxygen-bound hydrogen atom in the compound; the acceptance of hydrogen meant the presence of double or triple bonds. If the molecule was too complicated for analysis as a whole, it could be broken down into simpler portions by well-defined methods; the structures of the simpler portions could be worked out and the original molecule deduced from those.

Using the structural formula as a tool and guide, chemists could work out the structure of some useful naturally occurring organic compound (analysis) and then set about duplicating it or something like it in the laboratory (synthesis). One result was that something which was rare, expensive or difficult to obtain in nature might become cheaply available in quantity in the laboratory. Or, as in the case of the coal-tar dyes, the laboratory might create something that fulfilled a need better than did similar substances found in nature.

One startling case of a deliberate improvement on nature involves cocaine, found in the leaves of the coca plant, which is native to Bolivia and Peru but is now grown chiefly in Java. Like the compounds strychnine, morphine, and quinine, all mentioned earlier, cocaine is an example of an *alkaloid*, a nitrogen-containing plant product that, in small concentration, has profound physiological effects on man. Depending on the dose, alkaloids can cure or kill. The most famous alkaloid death of all times was that of Socrates, who was killed by *coniine*, an alkaloid in hemlock.

The molecular structure of the alkaloids is, in some cases, extraordinarily complicated, but that just sharpened chemical curiosity. The English chemist Robert Robinson tackled the alkaloids systematically. He worked out the structure of morphine (for all but one dubious atom) in 1925, and the structure of strychnine in 1946. He received the Nobel Prize for chemistry in 1947 as recognition of the value of his work.

Robinson had merely worked out the structure of alkaloids without using that structure as a guide to their synthesis. The American chemist Robert Burns Woodward took care of that. With his American colleague William von Eggers Doering, he synthesized quinine in 1944. It was the wild-goose chase after this particular compound by Perkin that had had such tremendous results. And, if you are curious, here is the structural formula of quinine:

No wonder it stumped Perkin.

That Woodward and von Doering solved the problem is not merely a tribute to their brilliance. They had at their disposal the new electronic theories of molecular structure and behavior worked out by men such as Pauling. Woodward went on to synthesize a variety of complicated molecules which had, before his time, represented hopeless challenges. In 1954, for instance, he synthesized strychnine.

Long before the structure of the alkaloids had been worked out, however, some of them—notably cocaine—were of intense interest to medical men. The South American Indians, it had been discovered, would chew coca leaves, finding it an antidote to fatigue and a source of happiness-sensation. The Scottish physician Robert Christison introduced the plant to Europe. (This is not the only gift to medicine on the part of the witch doctors and herb women of prescientific societies. There are also quinine and strychnine, already mentioned, as well as opium, digitalis, curare, atropine, strophanthidin, and reserpine. In addition, the smoking of tobacco, the chewing of betel nuts, the drinking of alcohol, and the taking of such drugs as marijuana and peyote are all inherited from primitive societies.)

Cocaine was not merely a general happiness-producer. Doctors discovered that it deadened the body, temporarily and locally, to sensations of pain. In 1884, the American physician Carl Koller discovered that cocaine could be used as a pain deadener when added to the mucous membranes around the eye. Eye operations could then be performed without pain. Cocaine could also be used in dentistry, allowing teeth to be extracted without pain.

This effect fascinated doctors, for one of the great medical victories of the nineteenth century had been that over pain. In 1799, Humphry Davy had prepared the gas *nitrous oxide* (N_2O) and studied its effects. He found that when it was inhaled, it released inhibitions so that anyone breathing it would laugh, cry, or otherwise act foolishly. Its common name is *laughing gas,* for that reason.

In the early 1840s, an American scientist, Gardner Quincy Cotton, discovered that nitrous oxide deadened the sensation of pain; and, in 1844, an American dentist, Horace Wells, used it in dentistry. By that time, something better had entered the field.

The American surgeon Crawford Williamson Long in 1842 had used ether to put a patient to sleep during tooth extractions. In 1846, the American dentist William Thomas Green Morton conducted a surgical operation under ether at the Massachusetts General Hospital. Morton usually gets the credit for the discovery, because Long did not describe his feat in the medical journals until after Morton's public demonstration, and Wells's earliest public demonstrations with nitrous oxide had been only indifferent successes.

The American poet and physician Oliver Wendell Holmes suggested that pain-deadening compounds be called *anesthetics* (from Greek words meaning "no feeling"). Some people at the time felt that anesthetics were a sacrilegious attempt to avoid pain inflicted on human beings by God; but if anything was needed to make anesthesia respectable, it was its use by the Scottish physician James Young Simpson for Queen Victoria of England during childbirth.

Anesthesia had finally converted surgery from torture-chamber butchery to something that was at least humane and, with the addition of antiseptic conditions, even life-saving. For that reason, any further advance in anesthesia was seized on with great interest. Cocaine's special interest was that it was a *local anesthetic,* deadening pain in a specific area without inducing general unconsciousness and lack of sensation, as in the case of such *general anesthetics* as ether.

There are several drawbacks to cocaine, however. In the first place, it can induce troublesome side effects and can even kill patients sensitive to it. Second, it can bring about addiction and has to be used skimpily and with caution. (Cocaine is one of the dangerous *drugs* that deaden not only pain but other unpleasant sensations and give a user the illusion of euphoria. The user may become habituated to the drug so that he may require increasing doses and, despite the actual bad effect upon his body, become so dependent on the illusions the drug carries with it that he cannot stop using it without developing painful *withdrawal symptoms.* Such *drug addiction* for cocaine and other drugs of this sort is an important social problem. Up to twenty tons of cocaine are produced illegally each year and sold with tremendous profits to a few and tremendous misery to many, despite worldwide efforts to stop the traffic.) Third, the molecule is fragile, and heating cocaine to sterilize it of any bacteria leads to changes in the molecule that interfere with its anesthetic effects.

The structure of the cocaine molecule is rather complicated:

$$\begin{array}{c} O \\ \parallel \\ C-O-CH_3 \\ / \\ CH-CH \\ CH_2 \quad / \qquad \backslash \\ \mid \quad N-CH_3 \quad CH-O-C \overset{O}{\underset{}{\diagup}} \\ CH_2 \quad \backslash \qquad / \qquad \quad C = CH \\ CH-CH_2 \qquad \mid \qquad CH \\ \qquad CH \qquad \parallel \\ \qquad \quad CH = CH \end{array}$$

The double ring on the left is the fragile portion, and that is the difficult one to synthesize. (The synthesis of cocaine was not achieved until 1923, when Richard Willstätter managed it.) However, it occurred to chemists that they might synthesize similar compounds in which the double ring was not closed, and so make the compound both easier to form and more stable. The synthetic substance might possess the anesthetic properties of cocaine, perhaps without the undesirable side effects.

For some twenty years, German chemists tackled the problem, turning out dozens of compounds, some of which were pretty good. The most successful modification was obtained in 1909, when a compound with the following formula was prepared:

$$\begin{array}{c} CH_2 \\ CH_3 \diagdown \mid \\ N \qquad CH_2-O-C \overset{O}{\underset{}{\diagup}} \\ CH_3 \diagup \mid \qquad \mid \qquad \quad C = CH \\ CH_2 \quad CH_2 \qquad \mid \qquad CH \\ \qquad \qquad CH \qquad \parallel \\ \qquad \qquad \quad CH \quad \diagup \quad C \\ \qquad \qquad \qquad NH_2 \end{array}$$

Compare this with the formula for cocaine, and you will see the similarity and also the important fact that the double ring no longer exists. This simpler molecule—stable, easy to synthesize, with good anesthetic properties and little in the way of side effects—does not exist in nature. It is a *synthetic substitute* far better than the real thing. It is called *procaine*, but is better known to the public by the trade-name Novocaine.

Perhaps the most effective and best-known of the general pain deadeners is *morphine*. Its very name is from the Greek word for "sleep." It is a purified derivative of the opium juice or *laudanum* used for centuries by peoples, both civilized and primitive, to combat the pains and tension of the workaday world. As a gift to the pain-wracked, it is heavenly, but it, too, carries the deadly danger of addiction. An attempt to find a substitute backfired. In 1898, a synthetic derivative, *diacetylmorphine*—better known as *heroin*—was introduced in the belief that it would be safer. Instead, it turned out to be the most dangerous drug of all.

Less dangerous *sedatives* (sleep inducers) are *chloral hydrate* and, particularly, the *barbiturates*. The first example of this latter group was introduced in 1902, and they are now the most common constituents of *sleeping pills*. Harmless enough when used properly, they can nevertheless induce addiction, and an overdose can cause death. In fact, because death comes quietly as the end product of a gradually deepening sleep, barbiturate overdosage is a rather popular method of suicide, or attempted suicide.

The most common sedative, and the longest in use, is, of course, alcohol. Methods of fermenting fruit juice and grain were known in prehistoric times. Distillation to produce stronger liquors than could be formed naturally was introduced in the Middle Ages. The value of light wines in areas where the water supply is nothing but a short cut to typhoid fever and cholera, and the social acceptance of drinking in moderation, make it difficult to treat alcohol as the drug it is, although it induces addiction as surely as morphine and, through sheer quantity of use, does much more harm. Legal prohibition of sale of liquor seems to be unhelpful; certainly the American experiment of Prohibition (1920–33) was a disastrous failure. Nevertheless, alcoholism is increasingly being treated as the disease it is rather than as a moral disgrace. The acute symptoms of alcoholism (*delirium tremens*) are probably not so much due to the alcohol itself as to the vitamin deficiencies induced in those who eat little while drinking much.

THE PROTOPORPHYRINS

Man now has at his disposal all sorts of synthetics of great potential use and misuse: explosives, poison gases, insecticides, weed-killers, antiseptics, disinfectants, detergents, drugs—almost no end of them, really. But synthesis is not merely the handmaiden of consumer needs. It can also be placed at the service of pure chemical research.

It often happens that a complex compound, produced either by living tissue or by the apparatus of the organic chemist, can be assigned only a tentative molecular structure, after all possible deductions have been drawn from the nature of the reactions it undergoes. In that case, a way out is to synthesize a compound by means of reactions designed to yield a molecular structure like the one that has been deduced. If the properties of the resulting compound are identical with the compound being investigated in the first place, a chemist can place confidence in the assigned structure.

An impressive case in point involves *hemoglobin*, the main component of the red blood cells and the pigment that gives the blood its red color. In 1831, the French chemist L. R. LeCanu split hemoglobin into two parts, of which the smaller portion, called *heme*, made up 4 percent of the mass of hemoglobin. Heme was found to have the empirical formula $C_{34}H_{32}O_4N_4Fe$. Since such compounds as heme were known to occur in other vitally important substances, in both the plant and animal kingdoms, the structure of the molecule was a matter of great moment to biochemists. For nearly a century after LeCanu's isolation of heme, however, all that could be done was to break it down into smaller molecules. The iron atom (Fe) was easily removed, and what was left then broke up into pieces roughly one-quarter the size of the original molecule. These fragments were found to be *pyrroles*—molecules built on rings of five atoms, of which four are carbon and one nitrogen. Pyrrole itself has the following structure:

$$CH - CH$$
$$CH \qquad CH$$
$$NH$$

The pyrroles actually obtained from heme possessed small groups of atoms containing one or two carbon atoms attached to the ring in place of one or more of the hydrogen atoms.

In the 1920s, the German chemist Hans Fischer tackled the problem further. Since the pyrroles were one-quarter the size of the original heme, he decided to try to combine four pyrroles and see what he got. What he finally succeeded in getting was a four-ring compound which he called *porphin* (from a Greek word meaning "purple," because of its purple color). Porphin would look like this:

$$CH \qquad CH \qquad CH$$
$$CH \quad C \qquad C \quad CH$$
$$C - NH \qquad N = C$$
$$CH \qquad CH$$
$$C - N \qquad NH - C$$
$$CH \quad C \qquad C \quad CH$$
$$CH \qquad CH \qquad CH$$

However, the pyrroles obtained from heme in the first place contained small *side chains* attached to the ring. These remained in place when the pyrroles were joined to form porphin. The prophin with various side chains attached make up a family of compounds called the *porphyrins*. Those compounds that possessed the particular side chains found in heme were *protoporphyrins*. It was obvious to Fischer, upon comparing the properties of heme with those of the porphyrins he had synthesized, that heme (minus its iron atom) was a protoporphyrin. But which one? No fewer than fifteen different protoporphyrins (each with a different arrangement of side chains) could be formed from the various pyrroles obtained from heme, according to Fischer's reasoning, and any one of those fifteen might be heme itself.

A straightforward answer could be obtained by synthesizing all fifteen and testing the properties of each one. Fischer put his students to work preparing, by painstaking chemical reactions that allowed only a particular structure to be built up, each of the fifteen possibilities. As each different protoporphyrin was formed, he compared its properties with those of the natural protoporphyrin of heme.

In 1928, he discovered that the ninth protoporphyrin in his series was the one he was after. The natural variety of protoporphyrin is therefore called *protoporphyrin IX* to this day. It was a simple procedure to convert protoporphyrin IX to heme by adding iron. Chemists at last felt confident that they knew the structure of that important compound. Here is the structure of heme, as worked out by Fischer:

For his achievement Fischer was awarded the Nobel Prize in chemistry in 1930.

NEW PROCESSES

All the triumphs of synthetic organic chemistry through the nineteenth century and the first half of the twentieth century, great as they were, were won by means of the same processes used by the alchemists of ancient times —mixing and heating substances. Heat was the one sure way of adding energy to molecules and making them interact, but the interactions were usually random in nature and took place by way of briefly existent, unstable intermediates, whose nature could only be guessed at.

What chemists needed was a more refined, more direct method for producing energetic molecules—a method that would produce a group of molecules all moving at about the same speed in about the same direction. This method would remove the random nature of interactions, for whatever one molecule would do, all would do. One way would be to accelerate ions in an electric field, much as subatomic particles are accelerated in cyclotrons.

In 1964, the German-American chemist Richard Leopold Wolfgang accelerated ions and molecules to high energies and, by means of what might be called a *chemical accelerator*, produced ion speeds that heat would produce only at temperatures of from 10,000° C to 100,000° C. Furthermore, the ions were all traveling in the same direction.

If the ions so accelerated are provided with a supply of electrons they can snatch up, they will be converted to neutral molecules which will still be traveling at great speeds. Such neutral beams were produced by the American chemist Leonard Wharton in 1969.

As to the brief intermediate stages of a chemical reaction, computers could help. It was necessary to work out the quantum-mechanical equations governing the state of the electrons in different atom-combinations and to work out the events that would take place on collision. In 1968, for instance, a computer guided by the Italian-American chemist Enrico Clementi *collided* ammonia and hydrochloric acid on closed-circuit television to make ammonium chloride, with the computer working out the events that must take place. The computer indicated that the ammonium chloride that was formed could exist as a high-pressure gas at 700° C. This possibility was not previously known but was proved experimentally a few months later.

In the last decade, chemists have developed brand-new tools, both theoretically and experimentally. Intimate details of reactions not hitherto available will be known, and new products—unattainable before or at least attainable only in small lots—will be formed. We may be at the threshold of unexpected wonders.

Polymers and Plastics

When we consider molecules like those of heme and quinine, we are approaching a complexity that even the modern chemist can cope with only with great difficulty. The synthesis of such a compound requires so many steps and such a variety of procedures that we can hardly expect to produce it in quantity without the help of some living organism (other than the chemist). This is nothing to make for an inferiority complex, however. Living tissue itself approaches the limit of its capacity at this level of complexity. Few molecules in nature are more complex than heme and quinine.

To be sure, there are natural substances composed of hundreds of thousands, even millions, of atoms, but these are not really individual molecules, constructed in one piece, so to speak. Rather, these large molecules are built up of units strung together like beads in a necklace. Living tissue usually synthesizes some small, fairly simple compound and then merely hooks the units together in chains. And that, as we shall see, the chemist also is capable of doing.

CONDENSATION AND GLUCOSE

In living tissue, this union of small molecules (*condensation*) is usually accompanied by the over-all elimination of two hydrogen atoms and an oxygen atom (which combine to form a water molecule) at each point of junction. Invariably, the process can be reversed (both in the body and in the test tube): by the addition of water, the units of the chain can be loosened and separated. This reverse of condensation is called *hydrolysis,* from Greek words meaning "loosening through water." In the test tube, the hydrolysis of these long chains can be hastened by a variety of methods, the most common being the addition of a certain amount of acid to the mixture.

The first investigation of the chemical structure of a large molecule dates back to 1812, when a Russian chemist, Gottlieb Sigismund Kirchhoff, found that boiling starch with acid produced a sugar identical in properties with glucose, the sugar obtained from grapes. In 1819, the French chemist Henri Braconnot also obtained glucose by boiling various plant products such as sawdust, linen, and bark, all of which contain a compound called *cellulose.* It was easy to guess that both starch and cellulose were built of glucose units, but the details of the molecular structure of starch and cellulose had to await knowledge of the molecular structure of glucose. At first, before the days of structural formulas, all that was known of glucose was its empirical formula, $C_6H_{12}O_6$. This proportion suggested that there was one water molecule, H_2O, attached to each of the six carbon atoms. Hence, glucose, and compounds similar to it in structure, were called *carbohydrates* ("watered carbon").

The structural formula of glucose was worked out in 1886 by the German chemist Heinrich Kiliani. He showed that its molecule consists of a chain of six carbon atoms, to which hydrogen atoms and oxygen-hydrogen groups are separately attached. There are no intact water combinations anywhere in the molecule.

Over the next decade or so, the German chemist Emil Fischer studied glucose in detail and worked out the exact arrangement of the oxygen-hydrogen groups around the carbon atoms, four of which were asymmetric. There are sixteen possible arrangements of these groups, and therefore sixteen possible optical isomers, each with its own properties. Chemists have, indeed, made all sixteen, only a few of which actually occur in nature. It was as a result of his work on the optical activity of these sugars that Fischer suggested the establishment of the L-series and D-series of compounds. For putting carbohydrate chemistry on a firm structural foundation, Fischer received the Nobel Prize in chemistry in 1902.

Here are the structural formulas of glucose and of two other common sugars, fructose and galactose:

$$
\begin{array}{ccc}
\text{CH} = \text{O} & \text{CH}_2 - \text{OH} & \text{CH} = \text{O} \\
| & | & | \\
\text{H} - \text{C} - \text{OH} & \text{C} = \text{O} & \text{H} - \text{C} - \text{OH} \\
| & | & | \\
\text{HO} - \text{C} - \text{H} & \text{HO} - \text{C} - \text{H} & \text{HO} - \text{C} - \text{H} \\
| & | & | \\
\text{H} - \text{C} - \text{OH} & \text{H} - \text{C} - \text{OH} & \text{HO} - \text{C} - \text{H} \\
| & | & | \\
\text{H} - \text{C} - \text{OH} & \text{H} - \text{C} - \text{OH} & \text{H} - \text{C} - \text{OH} \\
| & | & | \\
\text{CH}_2 - \text{OH} & \text{CH}_2 - \text{OH} & \text{CH}_2 - \text{OH} \\
\end{array}
$$

Glucose   Fructose   Galactose

These are the simplest structures that adequately present the asymmetries of the molecule; but in actual fact, the molecules are in the shape of nonplanar rings, each ring made up of five (sometimes four) carbon atoms and an oxygen atom.

Once chemists knew the structure of the simple sugars, it was relatively easy to work out the manner in which they are built up into more complex compounds. For instance, a glucose molecule and a fructose can be condensed to form sucrose—the sugar we use at the table. Glucose and galactose combine to form lactose, which occurs in nature only in milk.

There is no reason why such condensations cannot continue indefinitely, and in starch and cellulose they do. Each consists of long chains of glucose units, condensed in a particular pattern.

The details of the pattern are important, because although both compounds are built up of the same unit, they are profoundly different. Starch in one form or another forms the major portion of humanity's diet, while cellulose is completely inedible to human beings. The difference in the pattern of condensation, as painstakingly worked out by chemists, is analogous to the following: Suppose a glucose molecule is viewed as either right side up (when it may be symbolized as u) or upside down (symbolized as n). The starch molecule can then be viewed as consisting of a string of glucose molecules after this fashion ". . . uuuuuuuuu . . . ," while cellulose consists of ". . . unununun. . . ." The body's digestive juices possess the ability to hydrolyze the "uu" linkage of starch, breaking it up to glucose, which we can then absorb to obtain energy. Those same juices are helpless to touch the "un" or "nu" linkages of cellulose, and any cellulose we ingest travels through the alimentary canal and out.

Certain microorganisms can digest cellulose, though none of the higher animals can. Some of these microorganisms live in the intestinal tracts of ruminants and termites, for instance. It is thanks to these small helpers that cows, to our advantage, can live on grass, and that termites, often to our discomfiture, can live on wood. The microorganisms form glucose from cellulose in quantity, use what they need, and the host uses the overflow. The microorganisms supply the processed food, while the host supplies the raw material and the living quarters. This form of cooperation between two forms of life for mutual benefit is called *symbiosis,* from Greek words meaning "life together."

Christopher Columbus discovered South American natives playing with balls of a hardened plant juice. Columbus and the other explorers who visited South America over the next two centuries were fascinated by these bouncy balls (obtained from the sap of trees in Brazil). Samples were brought back to Europe eventually as a curiosity. About 1770, Joseph Priestley (soon to discover oxygen) found that a lump of this bouncy material would rub out pencil marks, so he invented the uninspired name of *rubber,* still the English word for the substance. The British called it *India rubber,* because it came from the "Indies" (the original name of Columbus's new world).

People eventually found other uses for rubber. In 1823, a Scotsman named Charles Macintosh patented garments made of a layer of rubber between two layers of cloth for use in rainy weather, and raincoats are still sometimes called *mackintoshes* (with an added *k*).

The trouble with rubber used in this way, however, was that in warm weather it became gummy and sticky, while in cold weather it was leathery and hard. A number of individuals tried to discover ways of treating rubber so as to remove these undesirable characteristics. Among them was an American named Charles Goodyear, who was innocent of chemistry but worked stubbornly along by trial and error. One day in 1839, he accidentally spilled a mixture of rubber and sulfur on a hot stove. He scraped it off as quickly as he could and found, to his amazement, that the heated rubber-sulfur mixture was dry even while it was still warm. He heated it and cooled it and found that he had a sample of rubber that did not turn gummy with heat or leathery with cold but remained soft and springy throughout.

This process of adding sulfur to rubber is now called *vulcanization* (after Vulcan, the Roman god of fire). Goodyear's discovery founded the rubber industry. It is sad to have to report that Goodyear himself never reaped a reward despite this multimillion-dollar discovery. He spent his life fighting for patent rights and died deeply in debt.

Knowledge of the molecular structure of rubber dates back to 1879, when a French chemist, Gustave Bouchardat, heated rubber in the absence of air and obtained a liquid called *isoprene.* Its molecule is composed of five carbon atoms and eight hydrogen atoms, arranged as follows:

$$CH_2 = \overset{\overset{\displaystyle CH_3}{\displaystyle |}}{C} - CH = CH_2$$

A second type of plant juice (*latex*), obtained from certain trees in southeast Asia, yields a substance called *gutta percha.* This lacks the elasticity of rubber; but, when heated in the absence of air, it, too, yields isoprene.

Both rubber and gutta percha are made up of thousands of isoprene units.

As in the case of starch and cellulose, the difference between them lies in the pattern of linkage. In rubber, the isoprene units are joined in the "... uuuuu ..." fashion and in such a way that they form coils, which can straighten out when pulled, thus allowing stretching. In gutta percha, the units join in the "... unununun ..." fashion, and these form chains that are straighter to begin with and therefore much less stretchable (figure 11.3).

Figure 11.3. The gutta percha molecule, a portion of which is shown here, is made up of thousands of isoprene units. The first five carbon atoms at the left (*black balls*) and the eight hydrogen atoms bonded to them make up an isoprene unit.

A simple sugar molecule, such as glucose, is a *monosaccharide* (Greek for "one sugar"); sucrose and lactose are *disaccharides* ("two sugars"); and starch and cellulose are *polysaccharides* ("many sugars"). Because two isoprene molecules join to form a well-known type of compound called *terpene* (obtained from turpentine), rubber and gutta percha are called *polyterpenes*.

The general term for such compounds was invented by Berzelius (a great inventor of names and symbols) as far back as 1830. He called the basic unit a *monomer* ("one part") and the large molecule a *polymer* ("many parts"). Polymers consisting of many units (say more than a hundred) are now called *high polymers*. Starch, cellulose, rubber, and gutta percha are all examples of high polymers.

Polymers are not clear-cut compounds but are complex mixtures of molecules of different sizes. The average molecular weight can be determined by several methods. One involves measurement of *viscosity* (the ease or difficulty with which a liquid flows under a given pressure). The larger the molecule and the more elongated it is, the more it contributes to the *internal friction* of a liquid and the more it makes it pour like molasses, rather than like water. The German chemist Hermann Staudinger worked out this method in 1930 as part of his general work on polymers; and in 1953, he was awarded the Nobel Prize for chemistry for his contribution toward the understanding of these giant molecules.

In 1913, two Japanese chemists discovered that natural fibers such as those of cellulose diffract X rays, just as a crystal does. The fibers are not crystals in the ordinary sense but are *microcrystalline* in character: that is, the long chains of units making up their molecules tend to run in parallel bundles for longer or shorter distances, here and there. Over the course of those parallel bundles, atoms are arranged in a repetitive order as they are in crystals, and X rays striking these sections of the fiber are diffracted.

So polymers have come to be divided into two broad classes—*crystalline* and *amorphous*.

In a crystalline polymer, such as cellulose, the strength of the individual chains is increased by the fact that parallel neighbors are joined together by chemical bonds. The resulting fibers have considerable tensile strength. Starch is crystalline, too, but far less so than cellulose, and therefore lacks the strength of cellulose and its capacity for fiber formation.

Rubber is an amorphous polymer. Since the individual chains do not line up, cross-links do not occur. If heated, the various chains can vibrate independently and slide freely over and around one another. Consequently, rubber or a rubberlike polymer will grow soft and sticky and eventually melt with heat. (Stretching rubber straightens the chains and introduces some microcrystalline character. Stretched rubber, therefore, has considerable tensile strength.) Cellulose and starch, in which the individual molecules are bound together here and there, cannot undergo the same independence of vibration, so there is no softening with heat. They remain stiff until the temperature is high enough to induce vibrations that shake the molecule apart so that charring and smoke emission take place.

At temperatures below the gummy, sticky stage, amorphous polymers are often soft and springy. At still lower temperatures, however, they become hard and leathery, even glassy. Raw rubber is dry and elastic only over a rather narrow temperature range. The addition of sulfur to the extent of 5 percent to 8 percent provides flexible sulfur links from chain to chain, which reduce the independence of the chains and thus prevent gumminess at moderate heat. They also increase the free play between the chains at moderately low temperatures; therefore the rubber does not harden. The addition of greater amounts of sulfur, up to 30 percent to 50 percent, will bind the chains so tightly that the rubber grows hard. It is then known as *hard rubber* or *ebonite*.

(Even vulcanized rubber will turn glassy if the temperature is lowered sufficiently. An ordinary rubber ball, dipped in liquid air for a few moments, will shatter if thrown against a wall. This is a favorite demonstration in introductory chemistry courses.)

Various amorphous polymers show different physical properties at a given temperature. At room temperature, natural rubber is elastic, various resins are glassy and solid, and chicle (from the sapodilla tree of South America) is soft and gummy (it is the chief ingredient of chewing gum).

CELLULOSE AND EXPLOSIVES

Aside from our food, which is mainly made up of high polymers, probably the one polymer that man has depended on longest is cellulose. It is the major component of wood, which has been indispensable as a fuel and a construction material. Cellulose is also used to make paper. In the pure fibrous forms of cotton and linen, cellulose has been man's most important textile material. And the organic chemists of the mid-nineteenth century naturally turned to cellulose as a raw material for making other giant molecules.

One way of modifying cellulose is by attaching the *nitrate group* of atoms (a nitrogen atom and three oxygen atoms) to the oxygen-hydrogen combinations (*hydroxyl groups*) in the glucose units. When this was done, by treating cellulose with a mixture of nitric acid and sulfuric acid, an explosive of until-then unparalleled ferocity was created. The explosive was discovered by accident in 1846 by a German-born Swiss chemist named Christian Friedrich Schönbein (who, in 1839, had discovered ozone). He had spilled an acid mixture in the kitchen (where he was forbidden to experiment, but he had taken advantage of his wife's absence to do just that) and snatched up his wife's cotton apron, so the story goes, to wipe up the mess. When he hung the apron over the fire to dry, it went poof!, leaving nothing behind.

Schönbein recognized the potentialities at once, as can be told from the name he gave the compound, which in English translation is *guncotton*. (It is also called *nitrocellulose*.) Shönbein peddled the recipe to several governments. Ordinary gunpowder was so smoky that it blackened the gunners, fouled the cannon, which then had to be swabbed between shots, and raised such a pall of smoke that, after the first volleys, battles had to be fought by dead reckoning. War offices therefore leaped at the chance to use an explosive that was not only more powerful but also smokeless. Factories for the manufacture of guncotton began to spring up. And almost as fast as they sprang up, they blew up. Guncotton was too eager an explosive; it would not wait for the cannon. By the early 1860s, the abortive guncotton boom was over, figuratively as well as literally.

Later, however, methods were discovered for removing the small quantities of impurities that encouraged guncotton to explode. It then became reasonably safe to handle. The English chemist Dewar (of liquefied gas fame) and a co-worker, Frederick Augustus Abel, introduced the technique, in 1889, of mixing it with *nitroglycerine*, and adding Vaseline to the mixture to make it moldable into cords (the mixture was called *cordite*). That, finally, was a useful smokeless powder. The Spanish-American War of 1898 was the last war of any consequence fought with ordinary gunpowder.

(The machine age added its bit to the horrors of gunnery. In the 1860s, the American inventor Richard Gatling produced the first *machine gun* for the rapid firing of bullets; and this was improved by another American inventor, Hiram Stevens Maxim, in the 1880s. The *Gatling gun* gave rise to the slang term *gat* for gun. It and its descendant, the *Maxim gun*, gave the unabashed imperialists of the late nineteenth century an unprecedented advantage over the "lesser breeds," to use Rudyard Kipling's offensive phrase, of Africa and Asia. "Whatever happens, we have got / The Maxim gun and they have not!" went a popular jingle.)

"Progress" of this sort continued in the twentieth century. The most important explosive in the First World War was *trinitrotoluene*, familiarly abbreviated as TNT. In the Second World War, an even more powerful explosive, *cyclonite*, came into use. Both contain the nitro group (NO_2) rather than the nitrate group (ONO_2). As lords of war, however, all chemical explosives gave way to nuclear bombs in 1945 (see chapter 10).

Nitroglycerine, by the way, was discovered in the same year as guncotton. An Italian chemist named Ascanio Sobrero treated glycerol with a mixture of nitric acid and sulfuric acid and knew he had something when he nearly killed himself in the explosion that followed. Sobrero, lacking Schönbein's promotional impulses, felt nitroglycerine to be too dangerous a substance to deal with and virtually suppressed information about it. But within ten years, a Swedish family, the Nobels, took to manufacturing it as a "blasting oil" for use in mining and construction work. After a series of accidents, including one that took the life of a member of the family, Alfred Bernhard Nobel, the brother of the victim, discovered a method of mixing nitroglycerine with an absorbent earth called *kieselguhr* or *diatomaceous earth* (kieselguhr consists largely of the tiny skeletons of one-celled organisms called diatoms). The mixture consisted of three parts of nitroglycerine to one of kieselguhr, but such was the absorptive power of the latter that the mixture was virtually a dry powder. A stick of this impregnated earth (*dynamite*) could be dropped, hammered, even burned, without explosion. When set off by a percussion cap (electrically, and from a distance), it displayed all the shattering force of pure nitroglycerine.

Percussion caps contain sensitive explosives that detonate by heat or by mechanical shock and are therefore called *detonators*. The strong shock of the detonation sets off the less sensitive dynamite. It might seem as though the danger were merely shifted from nitroglycerine to detonators, but it is not so bad as it sounds, since the detonator is only needed in tiny quantities. The detonators most used are mercury fulminate ($HgC_2N_2O_2$) and lead azide (PbN_6).

Sticks of dynamite eventually made it possible to carve the American West into railroads, mines, highways, and dams at a rate unprecedented in history. Dynamite, and other explosives he discovered, made a millionaire of the lonely and unpopular Nobel (who found himself, against his humanitarian will, regarded as a "merchant of death"). When he died in 1896, he left behind a fund out of which the famous Nobel Prizes were to be granted each year in five fields: chemistry, physics, medicine and physiology, literature, and peace. Aside from the fabulous honor, a cash reward of about forty thousand dollars was involved, and the amount has risen since then. The first prizes were awarded on 10 December 1901, the fifth anniversary of his death, and these have now become the greatest honor any scientist can receive.

Considering the nature of human society, explosives continued to take up a sizable fraction of the endeavor of great scientists. Since almost all explosives contain nitrogen, the chemistry of that element and its compounds was of key importance. (It is also, it must be admitted, of key importance to life as well.)

The German chemist Wilhelm Ostwald, who was interested in chemical theory rather than in explosives, studied the rates at which chemical reactions proceed. He applied to chemistry the mathematical principles associated with physics thus becoming one of the founders of *physical chemistry.* Toward the turn of the century, he worked out new methods for converting ammonia (NH_3) to nitrogen oxides, which could then be used to manufacture explosives. For his theoretical work, particularly on catalysis, Ostwald received the Nobel Prize for chemistry in 1909.

The ultimate source of usable nitrogen was, in the early decades of the twentieth century, the nitrate deposits in the desert of northern Chile. During the First World War, these fields were placed out of reach of Germany by the British Navy. However, the German chemist Fritz Haber had devised a method by which the molecular nitrogen of the air could be combined with hydrogen under pressure, to form the ammonia needed for the Ostwald process. This *Haber process* was improved by the German chemist Karl Bosch, who supervised the building of plants during the war for the manufacture of ammonia. Haber received the Nobel Prize for chemistry in 1918, and Bosch shared one in 1931. By the late 1960s, the United States alone was manufacturing 12 million tons of ammonia per year by the Haber process.

PLASTICS AND CELLULOID

But let us return to modified cellulose. Clearly, it was the addition of the nitrate group that made for explosiveness. In guncotton all of the available hydroxyl groups were nitrated. What if only some of them were nitrated? Would they not be less explosive? Actually, such partly nitrated cellulose proved not to be explosive at all. However, it did burn very readily; the material was eventually named *pyroxylin* (from Greek words meaning "firewood").

Pyroxylin can be dissolved in mixtures of alcohol and ether—as was discovered independently by the French scholar Louis Nicolas Ménard and an American medical student named J. Parkers Maynard (and an odd similarity in names that is). When the alcohol and ether evaporate, the pyroxylin is left behind as a tough, transparent film, which was named *collodion.* Its first use was as a coating over minor cuts and abrasions; it was called *new skin.* However, the adventures of pyroxylin were only beginning. Much more lay ahead.

Pyroxylin itself is brittle in bulk. But the English chemist Alexander Parkes found that if it was dissolved in alcohol and ether and mixed with a substance such as camphor, the evaporation of the solvent left behind a hard solid that became soft and malleable when heated. It could then be modeled into some desired shape which it would retain when cooled and hardened. So nitrocellulose was transformed into the first artificial *plastic,* in the year 1865. Camphor, which introduced the plastic properties into an otherwise brittle substance, was the first *plasticizer.*

What brought plastics to the attention of the public and made it more than a chemical curiosity was its dramatic introduction into the billiard parlor. Billiard balls were then made from ivory, a commodity that could be obtained only over an elephant's dead body—a point that naturally produced problems. In the early 1860s, a prize of 10,000 dollars was offered for the best substitute for ivory that would fulfill the billiard ball's manifold requirements of hardness, elasticity, resistance to heat and moisture, lack of grain, and so on. The American inventor John Wesley Hyatt was one of those who went out for the prize. He made no progress until he heard of Parkes's trick of plasticizing pyroxylin to a moldable material that would set as a hard solid. Hyatt set about working out improved methods of manufacturing the material, using less of the expensive alcohol and ether and more in the way of heat and pressure. By 1869, Hyatt was turning out cheap billiard balls of this material, which he called *celluloid*. It won him the prize.

Celluloid turned out to have significance away from the pool table. It was versatile indeed. It could be molded at the temperature of boiling water; it could be cut, drilled, and sawed at lower temperatures; it was strong and hard in bulk but could also be produced in the form of thin flexible films that served for shirt collars, baby rattles, and so on. In the form of still thinner and more flexible films, it could be used as a base for silver compounds in gelatin, and thus it became the first practical photographic film.

The one fault of celluloid was that, thanks to its nitrate groups, it had a tendency to burn with appalling quickness, particularly when in the form of thin film. It was the cause of a number of fire tragedies.

The substitution of acetate groups (CH_3COO^-) for nitrate groups led to the formation of another kind of modified cellulose called *cellulose acetate*. Properly plasticized, this has properties as good or almost as good as those of celluloid, plus the saving grace of being much less likely to burn. Cellulose acetate came into use just before the First World War; and after the war, it completely replaced celluloid in the manufacture of photographic film and many other items.

HIGH POLYMERS

Within half a century after the development of celluloid, chemists emancipated themselves from dependence on cellulose as the base for plastics. As early as 1872, Baeyer (who was later to synthesize indigo) had noticed that when phenols and aldehydes are heated together, a gooey, resinous mass results. Since he was interested only in the small molecules he could isolate from the reaction, he ignored this mess at the bottom of the flask (as nineteenth-century organic chemists typically tended to do when goo fouled up their glassware). Thirty-seven years later, the Belgian-born American chemist Leo Hendrik Baekeland, experimenting with formaldehyde, found that under certain conditions the reaction would yield a resin that on continued heating under pressure became first a soft solid, then a hard, insoluble substance. This resin could be molded while soft and then be allowed to set into a hard, permanent shape. Or, once hard, it could be powdered, poured into a mold and set into one piece by heat and pressure. Very complex forms could be cast easily and quickly. Furthermore, the product was inert and impervious to most environmental vicissitudes.

Baekeland named his product Bakelite, after his own name. Bakelite belongs to the class of *thermosetting plastics*, which, once they set on cooling, cannot be softened again by heating (though, of course, they can be destroyed by intense heat). Materials such as the cellulose derivatives, which can be softened again and again, are called *thermoplastics*. Bakelite has numerous uses —as insulator, adhesive, laminating agent, and so on. Although the oldest of the thermosetting plastics, it is still the most used.

Bakelite was the first production, in the laboratory, of a useful high polymer from small molecules. For the first time the chemist had taken over this particular task completely. It does not, of course, represent synthesis in the sense of that of heme or quinine, where chemists must place every last atom into just the proper position, almost one at a time. Instead, the production of high polymers requires merely that the small units of which they are composed be mixed under the proper conditions. A reaction is then set up in which the units form a chain automatically, without the specific point-to-point intervention of the chemist. The chemist can, however, alter the nature of the chain indirectly by varying the starting materials or the proportions among them, or by the addition of small quantities of acids, alkalies, or various substances that act as *catalysts* and tend to guide the precise nature of the reaction.

With the success of Bakelite, chemists naturally turned to other possible starting materials in search of more synthetic high polymers that might be useful plastics. And, as time went on, they succeeded many times over.

British chemists discovered in the 1930s, for instance, that the gas ethylene $(CH_2 = CH_2)$, under heat and pressure, would form very long chains. One of the two bonds in the double bond between the carbon atoms opens up and attaches itself to a neighboring molecule. With this happening over and over again, the result is a long-chain molecule called *polythene* in England and *polyethylene* in the United States.

The paraffin-wax molecule is a long chain made up of the same units, but the molecule of polyethylene is even longer. Polyethylene is therefore like wax, but more so. It has the cloudy whiteness of wax, the slippery feel, the electrical insulating properties, the waterproofness, and the lightness (it is about the only plastic that will float on water). It is, however, at its best, much tougher than paraffin and much more flexible.

As it was first manufactured, polyethylene required dangerous pressures, and the product had a rather low melting point—just above the boiling point of water. It softened to uselessness at temperatures below the melting point. Apparently this effect was due to the fact that the carbon chain had branches which prevented the molecules from forming close-packed, crystalline arrays. In 1953, a German chemist named Karl Ziegler found a way to produce unbranched polyethylene chains, without the need for high pressures. The result was a new variety of polyethylene, tougher and stronger than the old, and capable of withstanding boiling-water temperatures without softening too much. Ziegler accomplished this by using a new type of catalyst—a resin with ions of metals such as aluminum or titanium attached to negatively charged groups along the chain.

On hearing of Ziegler's development of metal-organic catalysts for polymer formation, the Italian chemist Giulio Natta began applying the technique to *propylene* (ethylene to which a small one-carbon methyl group, CH_{3-}, is attached). Within ten weeks, he had found that, in the resultant polymer, all the methyl groups face in the same direction, rather than (as was usual in polymer formation before that time) facing, in random fashion, in either direction. Such *isotactic polymers* (the name was proposed by Natta's wife) proved to have useful properties and can now be manufactured virtually at will. Chemists can design polymers, in other words, with greater precision than ever before. For their work in this field, Ziegler and Natta shared the 1963 Nobel Prize for chemistry.

The atomic-bomb project contributed another useful high polymer in the form of an odd relative of polyethylene. In the separation of uranium 235 from natural uranium, the nuclear physicists had to combine the uranium with fluorine to form uranium hexafluoride. Fluorine is the most active of all substances and will attack almost anything. Looking for lubricants and seals for their vessels that would be impervious to attack by fluorine, the physicists resorted to *fluorocarbons*—substances in which the carbon has already combined with fluorine (replacing hydrogen).

Until then, fluorocarbons had been only laboratory curiosities. The first (and simplest) of this type of molecule, *carbon tetrafluoride* (CF_4), had been obtained in pure form only in 1926. The chemistry of these interesting substances was now pursued intensively. Among the fluorocarbons studied was *tetrafluoroethylene* ($CF_2 = CF_2$), which had first been synthesized in 1933 and is, as you see, ethylene with its four hydrogens replaced by four fluorines. It was bound to occur to someone that tetrafluoroethylene might polymerize as ethylene itself did. After the war, Du Pont chemists produced a long-chain polymer which was as monotonously $CF_2CF_2CF_2$ . . . as polyethylene was $CH_2CH_2CH_2$. . . . Its trade name is Teflon, the *tefl* being an abbreviation of *tetrafluoro-*.

Teflon is like polyethylene, only more so. The carbon-fluorine bonds are stronger than the carbon-hydrogen bonds and offer even less opportunity for the interference of the environment. Teflon is insoluble in everything, unwettable by anything, an extremely good electrical insulator, and considerably more resistant to heat than is even the new and improved polyethylene. Teflon's best-known application, so far as the housewife is concerned, is as a coating upon frying pans, thus enabling food to be fried without fat, since food will not stick to the standoffish fluorocarbon polymer.

An interesting compound that is not quite a fluorocarbon is Freon (CF_2Cl_2), mentioned earlier in the book. It was introduced in 1932 as a refrigerant. It is more expensive than the ammonia or sulfur dioxide used in large-scale freezers; but, on the other hand, Freon is nonodorous, nontoxic, and nonflammable, so that accidental leakage introduces a minimum of danger. Midgley, its discoverer, demonstrated its harmlessness by taking in a deep lungful and letting it trickle out over a candle flame. The candle went out, but Midgley was unharmed. It is through Freon that room air conditioners have become a characteristic part of the American scene since the Second World War.

GLASS AND SILICONE

Plastic properties do not, of course, belong solely to the organic world. One of the most ancient of all plastic substances is glass. The large molecules of glass are essentially chains of silicon and oxygen atoms: that is, -Si-O-Si-O-Si-O-Si-, and so on indefinitely. Each silicon atom in the chain has two unoccupied bonds to which other groups can be added. The silicon atom, like the carbon atom, has four valence bonds. The silicon-silicon bond, however, is weaker than the carbon-carbon bond, so that only short silicon chains can be formed, and those (in compounds called *silanes*) are unstable. The silicon-oxygen bond is a strong one, however, and such chains are even more stable than those of carbon. In fact, since the earth's crust is half oxygen and a quarter silicon, the solid ground we stand upon may be viewed as essentially a silicon-oxygen chain.

Although the beauties and usefulness of glass (a kind of sand, made transparent) are infinite, it possesses the great disadvantage of being breakable. And in the process of breaking, it produces hard, sharp pieces which can be dangerous, even deadly. With untreated glass in the windshield of a car, a crash may convert the auto into a shrapnel bomb.

Glass can be prepared, however, as a double sheet between which is placed a thin layer of a transparent polymer, which hardens and acts as an adhesive. This is *safety glass,* for when it is shattered, even into powder, each piece is held firmly in place by the polymer. None goes flying out on death-dealing missions. Originally, as far back as 1905, collodion was used as the binder, but nowadays that has been replaced for the most part by polymers built of small molecules such as vinyl chloride. (Vinyl chloride is like ethylene, except that one of the hydrogen atoms is replaced by a chlorine atom.) The *vinyl resin* is not discolored by light, so safety glass can be trusted not to develop a yellowish cast with time.

Then there are the transparent plastics that can completely replace glass, at least in some applications. In the middle 1930s, Du Pont polymerized a small molecule called *methyl methacrylate* and cast the polymer that resulted (a *polyacrylic plastic*) into clear, transparent sheets. The trade names of these products are Plexiglas and Lucite. Such *organic glass* is lighter than ordinary glass, more easily molded, less brittle, and simply snaps instead of shattering when it does break. During the Second World War, molded transparent plastic sheets came into important use as windows and transparent domes in airplanes, where lightness and nonbrittleness are particularly useful. To be sure, the polyacrylic plastics have their disadvantages. They are affected by organic solvents, are more easily softened by heat than glass is, and are easily scratched. Polyacrylic plastics used in the windshields of cars, for instance, would quickly scratch under the impact of dust particles and become dangerously hazy. Consequently, glass is not likely ever to be replaced entirely. In fact, it is actually developing new versatility. Glass fibers have been spun into textile material that has all the flexibility of organic fibers and the inestimable further advantage of being absolutely fireproof.

In addition to glass substitutes, there is also what might be called a glass compromise. As I said, each silicon atom in a silicon-oxygen chain has two spare bonds for attachment to other atoms. In glass these other atoms are oxygen atoms, but they need not be. What if carbon-containing groups are attached instead of oxygen? You will then have an inorganic chain with organic offshoots, so to speak—a compromise between an organic and an inorganic material. As long ago as 1908, the English chemist Frederic Stanley Kipping formed such compounds, and they have come to be known as *silicones*.

During the Second World War, long-chain *silicone resins* came into prominence. Such silicones are essentially more resistant to heat than purely organic polymers. By varying the length of the chain and the nature of the side chains, one can obtain a list of desirable properties not possessed by glass itself. For instance, some silicones are liquid at room temperature and change very little in viscosity over large ranges of temperature: that is, they do not thin out with heat or thicken with cold. This is a particularly useful property for a hydraulic fluid—the type of fluid used to lower landing gear on airplanes, for instance. Other silicones form soft, puttylike sealers that do not harden or crack at the low temperatures of the stratosphere and are remarkably water-repellent. Still other silicones serve as acid-resistant lubricants, and so on.

Synthetic Fibers

In the story of organic synthesis, a particularly interesting chapter is that of the synthetic fibers. The first artificial fibers (like the first bulk plastics) were made from cellulose as the starting material. Naturally, the chemists began with cellulose nitrate, since it was available in reasonable quantity. In 1884, Hilaire Bernigaud de Chardonnet, a French chemist, dissolved cellulose nitrate in a mixture of alcohol and ether and forced the resulting thick solution through small holes. As the solution sprayed out, the alcohol and ether evaporated, leaving behind the cellulose nitrate as a thin thread of collodion. (This is essentially the manner in which spiders and silkworms spin their threads: they eject a liquid through tiny orifices, and this becomes a solid fiber on exposure to air.) The cellulose-nitrate fibers were too flammable for use, but the nitrate groups could be removed by appropriate chemical treatment, and the result was a glossy cellulose thread resembling silk.

De Chardonnet's process was expensive, of course, what with nitrate groups being first put on and then taken off, to say nothing of the dangerous interlude while they were in place and of the fact that the alcohol-ether mixture used as solvent was also dangerously flammable. In 1892, methods were discovered for dissolving cellulose itself. The English chemist Charles Frederick Cross, for instance, dissolved it in carbon disulfide and formed a thread from the resulting viscous solution (named *viscose*). The trouble was that carbon disulfide is flammable, toxic, and evil smelling. In 1903, a competing process employing acetic acid as part of the solvent, and forming a substance called *cellulose acetate*, came into use.

These artificial fibers were first called *artificial silk*, but were later named *rayon* because their glossiness reflects rays of light. The two chief varieties of rayon are usually distinguished as *viscose rayon* and *acetate rayon*.

Viscose, by the way, can be squirted through a slit to form a thin, flexible, waterproof, transparent sheet—*cellophane*—a process invented in 1908 by a French chemist, Jacques Edwin Brandenberger. Some synthetic polymers also can be extruded through a slit for the same purpose. Vinyl resins, for instance, yielded the covering material known as Saran.

It was in the 1930s that the first completely synthetic fiber was born.

Let me begin by saying a little about silk. Silk is an animal product made by certain caterpillars that are exacting in their requirements for food and care. The fiber must be tediously unraveled from their cocoons. For these reasons, silk is expensive and cannot be mass-produced. It was first produced in China more than 2,000 years ago, and the secret of its preparation was jealously guarded by the Chinese, so that it could be kept a lucrative monopoly for export. However, secrets cannot be kept forever, despite all security measures. The secret spread to Korea, Japan, and India. Ancient Rome received silk by the long overland route across Asia, with middlemen levying tolls every step of the way; thus, the fiber was beyond the reach of anyone except the most wealthy. In 550 A.D., silkworm eggs were smuggled into Constantinople, and silk production in Europe got its start. Nevertheless, silk has always remained more or less a luxury item. Moreover, until recently there was no good substitute for it. Rayon can imitate its glossiness but not its sheerness or strength.

After the First World War, when silk stockings became an indispensable item of the feminine wardrobe, the pressure for greater supplies of silk or of some adequate substitute became very strong. This was particularly true in the United States, where silk was used in greatest quantity and where relations with the chief supplier, Japan, were steadily deteriorating. Chemists dreamed of somehow making a fiber that could compare with it.

Silk is a protein (see chapter 12). Its molecule is built up of monomers called *amino acids*, which in turn contain *amino* ($-NH_2$) and *carboxyl* ($-COOH$) groups. The two groups are joined by a carbon atom between them; labeling the amino group *a* and the carboxyl group *c*, and symbolizing the intervening carbon by a hyphen, we can write an amino acid like this: a - c. These amino acids polymerize in head-to-tail fashion: that is, the amino group of one condenses with the carboxyl group of the next. Thus, the structure of the silk molecule runs like this: . . . a - c . a - c . a - c . a - c . . . .

In the 1930s, a Du Pont chemist named Wallace Hume Carothers was investigating molecules containing amine groups and carboxyl groups in the hope of discovering a good method of making them condense in such a way as to form molecules with large rings. (Such molecules are of importance in perfumery.) Instead, he found them condensing to form long-chain molecules.

Carothers had already suspected that long chains might be possible, and he was not caught napping. He lost little time in following up this development. He eventually formed fibers from adipic acid and hexamethylenediamine. The adipic-acid molecule contains two carboxyl groups separated by four carbon atoms, so it can be symbolized as: c----c. Hexamethylenediamene consists of two amine groups separated by six carbon atoms, thus: a------a. When Carothers mixed the two substances together, they condensed to form a polymer like this: . . . a------a . c----c . a------a . c----c . a------a . . . . The points at which condensation took place had the "c.a" configuration found in silk, you will notice.

At first the fibers produced were not much good; they were too weak. Carothers decided that the trouble lay in the presence of the water produced in the condensation process. The water set up a counteracting hydrolysis reaction which prevented polymerization from going very far. Carothers found a cure: he arranged to carry on the polymerization under low pressure, so that the water vaporized and was easily removed by letting it condense on a cooled glass surface held close to the reacting liquid and so slanted as to carry the water away (a *molecular still*). Now the polymerization could continue indefinitely. It formed nice long, straight chains; and in 1935, Carothers finally had the basis for a dream fiber.

The polymer formed from adipic acid and hexamethylenediamine was melted and extruded through holes. It was then stretched so that the fibers would lie side by side in crystalline bundles. The result was a glossy, silklike thread that could be used to weave a fabric as sheer and beautiful as silk, and even stronger. This first of the completely synthetic fibers was named *nylon*. Carothers did not live to see his discovery come to fruition, however. He died in 1937.

Du Pont announced the existence of the synthetic fiber in 1938 and began producing it commercially in 1939. During the Second World War, the United States Armed Forces took all the production of nylon for parachutes and for a hundred other purposes. But after the war nylon completely replaced silk for hosiery; indeed, women's stockings are now called *nylons*.

Nylon opened the way to the production of many other synthetic fibers. Acrylonitrile, or vinyl cyanide ($CH_2 = CHCN$), can be made to polymerize into a long chain like that of polyethylene but with cyanide groups (completely nonpoisonous in this case) attached to every other carbon. The result, introduced in 1950, is Orlon. If vinyl chloride ($CH_2 = CHCl$) is added, so that the eventual chain contains chlorine atoms as well as cyanide groups, Dynel results. Or the addition of acetate groups, through the use of vinyl acetate ($CH_2 = CHOOCCH_3$), produces Acrilan.

The British in 1941 made a *polyester* fiber, in which the carboxyl group of one monomer condenses with the hydroxyl group of another. The result is the usual long chain of carbon atoms, broken in this case by the periodic insertion of an oxygen in the chain. The British call it Terylene, but in the United States, it has appeared under the name of Dacron.

These new synthetic fibers are more water-repellent than most of the natural fibers; thus they resist dampness and are not easily stained. They are not subject to destruction by moths or beetles. Some are crease-resistant and can be used to prepare "wash-and-wear" fabrics.

Synthetic Rubber

It is a bit startling to realize that humans have been riding on rubber wheels for only about a hundred years. For thousands of years they rode on wooden or metal rims. When Goodyear's discovery made vulcanized rubber available, it occurred to a number of people that rubber rather than metal might be wrapped around wheels. In 1845, a British engineer, Robert William Thomson, went this idea one better: he patented a device consisting of an inflated rubber tube that would fit over a wheel. By 1890, *tires* were routinely used for bicycles; and in 1895, they were placed on horseless carriages.

Amazingly enough, rubber, though a soft, relatively weak substance, proved to be much more resistant to abrasion than wood or metal. This durability, coupled with its shock-absorbing qualities and the air-cushioning idea, introduced unprecedented riding comfort.

As the automobile increased in importance, the demand for rubber for tires grew astronomical. In half a century, the world production of rubber increased forty-two-fold. You can judge the quantity of rubber in use for tires today when I tell you that, in the United States, they leave no less than 200,000 tons of abraded rubber on the highways each year, in spite of the relatively small amount abraded from the tires of an individual car.

The increasing demand for rubber introduced a certain insecurity in the war resources of many nations. As war was mechanized, armies and supplies began to move on rubber, and rubber could be obtained in significant quantity only from the Malayan peninsula, far removed from the "civilized" nations most apt to engage in "civilized" warfare. (The Malayan peninsula is not the natural habitat of the rubber tree. The tree was transplanted there, with great success, from Brazil, where the original rubber supply steadily diminished.) The supply of the United States was cut off at the beginning of its entry into the Second

World War when the Japanese overran Malaya. American apprehensions in this respect were responsible for the fact that the very first object rationed during the war emergency, even before the attack on Pearl Harbor, was rubber tires.

Even in the First World War, when mechanization was just beginning, Germany was hampered by being cut off from rubber supplies by Allied sea power.

By that time, then, there was reason to consider the possibility of constructing a synthetic rubber. The natural starting material for such a synthetic rubber was isoprene, the building block of natural rubber. As far back as 1880, chemists had noted that isoprene, on standing, tended to become gummy and, if acidified, would set into a rubberlike material. Kaiser Wilhelm II eventually had the tires of his official automobile made of such material, as a kind of advertisement of Germany's chemical virtuosity.

However, there were two catches to the use of isoprene as the starting material for synthesizing rubber. First, the only major source of isoprene is rubber itself. Second, when isoprene polymerizes, it is most likely to do so in a completely random manner. The rubber chain possesses all the isoprene units oriented in the same fashion: ---uuuuuuuuu---. The gutta percha chain has them oriented in strict alternation: ----unununununun----. When isoprene is polymerized in the laboratory under ordinary conditions, however, the u's and the n's are mixed randomly, forming a material which is neither rubber nor gutta percha. Lacking the flexibility and resilience of rubber, it is useless for automobile tires (except possibly for imperial automobiles used on state occasions).

Eventually, catalysts like those that Ziegler introduced in 1953 for manufacturing polyethylene made it possible to polymerize isoprene to a product almost identical with natural rubber, but by that time many useful synthetic rubbers, very different chemically from natural rubber, had been developed.

The first efforts, naturally, concentrated on attempts to form polymers from readily available compounds resembling isoprene. For instance, during the First World War, under the pinch of the rubber famine, Germany made use of dimethylbutadiene:

$$CH_2 = C - C = CH_2$$
$$\quad\quad | \quad\ |$$
$$\quad\ CH_3\ CH_3$$

Dimethylbutadiene differs from isoprene only in containing a methyl group (CH_3) on both middle carbons of the four-carbon chain instead of on only one of them. The polymer built of dimethylbutadiene, called *methyl rubber*, could be formed cheaply and in quantity. Germany produced about 2,500 tons of it during the First World War. While it did not stand up well under stress, it was nonetheless the first of the usable synthetic rubbers.

About 1930, both Germany and the Soviet Union tried a new tack. They used as the monomer, butadiene, which has no methyl group at all:

$$CH_2 = CH - CH = CH_2$$

With sodium metal as a catalyst, they formed a polymer called Buna (from "*bu*tadiene" and *Na* for sodium).

Buna rubber was a synthetic rubber that could be considered satisfactory in a pinch. It was improved by the addition of other monomers, alternating with butadiene at intervals in the chain. The most successful addition was *styrene,* a compound resembling ethylene but with a benzene ring attached to one of the carbon atoms. This product was called Buna S. Its properties were very similar to those of natural rubber; and, in fact, thanks to Buna S, Germany's armed forces suffered no serious rubber shortage in the Second World War. The Soviet Union also supplied itself with rubber in the same way. The raw materials can be obtained from coal or petroleum.

The United States was later in developing synthetic rubber in commercial quantities, perhaps because it was in no danger of a rubber famine before 1941. But after Pearl Harbor, it took up synthetic rubber with a vengeance. It began to produce buna rubber and another type of synthetic rubber called *neoprene,* built up of *chloroprene:*

$$CH_2 = C - CH = CH_2$$
$$|$$
$$Cl$$

This molecule, as you see, resembles isoprene except for the substitution of a chlorine atom for the methyl group.

The chlorine atoms, attached at intervals to the polymer chain, confer upon neoprene certain resistances that natural rubber does not have. For instance, it is more resistant to organic solvents such as gasoline: it does not soften and swell nearly as much as would natural rubber. Thus neoprene is actually preferable to rubber for such uses as gasoline hoses. Neoprene first clearly demonstrated that in the field of synthetic rubbers, as in many other fields, the product of the test tube need not be a mere substitute for nature but could be an improvement.

Amorphous polymers with no chemical resemblance to natural rubber but with rubbery qualities have now been produced, and they offer a whole constellation of desirable properties. Since they are not actually rubbers, they are called *elastomers* (an abbreviation of *elastic polymer*).

The first rubber-unlike elastomer had been discovered in 1918. This was a *polysulfide rubber;* its molecule was a chain composed of pairs of carbon atoms alternating with groups of four sulfur atoms. The substance was given the name Thiokol, the prefix coming from the Greek word for "sulfur." The odor involved in its preparation held it in abeyance for a long time, but eventually it was put into commercial production.

Elastomers have also been formed from acrylic monomers, fluorocarbons, and silicones. Here, as in almost every field he or she touches, the organic chemist works as an artist, using materials to create new forms and improve upon nature.

Chapter 12

The Proteins

Amino Acids

Early in their study of living matter, chemists noticed a group of substances that behaved in a peculiar manner. Heating changed these substances from the liquid to the solid state, instead of the other way round. The white of eggs, a substance in milk (*casein*), and a component of the blood (*globulin*) were among the things that showed this property. In 1777, the French chemist Pierre Joseph Macquer put all the substances that coagulate on heating into a special class that he called *albuminous*, after *albumen*, the name the Roman encyclopedist Pliny had given to egg white.

When the nineteenth-century organic chemists undertook to analyze the albuminous substances, they found these compounds considerably more complicated than other organic molecules. In 1839, the Dutch chemist Gerardus Johannes Mulder worked out a basic formula, $C_{40}H_{62}O_{12}N_{10}$, which he thought the albuminous substances had in common. He believed that the various albuminous compounds were formed by the addition to this central formula of small sulfur-containing groups or phosphorus-containing groups. Mulder named his root formula *protein* (a word suggested to him by the inveterate word-coiner Berzelius), from a Greek word meaning "of first importance." Presumably the term was meant merely to signify that this core formula was of first importance in determining the structure of the albuminous substances; but as things turned out, it proved to be very apt for the substances themselves. The *proteins*, as they came to be known, were soon found to be of key importance to life.

Within a decade after Mulder's work, Justus von Liebig had established that proteins are even more essential for life than carbohydrates or fats: they supply not only carbon, hydrogen, and oxygen but also nitrogen, sulfur, and often phosphorus, which are absent from fats and carbohydrates.

The attempts of Mulder and others to work out complete empirical formulas for proteins were doomed to failure at the time they were made. The protein molecule is far too complicated to be analyzed by the methods then available. However, a start had already been made on another line of attack that was eventually to reveal, not only the composition, but also the structure of proteins. Chemists had begun to learn something about the building blocks of which they are made.

In 1820, Henri Braconnot, having succeeded in breaking down cellulose into its glucose units by heating the cellulose in acid (see chapter 11), decided to try the same treatment with gelatin, an albuminous substance. The treatment yielded a sweet, crystalline substance. Despite Braconnot's first suspicions, this turned out to be not a sugar but a nitrogen-containing compound, for ammonia (NH_3) could be obtained from it. Nitrogen-containing substances are conventionally given names ending in *-ine*, and the compound isolated by Braconnot is now called *glycine*, from the Greek word for "sweet."

Shortly afterward, Braconnot obtained a white, crystalline substance by heating muscle tissue with acid. He named this one *leucine*, from the Greek word for "white."

Eventually, when the structural formulas of glycine and leucine were worked out, they were found to have a basic resemblance:

Glycine   Leucine

Each compound, as you see, has at its ends an amine group (NH_2) and a carboxyl group (COOH). Because the carboxyl group gives acid properties to any molecule that contains it, molecules of this kind were named *amino acids*. Those that have the amine group and carboxyl group linked by a single carbon atom between them, as both these molecules have, are called *alpha-amino acids*.

As time went on, chemists isolated other alpha-amino acids from proteins. For instance, Liebig obtained one from the protein of milk (casein), which he called *tyrosine* (from the Greek word for "cheese"; *casein* itself comes from the Latin word for "cheese"):

$$
\begin{array}{c}
\text{OH} \\
| \\
\diagup \text{C} \diagdown \\
\text{CH} \qquad \text{CH} \\
| \qquad\quad || \\
\text{CH} \qquad \text{CH} \\
\diagdown \text{C} \diagup \\
| \\
\text{CH}_2 \\
| \qquad\qquad\; O \\
\qquad\qquad\qquad \diagup\!\!/ \\
\text{NH}_2 - \text{CH} - \text{C} \\
\qquad\qquad\qquad \diagdown \text{OH}
\end{array}
$$

Tyrosine

The differences among the various alpha-amino acids lie entirely in the nature of the atom grouping attached to that single carbon atom between the amine and the carboxyl groups. Glycine, the simplest of all the amino acids, has only a pair of hydrogen atoms attached there. The others all possess a carbon-containing *side chain* attached to that carbon atom.

I shall give the formula of just one more amino acid, which will be useful in connection with matters to be discussed later in the chapter. It is *cystine*, discovered in 1899 by the German chemist K. A. H. Mörner. This is a double-headed molecule containing two atoms of sulfur:

$$
\begin{array}{c}
\qquad\qquad\qquad\quad O \\
\qquad\qquad\qquad\; \diagup\!\!/ \\
\text{NH}_2 - \text{CH} - \text{C} \\
\qquad\qquad | \qquad\;\; \diagdown \text{OH} \\
\qquad\qquad \text{CH}_2 \\
\qquad\qquad | \\
\qquad\qquad \text{S} \\
\qquad\qquad | \\
\qquad\qquad \text{S} \\
\qquad\qquad | \\
\qquad\qquad \text{CH}_2 \\
\qquad\qquad | \qquad\qquad\;\; O \\
\qquad\qquad\qquad\qquad\;\; \diagup\!\!/ \\
\text{NH}_2 - \text{CH} - \text{C} \\
\qquad\qquad\qquad\;\;\; \diagdown \text{OH}
\end{array}
$$

Cystine

Actually, cystine had first been isolated in 1810 by the English chemist William Hyde Wollaston from a bladder stone; hence, its name from the Greek word for "bladder." What Mörner did was to show that this century-old compound is a component of protein as well as the substance in bladder stones.

Cystine is easily *reduced* (a term that, chemically, is the opposite of *oxidized*): that is, it will easily add on two hydrogen atoms, which fall into place at the S-S bond. The molecule then divides into two halves, each containing an -SH (*mercaptan*, or *thiol*) group. This reduced half is *cysteine* and is easily oxidized back to cystine.

The general fragility of the thiol group is such that it is important to the functioning of a number of protein molecules. A delicate balance and a capability of moving this way or that under slight impulse is the hallmark of the chemicals most important to life; the members of the thiol group are among the atomic combinations that contribute to this ability.

Altogether, nineteen important amino acids (that is, occurring in most proteins) have now been identified. The last of these was discovered in 1935 by the American chemist William Cumming Rose. It is unlikely that any other common ones remain to be found.

THE COLLOIDS

By the end of the nineteenth century, biochemists had become certain that proteins are giant molecules built up of amino acids, just as cellulose is constructed of glucose and rubber of isoprene units. But there is this important difference: whereas cellulose and rubber are made with just one kind of building block, a protein is built from a number of different amino acids. Hence, working out protein structure would pose special and subtle problems.

The first problem was to find out just how the amino acids are joined together in the protein-chain molecule. Emil Fischer made a start on the problem by linking amino acids together in chains, in such a way that the carboxyl group of one amino acid was always joined to the amino group of the next. In 1901, he achieved his first such condensation, linking one glycine molecule to another with the elimination of a molecule of water:

$$NH_2 - CH_2 - C \overset{O}{\underset{}{\overset{\parallel}{-}}} OH + NH_2 - CH_2 - C \overset{O}{\underset{OH}{\overset{\parallel}{\diagup}}}$$

$$\downarrow$$

$$NH_2 - CH_2 - C \overset{O}{\underset{}{\overset{\parallel}{-}}} NH - CH_2 - C \overset{O}{\underset{OH}{\overset{\parallel}{\diagup}}} + H_2O$$

This is the simplest condensation possible. By 1907, Fischer had synthesized a chain made up of eighteen amino acids, fifteen of them glycine and the remaining three leucine. This molecule did not show any of the obvious properties of proteins, but Fischer felt that was only because the chain was not long enough. He called his synthetic chains *peptides*, from a Greek word meaning "digest," because he believed that proteins broke down into such groups when they were digested. Fischer named the combination of the carboxyl's carbon with the amine group a *peptide link*.

In 1932, the German biochemist Max Bergmann (a pupil of Fischer's) devised a method of building up peptides from various amino acids. Using Bergmann's method, the Polish-American biochemist Joseph Stewart Fruton prepared peptides that could be broken down, by digestive juices, into smaller fragments. Since there was good reason to believe that digestive juices would *hydrolyze* (split by the addition of water) only one kind of molecular bond, the bond between the amino acids in the synthetic peptides must therefore be of the same kind as the one joining amino acids in true proteins. The demonstration laid to rest any lingering doubts about the validity of Fischer's peptide theory of protein structure.

Still, the synthetic peptides of the early decades of the twentieth century were very small and nothing like proteins in their properties. Fischer had made one consisting of eighteen amino acids, as I have said; in 1916, the Swiss chemist Emil Abderhalden went him one better by preparing a peptide with nineteen amino acids, but that held the record for thirty years. And chemists knew that such a peptide must be a tiny fragment indeed compared with the size of a protein molecule, because the molecular weights of proteins were enormous.

Consider, for instance, hemoglobin, a protein of the blood. Hemoglobin contains iron, making up just 0.34 percent of the weight of the molecule. Chemical evidence indicates that the hemoglobin molecule has four atoms of iron, so the total molecular weight must be about 67,000; four atoms of iron, with a total weight of 4×55.85, would come to 0.34 percent of such a molecular weight. Consequently, hemoglobin must contain about 550 amino acids (the average molecular weight of the amino acids being about 120). Compare that with Abderhalden's puny nineteen. And hemoglobin is only an average-sized protein.

The best measurement of the molecular weights of proteins has been obtained by whirling them in a *centrifuge,* a spinning device that pushes particles outward from the center by centrifugal force (figure 12.1). When the

Figure 12.1. Principle of the centrifuge.

DENSE PARTS
MOVE OUTWARD

LIGHT PARTS
CONCENTRATE
NEAR CENTER

SPINNING BOWL

centrifugal force is more intense than the earth's gravitational force, particles suspended in a liquid will settle outward away from the center at a faster rate than they would settle downward under gravity. For instance, red blood corpuscles will settle out quickly in such a centrifuge, and fresh milk will separate into two fractions, the fatty cream and the denser skim milk. These particular separations will take place slowly under ordinary gravitational forces, but centrifugation speeds them up.

Protein molecules, though very large for molecules, are not heavy enough to settle out of solution under gravity; nor will they settle out rapidly in an ordinary centrifuge. But in 1923, the Swedish chemist Theodor Svedberg developed an *ultracentrifuge* capable of separating molecules according to their weight. This high-speed device whirls at more than 10,000 revolutions per second and produces centrifugal forces up to 900,000 times as intense as the gravitational force at the earth's surface. For his contributions to the study of suspensions, Svedberg received the Nobel Prize in chemistry in 1926.

With the ultracentrifuge, chemists were able to determine the molecular weights of a number of proteins on the basis of their rate of sedimentation (measured in *svedbergs* in honor of the chemist). Small proteins turned out to have molecular weights of only a few thousand and to contain perhaps not more than fifty amino acids (still decidedly more than nineteen). Other proteins have molecular weights in the hundreds of thousands and even in the millions, which means that they must consist of thousands or tens of thousands of amino acids. The possession of such large molecules put proteins into a class of substances that have only been studied systematically from the mid-nineteenth century onward.

The Scottish chemist Thomas Graham was the pioneer in this field through his interest in *diffusion*—that is, in the manner in which the molecules of two substances, brought into contact, will intermingle. He began by studying the rate of diffusion of gases through tiny holes or fine tubes. By 1831, he was able to show that the rate of diffusion of a gas was inversely proportional to the square root of its molecular weight (*Graham's law*). (It was through the operation of Graham's law that uranium 235 was separated from uranium 238, by the way.)

In following decades, Graham passed to the study of the diffusion of dissolved substances. He found that solutions of such compounds as salt, sugar, or copper sulfate would find their way through a blocking sheet of parchment (presumably containing submicroscopic holes). On the other hand, solutions of such materials as gum arabic, glue, and gelatin would not. Clearly, the giant molecules of the latter group of substances would not fit through the holes in the parchment.

Graham called materials that could pass through parchment (and that happened to be easily obtained in crystalline form) *crystalloids*. Those that did not, such as glue (in Greek, *kolla*), he called *colloids*. The study of giant molecules (or giant aggregates of atoms, even where these do not form distinct molecules) thus came to be known as *colloid chemistry*. Because proteins and other key molecules in living tissue are of giant size, colloid chemistry is of particular importance to *biochemistry* (the study of the chemical reactions proceeding in living tissue).

Advantage can be taken of the giant size of protein molecules in a number of ways. Suppose that pure water is on one side of a sheet of parchment and a colloidal solution of protein on the other. The protein molecules cannot pass through the parchment; moreover, they block the passage of some of the water molecules, which might otherwise move through. For this reason, water moves more readily into the colloidal portion of the system than out of it. Fluid builds up on the side of the protein solution and sets up an *osmotic pressure.*

In 1877, the German botanist Wilhelm Pfeffer showed how one could measure this osmotic pressure and from it determine the molecular weight of a giant molecule. It was the first reasonably good method for estimating the size of such molecules.

Again, protein solutions could be placed in bags made of *semipermeable membranes* (membranes with pores large enough to permit the passage of small, but not large, molecules). If these were placed in running water, small molecules and ions would pass through the membrane and be washed away, while the large protein molecule would remain behind. This process of *dialysis* is the simplest method of purifying protein solutions.

Molecules of colloidal size are large enough to scatter light; small molecules cannot. Furthermore, light of short wavelength is more efficiently scattered than that of long wavelength. The first to note this effect, in 1869, was the Irish physicist John Tyndall; in consequence, it is called the *Tyndall effect.* The blue of the sky is explained now by the scattering effect of dust particles in the atmosphere upon the short-wave sunlight. At sunset, when light passes through a greater thickness of atmosphere rendered particularly dusty by the activity of the day, enough light is scattered to leave chiefly the red and the orange, thus accounting for the beautiful ruddy color of sunsets.

Light passing through a colloidal solution is scattered so that it can be seen as a visible cone of illumination when viewed from the side. Solutions of crystalloidal substances do not show such a visible cone of light when illuminated, and are *optically clear.* In 1902, the Austro-German chemist Richard Adolf Zsigmondy took advantage of this observation to devise an *ultramicroscope,* which viewed a colloidal solution at right angles, with individual particles (too small to be seen in an ordinary microscope) showing up as bright dots of light. For his endeavor, he received the Nobel Prize for chemistry in 1925.

The protein chemists naturally were eager to synthesize long, *polypeptide* chains, with the hope of producing proteins. But the methods of Fischer and Bergmann allowed only one amino acid to be added at a time—a procedure that seemed then to be completely impractical. What was needed was a procedure that would cause amino acids to join up in a kind of chain reaction, such as Baekeland had used in forming his high-polymer plastics. In 1947, both the Israeli chemist E. Katchalski and the Harvard chemist Robert Woodward (who had synthesized quinine) reported success in producing polypeptides through chain-reaction polymerization. Their starting material was a slightly modified amino acid. (The modification eliminated itself neatly during the reaction.) From this beginning, they built up synthetic polypeptides consisting of as many as a hundred or even a thousand amino acids.

These chains are usually composed of only one kind of amino acid, such as glycine or tyrosine, and are therefore called *polyglycine* or *polytyrosine*. It is also possible, by beginning with a mixture of two modified amino acids, to form a polypeptide containing two different amino acids in the chain. But these synthetic constructions resemble only the simplest kind of protein—for example, *fibroin*, the protein in silk.

THE POLYPEPTIDE CHAINS

Some proteins are as fibrous and crystalline as cellulose or nylon: for example, fibroin; keratin, the protein in hair and skin; and collagen, the protein in tendons and in connective tissue. The German physicist R. O. Herzog proved the crystallinity of these substances by showing that they diffract X rays. Another German physicist, Rudolf Brill, analyzed the pattern of the diffraction and determined the spacing of the atoms in the polypeptide chain. The British biochemist William Thomas Astbury and others in the 1930s obtained further information about the structure of the chain by means of X-ray diffraction. They were able to calculate with reasonable precision the distances between adjacent atoms and the angles at which adjacent bonds are set. And they learned that the chain of fibroin is fully extended: that is, the atoms are in as nearly a straight line as the angles of the bonds between them permit.

This full extension of the polypeptide chain is the simplest possible arrangement. It is called the *beta configuration*. When hair is stretched, its keratin molecule, like that of fibroin, takes up this configuration. (If hair is moistened, it can be stretched up to three times its original length.) But in its ordinary, unstretched state, keratin shows a more complicated arrangement, called the *alpha configuration*.

In 1951, Linus Pauling and Robert Brainard Corey of the California Institute of Technology suggested that, in the alpha configuration, polypeptide chains take a *helical* shape (like a spiral staircase). After building various models to see how the structure would arrange itself if all the bonds between atoms lay in their natural directions without strain, they decided that each turn of the helix would have the length of 3.6 amino acids, or 5.4 angstrom units.

What enables a helix to hold its structure? Pauling suggested that the agent is the so-called *hydrogen bond*. As we have seen, when a hydrogen atom is attached to an oxygen or a nitrogen atom, the latter holds the major share of the bonding electrons, so that the hydrogen atom has a slight positive charge and the oxygen or nitrogen a slight negative charge. In the helix, it appears, a hydrogen atom periodically occurs close to an oxygen or a nitrogen atom on the turn of the helix immediately above or below it. The slightly positive hydrogen atom is attracted to its slightly negative neighbor. This attraction has only 1/20 of the force of an ordinary chemical bond, but it is strong enough to hold the helix in place. However, a pull on the fiber easily uncoils the helix and thereby stretches the fiber.

We have considered so far only the "backbone" of the protein molecule—the chain that runs . . . CCNCCNCCNCCN. . . . But the various side chains of the amino acids also play an important part in protein structure.

All the amino acids except glycine have at least one asymmetric carbon atom—the one between the carboxyl group and the amine group. Thus each could exist in two optically active isomers. The general formulas of the two isomers are:

$$
\begin{array}{ccc}
\overset{\displaystyle O}{\underset{\displaystyle \|}{}} & & \overset{\displaystyle O}{\underset{\displaystyle \|}{}} \\
C - OH & & C - OH \\
| & & | \\
H - C - NH_2 & & NH_2 - C - H \\
| & & | \\
\text{side chain} & & \text{side chain} \\
\text{D-amino acid} & & \text{L-amino acid}
\end{array}
$$

However, it seems quite certain, from both chemical and X-ray analysis, that polypeptide chains are made up only of L-amino acids. In this situation, the side chains stick out alternately on one side of the backbone and then the other. A chain composed of a mixture of both isomers would not be stable, because, whenever an L-amino and a D-amino acid were next to each other, two side chains would be sticking out on the same side, which would crowd them and strain the bonds.

The side chains are important factors in holding neighboring peptide chains together. Wherever a negatively charged side chain on one chain is near a positively charged side chain on its neighbor, they will form an electrostatic link. The side chains also provide hydrogen bonds that can serve as links. And the double-headed amino acid cystine can insert one of its amine-carboxyl sequences in one chain and the other in the next. The two chains are then tied together by the two sulfur atoms in the side chain (the *disulfide link*). The binding together of polypeptide chains accounts for the strength of protein fibers. It explains the remarkable toughness of the apparently fragile spider web and the fact that keratin can form structures as hard as fingernails, tiger claws, alligator scales, and rhinoceros horns.

PROTEINS IN SOLUTION

All this nicely describes the structure of protein fibers. What about proteins in solution? What sort of structure do they have?

They certainly possess a definite structure, but it is extremely delicate. Gentle heating or stirring of a solution or the addition of a bit of acid or alkali or any of a number of other environmental stresses will *denature* a dissolved protein: that is, the protein loses its ability to perform its natural functions, and many of its properties change. Furthermore, denaturation usually is irreversible: for instance, a hard-boiled egg can never be un-hard-boiled again.

It seems certain that denaturation involves the loss of some specific configuration of the polypeptide backbone. Just what feature of the structure is destroyed? X-ray diffraction will not help us when proteins are in solution, but other techniques are available.

In 1928, for instance, the Indian physicist Chandrasekhara Venkata Raman found that light scattered by molecules in solution was, to some extent, altered in wavelength. From the nature of the alteration, deductions could be made about the structure of the molecule. For this discovery of the *Raman effect*, Raman received the 1930 Nobel Prize for physics. (The altered wavelengths of light are usually referred to as the *Raman spectrum* of the molecule doing the scattering.)

Another delicate technique was developed twenty years later, one based on the fact that atomic nuclei possess magnetic properties. Molecules exposed to a high intensity magnetic field will absorb certain frequencies of radio waves. From such absorption, referred to as *nuclear magnetic resonance* and frequently abbreviated NMR, information concerning the bonds between atoms can be deduced. In particular, NMR techniques can locate the position of the small hydrogen atoms within molecules, as X-ray diffraction cannot do. NMR techniques were worked out in 1946 by two teams, working independently: one under E. M. Purcell (later to be the first to detect the radio waves emitted by the neutral hydrogen atom in space; see chapter 2); and the other under the Swiss-American physicist Felix Bloch. Purcell and Bloch shared the Nobel Prize for physics in 1952 for this feat.

To return, then, to the question of the denaturation of proteins in solution. The American chemists Paul Mead Doty and Elkan Rogers Blout used light-scattering techniques on solutions of synthetic polypeptides and found them to have a helical structure. By changing the acidity of the solution, Doty and Blout could break down the helices into randomly curved coils; by readjusting the acidity, they could restore the helices. And they showed that the conversion of the helices to random coils reduced the amount of the solution's optical activity. It was even possible to show which way a protein helix is twisted: it runs in the direction of a right-handed screw thread.

All these findings suggest that the denaturation of a protein involves the destruction of its helical structure.

BREAKING DOWN A PROTEIN MOLECULE

So far I have taken an over-all look at the structure of the protein molecule —the general shape of the chain. What about the details of its construction? For instance, how many amino acids of each kind are there in a given protein molecule?

We might break down a protein molecule into its amino acids (by heating it with acid) and then determine how much of each amino acid is present in the mixture. Unfortunately, some of the amino acids resemble each other chemically so closely that it is almost impossible to get clear-cut separations by ordinary chemical methods. The amino acids can, however, be separated neatly by chromatography (see chapter 6). In 1941, the British biochemists Archer John Porter Martin and Richard Laurence Millington Synge pioneered the application of chromatography to this purpose. They introduced the use of starch as the packing material in the column. In 1948, the American biochemists Stanford Moore and William Howard Stein brought the starch chromatography of amino acids to a high pitch of efficiency and, as a result, shared the 1972 Nobel Prize in chemistry.

After the mixture of amino acids has been poured into the starch column, and all the amino acids have attached themselves to the starch particles, they are slowly washed down the column with fresh solvent. Each amino acid moves down the column at its own characteristic rate. As each emerges at the bottom separately, the drops of solution of that amino acid are caught in a container. The solution in each container is then treated with a chemical that turns the amino acid into a colored product. The intensity of the color is a measure of the amount of the particular amino acid present. This color intensity is measured by an instrument called a *spectrophotometer*, which indicates the intensity by means of the amount of light of the particular wavelength that is absorbed (figure 12.2).

Figure 12.2. A spectrophotometer. The beam of light is split into two, so that one beam passes through the specimen being analyzed and the other goes directly to the photocell. Since the weakened beam that has passed through the specimen liberates fewer electrons in the photocell than the unabsorbed beam does, the two beams create a difference in potential that measures the amount of absorption of the light by the specimen.

(Spectrophotometers can, by the way, be used for other kinds of chemical analysis. If light of successively increased wavelength is sent through a solution, the amount of absorption changes smoothly, rising to maxima at some wavelengths and falling to minima at others. The result is an *absorption spectrum*. A given atomic group has its own characteristic absorption peak or peaks. This is especially true in the region of the infrared, as was first shown by the American physicist William Weber Coblentz shortly after 1900. His instruments were too crude to make the technique practical then; but since the Second World War, the *infrared spectrophotometer*, designed to scan, automatically, the spectrum from 2 to 40 microns, and to record the results, has come into increasing use for analysis of the structure of complex compounds. *Optical methods* of chemical analysis, involving radio-wave absorption, light absorption, light scattering, and so on, are extremely delicate and nondestructive—the sample survives the inspection, in other words—and are completely replacing the classical analytical methods of Liebig, Dumas, and Pregl that were mentioned in the previous chapter.)

The measurement of amino acids with starch chromatography is quite satisfactory; but by the time this procedure was developed, Martin and Synge had worked out a simpler method of chromatography. It is called *paper chromatography* (figure 12.3). The amino acids are separated on a sheet of

Figure 12.3. Paper chromatography.

filter paper (an absorbent paper made of particularly pure cellulose). A drop or two of a mixture of amino acids is deposited near a corner of the sheet, and this edge of the sheet is then dipped into a solvent, such as butyl alcohol. The solvent slowly creeps up the paper through capillary action. (Dip the corner of a blotter into water and see it happen yourself.) The solvent picks up the molecules in the deposited drop and sweeps them along the paper. As in column chromatography, each amino acid moves up the paper at a characteristic rate. After a while the amino acids in the mixture become separated in a series of spots on the sheet. Some of the spots may contain two or three amino acids. To separate these, the filter paper, after being dried, is turned around ninety degrees from its first position, and the new edge is now dipped into a second solvent which will deposit the components in separate spots. Finally, the whole sheet, after once again being dried, is washed with chemicals that cause the patches of amino acids to show up as colored or darkened spots. It is a dramatic sight: all the amino acids, originally mixed in a single solution, are now spread out over the length and breadth of the paper in a mosaic of colorful spots. Experienced biochemists can identify each amino acid by the spot it occupies, and thus can read the composition of the original protein almost at a glance. By dissolving a spot, they can even measure how much of a particular amino acid was present in the protein. For their development of this technique, Martin and Synge received the 1952 Nobel Prize in chemistry.

(Martin, along with A. T. James, applied the principles of this technique to the separation of gases in 1952. Mixtures of gases or vapors may be passed through a liquid solvent or over an adsorbing solid by means of a current of inert *carrier gas,* such as nitrogen or helium. The mixture is pushed through and emerges at the other end separated. Such *gas chromatography* is particularly useful because of the speed of its separations and the great delicacy with which it can detect trace impurities.)

Chromatographic analysis yielded accurate estimates of the amino-acid contents of various proteins. For instance, the molecule of a blood protein called *serum albumin* was found to contain 15 glycines, 45 valines, 58 leucines, 9 isoleucines, 31 prolines, 33 phenylalanines, 18 tyrosines, 1 tryptophan, 22 serines, 27 threonines, 16 cystines, 4 cysteines, 6 methionines, 25 arginines, 16 histidines, 58 lysines, 46 aspartic acids, and 80 glutamic acids—a total of 526 amino acids of 18 different types built into a protein with a molecular weight of about 69,000. (In addition to these 18, there is one other common amino acid—alanine.)

The German-American biochemist Erwin Brand suggested a system of symbols for the amino acids which is now in general use. To avoid confusion with the symbols of the elements, he designated each amino acid by the first three letters of its name, instead of just the initial. There are a few special variations: cystine is symbolized CyS, to show that its two halves are usually incorporated in two different chains; cysteine is CySH, to distinguish it from cystine; and isoleucine is Ileu rather than Iso, for *iso* is the prefix of many chemical names.

In this shorthand, the formula of serum albumin can be written: Gly_{15} Val_{45} Leu_{58} $Ileu_9$ Pro_{31} Phe_{33} Tyr_{18} Try_1 Ser_{22} Thr_{27} CyS_{32} $CySH_4$ Met_6 Arg_{25} His_{16} Lys_{58} Asp_{46} Glu_{80} —more concise, you will admit, though certainly nothing to be rattled off.

ANALYZING THE PEPTIDE CHAIN

Discovering the empirical formula of a protein was only half the battle— in fact, much less than half. Now came the far more difficult task of deciphering the structure of a protein molecule. There was every reason to believe that the properties of every protein depend on exactly how—in what order—all those amino acids are arranged in the molecular chain. This assumption presents the biochemist with a staggering problem. The number of possible arrangements in which nineteen amino acids can be placed in a chain (even assuming that only one of each is used) comes to nearly 120 million billion. If you find this hard to believe, try multiplying out 19 times 18 times 17 times 16, and so on—the way the number of possible arrangements is calculated. And if you do not trust the arithmetic, get nineteen checkers, number them 1 to 19, and see in how many different orders you can arrange them. I guarantee you will not continue the game long.

When you have a protein of the size of serum albumin, composed of more than 500 amino acids, the number of possible arrangements comes out to something like 10^{600}—that is, 1 followed by 600 zeros. This is a completely fantastic number—far more than the number of subatomic particles in the entire known universe—or, for that matter, far more than the universe could hold if it were packed solid with such particles.

Nevertheless, although it may seem hopeless to find out which one of all those possible arrangements a serum albumin molecule actually possesses, this sort of problem has actually been tackled and solved.

In 1945, the British biochemist Frederick Sanger set out to determine the order of amino acids in a peptide chain. He started by trying to identify the amino acid at one end of the chain—the amine end.

Obviously, the amine group of this end amino acid (called the *N-terminal amino acid*) is free—that is, not attached to another amino acid. Sanger made use of a chemical that combines with a free amine group but not with an amine group bound to a carboxyl group and produces a DNP (*dinitrophenyl*) derivative of the peptide chain. With DNP he could label the N-terminal amino acid, and since the bond holding this combination together is stronger than the bonds linking the amino acids in the chain, he could break up the chain into its individual amino acids and isolate the one with the DNP label. As it happens, the DNP group has a yellow color, so this particular amino acid, with its DNP label, shows up as a yellow spot on a paper chromatogram.

Thus, Sanger was able to separate and identify the amino acid at the amine end of a peptide chain. In a similar way, he identified the amino acid at the other end of the chain—the one with a free carboxyl group, called the *C-terminal amino acid*. He was also able to peel off a few other amino acids one by one and identify the *end sequence* of a peptide chain in several cases.

Now Sanger proceeded to attack the peptide chain all along its length. He worked with *insulin*, a protein that has the merit of being very important to the functioning of the body and the added virtue of being rather small for a protein, having a molecular weight of only 6,000 in its simplest form. DNP treatment showed this molecule to consist of two peptide chains, for it contains two different N-terminal amino acids. The two chains are joined by cystine molecules. By a chemical treatment that broke the bond between the two sulfur atoms in the cystine, Sanger split the insulin molecule into its two peptide chains, each intact. One of the chains had glycine as the N-terminal amino acid (call it the *G-chain*), and the other had phenylalanine as the N-terminal amino acid (the *P-chain*). The two could now be worked on separately.

Sanger and a co-worker, Hans Tuppy, first broke up the chains into individual amino acids and identified the twenty-one amino acids that make up the G-chain and the thirty that compose the P-chain. Next, to learn some of the sequences, they broke the chains, not into individual amino acids, but into fragments consisting of two or three. This task could be done by partial hydrolysis, breaking only the weaker bonds in the chain, or by attacking the insulin with certain digestive substances which broke only certain links between amino acids and left the others intact.

By these devices Sanger and Tuppy broke each of the chains into many different pieces. For instance, the P-chain yielded 48 different fragments, 22 of which were made up of two amino acids (*dipeptides*), 14 of three, and 12 of more than three.

The various small peptides, after being separated, could then be broken down into their individual amino acids by paper chromatography. Now the investigators were ready to determine the order of the amino acids in these fragments. Suppose they had a dipeptide consisting of valine and isoleucine. The question would be: Was the order Val-Ileu or Ileu-Val? In other words, was valine or isoleucine the N-terminal amino acid? (The amine group, and consequently the N-terminal unit, is conventionally considered to be at the left end of a chain.) Here the DNP label could provide the answer. If it was present on the valine, that would be the N-terminal amino acid, and the arrangement in the dipeptide would then be established to be Val-Ileu. If it was present on the isoleucine, it would be Ileu-Val.

The arrangement in a fragment consisting of three amino acids also could be worked out. Say its components were leucine, valine, and glutamic acid. The DNP test could first identify the N-terminal amino acid. If it was, say, leucine, the order had to be either Leu-Val-Glu or Leu-Glu-Val. Each of these combinations was then synthesized and deposited as a spot on a chromatogram to see which would occupy the same place on the paper as did the fragment being studied.

As for peptides of more than three amino acids, these could be broken down to smaller fragments for analysis.

After thus determining the structures of all the fragments into which the insulin molecule had been divided, the next step was to put the pieces together in the right order in the chain—in the fashion of a jigsaw puzzle. There were a number of clues. For instance, the G-chain was known to contain only one unit of the amino acid alanine. In the mixture of peptides obtained from the breakdown of G-chains, alanine was found in two combinations: alanine-serine and cystine-alanine. Hence, in the intact G-chain, the order must be CyS-Ala-Ser.

By means of such clues, Sanger and Tuppy gradually put the pieces together. It took a couple of years to identify all the fragments definitely and arrange them in a completely satisfactory sequence; but by 1952, they had worked out the exact arrangement of all the amino acids in the G-chain and the P-chain. They then went on to establish how the two chains were joined. In 1953, their final triumph in deciphering the structure of insulin was announced. The complete structure of an important protein molecule had been worked out for the first time. For this achievement, Sanger was awarded the Nobel Prize in chemistry in 1958.

Biochemists immediately adopted Sanger's methods to determine the structure of other protein molecules. Ribonuclease, a protein molecule consisting of a single peptide chain with 124 amino acids, was conquered in 1959; and the protein unit of tobacco mosaic virus, with 158 amino acids, in 1960. In 1964, trypsin, a protein with 223 amino acids, was deciphered. By 1967, the technique was actually automated. The Swedish-Australian biochemist Pehr Edman devised a *sequenator* which could work on 5 milligrams of pure protein, peeling off and identifying the amino acids one by one. Sixty amino acids of the myoglobin chain were identified in this fashion in four days.

Ever longer peptide chains have been worked out in full detail; and by the 1980s, it was quite certain that the detailed structure of any protein, however large, could be determined. It was only necessary to take the trouble.

In general, such analyses have shown that most proteins have all the various amino acids (or almost all) well represented along the chain. Only a few of the simpler fibrous proteins, such as those found in silk or in tendons, are heavily weighted with two or three amino acids.

In those proteins made up of all nineteen amino acids, the individual amino acids are lined up in no obvious order; there are no easily spotted periodic repetitions. Instead, the amino acids are so arranged that when the chain folds up through the formation of hydrogen bonds here and there, various side chains make up a surface containing the proper arrangement of atomic groupings or of electric-charge pattern to enable the protein to do its work.

SYNTHETIC PROTEINS

Once the amino-acid order in a polypeptide chain was worked out, it became possible to attempt to put together amino acids in just that right order. Naturally, the beginning was a small one. The first protein to be synthesized in the laboratory was *oxytocin*, a hormone with important functions in the body. Oxytocin is extremely small for a protein molecule: it consists of only eight amino acids. In 1953, the American biochemist Vincent du Vigneaud succeeded in synthesizing a peptide chain exactly like that thought to represent the oxytocin molecule. And, indeed, the synthetic peptide showed all the properties of the natural hormone. Du Vigneaud was awarded the Nobel Prize in chemistry in 1955.

More complicated protein-molecules were synthesized as the years passed; but in order to synthesize a specific molecule with particular amino acids arranged in a particular order, the string had to be threaded, so to speak, one at a time. That was as difficult in the 1950s as it had been a half-century earlier in Fischer's time. Each time a particular amino acid was coupled to a chain, the new compound had to be separated from all the rest by tedious procedures, and then a new start had to be made to add one more particular amino acid. At each step, a good part of the material was lost in side reactions, and only small quantities of even simple chains were formed.

Beginning in 1959, however, a team under the leadership of the American biochemist Robert Bruce Merrifield, struck out in a new direction. An amino acid, the beginning of the desired chain, was bound to beads of polystyrene resin. These beads were insoluble in the liquid being used and could be separated from everything else by simple filtration. A new solution would be added containing the next amino acid, which would bind to the first. Again a filtration, then another. The steps between additions were so simple and quick that they could be automated with almost nothing lost. In 1965, the molecule of insulin was synthesized in this fashion; in 1969, it was the turn of the still longer chain of ribonuclease with all its 124 amino acids. Then, in 1970, the Chinese-American biochemist Cho Hao Li synthesized the 188-amino-acid chain of human-growth hormone. In principle, any protein can now be synthesized; it only requires that enough trouble be taken.

THE SHAPE OF THE PROTEIN MOLECULE

With the protein molecule understood, so to speak, as a string of amino acids, it became desirable to take a still more sophisticated view. What is the exact manner in which that amino acid chain bends and curves? What is the exact shape of the protein molecule?

Tackling this problem were the Austrian-English chemist Max Ferdinand Perutz and his English colleague John Cowdery Kendrew. Perutz took as his province hemoglobin, the oxygen-carrying protein of blood, containing something like 12,000 atoms. Kendrew took on myoglobin, a muscle protein similar in function to hemoglobin but only about a quarter the size. As their tool, they used X-ray diffraction studies.

Perutz used the device of combining the protein molecules with a massive atom, such as that of gold or mercury, which was particularly efficient in diffracting X rays. Thus, he got clues that allowed him more accurately to deduce the structure of the molecule without the massive atom. By 1959, myoglobin, and then hemoglobin, the year after, fell into place. It became possible to prepare three-dimensional models in which every single atom could be located in what seemed very likely to be the correct place. In both cases, the protein structure was clearly based upon the helix. As a result, Perutz and Kendrew shared the Nobel Prize in chemistry in 1962.

There is reason to think that the three-dimensional structures worked out by the Perutz-Kendrew techniques are after all determined by the nature of the string of amino acids. The amino-acid string has, so to speak, natural crease *points;* and when they bend, certain interconnections inevitably take place and keep it properly folded. It is possible to determine what these folds and interconnections are by working out all the interatomic distances and the angles at which the connecting bonds are placed, but it is a tedious job indeed. Here, too, computers have been called in to help, and these have not only made the calculation but thrown the results on a screen.

What with one thing or another, the list of protein molecules whose shapes are known in three-dimensional detail is growing rapidly. Insulin, which started the new forays into molecular biology, had its three-dimensional shape worked out by the English biochemist Dorothy Crowfoot Hodgkin in 1969.

Enzymes

Useful consequences follow from the complexity and almost infinite variety of protein molecules. Proteins have a multitude of different functions to perform in living organisms.

One major function is to provide the structural framework of the body. Just as cellulose serves as the framework of plants, so fibrous proteins act in the same capacity for the complex animals. Spiders spin gossamer threads, and insect larvae spin cocoon threads of protein fibers. The scales of fish and reptiles are made up mainly of the protein keratin. Hair, feathers, horns, hoofs, claws, and fingernails—all merely modified scales—also contain keratin. Skin owes its strength and toughness to its high content of keratin. The internal supporting tissues—cartilage, ligaments, tendons, even the organic framework of bones—are made up largely of protein molecules, such as collagen and elastin. Muscle is made of a complex fibrous protein called *actomyosin*.

In all these cases, the protein fibers are more than a cellulose substitute. They are an improvement; they are stronger and more flexible. Cellulose will do to support a plant, which is not called on for any motion more complex than swaying with the wind. But protein fibers must be designed for the bending and flexing of the appendages of the body, for rapid motions and vibrations, and so on.

The fibers, however, are among the simplest of the proteins, in form as well as function. Most of the other proteins have more subtle and more complicated jobs to do.

To maintain life in all its aspects, numerous chemical reactions must proceed in the body. These must go on at high speed and in great variety, each reaction meshing with all the others, for it is not upon any one reaction, but upon all together, that life's smooth workings must depend. Moreover, all the reactions must proceed under the mildest of environments—without high temperatures, strong chemicals, or great pressures. The reactions must be under strict yet flexible control and must be constantly adjusted to the changing characteristics of the environment and the changing needs of the body. The undue slowing down, or speeding up, of even one reaction out of the many thousands would more or less seriously disorganize the body.

All this is made possible by protein molecules.

CATALYSIS

Toward the end of the eighteenth century, chemists, following the leadership of Lavoisier, began to study reactions in a quantitative way—in particular, to measure the rates at which chemical reactions proceed. They quickly noted that reaction rates can be changed drastically by comparatively minor changes in the environment. For instance, when Kirchhoff found that starch could be converted to sugar in the presence of acid, he noticed that while the acid greatly speeded up this reaction, it was not itself consumed in the process. Other such examples were soon discovered. The German chemist Johann Wolfgang Döbereiner found that finely divided platinum (called *platinum black*) encouraged the combination of hydrogen and oxygen to form water— a reaction that, without this help, could take place only at a high temperature. Döbereiner even designed a self-igniting lamp in which a jet of hydrogen, played upon a surface coated with platinum black, caught fire.

Because the "hastened reactions" were usually in the direction of breaking down a complex substance to a simpler one, Berzelius named the phenomenon *catalysis* (from Greek words essentially meaning "break down"). Thus, platinum black came to be called a catalyst for the combination of hydrogen and oxygen, and acid a catalyst for the hydrolysis of starch to glucose.

Catalysis has proved of the greatest importance in industry. For instance, the best way of making sulfuric acid (the most important single inorganic chemical next to air, water, and, perhaps, salt) involves the burning of sulfur —first to sulfur dioxide (SO_2), then to sulfur trioxide (SO_3). The step from the dioxide to the trioxide would not proceed at more than a snail's pace without the help of a catalyst such as platinum black. Finely divided nickel (which has replaced platinum black in most cases, because it is cheaper) and such compounds as copper chromite, vanadium pentoxide, ferric oxide, and manganese dioxide also are important catalysts. In fact, a great deal of the success of an industrial chemical process depends on finding just the right catalyst for the reaction involved. It was the discovery of a new type of catalyst by Ziegler that revolutionized the production of polymers.

How is it possible for a substance, sometimes present only in very small concentrations, to bring about large quantities of reaction without itself being changed?

Well, one kind of catalyst does in fact take part in the reaction, but in a cyclic fashion, so that it is continually restored to its original form. An example is vanadium pentoxide (V_2O_5), which can catalyze the change of sulfur dioxide to sulfur trioxide. Vanadium pentoxide passes on one of its oxygen atoms to SO_2, forming SO_3 and changing itself to vanadyl oxide (V_2O_4). But the vanadyl oxide rapidly reacts with oxygen in the air and is restored to V_2O_5. The vanadium pentoxide thus acts as a middleman, handing an oxygen atom to sulfur dioxide, taking another from the air, handing that to sulfur dioxide, and so on. The process is so rapid that a small quantity of vanadium pentoxide will suffice to bring about the conversion of large quantities of sulfur dioxide; and in the end, the vanadium pentoxide appears unchanged.

In 1902, the German chemist George Lunge suggested that this sort of thing was the explanation of catalysis in general. In 1916, Irving Langmuir went a step farther and advanced an explanation for the catalytic action of substances, such as platinum, that are so nonreactive that they cannot be expected to engage in ordinary chemical reactions. Langmuir suggested that excess valence bonds at the surface of platinum metal would seize hydrogen and oxygen molecules. While held imprisoned in close proximity on the platinum surface, the hydrogen and oxygen molecules would be much more likely to combine to form water molecules than in their ordinary free condition as gaseous molecules. Once a water molecule was formed, it would be displaced from the platinum surface by hydrogen and oxygen molecules. Thus, the process of seizure of hydrogen and oxygen, their combination into water, release of the water, seizure of more hydrogen and oxygen, and formation of more water could continue indefinitely.

This process is called *surface catalysis*. Naturally, the more finely divided the metal, the more surface a given mass will provide, and the more effectively catalysis can proceed. Of course, if any extraneous substance attaches itself firmly to the surface bonds of the platinum, it will *poison* the catalyst.

All surface catalysts are more or less selective, or *specific*. Some easily absorb hydrogen molecules and will catalyze reactions involving hydrogen; others easily absorb water molecules and catalyze condensations or hydrolyses; and so on.

The ability of surfaces to add on layers of molecules (*adsorption*) is widespread and can be put to uses other than catalysis. Silicon dioxide prepared in spongy form (*silica gel*) will adsorb large quantities of water. Packed in with electronic equipment, whose performance would suffer under conditions of high humidity, it acts as a *dessicant*, keeping humidity low.

Again, finely divided charcoal (*activated carbon*) will adsorb organic molecules readily—the larger the organic molecule, the more readily. Activated carbon can be used to decolorize solutions, for it would adsorb the colored impurities (usually of high molecular weight), leaving behind the desired substance (usually colorless and of comparatively low molecular weight).

Activated carbon is also used in gas masks, a use foreshadowed by an English physician, John Stenhouse, who first prepared a charcoal air filter in 1853. The oxygen and nitrogen of air pass through such a mass unaffected, but the relatively large molecules of poison gases are adsorbed.

FERMENTATION

The organic world, too, has its catalysts. Indeed, some of them have been known for thousands of years, though not by that name. They are as old as the making of bread and the brewing of wine.

Bread dough, left to itself and kept from contamination by outside influences, will not rise. Add a lump of *leaven* (from a Latin word meaning "rise"), and bubbles begin to appear, lifting and lightening the dough. The common English word for leaven is *yeast*, possibly descended from a Sanskrit word meaning "to boil."

Yeast also hastens the conversion of fruit juices and grain to alcohol. Here again, the conversion involves the formation of bubbles, so the process is called *fermentation,* from a Latin word meaning "boil." The yeast preparation is often referred to as *ferment.*

It was not until the seventeenth century that the nature of leaven was discovered. In 1680, for the first time, a Dutch investigator, Anton van Leeuwenhoek, saw yeast cells. For the purpose, he made use of an instrument that was to revolutionize biology—the *microscope.* It was based on the bending and focusing of light by lenses. Instruments using combinations of lenses (*compound microscopes*) were devised as early as 1590 by a Dutch spectacle maker, Zacharias Janssen. The early microscopes were useful in principle, but the lenses were so imperfectly ground that the objects magnified were almost useless, fuzzy blobs. Van Leeuwenhoek ground tiny but perfect lenses that magnified quite sharply up to 200 times. He used single lenses (*simple microscope*).

With time, the practice of using good lenses in combinations (for a compound misroscope is, potentially at least, much stronger than a simple one) spread, and the world of the very little opened up further. A century and a half after Leeuwenhoek, a French physicist, Charles Cagniard de la Tour, using a good compound microscope, studied the tiny bits of yeast intently enough to catch them in the process of reproducing themselves. The little blobs were alive. Then, in the 1850s, yeast became a dramatic subject of study.

France's wine industry was in trouble. Aging wine was going sour and becoming undrinkable, and millions of francs were being lost. The problem was placed before the young dean of the Faculty of Sciences at the University of Lille, in the heart of the vineyard area. The young dean was Louis Pasteur, who had already made his mark by being the first to separate optical isomers in the laboratory.

Pasteur studied the yeast cells in the wine under the microscope. It was obvious to him that the cells were of varying types. All the wine contained yeast that brought about fermentation, but those wines that went sour contained another type of yeast in addition. It seemed to Pasteur that the souring action did not get under way until the fermentation was completed. Since there was no need for yeast after the necessary fermentation, why not get rid of all the yeast at that point and avoid letting the wrong kind make trouble?

He therefore suggested to a horrified wine industry that the wine be heated gently after fermentation, in order to kill all the yeast in it. Aging, he predicted, would then proceed without souring. The industry reluctantly tried his outrageous proposal and found, to its delight, that souring ceased, while the flavor of the wine was not in the least damaged by the heating. The wine industry was saved. Furthermore, the process of gentle heating (*pasteurization*) was later applied to milk, to kill any disease germs present.

Other organisms besides yeast hasten breakdown processes. In fact, a process analogous to fermentation takes place in the intestinal tract. The first man to study digestion scientifically was the French physicist René Antoine Ferchault de Réaumur. He used a hawk as his experimental subject and, in 1752, made it swallow small metal tubes containing meat; the tubes protected the meat from any mechanical grinding action, but they had openings, covered by gratings, so that chemical processes in the stomach could act on the meat. Réaumur found that when the hawk regurgitated these tubes, the meat was partly dissolved, and a yellowish fluid was present in the tubes.

In 1777, the Scottish physician Edward Stevens isolated fluid from the stomach (*gastric juice*) and showed that the dissolving process could be made to take place outside the body, thus divorcing it from the direct influence of life.

Clearly, the stomach juices contained something that hastens the breakdown of meat. In 1834, the German naturalist Theodor Schwann added mercuric chloride to the stomach juice and precipitated a white powder. After freeing the powder of the mercury compound, and dissolving what was left, he found he had a very concentrated digestive juice. He called the powder he had discovered *pepsin*, from the Greek word meaning "digest."

Meanwhile, two French chemists, Anselme Payen and Jean François Persoz, had found in malt extract a substance that could bring about the conversion of starch to sugar more rapidly than could acid. They called this *diastase*, from a Greek word meaning "to separate," because they had separated it from malt.

For a long time, chemists made a sharp distinction between living ferments such as yeast cells and nonliving, or *unorganized*, ferments such as pepsin. In 1878, the German physiologist Wilhelm Kühne suggested that the latter be called *enzymes*, from Greek words meaning "in yeast," because their activity was similar to that brought about by the catalyzing substances in yeast. Kühne did not realize how important, indeed universal, that term "enzyme" was to become.

In 1897, the German chemist Eduard Buchner ground yeast cells with sand to break up all the cells and succeeded in extracting a juice that he found could perform the same fermentative tasks that the original yeast cells could. Suddenly the distinction between the ferments inside and outside of cells vanished. It was one more breakdown of the vitalists' semimystical separation of life from nonlife. The term "enzyme" was now applied to all ferments.

For this discovery Buchner received the Nobel Prize in chemistry in 1907.

PROTEIN CATALYSTS

Now it was possible to define an enzyme simply as an organic catalyst. Chemists began to try to isolate enzymes and find out what sort of substances they were. The trouble was that the amount of enzyme in cells and natural juices is very small, and the extracts obtained were invariably mixtures in which it was hard to tell what was an enzyme and what was not.

Many biochemists suspected that enzymes were proteins, because enzyme properties could easily be destroyed, as proteins could be denatured, by gentle heating. But, in the 1920s, the German biochemist Richard Willstätter reported that certain purified enzyme solutions, from which he believed he had eliminated all protein, showed marked catalytic effects. He concluded that enzymes were not proteins but relatively simple chemicals, which might, indeed, utilize a protein as a *carrier molecule*. Most biochemists went along with Willstätter, who was a Nobel Prize winner and had great prestige.

However, the Cornell University biochemist James Batcheller Sumner produced strong evidence against this theory almost as soon as it was advanced. From jackbeans (the white seeds of a tropical American plant), Sumner isolated crystals that, in solution, showed the properties of an enzyme called *urease*, which catalyzes the breakdown of urea to carbon dioxide and ammonia. Sumner's crystals showed definite protein properties, and he could find no way to separate the protein from the enzyme activity. Anything that denatured the protein also destroyed the enzyme. All this seemed to show that what he had was an enzyme in pure and crystalline form, and that enzyme was a protein.

Willstätter's greater fame for a time minimized Sumner's discovery. But, in 1930, the chemist John Howard Northrop and his co-workers at the Rockefeller Institute clinched Sumner's case. They crystallized a number of enzymes, including pepsin, and found all to be proteins. Northrop, furthermore, showed that these crystals are pure proteins and retain their catalytic activity even when dissolved and diluted to the point where the ordinary chemical tests, such as those used by Willstätter, could no longer detect the presence of protein.

Enzymes were thus established to be *protein catalysts*. By now, some 2,000 different enzymes have been identified, and over 200 enzymes have been crystallized; all without exception are proteins.

For their work, Sumner and Northrop shared in the Nobel Prize in chemistry in 1946.

ENZYME ACTION

Enzymes are remarkable as catalysts in two respects—efficiency and specificity. There is an enzyme known as catalase, for instance, that catalyzes the breakdown of hydrogen peroxide to water and oxygen. Now the breakdown of hydrogen peroxide in solution can also be catalyzed by iron filings or manganese dioxide. However, weight for weight, catalase speeds up the rate of breakdown far more than any inorganic catalyst can. Each molecule of catalase can bring about the breakdown of 44,000 molecules of hydrogen peroxide per second at 0° C. The result is that an enzyme need be present only in small concentration to perform its function.

For this same reason, to put an end to life, it takes but small quantities of substances (*poisons*) capable of interfering with the workings of a key enzyme. Heavy metals, when administered in such forms as mercuric chloride or barium nitrate, react with thiol groups, which are essential to the working of many enzymes. The action of those enzymes stops, and the organism is poisoned. Compounds such as potassium cyanide or hydrogen cyanide place their cyanide group (–CN) in combination with the iron atom of other key enzymes and bring death quickly and, it is to be hoped, painlessly, for hydrogen cyanide is the gas used for execution in the gas chambers of some of our Western states.

Carbon monoxide is an exception among the common poisons. It does not act on enzymes primarily but ties up the hemoglobin molecule (a protein but not an enzyme), which ordinarily carries oxygen from lungs to cells but cannot do so with carbon monoxide hanging on to it. Animals that do not use hemoglobin are not harmed by carbon monoxide.

Enzymes, with catalase a good example, are highly specific: catalase breaks down hydrogen peroxide and nothing else; whereas inorganic catalysts, such as iron filings and manganese dioxide, may break down hydrogen peroxide but will also catalyze numerous other reactions.

What accounts for the remarkable specificity of enzymes? Lunge's and Langmuir's theories about the behavior of a catalyst as a middleman suggested an answer. Suppose we consider that an enzyme forms a temporary combination with the *substrate*—the substance whose reaction it catalyzes. The form, or configuration, of the particular enzyme may therefore play a highly important role. Plainly, each enzyme must present a very complicated surface, for it has a number of different side chains sticking out of the peptide backbone. Some of these side chains have a negative charge; some, positive; some, no charge. Some are bulky; some, small. One can imagine that each enzyme may have a surface that just fits a particular substrate. In other words, it fits the substrate as a key fits a lock. Therefore, it will combine readily with that substance, but only clumsily or not at all with others. Hence, the high specificity of enzymes: each has a surface made to order, so to speak, for combining with a particular compound. That being the case, no wonder that proteins are built of so many different units and are constructed by living tissue in such great variety.

This theory of enzyme action was first suggested by the work of an English physiologist, William Maddock Bayliss, working with a digestive enzyme named *trypsin.* In 1913, the theory was used by the German chemist, Leonor Michaelis, and his assistant, Maud Lenora Menten, to work out the *Michaelis-Menten equation* which described the manner in which enzymes carry out their functions, and robbed these catalysts of much of their mystery.

This lock-and-key view of enzyme action was borne out also by the discovery that the presence of a substance similar in structure to a given substrate will slow down or inhibit the substrate's enzyme-catalyzed reaction. The best-known case involves an enzyme called *succinic acid dehydrogenase,* which catalyzes the removal of two hydrogen atoms from succinic acid. That reaction will not proceed in the presence of a substance called *malonic acid,* which is very similar to succinic acid. The structures of succinic acid and malonic acid are:

$$\underset{\text{Succinic acid}}{\begin{array}{c} CH_2 - C \overset{\displaystyle O}{\diagdown}_{OH} \\ | \\ CH_2 - C \overset{\displaystyle O}{\diagdown}_{OH} \end{array}} \qquad \underset{\text{Malonic acid}}{\begin{array}{c} C \overset{\displaystyle O}{\diagdown}_{OH} \\ CH_2 \\ C \overset{\displaystyle O}{\diagdown}_{OH} \end{array}}$$

The only difference between these two molecules is that succinic acid has one more CH_2 group at the left. Presumably the malonic acid, because of its structural similarity to succinic acid, can attach itself to the surface of the enzyme. Once it has pre-empted the spot on the surface to which the succinic acid would attach itself, it remains jammed there, so to speak, and the enzyme is out of action. The malonic acid "poisons" the enzyme, so far as its normal function is concerned. This sort of action is called *competitive inhibition*.

The most positive evidence in favor of the enzyme-substrate-complex theory has come from spectrographic analysis. Presumably, if an enzyme combines with its substrate, there should be a change in the absorption spectrum: the combination's absorption of light should be different from that of the enzyme or the substrate alone. In 1936, the British biochemists David Keilin and Thaddeus Mann detected a change of color in a solution of the enzyme peroxidase after its substrate, hydrogen peroxide, was added. The American biophysicist Britton Chance made a spectral analysis and found that there were two progressive changes in the absorption pattern, one following the other. He attributed the first change in pattern to the formation of the enzyme-substrate complex at a certain rate, and the second to the decline of this combination as the reaction was completed. In 1964, the Japanese biochemist Kunio Yagi announced the isolation of an enzyme-substrate complex, made up of a loose union of the enzyme D-amino acid oxidase and its substrate alanine.

Now the question arises: Is the entire enzyme molecule necessary for catalysis, or would some part of it be sufficient? This is an important question from a practical as well as a theoretical standpoint. Enzymes are in wide use today; they have been put to work in the manufacture of drugs, citric acid, and many other chemicals. If the entire enzyme molecule is not essential and some small fragment of it would do the job, perhaps this active portion could be synthesized, so that the processes would not have to depend on the use of living cells, such as yeasts, molds, and bacteria.

Some promising advances toward this goal have been made. For instance, Northrop found that when a few acetyl groups (CH_3CO) were added to the side chains of the amino acid tyrosine in the pepsin molecule, the enzyme lost some of its activity. There was no loss, however, when acetyl groups were added to the lysine side chains in pepsin. Tyrosine, therefore, must contribute to pepsin's activity, while lysine obviously does not. This was the first indication that an enzyme might possess portions not essential to its activity.

Recently the *active region* of another digestive enzyme was pinpointed with more precision. This enzyme is chymotrypsin. The pancreas first secretes it in an inactive form called *chymotrypsinogen*. This inactive molecule is converted into the active one by the splitting of a single peptide link (accomplished by the digestive enzyme trypsin): that is, it looks as if the uncovering of a single amino acid endows chymotrypsin with its activity. Now it turns out that the attachment of a molecule known as DFP (*diisopropylfluorophosphate*) to chymotrypsin stops the enzyme's activity. Presumably, the DFP attaches itself to the key amino acid. Thanks to its tagging by DFP, that amino acid had been identified as serine. In fact, DFP has also been found to attach itself to serine in other digestive enzymes. In each case, the serine is in the same position in a sequence of four amino acids: glycine–aspartic acid–serine–glycine.

It turns out that a peptide consisting of those four amino acids alone will not display catalytic activity. In some way, the rest of the enzyme molecule plays a role, too. We can think of the four-acid sequence—the active center—as analogous to the cutting edge of a knife, which is useless without a handle.

Nor need the active center, or cutting edge, necessarily exist all in one piece in the amino-acid chain. Consider the enzyme ribonuclease. Now that the exact order of its 124 amino acids is known, it has become possible to devise methods for deliberately altering this or that amino acid in the chain and noting the effect of the change on the enzyme's action. It was discovered that three amino acids, in particular, are necessary for action, but that they are widely separated. They are a histidine in position 12, a lysine in position 41, and another histidine in position 119.

This separation, of course, exists only in the chain viewed as a long string. In the working molecule, the chain is coiled into a specific three-dimensional configuration, held in place by four cystine molecules, stretching across the loops. In such a molecule, the three necessary amino acids are brought together into a close-knit unit.

The matter of an active center was made even more specific in the case of lysozyme, an enzyme found in many places, including tears and nasal mucus. It brings about the dissolution of bacterial cells by catalyzing the breakdown of key bonds in some of the substances that make up the bacterial cell wall. It is as though it causes the wall to crack and the cell contents to leak away.

Lysozyme was the first enzyme whose structure was completely analyzed (in 1965) in three dimensions. Once this was done, it could be shown that the molecule of the bacterial cell wall that is subject to lysozyme's action fits neatly along a cleft in the enzyme structure. The key bond was found to lie between an oxygen atom in the side chain of glutamic acid (position 35) and another oxygen atom in the side chain of aspartic acid (position 52). The two positions were brought together by the folding of the amino-acid chain with just enough separation that the molecule to be attacked could fit in between. The chemical reaction necessary for breaking the bond could easily take place under those circumstances—and it is in this fashion that lysozyme is specifically organized to do its work.

Then, too, it happens sometimes that the cutting edge of the enzyme molecule is not a group of amino acids at all but an atom combination of an entirely different nature. A few such cases will be mentioned later in the book.

We cannot tamper with the cutting edge, but could we modify the handle without impairing the usefulness of the tool? The handle has its purposes, of course. It would seem that the enzyme in its natural state is "jiggly" and can take up several different shapes without much strain. When the substrate adds on to the active site, the enzyme adjusts itself to the shape of the substrate, thanks to the "give" of the non-active portion of the molecule so that the fit becomes tight and the catalytic action is highly efficient. Substrates of slightly different shape might not take advantage of the "give" quite as well and will not be affected—might, indeed, inhibit the enzyme.

Still, the enzyme might be simplified, perhaps, at the cost of the loss of some efficiency, but not all. The existence of different varieties of such protein as insulin, for instance, encourages us to believe that simplification might be possible. Insulin is a hormone, not an enzyme, but its function is highly specific. At a certain position in the G-chain of insulin there is a three-amino-acid sequence that differs in different animals: in cattle, it is alanine-serine-valine; in swine, threonine-serine-isoleucine; in sheep, alanine-glycine-valine; in horses, threonine-glycine-isoleucine; and so on. Yet any of these insulins can be substituted for any other and still perform the same function.

What is more, a protein molecule can sometimes be cut down drastically without any serious effect on its activity (as the handle of a knife or an ax might be shortened without much loss in effectiveness). A case in point is the hormone called ACTH (*adrenocorticotropic hormone*). This is a peptide chain made up of thirty-nine amino acids, whose order has now been fully determined. Up to fifteen of the amino acids have been removed from the C-terminal end without destroying the hormone's activity. On the other hand, the removal of one or two amino acids from the N-terminal end (the cutting edge, so to speak) kills activity at once.

The same sort of thing has been done to an enzyme called *papain*, from the fruit and sap of the papaya tree. Its enzymatic action is similar to that of pepsin. Removal of the pepsin molecule's 180 amino acids from the N-terminal end does not reduce its activity to any detectable extent.

So it is at least conceivable that enzymes may yet be simplified to the point where they will fall within the region of practical mass synthesis. Synthetic enzymes, in the form of fairly simple organic compounds, may then be made on a large scale for various purposes. This would be a form of chemical miniaturization.

Metabolism

An organism, such as the human body, is a chemical plant of great diversity. It breathes in oxygen and drinks water. It takes in as food carbohydrates, fats, proteins, minerals, and other raw materials. It eliminates various indigestible materials plus bacteria and the products of the putrefaction they bring about. It also excretes carbon dioxide via the lungs, gives up water by way of both the lungs and the sweat glands, and excretes urine, which carries off a number of compounds in solution, the chief of these being urea. These chemical reactions determine the body's *metabolism*.

By examining the raw materials that enter the body and the waste products that leave it, we can tell a few things about what goes on within the body. For instance, since protein supplies most of the nitrogen entering the body, we know that urea (NH_2CONH_2) must be a product of the metabolism of proteins. But between protein and urea lies a long, devious, complicated road. Each enzyme of the body catalyzes only a specific small reaction, rearranging perhaps no more than two or three atoms. Every major conversion in the body involves a multitude of steps and many enzymes. Even an apparently simple organism such as the tiny bacterium must make use of many thousands of separate enzymes and reactions.

All this may seem needlessly complex, but it is the very essence of life. The vast complex of reactions in tissues can be controlled delicately by increasing or decreasing the production of appropriate enzymes. The enzymes control body chemistry as the intricate movements of fingers on the strings control the playing of a violin; and without this intricacy, the body could not perform its manifold functions.

To trace the course of the myriads of reactions that make up the body's metabolism is to follow the outline of life. The attempt to follow it in detail, to make sense of the intermeshing of countless reactions all taking place at once, may indeed seem a formidable and even hopeless undertaking. Formidable it is, but not hopeless.

THE CONVERSION OF SUGAR TO ETHYL ALCOHOL

The chemists' study of metabolism began modestly with an effort to find out how yeast cells convert sugar to ethyl alcohol. In 1905, two British chemists, Arthur Harden and William John Young, suggested that this process involves the formation of sugars bearing phosphate groups. Harden and Young were the first to note that phosphorus plays an important role in metabolism (and phosphorus has been looming larger ever since). Harden and Young even found in living tissue a sugar-phosphate ester consisting of the sugar fructose with two phosphate groups (PO_3H_2) attached. This *fructose diphosphate* (still sometimes known as *Harden-Young ester*) was the first *metabolic intermediate* to be identified definitely—the first compound, that is, recognized to be formed momentarily, in the process of passing from the compounds as taken into the body to the compounds eliminated by it. Harden and Young had thus founded the study of *intermediary metabolism*, which concentrates on the nature of such intermediates and the reactions involving them. For this work and for further work on the enzymes involved in the conversion of sugar to alcohol by yeast (see chapter 15), Harden shared the Nobel Prize in chemistry in 1929.

What began by involving only the yeast cell became of far broader importance when the German chemist Otto Fritz Meyerhof demonstrated in 1918 that animal cells, such as those of muscle, break down sugar in much the same way as yeast does. The chief difference is that in animal cells the breakdown does not proceed so far in this particular route of metabolism. Instead of converting the six-carbon glucose molecule all the way down to the two-carbon ethyl alcohol (CH_3CH_2OH), they break it down only as far as the three-carbon lactic acid ($CH_3CHOHCOOH$).

Meyerhof's work made clear for the first time a general principle that has since become commonly accepted: with only minor differences, metabolism follows the same routes in all creatures, from the simplest to the most complex. For his studies on the lactic acid in muscle, Meyerhof shared the Nobel Prize in physiology and medicine in 1922 with the English physiologist Archibald Vivian Hill. The latter had tackled muscle from the standpoint of its heat production and had come to conclusions quite similar to those obtained from Meyerhof's chemical attack.

The details of the individual steps involved in the transition from sugar to lactic acid were evolved between 1937 and 1941 by Carl Ferdinand Cori and his wife Gerty Theresa Cori, working at Washington University in St. Louis. They used tissue extracts and purified enzymes to bring about changes in various sugar-phosphate esters, then put all the changes together like a jigsaw puzzle. The scheme of step-by-step changes that they presented has stood with little modification to this day, and the Coris were awarded a share in the Nobel Prize in physiology and medicine in 1947.

In the path from sugar to lactic acid, a certain amount of energy is produced and is utilized by the cells. The yeast cell lives on it when it is fermenting sugar, and so, when necessary, does the muscle cell. It is important to remember that this energy is obtained without the use of oxygen from the air. Thus, a muscle is capable of working even when it must expend more energy than can be replaced by reactions involving the oxygen brought to it at a relatively slow rate by the blood. As the lactic acid accumulates, however, the muscle grows weary, and eventually it must rest until oxygen breaks up the lactic acid.

METABOLIC ENERGY

Next comes the question: In what form is the energy from the sugar-to-lactic-acid breakdown supplied to the cells, and how do they use it? The German-born American chemist Fritz Albert Lipmann found an answer in researches beginning in 1941. He showed that certain phosphate compounds formed in the course of carbohydrate metabolism store unusual amounts of energy in the bond that connects the phosphate group to the rest of the molecule. This *high-energy phosphate bond* is transferred to energy carriers present in all cells. The best known of these carriers is *adenosine triphosphate* (ATP). The ATP molecule and certain similar compounds represent the small currency of the body's energy. They store the energy in neat, conveniently sized, readily negotiable packets. When the phosphate bond is hydrolyzed off, the energy is available to be converted into chemical energy for the building of proteins from amino acids, or into electrical energy for the transmission of a nerve impulse, or into kinetic energy via the contraction of muscle, and so on. Although the quantity of ATP in the body is small at any one time, there is always enough (while life persists), for as fast as the ATP molecules are used up, new ones are formed.

For his key discovery, Lipmann shared the Nobel Prize in physiology and medicine in 1953.

The mammalian body cannot convert lactic acid to ethyl alcohol (as yeast can); instead, by another route of metabolism, the body bypasses ethyl alcohol and breaks down lactic acid all the way to carbon dioxide (CO_2) and water. In so doing, it consumes oxygen and produces a great deal more energy than is produced by the non-oxygen-requiring conversion of glucose to lactic acid.

The fact that consumption of oxygen is involved offers a convenient means of tracing a metabolic process—that is, finding out what intermediate products are created along the route. Let us say that at a given step in a sequence of reactions a certain substance (for example, succinic acid) is suspected to be the intermediate substrate. We can mix this acid with living tissue (or in many cases with a single enzyme) and measure the rate at which the mixture consumes oxygen. If it shows a rapid uptake of oxygen, we can be confident that this particular substance can indeed further the process.

The German biochemist Otto Heinrich Warburg devised the key instrument used to measure the rate of uptake of oxygen. Called the *Warburg manometer*, it consists of a small flask (where the substrate and the tissue or enzyme are mixed) connected to one end of a thin U-tube, whose other end is open. A colored fluid fills the lower part of the U. As the mixture of enzyme and substrate absorbs oxygen from the air in the flask, a slight vacuum is created there, and the colored liquid in the U-tube rises on the side of the U connected to the flask. The rate at which the liquid rises can be used to calculate the rate of oxygen uptake (figure 12.4).

Warburg's experiments on the uptake of oxygen by tissues won him the Nobel Prize in physiology and medicine in 1931.

Warburg and another German biochemist, Heinrich Wieland, identified the reactions that yield energy during the breakdown of lactic acid. In the course of the series of reactions, pairs of hydrogen atoms are removed from intermediate substances by means of enzymes called *dehydrogenases*. These hydrogen atoms then combine with oxygen, with the catalytic help of enzymes called *cytochromes*. In the late 1920s, Warburg and Wieland argued strenu-

PARTIAL VACUUM CREATED BY
TISSUE'S OXYGEN CONSUMPTION

STOPCOCK

LIVING TISSUE

SUBSTRATE WHICH
INITIATES REACTION
WHEN ADDED TO
MAIN FLASK

FLUID RISES
IN PARTIAL
VACUUM

Figure 12.4. Warburg manometer.

ously over which of these reactions is the important one, Warburg contending that it is the uptake of oxygen, and Wieland that it is the removal of hydrogen. Eventually, David Keilin showed that both steps are essential.

The German biochemist Hans Adolf Krebs went on to work out the complete sequence of reactions and intermediate products from lactic acid to carbon dioxide and water. This is called the *Krebs cycle*, or the *citric-acid cycle*, citric acid being one of the key products formed along the way. For this achievement, completed in 1940, Krebs received a share in the Nobel Prize in physiology and medicine in 1953 (with Lipmann).

The Krebs cycle produces the lion's share of energy for those organisms that make use of molecular oxygen in respiration (which means all organisms except a few types of anaerobic bacteria that depend for energy on chemical reactions not involving oxygen). At different points in the Krebs cycle, a compound will lose two hydrogen atoms, which are eventually combined with oxygen to form water. This "eventually" hides a good deal of detail. The two hydrogen atoms are passed from one variety of cytochrome molecule to another, until the final one, *cytochrome oxidase,* passes it on to molecular oxygen. Along the line of cytochromes, molecules of ATP are formed and the body is supplied with its chemical "small change" of energy. All told, for every turn of the Krebs cycle, eighteen molecules of ATP are formed. The entire process, because it involves oxygen and the piling up of phosphate groups to form the ATP, is called *oxidative phosphorylation* and is a key reaction of living tissue. Any serious interference with it (as when one swallows potassium cyanide) brings death in minutes.

All the substances and all the enzymes that take part in oxidative phosphorylation are contained in tiny granules within the cytoplasm. These were first detected in 1898 by the German biologist C. Benda, who did not at that time, of course, understand their importance. He called them *mitochondria* ("threads of cartilage," which he wrongly thought they were), and the name stuck.

The average mitochondrion is football-shaped, about 1/10,000 of an inch long and 1/25,000 of an inch thick. An average cell might contain anywhere from several hundred to a thousand mitochondria. Very large cells may contain a couple of hundred thousand, while anaerobic bacteria contain none. After the Second World War, electron-microscopic investigation showed the mitochondrion to have a complex structure of its own, for all its tiny size. The mitochondrion has a double membrane, the outer one smooth and the inner one elaborately wrinkled to present a large surface. Along the inner surface of the mitochondrion are several thousand tiny structures called *elementary particles*. It is these that seem to represent the actual sites of oxidative phosphorylation.

THE METABOLISM OF FATS

Meanwhile biochemists also made headway in solving the metabolism of fats. It was known that the fat molecules are carbon chains, that they can be hydrolyzed to *fatty acids* (most commonly sixteen or eighteen carbon atoms long), and that the molecules are broken down two carbons at a time. In 1947, Fritz Lipmann discovered a rather complex compound, which plays a part in *acetylation*—that is, transfer of a two-carbon fragment from one compound to another. He called the compound *coenzyme A* (the A standing for "acetylation"). Three years later, the German biochemist Feodor Lynen found coenzyme A to be deeply involved in the breakdown of fats. Once it attaches itself to a fatty acid, there follows a series of four steps which end in lopping off the two carbons at the end of the chain to which the coenzyme A is attached. Then another coenzyme A molecule attaches itself to what is left of the fatty acid, chops off two more atoms, and so on. This is called the *fatty-acid oxidation cycle*. This and other work won Lynen a share in the 1964 Nobel Prize in physiology and medicine.

The breakdown of proteins obviously must be, in general, more complicated than that of carbohydrates or fats, because some twenty different amino acids are involved. In some cases it turns out to be rather simple: one minor change in an amino acid may convert it into a compound that can enter the citric-acid cycle (as the two-carbon fragments from fatty acids can). But mainly amino acids are decomposed by complex routes.

We can now go back to the conversion of protein into urea—the question that I considered in the section on enzymes. This conversion happens to be comparatively simple.

A group of atoms that is essentially the urea molecule forms part of a side chain of the amino acid arginine. This group can be chopped off by an enzyme called *arginase*, and it leaves behind a kind of truncated amino acid, called *ornithine*. In 1932, Krebs and a co-worker, K. Henseleit, while studying the formation of urea by rat-liver tissue, discovered that when they added arginine to the tissue, it produced a flood of urea—much more urea, in fact, than the splitting of every molecule of arginine they had added could have produced. Krebs and Henseleit decided that the arginine molecules must be acting as agents that produce urea over and over again. In other words, after an arginine molecule has its urea combination chopped off by arginase, the ornithine that is left picks up amine groups from other amino acids (plus carbon dioxide from the body) and forms arginine again. So the arginine molecule is repeatedly split, re-formed, split again, and so on, each time yielding a molecule of urea. This is called the *urea cycle*, the *ornithine cycle*, or the *Krebs-Henseleit cycle*.

After the removal of nitrogen, by way of arginine, the remaining *carbon* skeletons of the amino acids can, by various routes, be broken down to carbon dioxide and water, producing energy. (For the overall metabolism of carbohydrates, fats, and proteins, see figure 12.5.)

Tracers

The investigations of metabolism by all these devices still left biochemists in the position of being on the outside looking in, so to speak. They could work out general cycles, but to find out what was really going on in the living animal they needed some means of tracing, in fine detail, the course of events through the stages of metabolism—to follow the fate of particular molecules, as it were. Actually, techniques for doing this had been discovered early in the century, but the chemists were rather slow in making full use of them.

The first to pioneer along these lines was a German biochemist named Franz Knoop. In 1904, he conceived the idea of feeding labeled fat molecules to dogs to see what happened to the molecules. He labeled them by attaching a benzene ring at one end of the chain; he used the benzene ring because mammals possess no enzymes that can break it down. Knoop expected that what the benzene ring carried with it when it showed up in the urine might tell something about how the fat molecule broke down in the body—and he was right. The benzene ring invariably turned up with a two-carbon side chain attached. From this, he deduced that the body must split off the fat molecule's carbon atoms two at a time. (As we have seen, more than forty years later the work with coenzyme A confirmed his deduction.)

The carbon chains in ordinary fats all contain an even number of carbon atoms. What if you use a fat whose chain has an odd number of carbon atoms? In that case, if the atoms are chopped off two at a time, you should end up

Figure 12.5. The overall scheme of metabolism of carbohydrates, fats, and proteins.

with just one carbon atom attached to the benzene ring. Knoop fed this kind of fat molecule to dogs and did indeed end up with that result.

Knoop had employed the first *tracer* in biochemistry. In 1913, the Hungarian chemist Georg von Hevesy and his co-worker, the German chemist Friedrich Adolf Paneth, hit upon another way to tag molecules: radioactive isotopes. They began with radioactive lead, and their first biochemical experiment was to measure how much lead, in the form of a lead-salt solution, a plant would take up. The amount was certainly too small to be measured by any available chemical method, but if radiolead was used, it could easily be measured by its radioactivity. Hevesy and Paneth fed the radioactively tagged lead-salt solution to plants; and at periodic intervals, they would burn a plant and measure the radioactivity of its ash. In this way, they were able to determine the rate of absorption of lead by plant cells.

But the benzene ring and lead were very "unphysiological" substances to use as tags. They might easily upset the normal chemistry of living cells. It would be much better to use as tags atoms that actually take part in the body's ordinary metabolism—such atoms as oxygen, nitrogen, carbon, hydrogen, phosphorus.

Once the Joliot-Curies had demonstrated artificial radioactivity in 1934, Hevesy took this direction at once and began using phosphates containing radioactive phosphorus. With these he measured phosphate uptake in plants. Unfortunately, the radioisotopes of some of the key elements in living tissue

—notably, nitrogen and oxygen—are not usable, because they are very short-lived, having a half-life of only a few minutes at most. But the most important elements do have *stable* isotopes that can be used as tags. These isotopes are carbon 13, nitrogen 15, oxygen 18, and hydrogen 2. Ordinarily, they occur in very small amounts (about 1 percent or less); consequently, by "enriching" natural hydrogen, say, in hydrogen 2, one can make it to serve as a distinguishing tag in a hydrogen-containing molecule fed to the body. The presence of the heavy hydrogen in any compound can be detected by means of the mass spectograph, which separates it by virtue of its extra weight. Thus, the fate of the tagged hydrogen can be traced through the body.

Hydrogen, in fact, served as the first physiological tracer. It became available for this purpose when Harold Urey isolated hydrogen 2 (deuterium) in 1931. One of the first things brought to light by the use of deuterium as a tracer was that hydrogen atoms in the body are much less fixed to their compounds than had been thought. It turned out that they shuttle back and forth from one compound to another, exchanging places on the oxygen atoms of sugar molecules, water molecules, and so on. Since one ordinary hydrogen atom cannot be told from another, this shuttling had not been detected before the deuterium atoms disclosed it. What the discovery implied was that hydrogen atoms hop about throughout the body, and that if deuterium atoms were attached to oxygen, they would spread through the body regardless of whether the compounds involved underwent overall chemical change. Consequently, the investigator must make sure that a deuterium atom found in a compound got there by some definite enzyme-catalyzed reaction and not just by the shuttling, or exchange, process. Fortunately, hydrogen atoms attached to carbon do not exchange, so deuterium found along carbon chains has metabolic significance.

The roving habits of atoms were further emphasized in 1937 when the German-born American biochemist Rudolf Schoenheimer and his associates began to use nitrogen 15. They fed rats on amino acids tagged with nitrogen 15, killed the rats after a set period, and analyzed the tissues to see which compounds carried nitrogen 15. Here again, exchange was found to be important. After one tagged amino acid had entered the body, almost all the amino acids were shortly found to carry nitrogen 15. In 1942, Schoenheimer published a book entitled *The Dynamic State of Body Constituents.* That title describes the new look in biochemistry that the isotopic tracers brought about. A restless traffic in atoms goes on ceaselessly, quite aside from actual chemical changes.

Little by little the use of tracers filled in the details of the metabolic routes. It corroborated the general pattern of such things as sugar breakdown, the citric-acid cycle, and the urea cycle. It resulted in the addition of new intermediates, in the establishment of alternate routes of reaction, and so on.

Thanks to the nuclear reactor, over a hundred different radioactive isotopes became available in quantity after the Second World War, and tracer work went into high gear. Ordinary compounds could be bombarded by neutrons in a reactor and come out loaded with radioactive isotopes. Almost every biochemical laboratory in the United States (I might almost say in the world, for the United States soon made isotopes available to other countries for scientific use) started research programs involving radioactive tracers.

The stable tracers were now joined by radioactive hydrogen (tritium), radio-phosphorus (phosphorus 32), radiosulfur (sulfur 35), radiopotassium (potassium 42), radiosodium, radioiodine, radioiron, radiocopper, and most important of all, radiocarbon (carbon 14). Carbon 14 was discovered in 1940 by the American chemists Martin David Kamen and Samuel Ruben and, to their surprise, turned out to have a half-life of more than 5,000 years—unexpectedly long for a radioisotope among the light elements.

CHOLESTEROL

Carbon 14 solved problems that had defied chemists for years and against which they had seemed to be able to make no headway at all. One of the riddles to which it gave the beginning of an answer was the production of the substance known as *cholesterol*. Cholesterol's formula, worked out by many years of painstaking investigation by men such as Wieland (who received the 1927 Nobel Prize in chemistry for his work on compounds related to cholesterol), had been found to be:

The function of cholesterol in the body is not yet completely understood, but the substance is clearly of central importance. Cholesterol is found in a large quantity in the fatty sheaths around nerves, in the adrenal glands, and in combination with certain proteins. An excess of it can cause gallstones and atherosclerosis. Most significant of all, cholesterol is the prototype of the whole family of *steroids*, the steroid nucleus being the four-ring combination you see in the formula. The steroids are a group of solid, fatlike substances, which include the sex hormones and the adrenocortical hormones. All of them undoubtedly are formed from cholesterol. But how is cholesterol itself synthesized in the body?

Until tracers came to their help, biochemists had not the foggiest notion. The first to tackle the question with a tracer were Rudolf Schoenheimer and his co-worker David Rittenberg. They gave rats heavy water to drink and found that its deuterium turned up in the cholesterol molecules. This effect in itself was not significant, because the deuterium could have got there merely by exchanges. But, in 1942 (after Schoenheimer had tragically committed suicide), Rittenberg and another co-worker, the German-American biochemist, Konrad Emil Bloch, discovered a more definite clue. They fed rats acetate ion (a simple two-carbon group, CH_3COO-) with the deuterium tracer attached to the carbon atom in the CH_3 group. The deuterium again showed up in cholesterol molecules, and this time it could not have arrived there by exchange: it must have been incorporated in the molecule as part of the CH_3 group.

Two-carbon groups (of which the acetate ion is one version) seem to represent a general crossroads of metabolism. Such groups, then, might very well serve as the pool of material for building cholesterol. But just how do they form the molecule?

In 1950, when carbon 14 had become available, Bloch repeated the experiment, this time labeling the two carbons of the acetate ion, each with a different tag. He marked the carbon of the CH_3 group with the stable tracer carbon 13, and he labeled the carbon of the COO− group with radioactive carbon 14. Then, after feeding the compound to a rat, he analyzed its cholesterol to see where the two tagged carbons would appear in the molecule. The analysis was a task that called for delicate chemical artistry, and Bloch and a number of other experimenters worked at it for years, identifying the source of one after another of the cholesterol carbon atoms. The pattern that developed eventually suggested that the acetate groups probably first formed a substance called *squalene*, a rather scarce thirty-carbon compound in the body to which no one had ever dreamed of paying serious attention before. Now it appeared to be a way station on the road to cholesterol, and biochemists have begun to study it with intense interest. For this work, Bloch shared the 1964 Nobel Prize in physiology and medicine with Lynen.

THE PORPHYRIN RING OF HEME

In much the same way as they tackled the synthesis of cholesterol, biochemists have gone after the construction of the porphyrin ring of heme, a key structure in hemoglobin and in many enzymes. David Shemin of Columbia University fed ducks the amino acid glycine, labeled in various ways. Glycine (NH_2CH_2COOH) has two carbon atoms. When he tagged the CH_2 carbon with carbon 14, that carbon showed up in the porphyrin extracted from the ducks' blood. When he labeled the COOH carbon, the radioactive tracer did not appear in the porphyrin. In short, the CH_2 group entered into the synthesis of porphyrin but the COOH group did not.

Shemin, working with Rittenberg, found that the incorporation of glycine's atoms into porphyrin can take place just as well in red blood cells in the test tube as it can in living animals. This finding simplified matters, gave more clear-cut results, and avoided sacrificing or inconveniencing the animals.

He then labeled glycine's nitrogen with nitrogen 15 and its CH_2 carbon with carbon 14, then mixed the glycine with duck blood. Later, he carefully took apart the porphyrin produced and found that all four nitrogen atoms in the porphyrin molecule came from the glycine. So did an adjacent carbon atom in each of the four small pyrrole rings (see the formula in chapter 10), and also the four carbon atoms that serve as bridges between the pyrrole rings. This left twelve other carbon atoms in the porphyrin ring itself and fourteen in the various side chains. These were shown to arise from acetate ion, some from the CH_3 carbon and some from the COO– carbon.

From the distribution of the tracer atoms, it was possible to deduce the manner in which the acetate and the glycine enter into the porphyrin. First, they form a one-pyrrole ring; then two such rings combine, and finally two two-ring combinations join to form the four-ring porphyrin structure.

In 1952, a compound called *porphobilinogen* was isolated in pure form, as a result of an independent line of research by the English chemist R. G. Westall. This compound occurs in the urine of persons with defects in porphyrin metabolism, so it was suspected of having something to do with porphyrins. Its structure turned out to be just about identical with the one-pyrrole-ring structure that Shemin and his co-workers had postulated as one of the early steps in porphyrin synthesis. Porphobilinogen was a key way station.

It was next shown that *delta-aminolevulinic acid,* a substance with a structure like that of a porphobilinogen molecule split in half, could supply all the atoms necessary for incorporation into the porphyrin ring by the blood cells. The most plausible conclusion is that the cells first form delta-aminolevulinic acid from glycine and acetate (eliminating the COOH group of glycine as carbon dioxide in the process), that two molecules of delta-aminolevulinic acid then combine to form porphobilinogen (a one-pyrrole ring), and that the latter in turn combines first into a two-pyrrole ring and finally into the four-pyrrole ring of porphyrin.

Photosynthesis

Of all the triumphs of tracer research, perhaps the greatest has been the tracing of the complex series of steps that builds green plants—on which all life on this planet depends.

The animal kingdom could not exist if animals could feed only on one another, any more than a community of people can grow rich solely by taking in one another's washing or a man can lift himself by yanking upward on his belt buckle. A lion that eats a zebra or a man who eats a steak is consuming precious substance that has been obtained at great pains and with considerable attrition from the plant world. The second law of thermodynamics tells us that, at each stage of the cycle, something is lost. No animal stores all of the carbohydrate, fat, and protein contained in the food it eats, nor can it make use of all the energy available in the food. Inevitably a large part—indeed, most—of the energy is wasted in unusable heat. At each level of eating, then, some chemical energy is frittered away. Thus, if all animals were strictly carnivorous, the whole animal kingdom would die off in a very few generations. In fact, it would never have come into being in the first place.

The fortunate fact is that most animals are herbivorous. They feed on the grass of the field, on the leaves of trees, on seeds, nuts, and fruit, or on the seaweed and microscopic green plant cells that fill the upper layers of the oceans. Only a minority of animals can be supported in the luxury of being carnivorous.

As for the plants themselves, they would be in no better plight were they not supplied with an external source of energy. They build carbohydrates, fats, and proteins from simple molecules, such as carbon dioxide and water. This synthesis calls for an input of energy, and the plants get it from the most copious possible source: sunlight. Green plants convert the energy of sunlight into the chemical energy of complex compounds, and that chemical energy supports all life forms (except for certain bacteria). This process was first clearly pointed out in 1845 by the German physicist Julius Robert von Mayer, who was one of those who pioneered the law of conservation of energy, and was therefore particularly aware of the problem of energy balance. The process by which green plants make use of sunlight is called *photosynthesis*, from Greek words meaning "put together by light."

THE PROCESS

The first attempt at a scientific investigation of plant growth was made early in the seventeenth century by the Flemish chemist Jan Baptista Van Helmont. He grew a small willow tree in a tub containing a weighed amount of soil, and found, to everyone's surprise, that although the tree grew large, the soil weighed just as much as before. It had been taken for granted that plants derive their substance from the soil. (Actually plants do take some minerals and ions from the soil, but not in any easily weighable amount.) If they did not get it there, where did they get it from? Van Helmont decided that plants must manufacture their substance from water, with which he had supplied the soil liberally. He was only partly right.

A century later, the English physiologist Stephen Hales showed that plants build their substance in great part from a material more ethereal than water —namely, air. Half a century later, the Dutch physician Jan Ingen-Housz identified the nourishing ingredient in air as carbon dioxide. He also demonstrated that a plant does not absorb carbon dioxide in the dark; it needs light (the *photo* of photosynthesis). Meanwhile Priestley, the discoverer of oxygen, had learned that green plants give off oxygen. And, in 1804, the Swiss chemist Nicholas Théodore de Saussure proved that water is incorporated in plant tissue, as Van Helmont had suggested.

The next important contribution came in the 1850s, when the French mining engineer Jean Baptiste Boussingault grew plants in soil completely free of organic matter. He showed, in this way, that plants can obtain their carbon from atmospheric carbon dioxide only. On the other hand, plants would not grow in soil free of nitrogen compounds: hence, they derive their nitrogen from the soil, and atmospheric nitrogen is not utilized (except, as it turned out, by certain bacteria). From Boussingault's time, it became apparent that the service of soil as direct nourishment for the plant was confined to certain inorganic salts, such as nitrates and phosphates. It is these ingredients that organic fertilizers (such as manure) add to soil. Chemists began to advocate the addition of chemical fertilizers, which served the purpose excellently and eliminated noisome odors as well as decreasing the dangers of infection and disease, much of which could be traced to the farm's manure pile.

Thus, the skeleton of the process of photosynthesis was established. In sunlight, a plant takes up carbon dioxide and combines it with water to form its tissues, giving off "left-over" oxygen in the process. Hence, it became plain that green plants not only provide food but also renew the earth's oxygen supply. Were it not for this renewal, within a matter of centuries the oxygen would fall to a low level, and the atmosphere would be loaded with enough carbon dioxide to asphyxiate animal life.

The scale on which the earth's green plants manufacture organic matter and release oxygen is enormous. The Russian-American biochemist Eugene I. Rabinowitch, a leading investigator of photosynthesis, estimates that each year the green plants of the earth combine a total of 150 billion tons of carbon (from carbon dioxide) with 25 billion tons of hydrogen (from water) and liberate 400 billion tons of oxygen. Of this gigantic performance, the plants of the forests and fields on land account for only 10 percent; for 90 percent we have to thank the one-celled plants and seaweed of the oceans.

CHLOROPHYLL

We still have only the skeleton of the process. What about the details? Well, in 1817, Pierre Joseph Pelletier and Joseph Bienaimé Caventou of France—who were later to discover quinine, strychnine, caffeine, and several other specialized plant products—isolated the most important plant product of all—the one that gives the green color to green plants. They called the compound *chlorophyll*, from Greek words meaning "green leaf." Then, in 1865, the German botanist Julius von Sachs showed that chlorophyll is not distributed generally through plant cells (though leaves appear uniformly green), but is localized in small subcellular bodies, later called *chloroplasts*.

It became clear that photosynthesis takes place within the chloroplasts and that chlorophyll is essential to the process. Chlorophyll was not enough, however. Chlorophyll by itself, however carefully extracted, could not catalyze the photosynthetic reaction in a test tube.

Chloroplasts generally are considerably larger than mitochondria. Some one-celled plants possess only one large chloroplast per cell. Most plant cells, however, contain as many as 40 smaller chloroplasts, each from two to three times as long and as thick as the typical mitochondrion.

The structure of the chloroplast seems to be even more complex than that of the mitochondrion. The interior of the chloroplast is made up of many thin membranes stretching across from wall to wall. These are the *lamellae*. In most types of chloroplasts, these lamellae thicken and darken in places to produce *grana*, and it is within the grana that the chlorophyll molecules are found.

If the lamellae within the grana are studied under the electron microscope, they in turn seem to be made up of tiny units, just barely visible, that look like the neatly laid tiles of a bathroom floor. Each of these objects may be a photosynthesizing unit containing 250 to 300 chlorophyll molecules.

The chloroplasts are more difficult than mitochondria to isolate intact. It was not until 1954 that the Polish-American biochemist Daniel Israel Arnon, working with disrupted spinach-leaf cells, could obtain chloroplasts completely intact and was able to carry through the complete photosynthetic reaction.

The chloroplast contains not only chlorophyll but a full complement of enzymes and associated substances, all properly and intricately arranged. It even contains cytochromes by which the energy of sunlight, trapped by chlorophyll, can be converted into ATP through oxidative phosphorylation.

Meanwhile, though, what about the structure of chlorophyll, the most characteristic substance of the chloroplasts? For decades, chemists had tackled this key substance with every tool at their command, but it yielded only slowly. Finally, in 1906, Richard Willstätter of Germany (who was later to rediscover chromatography and to insist, incorrectly, that enzymes are not proteins) identified a central component of the chlorophyll molecule: the metal magnesium. (Willstätter received the Nobel Prize in chemistry in 1915 for this discovery and other work on plant pigments.) Willstätter and Hans Fischer went on to work on the structure of the molecule—a task that took a full generation to complete. By the 1930s, it had been determined that chlorophyll has a porphyrin ring structure basically like that of heme (a molecule that Fischer had deciphered). Where heme has an iron atom at the center of the porphyrin ring, chlorophyll has a magnesium atom.

Any doubt on this point was removed by R. B. Woodward. That master synthesist—who had put together quinine in 1945, strychnine in 1947, and cholesterol in 1951—now capped his previous efforts by putting together a molecule in 1960 that matched the formula worked out by Willstätter and Fischer, and, behold, it had all the properties of chlorophyll isolated from green leaves. Woodward received the 1965 Nobel Prize for chemistry as a result.

Exactly what reaction in a plant does chlorophyll catalyze? All that was known, up to the 1930s, was that carbon dioxide and water go in and oxygen comes out. Investigation was made more difficult by the fact that isolated chlorophyll cannot be made to bring about photosynthesis. Only intact plant cells or, at best, intact chloroplasts, would do; hence, the system under study was very complex.

As a first guess, biochemists assumed that the plant cells synthesize glucose

($C_6H_{12}O_6$) from the carbon dioxide and water and then go on to build from this the various plant substances, adding nitrogen, sulfur, phosphorus, and other inorganic elements from the soil.

On paper, it seemed as if glucose might be formed by a series of steps which first combine the carbon atom of carbon dioxide with water (releasing the oxygen atoms of CO_2), and then polymerize the combination, CH_2O (formaldehyde), into glucose. Six molecules of formaldehyde would make one molecule of glucose.

This synthesis of glucose from formaldehyde could indeed be performed in the laboratory, in a tedious sort of way. Presumably, the plant might possess enzymes that speed the reactions. To be sure, formaldehyde is a very poisonous compound, but the chemists assumed the formaldehyde to be turned into glucose so quickly that at no time does a plant contain more than a very small amount of it. This formaldehyde theory, first proposed in 1870 by Baeyer (the synthesizer of indigo), lasted for two generations, simply because there was nothing better to take its place.

A fresh attack on the problem began in 1938, when Ruben and Kamen undertook to probe the chemistry of the green leaf with tracers. By the use of oxygen 18, the uncommon stable isotope of oxygen, they made one clear-cut finding: it turned out that when the water given a plant is labeled with oxygen 18, the oxygen released by the plant carries this tag, but the oxygen does not carry the tag when only the carbon dioxide supplied to the plant is labeled. In short, the experiment showed that the oxygen given off by plants comes from the water molecule and not from the carbon dioxide molecule, as had been mistakenly assumed in the formaldehyde theory.

Ruben and his associates tried to follow the fate of the carbon atoms in the plant by labeling the carbon dioxide with the radioactive isotope carbon 11 (the only radiocarbon known at the time). But this attempt failed. For one thing, carbon 11 has a half-life of only 20.5 minutes. For another, they had no available method at the time for separating individual compounds in the plant cell quickly and thoroughly enough.

But, in the early 1940s, the necessary tools came to hand. Ruben and Kamen discovered carbon 14, the long-lived radioisotope, which made it possible to trace carbon through a series of leisurely reactions. And the development of paper chromatography provided a means of separating complex mixtures easily and cleanly. (In fact, radioactive isotopes allowed a neat refinement of paper chromatography: the radioactive spots on the paper, representing the presence of the tracer, would produce dark spots on a photographic film laid under it, so that the chromatogram would take its own picture—a technique called *autoradiography*.)

After the Second World War, another group, headed by the American biochemist Melvin Calvin, picked up the ball. They exposed microscopic one-celled plants (*chlorella*) to carbon dioxide containing carbon 14 for short periods, in order to allow the photosynthesis to progress only through its earliest stages. Then they mashed the plant cells, separated their substances on a chromatogram, and made an autoradiograph.

They found that even when the cells had been exposed to the tagged carbon dioxide for only 1½ minutes, the radioactive carbon atoms turned up in as many as fifteen different substances in the cell. By cutting down the exposure time, they reduced the number of substances in which radiocarbon was incorporated, and eventually they decided that the first, or almost the first, compound in which the cell incorporated the carbon-dioxide carbon was *glyceryl phosphate*. (At no time did they detect any formaldehyde, so the venerable formaldehyde theory passed quietly out of the picture.)

Glyceryl phosphate is a three-carbon compound. Evidently it must be formed by a roundabout route, for no one-carbon or two-carbon precursor could be found. Two other phosphate-containing compounds were located that took up tagged carbon within a very short time. Both were varieties of sugars: *ribulose diphosphate* (a five-carbon compound) and *sedoheptulose phosphate* (a seven-carbon compound). The investigators identified enzymes that catalyze reactions involving such sugars, studied those reactions, and worked out the travels of the carbon-dioxide molecule. The scheme that best fits all their data is the following.

First, carbon dioxide is added to the five-carbon ribulose diphosphate, making a six-carbon compound. This quickly splits in two, creating the three-carbon glyceryl phosphate. A series of reactions involving sedoheptulose phosphate and other compounds then puts two glyceryl phosphates together to form the six-carbon glucose phosphate. Meanwhile, ribulose diphosphate is regenerated and is ready to take on another carbon-dioxide molecule. You can imagine six such cycles turning. At each turn, each cycle supplies one carbon atom (from the carbon dioxide), and out of these a molecule of glucose phosphate is built. Another turn of the six cycles produces another molecule of glucose phosphate, and so on.

This is the reverse of the citric-acid cycle, from an energy standpoint. Whereas the citric-acid cycle converts the fragments of carbohydrate breakdown to carbon dioxide, the ribulose-diphosphate cycle builds up carbohydrates from carbon dioxide. The citric-acid cycle delivers energy to the organism; the ribulose-diphosphate cycle, conversely, has to consume energy.

Here the earlier results of Ruben and Kamen fit in. The energy of sunlight is used, thanks to the catalytic action of chlorophyll, to split a molecule of water into hydrogen and oxygen, a process called *photolysis* (from Greek words meaning "loosening by light"). This is the way that the radiant energy of sunlight is converted into chemical energy, for the hydrogen and oxygen molecules contain more chemical energy than did the water molecule from which they came.

In other circumstances, it takes a great deal of energy to break up water molecules into hydrogen—for instance, heating the water to something like 2,000° C or sending a strong electric current through it. But chlorophyll does the trick easily at ordinary temperatures. All it needs is the relatively weak energy of visible light. The plant uses the light-energy that it absorbs with an efficiency of at least 30 percent; some investigators believe its efficiency may approach 100 percent under ideal conditions. If we humans could harness energy as efficiently as the plants do, we would have much less to worry about with regard to our supplies of food and energy.

After the water molecules have been split, half of the hydrogen atoms find their way into the ribulose-diphosphate cycle, and half of the oxygen atoms are liberated into the air. The rest of the hydrogens and oxygens recombine into water. In doing so, they release the excess of energy that was given to them when sunlight split the water molecules, and this energy is transferred to high-energy phosphate compounds such as ATP. The energy stored in these compounds is then used to power the ribulose-diphosphate cycle. For his work in deciphering the reactions involved in photosynthesis, Calvin received the Nobel Prize in chemistry in 1961.

To be sure, some forms of life gain energy without chlorophyll. About 1880, *chemosynthetic bacteria* were discovered: bacteria that trap carbon dioxide in the dark and do not liberate oxygen. Some oxidized sulfur compounds to gain energy; some oxidized iron compounds; and some indulged in still other chemical vagaries.

Then, too, some bacteria have chlorophyll-like compounds (*bacteriochlorophyll*), which enable them to convert carbon dioxide to organic compounds at the expense of light-energy—even, in some cases, in the near infrared, where ordinary chlorophyll will not work. However, only chlorophyll itself can bring about the splitting of water and the conservation of the large energy store so gained; bacteriochlorophyll must make do with less energetic devices.

All methods of fundamental energy gain, other than that which uses sunlight by way of chlorophyll, are essentially dead-end, and exist only under rare and specialized conditions complicated than a bacterium has successfully made use of them. For almost all of life, chlorophyll and photosynthesis, directly or indirectly, are the basis of life.

Chapter 13

The Cell

Chromosomes

It is an odd paradox that until recent times, we humans have known very little about our own bodies. In fact, it was only some three hundred years ago that we learned about the circulation of the blood, and only within the last fifty years or so we have discovered the functions of many of the organs.

Prehistoric people, from cutting up animals for cooking and from embalming their own dead in preparation for afterlife, were aware of the existence of the large organs, such as the brain, liver, heart, lungs, stomach, intestines, and kidneys. This awareness was intensified through the frequent use of the appearance of the internal organs of a ritually sacrificed animal (particularly the appearance of its liver) in foretelling the future or estimating the extent of divine favor or disfavor. Egyptian papyri, dealing validly with surgical technique and presupposing some familiarity with body structure, can be dated earlier than 2000 B.C.

The ancient Greeks went so far as to dissect animals and an occasional human cadaver with the deliberate purpose of learning something about *anatomy* (from Greek words meaning "to cut up"). Some delicate work was done. Alcmaeon of Croton, about 500 B.C., first described the optic nerve and the Eustachian tube. Two centuries later, in Alexandria, Egypt (then the world center of science), a school of Greek anatomy started brilliantly with Herophilus and his pupil Erasistratus. They investigated the parts of the brain, distinguishing the cerebrum and the cerebellum, and studied the nerves and blood vessels as well.

Ancient anatomy reached its peak with Galen, a Greek physician who practiced in Rome in the latter half of the second century. Galen worked up theories of bodily functions which were accepted as gospel for fifteen hundred years afterward. But his notions about the human body were full of curious errors—understandably so, for the ancients obtained most of their information from dissecting animals. Inhibitions of one kind or another made people uneasy about dissecting the human body.

In denouncing the pagan Greeks, early Christian writers accused them of having practiced heartless vivisections on human beings. But this comes under the heading of polemical literature: not only is it doubtful that the Greeks did human vivisections, but obviously they did not even dissect enough dead bodies to learn much about the human anatomy. In any case, the Church's disapproval of dissection virtually put a stop to anatomical studies throughout the Middle Ages. As this period of history approached its end, anatomy began to revive in Italy. In 1316, an Italian anatomist, Mondino de Luzzi, wrote the first book to be devoted entirely to anatomy, and he is therefore known as the "restorer of anatomy."

The interest in naturalistic art during the Renaissance also fostered anatomical research. In the fifteenth century, Leonardo da Vinci performed some dissections by means of which he revealed new facts of anatomy, picturing them with the power of artistic genius. He showed the double curve of the spine and the sinuses that hollow the bones of the face and forehead. He used his studies to derive theories of physiology more advanced than Galen's. But Leonardo, though a genius in science as well as in art, had little influence on scientific thought in his time. Either from neurotic disinclination or from sober caution, he did not publish any of his scientific work but kept it hidden in coded notebooks. It was left for later generations to discover his scientific achievements when his notebooks were finally published.

The French physician Jean Fernel was the first modern to take up dissection as an important part of a physician's duties. He published a book on the subject in 1542. However, his work was almost completely overshadowed by a much greater work published in the following year. This was the famous *De Humani Corporis Fabrica* ("Concerning the Structure of the Human Body") of Andreas Vesalius, a Belgian who did most of his work in Italy. On the theory that the proper study of mankind was man, Vesalius dissected the appropriate subject and corrected many of Galen's errors. The drawings of the human anatomy in his book (which are reputed to have been made by Jan Stevenzoon van Calcar, a pupil of the artist Titian) are so beautiful and accurate that they are still republished today and will always stand as classics. Vesalius can be called the father of modern anatomy. His *Fabrica* was as revolutionary in its way as Copernicus's *De Revolutionibus Orbium Coelestium,* published in the very same year.

Just as the revolution initiated by Copernicus was brought to fruition by

Galileo, so the one initiated by Vesalius came to a head in the crucial discoveries of William Harvey. Harvey was an English physician and experimentalist, of the same generation as Galileo and William Gilbert, the experimenter with magnetism. Harvey's particular interest was that vital body juice—the blood. What does it do in the body, anyway?

It was known that there were two sets of blood vessels: the veins and the arteries. (Praxagoras of Cos, a Greek physician of the third century B.C., had provided the name *artery* from Greek words meaning "I carry air," because these vessels were found to be empty in dead bodies. Galen had later shown that in life they carry blood.) It was also known that the heartbeat drives the blood in some sort of motion, for when an artery was cut, the blood gushed out in pulses that synchronized with the heartbeat.

Galen had proposed that the blood seesawed to and fro in the blood vessels, traveling first in one direction through the body and then in the other. This theory required him to explain why the back-and-forth movement of the blood was not blocked by the wall between the two halves of the heart; Galen answered simply that the wall was riddled with invisibly small holes that let the blood through.

Harvey took a closer look at the heart. He found that each half was divided into two chambers, separated by a one-way valve that allows blood to flow from the upper chamber (*auricle*) to the lower (*ventricle*), but not vice versa. In other words, blood entering one of the auricles could be pumped into its corresponding ventricle and from there into blood vessels issuing from it, but there could be no flow in the opposite direction.

Harvey then performed some simple but beautifully clear-cut experiments to determine the direction of flow in the blood vessels. He would tie off an artery or a vein in a living animal to see on which side of this blockage the pressure within the blood vessel would build up. He found that when he stopped the flow in an artery, the vessel always bulged on the side between the heart and the block. Hence, the blood in arteries must flow in the direction away from the heart. When he tied a vein, the bulge was always on the other side of the block; therefore, the blood flow in veins must be toward the heart. Further evidence in favor of this one-way flow in veins rests in the fact that the larger veins contain valves that prevent blood from moving away from the heart. This mechanism had been discovered by Harvey's teacher, the Italian anatomist Hieronymus Fabrizzi (better known by his Latinized name, Fabricius). Fabricius, however, under the load of Galenic tradition, refused to draw the inevitable conclusion and left the glory to his English student.

Harvey went on to apply quantitative measurements to the blood flow (the first time anyone had applied mathematics to a biological problem). His measurements showed that the heart pumps out blood at such a rate that in twenty minutes its output equals the total amount of blood contained in the body. It did not seem reasonable to suppose that the body could manufacture new blood, or consume the old, at any such rate. The logical conclusion, therefore, was that the blood must be recycled through the body. Since it flows away from the heart in the arteries and toward the heart in the veins, Harvey decided that the blood is pumped by the heart into the arteries, then passes from them into the veins, then flows back to the heart, then is pumped into the arteries again, and so on. In other words, it circulates continuously in one direction through the heart-and-blood-vessel system.

Earlier anatomists, including Leonardo da Vinci, had hinted at such an idea, but Harvey was the first to state and investigate the theory in detail. He set forth his reasoning and experiments in a small, badly printed book entitled *De Motus Cordis* (Concerning the Motion of the Heart), which was published in 1628 and has stood ever since as one of the great classics of science.

The main question left unanswered by Harvey's work was: How does the blood pass from the arteries into the veins? Harvey said there must be connecting vessels of some sort, though they were too small to be seen. This was reminiscent of Galen's theory about small holes in the heart wall, but whereas Galen's holes in the heart were never found and do not exist, Harvey's connecting vessels were confirmed as soon as a microscope became available. In 1661, just four years after Harvey's death, an Italian physician named Marcello Malpighi examined the lung tissues of a frog with a primitive microscope, and, sure enough, there were tiny blood vessels connecting the arteries with the veins. Malpighi named them *capillaries*, from a Latin word meaning "hair-like." (For the circulatory system, see figure 13.1.)

The use of the microscope made it possible to see other minute structures as well. The Dutch naturalist Jan Swammerdam discovered the red blood corpuscles, while the Dutch anatomist Regnier de Graaf discovered tiny *ovarian follicles* in animal ovaries. Small creatures, such as insects, could be studied minutely.

Work in such fine detail encouraged the careful comparison of structures in one species with structures in others. The English botanist Nehemiah Grew was the first *comparative anatomist* of note. In 1675, he published his studies comparing the trunk structure of various trees, and in 1681 studies comparing the stomachs of various animals.

CELL THEORY

The coming of the microscope introduced biologists, in fact, to a more basic level of organization of living things—a level at which all ordinary structures could be reduced to a common denominator. In 1665, the English scientist Robert Hooke, using a compound microscope of his own design, discovered that cork, the bark of a tree, was built of extremely tiny compartments, like a superfine sponge. He called these holes *cells*, likening them to small rooms, such as the cells in a monastery. Other microscopists then found similar cells, but full of fluid, in living tissue.

Over the next century and a half, it gradually dawned on biologists that all living matter is made up of cells and that each cell is an independent unit of life. Some forms of life—certain microorganisms—consist of only a single cell; the larger organisms are composed of many cooperating cells. One of the

Figure 13.1. The circulatory system.

earliest to propose this view was the French physiologist René Joachim Henri
Dutrochet. His report, published in 1824, went unnoticed, however; and the
cell theory gained prominence only after Matthias Jakob Schleiden and Theodor Schwann of Germany independently formulated it in 1838 and 1839.

The colloidal fluid filling certain cells was named *protoplasm* ("first form")
by the Czech physiologist Jan Evangelista Purkinje in 1839, and the German
botanist Hugo von Mohl extended the term to signify the contents of all cells.
The German anatomist Max Johann Sigismund Schultze emphasized the
importance of protoplasm as the "physical basis of life" and demonstrated the

essential similarity of protoplasm in all cells, both plant and animal, and in both very simple and very complex creatures.

The cell theory is to biology about what the atomic theory is to chemistry and physics. Its importance in the dynamics of life was established when, around 1860, the German pathologist Rudolf Virchow asserted, in a succinct Latin phrase, that all cells arise from cells. He showed that the cells in diseased tissue had been produced by the division of originally normal cells.

By that time it had become clear that every living organism, even the largest, begins life as a single cell. One of the earliest microscopists, Johann Ham, an assistant of Leeuwenhoek, had discovered in seminal fluid tiny bodies that were later named *spermatozoa* (from Greek words meaning "animal seed"). Much later, in 1827, the German physiologist Karl Ernst von Baer had identified the *ovum*, or egg cell, of mammals (figure 13.2). Biologists came to

Figure 13.2. Human egg and sperm cells.

realize that the union of an egg and a spermatozoon forms a fertilized ovum from which the animal eventually develops by repeated divisions and redivisions.

Larger organisms, then, do not have larger cells than smaller organisms do; they simply have more of them. The cells remain small, almost always microscopic. The typical plant or animal's cell has a diameter of between 5 and 40 micrometers (a *micrometer* is equal to about 1/25,000 inch), and the human eye can just barely make out something that is 100 micrometers across.

Despite the fact that a cell is so small, it is by no means a featureless droplet of protoplasm. A cell has an intricate substructure that was made out, little by little, only in the course of the nineteenth century. It was to this substructure that biologists had to turn for the answers to many questions concerning life.

For instance, since organisms grow through the multiplication of their constituent cells, how do cells divide? The answer lies in a small globule of comparatively dense material within the cell, making up about a tenth its volume. It was first reported by Robert Brown (the discoverer of Brownian motion) in 1831 and named the *nucleus*. (To distinguish it from the nucleus of the atom, I shall refer to it from now on as the *cell nucleus*.)

If a one-celled organism was divided into two parts, one of which contained the intact cell nucleus, the part containing the cell nucleus was able to grow and divide, but the other part could not. (Later it was also learned that the red blood cells of mammals, lacking nuclei, are short-lived and have no capacity for either growth or division. For that reason, they are not considered true cells and are usually called *corpuscles*.)

Unfortunately, further study of the cell nucleus and the mechanism of division was thwarted for a long time by the fact that the cell is more or less transparent, so that its substructures cannot be seen. Then the situation was improved by the discovery that certain dyes would stain parts of the cell and not others. A dye called *hematoxylin* (obtained from logwood) stained the cell nucleus black and brought it out prominently against the background of the cell. After Perkin and other chemists began to produce synthetic dyes, biologists found themselves with a variety of dyes from which to choose.

In 1879, the German biologist Walther Flemming found that with certain red dyes he could stain a particular material in the cell nucleus which was distributed through it as small granules. He called this material *chromatin* (from the Greek word for "color"). By examining this material, Flemming was able to follow some of the changes in the process of cell division. To be sure, the stain killed the cell, but in a slice of tissue he would catch various cells at different stages of cell division. They served as still pictures, which he put together in the proper order to form a kind of "moving picture" of the progress of cell division.

In 1882, Flemming published an important book in which he described the process in detail. At the start of cell division, the chromatin material gathers itself together in the form of threads. The thin membrane enclosing the cell nucleus seems to dissolve; and at the same time, a tiny object just outside it divides in two. Flemming called this object the *aster*, from a Greek word for "star," because radiating threads give it a starlike appearance. After dividing, the two parts of the aster travel to opposite sides of the cell. Its trailing threads apparently entangle the threads of chromatin, which have meanwhile lined up in the center of the cell, and the aster pulls half the chromatin threads to one side of the cell, half to the other. As a result, the cell pinches in at the middle and splits into two cells. A cell nucleus develops in each, and the chromatin material enclosed by the nuclear membrane breaks up into granules again (see figure 13.3).

Flemming called the process of cell division *mitosis*, from the Greek word for "thread," because of the prominent part played in it by the chromatin threads. In 1888, the German anatomist Wilhelm von Waldeyer gave the chromatin thread the name *chromosome* (from the Greek for "colored body"), and that name has stuck. It should be mentioned, though, that chromosomes,

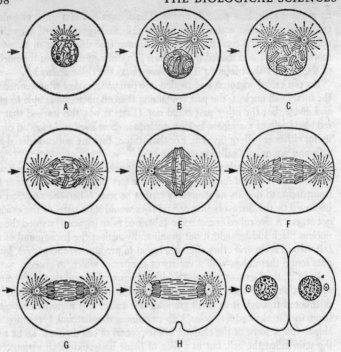

Figure 13.3. Division of a cell by mitosis.

despite their name, are colorless in their unstained natural state, in which of course they are quite difficult to make out against the very similar background. (Nevertheless, they had dimly been seen in flower cells as early as 1848 by the German amateur botanist Wilhelm Friedrich Benedict Hofmeister.)

Continued observation of stained cells showed that the cells of each species of plant or animal has a fixed and characteristic number of chromosomes. Before a cell divides in two during mitosis, the number of chromosomes is doubled, so that each of the two daughter cells after the division has the same number as the original mother cell.

The Belgian embryologist Eduard van Beneden discovered in 1885 that the chromosomes do *not* double in number when egg and sperm cells are being formed. Consequently each egg and each sperm cell has only half the number of chromosomes that ordinary cells of the organism possess. (The cell division that produces sperm and egg cells therefore is called *meiosis,* from a Greek word meaning "to make less.") When an egg and a sperm cell combine, however, the combination (the fertilized ovum) has a complete set of chromosomes, half contributed by the mother through the egg cell and half by the father through the sperm cell. This complete set is passed on by ordinary mitosis to all the cells that make up the body of the organism developing from the fertilized egg.

Even though the use of dyes makes the chromosomes visible, they do not make it easy to see one individual chromosome among the rest. Generally, they look like a tangle of stubby spaghetti. Thus, it was long thought that each human cell contained twenty-four pairs of chromosomes. It was not until 1956 that a more painstaking count of these cells showed twenty-three pairs to be the correct count.

Fortunately, this problem no longer exists. A technique has been devised whereby treatment with a low-concentration salt solution, in the proper manner, swells the cells and disperses the chromosomes. They can then be photographed, and that photograph can be cut into sections, each containing a separate chromosome. If these chromosomes are matched into pairs and then arranged in the order of decreasing length, the result is a *karyotype*, a picture of the chromosome content of the cell, consecutively numbered.

The karyotype offers a subtle tool in medical diagnosis, for separation of the chromosomes is not always perfect. In the process of cell division, a chromosome may be damaged or even broken. Sometimes the separation may not be even, so that one of the daughter cells gets an extra chromosome, while the other is missing one. Such abnormalities are sure to damage the working of the cell, often to such an extent that the cell cannot function. (This imperfection is what keeps the process of mitosis so accurate—not that it really is as accurate as it seems, but the mistakes are buried.)

Such imperfections are particularly dire when they take place in the process of meiosis, for then egg cells or sperm cells are produced with imperfections in the chromosome complement. If an organism can develop at all from such an imperfect start (and usually it cannot), every cell in its body has the imperfection: the result is a serious congenital disease.

The most frequent disease of this type involves severe mental retardation. It is called *Down's syndrome* (because it was first described in 1866 by the English physician John Langdon Haydon Down), and it occurs once in every thousand births. It is more commonly known as *mongolism*, because one of the symptoms is a slant to the eyelids that is reminiscent of the epicanthic fold of the peoples of eastern Asia. Since the syndrome has no more to do with the Asians, however, than with other ethnic groups, this is a poor name.

It was not until 1959 that the cause of Down's syndrome was discovered. In that year, three French geneticists—Jerome Jean Lejeune, Marthe Gautier, and Raymond Turpin—counted the chromosomes in cells from three cases and found that each had forty-seven chromosomes instead of forty-six. It turned out that the error was the possession of *three* members of chromosome pair 21. Then, in 1967, the mirror-image example of the disease was located. A mentally retarded three-year-old girl was found to have a single chromosome-21. She was the first discovered case of a living human being with a missing chromosome.

Cases of this sort involving other chromosomes seem less common but are turning up. Patients with a particular type of leukemia show a tiny extra chromosome fragment in their cells. This is called the *Philadelphia chromosome* because it was first located in a patient hospitalized in that city. Broken chromosomes, in general, turn up with greater than normal frequency in certain not very common diseases.

ASEXUAL REPRODUCTION

The formation of a new individual from a fertilized egg that contains half its chromosomes from each of two parents is *sexual reproduction*. It is the norm for human beings and for organisms generally that are at our level of complexity.

It is possible, however, for *asexual reproduction* to take place, with a new individual possessing a set of chromosomes derived from one parent only. A one-celled organism that divides in two, forming two independent cells, each with the same set of chromosomes as the original, offers an example of asexual reproduction.

Asexual reproduction is very common in the plant world, too. A twig of some plant can be placed in the ground, where it may take root and grow, producing a complete organism of the kind of which it was once only a twig. Or the twig can be grafted to the branch of another tree (of a different variety sometimes) where it can grow and flourish. Such a twig is called a *clone* from the Greek word for "twig"; and the term *clone* has come to be used for any one-parent organism of nonsexual origin.

Asexual reproduction can take place in multicellular animals as well. The more primitive the animal—that is, the less diversified and specialized its cells —the more likely it is that asexual reproduction can take place.

A sponge, or a freshwater hydra, or a flatworm, or a starfish, can, any of them, be torn into parts; and these parts, if kept in their usual environment, will each grow into a complete organism. The new organisms can be viewed as clones.

Even organisms as complex as insects can in some cases give birth to one-parent young and, in the case of aphids, for instance, do so as a matter of course. In such cases, an unfertilized egg cell, containing only a half-set of chromosomes, can do without a sperm cell. Instead, the egg cell's half-set merely duplicates itself, producing a full set all from the female parent; and the egg then proceeds to divide and become an independent organism, again a kind of clone.

Generally, in complex animals, however, no form of cloning takes place naturally, and reproduction is exclusively sexual. Yet human interference can bring about the cloning of vertebrates.

After all, a fertilized egg is capable of producing a complete organism; and as that egg divides and redivides, each new cell contains a complete set of chromosomes just like the original set. Why should not each new cell possess the capacity of producing a new individual if isolated and kept under conditions that allow the fertilized egg to develop?

Presumably, as the fertilized egg divides and redivides, the new cells *differentiate*, becoming liver cells, skin cells, nerve cells, muscle cells, kidney cells, and so on, and so on. Each has very different functions from any other; and, presumably, the chromosomes undergo subtle changes that make this differentiation possible. It is these subtle changes that make the differentiated cells incapable of starting from scratch and forming a new individual.

But are the chromosomes permanently and irreversibly changed? What if such chromosomes are restored to their original surroundings? Suppose, for instance, that one obtains an unfertilized egg cell of a particular species of animal and carefully removes its nucleus. One then obtains the nucleus of a skin cell from a developed individual of that species and inserts it into the egg cell. Under the influence of the egg cell, designed to promote the growth of a developed individual, might not the chromosomes within the skin-cell nucleus experience a "fountain of youth" effect which will restore them to their original function? Will the egg, *fertilized* in this fashion, develop to produce a new individual with the same chromosome set as the individual whose skin cell has been used for the purpose? Will not the new individual so obtained be a clone of the skin-cell donor?

This removal and substitution of nuclei within a cell is, of course, an excessively delicate operation, but it was successfully carried through in 1952 by the American biologists Robert William Briggs and Thomas J. King. Their work marked the beginning of the technique of *nuclear transplantation*.

In 1967, the British biologist John B. Gurdon successfully transplanted a nucleus from a cell from the intestine of a South African clawed frog to an unfertilized ovum of the same species and from that ovum developed a perfectly normal new individual—a clone of the first.

It would be enormously difficult to repeat this procedure in reptiles and birds whose egg cells are encased in hard shells—that is, to keep those egg cells alive and functioning after the shell is in some way broken for nuclear penetration.

What about mammalian egg cells? These are bare but are kept within the mother's body; they are particularly small and fragile, and microsurgical techniques must be further refined.

And yet nuclear transplantation has been successfully carried through in mice; and, in principle, cloning should be possible in any mammal, including the human being.

Genes

MENDELIAN THEORY

In the 1860s, an Austrian monk named Gregor Johann Mendel, who was too occupied with the affairs of his monastery to pay attention to the biologist's excitement about cell division, was quietly carrying through some experiments in his garden that were destined eventually to make sense out of chromosomes. Abbé Mendel, an amateur botanist, became particularly interested in the results of cross-breeding pea plants of varying characteristics. His great stroke of intuition was to study one clearly defined characteristic at a time.

He would cross plants with different seed colors (green or yellow), or smooth-seeded peas with wrinkle-seeded ones, or long-stemmed plants with short-stemmed ones, and then would follow the results in the offspring of the succeeding generations. Mendel kept a careful statistical record of his results, and his conclusions can be summarized essentially as follows:

1. Each characteristic is governed by *factors* that (in the cases that Mendel studied) can exist in one of two forms. One version of the factor for seed color, for instance, will cause the seeds to be green; the other form will make them yellow. (For convenience, let us use the present-day terms. The factors are now called *genes*, a term put forward in 1909 by the Danish biologist Wilhelm Ludwig Johannsen from a Greek word meaning "to give birth to"; and the different forms of a gene controlling a given characteristic are called *alleles*. Thus, the seed-color gene possesses two alleles, one for green seeds, the other for yellow seeds.)

2. Every plant has a pair of genes for each characteristic, one contributed by each parent. The plant transmits one of its pair to a germ cell (a general term used to include both egg cells and sperm cells), so that when the germ cells of two plants unite by pollination, the offspring has two genes for the characteristic once more. The two genes may be either identical or alleles.

3. When the two parent plants contribute alleles of a particular gene to the offspring, one allele may overwhelm the effect of the other. For instance, if a plant producing yellow seeds is crossed with one producing green seeds, all the members of the next generation will produce yellow seeds. The yellow allele of the seed-color gene is *dominant*, the green allele, *recessive*.

4. Nevertheless, the recessive allele is not destroyed. The green allele, in the case just cited, is still present, even though it produces no detectable effect. If two plants containing mixed genes (that is, each with one yellow and one green allele) are crossed, some of the offspring may have two green alleles in the fertilized ovum; in that case, those particular offspring will produce green seeds, and the offspring of such parents in turn will also produce green seeds. Mendel pointed out four possible ways of combining alleles from a pair of hybrid parents, each possessing one yellow and one green allele. A yellow allele from the first parent may combine with a yellow allele from the second; a yellow allele from the first may combine with a green allele from the second; a green allele from the first may combine with a yellow allele from the second; and a green allele from the first may combine with a green allele from the second. Of the four combinations, only the last will result in a plant that would produce green seeds. If all four combinations are equally probable, one-fourth of the plants of the new generation should produce green seeds—as Mendel indeed found to be so.

5. Mendel also found that characteristics of different kinds—for instance, seed color and flower color—to be inherited independently of each other: that is, red flowers are as likely to go with yellow seeds as with green seeds. The same is true of white flowers.

Mendel performed these experiments in the early 1860s, wrote them up carefully, and sent a copy of his paper to Karl Wilhelm von Nägeli, a Swiss botanist of great reputation. Von Nägeli's reaction was negative. Von Nägeli had, apparently, a predilection for all-encompassing theories (his own theoretical work was semimystical and turgid in expression), and he saw little merit in the mere counting of pea plants as a way to truth. Besides, Mendel was an unknown amateur.

It seems that Mendel allowed himself to be discouraged by von Nägeli's comments, for he turned to his monastery duties, grew fat (too fat to bend over in the garden), and abandoned his researches. He did, however, publish his paper in 1866 in a provincial Austrian journal, where it attracted no further attention for a generation.

But other scientists were slowly moving toward the same conclusions to which (unknown to them) Mendel had already come. One of the routes by which they arrived at an interest in genetics was the study of *mutations*—that is, of freak animals, or monsters, which had always been regarded as bad omens. (The word *monster* came from a Latin word meaning "warning.") In 1791, a Massachusetts farmer named Seth Wright took a more practical view of a sport that turned up in his flock of sheep. A lamb was born with abnormally short legs, and it occurred to the shrewd Yankee that short-legged sheep could not escape over the low stone walls around his farm. He therefore deliberately bred a line of short-legged sheep from his not unfortunate accident.

This practical demonstration stimulated other people to look for useful mutations. By the end of the nineteenth century, the American horticulturist Luther Burbank was making a successful career of breeding hundreds of new varieties of plants which were improvements over the old in one respect or another, not only by mutations, but by judicious crossing and grafting.

Meanwhile botanists tried to find an explanation of mutation. And in what is perhaps the most startling coincidence in the history of science, no fewer than three men, independently and in the very same year, came to precisely the same conclusions that Mendel had reached a generation earlier. They were Hugo De Vries of Holland, Karl Erich Correns of Germany, and Erich von Tschermak of Austria. None of them knew of each other's or Mendel's work. All three were ready to publish in 1900. All three, in a final check of previous publications in the field, came across Mendel's paper, to their own vast surprise. All three did publish in 1900, each citing Mendel's paper, giving Mendel full credit for the discovery, and advancing his own work only as confirmation.

GENETIC INHERITANCE

A number of biologists immediately saw a connection between Mendel's genes and the chromosomes seen under the microscope. The first to draw a parallel was an American cytologist named Walter Stanborough Sutton, in 1904. He pointed out that chromosomes, like genes, come in pairs, one of which is inherited from the father and one from the mother. The only trouble with this analogy was that the number of chromosomes in the cells of any organism is far smaller than the number of inherited characteristics. Man, for instance, has only twenty-three pairs of chromosomes and yet certainly possesses thousands of inheritable characteristics. Biologists therefore had to conclude that chromosomes are not genes. Each chromosome must be a collection of genes.

In short order, biologists discovered an excellent tool for studying specific genes. It was not a physical instrument but a new kind of laboratory animal. In 1906, the Columbia University zoologist Thomas Hunt Morgan, who was at first skeptical of Mendel's theories, conceived the idea of using fruit flies (*Drosophila melanogaster*) for research in genetics. (The term *genetics* was coined in 1902 by the British biologist William Bateson.)

Fruit flies had considerable advantages over pea plants (or any ordinary laboratory animal) for studying the inheritance of genes. They bred quickly and prolifically, could easily be raised by the hundreds on little food, had scores of inheritable characteristics that could be observed readily, and had a comparatively simple chromosomal setup—only four pairs of chromosomes per cell.

With the fruit fly, Morgan and his co-workers discovered an important fact about the mechanism of inheritance of sex. They found that the female fruit fly has four perfectly matched pairs of chromosomes so that all the egg cells, receiving one of each pair, are identical so far as chromosome makeup is concerned. However, in the male fruit fly, one of each of the four pairs consists of a normal chromosome, called the *X chromosome*, and a stunted one, the *Y chromosome*. Therefore, when sperm cells are formed, half have an X chromosome and half a Y chromosome. When a sperm cell with the X chromosome fertilizes an egg cell, the fertilized egg, with four matched pairs, naturally becomes a female. On the other hand, a sperm cell with a Y chromosome produces a male. Since both alternatives are equally probable, the number of males and females in the typical species of living things is roughly equal (figure 13.4). (In some creatures, notably various birds, it is the female that has a Y chromosome.)

This chromosomal difference explains why some disorders or mutations show up only in the male. If a defective gene occurs on one of a pair of X chromosomes, the other member of the pair is still likely to be normal and can salvage the situation. But in the male, a defect on the X chromosome paired with the Y chromosome generally cannot be compensated for, because the latter carries very few genes. Therefore the defect shows up.

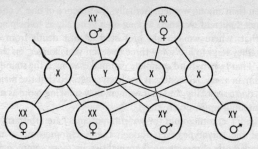

Figure 13.4. Combinations of X and Y chromosomes.

The most notorious example of such a *sex-linked disease* is *hemophilia,* a condition in which blood clots only with difficulty, if at all. Individuals with hemophilia run the constant risk of bleeding to death from slight causes or of suffering agonies from internal bleeding. A woman who carries a gene that will produce hemophilia on one of her X chromosomes is very likely to have a normal gene at the same position in the other X chromosome. She will therefore not show the disease. She will, however, be a *carrier.* Of the egg cells she forms, half will have the normal X chromosome and half the hemophiliac X chromosome. If the egg with the abnormal X chromosome is fertilized by sperm with an X chromosome from a normal male, the resulting child will be a girl who will not be hemophiliac but who will again be a carrier; if it is fertilized by sperm with a Y chromosome from a normal male, the hemophiliac gene in the ovum will not be counteracted by anything in the Y chromosome, and the result is a boy with hemophilia. By the laws of chance, half the sons of hemophilia carriers will be hemophiliacs; half the daughters will be, in their turn, carriers.

The most eminent hemophilia carrier in history was Queen Victoria of England. Only one of her four sons (the oldest, Leopold) was hemophiliac. Edward VII—from whom later British monarchs descended—escaped, so there is no hemophilia now in the British royal family. However, two of Victoria's daughters were carriers. One had a daughter (also a carrier) who married Czar Nicholas II of Russia. As a result, their only son was a hemophiliac; this circumstance helped alter the history of Russia and the world, for it was through his influence on the hemophiliac that the monk Gregory Rasputin gained power in Russia and helped bring on the discontent that eventually led to revolution. The other daughter of Victoria had a daughter (also a carrier) who married into the royal house of Spain, producing hemophilia there. Because of its presence among the Spanish Bourbons and the Russian Romanoffs, hemophilia was sometimes called the *royal disease,* but it has no particular connection with royalty, except for Victoria's misfortune.

A lesser sex-linked disorder is color-blindness, which is far more common among men than among women. Actually, the absence of one X chromosome may produce sufficient weakness among men generally as to help account for the fact that, where women are protected against death from childbirth infections, they tend to live some three to seven years longer, on the average, then men. That twenty-third complete pair makes women the sounder biological organism, in a way. (Recently, it has been suggested that the male's shorter life span is due to smoking, and that women, now smoking more as men smoke less, are catching up in death rate.)

The X and Y chromosomes (or *sex chromosomes*) are arbitrarily placed at the end of the karyotype, even though the X chromosome is among the longest. Apparently chromosome abnormalities are more common among the sex chromosomes than among the others. The reason may be not that the sex chromosomes are most likely to be involved in abnormal mitoses, but perhaps that sex-chromosome abnormalities are less likely to be fatal, so that more young manage to be born with them.

The type of sex-chromosome abnormality that has drawn the most attention is one in which a male ends up with an extra Y chromosome in his cells, so that he is XYY, so to speak. It turns out that XYY males are difficult to handle. They are tall, strong, and bright but are characterized by a tendency to rage and violence. Richard Speck, who killed eight nurses in Chicago in 1966, is supposed to have been an XYY. A murderer was acquitted in Australia in October 1968 on the grounds that he was an XYY and therefore not responsible for his action. Nearly 4 percent of the male inmates in a certain Scottish prison have turned out to be XYY, and there are some estimates that XYY combinations may occur in as many as 1 man in every 3,000.

There seems to be some reason for considering it desirable to run a chromosome check on everyone and certainly on every newborn child. As is the case of other procedures, simple in theory but tedious in practice, attempts are being made to computerize such a process.

CROSSING OVER

Research on fruit flies showed that traits are not necessarily inherited independently, as Mendel had thought. It happened that the seven characteristics of pea plants that he had studied were governed by genes on separate chromosomes. Morgan found that where two genes governing two different characteristics are located on the same chromosome, those characteristics are generally inherited together (just as a passenger in the front seat of a car and one in the back seat travel together).

This genetic linkage is not, however, unchangeable. Just as a passenger can change cars, so a piece of one chromosome occasionally switches to another, swapping places with a piece from it. Such *crossing over* may occur during the division of a cell (figure 13.5). As a result, linked traits are separated and reshuffled in a new linkage. For instance, there is a variety of fruit fly with scarlet eyes and curly wings. When it is mated with a white-eyed, miniature-winged fruit fly, the offspring will generally be either red-eyed and curly-

Figure 13.5. Crossing over in chromosomes.

winged or white-eyed and miniature-winged. But the mating may sometimes produce a white-eyed, curly-winged fly or a red-eyed, miniature-winged one as a result of crossing over. The new form will persist in succeeding generations unless another crossing over takes place.

Now picture a chromosome with a gene for red eyes at one end and a gene for curly wings at the other end. Let us say that, in the middle of the chromosome's length, there are two adjacent genes governing two other characteristics. Obviously, the probability of a break occurring at that particular point, separating these two genes, is smaller than the probability of a break coming at one of the many points along the length of the chromosome that would separate the genes at the opposite ends. By noting the frequency of separation of given pairs of linked characteristics by crossing over, Morgan and his co-workers, notably Alfred Henry Sturtevant, were able to deduce the relative locations of the genes in question and, in this way, worked out chromosome *maps* of gene locations for the fruit fly. The location, so determined, is the *locus* of a gene.

(But, as often happens when biological systems are studied, behavior does not follow the rules as rigidly as scientists are sometimes inclined to suppose. In the 1940s and afterward, the American biologist Barbara McClintock, carefully studying the genes in corn, and following them from generation to generation, came to the conclusion that some genes can shift rather easily and frequently from place to place on the chromosomes in the course of cell division. This idea seemed so out of line with the results obtained by Morgan and biologists who followed him that she was ignored—but she was right. When others began to find evidence of gene mobility, McClintock (now an octogenarian) received the Nobel Prize for physiology and medicine in 1983.)

From such chromosome maps—and from a study of giant chromosomes, many times the ordinary size, found in the salivary glands of the fruit fly—it has been established that the insect has a minimum of 10,000 genes in a chromosome pair. Hence, the individual gene must have a molecular weight of 60,000,000. Accordingly, the human's somewhat larger chromosomes may contain from 20,000 to 90,000 genes per chromosome pair, or up to 2,000,000 altogether.

For his work on the genetics of fruit flies, Morgan received the Nobel Prize in medicine and physiology in 1933.

Increasing knowledge of genes raises hopes that the genetic endowment of individual humans might someday be analyzed and modified: either preventing seriously anomalous conditions from developing, or correcting them if they slip by. Such *genetic engineering* would require human chromosome maps— clearly a tremendously larger job than in the case of the fruit fly. The task was made somewhat simpler in a startling way in 1967, when Howard Green of New York University formed hybrid cells containing both mouse and human chromosomes. Relatively few human chromosomes persisted after several cell divisions, and the effects due to their activity was more easily pinpointed.

Another step in the direction of gene knowledge and gene manipulation came in 1969, when the American biochemist Jonathan Beckwith and his co-workers isolated an individual gene for the first time in history. It was from an intestinal bacterium, and it controlled an aspect of sugar metabolism.

THE GENETIC LOAD

Every once in a while, with a frequency that can be calculated, a sudden change occurs in a gene. The *mutation* shows itself by some new and unexpected physical characteristic, such as the short legs of Farmer Wright's lamb. Mutations in nature are comparatively rare. In 1926, the geneticist Hermann Joseph Muller, who had been a member of Morgan's research team, discovered a way to increase the rate of mutations artificially in fruit flies so that the inheritance of such changes could be studied more easily. He found that X rays would do the trick—presumably by damaging the genes. The study of mutations made possible by Muller's discovery won him the Nobel Prize in medicine and physiology in 1946.

As it happens, Muller's researches have given rise to some rather disquieting thoughts concerning the future of the human species. While mutations are an important driving force in evolution, occasionally producing an improvement that enables a species to cope better with its environment, the beneficial mutation is very much the exception. Most mutations—at least 99 percent of them—are detrimental, some even lethal. Eventually, even those that are only slightly harmful die out, because their bearers do not get along as well and leave fewer descendants than healthy individuals do. But in the meantime a mutation may cause illness and suffering for many generations. Furthermore, new mutations keep cropping up continually, and every species carries a constant load of defective genes. Thus, more than 1,600 human diseases are thought to be the result of genetic defects.

The great number of different gene varieties—including large quantities of seriously harmful ones—in normal populations was clearly shown by the work of the Russian-American geneticist Theodosius Dobzhansky in the 1930s and 1940s. This diversity makes evolution march on as it does, but the number of deleterious genes (the *genetic load*) gives rise to fears justified anxiety.

Two modern developments seem to be adding steadily to this load. First, the advances in medicine and social care tend to compensate for the handicaps of people with detrimental mutations, at least so far as the ability to reproduce is concerned. Eyeglasses are available to individuals with defective vision; insulin keeps alive sufferers from diabetes (a hereditary disease), and so on. Thus they pass on their defective genes to future generations. The alternatives —allowing defective individuals to die young or sterilizing or imprisoning them—are, of course, unthinkable, except where the handicap is sufficiently great to make the individual less than human, as in idiocy or homicidal paranoia. Undoubtedly, the human species can still bear its load of negatively mutated genes, despite its humanitarian impulses.

But there is less excuse for the second modern hazard—namely, adding to the load by unnecessary exposure to radiation. Genetic research shows incontrovertibly that, for the population as a whole, even a slight increase in general exposure to radiation means a corresponding slight increase in the mutation rate. And since 1895 we have been exposed to types and intensities of radiation of which previous generations knew nothing. Solar radiation, the natural radioactivity of the soil, and cosmic rays have always been with us. Now, however, we use X rays in medicine and dentistry with abandon; we concentrate radioactive material; we form artificially radioactive isotopes of terrifying radiant potency; we even explode nuclear bombs. All of this increases the background radiation.

No one, of course, suggests that research in nuclear physics be abandoned, or that X rays never be used by doctor and dentist. There is, however, a strong recommendation that the danger be recognized and that exposure to radiation be minimized: that, for instance, X rays be used with discrimination and care, and that the sexual organs be routinely shielded during all such use. Another suggested precaution is that each individual keep a record of his or her total accumulated exposure to X rays so as to try to avoid exceeding a reasonable limit.

BLOOD TYPES

Of course, the geneticists could not be sure that the principles established by experiments on plants and insects necessarily applied to humans. After all, we are neither pea plants nor fruit flies. But direct studies of certain human characteristics showed that human genetics does follow the same rules. The best-known example is the inheritance of blood types.

Blood transfusion is a very old practice, and early physicians occasionally even tried to transfuse animal blood into persons weakened by loss of blood. But transfusions even of human blood often turned out badly, so that laws were sometimes passed forbidding transfusion. In the 1890s, the Austrian pathologist Karl Landsteiner finally discovered that human blood comes in different types, some of which are incompatible with each other. He found that sometimes when blood from one person was mixed with a sample of *serum* (the blood fluid remaining after the red cells and a clotting factor are removed) from another person, the red cells of the first person's whole blood would clump together. Obviously such a mixture would be very dangerous if it occurred in transfusion, and it might even kill the patient if the clumped cells blocked the blood circulation in key vessels. Landsteiner also found, however, that some bloods could be mixed without causing any deleterious clumping.

By 1902, Landsteiner was able to announce that there were four types of human blood, which he called A, B, AB, and O. Any given individual had blood of just one of these types. Of course, a particular type could be transferred without danger from one person to another having the same type. In addition, O blood could safely be transfused to a person possessing any of the other three types, and either A blood or B blood could be given to an AB patient. But red-cell clumping (*agglutination*) would result when AB blood was transfused to an A or a B individual, when A and B were mixed, or when an O individual received a transfusion of any blood other than O. (Nowadays, because of possible serum reactions, in good practice patients are given only blood of their own type.)

In 1930, Landsteiner (who by then had become a United States citizen) received the Nobel Prize in medicine and physiology.

Geneticists have established that these blood types (and all the others since discovered, including the Rh variations) are inherited in a strictly Mendelian manner. It seems that there are three gene alleles responsible, respectively, for A, B, and O blood. If both parents have O-type blood, all the children of that union will have O-type blood. If one parent is O-type and the other A-type, all the children may show A-type blood, for the A allele is dominant over the O. The B allele likewise is dominant over the O allele. The B allele and A allele, however, show no dominance with respect to each other, and an individual possessing both alleles has AB-type blood.

The Mendelian rules work out so strictly that blood groups can be (and are) used to test paternity. If an O-type mother has a B-type child, the child's father must be B-type, for that B allele must have come from somewhere. If the woman's husband happens to be A or O, it is clear that she has been unfaithful (or there has been a baby mix-up at the hospital). If an O-type woman with a B-type child accuses an A or an O man of being the parent, she is either mistaken or lying. On the other hand, while blood type can sometimes prove a negative, it can never prove a positive. If the woman's husband or the man accused is indeed a B-type, the case remains unproved. Any B-type man or any AB-type man could have been the father.

EUGENICS

The applicability of the Mendelian rules of inheritance to human beings has also been borne out by the existence of sex-linked traits. As I have said, color-blindness and hemophilia are found almost exclusively in males and are inherited in precisely the manner that sex-linked characteristics are inherited in the fruit fly.

Naturally, the thought will arise that by forbidding people with such afflictions to have children, the disorder can be wiped out. By directing proper mating, the human breed might even be improved, as breeds of cattle have been. This is by no means a new idea. The ancient Spartans believed this and tried to put it into practice 2,500 years ago. In modern times, the notion was revived by an English scientist, Francis Galton (a cousin of Charles Darwin). In 1883, he coined the word *eugenics* to describe his scheme. (The word derives from the Greek and means "good birth.")

Galton was not aware, in his time, of the findings of Mendel. He did not understand that characteristics might seem to be absent, yet be carried as recessives. He did not understand that groups of characteristics would be inherited intact, and that it might be difficult to get rid of an undesirable one without also getting rid of a desirable one. Nor was he aware that mutations would reintroduce undesirable characteristics in every generation.

Nevertheless, the desire to "improve" the human stock continues, and eugenics finds its supporters, even among scientists, to this day. Such support is almost invariably suspect, since those who are avid to show important genetic differences between recognizable groups of human beings are sure to find the groups to which they themselves belong to be "superior."

The English psychologist Cyril Lodowic Burt, for instance, reported studies of intelligence of different groups and claimed strong evidence for supposing men to be more intelligent than women, Christians to be more intelligent than Jews, Englishmen to be more intelligent than Irishmen, upper-class Englishmen to be more intelligent than lower-class Englishmen, and so on. Burt himself belonged, in every case, to the "superior" group. His results were, of course, accepted by many people who, like Burt, were in the "superior" group, and who were ready to believe that those who were worse off were the victims not of oppression and prejudice but, instead, of their own defects.

After Burt's death in 1971, however, doubts arose concerning his data. There were distinctly suspicious perfections about his statistics. The suspicions grew; and in 1978, the American psychologist D. D. Dorfman was able to show, rather conclusively, that Burt had simply fabricated his data, so anxious was he to prove a thesis that he deeply believed but that could not be proved honestly.

And yet, even so, Shockley, the co-inventor of the transistor, gained a certain notoriety for himself by maintaining that blacks are significantly less intelligent than whites, through genetic factors, so that attempts to better the lot of blacks by giving them equal opportunities are bound to fail. The German-British psychologist Hans J. Eysenck also maintains this view.

In 1980, Shockley laid himself open to some ill-natured jests when he incautiously revealed that he had contributed some of his then seventy-year-old sperm cells for preservation by freezing in a sperm bank designed for eventual use in the insemination of women volunteers of high intelligence.

My own belief is that human genetics is an enormously complicated subject that is not likely to be completely or neatly worked out in the foreseeable future. Because we breed neither as frequently nor as prolifically as the fruit fly; because our matings cannot be subjected to laboratory control for experimental purposes; because we have many more chromosomes and many more inherited characteristics than the fruit fly; because the human characteristics in which we are most interested—such as creative genius, intelligence, and moral strength—are extremely complex, involving the interplay of numerous genes plus environmental influences—for all these reasons, geneticists cannot deal with human genetics with the same confidence with which they study fruit-fly genetics.

Eugenics remains a dream, therefore, made hazy and insubstantial by lack of knowledge, and vicious because of the ease with which it can be exploited by racists and bigots.

CHEMICAL GENETICS

Just how does a gene bring the physical characteristic for which it is responsible into being? What is the mechanism whereby it gives rise to yellow seeds in pea plants, or curled wings in fruit flies, or blue eyes in human beings?

Biologists are now certain that genes exert their effects by way of enzymes. One of the clearest cases in point involves the color of eyes, hair, and skin. The color (blue or brown, yellow or black, pink or brown, or shades in between) is determined by the amount of pigment, called *melanin* (from the Greek word for "black"), that is present in the eye's iris, the hair, or the skin. Now melanin is formed from an amino acid, tyrosine, by way of a number of steps, most of which have now been worked out. A number of enzymes are involved, and the amount of melanin formed will depend upon the quantity of these enzymes. For instance, one of the enzymes, which catalyzes the first two steps, is tyrosinase. Presumably some particular gene controls the production of tyrosinase by the cells and, in that way, will control the coloring of the skin, hair, and eyes. And since the gene is transmitted from generation to generation, children will naturally resemble their parents in coloring. If a mutation happens to produce a defective gene that cannot form tyrosinase, there will be no melanin, and the individual will be an *albino*. The absence of a single enzyme (and hence the deficiency of a single gene) will thus suffice to bring about a major change in personal characteristics.

Granted that an organism's characteristics are controlled by its enzyme make-up, which in turn is controlled by genes, the next question is: How do the genes work? Unfortunately, even the fruit fly is much too complex an organism to trace out the matter in detail. But, in 1941, the American biologists George Wells Beadle and Edward Lawrie Tatum began such a study with a simple organism which they found admirably suited to this purpose: the common pink bread mold (scientific name, *Neurospora crassa*).

Neurospora is not very demanding in its diet. It will grow very well on sugar plus inorganic compounds that supply nitrogen, sulfur, and various minerals. Aside from sugar, the only organic substance that has to be supplied to it is a vitamin called *biotin*.

At a certain stage in its life cycle, the mold produces eight spores, all identical in genetic constitution. Each spore contains seven chromosomes; as in the sex cell of a higher organism, its chromosomes come singly, not in pairs.

Consequently, if one of its chromosomes is changed, the effect can be observed, because there is no normal partner present to mask the effect. Beadle and Tatum, therefore, were able to create mutations in *Neurospora* by exposing the mold to X rays and then to follow the specific effects in the behavior of the spores.

If, after the mold had received a dose of radiation, the spores still thrived on the usual medium of nutrients, clearly no mutation had taken place, at least so far as the organism's nutritional requirements for growth were concerned. If the spores would not grow on the usual medium, the experimenters proceeded to determine whether they were alive or dead, by feeding them a complete medium containing all the vitamins, amino acids, and other items they might possibly need. If the spores grew on this, the conclusion was that the X rays had produced a mutation that had changed *Neurospora*'s nutritional requirements. Apparently it now needed at least one new item in its diet. To find out what that was, the experimenters tried the spores on one diet after another, each time with some items of the complete medium missing. They might omit all the amino acids, or all the various vitamins, or all but one or two amino acids or one or two vitamins. In this way, they narrowed down the requirements until they identified just what the spore now needed in its diet because of the mutation.

It turned out sometimes that the mutated spore required the amino acid arginine. The normal *wild strain* had been able to manufacture its own arginine from sugar and ammonium salts. Now, thanks to the genetic change, it could no longer synthesize arginine; and unless this amino acid was supplied in its diet, it could not make protein and therefore could not grow.

The clearest way to account for such a situation was to suppose that the X rays had disrupted a gene responsible for the formation of an enzyme necessary for manufacturing arginine. For lack of the normal gene, *Neurospora* could no longer make the enzyme. No enzyme, no arginine.

Beadle and his co-workers went on to use this sort of information to study the relation of genes to the chemistry of metabolism. There was a way to show, for instance, that more than one gene is involved in the making of arginine. For simplicity's sake, let us say there are two—gene A and gene B—responsible for the formation of two different enzymes, both of which are necessary for the synthesis of arginine. Then a mutation of either gene A or gene B will rob *Neurospora* of the ability to make the amino acid. Suppose we irradiate two batches of *Neurospora* and produce an arginineless strain in each one. If we are lucky, one mutant may have a defective A gene and a normal B gene; the other, a normal A and defective B. To see if that has happened, let us cross the two mutants at the sexual stage of their life cycle. If the two strains do indeed differ in this way, the recombination of chromosomes may produce some spores whose A and B genes are both normal. In other words, from two mutants that are incapable of making arginine, we will get some offspring that *can* make it. Sure enough, exactly that sort of thing happened when the experiments were performed.

It was possible to explore the metabolism of *Neurospora* in finer detail than this. For instance, here were three different mutant strains incapable of making arginine on an ordinary medium. One would grow only if it was supplied with arginine itself. The second would grow if it received either arginine or a very similar compound called *citrulline*. The third could grow on arginine or citrulline or still another similar compound called *ornithine*.

What conclusion would you draw from all this? Well, we can guess that these three substances are steps in a sequence of which arginine is the final product. Each requires an enzyme. First, ornithine is formed from some simpler compound with the help of an enzyme; then, another enzyme converts ornithine to citrulline; and finally, a third enzyme converts citrulline to arginine. Now a *Neurospora* mutant that lacks the enzyme for making ornithine but possesses the other enzymes can get along if it is supplied with ornithine, for from it the spore can make citrulline and then the essential arginine. Of course, it can also grow on citrulline, from which it can make arginine, and on arginine itself. By the same token, we can reason that the second mutant strain lacks the enzyme needed to convert ornithine to citrulline. This strain therefore must be provided with citrulline, from which it can make arginine, or with arginine itself. Finally, we can conclude that the mutant that will grow only on arginine has lost the enzyme (and gene) responsible for converting citrulline to arginine.

By analyzing the behavior of the various mutant strains they were able to isolate, Beadle and his co-workers founded the science of *chemical genetics*. They worked out the course of synthesis of many important compounds by organisms. Beadle proposed what has become known as the *one-gene-one-enzyme theory*—that is, that every gene governs the formation of a single enzyme—a suggestion that is now generally accepted by geneticists. For their pioneering work, Beadle and Tatum shared in the Nobel Prize in medicine and physiology in 1958.

ABNORMAL HEMOGLOBIN

Beadle's discoveries put biochemists on the *qui vive* for evidence of gene-controlled changes in proteins—particularly in human mutants, of course. A case turned up, unexpectedly, in connection with the disease called *sickle-cell anemia*, one of the more than 1600 genetic diseases in human beings.

This disease had first been reported in 1910 by a Chicago physician named James Bryan Herrick. Examining a sample of blood from a black teenage patient under the microscope, Herrick found that the red cells, normally round, had odd, bent shapes, many of them resembling the crescent shape of a sickle. Other physicians began to notice the same peculiar phenomenon, almost always in black patients. Eventually investigators decided that sickle-cell anemia is a hereditary disease. It follows the Mendelian laws of inheritance: apparently there is a sickle-cell gene that, when inherited in double dose from both parents, produces these distorted red cells. Such cells are unable to carry oxygen properly and are exceptionally short-lived, so there is a shortage of red cells in the blood. Those who inherit the double dose tend to die of the disease in childhood. On the other hand, when a person has only one sickle-cell gene, from one of his parents, the disease does not appear. Sickling of his red cells shows up only when the person is deprived of oxygen to an unusual degree, as at high altitudes. Such people are considered to have the *sickle-cell trait,* but not the disease.

It was found that about 9 percent of the black people in America have the trait, and 0.25 percent have the disease. In some localities in Central Africa, as much as a quarter of the black population shows the trait. Apparently the sickle-cell gene arose as a mutation in Africa and has been inherited ever since by individuals of African descent. If the disease is fatal, why has the defective gene not died out? Studies in Africa during the 1950s turned up the answer. It seems that people with the sickle-cell trait tend to have greater immunity to malaria than do normal individuals. The sickle cells are somehow inhospitable to the malarial parasite. It is estimated that, in areas infested with malaria, children with the trait have a 25 percent better chance of surviving to child-bearing age than have those without the trait. Hence, possessing a single dose of the sickle-cell gene (but not the anemia-causing double dose) confers an advantage. The two opposing tendencies—promotion of the defective gene by the protective effect of the single dose, and elimination of the gene by its fatal effect in double dose—tend to produce an equilibrium that maintains the gene at a certain level in the population.

In regions where malaria is not an acute problem, the gene does tend to die out. In America, the incidence of sickle-cell genes among blacks may have started as high as 25 percent. Even allowing for a reduction to an estimated 15 percent by admixture with non-black individuals, the present incidence of only 9 percent shows that the gene is dwindling away. In all probability it will continue to do so. If Africa is freed of malaria, the gene will presumably dwindle there, too.

The biochemical significance of the sickle-cell gene suddenly came into prominence in 1949 when Linus Pauling and his co-workers at California Institute of Technology (where Beadle also was working) showed that the gene affects the hemoglobin of the red blood cells: persons with a double dose of the sickle-cell gene are unable to make normal hemoglobin. Pauling proved this by means of the technique called *electrophoresis,* a method that uses an electric current to separate proteins by virtue of differences in the net electric charge on the various protein molecules. (The electrophoretic technique was developed by the Swedish chemist Arne Wilhelm Kaurin Tiselius, who received the Nobel Prize in chemistry in 1948 for this valuable contribution.) Pauling, by electrophoretic analysis, found that patients with sickle-cell anemia had an abnormal hemoglobin (named *hemoglobin S*), which could be separated from normal hemoglobin. The normal kind was given the name *hemoglobin A* (for "adult") to distinguish it from a hemoglobin in fetuses, called *hemoglobin F.*

Since 1949, biochemists have discovered other abnormal hemoglobins besides the sickle-cell one, and they are lettered from hemoglobin C to hemoglobin M. Apparently, the gene responsible for the manufacture of hemoglobin has been mutated into many defective alleles, each giving rise to a hemoglobin that is inferior for carrying out the functions of the molecule in ordinary circumstances but perhaps helpful in some unusual condition. Thus, just as hemoglobin S in a single dose improves resistance to malaria, so hemoglobin C in a single dose improves the ability of the body to get along on marginal quantities of iron.

Since the various abnormal hemoglobins differ in electric charge, they must differ somehow in the arrangement of amino acids in the peptide chain, for the amino-acid make-up is responsible for the charge pattern of the molecule. The differences must be very small, because the abnormal hemoglobins all function as hemoglobin after a fashion. The hope of locating the difference in a huge molecule of some 600 amino acids was correspondingly small. Nevertheless, the German-American biochemist Vernon Martin Ingram and co-workers tackled the problem of the chemistry of the abnormal hemoglobins.

They first broke down hemoglobin A, hemoglobin S, and hemoglobin C into peptides of various sizes by digesting them with a protein-splitting enzyme. Then they separated the fragments of each hemoglobin by *paper electrophoresis*—that is, using the electric current to convey the molecules along a moistened piece of filter paper instead of through a solution. (We can think of this as a kind of electrified paper chromatography.) When the investigators had done this with each of the three hemoglobins, they found that the only difference among them was that a single peptide turned up in a different place in each case.

They proceeded to break down and analyze this peptide. Eventually they learned that it was composed of nine amino acids, and that the arrangement of these nine was exactly the same in all three hemoglobins except at one position. The respective arrangements were:

Hemoglobin A: His-Val-Leu-Leu-Thr-Pro-Glu-Glu-Lys
Hemoglobin S: His-Val-Leu-Leu-Thr-Pro-Val-Glu-Lys
Hemoglobin C: His-Val-Leu-Leu-Thr-Pro-Lys-Glu-Lys

As far as could be told, the only difference among the three hemoglobins lay in that single amino acid in the seventh position in the peptide: it was glutamic acid in hemoglobin A, valine in hemoglobin S, and lysine in hemoglobin C. Since glutamic acid gives rise to a negative charge, lysine to a positive charge, and valine to no charge at all, it is not surprising that the three proteins behave differently in electrophoresis. Their charge pattern is different.

But why should so slight a change in the molecule result in so drastic a change in the red cell? Well, the normal red cell is one-third hemoglobin A. The hemoglobin A molecules are packed so tight in the cell that they barely have room for free movement. In short, they are on the point of precipitating out of solution. Part of the influence that determines whether a protein is to precipitate out is the nature of its charge. If all the proteins have the same net charge, they repel one another and keep from precipitating. The greater the charge (that is, the *repulsion*), the less likely the proteins are to precipitate. In hemoglobin S the intermolecular repulsion may be slightly less than in hemoglobin A, and hemoglobin S is correspondingly less soluble and more likely to precipitate. When a sickle cell is paired with a normal gene, the latter may form enough hemoglobin A to keep the hemoglobin S in solution, though it is a near squeak. But when both of the genes are sickle-cell mutants, they will produce only hemoglobin S. This molecule cannot remain in solution. It precipitates out into crystals, which distort and weaken the red cell.

This theory would explain why the change of just one amino acid in each half of a molecule made up of nearly 600 is sufficient to produce a serious disease and the near-certainty of an early death.

METABOLIC ABNORMALITY

Albinism and sickle-cell anemia are not the only human defects that have been traced to the absence of a single enzyme or the mutation of a single gene. There is *phenylketonuria,* a hereditary defect of metabolism, which often causes mental retardation and results from the lack of an enzyme needed to convert the amino acid phenylalanine to tyrosine. There is *galactosemia,* a disorder causing eye cataracts and damage to the brain and liver, which has been traced to the absence of an enzyme required to convert a galactose phosphate to a glucose phosphate. There is a defect, involving the lack of one or another of the enzymes that control the breakdown of *glycogen* (a kind of starch) and its conversion to glucose, which results in abnormal accumulations of glycogen in the liver and elsewhere and usually leads to early death. These are examples of *inborn errors of metabolism,* a congenital lack of the capacity to form some more or less vital enzyme found in normal human beings. This concept was first introduced to medicine by the English physician Archibald Edward Garrod in 1908, but it lay disregarded for a generation until, in the mid-1930s, the English geneticist John Burdon Sanderson Haldane brought the matter to the attention of scientists once more.

Such disorders are generally governed by a recessive allele of the gene that produces the enzyme involved. When only one of a pair of genes is defective, the normal one can carry on, and the individual is usually capable of leading a normal life (as in the case of possessor of the sickle-cell trait). Trouble generally comes only when two parents happen to have the same unfortunate gene and have the further bad luck of combining those two in a fertilized egg. Their child, then, is the victim of a kind of Russian roulette. Probably all of us carry our load of abnormal, defective, even dangerous genes, usually masked by normal ones. You can understand why the human geneticists are so concerned about radiation or anything else that may increase the mutation rate and add to the load.

Nucleic Acids

The really remarkable thing about heredity is not these spectacular, comparatively rare aberrations, but the fact that, by and large, inheritance runs so strictly true to form. Generation after generation, millennium after millennium, the genes go on reproducing themselves in exactly the same form and generating exactly the same enzymes, with only an occasional accidental variation of the blueprint. They rarely fail by so much as the introduction of a single wrong amino acid in a large protein molecule. How do they manage to make true copies of themselves over and over again with such astounding faithfulness?

The answer must lie in the chemistry of the long strings of genes that we call chromosomes. One major portion of the chromosomes, about half of its mass, is made up of proteins. This is no surprise. As the twentieth century wore on, biochemists expected any complex bodily function to involve proteins. Proteins seemed to be *the* complex molecules of the body, the only ones complex enough to represent the versatility and sensitivity of life.

And yet, a major portion of chromosomal proteins belonged to a class called *histone*, whose molecules are rather small for a protein and (worse yet) made up of a surprisingly simple mix of amino acids. They did not seem nearly complicated enough to be responsible for the delicacies and intricacies of genetics. To be sure, there were nonhistone protein components that were made up of much larger and more complex molecules, but they amounted to but a minor portion of the whole.

Nevertheless, biochemists were stuck with the proteins. Surely, the mechanism of heredity could involve nothing else. About half the chromosome consisted of material that was not protein at all, but it did not seem possible that anything that was not protein would suit. Still, it is to this nonprotein constituent of chromosomes that we must turn.

GENERAL STRUCTURE

In 1869, a Swiss biochemist named Friedrich Miescher, while breaking down the protein of cells with pepsin, discovered that the pepsin did not break up the cell nucleus. The nucleus shrank a bit, but remained intact. By chemical analysis, Miescher then found that the cell nucleus consisted largely of a phosphorus-containing substance whose properties did not at all resemble protein. He called the substance *nuclein*. It was renamed *nucleic acid* twenty years later when it was found to be strongly acid.

Miescher devoted himself to a study of this new material and eventually discovered sperm cells (which consist almost entirely of nuclear material) to be particularly rich in nucleic acid. Meanwhile, the German chemist Felix Hoppe-Seyler, in whose laboratories Miescher had made his first discovery, and who had personally confirmed the young man's work before allowing it to be published, isolated nucleic acid from yeast cells. This seemed different in properties from Miescher's material, so Miescher's variety was named *thymus nucleic acid* (because it could be obtained with particular ease from the thymus gland of animals), and Hoppe-Seyler's, naturally, was called *yeast nucleic acid*. Since thymus nucleic acid was at first derived only from animal cells and yeast nucleic acid only from plant cells, it was thought for a while that this might represent a general chemical distinction between animals and plants.

The German biochemist Albrecht Kossel, another pupil of Hoppe-Seyler, was the first to make a systematic investigation of the structure of the nucleic-acid molecule. By careful hydrolysis, he isolated from it a series of nitrogen-containing compounds, which he named *adenine, guanine, cytosine,* and *thymine.* Their formulas are now known to be:

Adenine   Guanine   Cytosine

Thymine

The double-ring formation in the first two compounds is called the *purine ring,* and the single ring in the other two is the *pyrimidine ring.* Therefore, adenine and guanine are referred to as *purines,* and cytosine and thymine are *pyrimidines.*

For these researches, which started a fruitful train of discoveries, Kossel received the Nobel Prize in medicine and physiology in 1910.

In 1911, the Russian-born American biochemist Phoebus Aaron Theodore Levene, a pupil of Kossel, carried the investigation a stage further. Kossel had discovered, in 1891, that nucleic acids contain carbohydrate, but now Levene showed that the nucleic acids contain five-carbon sugar molecules. (This was, at the time, an unusual finding: the best-known sugars, such as glucose, contain six carbons.) Levene followed this discovery by showing that the two varieties of nucleic acid differ in the nature of the five-carbon sugar. Yeast nucleic acid contains *ribose,* while thymus nucleic acid contains a sugar that is very much like ribose except for the absence of one oxygen atom, and so was called *deoxyribose.* Their formulas are:

$$O{\Large\diagdown}CH$$
$$|$$
$$H-C-OH$$
$$|$$
$$H-C-OH$$
$$|$$
$$H-C-OH$$
$$|$$
$$CH_2-OH$$

Ribose

$$O{\Large\diagdown}CH$$
$$|$$
$$H-C-H$$
$$|$$
$$H-C-OH$$
$$|$$
$$H-C-OH$$
$$|$$
$$CH_2-OH$$

Deoxyribose

In consequence, the two varieties of nucleic acid came to be called *ribonucleic acid* (RNA) and *deoxyribonucleic acid* (DNA).

Besides the difference in their sugars, the two nucleic acids also differ in one of the pyrimidines. RNA has *uracil* in place of thymine. Uracil is very like thymine, however, as you can see from the formula:

$$OH$$
$$|$$
$$C$$
$$N{\diagup}\quad{\diagdown}CH$$
$$|\qquad\qquad\|$$
$$C\qquad CH$$
$$HO{\diagup}{\Large\diagdown}N{\diagdown}$$

Uracil

By 1934, Levene was able to show that the nucleic acids could be broken down to fragments that contain a purine or a pyrimidine, either the ribose or the deoxyribose sugar, and a phosphate group. This combination is called a *nucleotide*. Levene proposed that the nucleic-acid molecule is built up of nucleotides as a protein is built up of amino acids. His quantitative studies suggested to him that the molecule consists of just four nucleotide units, one containing adenine, one guanine, one cytosine, and one either thymine (in DNA) or uracil (in RNA).

This proposal seemed to make sense. The material in chromosomes and elsewhere was thought of as *nucleoprotein*, which in turn consisted of a large protein molecule to which were attached one or more of these *tetranucleotide* groups, which served some unknown but, presumably, subsidiary purpose.

It turned out, however, that what Levene had isolated were not nucleic-acid molecules but pieces of them; and by the middle 1950s, biochemists found that the molecular weights of nucleic acids ran as high as 6 million. Nucleic acids are thus certainly equal and very likely superior to proteins in molecular size.

The exact manner in which nucleotides are built up and interconnected was confirmed by the British biochemist Alexander Robertus Todd, who built up a variety of nucleotides out of simpler fragments and carefully bound nucleotides together under conditions that allowed only one variety of bonding. He received the Nobel Prize in chemistry in 1957 for this work.

As a result, the general structure of the nucleic acid could be seen to be somewhat like the general structure of protein. The protein molecule is made up of a polypeptide backbone out of which jut the side chains of the individual amino acids. In nucleic acids, the sugar portion of one nucleotide is bonded to the sugar portion of the next by means of a phosphate group attached to both. Thus, a *sugar-phosphate backbone* runs the length of the molecule, and from it extend purines and pyrimidines, one to each nucleotide.

Nucleoproteins, it became clear, consist of two parts that are each large *macromolecules*. The question of the function of the nucleic-acid portion became more urgent.

DNA

By the use of cell-staining techniques, investigators began to pin down the location of nucleic acids in the cell. The German chemist Robert Feulgen, employing a red dye that stained DNA but not RNA, found DNA located in the cell nucleus, specifically in the chromosomes. He detected it not only in animal cells but also in plant cells. In addition, by staining RNA, he showed that this nucleic acid, too, occurs in both plant and animal cells. In short, the nucleic acids are universal materials existing in all living cells.

The Swedish biochemist Torbjörn Caspersson studied the subject further by removing one of the nucleic acids (by means of an enzyme that reduced it to soluble fragments that could be washed out of the cell) and concentrating on the other. He would photograph the cell in ultraviolet light; since a nucleic acid absorbs ultraviolet much more strongly than do other cell materials, the location of the DNA or the RNA—whichever he had left in the cell—showed up clearly. By this technique, DNA showed up only in the chromosomes. RNA made its appearance mainly in certain particles in the cytoplasm. Some RNA also showed up in the *nucleolus,* a structure within the nucleus. (In 1948, the Rockefeller Institute biochemist Alfred Ezra Mirsky showed that small quantities of RNA are present even in the chromosomes, while Ruth Sager showed that DNA can occur in the cytoplasm, notably in the chloroplasts of plants. In 1966, DNA was located in the mitochondria, too.)

Caspersson's pictures disclosed that the DNA lies in localized bands in the chromosomes. Was it possible that DNA molecules are none other than the genes, which up to this time had had a rather vague and formless existence?

Through the 1940s, biochemists pursued this lead with growing excitement. They found it particularly significant that the amount of DNA in the cells of an organism was always rigidly constant, except that the sperm and egg cells had only half this amount—as expected, since they had only half the chromosome supply of normal cells. The amount of RNA and of the protein in chromosomes might vary all over the lot, but the quantity of DNA remained fixed. This certainly seemed to indicate a close connection between DNA and genes.

The nucleic acid tail was beginning to wag the protein dog, and then some

582   THE BIOLOGICAL SCIENCES

remarkable observations were reported that seemed to show that the tail *was* the dog.

Bacteriologists had long studied two different strains of pneumococci grown in the laboratory: one with a smooth coat made of a complex carbohydrate; the other lacking this coat and therefore rough in appearance. Apparently the rough strain lacked some enzyme needed to make the carbohydrate capsule. But an English bacteriologist named Fred Griffith had discovered that, if killed bacteria of the smooth variety were mixed with live ones of the rough strain and then injected into a mouse, the tissues of the infected mouse would eventually contain live pneumococci of the smooth variety! How could this happen? The dead pneumococci had certainly not been brought to life. Something must have transformed the rough pneumococci so that they were now capable of making the smooth coat. What was that something? Evidently it was some factor contributed by the dead bacteria of the smooth strain.

In 1944, three American biochemists—Oswald Theodore Avery, Colin Munro Macleod, and Maclyn McCarty—identified the transforming principle. It was DNA. When they isolated pure DNA from the smooth strain and gave it to rough pneumococci, that alone sufficed to transform the rough strain to a smooth.

Investigators went on to isolate other transforming principles, involving other bacteria and other properties, and in every case the principle turned out to be a variety of DNA. The only plausible conclusion was that DNA could act like a gene. In fact, various lines of research, particularly with viruses (see chapter 14), showed that the protein associated with DNA is almost superfluous from a genetic point of view: DNA can produce genetic effects all by itself, either in the chromosome or—in the case of nonchromosomal inheritance—in cytoplasmic bodies such as the chloroplasts and mitochondria.

THE DOUBLE HELIX

If DNA is the key to heredity, it must have a complex structure, because it has to carry an elaborate pattern, or code of instructions (the *genetic code*), for the synthesis of specific enzymes. If it is made up of the four kinds of nucleotide, they cannot be strung in a regular arrangement, such as 1, 2, 3, 4, 1, 2, 3, 4, 1, 2, 3, 4. . . . Such a molecule would be far too simple to carry a blueprint for enzymes. In fact, the American biochemist Erwin Chargaff and his co-workers found definite evidence, in 1948, that the composition of nucleic acids was more complicated than had been thought. Their analysis showed that the various purines and pyrimidines are not present in equal amounts, and that the proportions vary in different nucleic acids.

Everything seemed to show that the four purines and pyrimidines were distributed along the DNA backbone as randomly as the amino acid side chains were distributed along the peptide backbone. Yet some regularities did seem to exist. In any given DNA molecule, the total number of purines seemed always to be equal to the total number of pyrimidines. In addition, the number of adenines (one purine) was always equal to the number of thymines (one pyrimidine), while the number of guanines (the other purine) was always equal to the number of cytosines (the other pyrimidine).

We could symbolize adenine as A, guanine as G, thymine as T, and cytosine as C. The purines would then be A + G and the pyrimidines T + C. The findings concerning any given DNA molecule could then be summarized as:

$$A = T$$
$$G = C$$
$$A + G = T + C$$

More general regularities also emerged. As far back as 1938, Astbury had pointed out that nucleic acids scatter X rays in diffraction patterns, a good sign of the existence of structural regularities in the molecule. The New Zealand-born British biochemist Maurice Hugh Frederick Wilkins calculated that these regularities repeat themselves at intervals considerably greater than the distance from nucleotide to nucleotide. One logical conclusion was that the nucleic-acid molecule takes the form of a helix, with the coils of the helix forming the repetitive unit noted by the X rays. This thought seemed the more attractive because Linus Pauling was at that time demonstrating the helical structure of certain protein molecules.

Wilkins's conclusions were based largely on the X-ray diffraction work of his associate, Rosalind Elsie Franklin, whose role in the studies was consistently underplayed in part because of the anti-feminist attitudes of the British scientific establishment.

In 1953, the English physicist Francis Harry Compton Crick and his co-worker, the American biochemist (and one-time Quiz Kid) James Dewey Watson, put all the information together—making use of a key photograph taken by Franklin, apparently without her permission—and came up with a revolutionary model of the nucleic-acid molecule. This model represented it not merely as a helix but (and this was the key point) as a double helix—two sugar-phosphate backbones winding like a double-railed spiral staircase up the same vertical axis (figure 13.6). From each sugar-phosphate chain, purines and peptides extended inward toward each other, meeting as though to form the steps of this double-railed spiral staircase.

Just how might the purines and pyrimidines be arrayed along these parallel chains? To make a good uniform fit, a double-ring purine on one side should always face a single-ring pyrimidine on the other, to make a three-ring width altogether. Two pyrimidines could not stretch far enough to cover the space; while two purines would be too crowded. Furthermore, an adenine from one chain would always face a thymine on the other, and a guanine on one chain would always face a cytosine on the other. In this way, one could explain the finding that A = T, G = C, and A + T = G + C.

This *Watson-Crick model* of nucleic-acid structure has proved to be extraordinarily fruitful; and Wilkins, Crick, and Watson shared the 1962 Nobel Prize in medicine and physiology as a result. (Franklin had died in 1958, so the question of her contribution did not arise.)

Figure 13.6. Model of the nucleic-acid molecule. The drawing at the left shows the double helix; in the center, a portion of it is shown in detail (omitting the hydrogen atoms); at the right is a detail of the nucleotide combinations.

The Watson-Crick model makes it possible, for instance, to explain just how a chromosome may duplicate itself in the process of cell division. Consider the chromosome as a string of DNA molecules. The molecules can first divide by a separation of the two helices making up the double helix; the two chains unwind themselves fro n each other, so to speak. This can be done because opposing purines and pyrimidines are held by hydrogen bonds, weak enough to be easily broken. Each chain is a half-molecule that can bring about the synthesis of its own missing complement. Where it has a thymine, it attaches an adenine; where it has a cytosine, it attaches a guanine; and so on. All the raw materials for making the units, and the necessary enzymes, are on hand in the cell. The half-molecule simply plays the role of a *template*, or mold, for putting the units together in the proper order. The units eventually will fall into the appropriate places and stay there because that is the most stable arrangement.

To summarize, then, each half-molecule guides the formation of its own complement, held to itself by hydrogen bonds. In this way, it rebuilds the complete, double-helix DNA molecule, and the two half-molecules into which the original molecule divided thus form two molecules where only one existed before. Such a process, carried out by all the DNAs down the length of a chromosome, will create two chromosomes that are exactly alike and perfect copies of the original mother chromosome. Occasionally something may go wrong: the impact of a subatomic particle or of energetic radiation, or the intervention of certain chemicals, may introduce an imperfection somewhere or other in the new chromosome. The result is a mutation.

Evidence in favor of this mechanism of replication has been piling up.

Tracer studies, employing heavy nitrogen to label chromosomes and following the fate of the labeled material during cell division, have tended to bear out the theory. In addition, some of the important enzymes involved in replication have been identified.

In 1955, the Spanish-American biochemist Severo Ochoa isolated from a bacterium (*Aztobacter vinelandii*) an enzyme that proved capable of catalyzing the formation of RNA from nucleotides. In 1956, a former pupil of Ochoa's, Arthur Kornberg, isolated another enzyme (from the bacterium *Escherichia coli*), which could catalyze the formation of DNA from nucleotides. Ochoa proceeded to synthesize RNA-like molecules from nucleotides, and Kornberg did the same for DNA. (The two men shared the Nobel Prize in medicine and physiology in 1959.) Kornberg also showed that his enzyme, given a bit of natural DNA to serve as a template, could catalyze the formation of a molecule that seemed to be identical with natural DNA. In 1965, Sol Spiegelman of the University of Illinois used RNA from a living virus (the simplest class of living things) and produced additional molecules of that sort. Since these additional molecules showed the essential properties of the virus, this was the closest approach yet to producing test-tube life. In 1967, Kornberg and others did the same, using DNA from a living virus as template.

The amount of DNA associated with the simplest manifestations of life is small—a single molecule in a virus—and can be made smaller. In 1967, Spiegelman allowed the nucleic acid of a virus to replicate and selected samples after increasingly shorter intervals for further replication. In this way, he selected molecules that completed the job unusually quickly—because they were smaller than average. In the end, he had reduced the virus to one-sixth its normal size and multiplied replication speed fifteenfold.

Although it is DNA that replicates in cells, many of the simpler viruses contain RNA only. RNA molecules in double strands replicate in such viruses. The RNA in cells is single-stranded and does not replicate.

Nevertheless, a single-stranded structure and replication are not mutually exclusive. The American biophysicist Robert Louis Sinsheimer discovered a strain of virus that contained DNA made up of a single strand. That DNA molecule had to replicate itself; but how could that be done with but a single strand? The answer was not difficult. The single strand brought about the production of its own complement, and the complement then brought about the production of the "complement to the complement"—that is, a replica of the original strand.

It is clear that the single-strand arrangement is less efficient than the double-strand arrangement (which is probably why the former exists only in certain very simple viruses and the latter in all other living creatures). For one thing, a single strand must replicate itself in two successive steps, whereas the double strand does so in a single step. Second, it now seems that only one strand of the DNA molecule is the important working structure—the cutting edge of the molecule, so to speak. Its complement may be thought of as a protecting scabbard for that cutting edge. The double strand represents the cutting edge protected within the scabbard except when actually in use; the single strand is the cutting edge always exposed and continually subjected to blunting by accident.

GENE ACTIVITY

Replication, however, merely keeps a DNA molecule in being. How does it accomplish its work of bringing about the synthesis of a specific enzyme— that is, of a specific protein molecule? To form a protein, the DNA molecule has to direct the placement of amino acids in a certain specific order in a molecule made up of hundreds or thousands of units. For each position it must choose the correct amino acid from some twenty different amino acids. If there were twenty corresponding units in the DNA molecule, it would be easy. But DNA is made up of only four different building blocks—the four nucleo- tides. Thinking about this, the astronomer George Gamow suggested in 1954 that the nucleotides, in various combinations, might be used as what we now call a *genetic code* (just as the dot and dash of the Morse code can be combined in various ways to represent the letters of the alphabet, numerals, and so on).

If you take the four different nucleotides (A, G, C, T), two at a time, there are 4×4, or 16 possible combinations (AA, AG, AC, AT, GA, GG, GC, GT, CA, CG, CC, CT, TA, TG, TC, and TT)—still not enough. If you take them three at a time, there are $4 \times 4 \times 4$, or 64 different combinations— more than enough. (You may amuse yourself trying to list the different combi- nations and see if you can find a sixty-fifth.)

It seemed as though each different *nucleotide triplet* or *codon* represented a particular amino acid. In view of the great number of different codons possible, it could well be that two or even three different codons represented one particular amino acid. In this case, the genetic code would be what cryptographers call *degenerate*.

This left two chief questions: Which codon (or codons) correspond to which amino acid? And how does the codon information (which is securely locked in the nucleus where the DNA is to be found) reach the sites of enzyme formation in the cytoplasm?

To take the second problem first, suspicion soon fell upon RNA as the substance serving as go-between—as the French biochemists François Jacob and Jacques Lucien Monod were the first to suggest. The structure of such RNA would have to be very like DNA with such differences as existed not affecting the genetic code. RNA had ribose in place of deoxyribose (one extra oxygen atom per nucleotide) and uracil in place of thymine (one missing methyl group, CH_3, per nucleotide). Furthermore, RNA was present chiefly in the cytoplasm, but also, to a small extent, in the chromosomes themselves.

It was not hard to see, and then demonstrate, what was happening. Every once in a while, when the two coiled strands of the DNA molecule unwound, one of those strands (always the same one, the cutting edge) replicates its structure, not on nucleotides that form a DNA molecule, but on nucleotides that form an RNA molecule. In this case, the adenine of the DNA strand attaches not thymine nucleotides to itself but uracil nucleotides instead. The resulting RNA molecule, carrying the genetic code imprinted on its nucleotide pattern, can then leave the nucleus and enter the cytoplasm.

Since it carries the DNA *message*, it has been named *messenger-RNA*, or more simply, *mRNA*.

The Rumanian-American biochemist George Emil Palade, thanks to careful work with the electron microscope, demonstrated, in 1956, the site of enzyme manufacture in the cytoplasm to be tiny particles, about 2 millionths of a centimeter in diameter. They were rich in RNA and were therefore named *ribosomes*. There are as many as 15,000 ribosomes in a bacterial cell, perhaps ten times as many in a mammalian cell. They are the smallest of the subcellular particles or *organelles*. It was soon determined that the messenger-RNA—carrying the genetic code on its structure—makes its way to the ribosomes and layers itself onto one or more of them, and that the ribosomes are the site of protein synthesis.

The next step was taken by the American biochemist Mahlon Bush Hoagland, who had also been active in working out the notion of mRNA. He showed that in the cytoplasm are a variety of small RNA molecules, which might be called *soluble-RNA* or *sRNA*, because their small size enables them to dissolve freely in the cytoplasmic fluid.

At one end of each sRNA molecule was a particular triplet of nucleotides that just fitted a complementary triplet somewhere on the mRNA chain: that is, if the sRNA triplet were AGC, it would fit tightly to a UCG triplet on the mRNA and only there. At the other end of the sRNA molecule was a spot where it would combine with one particular amino acid and none other. On each sRNA molecule, the triplet at one end meant a particular amino acid on the other. Therefore, a complementary triplet on the mRNA meant that only a certain sRNA molecule carrying a certain amino acid molecule would affix itself there. A large number of sRNA molecules would affix themselves one after the other, right down the line, to the triplets making up the mRNA structure (triplets that had been molded right on the DNA molecule of a particular gene). All the amino acids properly lined up could then easily be hooked together to form an enzyme molecule.

Because the information from the messenger-RNA is, in this way, transferred to the protein molecule of the enzyme, sRNA has come to be called *transfer-RNA*, and this name is now well established.

In 1964, the molecule of alanine-transfer-RNA (the transfer-RNA that attaches itself to the amino acid alanine) was completely analyzed by a team headed by the American biochemist, Robert William Holley. This analysis was done by the Sanger-method of breaking down the molecule into small fragments by appropriate enzymes, then analyzing the fragments and deducing how they must fit together. The alanine-transfer-RNA, the first naturally occurring nucleic acid to be completely analyzed, was found to be made up of a chain of seventy-seven nucleotides. These include not only the four nucleotides generally found in RNA (A, G, C, and U) but also several of seven others closely allied to them.

It had been supposed at first that the single chain of a transfer-RNA would be bent like a hairpin at the middle and the two ends would twine about each other in a double helix. The structure of alanine theory transfer-RNA did not lend itself to this theory. Instead, it seemed to consist of three loops, so that it looked rather like a lopsided three-leaf clover. In subsequent years, other transfer-RNA molecules were analyzed in detail, and all seemed to have the same three-leaf-clover structure. For his work, Holley received a share of the 1968 Nobel Prize for medicine and physiology.

In this way, the structure of a gene controls the synthesis of a specific enzyme. Much, of course, remained to be worked out, for genes do not simply organize the production of enzymes at top speed at all times. The gene may be working efficiently now, slowly at another time, and not at all at still another time. Some cells manufacture protein at great rates, with an ultimate capacity of combining some 15 million amino acids per chromosome per minute; some only slowly; some scarcely at all—yet all the cells in a given organism have the same genic organization. Then, too, each type of cell in the body is highly specialized, with characteristic functions and chemical behavior of its own. An individual cell may synthesize a given protein rapidly at one time, slowly at another. And, again, all have the same genic organization all the time.

It is clear that cells have methods for blocking and unblocking the DNA molecules of the chromosomes. Through the pattern of blocking and unblocking, different cells with identical gene patterns can produce different combinations of proteins, while a particular cell with an unchanging gene pattern can produce different combinations from time to time.

In 1961, Jacob and Monod suggested that each gene has its own repressor, coded by a *regulator gene*. This repressor—depending on its geometry, which can be altered by delicate changes in circumstances within the cell—will block or release the gene. In 1967, such a repressor was isolated and found to be a small protein. Jacob and Monod, together with a co-worker, André Michael Lwoff, received the 1965 Nobel Prize for medicine and physiology as a result.

Through laborious work since 1973, it would seem that the long double helix of the DNA twists to form a second helix (a *superhelix*) about a core of a string of histone molecules, so that there is a succession of units called *nucleosomes*. In such nucleosomes, depending upon the detailed structure, some genes may be repressed, and others active; and the histones may have something to do with which active gene becomes repressed from time to time or is activated. (As usual, biological systems always seem more complex than expected once one probes deeply into the details.)

Nor is the flow of information entirely one way, from gene to enzyme. There is "feedback" as well. Thus, there is a gene that brings about the formation of an enzyme that catalyzes a reaction that converts the amino acid threonine to another amino acid, isoleucine. Isoleucine, by its presence, somehow serves to activate the repressor, which begins to shut down the very gene that produces the particular enzyme that led to that presence. In other words, as isoleucine concentration goes up, less is formed; if the concentration declines, the gene is unblocked, and more isoleucine is formed. The chemical machinery within the cell—genes, repressors, enzymes, end-products—is enormously complex and intricately interrelated. The complete unraveling of the pattern is not likely to take place rapidly.

But meanwhile, what of the other question: Which codon goes along with which amino acid? The beginning of an answer came in 1961, thanks to the work of the American biochemists Marshall Warren Nirenberg and J. Heinrich Matthaei. They began by making use of a synthetic nucleic acid, built up according to Ochoa's system from uracil nucleotides only. This *polyuridylic acid* was made up of a long chain of . . . UUUUUUUU . . . and could only possess one codon, UUU.

Nirenberg and Matthaei added this polyuridylic acid to a system that contained various amino acids, enzymes, ribosomes, and all the other components necessary to synthesize proteins. Out of the mixture tumbled a protein made up only of the amino acid phenylalanine. This meant that UUU was equivalent to phenylalanine. The first item in the *codon dictionary* was worked out.

The next step was to prepare a nucleotide made out of a preponderance of uridine nucleotides with a small quantity of adenine nucleotides added; thus, along with the UUU codon, an occasional UUA, or AUU, or UAU codon might appear. Ochoa and Nirenberg showed that, in such a case, the protein formed is mainly phenylalanine but also contains an occasional leucine, isoleucine, and tyrosine, three other amino acids.

Slowly, by methods such as these, the dictionary was extended. It was found that the code is indeed degenerate, and that GAU and GAC might each stand for aspartic acid, for instance, and that GUU, GAU, GUC, GUA, and GUG, all stand for glycine. In addition, there was some punctuation. The codon AUG not only stood for the amino acid methionine but apparently signified the beginning of a chain. It was a *capital letter*, so to speak. Then, too, UAA and UAG signaled the end of a chain: they were *periods*, or *full stops*.

By 1967, the dictionary was complete (see table 13.1). Nirenberg and his collaborator, the Indian-American chemist Har Gobind Khorana, were awarded shares (along with Holley) in the 1968 Nobel Prize for medicine and physiology.

The working out of the genetic code is not, however, a "happy ending" in the sense that now all mysteries are explained. (There are, perhaps, no happy endings of this sort in science—and a good thing, too, for a universe without mysteries would be unbearably dull.)

The genetic code was worked out largely through experiments on bacteria, where the chromosomes are packed tight with working genes that code the formation of proteins. Bacteria are *prokaryotes* (from Greek words meaning "before the nucleus"), since they lack cell nuclei but have chromosomal material distributed throughout their tiny cells.

TABLE 13.1

The genetic code. In the left-hand column are the initials of the four RNA bases (uracil, cytosine, adenine, guanine) representing the first "letter" of the codon triplet; the second letter is represented by the initials across the top, while the third but less important letter appears in the final column. For example, tyrosine (Tyr) is coded for by either UAU or UAC. Amino acids coded by each codon are shown abbreviated as follows: Phe— phenylalanine; Leu—leucine; Ileu—isoleucine; Met—methionine; Val—valine; Ser— serine; Pro—proline; Thr—threonine; Ala—alanine; Tyr—tyrosine; His—histidine; Glun—glutamine; Aspn—asparagine; Lys—lysine; asp—aspastic acid; Glu—glutamic acid; Cys—cysteine; Tryp—tryptophan; Arg—arginine; Gly—glycine.

First Position	Second Position				Third Position
	U	C	A	G	
U	Phe	Ser	Tyr	Cys	U
	Phe	Ser	Tyr	Cys	C
	Leu	Ser	(normal "full stop")	"full stop"	A
	Leu	Ser	(less common "full stop")	Tryp	G
C	Leu	Pro	His	Arg	U
	Leu	Pro	His	Arg	C
	Leu	Pro	Glun	Arg	A
	Leu	Pro	Glun	Arg	G
A	Ileu	Thr	Aspn	Ser	U
	Ileu	Thr	Aspn	Ser	C
	Ileu?	Thr	Lys	Arg	A
	Met ("capital letter")	Thr	Lys	Arg	G
G	Val	Ala	Asp	Gly	U
	Val	Ala	Asp	Gly	C
	Val	Ala	Glu	Gly	A
	Val ("capital letter")	Ala	Glu	Gly	G

As for the *eukaryotes*, which have a cell nucleus (and include all cells but those of bacteria and blue-green algae), the case is different. The length of nucleic acid is *not* solidly packed with working genes. Instead, those portions of the nucleotide chain that are used to encode messenger-RNA and, eventually, proteins (*exons*) are interspersed by sections of chain (*introns*) that may be described as gibberish. A single gene that controls the production of a single enzyme may consist of a number of exons separated by introns, and the nucleotide chain coils in such a way as to bring the exons together for the encoding of messenger-RNA. Thus, the estimate, given earlier in the chapter, of the existence of 2 million genes in the human cell is far too high, if one is referring to working genes.

Why eukaryotes should carry such a load of what seems dead weight is puzzling. Perhaps that is how genes developed in the first place; and in prokaryotes, the introns were disposed of in order to make shorter nucleotide chains that could be more swiftly replicated in the interest of faster growth and reproduction. In eukaryotes, the introns are not excised, perhaps because they offer some advantage that is not immediately visible. No doubt the answer, when it comes, will be illuminating.

And meanwhile scientists have found methods of participating directly in gene activity. In 1971, the American microbiologists Daniel Nathans and Hamilton Othanel Smith worked with *restriction enzymes* which were capable of cutting the DNA chain in specific fashion at a particular nucleotide junction and no other. There is another type of enzyme, *DNA ligase,* which is capable of uniting two strands of DNA. The American biochemist Paul Berg cut DNA strands by restriction enzymes and then recombined strands in fashions *other* than had originally existed. A molecule of *recombinant-DNA* was thus formed that was not like the original or, perhaps, not like any that had ever existed.

It became possible, as a result of such work, to modify genes or to design new ones: to insert them into bacterial cells (or into the nuclei of eukaryotic cells) and thus to form cells with new biochemical properties. As a result, Nathans and Smith were awarded shares of the 1978 Nobel Prize for physiology and medicine, while Berg received a share of the 1980 Nobel Prize for chemistry.

Recombinant-DNA work had its areas of apparent danger. What if, either deliberately or inadvertently, a bacterial cell was produced, or a virus with the ability to produce a toxin to which human beings had no natural immunity? If such a new microorganism escaped from the laboratory, it might inflict an indescribably disastrous epidemic upon humanity. With such thoughts in mind, Berg and others, in 1974, called on scientists for a voluntary adherence to strict controls in work on recombinant-DNA.

As it happened, though, further experience showed that there was little danger of anything untoward happening. Precautions were extreme, and the new genes placed in microorganisms produced strains that were so weak (an *unnatural* gene is not easy to live with) that they could barely be kept alive under the most favorable conditions.

Then, too, recombinant-DNA work involves the possibility of great benefits. Aside from the possibility of the advancement of knowledge concerning the fine details of the workings of cells and of the mechanism of inheritance in particular, there are more immediate benefits. By appropriately modifying a gene, or by inserting a foreign gene, a bacterial cell might become a tiny factory that is manufacturing molecules of something needed by human beings rather than by itself.

Thus, bacterial cells, in the 1980s, have been so modified as to manufacture human insulin, with the unattractive name of *humulin.* Hence, in time, diabetics will no longer be dependent upon the necessarily limited supplies available from the pancreases of slaughtered animals and will not have to use the adequate, but not ideal, insulin varieties of cattle and swine.

Other proteins that can be made available by appropriately modified microorganisms are interferon and growth hormone—with, on the horizon, unlimited possibilities. It is not surprising that the question has now arisen whether new forms of life can be patented.

The Origin of Life

Once we get down to the nucleic-acid molecules, we are as close to the basis of life as we can get. Here, surely, is the prime substance of life itself. Without DNA, living organisms could not reproduce, and life as we know it could not have started. All the substances of living matter—enzymes and all the others, whose production is catalyzed by enzymes—depend in the last analysis on DNA. How, then, did DNA, and life, start?

This is a question that science has always hesitated to ask, because the origin of life has been bound up with religious beliefs even more strongly than has the origin of the earth and the universe. It is still dealt with only hesitantly and apologetically. A book entitled *The Origin of Life*, by the Russian biochemist Aleksandr Ivanovich Oparin, brought the subject to the fore. The book was published in the Soviet Union in 1924 and in English translation in 1936. In it the problem of life's origin for the first time was dealt with in detail from a completely materialistic point of view. Since the Soviet Union is not inhibited by the religious scruples to which the Western nations feel bound, this, perhaps, is not surprising.

EARLY THEORIES

Most early cultures developed myths telling of the creation of the first human beings (and sometimes of other forms of life as well) by gods or demons. However, the formation of life itself was rarely thought of as being entirely a divine prerogative. At least the lower forms of life might arise spontaneously from nonliving material without supernatural intervention. Insects and worms might, for instance, arise from decaying meat, frogs from mud, mice from rotting wheat. This idea was based on actual observation, for decaying meat, to take the most obvious example, did indeed suddenly give rise to maggots. It was only natural to assume that the maggots were formed from the meat.

Aristotle believed in the existence of *spontaneous generation*. So did the great theologians of the Middle Ages, such as Thomas Aquinas. So did William Harvey and Isaac Newton. After all, the evidence of one's own eyes is hard to refute.

The first to put this belief to the test of experimentation was the Italian physician Francesco Redi. In 1668, he decided to check on whether maggots really formed out of decaying meat. He put pieces of meat in a series of jars and then covered some of them with fine gauze and left others uncovered. Maggots developed only in the meat in the uncovered jars, to which flies had had free access. Redi concluded that the maggots had arisen from microscopically small eggs laid on the meat by the flies. Without flies and their eggs, he insisted, meat could never produce maggots, however long it decayed and putrefied.

Experimenters who followed Redi confirmed this finding, and the belief that visible organisms arise from dead matter died. But when microbes were discovered, shortly after Redi's time, many scientists decided that these forms of life at least must come from dead matter. Even in gauze-covered jars, meat would soon begin to swarm with microorganisms. For two centuries after Redi's experiments, belief in the possibility of the spontaneous generation of microorganisms remained very much alive.

It was another Italian, the naturalist Lazzaro Spallanzani, who first cast serious doubt on this notion. In 1765, he set out two sets of vessels containing a broth. One he left open to the air. The other, which he had boiled to kill any organisms already present, he sealed up to keep out any organisms that might be floating in the air. The broth in the first vessels soon teemed with microorganisms, but the boiled and sealed-up broth remained sterile. This proved to Spallanzani's satisfaction that even microscopic life could not arise from inanimate matter. He even isolated a single bacterium and witnessed its division into two bacteria.

The proponents of spontaneous generation were not convinced. They maintained that boiling destroyed some *vital principle* and that, as a result, no microscopic life developed in Spallanzani's boiled, sealed flasks. It remained for Pasteur to settle the question, in 1862, seemingly once and for all. He devised a flask with a long swan neck in the shape of a horizontal S (figure 13.7). With the opening unstoppered, air could percolate into the flask, but dust particles and microorganisms could not, for the curved neck would serve as a trap, like the drain trap under a sink. Pasteur put some broth in the flask,

---

Figure 13.7. Pasteur's flask for the experiment on spontaneous generation.

---

attached the S-shaped neck, boiled the broth until it steamed (to kill any microorganisms in the neck as well as in the broth), and waited for developments. The broth remained sterile. There was no vital principle in air. Pasteur's demonstration apparently laid the theory of spontaneous generation to rest permanently.

All this left a germ of embarrassment for scientists. How had life arisen, after all, if not through divine creation or through spontaneous generation?

Toward the end of the nineteenth century some theorists went to the other extreme and made life eternal. The most popular theory was advanced by Svante Arrhenius (the chemist who had developed the concept of ionization). In 1907, he published a book entitled *Worlds in the Making*, picturing a universe in which life had always existed and migrated across space, continually colonizing new planets. Life traveled in the form of spores that escaped from the atmosphere of a planet by random movement and then were driven through space by the pressure of light from the sun.

Such light pressure is by no means to be sneered at as a possible driving force. The existence of radiation pressure had been predicted in the first place by Maxwell, on theoretical grounds and, in 1899, had been demonstrated experimentally by the Russian physicist Peter Nicolaevich Lebedev.

Arrhenius's views held, then, that spores traveled on and on through interstellar space, driven by light radiation this way and that, until they died or fell on some planet, where they would spring into active life and compete with life forms already present, or inoculate the planet with life if it was uninhabited but habitable.

At first blush, this theory looks attractive. Bacterial spores, protected by a thick coat, are very resistant to cold and dehydration and might conceivably last a long time in the vacuum of space. Also, they are of just the proper size to be more affected by the outward pressure of a sun's radiation than by the inward pull of its gravity. But Arrhenius's suggestion fell before the onslaught of ultraviolet light. In 1910, experimenters showed that ultraviolet light quickly kills bacterial spores; and in interplanetary space, the sun's ultraviolet light is intense—not to speak of other destructive radiations, such as cosmic rays, solar X rays, and zones of charged particles like the Van Allen belts around the earth. Conceivably, there may be spores somewhere that are resistant to radiation, but spores made of protein and nucleic acid, as we know them, could not make the grade. To be sure, some particularly resistant microorganisms were exposed to the radiation of outer space on board the *Gemini 9* capsule in 1966 and survived six hours of harsh unfiltered sunlight. But we are talking of exposures not of hours, but of months and years.

Besides, if we suppose Earth to bear life only because it was seeded by bits of life that originated elsewhere, we would have to wonder how it originated elsewhere. Thus, such seeding is not a solution to the problem but only shifts the problem elsewhere.

CHEMICAL EVOLUTION

Although some scientists, even today, find the possibility of seeding attractive, the large majority feel it appropriate to work out reasonable mechanisms for the origin of life right here on Earth.

They are back to spontaneous generation, but with a difference. The pre-Pasteur view of spontaneous generation was of something taking place *now* and *quickly.* The modern view is that it took place long ago and very slowly.

It could not take place now, for anything that even approached the complexity required of the simplest conceivable form of life would promptly be incorporated, as food, into one of the innumerable bits of life that already exist. Spontaneous generation, therefore, had to take place only on a planet on which life did not already exist. On Earth, that would be over three and a half billion years ago.

Then, too, life could not take place in an atmosphere rich in oxygen. Oxygen is an active element that would unite with the chemicals that were building up into near-life, and break them down again. However, as I said in chapter 5, scientists believe that Earth's primordial atmosphere was a reducing one and did not contain free oxygen. In fact, one possibility is that Earth's original atmosphere was composed of hydrogen-containing gases such as methane (CH_4), ammonia (NH_3) and water vapor (H_2O), with perhaps some hydrogen (H_2) as well.

Such a highly hydrogenated atmosphere we might call Atmosphere I. Through photodissociation, this would slowly turn into an atmosphere of carbon dioxide and nitrogen (see chapter 5), or Atmosphere II. After that an ozone layer would form in the upper atmosphere, and spontaneous change would halt. Can life then have formed in one or the other of the early atmospheres?

H. C. Urey felt life started in Atmosphere I. In 1952, Stanley Lloyd Miller, then a graduate student in Urey's laboratories, circulated water, plus ammonia, methane and hydrogen, past an electric discharge (to simulate the ultraviolet radiation of the sun). At the end of a week, he analyzed his solution by paper chromatography and found that, in addition to the simple substances without nitrogen atoms, he also had glycine and alanine, the two simplest of the amino acids, plus some indication of one or two more complicated ones.

Miller's experiment was significant in several ways. In the first place, these compounds had formed quickly and in surprisingly large quantities. One-sixth of the methane with which he had started had gone into the formation of more complex organic compounds; yet the experiment had only been in operation for a week.

Then, too, the kind of organic molecules formed in Miller's experiments were just those present in living tissue. The path taken by the simple molecules, as they grew more complex, seemed pointed directly toward life. This pointing-toward-life continued consistently in later, more elaborate experiments. At no time were molecules formed in significant quantity that seemed to point in an unfamiliar nonlife direction.

Thus, Philip Abelson followed Miller's work by trying a variety of similar experiments with starting materials made up of different gases in different combinations. It turned out that as long as he began with molecules that included atoms of carbon, hydrogen, oxygen, and nitrogen, amino acids of the kind normally found in proteins were formed. Nor were electric discharges the only source of energy that would work. In 1959, two German scientists, Wilhelm Groth and H. von Weyssenhoff, designed an experiment in which ultraviolet light could be used instead, and they also got amino acids.

If there was any doubt that the direction-toward-life was the line of least resistance, there was the fact that, in the late 1960s, more and more complicated molecules, representing the first stages of that direction, were found in gas clouds of outer space (see chapter 2). It may be, then, that at the time the earth was formed out of clouds of dust and gas, the first stages of building up complex molecules had already taken place.

The earth, at its first formation, may have had a supply of amino acids. Evidence in favor of this theory came in 1970. The Ceylon-born biochemist Cyril Ponnamperuma studied a meteorite that had fallen in Australia on 28 September 1969. Careful analyses showed the presence of small traces of five amino acids: glycine, alanine, glutamic acid, valine, and proline. There was no optical activity in these amino acids, so they were formed not by life processes (hence their presence was not the result of earthly contamination) but by the nonliving chemical processes of the type that took place in Miller's flask.

In fact, Fred Hoyle and an Indian colleague, Chandra Wickramasinghe, are so impressed by this finding that they feel that the syntheses may go far beyond what has been detected. Very small quantities of microscopic bits of life may be formed, they feel—not enough to be detected at astronomical distances, but large in an absolute sense; and these may be formed not only in distant gas clouds but in comets of our own solar system. Life on Earth may therefore have originated when spores were carried to Earth by comet tails. (It is only fair to say that almost no one takes this speculation seriously.)

Could chemists in the laboratory progress beyond the amino acid stage? One way of doing so would be to start with larger samples of raw materials and subject them to energy for longer periods. This process would produce increasing numbers of ever more complicated products, but the mixtures of these products would become increasingly complex and would be increasingly difficult to analyze.

Another possibility would be for chemists to begin at a later stage. The products formed in earlier experiments would be used as new raw materials. Thus, one of Miller's products was hydrogen cyanide. The Spanish-American biochemist Juan Oro added hydrogen cyanide to the starting mixture in 1961. He obtained a richer mixture of amino acids and even a few short peptides. He also formed purines—in particular, adenine, a vital component of nucleic acids. In 1962, Oro used formaldehyde as one of his raw materials and produced ribose and deoxyribose, also components of nucleic acids.

In 1963, Ponnamperuma also performed experiments similar to those of Miller, using electron beams as a source of energy, and found that adenine was formed. Together with Ruth Mariner and Carl Sagan, he went on to add adenine to a ribose solution; and under ultraviolet light, *adenosine*, a molecule formed of adenine and ribose linked together, was formed. If phosphate was also present, it, too, was hooked on to form the adenine nucleotide. Indeed, three phosphate groups could be added to form adenosine triphosphate (ATP), which, as was explained in chapter 12, is essential to the energy-handling mechanisms of living tissue. In 1965, he formed a *dinucleotide*, two nucleotides bound together. Additional products can be built up if substances such as cyanamide ($CNNH_2$) and ethane (CH_3CH_3)—substances which may well have been present in the primordial era—are added to the mixtures employed by various experimenters in this field. There is no question, then, but that normal chemical and physical changes in the primordial ocean and atmosphere could have acted in such a way as to build up proteins and nucleic acids.

Any compound that formed in the lifeless ocean would tend to endure and accumulate. There were no organisms, either large or small, to consume them or cause them to decay. Moreover, in the primeval atmosphere there was no free oxygen to oxidize and break down the molecules. The only important factors tending to break down complex molecules would have been the very ultraviolet and radioactive energies that built them up. But ocean currents might have carried much of the material to a safe haven at mid-levels in the sea, away from the ultraviolet-irradiated surface and the radioactive bottom. Indeed, Ponnamperuma and his co-workers have estimated that fully 1 percent of the primordial ocean may have been made up of these built-up organic compounds. If so, this would represent a mass of over a million billion tons. This is certainly an ample quantity for natural forces to play with; and in such a huge mass, even substances of most unlikely complexity are bound to be built up in not too long a period (particularly considering a billion years are available for the purpose).

There is no logical barrier, then, to supposing that out of the simple compounds in the primordial ocean and atmosphere there appeared, with time, ever higher concentrations of the more complicated amino acids, as well as simple sugars; that amino acids combined to form peptides; that purines, pyrimidines, sugar, and phosphate combined to form nucleotides; and that, gradually over the ages, proteins and nucleic acids were created. Then, eventually, must have come the key step—the formation, through chance combinations, of a nucleic acid molecule capable of inducing replication. That moment marked the beginning of life.

Thus a period of *chemical evolution* preceded the evolution of life itself.

A single living molecule, it seems, might well have been sufficient to get life under way and give rise to the whole world of widely varying living things, as a single fertilized cell can give rise to an enormously complex organism. In the organic "soup" that constituted the ocean at that time, the first living molecule could have replicated billions and billions of molecules like itself in short order. Occasional mutations would create slightly changed forms of the molecule, and those that were in some way more efficient than the others would multiply at the expense of their neighbors and replace the old forms. If one group was more efficient in warm water and another group in cold water, two varieties would arise, each restricted to the environment it fitted best. In this fashion, the course of *organic evolution* would be set in motion.

Even if several living molecules came into existence independently at the beginning, it is very likely that the most efficient one would have outbred the others, so that all life today may very well be descended from a single original molecule. In spite of the great present diversity of living things, all have the same basic ground plan. Their cells all carry out metabolism in pretty much the same way. Furthermore, it seems particularly significant that the proteins of all living things are composed of L-amino acids rather than amino acids of the D type. It may be that the original nucleoprotein from which all life is descended happened to be built from L-amino acids by chance; and since D could not be associated with L in any stable chain, what began as chance persisted by replication into grand universality. (This is not to imply that D-amino acids are totally absent in nature. They occur in the cell walls of some bacteria and in some antibiotic compounds. These, however, are exceptional cases.)

THE FIRST CELLS

Of course, the step from a living molecule to the kind of life we know today is still an enormous one. Except for the viruses, all life is organized into cells; and a cell, however small it may seem by human standards, is enormously complex in its chemical structure and interrelationships. How did that start?

The question of the origin of cells was illuminated by the researches of the American biochemist Sidney Walter Fox. It seemed to him that the early earth must have been quite hot, and that the energy of heat alone could be sufficient to form complex compounds out of simple ones. In 1958, to test this theory, Fox heated a mixture of amino acids and found they formed long chains that resembled those in protein molecules. These *proteinoids* were digested by enzymes that digested ordinary proteins, and could be used as food by bacteria.

Most startling of all, when Fox dissolved the proteinoids in hot water and let the solution cool, he found they would cling together in little *microspheres* about the size of small bacteria. These microspheres were not alive by the usual standards but behaved as cells do, in some respects at least (they are surrounded by a kind of membrane, for instance). By adding certain chemicals to the solution, Fox could make the microspheres swell or shrink, much as ordinary cells do. They can produce buds, which sometimes seem to grow larger and then break off. Microspheres can separate, divide in two, or cling together in chains.

Perhaps in primordial times, such tiny not-quite-living aggregates of materials formed in several varieties. Some were particularly rich in DNA and were very good at replicating, though only moderately successful at storing energy. Other aggregates could handle energy well but replicated only limpingly. Eventually, collections of such aggregates might have cooperated, each supplying the deficiencies of the other, to form the modern cell, which was much more efficient than any of its parts alone. The modern cell still has the nucleus —rich in DNA but unable of itself to handle oxygen—and numerous mitochondria—which handle oxygen with remarkable efficiency but cannot reproduce in the absence of nuclei. (That mitochondria may once have been independent entities is indicated by the fact that they still possess small quantities of DNA.)

To be sure, in the last few years, there is an increasing tendency to suspect that Atmosphere I did not last very long, and that Atmosphere II was present almost at the beginning. Both Venus and Mars have Atmosphere II (carbon dioxide and nitrogen), for instance; and Earth may have had one, too, at a time when, like Venus and Mars, it bore no life.

This is not a fatal change. Simple compounds can still be built up from carbon dioxide, water vapor, and nitrogen. The nitrogen could be converted to nitrogen oxides or cyanide or ammonia by combination with carbon dioxide or water, or both, under the influence of lightning discharges perhaps; and molecular changes would then continue upward toward life under the lash of sunlight and other energy sources.

ANIMAL CELLS

Throughout the existence of Atmospheres I and II, primitive life forms could only exist at the cost of breaking down complex chemical substances into simpler ones and storing the energy evolved. The complex substances were rebuilt by the action of the ultraviolet radiation of the sun. Once Atmosphere II was completely formed and the ozone layer was in place, the danger of starvation set in, for the ultraviolet supply was cut off.

By then, though, some mitochondrialike aggregate was formed which contained chlorophyll—the ancestor of the modern chloroplast. In 1966, the Canadian biochemists G. W. Hodson and B. L. Baker began with pyrrole and paraformaldehyde (both of which can be formed from still simpler substances in Miller-type experiments) and demonstrated the formation of porphyrin rings, the basic structure of chlorophyll, after merely three hours gentle heating.

Even the inefficient use of visible light by the first primitive chlorophyll-containing aggregates must have been much preferable to the slow starvation of nonchlorophyll systems at the time when the ozone layer was forming. Visible light could easily penetrate the ozone layer, and the lower energy of visible light (compared with ultraviolet) was enough to activate the chlorophyll system.

The first chlorophyll-using organisms may have been no more complicated than individual chloroplasts today. There are, in fact, 2,000 species of a group of one-celled photosynthesizing organisms called *blue-green algae* (they are not all blue-green, but the first ones studied were). These are very simple cells, prokaryotes, rather bacterialike in structure, except that they contain chlorophyll and bacteria do not. Blue-green algae may be the simplest descendants of the original chloroplast, while bacteria may be the descendants of chloroplasts that lost their chlorophyll and took to parasitism or to foraging on dead tissue and its components.

As chloroplasts multiplied in the ancient seas, carbon dioxide was gradually consumed and molecular oxygen took its place. The present Atmosphere III was formed. Plant cells grew steadily more efficient, each one containing numerous chloroplasts. At the same time, elaborate cells without chlorophyll could not exist on the previous basis, for new food did not form in the ocean except within plant cells. However, cells without chlorophyll but with elaborate mitochondrial equipment that could handle complex molecules with great efficiency and store the energy of their breakdown, could live by ingesting the plant cells and stripping the molecules the latter had painstakingly built up. Thus originated the animal cell. Eventually, organisms grew complex enough to begin to leave the fossil record (plant and animal) that we have today.

Meanwhile, the earth environment had changed fundamentally, from the standpoint of creation of new life. Life could no longer originate and develop from purely chemical evolution. For one thing, the forms of energy that had brought it into being in the first place—ultraviolet and radioactive energy— were effectively gone or at least seriously diminished. For another, the well-established forms of life would quickly consume any organic molecules that arose spontaneously. For both these reasons, there is virtually no chance of any new and independent breakthrough from nonlife into life (barring some future human intervention, if we learn to turn the trick). Spontaneous generation today is so highly improbable that it can be regarded as essentially impossible.

Life in Other Worlds

If we accept the view that life arose simply from the workings of physical and chemical laws, it follows that in all likelihood life is not confined to the earth. What are the possibilities of life elsewhere in the universe?

When it was first recognized that the planets of the solar system were worlds, it was taken for granted that they were the abode of life, even intelligent life. It was with certain shock that the moon was recognized as lacking air and water and, therefore, probably lacking life as well.

In the modern age of rockets and probes (see chapter 3), scientists are pretty well convinced that there is no life on the moon or on any of the other worlds of the inner solar system, except for Earth itself.

Nor is there much chance for the outer solar system. To be sure, Jupiter has a deep and complex atmosphere with a temperature very low at the visible cloud layer and very high within. Somewhere at moderate depths and moderate temperatures, and with the known presence of water and organic compounds, it is conceivable (as Carl Sagan suggests) that life may exist. If true of Jupiter, it may be true of the three other gas giants as well.

Then, too, the Jovian satellite of Europa has a world-girdling glacier; but beneath it, may be a water ocean warmed by Jupiter's tidal influence. Titan has an atmosphere of methane and nitrogen and may have liquid nitrogen and solid organic compounds on the surface—and so may Neptune's satellite Triton as well. On all three satellites, it is conceivable that some form of life may exist.

These are all long shots, however. We can hope, wistfully, but we cannot honestly expect much; and it is only fair to suppose that as far as the solar system is concerned, the earth, and only the earth, seems to be an abode of life. But the solar system is not all there is. What are the possibilities of life elsewhere in the universe?

The total number of stars in the known universe is estimated to be at least 10,000,000,000,000,000,000,000 (10 billion trillion). Our own galaxy contains well in excess of 200,000,000,000 stars. If all the stars developed by the same sort of process as the one that is believed to have created our own solar system (that is, the condensing of a large cloud of dust and gas), then it is likely that no star is solitary, but each is part of a local system containing more than one body. We know that there are many double stars, revolving around a common center, and it is estimated that at least half of the stars belong to a system containing two or more stars.

What we really want, though, is a multiple system in which a number of members are too small to be self-luminous and are planets rather than stars. Though we have no means (so far) of detecting directly any planet beyond our own solar system, even for the nearest stars, we can gather indirect evidence. This has been done at the Sproul Observatory of Swarthmore College under the guidance of the Dutch-American astronomer Peter Van de Kamp.

In 1943, small irregularities of one of the stars of the double-star system 61 Cygni showed that a third component, too small to be self-luminous, must exist. This third component, 61 Cygni C, had to be about eight times the mass of Jupiter and therefore (assuming the same density) about twice the diameter. In 1960, a planet of similar size was located circling about the small star Lalande 21185 (located, at least, in the sense that its existence was the most logical way of accounting for irregularities in the star's motion). In 1963, a close study of Barnard's star indicated the presence of a planet there, too— one that was only one and one-half times the mass of Jupiter.

Barnard's star is second closest to ourselves; Lalande 21185, third closest; and 61 Cygni, twelfth closest. That three planetary systems should exist in close proximity to ourselves would be extremely unlikely unless planetary systems were very common generally. Naturally, at the vast distances of the stars, only the largest planets could be detected and even then with difficulty.

Where super-Jovian planets exist, it seems quite reasonable (and even inevitable) to suppose that smaller planets also exist.

Unfortunately the observations that yield the supposition that these extrasolar planets exist are anything but clear-cut and are close to the limits that can be observed. There is considerable doubt among astronomers generally that the existence of such planets has really been demonstrated.

But, then, a new kind of evidence made its appearance. In 1983, an Infrared Astronomy Satellite (IRAS) was orbiting Earth. It was designed to detect and study infrared sources in the sky. In August, the astronomers Harmut H. Aumann and Fred Gillett turned its detecting system toward the star Vega and found, to their surprise, that Vega was much brighter in the infrared than seemed reasonable. A closer study showed the infrared radiation was coming not from Vega itself but from its immediate surroundings.

Apparently Vega was surrounded by a cloud of matter stretching outward twice as far as Pluto's orbit from our sun. Presumably, the cloud consisted of particles larger than dust grains (or it would long since have been gathered up by Vega). Vega is much younger than our sun, for it is less than a billion years old, and, being 60 times as luminous as the sun has a much stronger stellar wind which can act to keep the particles from coalescing. For both these reasons, Vega might be expected to have a planetary system still in the process of formation. Included among the vast cloud of gravel may already be planet-sized objects that are gradually sweeping their orbits clean.

This discovery strongly favors the supposition that planetary systems are common in the universe, perhaps as common as stars are.

But even assuming that all or most stars have planetary systems and that many of the planets are earthlike in size, we must know the criteria such planets must fulfill to be habitable. The American space scientist Stephen H. Dole made a particular study of this problem in his book *Habitable Planets for Man* (1964) and reached certain conclusions, admittedly speculative, but reasonable.

He pointed out, in the first place, that a star must be of a certain size in order to possess a habitable planet. The larger the star, the shorter-lived it is; and, if it is larger than a certain size, it will not live long enough to allow a planet to go through the long stage of chemical evolution prior to the development of complex life forms. A star that is too small cannot warm a planet sufficiently, unless that planet is so close that it will suffer damaging tidal effects. Dole concluded that only stars of spectral classes F2 to K1 are suitable for the nurturing of planets that are comfortably habitable for human beings: planets that we can colonize (if travel between the stars ever becomes practicable) without undue effort. There are, Dole estimated, 17 billion such stars in our galaxy.

Such a star might be capable of possessing a habitable planet and yet might not possess one. Dole estimated the probabilities that a star of suitable size might have a planet of the right mass and at the right distance, with an appropriate period of rotation and an appropriately regular orbit; and by making what seem to him to be reasonable estimates, he concluded that there are likely to be 600 million habitable planets in our galaxy alone, each of them already containing some form of life.

If these habitable planets are spread more or less evenly throughout the galaxy, Dole estimated that there is one habitable planet per 80,000 cubic light-years. Hence, the nearest habitable planet to ourselves may be some twenty-seven light-years away; and within one hundred light-years of ourselves, there may be a total of fifty habitable planets.

Dole listed fourteen stars within twenty-two light-years of ourselves that might possess habitable planets, and weighed the probabilities that this might be true in each case. He concluded that habitable planets are most likely to be found in the stars closest to us—the two sun-like stars of the Alpha Centauri system, Alpha Centauri A and Alpha Centauri B. These two companion stars, taken together, have, Dole estimates, 1 chance in 10 of possessing habitable planets. The total probability for all fourteen neighboring stars is about 2 chances in 5.

If life is the consequence of the chemical reactions described in the previous section, its development should prove inevitable on any earthlike planet. Of course, a planet may possess life and yet not possess intelligent life. We have no way of making even an intelligent guess as to the likelihood of the development of intelligence on a planet; and Dole, for instance, was careful to make none. After all, our own Earth, the only habitable planet we can study, existed for more than 3 billion years with a load of life, but not intelligent life.

It is possible that the porpoises and some of their relatives are intelligent, but, as sea creatures, they lack limbs and could not develop the use of fire; consequently, their intelligence, if it exists, could not be bent in the direction of a developed technology. If land life alone is considered, then it is only for about a million years or so that the earth has been able to boast a living creature with intelligence greater than that of an ape.

Still, this means that the earth has possessed intelligent life for 1/3500 of the time it has possessed life of any kind (at a rough guess). If we can say that of all life-bearing planets, 1 out of 3,500 bears intelligent life, then out of the 640 million habitable planets Dole wrote of, there may be 180,000 intelligences. We may well be far from alone in the universe.

This view of a universe rich in intelligent life forms, which Dole, Sagan (and I) favor, is not held unanimously by astronomers. Since Venus and Mars have been studied in detail and found to be hostile to life, there is the pessimistic view that the limits within which we may expect life to form and to be maintained for billions of years are very narrow, and that Earth is extraordinarily fortunate to be within those limits. A slight change in this direction or that in any of a number of properties, and life would not have formed or, if formed, would not have remained in existence long. In this view, there may not be more than one or two life-bearing planets per galaxy, and there may only be one or two technological civilizations in the entire universe.

Francis Crick takes the view that there may be a sizable number of planets in each galaxy that are habitable but do not have the much narrower range of properties required for life to originate. It may then be that life originates on one particular planet and, once a civilization arises that can manage interstellar flights, spreads elsewhere. Clearly, Earth has not yet developed interstellar flights, and it may be that some travelers from far off billions of years ago unwittingly (or deliberately) infected Earth with life on a visit here.

Both these views, the optimistic and the pessimistic—a universe full of life and a universe nearly empty of life—are *a priori* views. Both involve reasoning from certain assumptions, and neither has observational evidence.

Can such evidence be obtained? Is there any way of telling, at a distance, whether life exists somewhere in the vicinity of a distant star? It can be reasoned that any form of life intelligent enough to have developed a high-technological civilization comparable, or superior, to our own will certainly have developed radio astronomy and will certainly be capable of sending out radio signals—or will, as we ourselves do, send them out inadvertently as we go about our radio-filled lives.

American scientists took such a possibility seriously enough to set up an enterprise, under the leadership of Frank Donald Drake, called Project Ozma (deriving its name from one of the *Oz* books for children) to listen for possible radio signals from other worlds. The idea is to look for some pattern in radio waves coming in from space. If they detect signals in a complexly ordered pattern, as opposed to the random, formless broadcasts from radio stars or excited matter in space, or from the simple periodicity of pulsars it may be assumed that such signals will represent messages from some extraterrestrial intelligence. Of course, even if such messages were received, communication with the distant intelligence would still be a problem. The messages would have been many years on the way, and a reply also would take many years to reach the distant broadcasters, since the nearest potentially habitable planet is 4⅓ light-years away.

The sections of the heavens listened to at one time or another in the course of Project Ozma included the directions in which lie Epsilon Eridani, Tau Ceti, Omicron-2 Eridani, Epsilon Indi, Alpha Centauri, 70 Ophiuchi, and 61 Cygni. After two months of negative results, however, the project was suspended.

Other attempts of this sort were still briefer and less elaborate. Scientists dream of something better, however.

In 1971, a NASA group under Bernard Oliver suggested what has come to be called Project Cyclops. This would be a large array of radio telescopes, each 100 meters (109 yards) in diameter; all arranged in rank and file; all steered in unison by a computerized electronic system. The entire array, working together, would be equivalent to a single radio telescope some 10 kilometers (6.2 miles) across. Such an array would detect radio beams of the kind Earth is inadvertently leaking at a distance of a hundred light-years, and should detect a deliberately aimed radio-wave beacon from another civilization at a distance of a thousand light-years.

To set up such an array might take twenty years and cost 100 billion dollars. (Before exclaiming at the expense, think that the world is spending 500 billion dollars—five times as much—*each year* on war and preparations for war.)

But why make the attempt? There seems little chance we will succeed, and even if we do, so what? Is there any possible chance we can understand an interstellar message? Yet there are reasons for trying.

First, the mere attempt will advance the art of radio telescopy to the great advantage of humanity in understanding the universe. Second, if we search the sky for messages and find none, we may still find a great deal of interest. But, what if we actually do detect a message and do not understand it? What good will it do us?

Well, there is another argument against intelligent life existing on other planets. It goes as follows: If they exist, and are superior to us, why haven't *they* discovered *us?* Life has existed on Earth for billions of years without being disturbed by outside influences (as far as we can tell), and that is indication enough that there are no outside influences to begin with.

Other arguments can be used to counter this. It may be that the civilizations that exist are so far away that there is no convenient way to reach us; that interstellar travel is never developed by any civilization; and that every one of us is isolated so that we can reach each other only by long-distance messages. It may also be that they *have* reached us but, realizing that we are a planet that is in the process of developing life and an eventual civilization, are deliberately refraining from interfering with us.

Both are weak arguments. There is another, stronger, and very frightening one. It is possible that intelligence is a self-limiting property. Perhaps as soon as a species develops a sufficiently high technology, it destroys itself—as we, with our mounting stores of nuclear weapons and our penchant for over-populating and for destroying the environment, seem to be doing. In that case, it is not that there are no civilizations, and nothing more. There may be many civilizations not yet at the point of being able to send or receive messages, and many civilizations that are destroyed, and only one or two that have just reached the point of message sending and are about to destroy themselves but have not yet done so.

In that case, if we receive a message—*one* message—the one fact it may reveal to us is that somewhere *one* civilization anyway has reached a high level of technology (beyond ours, in all likelihood) and has *not* destroyed itself.

And if it has managed to survive, might we not as well?

It is the kind of encouragement that humanity badly needs at this stage in its history and something that I, for one, would gladly welcome.

Chapter 14

The

Microorganisms

Bacteria

Before the seventeenth century, the smallest known living creatures were tiny insects. It was taken for granted, of course, that no smaller organisms existed. Living beings might be made invisible by a supernatural agency (all cultures believed that in one way or another), but no one supposed there to be creatures in nature too small to be seen.

MAGNIFYING DEVICES

Had anyone suspected such a thing, people might have come much sooner to the deliberate use of magnifying devices. Even the Greeks and Romans knew that glass objects of certain shapes would focus sunlight on a point and would magnify objects seen through the glass. A hollow glass sphere filled with water would do so, for instance. Ptolemy discussed the optics of burning glasses; and Arabic writers such as Alhazen, about 1000 A.D., extended his observations.

It was Robert Grosseteste—English bishop, philosopher, and keen amateur scientist—who, early in the thirteenth century, first suggested a use for this. He pointed out that *lenses* (so named because they were shaped like lentils) might be useful in magnifying objects too small to be seen conveniently. His pupil Roger Bacon acted on this suggestion and devised spectacles to improve poor vision.

At first only convex lenses, to correct farsightedness, were made. Concave lenses, to correct nearsightedness, were not developed until about 1400. The invention of printing brought more and more demand for spectacles; and by the sixteenth century spectaclemaking was a skilled profession. It became a particular specialty in the Netherlands.

(*Bifocals*, serving for both far and near vision, were invented by Benjamin Franklin in 1760. In 1827, the British astronomer George Biddell Airy designed the first lenses to correct astigmatism, from which he suffered himself. And in 1887, a German physician, Adolf Eugen Fick, introduced the idea of contact lenses, which may some day make ordinary spectacles more or less obsolete.)

Let us get back to the Dutch spectaclemakers. In 1608, so the story goes, an apprentice to a spectaclemaker named Hans Lippershey, amused himself during an idle hour by looking at objects through two lenses held one behind the other. The apprentice was amazed to find that, when he held them a certain distance apart, far-off objects appeared close at hand. The apprentice promptly told his master about it, and Lippershey proceeded to build the first *telescope*, placing the two lenses in a tube to hold them at the proper spacing. Prince Maurice of Nassau, commander of the Dutch forces in rebellion against Spain, saw the military value of the instrument and endeavored to keep it secret.

He reckoned without Galileo, however. Hearing rumors of the invention of a far-seeing glass, and knowing no more than that it was made with lenses, Galileo soon discovered the principle and built his own telescope; his was completed within six months after Lippershey's.

By rearranging the lenses of his telescope, Galileo found that he could magnify close objects, so that it was in effect a *microscope*. Over the next decades, several scientists built microscopes. An Italian naturalist named Francesco Stelluti studied insect anatomy with one; Malpighi discovered the capillaries; and Hooke discovered the cells in cork.

But the importance of the microscope was not really appreciated until Anton van Leeuwenhoek, a merchant in the city of Delft, took it up. Some of van Leeuwenhoek's lenses could enlarge up to 200 times.

Van Leeuwenhoek looked at all sorts of objects quite indiscriminately, describing what he saw in lengthy detail in letters to the Royal Society in London. It was rather a triumph for the democracy of science that the tradesman was elected a fellow of the gentlemanly Royal Society. Before he died, the Queen of England and Peter the Great, czar of all the Russias, visited the humble microscope maker of Delft.

Through his lenses van Leeuwenhoek discovered sperm cells and actually saw blood moving through capillaries in the tail of a tadpole. More important, he was the first to see living creatures too small to be seen by the unaided eye. He discovered these *animalcules* in stagnant water in 1675. He also resolved the tiny cells of yeast and, at the limit of his lenses' magnifying power, finally, in 1676, came upon *germs*, which today we know as *bacteria.*

Microscopes improved only slowly, and it took a century and a half before objects the size of germs could be studied with ease. For instance, it was not until 1830 that the English optician Joseph Jackson Lister devised an *achromatic microscope,* which eliminated the rings of color that limited the sharpness of the image. Lister found that red-blood corpuscles (first detected as featureless blobs by the Dutch physician Jan Swammerdam in 1658) were biconcave disks—like tiny doughnuts with dents instead of a hole. The achromatic microscope was a great advance; and in 1878, the German physicist Ernst Abbé began a series of improvements that resulted in what might be called the modern optical microscope.

NAMING THE BACTERIA

The members of the new world of microscopic life gradually received names. Van Leeuwenhoek's animalcules actually were animals, feeding on small particles and moving about by means of small whips (*flagellae*) or hairlike *cilia* or advancing streams of protoplasm (*pseudopods*). These animals were given the name *protozoa* (Greek for "first animals"), and the German zoologist Karl Theodor Ernst Siebold identified them as single-celled creatures.

Germs were something else—much smaller than protozoa and much simpler. Although some germs could move about, most lay quiescent and merely grew and multiplied. Except for their lack of chlorophyll, they showed none of the properties associated with animals. For that reason, they were usually classified among the *fungi*—plants that lack chlorophyll and live on organic matter. Nowadays most biologists tend to consider germs as neither plant nor animal but as a class by themselves. *Germ* is a misleading name for them. The same term may apply to the living part of a seed (as in "wheat germ"), or to sex cells ("germ cells"), or to embryonic organs ("germ layers"), or, in fact, to any small object possessing the potentiality of life.

The Danish microscopist Otto Frederik Müller managed to see the little creatures well enough in 1773 to distinguish two types: *bacilli* (from a Latin word meaning "little rods") and *spirilla* (for their spiral shape). With the advent of achromatic microscopes, the Austrian surgeon Theodor Billroth saw still smaller varieties to which he applied the term *coccus* (from the Greek word for "berry"). It was the German botanist Ferdinand Julius Cohn who finally coined the name *bacterium* (also from a Latin word meaning "little rod"). (See figure 14.1.)

Pasteur popularized the general term *microbe* ("small life") for all forms of microscopic life—plant, animal, and bacterial. But this word was soon adopted

Figure 14.1. Types of bacteria: cocci *(A)*, bacilli *(B)*, and spirilla *(C)*. Each type has a number of varieties.

for the bacteria, just then coming into notoriety. Today the general term for microscopic forms of life is *microorganism*.

The larger microorganisms are eukaryotes, as are the cells of multicellular animals and plants (including our own). The protozoa have nuclei and mitochondria, together with other organelles. Indeed, many of the protozoan cells are larger and more complex than the cells of our own body, for instance, since the protozoan cell must perform all the functions inseparable from life, whereas the cells of multicellular organisms may specialize and depend on other cells to perform functions and supply products they themselves cannot.

One-celled plant cells, called *algae* are, again, as complex as the cells of multicellular plants or more so. The algae contain nuclei, chloroplasts, and so on.

Bacteria, however, are prokaryotes and do not contain a nucleus or other organelles. The genetic material, ordinarily confined within the nucleus in eukaryotes, is spread throughout the bacterial cell. Bacteria are also unique in possessing a cell wall made up chiefly of a polysaccharide and protein in combination. Bacteria, which range from 1 to 10 micrometers in diameter (averaging, in other words, about 1/10,000th of an inch in diameter), are much smaller than eukaryotic cells in general.

Another large group of prokaryotes are the blue-green algae, which differ from bacteria chiefly in possessing chlorophyll and being able to carry through photosynthesis. Sometimes these are simply called *blue-greens*, leaving the term *algae* for eukaryotic one-celled plants.

One must not be overwhelmed by the apparent simplicity of bacteria. Although they do not have nuclei and do not seem to transfer chromosomes in the fashion of sexual reproduction, they nevertheless do indulge in a kind of primitive sex. In 1946, Edward Tatum and his student Joshua Lederberg began a series of observations that showed that bacteria do, on occasion, transfer sections of nucleic acid from one individual to another. Lederberg called the process *conjugation,* and he and Tatum shared in the 1958 Nobel Prize for physiology and medicine as a result.

In the study of conjugation, it appeared that the portions of the nucleic acid that underwent transfer were molecules that formed rings rather than straight lines. In 1952, Lederberg named these nucleic acid rings *plasmids.* The plasmids are the nearest thing to organelles that bacteria have. They possess genes, control the formation of certain enzymes, and can transfer properties from cell to cell.

THE GERM THEORY OF DISEASE

It was Pasteur who first definitely connected microorganisms with disease, thus founding the modern science of *bacteriology* or, to use a more general term, *microbiology.* This came about through Pasteur's concern with something that seemed an industrial problem rather than a medical one. In the 1860s, the French silk industry was being ruined by a disease of the silkworms. Pasteur, having already rescued France's wine makers, was put to work on this problem, too. Again making inspired use of the microscope, as he had in studying asymmetric crystals and varieties of yeast cells, Pasteur found microorganisms infecting the sick silkworms and the mulberry leaves on which they fed. He recommended that all infected worms and leaves be destroyed and a fresh start be made with the uninfected worms and leaves that remained. This drastic step was taken, and it worked.

Pasteur did more with these researches than merely to revive the silk industry. He generalized his conclusions and enunciated the *germ theory of disease*—without question the greatest single medical discovery ever made (and it was made not by a physician but by a chemist, as chemists such as myself delight in pointing out).

Before Pasteur, doctors had been able to do little more for their patients than recommend rest, good food, fresh air, and clean surroundings and, occasionally, handle a few types of emergency. This much had been advocated by the Greek physician Hippocrates of Cos (the "father of medicine") as long ago as 400 B.C. It was Hippocrates who introduced the rational view of medicine, turning away from the arrows of Apollo and demonic possession to proclaim that even epilepsy, called the "sacred disease," was not the result of being affected by some god's influence, but was a mere physical disorder to be treated as such. The lesson was never entirely forgotten by later generations.

However, medicine progressed surprisingly little in the next two millennia. Doctors could lance boils, set broken bones, and prescribe a few specific remedies that were simply products of folk wisdom: such drugs as quinine from the bark of the cinchona tree (originally chewed by the Peruvian Indians to cure themselves of malaria), and digitalis from the foxglove plant (an old herbwomen's remedy to stimulate the heart). Aside from these few treatments (and the smallpox vaccine, which I shall discuss later), many of the medicines and treatments dispensed by physicians after Hippocrates tended to heighten the death rate rather than lower it.

One of the interesting advances made in the first two and a half centuries of the Age of Science was the invention, in 1819, of the *stethoscope* by the French physician, René Théophile Hyacinthe Laennec. In its original form, it was little more than a wooden tube designed to help the doctor hear and interpret the sounds of the beating heart. Improvements since then have made it as characteristic and inevitable an accompaniment of the physician as the pocket computer is of an engineer.

It is not surprising, then, that up to the nineteenth century, even the most civilized countries were periodically swept by plagues, some of which had a profound effect on history. The plague in Athens that killed Pericles, at the time of the Peloponnesian War, was the first step in the ultimate ruin of Greece. Rome's downfall probably began with the plagues that fell upon the empire during the reign of Marcus Aurelius. The Black Death of the four-teenth century is estimated to have killed off a fourth of the population of Europe; this plague and gunpowder combined to destroy the social structure of the Middle Ages.

To be sure, plagues did not end when Pasteur discovered that infectious diseases are caused and spread by microorganisms. In India, cholera has long been endemic, and other underdeveloped countries suffer severely from epi-demics. Disease has remained a major hazard of wartime. Virulent new organ-isms arise from time to time and sweep over the world; indeed, the influenza pandemic of 1918 killed an estimated 15 million people, a larger number of people than died in any other plague in human history, and nearly twice as many as were killed in the then just-completed world war.

Nevertheless, Pasteur's discovery was a great turning point. The death rate in Europe and the United States began to fall markedly, and life expectancy steadily rose. Thanks to the scientific study of disease and its treatment, which began with Pasteur, men and women in the more advanced regions of the world can now expect to live an average of over seventy years; whereas before Pasteur, the average was only forty years under the most favorable conditions and perhaps only twenty-five years under unfavorable conditions. Since the Second World War, life expectancy has been zooming upward even in the less advanced regions of the world.

IDENTIFYING BACTERIA

Even before Pasteur advanced the germ theory in 1865, a Viennese physi-cian named Ignaz Philipp Semmelweiss had made the first effective attack on bacteria, without, of course, knowing what he was fighting. He was working in the maternity ward of one of Vienna's hospitals, where 12 percent or more of the new mothers died of something called *puerperal fever* (in plain English, "childbed fever"). Semmelweiss noted uneasily that women who bore their babies at home, with only the services of ignorant midwives, practically never got puerperal fever. His suspicions were further aroused by the death of a doctor in the hospital with symptoms that strongly resembled those of puerperal fever, after the doctor had cut himself while dissecting a cadaver. Were the doctors and students who came in from the dissection wards somehow transmitting this disease to the women whose delivery they attended? Semmelweiss insisted that the doctors wash their hands in a solution of chlorinated lime. Within a year, the death rate in the maternity wards fell from 12 percent to 1.5 percent.

But the veteran doctors were livid. Resentful of the implication that they had been murderers, and humiliated by all the hand washing, they drove Semmelweiss out of the hospital. (They were helped by the fact that he was a Hungarian, and Hungary was in revolt against its Austrian rulers.) Semmelweiss went to Budapest, where he reduced the maternal death rate; while in Vienna, the hospitals reverted to death traps for another decade or so. But Semmelweiss himself died of puerperal fever from an accidental infection (at the age of forty-seven) in 1865—just too soon to see the scientific vindication of his suspicions about the transmission of disease. That was the year when Pasteur discovered microorganisms in the diseased silkworms, and when an English surgeon named Joseph Lister (the son of the inventor of the achromatic microscope) independently introduced the chemical attack upon germs.

Lister resorted to the drastic substance *phenol* (carbolic acid). He used it first in dressings for a patient with a compound fracture. Up to that time, any serious wound almost invariably led to infection. Of course, Lister's phenol killed the tissues around the wound, but it did kill the bacteria. The patient made a remarkably untroubled recovery.

Lister followed up this success with the practice of spraying the operating room with phenol. It must have been hard on people who had to breathe it, but it began to save lives. As in Semmelweiss's case, there was opposition, but Pasteur's experiments had created a rationale for antisepsis, and Lister easily won the day.

Pasteur himself had somewhat harder going in France (unlike Lister, he lacked the union label of the M.D.), but he prevailed on surgeons to boil their instruments and steam their bandages. Sterilization with steam *à la* Pasteur replaced Lister's unpleasant phenol spray. Milder antiseptics, which could kill bacteria without unduly damaging tissue, were sought and found. The French physician Casimir Joseph Davaine reported on the antiseptic properties of iodine in 1873, and *tincture of iodine* (that is, iodine dissolved in a mixture of alcohol and water) came into common use in the home. It and similar products are automatically applied to every scratch. The number of infections prevented in this way is undoubtedly enormous.

In fact, the search for protection against infection leaned more and more in the direction of preventing germ entry (*asepsis*) rather than of destroying germs after they had gained a foothold, as was implied in antisepsis. In 1890, the American surgeon William Stewart Halstead introduced the practice of using sterilized rubber gloves during operations; by 1900, the British physician William Hunter had added the gauze mask to protect the patient against the germs in the physician's breath.

Meanwhile the German physician Robert Koch had begun to identify the specific bacteria responsible for various diseases, by introducing a vital improvement in the nature of *culture media*—that is, the food supply in which bacteria are grown. Where Pasteur used liquid media, Koch introduced solid media. He planted isolated samples on gelatin (for which *agar*, a gelatinlike substance obtained from seaweed, was substituted later). If a single bacterium is deposited (with a fine needle) in a spot on this medium, a pure colony will grow around the spot, because on the solid surface of the agar the bacteria lacks the ability to move or drift away from the original parent, as they would do in a liquid. An assistant of Koch, Julius Richard Petri, introduced the use of shallow glass dishes with covers, to protect the cultures from contamination by bacterial spores floating in air; such *Petri dishes* have been used for the purpose ever since.

In this way, individual bacteria give rise to colonies which can then be cultured separately and tested to see what disease they produce in an experimental animal. The technique not only made it possible to identify a given infection but also permitted experiments with various possible treatments to kill specific bacteria.

With his new techniques, Koch isolated a bacillus that causes anthrax and, in 1882, another that causes tuberculosis. In 1884, he also isolated the bacterium that causes cholera. Others followed in Koch's path. In 1883, for instance, the German pathologist Edwin Klebs isolated the bacterium that causes diphtheria. In 1905, Koch received the Nobel Prize in medicine and physiology.

Chemotherapy

Once bacteria had been identified, the next task was to find drugs that would kill a bacterium without killing the patient as well. To such a search, the German physician and bacteriologist Paul Ehrlich, who had worked with Koch, now addressed himself. He thought of the task as looking for a "magic bullet" which would not harm the body but strike only the bacteria.

Ehrlich was interested in dyes that stain bacteria—an area that had an important relationship to cell research. The cell in its natural state is colorless and transparent so that little detail within it could be seen. Early microscopists had tried to use dyes to color the cells, but it was only after Perkin's discovery of aniline dyes (see chapter 11) that the technique became practical. Though Ehrlich was not the first to use synthetic dyes in staining, he worked out the techniques in detail in the late 1870s and thus led the way to Flemming's study of mitosis and Feulgen's study of DNA in the chromosomes (see chapter 13).

But Ehrlich had other game in mind, too. He turned to these dyes as possible bactericides. A stain that reacted with bacteria more strongly than with other cells might well kill the bacteria, even when it was injected into the blood in a concentration low enough not to harm the cells of the patient. By 1907, Ehrlich had discovered a dye, called *trypan red*, which would stain trypanosomes, the organisms responsible for the dreaded African sleeping sickness, transmitted via the tsetse fly. Trypan red, when injected in the blood in proper doses, can kill trypanosomes without killing the patient.

Ehrlich was not satisfied: he wanted a surer kill of the microorganisms. Assuming that the toxic part of the trypan-red molecule was the *azo* combination—that is, a pair of nitrogen atoms ($-N=N-$)—he wondered what a similar combination of arsenic atoms ($-As=As-$) might accomplish. Arsenic is chemically similar to nitrogen but much more toxic. Ehrlich began to test arsenic compounds one after the other almost indiscriminately, numbering them methodically as he went. In 1909, a Japanese student of Ehrlich's, Sahachiro Hata, tested compound 606, which had failed against the trypanosomes, on the bacterium that causes syphilis. It proved deadly against this microbe (called a *spirochete* because it is spiral-shaped).

At once Ehrlich realized he had stumbled on something more important than a cure for trypanosomiasis, which after all was a limited disease confined to the tropics. Syphilis had been a hidden scourge of Europe for more than 400 years, ever since Columbus's time. (Columbus's men are supposed to have brought it back from the Caribbean Indians; in return, Europe donated smallpox to the Indians.) Not only was there no cure for syphilis, but prudishness had clothed the disease in a curtain of silence that let it spread unchecked.

Ehrlich devoted the rest of his life (he died in 1915) to the attempt to combat syphilis with compound 606, or, as he called it, Salvarsan ("safe arsenic"; its chemical name is arsphenamine). It could cure the disease, but its use was not without risk, and Ehrlich had to bully hospitals into using it correctly.

With Ehrlich, a new phase of chemotherapy came into being. *Pharmacology*, the study of the action of chemicals other than foods (that is "drugs") upon organisms, finally came into its own as a twentieth-century adjunct of medicine. Arsphenamine was the first synthetic drug, as opposed to the plant remedies such as quinine or the mineral remedies of Paracelsus and those who imitated him.

SULFA DRUGS

Naturally, the hope at once arose that every disease might be fought with a little tailored antidote all its own. But for a quarter of a century after Ehrlich's discovery, the concocters of new drugs had little luck. About the only success of any sort was the synthesis by German chemists of *plasmochin* in 1924 and of atabrine in 1930; they could be used as substitutes for quinine against malaria. (These drugs were very helpful to Western troops in jungle areas during the Second World War, when the Japanese held Java, the source of the world supply of quinine, which, like rubber, had moved from South America to Southeast Asia.)

In 1932 came a breakthrough. A German chemist named Gerhard Domagk had been injecting various dyes into infected mice. He tried a new red dye called Prontosil on mice infected with the deadly hemolytic streptococcus. The mice survived! He used it on his own daughter, who was dying of streptococcal blood poisoning. She survived also. Within three years, Prontosil had gained worldwide renown as a drug that could stop the strep infection in man.

Oddly, Prontosil did not kill streptococci in the test tube—only in the body. At the Pasteur Institute in Paris, Jacques Trefouël and his co-workers decided that the body must change Prontosil into some other substance that takes effect on the bacteria. They proceeded to break down Prontosil to the effective fragment, named *sulfanilamide*. This compound had been synthesized in 1908, reported perfunctorily, and forgotten. Sulfanilamide's structure is:

$$
\begin{array}{c}
NH_2 \\
| \\
C \\
CH \quad\quad CH \\
| \quad\quad\quad || \\
CH \quad\quad CH \\
C \\
| \\
O = S = O \\
| \\
NH_2
\end{array}
$$

It was the first of the "wonder drugs." One after another bacterium fell before it. Chemists found that, by substituting various groups for one of the hydrogen atoms on the sulfur-containing group, they could obtain a series of compounds, each of which had slightly different antibacterial properties. *Sulfapyridine* was introduced in 1937; *sulfathiazole*, in 1939; and *sulfadiazine*, in 1941. Physicians now could choose from a whole platoon of *sulfa drugs* for various infections. In the medically advanced countries, the death rates from bacterial diseases—notably, pneumococcal pneumonia—dropped dramatically.

Domagk was awarded the Nobel Prize in medicine and physiology in 1939. When he wrote the usual letter of acceptance, he was promptly arrested by the Gestapo; the Nazi government, for peculiar reasons of its own, refused to have anything to do with the Nobel Prizes. Domagk felt it the better part of valor to refuse the prize. After the Second World War, when he was at last free to accept the honor, Domagk went to Stockholm to receive it officially.

THE ANTIBIOTICS

The sulfa drugs had only a brief period of glory, for they were soon put in the shade by the discovery of a far more potent kind of antibacterial weapon —the antibiotics.

All living matter (including human beings) eventually returns to the soil to decay and decompose. With the dead matter and the wastes of living creatures go the germs of the many diseases that infect those creatures. Why is it, then, that the soil is usually so remarkably clean of infectious germs? Very few of them (the anthrax bacillus is one of the few) survive in the soil. A number of years ago, bacteriologists began to suspect that the soil harbors microorganisms or substances that destroy bacteria. As early as 1877, for instance, Pasteur had noticed that some bacteria died in the presence of others. And if the suspicion were correct, the soil would offer a large variety of organisms that might bring death to others of their kind. It is estimated that each acre of soil contains about 2,000 pounds of molds, 1,000 pounds of bacteria, 200 pounds of protozoa, 100 pounds of algae, and 100 pounds of yeast.

One of those who conducted a deliberate search for bactericides in the soil was the French-American microbiologist René Jules Dubos. In 1939, he isolated from a soil microorganism called *Bacillus brevis* a substance, *tyrothricin*, from which he isolated two bacteria-killing compounds that he named *gramicidin* and *tyrocidin*. They turned out to be peptides containing D-amino acids—the mirror images of the ordinary L-amino acids that make up most natural proteins.

Gramicidin and tyrocidin were the first antibiotics produced as such. But an antibiotic that was to prove immeasurably more important had been discovered—and merely noted in a scientific paper—twelve years earlier.

The British bacteriologist Alexander Fleming one morning found that some cultures of staphylococcus (the common pus-forming bacterium), which he had left on a bench, were contaminated with something that had killed the bacteria. There were little clear circles where the staphylococci had been destroyed in the culture dishes. Fleming, being interested in antisepsis (he had discovered that an enzyme in tears, called *lysozome*, had antiseptic properties), at once investigated to see what had killed the bacteria, and discovered that it was a common bread mold, *Penicillium notatum*. Some substance, which he named *penicillin*, produced by the mold was lethal to germs. Fleming dutifully published his results in 1929, but no one paid much attention at the time.

Ten years later the British biochemist Howard Walter Florey and his German-born associate, Ernst Boris Chain, became intrigued by the almost forgotten discovery and set out to try to isolate the antibacterial substance. By 1941, they had obtained an extract that proved effective clinically against a number of *gram-positive* bacteria (bacteria that retain a dye developed in 1884 by the Danish bacteriologist Hans Christian Joachim Gram).

Because wartime Britain was in no position to produce the drug, Florey went to the United States and helped to launch a program that developed methods of purifying penicillin and speeding up its production by the mold. In 1943, five hundred cases were treated with penicillin; and, by the war's end, large-scale production and use of penicillin were under way. Not only did penicillin pretty much supplant the sulfa drugs, but it became (and still is) one of the most important drugs in the entire practice of medicine. It is effective against a wide range of infections, including pneumonia, gonorrhea, syphilis, puerperal fever, scarlet fever, and meningitis. (The range of effectivity is called the *antibiotic spectrum.*) Furthermore, it has practically no toxicity or undesirable side effects, except in penicillin-sensitive individuals.

In 1945, Fleming, Florey, and Chain shared the Nobel Prize in medicine and physiology.

Penicillin set off an almost unbelievably elaborate hunt for other antibiotics. (The word was coined in 1942 by the Rutgers University bacteriologist Selman Abraham Waksman.)

In 1943, Waksman isolated from a soil mold of the genus *Streptomyces* the antibiotic known as *streptomycin.* Streptomycin hit the *gram-negative* bacteria (those that easily lose the Gram stain). Its greatest triumph was against the tubercle bacillus. But streptomycin, unlike penicillin, is rather toxic and must be used with caution.

For the discovery of streptomycin, Waksman received the Nobel Prize in medicine and physiology in 1952.

Another antibiotic, *chloramphenicol,* was isolated from molds of the genus *Streptomyces* in 1947. Chloramphenicol attacks not only gram-positive and gram-negative bacteria but also certain smaller organisms—notably those causing typhus fever and psittacosis (*parrot fever*). But its toxicity calls for care in its use.

Then came a whole series of *broad-spectrum* antibiotics, found after painstaking examination of many thousands of soil samples—Aureomycin, Terramycin, Achromycin, and so on. The first of these, Aureomycin, was isolated by Benjamin Minge Duggar and his co-workers in 1944 and was placed on the market in 1948. These antibiotics are called *tetracyclines,* because in each case the molecule is composed of four rings side by side. They are effective against a wide range of infections with the result that infectious diseases have fallen to cheeringly low levels. (Of course human beings left alive by our continuing mastery over infectious disease have a much greater chance of succumbing to a metabolic disorder. Thus, in the last eighty years, the incidence of diabetes, the most common such disorder, has increased tenfold.)

RESISTANT BACTERIA

The chief disappointment in the development of chemotherapy has been the speedy rise of resistant strains of bacteria. In 1939, for instance, all cases of meningitis and pneumococcal pneumonia showed a favorable response to the administration of sulfa drugs. Twenty years later, only half the cases did. The various antibiotics also became less effective with time. It is not that the bacteria "learn" to resist but that resistant mutants among them flourish and multiply when the "normal" strains are killed off. Mutation is, ordinarily, a slow process; and in the eukarotes, the multicellular ones particularly, variation and change are brought about much more quickly by the constant shuffling of genes and chromosomes in each generation. Bacterial transformation is swift through mutation alone, however, because bacteria multiply so quickly. Uncounted numbers arise from a few progenitors; and though the percentage of useful mutations, such as those capable of giving rise to an enzyme that destroys an otherwise effective chemotherapeutic agent, is very low, the absolute numbers of such mutations is fairly high.

Furthermore, the genes necessary for the production of such enzymes are often found in plasmids and are transferred from bacterium to bacterium, making the spread of resistance even more rapid.

The danger of resistant strains of bacteria is greatest in hospitals, where antibiotics are used constantly, and where the patients naturally have below-normal resistance to infection. Certain new strains of staphylococci resist antibiotics with particular stubbornness. This *hospital staph* is a serious worry in maternity wards, for instance, and attained headline fame in 1961, when an attack of pneumonia, sparked by such resistant bacteria, nearly ended the life of screen star Elizabeth Taylor.

Fortunately, where one antibiotic fails another may still attack a resistant strain. New antibiotics, and synthetic modifications of the old, may hold the line in the contest against mutations. The ideal thing would be to find an antibiotic to which no mutants are immune. Then there would be no survivors of that particular bacterium to multiply. A number of such candidates have been produced. For instance, a modified penicillin, called Staphcyllin, was developed in 1960. It is partly synthetic; and because its structure is strange to bacteria, its molecule is not split and its activity ruined by enzymes such as *penicillinase* (first discovered by Chain), which resistant strains use against ordinary penicillin. Consequently, Staphcyllin is death to otherwise resistant strains; it was used to save Taylor's life, for instance. Yet strains of staphylococcus, resistant to synthetic penicillins, have also turned up. Presumably, the merry-go-round will go on forever.

Additional allies against resistant strains are various other new antibiotics and modified versions of old ones. One can only hope that the stubborn versatility of chemical science will manage to keep the upper hand over the stubborn versatility of the disease germs.

PESTICIDES

The same problem of the development of resistant strains arises in the battle with our larger enemies, the insects, which not only compete dangerously for food but also spread disease. The modern chemical defenses against insects arose in 1939, with the development by the Swiss chemist Paul Müller of the chemical *dichlorodiphenyltrichloroethane,* better known by its initials, DDT. Müller was awarded the Nobel Prize in medicine and physiology for this feat in 1948.

By then, DDT had come into large-scale use, and resistant strains of house-flies had developed. Newer *insecticides*—or, to use a more general term that will cover chemicals used against rats or against weeds, *pesticides*—must continually be developed.

In addition, there are critics of the overchemicalization of our battle against other, nonhuman forms of life. Some critics are concerned lest science make it possible for an increasingly large segment of the population to remain alive only through the grace of chemistry; they fear that if ever our technological organization falters, even temporarily, great carnage will result as populations fall prey to the infections and diseases from which they have been kept safe by chemical fortification and to which they lack natural resistance.

As for the pesticides themselves, the American science-writer Rachel Louise Carson published a book, *Silent Spring,* in 1962 that dramatically brought to the fore the possibility that our indiscriminate use of chemicals might kill harmless and even useful species along with those we are actually attempting to destroy. Furthermore, Carson maintained that to destroy living things without due consideration might lead to a serious upsetting of the intricate system whereby one species depends on another and, in the end, hurt more than it helps humanity. The study of this interlinking of species is termed *ecology,* and there is no question but that Carson's book encouraged a new hard look at this branch of biology.

The answer, of course, is not to abandon technology and give up all attempts to control insects (the price in disease and starvation would be too high) but to find methods that are more specific and less damaging to the ecological structure generally.

For instance, insects have their enemies. Those enemies, whether insect-parasites or insect eaters, might be encouraged. Sounds and odors might be used to repel insects or to lure them to their death. Insects might be sterilized through radiation. In each case, every effort should be made to zero in on the insect being fought.

One hopeful line of attack, led by the American biologist Carroll Milton Williams, is to make use of the insects' own hormones. Insects molt periodi-cally and pass through two or three well-defined stages: larva, pupa, and adult. The transitions are complex and are controlled by hormones. Thus, one called *juvenile hormone* prevents formation of the adult stage until an appropriate time. By isolating and applying the juvenile hormone, the adult stage is held back to the point where the insect is killed. Each insect has its own juvenile hormone and is affected only by its own. A particular juvenile hormone might thus be used to attack one particular species of insect without affecting any other organism in the world. Guided by the structure of the hormone, biolo-gists may even prepare synthetic substitutes that will be much cheaper and do the job as well.

In short, the answer to the fact that scientific advance may sometimes have damaging side effects, is not to abandon scientific advance, but to substitute still more advance—intelligently and cautiously applied.

HOW CHEMOTHERAPY WORKS

As to how the chemotherapeutic agents work, the best guess seems to be that each drug inhibits some key enzyme in the microorganism in a competitive way. This action is best established in the case of the sulfa drugs. They are very similar to *para aminobenzoic acid* (generally written "*p*-aminobenzoic acid"), which has this structure:

$$
\begin{array}{c}
NH_2 \\
| \\
C \\
\diagup \diagdown \\
CH \qquad CH \\
| \qquad\quad \| \\
CH \qquad CH \\
\diagdown \diagup \\
C \\
| \\
C = O \\
| \\
OH
\end{array}
$$

P-aminobenzoic acid is necessary for the synthesis of *folic acid*, a key substance in the metabolism of bacteria as well as other cells. A bacterium that picks up a sulfanilamide molecule instead of *p*-aminobenzoic acid can no longer produce folic acid, because the enzyme needed for the process is put out of action. Consequently, the bacterium ceases to grow and multiply. The cells of the human patient, on the other hand, are not disturbed; they obtain folic acid from food and do not have to synthesize it. There are no enzymes in human cells to be inhibited by moderate concentrations of the sulfa drugs in this fashion.

Even where a bacterium and the human cell possess similar enzymes, there are other ways of attacking the bacterium selectively. The bacterial enzyme may be more sensitive to a given drug than the human enzyme is, so that a certain dose will kill the bacterium without seriously disturbing the human cells. Or a drug of the proper design may be able to penetrate the bacterial cell membrane but not the human cell membrane. Penicillin, for instance, interferes with the manufacture of cell walls, which bacteria possess but animals cells do not.

Do the antibiotics also work by competitive inhibition of enzymes? Here the answer is less clear. But there is good ground for believing that at least some of them do.

Gramicidin and tyrocidin, as I mentioned earlier, contain the "unnatural" D-amino acids. Perhaps these jam up the enzymes that form compounds from the natural L-amino acids. Another peptide antibiotic, bacitracin, contains ornithine; this may inhibit enzymes from making use of arginine, which ornithine resembles. There is a similar situation in streptomycin: its molecule contains an odd variety of sugar which may interfere with some enzyme acting on one of the normal sugars of living cells. Again, chloramphenicol resembles the amino acid phenylalanine; likewise, part of the penicillin molecule resembles the amino acid cysteine. In both of these cases the possibility of competitive inhibition is strong.

The clearest evidence so far of competitive action by an antibiotic involves *puromycin*, a substance produced by a *Streptomyces* mold. This compound has a structure much like that of nucleotides (the building units of nucleic acids), and Michael Yarmolinsky and his co-workers at Johns Hopkins University have shown that puromycin, competing with transfer-RNA, interferes with the synthesis of proteins. Again, streptomycin interferes with transfer-RNA, forcing the misreading of the genetic code and the formation of useless protein. Unfortunately, this form of interference makes it toxic to other cells besides bacteria, because it prevents their normal production of necessary proteins. Thus puromycin is too dangerous a drug to use, and streptomycin is nearly so.

BENEFICENT BACTERIA

Naturally, human attention focuses on those bacteria that are pathogenic and (in our selfish judgment) do harm. These are, however, a minute fraction of the total number. It has been estimated that, for every harmful bacterium, there are 30,000 that are harmless, useful, or even necessary. If we go by species, then out of 1,400 identified species of bacteria, only about 150 cause disease in human beings or in those plants and animals that we have cultivated or domesticated.

Consider, for instance, the fact that, at each moment, countless organisms are dying, and that relatively few of them serve as food for other organisms at the level of ordinary animals. Less than 10 percent of fallen leaves, and less than 1 percent of dead wood, are eaten by animals. The rest falls prey to fungi and bacteria. Were it not for these *decomposers*, especially the popularly named *decay bacteria*, the world of life would choke on the ever-increasing accumulation of indigestible fragments which would contain within themselves an ever-increasing fraction of those elements necessary for life. And, in the not distant end, there would be no life at all.

Cellulose, in particular, is indigestible to multicellular animals and is the most common of all the structures produced by life. Even though animals such as cattle and termites seem to live on cellulose-rich food such as grass and wood, they do so only through innumerable bacteria that live in their digestive tracts. It is these bacteria that decompose the cellulose and restore it to an active role in the overall life cycle.

Again, all plant life requires nitrogen, out of which to build up amino acids and proteins. Animal life also requires nitrogen and obtains it (already built up into amino acids and proteins) from the plant world. Plant life obtains it from nitrates in the soil. The nitrates, however, are inorganic salts which are soluble in water. If it were merely a question of nitrates, these would be leached out of the soil by rainfall, and the land would become unproductive.

On land at least, plant life would be impossible, and only such animals could exist as fed on sea life.

Where do the nitrates come from, then, since there are always some present in the soil despite the action of millions of years of rainfall? The obvious source is the nitrogen of the atmosphere, but plants and animals have no means of making use of gaseous nitrogen (which is quite inert chemically) and of "fixing" it in the form of compounds. There are, however *nitrogen-fixing bacteria* that are capable of converting atmospheric nitrogen into ammonia. Once that is formed, it is easily converted into nitrates by *nitrifying bacteria*. Without such activity by bacteria (and by blue-green algae), land life would be impossible.

(Of course, human beings—thanks to modern technology, such as the Haber process, as described in chapter 11—are also capable of fixing atmospheric nitrogen, but were able to do so only after land life had existed for some hundreds of millions of years. By now, the industrial fixation of nitrogen has reached such a point that there is some concern about whether natural processes of denitrification—the reconversion of nitrates to gaseous nitrogen by still other bacteria—can keep pace. The overaccumulation of nitrates in rivers and lakes can encourage the growth of algae and the death of higher organisms such as fish, to the overall detriment of a balanced ecological system.)

Microorganisms of various sorts (including bacteria) have been of direct use to human beings from prehistoric times. Various yeasts (single-celled fungi that are eukaryotic) readily convert sugars and starches to alcohol and carbon dioxide and have therefore been used, from remote antiquity, to ferment fruit and grain into wine and beer. The production of carbon dioxide has been used to convert wheat flour into the soft and puffy breads and pastries we are accustomed to.

Molds and bacteria produce other changes that convert milk into yogurt or into any of a myriad of cheeses.

In modern times, we have *industrial microbiology* where specific strains of molds and bacteria are cultivated in order to produce substances of pharmaceutical value—such as antibiotics, vitamins, or amino acids—or of industrial value, such as acetone, butyl alcohol, or citric acid.

With the use of genetic engineering (as mentioned in the previous chapter), bacteria and other microorganisms might make more efficient capacities they already possess—such as nitrogen fixation—or develop new capacities—such as the ability to oxidize hydrocarbon molecules under the proper conditions and thus clean up oil spills. They might also gain the capacity to produce desirable substances such as various blood fractions and hormones.

Viruses

To most people it may seem mystifying that the wonder drugs have had so much success against the bacterial diseases and so little success against the virus diseases. Since viruses, after all, can cause disease only if they reproduce themselves, why should it not be possible to jam the virus's machinery just as we jam the bacterium's? The answer is simple and, indeed, obvious once you realize how a virus reproduces itself. As a complete parasite, incapable of multiplying anywhere except inside a living cell, the virus has very little, if any, metabolic machinery of its own. To make copies of itself, it depends entirely on materials supplied by the cell it invades—as it can do with great efficiency. One virus within a cell can become 200 in 25 minutes. And it is therefore difficult to deprive the virus of those materials or jam the machinery without destroying the cell itself.

Biologists discovered the viruses only recently, after a series of encounters with increasingly simple forms of life. Perhaps as good a place as any to start this story is the discovery of the cause of malaria.

NONBACTERIAL DISEASE

Malaria has, year in and year out, probably killed more people in the world than any other infectious ailment, since until recently about 10 percent of the world's population suffered from the disease, which caused 3 million deaths a year. Until 1880, it was thought to be caused by the bad air (*mala aria* in Italian) of swampy regions. Then a French bacteriologist, Charles Louis Alphonse Laveran, discovered that the red-blood cells of malaria-stricken individuals were infested with parasitic protozoa of the genus *Plasmodium*. (For this discovery, Laveran was awarded the Nobel Prize in medicine and physiology in 1907.)

In 1894, a British physician named Patrick Manson, who had conducted a missionary hospital in Hong Kong, pointed out that swampy regions harbor mosquitoes as well as dank air, and he suggested that mosquitoes might have something to do with the spread of malaria. A British physician in India, Ronald Ross, pursued this idea and, in 1897, was able to show that the malarial parasite does indeed pass part of its life cycle in mosquitoes of the genus *Anopheles* (see figure 14.2). The mosquito picks up the parasite in sucking the blood of an infected person and then passes it on to any person it bites.

For his work, bringing to light for the first time the transmission of a disease by an insect *vector*, Ross received the Nobel Prize in medicine and physiology in 1902. It was a crucial discovery of modern medicine, for it showed that a disease might be stamped out by killing off the insect carrier. Drain the swamps that breed mosquitoes; eliminate stagnant water; destroy the mosquitoes with insecticides—and you can stop the disease. Since the Second World War, large areas of the world have been freed of malaria in just this

CYCLE IN MOSQUITO

CYCLE IN MAN

Figure 14.2. Life cycle of the malarial microorganism.

way, and the total number of deaths from malaria has declined by at least one third from its maximum.

Malaria was the first infectious disease traced to a nonbacterial microorganism (a protozoan in this case). Very shortly afterward, another nonbacterial disease was tracked down in a similar way. It was the deadly yellow fever, which as late as 1898, during an epidemic in Rio de Janeiro, killed nearly 95 percent of those it struck. In 1899, when an epidemic of yellow fever broke out in Cuba, a United States board of inquiry, headed by the bacteriologist Walter Reed, went to Cuba to investigate the causes of the disease.

Reed suspected a mosquito vector, such as had just been exposed as the transmitter of malaria. He first established that the disease could not be transmitted by direct contact between the patients and doctors or by way of the patient's clothing or bedding. Then some of the doctors deliberately let themselves be bitten by mosquitoes that had previously bitten a man sick with yellow fever. They got the disease, and one of the courageous investigators, Jesse William Lazear, died. But the culprit was identified as the *Aedes aegypti* mosquito. The epidemic in Cuba was checked, and yellow fever is no longer a serious disease in the medically advanced parts of the world. The cause of yellow fever is non-bacterial, but non-protozoan, too. The disease agent is something even smaller than a bacterium.

As a third example of a nonbacterial disease, there is typhus fever. This infection is endemic in North Africa and was brought into Europe via Spain during the long struggle of the Spaniards against the Moors of North Africa. Commonly known as *plague*, it is very contagious and has devastated nations. In the First World War, the Austrian armies were driven out of Serbia by the typhus when the Serbian army itself was unequal to the task. The ravages of typhus in Poland and Russia during that war and its aftermath (some 3 million persons died of the disease) did as much as military action to ruin those nations.

At the turn of the twentieth century, the French bacteriologist Charles Nicolle, then in charge of the Pasteur Institute in Tunis, noticed that although typhus was rife in the city, no one caught it in the hospital. The doctors and nurses were in daily contact with typhus-ridden patients, and the hospital was crowded; yet there was no spread of the disease there. Nicolle considered what happened when a patient came into the hospital, and it struck him that the most significant change was a thorough washing of the patient and removal of his lice-infested clothing. Nicolle decided that the body louse must be the vector of typhus. He proved the correctness of his guess by experiments. He received the Nobel Prize in medicine and physiology in 1928 for his discovery. Thanks to his finding, and the discovery of DDT, typhus fever did not repeat its deadly carnage in the Second World War. In January 1944, DDT was brought into play against the body louse. The population of Naples was sprayed *en masse*, and the lice died. For the first time in history, a winter epidemic of typhus (when the multiplicity of clothes, not removed very often, made louse-infestation almost certain and almost universal) was stopped in its tracks. A similar epidemic was stopped in Japan in late 1945 after the American occupation. The Second World War became almost unique among history's wars in possessing the dubious merit of killing fewer people by disease than by guns and bombs.

Typhus, like yellow fever, is caused by an agent smaller than a bacterium, and we must now enter the strange and wonderful realm populated by subbacterial organisms.

SUBBACTERIA

To get some idea of the dimensions of objects in this world, let us look at them in order of decreasing size. The human ovum is about 100 micrometers (100 millionths of a meter, or about 1/250 inch) in diameter and is just barely visible to the naked eye. The paramecium, a large protozoan which in bright light can be seen moving about in a drop of water, is about the same size. An ordinary human cell is only 1/10 as large (about 10 micrometers in diameter) and is quite invisible without a microscope. Smaller still is the red-blood corpuscle—some 7 micrometers in maximum diameter. The bacteria, starting with species as large as ordinary cells, drop down to a tinier level: the average rod-shaped bacterium is only 2 micrometers long, and the smallest bacteria are spheres perhaps no more than 0.4 micrometers in diameter. They can barely be seen in ordinary microscopes.

At this level, organisms apparently have reached the smallest possible volume into which can be crowded all the metabolic machinery necessary for an independent life. Any smaller organism cannot be a self-sufficient cell and must live as a parasite. It must shed most of the enzymatic machinery as excess baggage, so to speak. It is unable to grow or multiply on any artificial supply of food, however ample; hence it cannot be cultured, as bacteria can, in the test tube. The only place it can grow is in a living cell, which supplies the enzymes that it lacks. Such a parasite grows and multiplies, naturally, at the expense of the host cell.

The first subbacteria were discovered by a young American pathologist named Howard Taylor Ricketts. In 1909, he was studying a disease called *Rocky Mountain spotted fever,* which is spread by ticks (blood-sucking arthropods, related to the spiders rather than to insects). Within the cell's infected hosts, he found *inclusion bodies* which turned out to be very tiny organisms, now called *rickettsia* in his honor. Ricketts and others soon found that typhus also is a rickettsial disease. In the process of establishing a proof of this fact, Ricketts himself caught typhus, and died in 1910 at the age of thirty-nine.

The rickettsia are still big enough to be attacked by antibiotics such as chloramphenicol and the tetracyclines. They range in diameter from about 0.8 to 0.2 micrometers. Apparently they possess enough metabolic machinery of their own to differ from the host cells in their reaction to drugs. Antibiotic therapy has therefore considerably reduced the danger of rickettsial diseases.

At the lowest end of the scale, finally, come the viruses. They overlap the rickettsia in size; in fact, there is no actual dividing line between rickettsia and viruses. But the smallest viruses are small indeed. The virus of yellow fever, for instance, is only 0.02 micrometers in diameter. The viruses are much too small to be detected in a cell or to be seen under any optical microscope. The average virus is only 1/1,000 the size of the average bacterium.

A virus is stripped practically clean of metabolic machinery. It depends almost entirely upon the enzyme equipment of the host cell. Some of the largest viruses are affected by certain antibiotics; but against the run-of-the-mill viruses, drugs are helpless.

The existence of viruses was suspected many decades before they were finally seen. Pasteur, in his studies of hydrophobia, could find no organism in the body that could reasonably be suspected of causing the disease. Rather than decide that his germ theory of disease was wrong, Pasteur suggested that the germ in this case was simply too small to be seen. He was right.

In 1892, a Russian bacteriologist, Dmitri Iosifovich Ivanovski, was studying *tobacco-mosaic disease,* a disease that gives the leaves of the tobacco plant a mottled appearance. He found that the juice of infected leaves could transmit the disease when placed on the leaves of healthy plants. In an effort to trap the germs, he passed the juice through porcelain filters with holes so fine that not even the smallest bacterium could pass through. Yet the filtered juice still infected tobacco plants. Ivanovski decided that his filters must be defective and were actually letting bacteria through.

A Dutch bacteriologist, Martinus Willem Beijerinck, repeated the experiment in 1897 and came to the decision that the agent of the disease was small enough to pass through the filter. Since he could see nothing in the clear, infective fluid under any microscope and was unable to grow anything from it in a test-tube culture, he thought the infective agent might be a small molecule, perhaps about the size of a sugar molecule. Beijerinck called the infective agent a *filtrable virus* (*virus* being a Latin word meaning "poison").

In the same year, a German bacteriologist, Friedrich August Johannes Löffler, found that the agent causing hoof-and-mouth disease in cattle could also pass through a filter. And, in 1901, Walter Reed, in the course of his yellow-fever researches, found that the infective agent of that disease also was a filtrable virus. In 1914, the German bacteriologist Walther Kruse demonstrated the common cold to be virus-produced.

By 1931, some forty diseases (including measles, mumps, chicken pox, influenza, smallpox, poliomyelitis, and hydrophobia) were known to be caused by viruses, but the nature of viruses was still a mystery. Then an English bacteriologist, William Joseph Elford, finally began to trap some in filters and to prove that at least they were material particles of some kind. He used fine collodion membranes, graded to keep out smaller and smaller particles, and he worked his way down to membranes fine enough to remove the infectious agent from a liquid. From the fineness of the membrane that could filter out the agent of a given disease, he was able to judge the size of that virus. He found that Beijerinck had been wrong: even the smallest virus was larger than most molecules. The largest viruses approached the rickettsia in size.

For some years afterward, biologists debated whether viruses were living or dead particles. Their ability to multiply and transmit disease certainly suggested that they were alive. But in 1935, the American biochemist Wendell Meredith Stanley produced a piece of evidence that seemed to speak forcefully in favor of "dead." He mashed up tobacco leaves heavily infected with the tobacco-mosaic virus and set out to isolate the virus in as pure and concentrated a form as he could, using protein-separation techniques for the purpose. Stanley succeeded beyond his expectations, for he obtained the virus in crystalline form! His preparation was just as crystalline as a crystallized molecule, yet the virus evidently was still intact; when he redissolved it in liquid, it was just as infectious as before.

For his crystallization of the virus, Stanley shared the 1946 Nobel Prize in chemistry with Summer and Northrop, the crystallizers of enzymes (see chapter 12).

Still, for twenty years after Stanley's feat, the only viruses that could be crystallized were the very simple *plant viruses* (those infesting plant cells). Not until 1955 was the first *animal virus* crystallized. In that year, Carlton E. Schwerdt and Frederick L. Schaffer crystallized the poliomyelitis virus.

The fact that viruses could be crystallized seemed to many, including Stanley himself, to be proof that they were merely dead protein. Nothing living had ever been crystallized, and life and crystallinity seemed to be mutually contradictory. Life was flexible, changeable, dynamic; a crystal was rigid, fixed, strictly ordered.

Yet the fact remained that viruses are infective, thay they can grow and multiply even after having been crystallized. And growth and reproduction have always been considered the essence of life.

The turning point came in 1936 when two British biochemists, Frederick Charles Bawden and Norman Wingate Pirie, showed that the tobacco mosaic virus contains ribonucleic acid! Not much, to be sure: the virus is 94 percent protein and only 6 percent RNA; but it is nonetheless definitely a nucleoprotein. Furthermore, all other viruses proved to be nucleoprotein, containing RNA or DNA or both.

The difference between being nucleoprotein and being merely protein is practically the difference between being alive and dead. Viruses turned out to be composed of the same stuff as genes, and the genes are the very essence of life. The larger viruses give every appearance of being chromosomes on the loose, so to speak. Some contain as many as seventy-five genes, each of which controls the formation of some aspect of its structure—a fiber here, a folding there. By producing mutations in the nucleic acid, one gene or another may be made defective, and through this means, its function and even its location can be determined. The total gene analysis (both structural and functional) of a virus is within reach, though of course this represents but a small step toward a similar total analysis for cellular organisms, with their much more elaborate genic equipment.

We can picture viruses in the cell as raiders that, pushing aside the supervising genes, take over the chemistry of the cell in their own interests, often causing the death of the cell or of the entire host organism in the process. Sometimes, a virus may even replace a gene, or series of genes, with its own, introducing new characteristics that can be passed along to daughter cells. A virus may also pick up DNA from a bacterial cell it has infected, and carry it to a new cell which it then infects. This phenomenon is called *transduction* —a name given it by Lederberg, who discovered the phenomenon in 1952.

If the genes carry the "living" properties of a cell, then viruses are living things. Of course, a lot depends on how one defines life. I, myself, think it fair to consider any nucleoprotein molecule capable of replication to be living. By that definition, viruses are as alive as elephants and human beings.

No amount of indirect evidence of the existence of viruses is as good as seeing one. Apparently the first man to lay eyes on a virus was a Scottish physician named John Brown Buist. In 1887, he reported that, in the fluid from a vaccination blister, he had managed to make out some tiny dots under the microscope. Presumably they were the cowpox virus, the largest known virus.

To get a good look—or any look at all—at a typical virus, something better than an ordinary microscope was needed. The something better was finally invented in the late 1930s: the electron microscope, which can reach magnifications as high as 100,000 and resolve objects as small as 0.001 micrometers in diameter.

The electron microscope has its drawbacks. The object has to be placed in a vacuum, and the inevitable dehydration may change its shape. An object such as a cell must be sliced extremely thin. The image is only two-dimensional; furthermore, the electrons tend to go right through a biological material, so that it does not stand out against the background.

In 1944, the American astronomer and physicist Robley Cook Williams and the electron microscopist Ralph Walter Graystone Wyckoff jointly worked out an ingenious solution of these last difficulties. It occurred to Williams, as an astronomer, that just as the craters and mountains of the moon are brought into relief by shadows when the sun's light falls on them obliquely, so viruses might be seen in three dimensions in the electron microscope if they could somehow be made to cast shadows. The solution the experimenters hit upon was to blow vaporized metal obliquely across the virus particles set up on the stage of the microscope. The metal stream left a clear space—a "shadow"—behind each virus particle. The length of the shadow indicated the height of the blocking particle. And the metal, condensing as a thin film, also defined the virus particles sharply against the background.

The shadow pictures of various viruses then disclosed their shapes (figure 14.3). The cowpox virus was found to be shaped something like a barrel. It turned out to be about 0.25 micrometers thick—about the size of the smallest rickettsia. The tobacco-mosaic virus proved to be a thin rod 0.28 micrometers long by 0.015 micrometers thick. The smallest viruses, such as those of poliomyelitis, yellow fever, and hoof-and-mouth disease, were tiny spheres ranging in diameter from 0.025 down to 0.020 micrometers—considerably smaller than the estimated size of a single human gene. The weight of these viruses is only about 100 times that of an average protein molecule. The brome-grass mosaic virus has a molecular weight of 4.5 million. It is only one-tenth the size of the tobacco-mosaic virus.

In 1959, the Finnish cytologist Alvar P. Wilska designed an electron microscope using comparatively low-speed electrons. Because they are less penetrating than high-speed electrons, they can define some of the internal detail in the structure of viruses. And in 1961, the French cytologist Gaston DuPouy devised a way of placing bacteria in air-filled capsules and taking electron microscope views of living cells in this way. In the absence of metal-shadowing, however, detail was lacking.

The ordinary electron microscope is a *transmission* device because the electrons pass through the thin slice and are recorded on the other side. It is

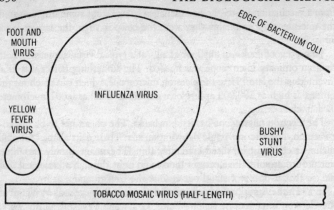

Figure 14.3. Relative sizes of simple substances and proteins and of various particles and bacteria. (An inch and a half on this scale = 1/10,000 of a millimeter in life.)

possible to use a low-energy electron beam that scans the object to be viewed, much as an electron beam scans a television tube. The electron beam causes material on the surface to emit electrons of their own. It is these emitted electrons that are studied. In such a *scanning electron microscope,* a great deal of surface detail can be made out. Such a device was suggested by the British scientist C. W. Oatley in 1948; and by 1958, such electron microscopes were in use.

THE ROLE OF NUCLEIC ACID

Virologists have actually begun to take viruses apart and put them together again. For instance, at the University of California, the German-American biochemist Heinz Fraenkel-Conrat, working with Robley Williams, found that gentle chemical treatment broke down the protein of the tobacco-mosaic virus into some 2,200 fragments, consisting of peptide chains made up of 158 amino acids apiece, and individual molecular weights of 18,000. The exact amino-acid constitution of these virus-protein units was completely worked out in 1960. When such units are dissolved, they tend to coalesce once more into the long, hollow rod (in which form they exist in the intact virus). The units are held together by calcium and magnesium atoms.

In general, virus-protein units make up geometric patterns when they combine. Those of tobacco-mosaic virus, just discussed, form segments of a helix.

The sixty subunits of the protein of the poliomyelitis virus are arranged in twelve pentagons. The twenty subunits of the *Tipula* iridescent virus are aranged in a regular twenty-sided solid, an icosahedron.

The protein of the virus is hollow. The protein helix of tobacco-mosaic virus, for instance, is made up of 130 turns of the peptide chain, producing a long, straight cavity within. Inside the protein cavity is the nucleic-acid portion of the virus. This may be DNA or RNA, but, in either case, it is made up of about 6,000 nucleotides, although Sol Spiegelman has detected an RNA molecule with as few as 470 nucleotides that is capable of replication.

Fraenkel-Conrat separated the nucleic acid and protein portions of tobacco-mosaic viruses and tried to find out whether each portion alone could infect a cell. It developed that separately they could not, as far as he could tell. But when he mixed the protein and nucleic acid together again, as much as 50 percent of the original infectiousness of the virus sample could eventually be restored!

What had happened? The separated virus protein and nucleic acid had seemed dead, to all intents and purposes; yet, mixed together again, some at least of the material seemed to come to life. The public press hailed Fraenkel-Conrat's experiment as the creation of a living organism from nonliving matter. The stories were mistaken, as we shall see in a moment.

Apparently, some recombination of protein and nucleic acid had taken place. Each, it seemed, had a role to play in infection. What were the respective roles of the protein and the nucleic acid, and which was more important?

Fraenkel-Conrat performed a neat experiment that answered the question. He mixed the protein part of one strain of the virus with the nucleic-acid portion of another strain. The two parts combined to form an infectious virus with a mixture of properties! In virulence (that is, the degree of its power to infect tobacco plants), it was the same as the strain of virus that had contributed the protein; in the particular disease produced (that is, the nature of the mosaic pattern on the leaf), it was identical with the strain of virus that had supplied the nucleic acid.

This finding fitted well with what virologists already suspected about the respective functions of the protein and the nucleic acid. It seems that when a virus attacks a cell, its protein shell, or coat, serves to attach itself to the cell and to break open an entrance into the cell. Its nucleic acid then invades the cell and engineers the production of virus particles.

After Fraenkel-Conrat's hybrid virus had infected a tobacco leaf, the new generation of virus that it bred in the leaf's cells turned out to be not a hybrid but just a replica of the strain that had contributed the nucleic acid. It copied that strain in degree of infectiousness as well as in the pattern of disease produced. In other words, the nucleic acid had dictated the construction of the new virus's protein coat. It had produced the protein of its own strain, not that of the strain with which it had been combined in the hybrid.

This reinforced the evidence that the nucleic acid is the "live" part of a virus, or, for that matter, of any nucleoprotein. Actually, Fraenkel-Conrat found in further experiments that pure virus nucleic acid alone could produce a little infection in a tobacco leaf—about 0.1 percent as much as the intact virus. Apparently once in a while the nucleic acid somehow managed to breach an entrance into a cell all by itself.

So putting virus nucleic acid and protein together to form a virus is not creating life from nonlife; the life is already there, in the shape of the nucleic acid. The protein merely serves to protect the nucleic acid against the action of hydrolyzing enzymes (*nucleases*) in the environment and to help it go about the business of infection and reproduction more efficiently. We might compare the nucleic-acid fraction to a man and the protein fraction to an automobile. The combination makes easy work of traveling from one place to another. The automobile by itself could never make the trip. The man could make it on foot (and occasionally does), but the automobile is a big help.

The clearest and most detailed information about the mechanism by which viruses infect a cell has come from studies of the viruses called *bacteriophages*, first discovered by the English bacteriologist Frederick William Twort in 1915 and, independently, by the Canadian bacteriologist Félix Hubert d' Hérelle in 1917. Oddly enough, these viruses are germs that prey on germs —namely, bacteria. D'Hérelle gave them the name *bacteriophage*, from Greek words meaning "bacteria eater."

The bacteriophages are beautifully convenient things to study, because they can be cultured with their hosts in a test tube. The process of infection and multiplication goes about as follows:

A typical bacteriophage (usually called *phage* by those who work with it) is shaped like a tiny tadpole, with a blunt head and a tail. Under the electron microscope, investigators have been able to see that the phage first lays hold of the surface of a bacterium with its tail. The best guess about how it does this is that the pattern of electric charge on the tip of the tail (determined by charged amino acids) just fits the charge pattern on certain portions of the bacterium's surface. The configurations of the opposite, and attracting, charges on the tail and on the bacterial surface match so neatly that they come together with something like the click of perfectly meshing gear teeth. Once the virus has attached itself to its victim by the tip of its tail, it cuts a tiny opening in the cell wall, perhaps by means of an enzyme that cleaves the molecules at that point. As far as the electron-microscope pictures show, nothing whatever is happening. The phage, or at least its visible shell, remains attached to the outside of the bacterium. Inside the bacterial cell, there is no visible activity. But, within half an hour, the cell bursts open, and hundreds of full-grown viruses pour out.

Evidently only the protein shell of the attacking virus stays outside the cell. The nucleic acid within the virus's shell must pour into the bacterium through the hole in its wall made by the protein. That the invading material is just nucleic acid, without any detectable admixture of protein, was proved by the American bacteriologist Alfred Day Hershey by means of radioactive tracers. He tagged phages with radioactive phosphorus and radioactive sulfur atoms (by growing them in bacteria that had incorporated these radioisotopes from their nutritive medium). Now phosphorus occurs both in proteins and in nucleic acids, but sulfur will turn up only in proteins, because there is no sulfur in a nucleic acid. Therefore if a phage labeled with both tracers invaded a bacterium and its progeny turned up with radiophosphorus but no radiosulfur, the experiment would indicate that the parent virus's nucleic acid had entered the cell but its protein had not. The absence of radiosulfur would suggest that all the protein in the virus progeny was supplied by the host bacterium. The experiment, in fact, turned out just this way: the new viruses contained radiophosphorus (contributed by the parent) but no radiosulfur.

Once more, the dominant role of nucleic acid in the living process was demonstrated. Apparently, only the phage's nucleic acid went into the bacterium, and there is superintended the construction of new viruses—protein and all—from the material in the cell.

Indeed, the infectious agent that causes spindle-tuber disease in potatoes was found to be an unusually small virus. In 1967, in fact, the microbiologist T. O. Diener suggested the virus in question was a naked strand of RNA. Such infectious bits of nucleic acid (minus protein) he called *viroids,* and some half dozen plant diseases have now been attributed to viroid infection.

The molecular weight of a viroid has been estimated at 130,000, only 1/300 that of a tobacco-mosaic virus. A viroid might consist of only 400 nucleotides in the string, yet that is enough for replication and, apparently, life. The viroids may be the smallest known living things.

Such viroids may conceivably be involved with certain little-understood degenerative diseases in animals which, if virus-caused, are brought about by *slow viruses* that take a long time to produce the symptoms. This may be the result of the low infectivity rates of short strings of uncoated nucleic acid.

Immune Reactions

Viruses are our most formidable living enemy (except human beings themselves). By virtue of their intimate association with the body's own cells, viruses have been all but invulnerable to attack by drugs or any other artificial weapon. And yet we have been able to hold our own against them, even under the most unfavorable conditions. The human organism is endowed with impressive natural defenses against disease.

Consider the Black Death, the great plague of the fourteenth century. It attacked a Europe living in appalling filth, without any modern conception of cleanliness and hygiene, without plumbing, without any form of reasonable medical treatment—a crowded and helpless population. To be sure, people could flee from the infected villages, but the fugitive sick only spread the epidemics faster and farther. Nonetheless, three fourths of the population successfully resisted the infections. Under the circumstances, the marvel is not that one out of four died, but that three out of four survived.

There is clearly such a thing as natural resistance to any given disease. Of people exposed to a serious contagious disease, some will have a relatively mild case, some will be very sick, some will die. There is also such a thing as complete immunity—sometimes inborn, sometimes acquired. A single attack of measles, mumps, or chickenpox, for instance, will usually make one immune to that particular disease for the rest of one's life.

All three of these diseases, as it happens, are caused by viruses. Yet they are comparatively minor infections, seldom fatal. Measles, the most dangerous of the three, usually produces only mild symptoms, at least in a child. How does the body fight off these viruses and then fortify itself so that the virus it has defeated never troubles it again? The answer to that question forms a thrilling episode in modern medical science, and for the beginning of the story we must go back to the conquest of smallpox.

SMALLPOX

Up to the end of the eighteenth century, smallpox was a particularly dreaded disease, not only because it was often fatal but also because those who recovered were permanently disfigured. A light case would leave the skin pitted; a severe attack could destroy all traces of beauty and almost of humanity. A very large proportion of the population bore the marks of smallpox on their faces. And those who had not yet caught it lived in fear of its striking.

In the seventeenth century, people in Turkey began to infect themselves deliberately with mild forms of smallpox, in the hope of making themselves immune to severe attack. They would have themselves scratched with the serum from blisters of a person who had a mild case. Some people developed only a light infection; others suffered the very disfigurement or death they had sought to avoid. It was risky business, but it is a measure of the horror of the disease that people were willing to risk the horror itself in order to escape from it.

In 1718, the famous beauty Lady Mary Wortley Montagu learned about this practice when she went to Turkey with her husband, sent there briefly as the British ambassador, and she had her own children inoculated. They escaped without harm. But the idea did not catch on in England, perhaps partly because Lady Montagu was considered a notorious eccentric. A similar case, across the ocean, was that of Zabdiel Boylston, an American physician. During a smallpox epidemic in Boston, he inoculated 241 people, of whom 6 died. He underwent considerable criticism for this.

Certain country folk in Gloucestershire had their own idea about how to avoid smallpox. They believed that a case of cowpox, a disease that attacked cows and sometimes people, would make a person immune to both cowpox and smallpox. This was wonderful, if true, for cowpox produced hardly any blisters and left hardly any marks. A Gloucestershire doctor, Edward Jenner, decided that there might be some truth in this folk "superstition." Milkmaids, he noticed, were particularly likely to catch cowpox and apparently also particularly likely not to be pockmarked by smallpox. (Perhaps the eighteenth-century vogue of romanticizing the beautiful milkmaid was based on the fact that milkmaids, having clear complexions, were indeed beautiful in a pockmarked world.)

Was it possible that cowpox and smallpox were so alike that a defense formed by the body against cowpox would also protect against smallpox? Very cautiously Dr. Jenner began to test this notion (probably experimenting on his own family first). In 1796, he decided to chance the supreme test. First he inoculated an eight-year-old boy named James Phipps with cowpox, using fluid from a cowpox blister on a milkmaid's hand. Two months later came the crucial and desperate part of the test. Jenner deliberately inoculated young James with smallpox itself.

The boy did not catch the disease. He was immune.

Jenner called the process *vaccination,* from *vaccinia,* the Latin name for cowpox. Vaccination spread through Europe like wildfire. It is one of the rare cases of a revolution in medicine that was adopted easily and almost at once —a true measure of the deadly fear inspired by smallpox and of the eagerness of the public to try anything that promised escape. Even the medical profession put up only weak opposition to vaccination—though its leaders put up such stumbling blocks as they could. When Jenner was proposed for election to the Royal College of Physicians in London in 1813, he was refused admission, on the ground that he was not sufficiently up on Hippocrates and Galen.

Today smallpox seems to have been wiped out as an active disease for lack of enough people who have not been rendered immune by vaccination and can serve as hosts. There has not been a single case of smallpox in the United States since 1949 or anywhere in the world since 1977. Samples of the virus still exist in some laboratories for purposes of research, and accidents may yet happen.

VACCINES

Attempts to discover similar inoculations for other severe diseases got nowhere for more than a century and a half. It was Pasteur who made the next big step forward. He discovered, more or less by accident, that he could change a severe disease into a mild one by weakening the microbe that produced it.

Pasteur was working with a bacterium that caused cholera in chickens. He concentrated a preparation so virulent that a little injected under the skin of a chicken would kill it within a day. On one occasion, he used a culture that had been standing for a week. This time the chickens became only slightly sick and recovered. Pasteur decided that the culture was spoiled and prepared a virulent new batch. But his fresh culture failed to kill the chickens that had recovered from the dose of "spoiled" bacteria. Clearly, the infection with the weakened bacteria had equipped the chickens with a defense against the fully potent ones.

In a sense, Pasteur had produced an artificial "cowpox" for this particular "smallpox." He recognized the philosophical debt he owed to Jenner by calling his procedure *vaccination,* too, although it had nothing to do with vaccinia. Since then, the term has been used quite generally to mean inoculations against any disease, and the preparation used for the purpose is called a *vaccine.*

Pasteur developed other methods of weakening (or *attenuating*) disease agents. For instance, he found that culturing anthrax bacteria at a high temperature produced a weakened strain that would immunize animals against the disease. Until then, anthrax had been so hopelessly fatal and contagious that as soon as one member of a herd came down with it, the whole herd had to be slaughtered and burned.

Pasteur's most famous victory, however, was over the virus disease called *hydrophobia,* or *rabies* (from a Latin word meaning "to rave," because the disease attacks the nervous system and produces symptoms akin to madness). A person bitten by a rabid dog would, after an incubation period of a month or two, be seized by violent symptoms and almost invariably die an agonizing death.

Pasteur could find no visible microbe as the agent of the disease (of course, he knew nothing of viruses), so he had to use living animals to cultivate it. He would inject the infectious fluid into the brain of a rabbit, let it incubate, mash up the rabbit's spinal cord, inject the extract into the brain of another rabbit, and so on. Pasteur attenuated his preparations by aging and testing them continuously until the extract could no longer cause noticeable disease in a rabbit. He then injected the rabbit with hydrophobia in full strength and found the animal immune.

In 1885, Pasteur got his chance to try the cure on a human being. A nine-year-old boy, Joseph Meister, who had been severely bitten by a rabid dog, was brought to him. With considerable hesitation and anxiety, Pasteur treated the boy with inoculations of successively less and less attenuated virus, hoping to build up resistance before the incubation period had elapsed. He succeeded. At least, the boy survived. (Meister became the gatekeeper of the Pasteur Institute and, in 1940, committed suicide when the Nazi army in Paris ordered him to open Pasteur's crypt.)

In 1890, a German army doctor named Emil von Behring, working in Koch's laboratory, tried another idea. Why take the risk of injecting the microbe itself, even in attenuated form, into a human being? Assuming that the disease agent causes the body to manufacture some defensive substance, would it not serve just as well to infect an animal with the agent, extract the defense substance that it produces, and inject that substance into the human patient?

Von Behring found that this scheme did indeed work. The defensive substance turned up in the blood serum, and von Behring called it *antitoxin.* He caused animals to produce antitoxins against tetanus and diphtheria. His first use of the diphtheria antitoxin on a child with the disease was so dramatically successful that the treatment was adopted immediately and proceeded to cut the death rate from diphtheria drastically.

Paul Ehrlich (who later was to discover the "magic bullet" for syphilis) worked with von Behring and probably calculated the appropriate antitoxin dosages. Later he broke with von Behring (Ehrlich was an irascible individual who found it easy to break with anyone) and alone went on to work out the rationale of serum therapy in detail. Von Behring received the Nobel Prize in medicine and physiology in 1901, the first year in which it was awarded. Ehrlich also was awarded that Nobel Prize, sharing it with a Russian biologist in 1908.

The immunity conferred by an antitoxin lasts only as long as the antitoxin remains in the blood. But the French bacteriologist Gaston Ramon found that, by treating the toxin of diphtheria or tetanus with formaldehyde or heat, he was able to change its structure in such a way that the new substance (called toxoid) could safely be injected in a human patient. The antitoxin then made by the patient himself lasts longer than that from an animal; furthermore, new doses of the toxoid can be injected when necessary to renew immunity. After toxoid was introduced in 1925, diphtheria lost most of its terrors.

Serum reactions were also used to detect the presence of disease. The best-known example of this is the Wasserman test, introduced by the German bacteriologist August von Wasserman, in 1906, for the detection of syphilis. This was based on techniques first developed by a Belgian bacteriologist, Jules Bordet, who worked with serum fractions that came to be called complement. This has turned out to be a complex system made up of a number of interrelated enzymes. For his work, Bordet received the Nobel Prize in medicine and physiology in 1919.

Pasteur's laborious wrestle with the virus of rabies showed the difficulty of dealing with viruses. Bacteria can be cultured, manipulated, and attenuated on artificial media in the test tube. Viruses cannot; they can be grown only in living tissue. In the case of smallpox, the living hosts for the experimental material (the cowpox virus) were cows and milkmaids. In the case of rabies, Pasteur used rabbits. But living animals are, at best, an awkward, expensive, and time-consuming medium for culturing microorganisms.

In the first quarter of this century, the French biologist Alexis Carrel won considerable fame with a feat that was to prove immensely valuable to medical research—keeping bits of tissue alive in the test tube. Carrel had become interested in this sort of thing through his work as a surgeon. He had developed new methods of transplanting animals' blood vessels and organs, for which he received the Nobel Prize in medicine and physiology in 1912. Naturally, he had to keep the excised organ alive while he was getting ready to transplant it. He worked out a way to nourish it, by perfusing the tissue with blood and supplying the various extracts and ions. As an incidental dividend, Carrel, with the help of Charles Augustus Lindbergh, developed a crude mechanical heart to pump the blood through the tissue.

Carrel's devices were good enough to keep a piece of embryonic chicken heart alive for thirty-four years—much longer than a chicken's lifetime. Carrel even tried to use his tissue cultures to grow viruses—and he succeeded in a way. The only trouble was that bacteria also grew in the tissues; and in order to keep the virus pure, such tedious aseptic precautions had to be taken that it was easier to use animals.

The chick-embryo idea, however, was in the right ball park, so to speak. Better than just a piece of tissue would be the whole thing—the chick embryo itself. A chick embryo is a self-contained organism, protected by the egg shell, equipped with its own natural defenses against bacteria, and cheap and easy to come by in quantity. And, in 1931, the pathologist Ernest William Goodpasture and his co-workers at Vanderbilt University succeeded in transplanting a virus into a chick embryo. For the first time, pure viruses could be cultured almost as easily as bacteria.

The first great medical victory by means of the culture of viruses in fertile eggs came in 1937. At the Rockefeller Institute, bacteriologists were still hunting for further protection against the yellow-fever virus. It was impossible to eradicate the mosquito completely, after all, and infected monkeys maintained a constantly threatening reservoir of the disease in the tropics. The South-African bacteriologist Max Theiler at the institute set out to produce an attenuated yellow-fever virus. He passed the virus through 200 mice and 100 chick embryos until he had a mutant that caused only mild symptoms yet gave rise to complete immunity against yellow fever. For this achievement Theiler received the 1951 Nobel Prize in medicine and physiology.

When all is said and done, nothing can beat culture in glassware for speed, control of the conditions, and efficiency. In the late 1940s John Franklin Enders, Thomas Huckle Weller, and Frederick Chapman Robbins at the Harvard Medical School went back to Carrel's approach. (He had died in 1944 and was not to see their success.) This time they had a new and powerful weapon against bacteria contaminating the tissue culture—the antibiotics. They added penicillin and streptomycin to the supply of blood that kept the tissues alive, and found that they could grow viruses without trouble. On impulse, they tried the poliomyelitis virus. To their delight, it flourished in this medium. It was the breakthrough that was to conquer polio, and the three men received the Nobel Prize in medicine and physiology in 1954.

The poliomyelitis virus could now be bred in the test tube, instead of solely in monkeys (which are expensive and temperamental laboratory subjects). Large-scale experimentation with the virus became possible. Thanks to the tissue-culture technique, Jonas Edward Salk of the University of Pittsburgh was able to experiment with chemical treatment of the virus, to learn that polio viruses killed by formaldehyde could still produce immune reactions in the body, and to develop his now-famous *Salk vaccine*.

Polio's sizable death rate, its dreaded paralysis, its partiality for children (so that it has the alternate name of *infantile paralysis*), the fact that it seems to be a modern scourge with no epidemics on record prior to 1840, and particularly the interest attracted to the disease by its eminent victim, Franklin Delano Roosevelt, made its conquest one of the most celebrated victories over a disease in all human history. Probably no medical announcement ever received such a Hollywood-premiere type of reception as did the report, in

1955, of the evaluating committee that found the Salk vaccine effective. Of course, the event merited such a celebration—more than do most of the performances that arouse people to throw ticker tape and trample one another. But science does not thrive on furor or wild publicity. The rush to respond to the public pressure for the vaccine apparently resulted in a few defective, disease-producing samples of the vaccine slipping through, and the subsequent counterfuror set back the vaccination program against the disease.

The setback was, however, made up, and the Salk vaccine was found effective and, properly prepared, safe. In 1957, the Polish-American microbiologist Albert Bruce Sabin went a step further. He made use not of dead virus (which, when not entirely dead, could be dangerous) but of strains of living virus, incapable of producing the disease itself, but capable of bringing about the production of appropriate antibodies. Such a *Sabin vaccine* could be taken by mouth, moreover, and did not require the hypodermic. The Sabin vaccine gained popularity first in the Soviet Union and then in eastern European countries; but by 1960, it came into use in the United States as well, and the fear of poliomyelitis has lifted.

ANTIBODIES

What does a vaccine do, exactly? The answer to this question may some day give us the chemical key to immunity.

For more than half a century, biologists have known the body's main defenses against infection as *antibodies*. (Of course, there are also the white-blood cells called *phagocytes*, which devour bacteria—as was discovered in 1883 by the Russian biologist Ilya Ilitch Mechnikov, who later succeeded Pasteur as the head of the Pasteur Institute in Paris and shared the 1908 Nobel Prize in medicine and physiology with Ehrlich. But phagocytes are no help against viruses and seem not to be involved in the immunity process I am considering.) A virus, or indeed almost any foreign substance entering into the body's chemistry, is called an *antigen*. The antibody is a substance manufactured by the body to fight the specific antigen: it puts the antigen out of action by combining with it.

Long before the chemists actually ran down an antibody, they were pretty sure the antibodies must be proteins. For one thing, the best-known antigens were proteins, and presumably it would take a protein to catch a protein. Only a protein could have the subtlety of structure necessary to single out and combine with a particular antigen.

Early in the 1920s, Landsteiner (the discoverer of blood groups) carried out a series of experiments that clearly showed antibodies to be very specific indeed. The substances he used to generate antibodies were not antigens but much simpler compounds whose structure was well known. They were arsenic-containing compounds called *arsanilic acids*. In combination with a simple protein, such as the albumin of egg white, an arsanilic acid acted as an antigen: when injected into an animal, it gave rise to an antibody in the blood serum. Furthermore, this antibody was specific for the arsanilic acid; the blood serum of the animal would clump only the arsanilic-albumin combination, not albumin alone. Indeed, sometimes the antibody could be made to react with just an arsanilic acid, not combined with albumin. Landsteiner also showed that very small changes in the structure of the arsanilic acid would be reflected in the antibody. An antibody evoked by one variety of arsanilic acid would not react with a slightly altered variety.

Landsteiner coined the name *haptens* (from a Greek word meaning "to bind") for compounds, such as the arsanilic acids, that can give rise to antibodies when they are combined with protein. Presumably, each natural antigen has a specific region in its molecule that acts as a hapten. On that theory, a germ or a virus that can serve as a vaccine is one that has had its structure changed sufficiently to reduce its ability to damage cells but still has its hapten group intact, so that it can cause the formation of a specific antibody. In the early 1980s, a group headed by Richard A. Lerner prepared a synthetic vaccine by using a synthetic protein modeled on the flu virus. This synthetic vaccine immunized guinea pigs against the disease.

It would be interesting to learn the chemical nature of the natural haptens. If that could be determined, it might be possible to use a hapten, perhaps in combination with some harmless protein, to serve as a vaccine giving rise to antibodies for a specific antigen. That would avoid the necessity of resorting to toxins or attenuated viruses, which always carries some small risk.

Just how does an antigen evoke an antibody? Ehrlich believed that the body normally contains a small supply of all the antibodies it may need, and that when an invading antigen reacts with the appropriate antibody, this stimulates the body to produce an extra supply of that particular antibody. Some immunologists still adhere to that theory or to modifications of it. Yet it seems highly unlikely that the body is prepared with specific antibodies for all the possible antigens, including unnatural substances such as the arsanilic acids.

The alternate suggestion is that the body has some generalized protein molecule that can be molded to fit any antigen. The antigen, then, acts as a template to shape the specific antibody formed in response to it. Pauling proposed such a theory in 1940. He suggested that the specific antibodies are varying versions of the same basic molecule, merely folded in different ways. In other words, the antibody is molded to fit its antigen as a glove fits the hand.

By 1969, however, the advance of protein analysis had made it possible for a team under Gerald Maurice Edelman to work out the amino-acid structure of a typical antibody made up of well over 1,000 amino acids. For this work, he received a share of the 1972 Nobel Prize for physiology and medicine.

J. Donald Capra went on to show that there were *hypervariable* regions in the amino-acid chain. Apparently, the relatively constant sections of the chain serve to form a three-dimensional structure that holds the hypervariable region, which can itself be designed to fit a particular antigen by a combination of changes in particular amino acids within the chain and by changes in geometrical configuration.

By the act of combination, an antibody can neutralize a toxin and make it impossible for it to participate in whatever reactions serve to harm the body. An antibody might also combine with regions on the surface of a virus or a bacterium. If it has the capacity to combine with two different spots, and one is on the surface of one microorganism and the other on the surface of a second, an antibody can initiate the process of agglutination in which the microorganisms stick together and lose their ability to multiply or to enter cells.

The antibody combination may serve to label cells it involves, so that phagocytes more easily engulf it. The antibody combination may also serve to activate the complement system which can then utilize its enzyme components to puncture the wall of the intruding cell and thus destroy it.

The very specificity of antibodies is a disadvantage in some ways. Suppose a virus mutates so that its protein has a slightly different structure. The old antibody for the virus often will not fit the new structure. It follows that immunity against one strain of virus is no safeguard against another strain. The virus of influenza and of the common cold are particularly susceptible to minor mutations—one reason that we are plagued by frequent recurrences of these diseases. Influenza, in particular, will occasionally develop a mutant of extraordinary virulence, which may then sweep a surprised and nonimmune world— as happened in 1918 and, with much less fatal result, in the *Asian flu* pandemic of 1957.

A still more annoying effect of the body's oversharp efficiency in forming antibodies is its tendency to produce them even against a harmless protein that happens to enter the body. The body then becomes *sensitized* to that protein and may react violently to any later incursion of the originally innocent protein. The reaction may take the form of itching, tears, production of mucus in the nose and throat, asthma, and so on. Such *allergic reactions* are evoked by the pollen of certain plants (causing hay fever), by certain foods, by the fur or dandruff of animals, and so on. An allergic reaction may be acute enough to cause serious disablement or even death. The discovery of such *anaphylactic shock* won for the French physiologist Charles Robert Richet the Nobel Prize in medicine and physiology in 1913.

In a sense, every human being is more or less allergic to every other human being. A transplant, or graft, from one individual to another will not take, because the receiver's body treats the transplanted tissue as foreign protein and manufactures antibodies against it. The person-to-person graft that will work best is from one identical twin to the other. Since their identical heredity gives them exactly the same proteins, they can exchange tissues or even a whole organ, such as a kidney.

The first successful kidney transplant took place in December 1954 in Boston, from one identical twin to another. The receiver died in 1962 at the age of thirty of coronary artery disease. Since then, hundreds of individuals have lived for months and even years with kidneys transplanted from *other* than identical twins.

Attempts at transplanting other organs, such as the lungs or the liver, have been made, but what most caught the public fancy was the heart transplant. The first reasonably successful heart transplants were conducted in December 1967 by the South African surgeon Christiaan Barnard. The fortunate receiver —Philip Blaiberg, a retired South African dentist—lived for many months on someone else's heart.

For a while afterward, heart transplants became the rage, but the furor by late 1969 had died down. Few receivers lived very long, for the problems of tissue rejection seemed overwhelming, despite massive attempts to solve the reluctance of the body to incorporate any tissue but its own.

The Australian bacteriologist Macfarlane Burnet had suggested that embryonic tissues might be *immunized* to foreign tissues and that the free-living animal might then tolerate grafts of that tissue. The British biologist Peter Medawar demonstrated this to be so, using mouse embryos. The two men shared in the 1960 Nobel Prize in medicine and physiology as a result.

In 1962, a French-Australian immunologist, Jacques Francis Albert Pierre Miller, working in England, went even further and discovered what may be the reason for this ability to work with embryos in order to make future toleration possible. He discovered that the thymus gland (a piece of tissue which until then had had no known use) was the tissue capable of forming antibodies. If the thymus gland was removed from mice at birth, those mice died after three or four months out of sheer incapacity to protect themselves against the environment. If the thymus was allowed to remain in the mice for three weeks, it already had time to bring about the development of antibody-producing cells in the body, and might then be removed without harm. Embryos in which the thymus has not yet done its work may be so treated as to "learn" to tolerate foreign tissue; the day may yet come when, by the way of the thymus, we may improve tissue toleration, when desirable, perhaps even in adults.

And yet, even if the problem of tissue rejection were surmounted, there would remain serious problems. After all, every person who receives a living organ must receive it from someone who is giving it up, and the question arises when the prospective donor may be considered dead enough to yield up his or her organs.

In that respect it might prove better if mechanical organs were prepared which would involve neither tissue rejection nor knotty ethical issues. Artificial kidneys became practical in the 1940s, and it is possible for patients without natural kidney function to visit a hospital once or twice a week and have their blood cleansed of wastes. It makes for a restricted life even for those fortunate enough to be serviced, but it is preferable to death.

In the 1940s, researchers found that allergic reactions are brought about by the liberation of small quantities of a substance called *histamine* into the blood-stream. This led to the successful search for neutralizing *antihistamines,* which can relieve the allergic symptoms but, of course, do not remove the allergy. The first successful antihistamine was produced at the Pasteur Institute in Paris in 1937 by the Swiss-born chemist Daniel Bovet, who for this and subsequent researches in chemotherapy was awarded the Nobel Prize in physiology and medicine in 1957.

Noting that sniffling and other allergic symptoms were much like those of the common cold, pharmaceutical firms decided that what works for one ought to work for the other, and, in 1949 and 1950, flooded the country with antihistamine tablets. (The tablets turned out to do little or nothing for colds, and their vogue diminished.)

Allergies do their maximum harm when the body becomes allergic to one or another of its own proteins. Ordinarily, the body adjusts to its own proteins in the course of its development from a fertilized egg; but on occasion, some of this adjustment is lost. The reason may be that the body manufactures antibodies against a foreign protein that, in some respects, is uncomfortably close in structure to one of the body's own; or it may be that with age enough changes take place in the surface of certain cells that they begin to seem foreign to the antibody cells; or certain obscure viruses which, on infection, do little or no harm to the cells ordinarily, may produce subtle changes in the surface. The result is *autoimmune disease.*

In any case, autoimmune responses figure more commonly in human disorders than had been realized until recently. While most autoimmune diseases are uncommon, it may be that rheumatoid arthritis is one. Treatment for such diseases is difficult, but hope naturally improves if we know the cause and, therefore, the direction in which to look for effective treatment.

In 1937, thanks to the protein-isolating techniques of electrophoresis, biologists finally tracked down the physical location of antibodies in the blood. The antibodies were located in the blood fraction called *gamma globulin.*

Physicians have long been aware that some children are unable to form antibodies and therefore are easy prey to infection. In 1951, doctors at the Walter Reed Hospital in Washington made an electrophoretic analysis of the plasma of an eight-year-old boy suffering from a serious *septicemia* ("blood poisoning") and, to their astonishment, discovered that his blood had no gamma globulin at all. Other cases were quickly discovered. Investigators established that this lack is due to an inborn defect of metabolism which deprives the person of the ability to make gamma globulin; it is called *agammaglobulinemia.* Such persons cannot develop immunity to bacteria. They can now be kept alive, however, by antibiotics. Surprisingly enough, they *are* able to become immune to virus infections, such as measles and chickenpox, after having the disease once. Apparently, antibodies are not the body's only defense against viruses.

In 1957, a group of British bacteriologists, headed by Alick Isaacs, showed that cells, under the stimulus of a virus invasion, liberates a protein that has broad antiviral properties. It counters not only the virus involved in the immediate infection but other viruses as well. This protein, named *interferon,* is produced more quickly than antibodies are and may explain the antivirus defenses of those with agammaglobulinemia. Apparently its production is stimulated by the presence of RNA in the double-stranded variety found in viruses. Interferon seems to direct the synthesis of a messenger-RNA that produces an antivirus protein that inhibits production of virus protein but not of other forms of protein. Interferon seems to be as potent as antibiotics and does not activate resistance. It is, however, fairly species-specific. Only interferon from humans and from other primates will work on human beings.

The fact that human, or near-human, interferon is required and that human cells produce it in only trace quantities, has made it impossible for a long time to obtain the material in amounts sufficient to make it clinically useful.

Beginning in 1977, however, Sydney Pestka at the Roche Institute of Molecular Biology worked on methods for purifying interferon. This was done, and interferon was found to exist as several closely allied proteins. The first *alpha-interferon* to be purified had a molecular weight of 17,500 and consisted of a chain of 166 amino acids. The amino-acid sequence of a dozen different interferon species were worked out, and there were only relatively minor differences among them.

The gene responsible for the formation of interferon was located and, by means of recombinant-DNA techniques, was inserted into the common bacterium *Escherichia coli*. A colony of these bacteria was thus induced to form human interferon in very pure form, so that it could be isolated and crystallized. The crystals could be analyzed by X rays, and the three-dimensional structure determined.

By 1981, enough interferon was on hand for clinical trials. No miracles resulted but it takes time to work out appropriate procedures.

New infectious diseases occasionally make their appearance. The 1980s saw a frightening one called *acquired immune deficiency syndrome* (AIDS) in which the immune mechanism breaks down and a simple infection can kill. The disease, attacking chiefly male homosexuals, Haitians, and those receiving blood transfusions, is spreading rapidly and is usually fatal. So far it is incurable, but in 1984, the virus causing it was isolated in France and in the United States and that is a first step forward.

Cancer

As the danger of infectious diseases diminishes, the incidence of other types of disease increases. Many people who a century ago would have died young of tuberculosis or diphtheria or pneumonia or typhus now live long enough to die of heart disease or cancer. That is one reason heart disease and cancer have become, respectively, the number-one and the number-two killers in the Western world. Cancer, in fact, has succeeded plague and smallpox as a universal fear. It is a nightmare hanging over all of us, ready to strike anyone without warning or mercy. Three hundred thousand Americans die of it each year, while 10,000 new cases are recorded each week. The incidence has risen 50 percent since 1900.

Cancer is actually a group of many diseases (about 200 types are known), affecting various parts of the body in various fashions. But the primary disorder is always the same: disorganization and uncontrolled growth of the affected tissues. The name *cancer* (the Latin word for "crab") comes from the fact that Hippocrates and Galen fancied the disease spreading its ravages through diseased veins like the crooked, outstretched claws of a crab.

Tumor (from the Latin word meaning "grow") is by no means synonymous with cancer but applies to harmless growths such as warts and moles (*benign tumors*) as well as to cancers (*malignant tumors*). The cancers are variously named according to the tissues affected. Cancers of the skin or the intestinal linings (the most common malignancies) are called *carcinomas* (from the Greek word for "crab"); cancers of the connective tissues are *sarcomas;* of the liver, *hepatoma;* of glands generally, *adenomas;* of the white blood cells, *leukemia;* and so on.

Rudolf Virchow of Germany, the first to study cancer tissue under the microscope, believed that cancer was caused by the irritations and shocks of the outer environment. This is a natural thought, for it is just those parts of the body most exposed to the outer world that are most subject to cancer. But when the germ theory of disease became popular, pathologists began to look for some microbe as the cause of cancer. Virchow, a staunch opponent of the germ theory of disease, stubbornly insisted on the irritation theory. (He quit pathology for archaeology and politics when it turned out that the germ theory of disease was going to win out. Few scientists in history have gone down with the ship of mistaken beliefs in quite so drastic a fashion.)

If Virchow was stubborn for the wrong reason, he may have been so in the right cause. There has been increasing evidence that some environments are particularly conducive to cancer. In the eighteenth century, chimney sweeps were found to be more prone to cancer of the scrotum than other people were. After the coal-tar dyes were developed, workers in the dye industries showed an above-average incidence of cancers of the skin or bladder. It seemed that something in soot and in the aniline dyes must be capable of causing cancer. Then, in 1915, two Japanese scientists, K. Yamagiwa and K. Ichikawa, discovered that a certain coal-tar fraction could produce cancer in rabbits when applied to the rabbits' ears for long periods. In 1930, two British chemists induced cancer in animals with a synthetic chemical called *dibenzanthracene* (a hydrocarbon with a molecule made up of five benzene rings). This does not occur in coal tar; but three years later, it was discovered that *benzpyrene* (also containing five benzene rings but in a different arrangement), a chemical that *does* occur in coal tar, can cause cancer.

Quite a number of *carcinogens* (cancer producers) have now been identified. Many are hydrocarbons made up of numerous benzene rings, like the first two discovered. Some are molecules related to the aniline dyes. In fact, one of the chief concerns about using artificial dyes in foods is the possibility that in the long run such dyes may be carcinogenic.

Many biologists believe that human beings have introduced a number of new cancer-producing factors into the environment within the last two or three centuries. There is the increased use of coal; there is the burning of oil on a large scale, particularly in gasoline engines; there is the growing use of synthetic chemicals in food, cosmetics, and so on. The most clearly guilty of the suspects, of course, is cigarette smoking, which is accompanied by a relatively high rate of incidence of lung cancer.

THE EFFECTS OF RADIATION

One other environmental factor that is certainly carcinogenic is energetic radiation, to which human beings have been exposed in increasing measure since 1895.

On 5 November 1895, the German physicist Wilhelm Konrad Roentgen performed an experiment to study the luminescence produced by cathode rays. The better to see the effect, he darkened the room. His cathode-ray tube was enclosed in a black cardboard box. When he turned on the cathode-ray tube, he was startled to catch a flash of light from something across the room. The flash came from a sheet of paper coated with barium platinocyanide, a luminescent chemical. Was it possible that radiation from the closed box had made it glow? Roentgen turned off his cathode-ray tube, and the glow stopped. He turned it on again—the glow returned. He took the paper into the next room, and it still glowed. Clearly, the cathode-ray tube was producing some form of radiation which could penetrate cardboard and walls.

Roentgen, having no idea what kind of radiation this might be, called it simply X rays. Other scientists tried to change the name to *Roentgen rays*, but this was so hard for anyone but Germans to pronounce that *X rays* stuck. (We now know that the speeding electrons making up the cathode rays are strongly decelerated on striking a metal barrier. The kinetic energy lost is converted into radiation that is called *Bremsstrahlung*—German for "braking radiation." X rays are an example of such radiation.)

The X rays revolutionized physics: they captured the imagination of physicists, started a typhoon of experiments, led within a few months to the discovery of radioactivity, and opened up the inner world of the atom. When the award of Nobel Prizes began in 1901, Roentgen was the first to receive the prize in physics.

The hard X radiation also started something else—exposure of human beings to intensities of energetic radiation such as they had never experienced before. Four days after the news of Roentgen's discovery reached the United States, X rays were used to locate a bullet in a patient's leg. They were a wonderful means of exploring the interior of the body. X rays pass easily through the soft tissues (consisting chiefly of elements of low atomic weight) and tend to be stopped by elements of higher atomic weight, such as make up the bones (composed largely of phosphorus and calcium). On a photographic plate placed behind the body, bone shows up as a cloudy white, in contrast to the black areas where X rays have come through in greater intensity because they have been much less absorbed by the soft tissues. A lead bullet shows up as pure white; it stops the X rays completely.

X rays are obviously useful for showing bone fractures, calcified joints, cavities in the teeth, foreign objects in the body, and so on. But it is also a simple matter to outline the soft tissues by introducing an insoluble salt of a heavy element. Barium sulfate, when swallowed, will outline the stomach or intestines. An iodine compound injected into the blood will travel to the kidneys and the ureter and outline those organs, for iodine has a high atomic weight and therefore is opaque to X rays.

Even before X rays were discovered, a Danish physician, Niels Ryberg Finsen, had found that high-energy radiation could kill microorganisms; he used ultraviolet light to destroy the bacteria causing *lupus vulgaris*, a skin disease. (For this he was awarded the Nobel Prize in physiology and medicine in 1903.) The X rays turned out to be far more deadly: they could kill the fungus of ringworm; they could damage or destroy human cells and were eventually used to kill cancer cells beyond reach of the surgeon's knife.

What was also discovered—the hard way—was that high-energy radiation could *cause* cancer. At least one hundred of the early workers with X rays and radioactive materials died of cancer, the first death taking place in 1902. As a matter of fact, both Marie Curie and her daughter, Irène Joliot-Curie, died of leukemia, and it is easy to believe that radiation was a contributing cause in both cases. In 1928, a British physician, George William Marshall Findlay, found that even ultraviolet radiation was energetic enough to cause skin cancer in mice.

It is certainly reasonable to suspect that the increasing exposure of human beings to energetic radiation (in the form of medical X rays, nuclear experimentation, and so on) may be responsible for part of the increased incidence of cancer.

MUTAGENS AND ONCOGENES

What can all the various carcinogens—chemicals, radiation, and so on—possibly have in common? One reasonable thought is that all of them may cause genetic mutations, and that cancer may be the result of mutations in body cells. This notion was first suggested by the German zoologist Theodor Boveri in 1914.

After all, suppose that some gene is changed so that it no longer can produce a key enzyme needed in the process that controls the growth of cells. When a cell with such a defective gene divides, it will pass on the defect. With the control mechanism not functioning, further division of these cells may continue indefinitely, without regard to the needs of the body as a whole or even to the needs of the tissue involved (for example, the specialization of cells in an organ). The tissue is disorganized. It is, so to speak, a case of anarchy in the body.

That energetic radiation can produce mutations is well established. What about the chemical carcinogens? Well, mutation by chemicals also has been demonstrated. The *nitrogen mustards* are a clear example. These compounds, like the *mustard gas* of the First World War, produce on the skin burns and blisters resembling those caused by X rays, and can also damage the chromosomes and increase the mutation rate. Moreover, a number of other chemicals have been found to imitate energetic radiation in the same way.

The chemicals that can induce mutations are called *mutagens*. Not all mutagens have been shown to be carcinogens, and not all carcinogens have been shown to be mutagens. But there are enough cases of compounds that are both carcinogenic and mutagenic to arouse suspicion that their relationship is more than coincidental.

Beginning in 1960, scientists began to search for nonrandom changes in chromosomes in tumor cells, as compared with normal ones. Changes were indeed found and were pinpointed more surely when techniques were developed to form hybrid mouse/human cells. Such hybrid cells would contain relatively few of the human chromosomes; and if one of those suspected in activating tumors was included, it would give rise to a tumor when the hybrid cell was injected into a mouse.

Further investigations pinned the cancerous change to a single gene on such a chromosome, when a group at the Massachusetts Institute of Technology, under Robert A. Weinberg, successfully produced tumors in mice, in 1978, by the transfer of individual genes. These were called *oncogenes* (the prefix *onco-*, from a Greek word meaning "mass," is commonly used in medical terminology for "tumor").

The oncogene was found to be very similar to a normal gene. The two might differ, in fact, in a single amino acid along the chain. The picture therefore arises of a *proto-oncogene*, a normal gene, which exists in cells and is passed along with every generation of cell division, and which, through any number of different influences may, at any time, undergo some small change that will make it an active oncogene. (One may well wonder about the purpose of having a proto-oncogene hanging around a cell when the potential danger is so great. There is no answer yet, but at least we have a new direction of investigation and attack, and that is no small thing.)

THE VIRUS THEORY

Meanwhile, the notion that microorganisms may have something to do with cancer is far from dead. With the discovery of viruses, this suggestion of the Pasteur era was revived. In 1903, the French bacteriologist Amédée Borrel suggested that cancer might be a virus disease; and, in 1908, two Danes, Wilhelm Ellerman and Olaf Bang, showed that fowl leukemia was indeed caused by a virus. However, leukemia was not at the time recognized as a form of cancer, and the issue hung fire. In 1909, however, the American physician Francis Peyton Rous ground up a chicken tumor, filtered it, and injected the clear filtrate into other chickens. Some of them developed tumors. The finer the filter, the fewer the tumors. It certainly looked as if particles of some kind were responsible for the initiation of tumors, and it seemed that these particles were the size of viruses.

The *tumor viruses* have had a rocky history. At first, the tumors pinned down to viruses turned out to be uniformly benign; for instance, viruses were shown to cause such things as rabbits' papillomas (similar to warts). In 1936, John Joseph Bittner, working in the famous mouse-breeding laboratory at Bar Harbor, Maine, came on something more exciting. Maude Slye of the same laboratory had bred strains of mice that seemed to have an inborn resistance to cancer, and other strains that seemed cancer-prone. The mice of some strains rarely developed cancer; those of other strains almost invariably did, after reaching maturity. Bittner tried the experiment of switching mothers on the newborn mice so that they would suckle at the opposite strain. He discovered that when baby mice of a "cancer-resistant" strain suckled at mothers of a "cancer-prone" strain, they usually developed cancer. On the other hand, supposedly cancer-prone baby mice that were fed by cancer-resistant mothers did not develop cancer. Bittner concluded that the cancer cause, whatever it was, was not inborn but was transmitted in the mother's milk. He called it the *milk factor*.

Naturally, Bittner's milk factor was suspected to be a virus. Eventually the Columbia University biochemist Samuel Graff identified the factor as a particle containing nucleic acids. Other tumor viruses, causing certain types of mouse tumors and animal leukemias, have been found, and all of them contain nucleic acids. No viruses have been detected in connection with human cancers, but research on human cancer is obviously limited.

Now the mutation and virus theories of cancer begin to converge. Perhaps the seeming contradiction between the two notions is not a contradiction after all. Viruses and genes are very similar in structure; and some viruses, on invading a cell, may become part of the cell's permanent equipment and may play the role of an oncogene.

To be sure, tumor viruses seem to possess RNA every time, while the human gene possesses DNA. As long as it was taken for granted that information always flows from DNA to RNA, it was hard to see how tumor viruses could play the role of genes. It is now known, however, that there are occasions when RNA can bring about the production of DNA that carries the RNA pattern of nucleotides. A tumor virus, therefore, may not *be* an oncogene, but it might *form* an oncogene.

For that matter, a virus may be less direct in its attack. It may merely play an important role in bringing about the conversion of the proto-oncogene to the oncogene.

It was not till 1966, however, that the virus hypothesis was deemed fruitful enough to be worth a Nobel Prize. Fortunately, Peyton Rous, who had made his discovery fifty-five years before, was still alive and received a share of the 1966 Nobel Prize for medicine and physiology. (He lived on to 1970, dying at the age of ninety, active in research nearly to the end.)

POSSIBLE CURES

What goes wrong in the metabolic machinery when cells grow unrestrainedly? This question has as yet received no answer. One strong suspicion rests on some of the hormones, especially the sex hormones.

For one thing, the sex hormones are known to stimulate rapid, localized growth in the body (as in the breasts of an adolescent girl). For another, the tissues of sexual organs—the breasts, cervix, and ovaries in a woman; the testes and prostate in a man—are particularly vulnerable to cancer. Strongest of all is the chemical evidence. In 1933, the German biochemist Heinrich Wieland (who had won the Nobel Prize in chemistry in 1927 for his work with bile acids) managed to convert a bile acid into a complex hydrocarbon called *methylcholanthrene*, a powerful carcinogen. Now methylcholanthrene (like the bile acids) has the four-ring structure of a steroid, and it so happens that all the sex hormones are steroids. Could a misshapen sex-hormone molecule act as a carcinogen? Or might even a correctly shaped hormone be mistaken for a carcinogen, so to speak, by a distorted gene pattern in a cell, and so stimulate uncontrolled growth? It is anyone's guess, but these are interesting speculations.

Curiously enough, changing the supply of sex hormones sometimes checks cancerous growth. For instance, castration, to reduce the body's manufacture of male sex hormone, or the administration of neutralizing female sex hormone, has a mitigating effect on cancer of the prostate. As a treatment, this is scarcely something to shout about, and it is a measure of the desperation regarding cancer that such devices are resorted to.

The main line of attack against cancer still is surgery. And its limitations are still what they have always been: sometimes the cancer cannot be cut out without killing the patient; often the knife frees bits of malignant tissue (since the disorganized cancer tissue has a tendency to fragment), which are then carried by the bloodstream to other parts of the body where they take root and grow.

The use of energetic radiation to kill the cancer tissue also has its drawbacks. Artificial radioactivity has added new weapons to the traditional X rays and radium. One of them is cobalt 60, which yields high-energy gamma rays and is much less expensive than radium; another is a solution of radioactive iodine (the "atomic cocktail"), which concentrates in the thyroid gland and thus attacks a thyroid cancer. But the body's tolerance of radiation is limited, and there is always the danger that the radiation will start more cancers than it stops.

The increasing knowledge of the last decade offers hope for some method of treatment that would be less drastic, more subtle, and more effective.

For instance, if viruses are involved in some fashion in the initiation of a cancer, any agent that would inhibit virus action might cut down the incidence of cancer or stop the growth of a cancer once it has started. The obvious possibility here is interferon; and now that interferon is available pure and in quantity, it has been tried on cancer patients. So far it has not been markedly successful; but because it is an experimental procedure, it has been tried only on patients who are far along in the disease and may be beyond help. Then, too, there may be subtleties to the method of use that have not yet been worked out.

Another approach is this: Oncogenes differ so slightly from normal genes that it seems reasonable to suppose that they are forming frequently, and that the production of a cancerous cell is more common than we suppose. Such a cell would have to be different in some ways from a normal one, and perhaps the body's immune system may recognize it early on and dispose of it. It may be, then, that the development of cancer means not that a cancer cell has formed, but that a cancer cell, having formed, has not been stopped. Perhaps cancer is the result of a failure of the immune system—in a way, the opposite of autoimmune disease, where the immune system works too well.

Prevention and cure may rest then with our increased understanding of the manner in which the immune system works. Or, until such an end is achieved, perhaps the body may be given artificial aid in the form of compounds that will distinguish between normal cells and cancer cells.

Some plants, for instance, produce substances that react with certain sugars, as antibodies react with certain proteins. (The purpose of such sugar-distinguishing substances in plants is not yet known.)

The membranes that enclose cells are made up of proteins and fatty substances, but the proteins usually incorporate into their structures certain moderately complex sugar molecules. Because the nature of the sugars in the membranes are different, blood cells are of several different types that can be distinguished by the fact that some types agglutinate under some conditions and some under others.

The American biochemist William Clouser Boyd wondered whether there might be plant substances that can distinguish between one blood group and another. In 1954, rather to his own surprise, he found such a substance in lima beans, which was among the first plants he tried. He named such substances *lectins,* from a Latin word meaning "to choose."

If a lectin can choose between one kind of red corpuscle and another, on the basis of subtle differences in surface chemistry, some lectins might be found that can distinguish between a tumor cell and its normal cell of origin, agglutinating the tumor cells and not the normal ones. Thus, these lectins might put the tumor cells out of action and slow or reverse the growth of cancer. Some preliminary investigations have yielded hopeful results.

Finally, the more we learn about oncogenes and their method of production, the greater the chance of our learning ways of preventing their appearance or of encouraging their disappearance.

Meanwhile, though, the fear of cancer and the apparent hopelessness of cure frequently causes the public to yearn after pseudo-scientific cures such as those attributed to the substances called *krebiozen* and *laetrile.* One can scarcely blame people who clutch at straws, but so far these substances have never helped and have sometimes prevented patients from seeking out more hopeful treatment.

Chapter 15

The Body

Food

Perhaps the first great advance in medical science was the recognition by physicians that good health calls for a simple, balanced diet. The Greek philosophers recommended moderation in eating and drinking, not only for philosophical reasons but also because those who followed this rule were more comfortable and lived longer. That was a good start, but biologists eventually learned that moderation alone is not enough. Even if one has the good fortune to avoid eating too little and the good sense to avoid eating too much, one will still do poorly on a diet that happens to be shy of certain essential ingredients, as is actually the case for large numbers of people in some parts of the world.

The human body is rather specialized (as organisms go) in its dietary needs. A plant can live on just carbon dioxide, water, and certain inorganic ions. Some of the microorganisms likewise get along without any organic food; they are called *autotrophic* ("self-feeding"), which means that they can grow in environments in which there is no other living thing. The bread mold *Neurospora* begins to get a little more complicated: in addition to inorganic substances, it has to have sugar and the vitamin biotin. And as the forms of life become more and more complex, they seem to become more and more dependent on their diet to supply the organic building blocks necessary for building living tissue. The reason is simply that they have lost some of the enzymes that primitive organisms possess. A green plant has a complete supply of enzymes for making all the necessary amino acids, proteins, fats, and carbohydrates from inorganic materials. *Neurospora* has all the enzymes except one or more of those needed to make sugar and biotin. By the time we get to humans, we find that they lack the enzymes required to make many of the amino acids, the vitamins, and various other necessities, and thus must get these ready-made in food.

This may seem a kind of degeneration—a growing dependence on the environment which puts the organism at a disadvantage. Not so. If the environment supplies the building blocks, why should the cell carry the elaborate enzymatic machinery needed to make them? By dispensing with this machinery, the cell can use its energy and space for more refined and specialized purposes.

For human beings (or other animals) to get the food they need, they must depend on physically ingesting other organisms. It is the organic constituents of those organisms that make up food. In the intestines of the eater, the small molecules of the eaten can be absorbed directly. The large molecules of starch, protein and so on are broken down, or *digested,* by means of enzyme action; and the fragments (amino acids, glucose, and so on) are absorbed. Within the eater's body, those fragments are further broken down for the production of energy, or put together again to form the large molecules characteristic of the eater, rather than of the eaten. In a sense, animal life is an endless burglary-by-force.

Some animals are *carnivorous* and eat only other animals. If all animals ate in this fashion, animal life would not long endure, for the transfer of energy and of tissue components from the eaten to the eater is very inefficient. As a general rule of thumb, it takes 10 pounds of the eaten to support 1 pound of the eater.

There are animals that are *herbivorous* and eat plants. Plant life is much more common than animal life, so that the total mass of herbivorous animals is far higher than that of carnivorous animals, and the former can better support the latter. (Some animals—like human beings, bears, and swine—are *omnivorous* and eat both plants and animals.)

The transfer of energy and of tissue components from plants to the animals that eat them is also highly inefficient, and life would soon dwindle to nothing if plants could not somehow renew themselves as fast as they were eaten. This they do by making use of solar energy in the process of photosynthesis (see chapter 12). In this way, plants live on the nonliving and keep virtually all of life going—and have been doing so for all the time they have existed.

To be sure photosynthesis is even less efficient than the eating processes of animals. It is estimated that less than one-tenth of 1 percent of all the solar energy that bathes the earth is trapped by plants and converted into tissue, but this is still enough to produce between 150 billion and 200 billion tons of dry organic matter every year the world over. Naturally, this process can only last on Earth in its present form as long as the sun remains in essentially its present form—a matter of some billions of years.

THE ORGANIC FOODS

If energy were all that were required of food, we would not really need very much food. Half a pound of butter would supply me with all I would need for a day's worth of energy at my sedentary occupation. However, food does not supply energy alone but is also a source of the building blocks I need to repair and rebuild my tissues, and these are found in a wide variety of places. Butter alone would not supply my needs in those respects.

It was the English physician William Prout (the same Prout who was a century ahead of his time in suggesting that all the elements are built from hydrogen) who first suggested that the organic foods could be divided into three types of substances, later named *carbohydrates, fats,* and *proteins.*

The chemists and biologists of the nineteenth century, notably Justus von Liebig of Germany, gradually worked out the nutritive properties of these foods. Protein, they found, is the most essential, and the organism could get along on it alone. The body cannot make protein from carbohydrate and fat, because these substances have no nitrogen, but it can make the necessary carbohydrates and fats from materials supplied by protein. Since protein is comparatively scarce in the environment, however, it would be wasteful to live on an all-protein diet—like stoking the fire with furniture when firewood is available.

Throughout history, and in many places even today, people have difficulty getting enough food. Either there is literally insufficient food to go around, as during the famine that follows a bad harvest; or else there are flaws in the distribution, either physical or economic, so that there are people who cannot obtain the food or who cannot afford to buy it.

Even when there seems to be enough food to chew on, the protein content may be too low, so that though there is no *undernutrition* in the broad sense of the word, there is *malnutrition.* Children, especially, may suffer from protein deficiency, since they need protein not merely for replacement but for the building of new tissue, for growth. In Africa, such protein deficiency is particularly common among children forced to exist on a monotonous diet of cornmeal. (Any monotonous diet is dangerous, for few items of food have everything one needs. In variety, there is safety.)

There have always been a minority of people who can eat freely, and who therefore do and take in more of everything than they need. The body stores the excess as fat (the most economical way of packing calories into as little space as possible), and this is useful in many ways—as a store of calories to tide one over during a period when little food is available. If, however, there are no such periods, the fat remains, and a person is overweight or, in the extreme, obese. This state has its evils and discomforts and is associated with a greeater incidence of degenerative and metabolic diseases, such as diabetes, atherosclerosis, and so on. (And even overweight does not ensure against deficiencies in needed nutrients if one's diet is not properly balanced.)

In a country like the United States, where the standard of living is unusually high, and where fatness is considered unesthetic, overweight is a serious problem. The only rational way to prevent this is to cut down food intake or increase activity (or both), and people who refuse to do either are doomed to remain overweight, regardless of what tricks they try.

PROTEINS

On the whole, foods high in protein tend to be more expensive and in shorter supply (those two characteristics usually go together) than those low in protein. In general, animal food tends to be higher in protein than plant food does.

This creates a problem for those human beings who choose to be vegetarians. While vegetarianism has its points, those practicing it have to try a little harder to make sure they maintain an adequate protein intake. It can be done, for as little as 2 ounces of protein per day might be enough for the average adult. Children and pregnant or nursing mothers need somewhat more.

Of course, a lot depends on the proteins you choose. Nineteenth-century experimenters tried to find out whether the population could get along, in times of famine, on *gelatin*—a protein material obtained by heating bones, tendons, and other otherwise inedible parts of animals. But the French physiologist François Magendie demonstrated that dogs lost weight and died when gelatin was their sole source of protein. This does not mean there is anything wrong with gelatin as a food, but it simply does not supply all the necessary building blocks when it is the only protein in the diet. Again, in variety lies safety.

The key to the usefulness of a protein lies in the efficiency with which the body can use the nitrogen it supplies. In 1854, the English agriculturists John Bennet Lawes and Joseph Henry Gilbert fed pigs protein in two forms—lentil meal and barley meal. They found that the pigs retained much more of the nitrogen in barley than of that in lentils. These were the first *nitrogen balance* experiments.

A growing organism gradually accumulates nitrogen from the food it ingests (*positive nitrogen balance*). If it is starving or suffering a wasting disease, and gelatin is the sole source of protein, the body continues to starve or waste away, from a nitrogen-balance standpoint (a situation called *negative nitrogen balance*). It keeps losing more nitrogen than it takes in, regardless of how much gelatin it is fed.

Why so? The nineteenth-century chemists eventually discovered that gelatin is an unusually simple protein. It lacks tryptophan and other amino acids present in most proteins. Without these building blocks, the body cannot build the proteins it needs for its own substance. Therefore, unless it gets other protein in its food as well, the amino acids that do occur in the gelatin are useless and have to be excreted. It is as if housebuilders found themselves with plenty of lumber but no nails. Not only could they not build the house, but the lumber would just be in the way and eventually would have to be disposed of. Attempts were made in the 1890s to make gelatin a more efficient article of diet by adding some of those amino acids in which it was deficient, but without success. Better results were obtained with proteins not as drastically limited as gelatin.

In 1906, the English biochemists Frederick Gowland Hopkins and Edith Gertrude Willcock fed mice a diet in which the only protein was *zein*, found in corn. They knew that this protein had very little of the amino acid tryptophan. The mice died in about fourteen days. (It is the lack of tryptophan that is the chief cause of the protein-deficiency disease *Kwashiorkor*, common among African children.) The experimenters then tried mice on zein plus tryptophan. This time the mice survived twice as long. It was the first hard evidence that amino acids, rather than protein, might be the essential components of the diet. (Although the mice still died prematurely, this was probably due mainly to a lack of certain vitamins not known at the time.)

In the 1930s, the American nutritionist William Cumming Rose got to the bottom of the amino-acid problem. By that time the major vitamins were known, so he could supply the animals with those needs and focus on the amino acids. Rose fed rats a mixture of amino acids instead of protein. The rats did not live long on this diet. But when he fed rats on the milk protein casein, they did well. Apparently there was something in casein—some undiscovered amino acid, in all probability—which was not present in the amino-acid mixture he was using. Rose broke down the casein and tried adding various of its molecular fragments to his amino-acid mixture. In this way he tracked down the amino acid *threonine*, the last of the major amino acids to be discovered. When he added the threonine from casein to his amino-acid mixture, the rats grew satisfactorily, without any intact protein in the diet.

Rose proceeded to remove the amino acids from their diet one at a time. By this method he eventually identified ten amino acids as indispensable items in the diet of the rat: lysine, tryptophan, histidine, phenylalanine, leucine, isoleucine, threonine, methionine, valine, and arginine. If supplied with ample quantities of these, the rat could manufacture all it needed of the others, such as glycine, proline, aspartic acid, alanine, and so on.

In the 1940s, Rose turned his attention to human requirements of amino acids. He persuaded graduate students to submit to controlled diets in which a mixture of amino acids was the only source of nitrogen. By 1949, he was able to announce that the adult male required only eight amino acids in the diet: phenylalanine, leucine, isoleucine, methionine, valine, lysine, tryptophan, and threonine. Since arginine and histidine, indispensable to the rat, are dispensable in the human diet, it would seem that in this respect the human being is less specialized than the rat, or, indeed, than any other mammal that has been tested in detail.

Potentially a person could get along on the eight dietarily essential amino acids; given enough of these, he can make not only all the other amino acids he needs but also all the carbohydrates and fats. Actually a diet made up only of amino acids would be much too expensive, to say nothing of its flatness and monotony. But it is enormously helpful to have a complete blueprint of our amino-acid needs so that we can reinforce natural proteins when necessary for maximum efficiency in absorbing and utilizing nitrogen.

FATS

Fats, too, can be broken down to simpler building blocks, of which the chief are *fatty acids*. Fatty acids can be *saturated*, with the molecules containing all the hydrogen atoms they can carry; or *unsaturated*, with one or more pairs of hydrogen atoms missing. If more than one pair are missing, they are *polyunsaturated*.

Fats containing unsaturated fatty acids tend to melt at lower temperatures than those containing saturated fatty acids. In the organism, it is desirable to have fat in a liquid state; thus, plants and cold-blooded animals tend to have fat that is more unsaturated than that fat in birds and mammals, which are warm-blooded. The human body cannot make polyunsaturated fats out of saturated ones, and so the polyunsaturated fatty acids are *essential fatty acids*. In this respect, vegetarians have an advantage and are less likely to suffer a deficiency.

Vitamins

Food fads and superstitions unhappily still delude too many people—and spawn too many cure-everything best sellers—even in these enlightened times. In fact, it is perhaps because these times are enlightened that food faddism is possible. Through most of human history, people's food consisted of whatever could be produced in the vicinity, of which there usually was not very much. It was eat what there was to eat or starve; no one could afford to be picky, and without pickiness there can be no food faddism.

Modern transportation has made it possible to ship food from any part of the earth to any other, particularly with the use of large-scale refrigeration, and thus reduced the threat of famine. Before modern times, famine was invariably local; neighboring provinces could be loaded with food that could not be transported to the famine area.

Home storage of a variety of foods became possible as early humans learned to preserve foods by drying, salting, increasing the sugar content, fermenting, and so on. It became possible to preserve food in states closer to the original when methods of storing cooked food in vacuum were developed. (The cooking kills microorganisms, and the vacuum prevents others from growing and reproducing.) Vacuum storage was first made practical by a French chef, François Appert, who developed the technique in response to a prize offered by Napoleon I for a way of preserving food for his armies. Appert made use of glass jars; but nowadays, tin-lined steel cans (inappropriately called *tin cans* or, in Great Britain, just *tins*) are used for the purpose. Since the Second World War, fresh-frozen food has become popular, and the growing number of home freezers has further increased the general availability and variety of fresh foods. Each broadening of food availability has increased the practicality of food faddism.

DEFICIENCY DISEASES

All this is not to say that a shrewd choice of food may not be useful. There are certain cases in which specific foods will definitely cure a particular disease. In every instance, these are *deficiency diseases*—diseases that occur when food, and even protein, is ample. They are produced by the lack in the diet of some substance essential to the body's chemical machinery in tiny amounts —yet where these tiny amounts are not present in the diet. Such deficiency diseases arise almost invariably when a person is deprived of a normal, balanced diet—one containing a wide variety of foods.

To be sure, the value of a balanced and variegated diet was understood by a number of medical practitioners of the nineteenth century and before, when the chemistry of food was still a mystery. A famous example is that of Florence Nightingale, the heroic English nurse of the Crimean War who pioneered the adequate feeding of soldiers, as well as decent medical care. And yet *dietetics* (the systematic study of diet) had to await the end of the century and the discovery of trace substances in food, essential to life.

The ancient world was well acquainted with *scurvy,* a disease in which the capillaries become increasingly fragile, gums bleed and teeth loosen, wounds heal with difficulty if at all, and the patient grows weak and eventually dies. It was particularly prevalent in besieged cities and on long ocean voyages. (It first made its appearance on shipboard during Vasco da Gama's voyage around Africa to India in 1497; and Magellan's crew, during the first circumnavigation of the world a generation later, suffered more from scurvy than from general undernourishment.) Ships on long voyages, lacking refrigeration, had to carry nonspoilable food, which meant hardtack and salt pork. Nevertheless, physicians for many centuries failed to connect scurvy with diet.

In 1536, while the French explorer Jacques Cartier was wintering in Canada, 110 of his men were stricken with scurvy. The native Indians knew and suggested a remedy: drinking water in which pine needles had been soaked. Cartier's men in desperation followed this seemingly childish suggestion. It cured them of their scurvy.

Two centuries later, in 1747, the Scottish physician James Lind took note of several incidents of this kind and experimented with fresh fruits and vegetables as a cure. Trying his treatments on scurvy-ridden sailors, he found that oranges and lemons brought about improvement most quickly. Captain Cook, on a voyage of exploration across the Pacific from 1772 to 1775, kept his crew scurvy-free by enforcing the regular eating of sauerkraut. Nevertheless, it was not until 1795 that the brass hats of the British navy were sufficiently impressed by Lind's experiments (and by the fact that a scurvy-ridden flotilla could lose a naval engagement with scarcely a fight) to order daily rations of lime juice for British sailors. (They have been called *limeys* ever since, and the Thames area in London where the crates of limes were stored is still called Limehouse.) Thanks to the lime juice, scurvy disappeared from the British navy.

A century later, in 1884, Admiral Kanehiro Takaki of the Japanese navy similarly introduced a broader diet into the rice monotony of his ships. The scourge of a disease known as *beri-beri* came to an end in the Japanese navy as a result.

In spite of occasional dietary victories of this kind (which no one could explain), nineteenth-century biologists refused to believe that a disease could be cured by diet, particularly after Pasteur's germ theory of disease came into its own. In 1896, however, a Dutch physician named Christiaan Eijkman convinced them almost against his own will.

Eijkman was sent to the Dutch East Indies to investigate beri-beri, which was endemic in those regions (and which, even today, when medicine knows its cause and cure, still kills 100,000 people a year). Takaki had stopped beri-beri by dietary measures; but the West, apparently, placed no stock in what might have seemed merely the mystic lore of the Orient.

Supposing beri-beri to be a germ disease, Eijkman took along some chickens as experimental animals in which to establish the germ. A highly fortunate piece of skulduggery upset his plans. Without warning, most of his chickens came down with a paralytic disease from which some died; but after about four months, those still surviving regained their health. Eijkman, mystified by failing to find any germ responsible for the attack, finally investigated the chickens' diet. He discovered that the person originally in charge of feeding the chickens had economized (and no doubt profited) by using scraps of leftover food, mostly polished rice, from the wards of the military hospital. It happened that after a few months a new cook had arrived and taken over the feeding of the chickens; he had put a stop to the petty graft and supplied the animals with the usual chicken feed, containing unhulled rice. It was then that the chickens had recovered.

Eijkman experimented. He put chickens on a polished-rice diet, and they fell sick. Back on the unhulled rice, they recovered. It was the first case of a deliberately produced dietary-deficiency disease. Eijkman decided that this *polyneuritis* that afflicted fowls was similar in symptoms to human beri-beri. Did human beings get beri-beri because they ate only polished rice?

For human consumption, rice was stripped of its hulls mainly so that it would keep better, for the rice germ removed with the hulls contains oils that go rancid easily. Eijkman and a co-worker, Gerrit Grijns, set out to see what it was in rice hulls that prevented beri-beri. They succeeded in dissolving the crucial factor out of the hulls with water, and found that it would pass through membranes that would not pass proteins. Evidently the substance in question must be a fairly small molecule. They could not, however, identify it.

Meanwhile other investigators were coming across other mysterious factors that seemed to be essential for life. In 1905, a Dutch nutritionist, Cornelis Adrianis Pekelharing, found that all his mice died within a month on an artificial diet that seemed ample as far as fats, carbohydrates, and proteins were concerned. But mice did fine when he added a few drops of milk to this diet. And in England, the biochemist Frederick Hopkins, who was demonstrating the importance of amino acids in the diet, carried out a series of experiments in which he, too, showed that something in the casein of milk would support growth if added to an artificial diet. This something was soluble in water. Even better than casein as the dietary supplement was a small amount of a yeast extract.

For their pioneer work in establishing that trace substances in the diet were essential to life, Eijkman and Hopkins shared the Nobel Prize in medicine and physiology in 1929.

ISOLATING VITAMINS

The next task was to isolate these vital trace factors in food. By 1912, three Japanese biochemists—Umetaro Suzuki, T. Shimamura, and S. Ohdake—had extracted from rice hulls a compound that was very potent in combating beri-beri. Doses of 5 to 10 milligrams sufficed to effect a cure in fowl. In the same year, the Polish-born biochemist Casimir Funk (then working in England and later to come to the United States) prepared the same compound from yeast.

Because the compound proved to be an amine (that is, one containing the amine group, NH_2), Funk called it a *vitamine*, Latin for "life amine." He made the guess that beri-beri, scurvy, pellagra, and rickets all arise from deficiencies of "vitamines." Funk's guess was correct as far as his identification of these diseases as dietary-deficiency diseases was concerned. But it turned out that not all "vitamines" were amines.

In 1913, two American biochemists, Elmer Vernon McCollum and Marguerite Davis, discovered another trace factor vital to health in butter and egg yolk. This one was soluble in fatty substances instead of water. McCollum called it *fat-soluble A*, to contrast it with *water-soluble B*, which was the name he applied to the antiberi-beri factor. In the absence of chemical information about the nature of the factors, this seemed fair enough, and it started the custom of naming them by letters. In 1920, the British biochemist Jack Cecil Drummond changed the names to *vitamin A* and *vitamin B*, dropping the final *e* of *vitamine* as a gesture toward taking *amine* out of the name. He also suggested that the antiscurvy factor was still a third such substance, which he named *vitamin C*.

Vitamin A was quickly identified as a food factor required to prevent the development of abnormal dryness of the membranes around the eye, called *xerophthalmia*, from Greek words meaning "dry eyes." In 1920, McCollum and his associates found that a substance in cod-liver oil, which was effective in curing both xerophthalmia and a bone disease called *rickets*, could be so treated as to cure rickets only. They decided the antirickets factor must represent a fourth vitamin, which they named *vitamin D*. Vitamins D and A are fat-soluble; C and B are water-soluble.

By 1930, it had become clear that vitamin B was not a simple substance but a mixture of compounds with different properties. The food factor that cured beri-beri was named *vitamin B_1*, a second factor was called *vitamin B_2*, and so on. Some of the reports of new factors turned out to be false alarms, so that one does not hear of B_3, B_4, or B_5 any longer. However, the numbers worked their way up to B_{14}. The whole group of vitamins (all water-soluble) is frequently referred to as the *B-vitamin complex*.

New letters also were added. Of these, *vitamins E* and *K* (both fat-soluble) remain as veritable vitamins; but *vitamin F* turned out to be not a vitamin, and *vitamin H* turned out to be one of the B-vitamin complex.

Nowadays, with their chemistry identified, the letters of even the true vitamins are going by the board, and most of them are known by their chemical names, though the fat-soluble vitamins, for some reason, have held on to their letter designations more tenaciously than the water-soluble ones.

CHEMICAL COMPOSITION AND STRUCTURE

It was not easy to work out the chemical composition and structure of the vitamins, for these substances occur only in minute amounts. For instance, a ton of rice hulls contains only about 5 grams (a little less than one-fifth of an ounce) of vitamin B_1. Not until 1926 did anyone extract enough of the reasonably pure vitamin to analyze it chemically. Two Dutch biochemists, Barend Coenraad Petrus Jansen and William Frederick Donath, worked up a composition for vitamin B, from a tiny sample, but it turned out to be wrong. In 1932, Ohdake tried again on a slightly larger sample and got it almost right. He was the first to detect a sulfur atom in a vitamin molecule.

Finally, in 1934, Robert Runnels Williams, then director of chemistry at the Bell Telephone Laboratories, climaxed twenty years of research by painstakingly separating vitamin B_1 from tons of rice hulls until he had enough to work out a complete structural formula. The formula follows:

Since the most unexpected feature of the molecule was the atom of sulfur (*theion* in Greek), the vitamin was named *thiamine*.

Vitamin C was a different sort of problem. Citrus fruits furnish a comparatively rich source of this material, but one difficulty was finding an experimental animal that does not make its own vitamin C. Most mammals, aside from humans and the other primates, have retained the capacity to form this vitamin. Without a cheap and simple experimental animal that would develop scurvy, it was difficult to follow the location of vitamin C among the various fractions into which the fruit juice was broken down chemically.

In 1918, the American biochemists B. Cohen and Lafayette Benedict Mendel solved this problem by discovering that guinea pigs cannot form the vitamin. In fact, guinea pigs develop scurvy much more easily than humans do. But another difficulty remained. Vitamin C was found to be very unstable (it is the most unstable of the vitamins), so it was easily lost in chemical procedures to isolate it. A number of research workers ardently pursued the vitamin without success.

As it happened, vitamin C was finally isolated by someone who was not particularly looking for it. In 1928, the Hungarian-born biochemist Albert Szent-Györgi, then working in London in Hopkins's laboratory and interested mainly in finding out how tissues make use of oxygen, isolated from cabbages a substance that helped transfer hydrogen atoms from one compound to another. Shortly afterward, Charles Glen King and his co-workers at the University of Pittsburgh, who *were* looking for vitamin C, prepared some of the substance from cabbages and found that it was strongly protective against scurvy. Furthermore, they found it identical with crystals they had obtained from lemon juice. King determined its structure in 1933, and it turned out to be a sugar molecule of six carbons, belonging to the L-series instead of the D-series:

$$O = C \underset{\underset{OH}{\diagup}}{\overset{\overset{O}{\diagdown}}{\underset{\diagdown}{C}}} = \underset{\overset{}{\underset{OH}{C}}}{\overset{O}{\diagup}} CH - \underset{\overset{|}{OH}}{CH} - CH_2OH$$

It was named *ascorbic acid* (from Greek words meaning "no scurvy").

As for vitamin A, the first hint about its structure came from the observation that the foods rich in vitamin A are often yellow or orange (butter, egg yolk, carrots, fish-liver oil, and so on). The substance largely responsible for this color was found to be a hydrocarbon named *carotene;* and in 1929, the British biochemist Thomas Moore demonstrated that rats fed on diets containing carotene stored vitamin A in the liver. The vitamin itself was not colored yellow, so the deduction was that though carotene is not itself vitamin A, the liver converts it into something that is vitamin A. (Carotene is now considered an example of a *provitamin.*)

In 1937, the American chemists Harry Nicholls Holmes and Ruth Elizabeth Corbet isolated vitamin A as crystals from fish-liver oil. It turned out to be a 20-carbon compound—half of the carotene molecule with a hydroxyl group added:

$$\begin{array}{c}
CH_3 \quad CH_3 \\
\backslash \; C \; / \\
CH_2 \quad\; C - CH = CH - \overset{CH_3}{\underset{\|}{C}} = CH - CH = CH - \overset{CH_3}{\underset{\|}{C}} = CH - CH_2 - OH \\
| \quad\quad \| \\
CH_2 \quad\; C - CH_3 \\
\backslash \; CH_2 \; /
\end{array}$$

The chemists hunting for vitamin D found their best chemical clue by means of sunlight. As early as 1921, the McCollum group (who first demonstrated the existence of the vitamin) showed that rats do not develop rickets on a diet lacking vitamin D if they are exposed to sunlight. Biochemists guessed that the energy of sunlight converts some provitamin in the body into vitamin D. Since vitamin D is fat-soluble, they went searching for the provitamin in the fatty substances of food.

By breaking down fats into fractions and exposing each fragment separately to sunlight, they determined that the provitamin that sunlight converts into vitamin D is a steroid. What steroid? They tested cholesterol, the most common steroid of the body, and that was not it. Then, in 1926, the British biochemists Otto Rosenheim and T. A. Webster found that sunlight would convert a closely related sterol, *ergosterol* (so named from the fact that it was first isolated from ergot-infested rye), into vitamin D. The German chemist Adolf Windaus made this discovery independently at about the same time. For this and other work in steroids, Windaus received the Nobel Prize in chemistry in 1928.

The difficulty in producing vitamin D from ergosterol rested on the fact that ergosterol does not occur in animals. Eventually the human provitamin was identified as *7-dehydrocholesterol,* which differs from cholesterol only in having two hydrogen atoms fewer in its molecule. The vitamin D formed from it has this formula:

Vitamin D in one of its forms is called *calciferol,* from Latin words meaning "calcium-carrying," because it is essential to the proper laying down of bone structure.

Not all the vitamins show their absence by producing an acute disease. In 1922, Herbert McLean Evans and K. J. Scott at the University of California implicated a vitamin as a cause of sterility in animals. Evans and his group did not succeed in isolating this one, vitamin E, until 1936. It was then given the name *tocopherol* (from Greek words meaning "to bear children").

Unfortunately, whether human beings need vitamin E, or how much, is not yet known. Obviously, dietary experiments designed to bring about sterility cannot be tried on human subjects. And even in animals, the fact that they can be made sterile by withholding vitamin E does not necessarily mean that natural sterility arises in this way.

In the 1930s, the Danish biochemist Carl Peter Henrik Dam discovered by experiments on chickens that a vitamin is involved in the clotting of blood. He named it *Koagulationsvitamine,* and this was eventually shortened to *vitamin K.* Edward Doisy and his associates at St. Louis University then isolated vitamin K and determined its structure. Dam and Doisy shared the Nobel Prize in medicine and physiology in 1943.

Vitamin K is not a major vitamin or a nutritional problem. Normally a more than adequate supply of this vitamin is manufactured by the bacteria in the intestines. In fact, they make so much of it that the feces may be richer in vitamin K than the food is. Newborn infants are the most likely to run a danger of poor blood clotting and consequent hemorrhage because of vitamin-K deficiency. In the hygienic modern hospital, it takes infants three days to accumulate a reasonable supply of intestinal bacteria, and they are protected by injections of the vitamin into themselves directly or into the mother shortly before birth. In the old days, the infants picked up the bacteria almost at once; and though they might die of various infections and disease, they were at least safe from the dangers of hemorrhage.

In fact, one might wonder whether organisms could live at all in the complete absence of intestinal bacteria, or whether the symbiosis had not become too intimate to abandon. However, germ-free animals have been grown from birth under completely sterile conditions and have even been allowed to reproduce under such conditions. Mice have been carried through twelve generations in this fashion. Experiments of this sort have been conducted at the University of Notre Dame since 1928.

During the late 1930s and early 1940s, biochemists identified several additional B vitamins, which now go under the names of *biotin, pantothenic acid, pyridoxine, folic acid,* and *cyanocobalamine.* These vitamins are all made by intestinal bacteria; moreover, they are present so universally in foodstuffs that no cases of deficiency diseases have appeared. In fact, investigators have had to feed animals an artificial diet deliberately excluding them, and even to add *antivitamins* to neutralize those made by the intestinal bacteria, in order to see what the deficiency symptoms are. (Antivitamins are substances similar to the vitamin in structure. They immobilize the enzyme making use of the vitamin by means of competitive inhibition.)

VITAMIN THERAPY

The determination of the structure of each of the various vitamins was usually followed speedily (or even preceded) by synthesis of the vitamin. For instance, Williams and his group synthesized thiamine in 1937, three years after they had deduced its structure. The Polish-born Swiss biochemist Tadeus

Reichstein and his group synthesized ascorbic acid in 1933, somewhat before the structure was completely determined by King. Vitamin A, for another example, was synthesized in 1936 (again somewhat before the structure was completely determined) by two different groups of chemists.

The use of synthetic vitamins has made it possible to fortify food (milk was first vitamin-fortified as early as 1924) and to prepare vitamin mixtures at reasonable prices and sell them over the drugstore counter. The need for vitamin pills varies with individual cases. Of all the vitamins, the one most likely to be deficient in supply is vitamin D. Young children in northern climates, where sunlight is weak in winter time, run the danger of rickets and may require irradiated foods or vitamin supplements. But the dosage of vitamin D (and of vitamin A) should be carefully controlled, because an overdose of these vitamins can be harmful. As for the B vitamins, anyone eating an ordinary, rounded diet does not need to take pills for them. The same is true of vitamin C, which in any case should not present a problem, for there are few people who do not enjoy orange juice or who do not drink it regularly in these vitamin-conscious times.

On the whole, the wholesale use of vitamin pills, while redounding chiefly to the profit of drug houses, usually does people no harm and may be partly responsible for the fact that the current generation of Americans is taller and heavier than previous generations.

During the 1970s, schemes for *megavitamin therapy* were advanced. There were suggestions that minimum quantities of vitamins that were sufficient to stave off deficiency diseases were not necessarily enough for optimum working of the body or were not enough to stave off some other diseases. It was maintained, for instance, that large doses of some B vitamins might ameliorate schizophrenic conditions.

The most important exponent of megavitamin therapy is Linus Pauling who, in 1970, maintained that large daily doses of vitamin C would prevent colds and would have other beneficial effects on health. He has not convinced the medical profession generally; but the general public, which always accentuates the positive in connection with vitamins (especially since they are readily available and quite cheap), stripped the druggists' shelves of vitamin C in their eagerness to gulp it down.

Taking more than enough of the water-soluble vitamins such as the B-complex and C, is not likely to do positive harm since they are not stored by the body and are easily excreted. Therefore, if a large dose is not actually needed by the body, the excess merely serves to enrich the urine.

The case is otherwise with the fat-soluble vitamins, particularly A and D. These tend to dissolve in the body fat and to be stored there, and are then relatively immobile, as is the fat itself. Too great a supply may therefore overload the body and disturb its workings, giving rise to *hypervitaminoses,* as the condition is called. Since vitamin A is stored in the liver, particularly in fish and in animals that live on fish (a whole generation of youngsters had their life made hideous by regular doses of *cod-liver oil*), there have been horror tales of Arctic explorers who were rendered seriously ill or even killed by dining on polar-bear liver—poisoned by vitamin A.

VITAMINS AS ENZYMES

Biochemists naturally were curious to find out how the vitamins, present in the body in such tiny quantities, exert such important effects on the body chemistry. The obvious guess was that they have something to do with enzymes, also present in small quantities.

The answer finally came from detailed studies of the chemistry of enzymes. Protein chemists had known for a long time that some proteins are not made up solely of amino acids, and that nonamino-acid prosthetic groups might exist, such as the heme in hemoglobin (see chapter 11). In general, these prosthetic groups tended to be tightly bound to the rest of the molecule. With enzymes, however, there were in some cases nonamino-acid portions that were quite loosely bound and might be removed with little trouble.

This was first discovered in 1904 by Arthur Harden (who was soon to discover phosphorus-containing intermediates; see chapter 12). Harden worked with a yeast extract capable of bringing about the fermentation of sugar. He placed it in a bag made of a semipermeable membrane and placed that bag in fresh water. Small molecules could penetrate the membrane, but the large protein molecule could not. After this *dialysis* had progressed for a while, Harden found that the activity of the extract was lost. Neither the fluid within nor that outside the bag would ferment sugar. If the two fluids were combined, activity was regained.

Apparently, the enzyme was made up not only of a large protein molecule, but also of a *coenzyme* molecule, small enough to pass through the pores of the membrane. The coenzyme was essential to enzyme activity (it was the cutting edge, so to speak).

Chemists at once tackled the problem of determining the structure of this coenzyme (and of similar adjuncts to other enzymes). The German-Swedish chemist Hans Karl August Simon von Euler-Chelpin was the first to make real progress in this respect. As a result, he and Harden shared the Nobel Prize in chemistry in 1929.

The coenzyme of the yeast enzyme studied by Harden proved to consist of a combination of an adenine molecule, two ribose molecules, two phosphate groups, and a molecule of *nicotinamide*. Now this last was an unusual thing to find in living tissue, and interest naturally centered on the nicotinamide. (It is called *nicotinamide* because it contains an amide group, $CONH_2$, and can be formed easily from nicotinic acid. Nicotinic acid is structurally related to the tobacco alkaloid *nicotine*, but they are utterly different in properties; for one thing, nicotinic acid is necessary to life, whereas nicotine is a deadly poison.) The formulas of nicotinamide and nicotinic acid are:

$$
\begin{array}{ccc}
& \text{CH} & \\
\text{HC} & \diagup \quad \diagdown & \text{C} - \text{C} \diagup\!\!^{\text{O}} \\
\parallel & \text{C} & \diagdown \text{OH} \\
\text{HC} & \diagdown \quad \diagup & \text{CH} \\
& \text{N} &
\end{array}
\qquad
\begin{array}{ccc}
& \text{CH} & \\
\text{HC} & \diagup \quad \diagdown & \text{C} - \text{C} \diagup\!\!^{\text{O}} \\
\parallel & \text{C} & \diagdown \text{NH}_2 \\
\text{HC} & \diagdown \quad \diagup & \text{CH} \\
& \text{N} &
\end{array}
$$

Nicotinic acid   Nicotinamide

Once the formula of Harden's coenzyme was worked out, it was promptly renamed *diphosphopyridine nucleotide* (DPN): *nucleotide* from the characteristic arrangement of the adenine, ribose, and phosphate, similar to that of the nucleotides making up nucleic acid; and *pyridine* from the name given to the combination of atoms making up the ring in the nicotinamide formula.

Soon a similar coenzyme was found, differing from DPN only in the fact that it contained three phosphate groups rather than two. This, naturally, was named *triphosphopyridine nucleotide* (TPN). Both DPN and TPN proved to be coenzymes for a number of enzymes in the body, all serving to transfer hydrogen atoms from one molecule to another. (Such enzymes are called *dehydrogenases.*) It was the coenzyme that does the actual job of hydrogen transfer; the enzyme proper in each case selects the particular substrate on which the operation is to be performed. The enzyme and the coenzyme each have a vital function; and if either were deficient in supply, the release of energy from foodstuffs via hydrogen transfer would slow to a limp.

What was immediately striking about all this was that the nicotinamide group represents the only part of the enzyme the body cannot manufacture itself. It can make all the protein it needs and all the ingredients of DPN and TPN except the nicotinamide: that it must find ready-made (or at least in the form of nicotinic acid) in the diet. If not, then the manufacture of DPN and TPN stops, and all the hydrogen-transfer reactions they control slow down.

Was nicotinamide or nicotinic acid a vitamin? As it happened, Funk (who coined the word *vitamine*) had isolated nicotinic acid from rice hulls. Nicotinic acid was not the substance that cured beri-beri, so he had ignored it. But on the strength of nicotinic acid's appearance in connection with coenzymes, the University of Wisconsin biochemist Conrad Arnold Elvehjem and his co-workers tried it on another deficiency disease.

In the 1920s, the American physician Joseph Goldberger had studied *pellagra*, a disease endemic in the Mediterranean area and almost epidemic in the southern United States in the early part of this century. Pellagra's most noticeable symptoms are a dry, scaly skin, diarrhea, and an inflamed tongue; it sometimes leads to mental disorders. Goldberger noticed that the disease struck people who lived on a limited diet (for example, mainly cornmeal) and spared families that owned a milch cow. He began to experiment with artificial diets, feeding them to animals and inmates of jails (where pellagra seemed to blossom). He succeeded in producing *blacktongue* (a disease analogous to pellagra) in dogs and in curing this disease with a yeast extract. He found he could cure jail inmates of pellagra by adding milk to their diet. Goldberger decided that a vitamin must be involved, and he named it the P-P (*pellagra-preventive*) factor.

It was pellagra, then, that Elvehjem chose for the test of nicotinic acid. He fed a tiny dose to a dog with blacktongue, and the dog responded with remarkable improvement. A few more doses cured it. Nicotinic acid was a vitamin, all right; it was the P-P factor.

The American Medical Association, worried that the public might get the impression there were vitamins in tobacco, urged that the vitamin not be called nicotinic acid and suggested instead the name *niacin* (an abbreviation of "*ni*cotine *ac*id") or *niacinamide*. Niacin has caught on fairly well.

Gradually, it became clear that the various vitamins were merely portions of coenzymes, each consisting of a molecular group that an animal or a human being cannot make for itself. In 1932, Warburg had found a yellow coenzyme that catalyzed the transfer of hydrogen atoms. The Austrian chemist Richard Kuhn and his associates shortly afterward isolated vitamin B_2, which proved to be yellow, and worked out its structure:

The carbon chain attached to the middle ring is like a molecule called *ribitol*, so vitamin B_2 was named *riboflavin, flavin* coming from a Latin word meaning "yellow." Since examination of its spectrum showed riboflavin to be very similar in color to Warburg's yellow coenzyme, Kuhn tested the coenzyme, for riboflavin activity in 1935 and found such activity to be there. In the same year, the Swedish biochemist Hugo Theorell worked out the structure of Warburg's yellow coenzyme and showed it to be riboflavin with a phosphate group added. (In 1954, a second and more complicated coenzyme also was shown to have riboflavin as part of its molecule.)

Kuhn was awarded the 1938 Nobel Prize in chemistry, and Theorell received the 1955 Nobel Prize in medicine and physiology. Kuhn, however, was unfortunate enough to be selected for his prize shortly after Austria had been absorbed by Nazi Germany, and (like Gerhard Domagk) was compelled to refuse it.

Riboflavin was synthesized independently by the Swiss chemist Paul Karrer. For this and other work on vitamins, Karrer was awarded a share of the 1937 Nobel Prize in chemistry. (He shared it with the English chemist Walter Norman Haworth, who had worked on the ring structure of carbohydrate molecules.)

In 1937, the German biochemists Karl Heinrich Adolf Lohmann and P. Schuster discovered an important coenzyme that contains thiamine as part of its structure. Through the 1940s other connections were found between B vitamins and coenzymes. Pyridoxine, pantothenic acid, folic acid, biotin— each in turn was found to be tied to one or more groups of enzymes.

The vitamins beautifully illustrate the economy of the human body's chemical machinery. The human cell can dispense with making them because they serve only one special function, and the cell can take the reasonable risk of finding the necessary supply in the diet. There are many other vital substances that the body needs only in trace amounts but must make for itself. ATP, for instance, is formed from much the same building blocks that make up the indispensable nucleic acids. It is inconceivable that any organism could lose any enzyme necessary for nucleic-acid synthesis and remain alive, for nucleic acid is needed in such quantities that the organism dare not trust to the diet for its supply of the necessary building blocks. And the ability to make nucleic acid automatically implies the ability to make ATP. Consequently, no organism is known that is incapable of manufacturing its own ATP, and in all probability no such organism will ever be found.

To make such special products as vitamins would be like setting up a special machine next to an assembly line to turn out nuts and bolts for the automobiles. The nuts and bolts can be obtained more efficiently from a parts supplier, without any loss to the apparatus for assembling the automobiles; by the same token the organism can obtain vitamins in its diet, with a saving in space and material.

The vitamins illustrate another important fact of life. As far as is known, all living cells require the B vitamins. The coenzymes are an essential part of the cell machinery of every cell alive—plant, animal, or bacterial. Whether the cell gets the B vitamins from its diet or makes them itself, it must have them if it is to live and grow. This universal need for a particular group of substances is an impressive piece of evidence for the essential unity of all life and its descent (possibly) from a single original scrap of life formed in the primeval ocean.

VITAMIN A

While the roles of the B vitamins are now well known, the chemical functions of the other vitamins have proved rather hard nuts to crack. The only one on which any real advance has been made is vitamin A.

In 1925, the American physiologists L. S. Fridericia and E. Holm found that rats fed on a diet deficient in vitamin A had difficulty performing tasks in dim light. An analysis of their retinas showed that they were deficient in a substance called *visual purple*.

There are two kinds of cell in the retina of the eye—*rods* and *cones*. The rods specialize in vision in dim light, and they contain the visual purple. A shortage of visual purple therefore hampers only vision in dim light and results in what is known as *night-blindness*.

In 1938, the Harvard biologist George Wald began to work out the chemistry of vision in dim light. He showed that light causes visual purple, or *rhodopsin*, to separate into two components: the protein *opsin* and a non-protein called *retinene*. Retinene proved to be very similar in structure to vitamin A.

The retinene always recombines with the opsin to form rhodopsin in the dark. But during its separation from opsin in the light, a small percentage of it breaks down, because it is unstable. However, the supply of retinene is replenished from vitamin A, which is converted to retinene by the removal of two hydrogen atoms with the aid of enzymes. Thus vitamin A acts as a stable reserve for retinene. If vitamin A is lacking in the diet, eventually the retinene supply and the amount of visual purple decline, and night-blindness is the result. For his work in this field, Wald shared in the 1967 Nobel Prize for medicine and physiology.

Vitamin A must have other functions as well, for a deficiency causes dryness of the mucous membranes and other symptoms which cannot very well be traced to troubles in the retina of the eye. But the other functions are still unknown.

The same has to be said about the chemical functions of vitamins C, D, E, and K.

Minerals

It is natural to suppose that the materials making up anything as wonderful as living tissue must themselves be something pretty exotic. Wonderful the proteins and nucleic acids certainly are, but it is a little humbling to realize that the elements making up the human body are as common as dirt, and the whole lot could be bought for a few dollars. (It used to be cents, but inflation has raised the price of everything.)

In the early nineteenth century, when chemists were beginning to analyze organic compounds, it became clear that living tissue is made up, in the main, of carbon, hydrogen, oxygen, and nitrogen. These four elements alone constitute about 96 percent of the weight of the human body. Then there is also a little sulfur in the body. If you burned off these five elements, you would be left with a bit of white ash, mostly the residue from the bones. The ash would be a collection of minerals.

It would not be surprising to find common salt, sodium chloride, in the ash.

After all, salt is not a mere condiment to improve the taste of food—as dispensable as, say, basil, rosemary, or thyme. It is a matter of life and death. You need only taste blood to realize that salt is a basic component of the body. Herbivorous animals, which presumably lack sophistication as far as the delicacies of food preparation are concerned, will undergo much danger and privation to reach a *salt lick,* where they can make up the lack of salt in their diet of grass and leaves.

As early as the mid-eighteenth century, the Swedish chemist Johann Gottlieb Gahn had shown that bones are made up largely of calcium phosphate; and an Italian scientist, Vincenzo Antonio Menghini, had established that the blood contains iron. In 1847, Justus von Liebig found potassium and magnesium in the tissues. By the mid-nineteenth century, then, the mineral constituents of the body were known to include calcium, phosphorus, sodium, potassium, chlorine, magnesium, and iron. Furthermore, these are as active in life processes as any of the elements usually associated with organic compounds.

The case of iron is the clearest. If it is lacking in the diet, the blood becomes deficient in hemoglobin and transports less oxygen from the lungs to the cells. The condition is known as *iron-deficiency anemia.* The patient is pale for lack of the red pigment and tired for lack of oxygen.

In 1882, the English physician Sidney Ringer found that a frog heart could be kept alive and beating outside its body in a solution (called *Ringer's solution* to this day) containing, among other things, sodium, potassium, and calcium in about the proportions found in the frog's blood. Each is essential for functioning of muscle. An excess of calcium causes the muscle to lock in permanent contraction (*calcium rigor*) whereas an excess of potassium causes it to unlock in permanent relaxation (*potassium inhibition*). Calcium, moreover, is vital to blood clotting. In its absence blood would not clot, and no other element can substitute for calcium in this respect.

Of all the minerals, phosphorus was eventually discovered to perform the most varied and crucial functions in the chemical machinery of life (see chapter 13).

Calcium, a major component of bone, makes up 2 percent of the body; phosphorus, 1 percent. The other minerals I have mentioned come in smaller proportions, down to iron, which makes up only 0.004 percent of the body. (That still leaves the average adult male 1/10 ounce of iron in his tissues.) But we are not at the end of the list; there are other minerals that, though present in tissue only in barely detectable quantities, are yet essential to life.

The mere presence of an element is not necessarily significant; it may be just an impurity. In our food we take in at least traces of every element in our environment, and some small amount of each finds its way into our tissues. But elements such as titanium and nickel, for instance, contribute nothing. On the other hand, zinc is vital. How does one distinguish an essential mineral from an accidental impurity?

The best way is to show that some necessary enzyme contains the trace element as an essential component. (Why an enzyme? Because in no other way can any trace component possibly play an important role.) In 1939, David Keilin and Thaddeus Robert Rudolph Mann of England showed that zinc is an integral part of the enzyme carbonic anhydrase. Now carbonic anhydrase is essential to the body's handling of carbon dioxide, and the proper handling of that important waste material, in turn, is essential to life. It follows in theory that zinc is indispensable to life, and experiment shows that it actually is. Rats fed on a diet low in zinc stop growing, lose hair, suffer scaliness of the skin, and die prematurely for lack of zinc as surely as for lack of a vitamin.

In the same way it has been shown that copper, manganese, cobalt, and molybdenum are essential to animal life. Their absence from the diet gives rise to deficiency diseases. Molybdenum is a constituent of an enzyme called *xanthine oxidase*. The importance of molybdenum was first noticed in the 1940s in connection with plants, when soil scientists found that plants would not grow well in soils deficient in the element. It seems that molybdenum is a component of certain enzymes in soil microorganisms that catalyze the conversion of the nitrogen of the air into nitrogen-containing compounds. Plants depend on this help from microorganisms because they cannot themselves take nitrogen from the air. (This is only one of an enormous number of examples of the close interdependence of all life on our planet. The living world is a long and intricate chain which may suffer hardship or even disaster if any link is broken.)

Not all trace elements are universally essential. Boron seems to be essential in traces to plant life but not, apparently, to animals. Certain tunicates gather vanadium from sea water and use it in their oxygen-transporting compound, but few, if any, other animals require vanadium for any reason. Some elements, such as selenium and chromium, are suspected of being essential, but their exact role has not been determined.

It is now realized that there are trace-element deserts, just as there are waterless deserts; the two usually go together but not always. In Australia soil, scientists have found that 1 ounce of molybdenum in the form of some appropriate compound spread over 16 acres of molybdenum-deficient land results in a considerable increase in fertility. Nor is this a problem of exotic lands only. A survey of American farmland in 1960 showed areas of boron deficiency in forty-one states, of zinc deficiency in twenty-nine states, and of molybdenum deficiency in twenty-one states. The dosage of trace elements is crucial. Too much is as bad as too little, for some substances that are essential for life in small quantities (such as, copper) become poisonous in larger quantities.

This, of course, carries to its logical extreme the much older custom of using *fertilizers* for soil. Until modern times, fertilization was through the use of animal excreta, manure or guano, which restored nitrogen and phosphorus to the soil. While this worked, it was accompanied by foul odors and by the ever-present possibility of infection. The substitution of chemical fertilizers, clean and odor-free, was through the work of Justus von Liebig in the early nineteenth century.

COBALT

One of the most dramatic episodes in the discovery of mineral deficiencies has to do with cobalt. It involves the once incurably fatal disease called *pernicious anemia.*

In the early 1920s, the University of Rochester pathologist George Hoyt Whipple was experimenting on the replenishment of hemoglobin by means of various food substances. He would bleed dogs to induce anemia and then feed them various diets to see which would permit the dogs to replace the lost hemoglobin most rapidly. He did this not because he was interested in pernicious anemia, or in any kind of anemia, but because he was investigating bile pigments, compounds produced by the body from hemoglobin. Whipple discovered that liver was the food that enabled the dogs to make hemoglobin most quickly.

In 1926, two Boston physicians, George Richards Minot and William Parry Murphy, considered Whipple's results, decided to try liver as a treatment for pernicious-anemia patients. The treatment worked. The incurable disease was cured, so long as the patients ate liver as an important portion of their diet. Whipple, Minot, and Murphy shared the Nobel Prize in physiology and medicine in 1934.

Unfortunately liver, although it is a great delicacy when properly cooked, then chopped, and lovingly mixed with such things as eggs, onions, and chicken fat, becomes wearing as a steady diet. (After a while, a patient might be tempted to think pernicious anemia was preferable.) Biochemists began to search for the curative substance in liver; and by 1930, Edwin Joseph Cohn and his co-workers at the Harvard Medical School had prepared a concentrate a hundred times as potent as liver itself. To isolate the active factor, however, further purification was needed. Fortunately, chemists at the Merck Laboratories discovered in the 1940s that the concentrate from liver could accelerate the growth of certain bacteria. This provided an easy test of the potency of any preparation from it, so the biochemists could proceed to break down the concentrate into fractions and test them in quick succession. Because the bacteria reacted to the liver substance in much the same way that they reacted to, say, thiamine or riboflavin, the investigators now suspected strongly that the factor they were hunting for was a B vitamin. They called it *vitamin B_{12}.*

By 1948, using bacterial response and chromatography, Ernest Lester Smith in England and Karl August Folkers at Merck succeeded in isolating pure samples of vitamin B_{12}. The vitamin proved to be a red substance, and both scientists thought it resembled the color of certain cobalt compounds. It was known by this time that a deficiency of cobalt caused severe anemia in cattle and sheep. Both Smith and Folkers burned samples of vitamin B_{12}, analyzed the ash, and found that it did indeed contain cobalt. The compound has now been named *cyanocobalamine.* So far it is the only cobalt-containing compound that has been found in living tissue.

By breaking it up and examining the fragments, chemists quickly decided that vitamin B_{12} was an extremely complicated compound, and they worked out an empirical formula of $C_{63}H_{88}O_{14}N_{14}PCo$. Then a British chemist, Dorothy Crowfoot Hodgkin, determined its over-all structure by means of X rays. The diffraction pattern given by crystals of the compound allowed her to build up a picture of the *electron densities* along the molecule—that is, those regions where the probability of finding an electron is high and those where it is low. If lines are drawn through regions of equal probability, a kind of skeletal picture is built up of the shape of the molecule as a whole.

This is not as easy as it sounds. Complicated organic molecules can produce an X-ray scattering truly formidable in its complexity. The mathematical operations required to translate that scattering into electron densities are tedious in the extreme. By 1944, electronic computers had been called in to help work out the structural formula of penicillin. Vitamin B_{12} was much more complicated, and Hodgkin had to use a more advanced computer—the National Bureau of Standards Western Automatic Computer (SWAC)—and do some heavy spadework. It eventually earned for her, however, the 1964 Nobel Prize for chemistry.

The molecule of vitamin B_{12}, or cyanocobalamine, turned out to be a lopsided porphyrin ring, with one of the carbon bridges connecting two of the smaller pyrrole rings missing, and with complicated side chains on the pyrrole rings. It resembles the somewhat simpler heme molecule, with this key difference: where heme has an iron atom at the center of the porphyrin ring, cyanocobalamine has a cobalt atom.

Cyanocobalamine is active in very small quantities when injected into the blood of pernicious-anemia patients. The body can get along on only 1/1,000 as much of this substance as it needs of the other B vitamins. Any diet, therefore, ought to have enough cyanocobalamine for our needs. Even if it did not, the bacteria in the intestines manufacture quite a bit of it. Why, then, should anyone ever have pernicious anemia?

Apparently, the sufferers from this disease are simply unable to absorb enough of the vitamin into the body through the intestinal walls. Their feces are actually rich in the vitamin (for want of which they are dying). From feedings of liver, providing a particularly abundant supply, such a patient manages to absorb enough cyanocobalamine to stay alive. But he needs 100 times as much of the vitamin if he takes it by mouth as he does when it is injected directly into the blood.

Something must be wrong with the patient's intestinal apparatus, preventing the passage of the vitamin through the walls of the intestines. It has been known since 1929, thanks to the researches of the American physician William Bosworth Castle, that the answer lies somewhere in the gastric juice. Castle called the necessary component of gastric juice *intrinsic factor*. And in 1954 investigators found a product, from the stomach linings of animals, that assists the absorption of the vitamin and proved to be Castle's intrinsic factor. Apparently this substance is missing in pernicious-anemia patients. When a small amount of it is mixed with cyanocobalamine, the patient has no difficulty in absorbing the vitamin through the intestines. The intrinsic factor has proved to be a *glycoprotein* (a sugar-containing protein) that binds a molecule of cyanocobalamine and carries it into the intestinal cells.

IODINE

Getting back to the trace elements. . . . The first one discovered was not a metal; it was iodine, an element with properties like those of chlorine. This story begins with the thyroid gland.

In 1896, a German biochemist, Eugen Baumann, discovered that the thyroid was distinguished by containing iodine, practically absent from all other tissues. In 1905, a physician named David Marine, who had just set up practice in Cleveland, was amazed to find how widely prevalent goiter was in that area. Goiter is a conspicuous disease, sometimes producing grotesque enlargement of the thyroid and causing its victims to become either dull and listless or nervous, overactive, and pop-eyed. For the development of surgical techniques in the treatment of abnormal thyroids with resulting relief from goitrous conditions, the Swiss physician Emil Theodor Kocher earned the 1909 Nobel Prize in medicine and physiology.

But Marine wondered whether the enlarged thyroid might not be the result of a deficiency of iodine, the one element in which the thyroid specializes, and whether goiter might not be treated more safely and expeditiously by chemicals rather than by the knife. Iodine deficiency and the prevalence of goiter in the Cleveland area might well go hand in hand, at that, for Cleveland, being inland, might lack the iodine that was plentiful in the soil near the ocean and in the seafood that is an important article of diet there.

The doctor experimented on animals and, after ten years, felt sure enough of his ground to try feeding iodine-containing compounds to goiter patients. He was probably not too surprised to find that it worked. Marine then suggested that iodine-containing compounds be added to table salt and to the water supply of inland cities where the soil was poor in iodine. There was strong opposition to his proposal, however; and it took another ten years to get water iodination and iodized salt generally accepted. Once the iodine supplements became routine, simple goiter declined in importance as a human woe.

FLUORIDES

A half-century later American researchers (and the public) were engaged in studies and discussion of a similar health question—the fluoridation of water to prevent tooth decay. This issue was a matter of bitter controversy in the nonscientific and political arena—with the opposition far more stubborn than in the case of iodine. Perhaps one reason is that cavities in the teeth do not seem nearly as serious as the disfigurement of goiter.

In the early decades of this century dentists noticed that people in certain areas in the United States (for example, some localities in Arkansas) tended to have darkened teeth—a mottling of the enamel. Eventually the cause was traced to a higher-than-average content of fluorine compounds (*fluorides*) in the natural drinking water of those areas. With the attention of researchers directed to fluoride in the water, another interesting discovery turned up. Where the fluoride content of the water was above average, the population had an unusually low rate of tooth decay. For instance, the town of Galesburg in Illinois, with fluoride in its water, had only one-third as many cavities per youngster as the nearby town of Quincy, whose water contained practically no fluoride.

Tooth decay is no laughing matter, as anyone with a toothache will readily agree. It costs the people of the United States more than a billion and a half dollars a year in dental bills; and by the age of thirty-five, two thirds of all Americans have lost at least some of their teeth. Dental researchers succeeded in getting support for large-scale studies to find out whether fluoridation of water would be safe and would really help to prevent tooth decay. They found that one part per million of fluoride in the drinking water, at an estimated cost of 5 to 10 cents per person per year, did not mottle teeth and yet showed an effect in decay prevention. They therefore adopted one part per million as a standard for testing the results of fluoridation of community water supplies.

The effect is, primarily, on those whose teeth are being formed—that is, on children. The presence of fluoride in the drinking water ensures the incorporation of tiny quantities of fluoride into the tooth structure; it is this, apparently, that makes the tooth mineral unpalatable to bacteria. (The use of small quantities of fluoride in pill form or in toothpaste has also shown some protective effect against tooth decay.)

The dental profession is now convinced, on the basis of a quarter of a century of research, that for a few pennies per person per year, tooth decay can be reduced by about two thirds, with a saving of at least a billion dollars a year in dental costs and a relief of pain and of dental handicaps that cannot be measured in money.

Two chief arguments have been employed by the opponents of fluoridation with the greatest effect. One is that fluorine compounds are poisonous. So they are, but not in the doses used for fluoridation! The other is that fluoridation is compulsory medication, infringing the individual's freedom. That may be so, but it is questionable whether the individual in any society should have the freedom to expose others to preventable sickness. If compulsory medication is evil, then we have a quarrel not only with fluoridation but also with chlorination, iodination, and, for that matter, with all the forms of inoculation, including vaccination against smallpox, that are compulsory in most civilized countries today.

Hormones

Enzymes, vitamins, trace elements—how potently these sparse substances decide life-or-death issues for the organism! But there is a fourth group of substances that, in a way, are even more potent. They conduct the whole performance; they are like a master switch that awakens a city to activity, or the throttle that controls an engine, or the red cape that excites the bull.

At the turn of the century, two English physiologists, William Maddock Bayliss and Ernest Henry Starling, became intrigued by a striking little performance in the digestive tract. The gland behind the stomach known as the pancreas releases its digestive fluid into the upper intestines at just the moment when food leaves the stomach and enters the intestine. How does the pancreas get the message? What tells it that the right moment has arrived? The obvious guess was that the information must be transmitted via the nervous system, which was then the only known means of communication in the body. Presumably, the entry of food into the intestines from the stomach stimulated nerve endings that relayed the message to the pancreas by way of the brain or the spinal cord.

To test this theory, Bayliss and Starling cut every nerve to the pancreas. Their maneuver failed! The pancreas still secreted juice at precisely the right moment.

The puzzled experimenters went hunting for an alternate signaling system. In 1902, they tracked down a *chemical messenger*. It was a substance secreted by the walls of the intestine. When they injected this into an animal's blood, it stimulated the secretion of pancreatic juice even though the animal was not eating. Bayliss and Starling concluded that, in the normal course of events, food entering the intestines stimulates their linings to secrete the substance, which then travels via the bloodstream to the pancreas and triggers the gland to start giving forth pancreatic juice. The two investigators named the substance secreted by the intestines *secretin*, and they called it a *hormone*, from a Greek word meaning "rouse to activity." Secretin is now known to be a small protein molecule.

Several years earlier, physiologists had discovered that an extract of the adrenals (two small organs just above the kidneys) could raise blood pressure if injected into the body. The Japanese chemist Jokichi Takamine, working in the United States, isolated the responsible substance in 1901 and named it *adrenalin*. (This later became a trade name; the chemists' name for it now is *epinephrine*.) Its structure proved to resemble that of the amino acid tyrosine, from which it is derived in the body.

Plainly, adrenalin, too, is a hormone. As the years went on, the physiologists found that a number of other *glands* in the body secrete hormones. (The word *gland* comes from the Greek word for "acorn" and was originally applied to any small lump of tissue in the body, but it became customary to give the name to any tissue that secretes a fluid, even large organs such as the liver and the mammaries. Small organs that do not secrete fluids gradually lost this name, so that the *lymph glands*, for instance, were renamed the *lymph nodes*. Even so, when lymph nodes in the throat or the armpit become enlarged during infections, physicians and mothers alike still refer to them as "enlarged glands.")

Many of the glands—such as those along the alimentary canal, the sweat glands, and the salivary glands—discharge their fluids through ducts. Some, however, are *ductless;* they release substances directly into the bloodstream,

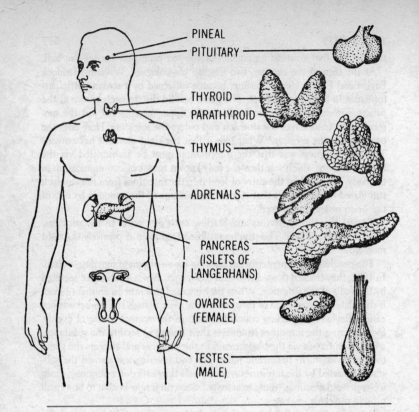

Figure 15.1. The endocrine glands.

which then circulates the secretions through the body. It is the secretions of these ductless, or *endocrine*, glands that contain hormones (see figure 15.1). The study of hormones is for this reason termed *endocrinology*.

Naturally, biologists are most interested in hormones that control functions of the mammalian body and, in particular, the human one. However, I should like at least to mention the fact that there are *plant hormones* that control and accelerate plant growth, *insect hormones* that control pigmentation and molting, and so on.

When biochemists found that iodine was concentrated in the thyroid gland, they made the reasonable guess that the element was part of a hormone. In 1915, Edward Calvin Kendall of the Mayo Foundation in Minnesota isolated from the thyroid an iodine-containing amino acid which behaved like a hormone, and named it *thyroxine*. Each molecule of thyroxine contained four atoms of iodine. Like adrenalin, thyroxine has a strong family resemblance to tyrosine and is manufactured from it in the body. (Many years later, in 1952, the biochemist Rosalind Pitt-Rivers and her associates isolated another thyroid hormone—*triiodothyronine,* so named because its molecule contains three atoms of iodine rather than four. It is less stable than thyroxine but three to five times as active.)

The thyroid hormones control the overall rate of metabolism in the body: they arouse all the cells to activity. People with an underactive thyroid are sluggish, torpid, and after a time may become mentally retarded, because the various cells are running in low gear. Conversely, people with an overactive thyroid are nervous and jittery, because their cells are racing. Either an underactive or an overactive thyroid can produce goiter.

The thyroid controls the body's *basal metabolism:* that is, its rate of consumption of oxygen at complete rest in comfortable environmental conditions —the "idling rate," so to speak. If a person's basal metabolism is above or below the norm, suspicion falls upon the thyroid gland. Measurement of the basal metabolism was at one time a tedious affair, for the subject had to fast for a period in advance and lie still for half an hour while the rate is measured, to say nothing of an even longer period beforehand. Instead of going through this troublesome procedure, why not go straight to the horse's mouth: that is, measure the amount of rate-controlling hormone that the thyroid is producing? In recent years researchers have developed a method of measuring the amount of *protein-bound iodine* (PBI) in the bloodstream; this indicates the rate of thyroid-hormone production and so has provided a simple, quick blood test to replace the basal-metabolism determination.

INSULIN AND DIABETES

The best-known hormone is insulin, the first protein whose structure was fully worked out (see chapter 12). Its discovery was the culmination of a long chain of events.

Diabetes is the name of a whole group of diseases, all characterized by unusual thirst and, in consequence, an unusual output of urine. It is the most common of the inborn errors of metabolism. There are 1,500,000 diabetics in the United States, 80 percent of whom are over forty-five. It is one of the few diseases to which the female is more subject than the male: women diabetics outnumber men four to three.

The name comes from a Greek word meaning "syphon" (apparently the coiner pictured water syphoning endlessly through the body). The most serious form of the disease is *diabetes mellitus. Mellitus* comes from the Greek word for "honey" and refers to the fact that, in advanced stages of certain cases of the disease, the urine has a sweet taste. (This may have been determined directly by some heroic physician, but the first indication was rather indirect: diabetic urine tended to gather flies.) In 1815, the French chemist Michel Eugène Chevreul was able to show that the sweetness is due to the presence of the simple sugar glucose. This waste of glucose plainly indicates that the body is not utilizing its food efficiently. In fact, the diabetic patient, despite an increase in appetite, may steadily lose weight as the disease advances. Up to a generation ago there was no helpful treatment for the disease.

In the nineteenth century, the German physiologists Joseph von Mering and Oscar Minkowski found that removal of the pancreas gland from a dog produced a condition just like human diabetes. After Bayliss and Starling discovered the hormone secretin, it began to appear that a hormone of the pancreas might be involved in diabetes. But the only known secretion from the pancreas was the digestive juice. Where did the hormone come from? A significant clue turned up. When the duct of the pancreas was tied off, so that it could not pour out its digestive secretions, the major part of the gland shriveled, but the groups of cells known as the *islets of Langérhans* (after the German physician Paul Langerhans, who had discovered them in 1869) remained intact.

In 1916, a Scottish physician, Albert Sharpey-Schafer, suggested, therefore, that the islets must be producing the antidiabetes hormone. He named the assumed hormone *insulin*, from the Latin word for "island."

Attempts to extract the hormone from the pancreas at first failed miserably. As we now know, insulin is a protein, and the protein-splitting enzymes of the pancreas destroyed it even while the chemists were trying to isolate it. In 1921, the Canadian physician Frederick Grant Banting and the physiologist Charles Herbert Best (working in the laboratories of John James Rickard MacLeod at the University of Toronto) tried a new approach. First they tied off the duct of the pancreas. The enzyme-producing portion of the gland shriveled, the production of protein-splitting enzymes stopped, and the scientists were then able to extract the intact hormone from the islets. It proved indeed effective in countering diabetes, and it is estimated that in the next fifty years it saved the lives of some 20 million to 30 million diabetics. Banting called the hormone *isletin*, but the older and more Latinized form proposed by Sharpey-Schafer won out. Insulin it became and still is.

In 1923, Banting and, for some reason, MacLeod (whose chief service to the discovery of insulin was to allow the use of his laboratory over the summer while he was on vacation) received the Nobel Prize in physiology and medicine.

The effect of insulin within the body shows most clearly in connection with the level of glucose concentration in the blood. Ordinarily the body stores most of its glucose in the liver, in the form of a kind of starch called *glycogen* (discovered in 1856 by the French physiologist Claude Bernard), keeping only a small quantity of glucose in the bloodstream to serve the immediate energy needs of the cells. If the glucose concentration in the blood rises too high, the pancreas is stimulated to increase its production of insulin, which pours into the bloodstream and brings about a lowering of the glucose level. On the other hand, when the glucose level falls too low, the lowered concentration inhibits the production of insulin by the pancreas, so that the sugar level rises. Thus a balance is achieved. The production of insulin lowers the level of glucose, which lowers the production of insulin, which raises the level of glucose, which raises the production of insulin, which lowers the level of glucose—and so on. This is an example of what is called *feedback*. The thermostat that controls the heating of a house works in the same fashion.

Feedback is probably the customary device by which the body maintains a constant internal environment. Another example involves the hormone produced by the parathyroid glands, four small bodies embedded in the thyroid gland. The hormone *parathormone* was finally purified in 1960 by the American biochemists Lyman Creighton Craig and Howard Rasmussen after five years of work.

The molecule of parathormone is somewhat larger than that of insulin, being made up of eighty-three amino acids and possessing a molecular weight of 9,500. The action of the hormone is to increase calcium absorption in the intestine and decrease calcium loss through the kidneys. Whenever calcium concentration in the blood falls slightly below normal, secretion of the hormone is stimulated. With more calcium coming in and less going out, the blood level soon rises; this rise inhibits the secretion of the hormone. This interplay between calcium concentration in the blood and parathyroid hormone flow keeps the calcium level close to the needed level at all times. (And a good thing, too, for even a small departure of the calcium concentration from the proper level can lead to death. Thus, removal of the parathyroids is fatal. At one time, doctors, in their anxiety to snip away sections of thyroid to relieve goiter, thought nothing of tossing away the much smaller and less prominent parathyroids. The death of the patient taught them better.)

At some times, the action of feedback is refined by the existence of two hormones working in opposite directions. In 1961, for instance, D. Harold Copp, at the University of British Columbia, demonstrated the presence of a thyroid hormone he called *calcitonin*, which acted to depress the level of calcium in the blood by encouraging the deposition of its ions in bone. With parathormone pulling in one direction and calcitonin in the other, the feedback produced by calcium levels in the blood can be all the more delicately controlled. (The calcitonin molecule is made up of a single polypeptide chain that is thirty-two amino acids long.)

Then, too, in the case of blood-sugar concentration, where insulin is involved, a second hormone, also secreted by the islets of Langerhans, cooperates. The islets are made up of two distinct kinds of cells, *alpha* and *beta*. The beta cells produce insulin, while the alpha cells produce *glucagon*. The existence of glucagon was first suspected in 1923, and it was crystallized in 1955. Its molecule is made up of a single chain of twenty-nine acids, and, by 1958, its structure had been completely worked out.

Glucagon opposes the effect of insulin, so the two hormonal forces push in opposite directions, and the balance shifts very slightly this way and that under the stimulus of the glucose concentration in blood. Secretions from the pituitary gland (which I shall discuss shortly) also have a countereffect on insulin activity. For the discovery of this effect, the Argentinian physiologist Bernardo Alberto Houssay shared in the 1947 Nobel Prize for medicine and physiology.

Now the trouble in diabetes is that the islets have lost the ability to turn out enough insulin. The glucose concentration in the blood therefore drifts upward. When the level rises to about 50 percent higher than normal, it crosses the *renal threshold:* that is, glucose spills over into the urine. In a way, this loss of glucose into the urine is the lesser of two evils, for if the glucose concentration were allowed to build up any higher, the resulting rise in viscosity of the blood would cause undue heartstrain. (The heart is designed to pump blood, not molasses.)

The classic way of checking for the presence of diabetes is to test the urine for sugar. For instance, a few drops of urine can be heated with *Benedict's solution* (named for the American chemist Francis Gano Benedict). The solution contains copper sulfate, which gives it a deep blue color. If glucose is not present in the urine, the solution remains blue. If glucose is present, the copper sulfate is converted to *cuprous oxide.* Cuprous oxide is a brick-red, insoluble substance. A reddish precipitate at the bottom of the test tube therefore is an unmistakable sign of sugar in the urine, which usually means diabetes.

Nowadays an even simpler method is available. Small paper strips about two inches long are impregnated with two enzymes, glucose dehydrogenase and peroxidase, plus an organic substance called *orthotolidine.* The yellowish strip is dipped into a sample of the patient's urine and then exposed to the air. If glucose is present, it combines with oxygen from the air with the catalytic help of the glucose dehydrogenase. In the process, hydrogen peroxide is formed. The peroxidase in the paper then causes the hydrogen peroxide to combine with the orthotolidine to form a deep blue compound. In short, if the yellowish paper is dipped into urine and turns blue, diabetes can be strongly suspected.

Once glucose begins to appear in the urine, diabetes mellitus is fairly far along in its course. It is better to catch the disease earlier by checking the glucose level in the blood before it crosses the renal threshold. The *glucose tolerance test,* now in general use, measures the rate of fall of the glucose level in the blood after it has been raised by feeding a person glucose. Normally, the pancreas responds with a flood of insulin. In a healthy person the sugar level will drop to normal within two hours. If the level stays high for three hours or more, it shows a sluggish insulin response, and the person is likely to be in the early stages of diabetes.

It is possible that insulin has something to do with controlling appetite.

To begin with, we are all born with what some physiologists call an *appestat,* which regulates appetite as a thermostat regulates a furnace. If one's appestat is set too high, one finds oneself continually taking in more calories than one expends, unless one exerts a strenuous self-control which sooner or later wears the individual out.

In the early 1940s, a physiologist, Stephen Walter Ranson, showed that animals grew obese after destruction of a portion of the *hypothalamus* (located in the lower part of the brain). This seems to fix the location of the appestat. What controls its operation? *Hunger pangs* spring to mind. An empty stomach contracts in waves, and the entry of food ends the contractions. Perhaps it is these contractions that signal to the appestat. Not so: surgical removal of the stomach has never interfered with appetite control.

The Harvard physiologist Jean Mayer has advanced a more subtle suggestion. He believes that the appestat responds to the level of glucose in the blood. After food has been digested, the glucose level in the blood slowly drops. When it falls below a certain level, the appestat is turned on. If, in response to the consequent urgings of the appetite, one eats, the glucose level in one's blood momentarily rises, and the appestat is turned off.

THE STEROID HORMONES

The hormones I have discussed so far are all either proteins (as insulin, glucagon, secretin, parathormone) or modified amino acids (as thyroxine, triiodothyronine, adrenalin). We come now to an altogether different group —the steroid hormones.

The story of these begins in 1927, when two German physiologists, Bernhard Zondek and Selmar Aschheim, discovered that extracts of the urine of pregnant women, when injected into female mice or rats, aroused them to sexual heat. (This discovery led to the first early test for pregnancy.) It was clear at once that Zondek and Aschheim had found a hormone—specifically, a *sex hormone*.

Within two years pure samples of the hormone were isolated by Adolf Butenandt in Germany and by Edward Adelbert Doisy at St. Louis University. It was named *estrone*, from *estrus*, the term for sexual heat in females. Its structure was quickly found to be that of a steroid, with the four-ring structure of cholesterol. For his part in the discovery of sex hormones, Butenandt was awarded the Nobel Prize for chemistry in 1939. He, like Domagk and Kuhn, was forced to reject it and could only accept the honor in 1949 after the destruction of the Nazi tyranny.

Estrone is now one of a group of known female sex hormones, called *estrogens* ("giving rise to estrus"). In 1931, Butenandt isolated the first male sex hormone, or *androgen* ("giving rise to maleness"). He called it *androsterone*.

It is the production of sex hormones that governs the changes that take place during adolescence: the development of facial hair in the male and of enlarged breasts in the female, for instance. The complex menstrual cycle in females depends on the interplay of several estrogens.

The female sex hormones are produced, in large part, in the ovaries; the male sex hormones, in the testes.

The sex hormones are not the only steroid hormones. The first nonsexual chemical messenger of the steroid type was discovered in the adrenals. These, as a matter of fact, are double glands, consisting of an inner gland called the adrenal *medulla* (the Latin word for "marrow") and an outer gland called the adrenal *cortex* (the Latin word for "bark"). It is the medulla that produces adrenalin. In 1929, investigators found that extracts from the cortex could keep animals alive after their adrenal glands had been removed—a 100 percent fatal operation. Naturally, a search immediately began for *cortical hormones*.

The search had a practical medical reason behind it. The well-known affliction called *Addison's disease* (first described by the English physician Thomas Addison in 1855) had symptoms like those resulting from the removal of the adrenals. Clearly, the disease must be caused by a failure in hormone production by the adrenal cortex. Perhaps injections of cortical hormones might deal with Addison's disease as insulin dealt with diabetes.

Two men were outstanding in this search: Tadeus Reichstein (who was later to synthesize vitamin C) and Edward Kendall (who had first discovered the thyroid hormone nearly twenty years before). By the late 1930s, the researchers had isolated more than two dozen different compounds from the adrenal cortex. At least four showed hormonal activity. Kendall named the substances Compound A, Compound B, Compound E, Compound F, and so on. All the cortical hormones proved to be steroids.

Now the adrenals are very tiny glands, and it would take the glands of countless numbers of animals to provide enough cortical extracts for general use. Apparently, the only reasonable solution was to try to synthesize the hormones.

A false rumor drove cortical-hormone research forward under full steam during the Second World War. It was reported that the Germans were buying up adrenal glands in Argentine slaughterhouses to manufacture cortical hormones that improved the efficiency of their airplane pilots in high-altitude flight. There was nothing to it, but the rumor had the effect of stimulating the United States Government to place a high priority on research into methods for the synthesis of the cortical hormones; the priority was even higher than that given to the synthesis of penicillin or the antimalarials.

Compound A was synthesized by Kendall in 1944; and by the following year, Merck & Co. had begun to produce it in substantial amounts. It proved of little value for Addison's disease, to the disappointment of all. After prodigious labor, the Merck biochemist Lewis H. Sarrett then synthesized, by a process involving thirty-seven steps, Compound E, which was later to become known as *cortisone.*

The synthesis of Compound E created little immediate stir in medical circles. The war was over; the rumor of cortical magic worked on German pilots had proved untrue; and Compound A had fizzled. Then, in an entirely unexpected quarter, Compound E suddenly came to life.

For twenty years, the Mayo Clinic physician Philip Showalter Hench had been studying rheumatoid arthritis, a painful, sometimes paralytic disease. Hench suspected that the body possessed natural mechanisms for countering this disease, because the arthritis was often relieved during pregnancy or during attacks of jaundice. He could not think of any biochemical factor that jaundice and pregnancy held in common. He tried injections of bile pigments (involved in jaundice) and sex hormones (involved in pregnancy) but neither helped his arthritic patients.

However, various bits of evidence pointed toward cortical hormones as a possible answer; and, in 1949, with cortisone available in reasonable quantity, Hench tried that. It worked! It did not cure the disease, any more than insulin cures diabetes, but it seemed to relieve the symptoms, and to an arthritic that alone is manna from heaven. What was more, cortisone later proved to be helpful as a treatment for Addison's disease, where Compound A had failed.

For their work on the cortical hormones, Kendall, Hench, and Reichstein shared the Nobel Prize in medicine and physiology in 1950.

Unfortunately, the influences of the cortical hormones on the body's workings are so multiplex that there are always side effects, sometimes serious. Physicians are reluctant to use cortical-hormone therapy unless the need is clear and urgent. Synthetic substances related to cortical hormones (some with a fluorine atom inserted in the molecule) are being used in an attempt to avoid the worst of the side effects, but nothing approaching a reasonable ideal has yet been found. One of the most active of the cortical hormones discovered so far is *aldosterone*, isolated in 1953 by Reichstein and his co-workers.

THE PITUITARY AND THE PINEAL GLANDS

What controls all the varied and powerful hormones? All of them (including a number I have not mentioned), can exert more or less drastic effects in the body. Yet they are tuned together so harmoniously that they keep the body functioning smoothly without a break in the rhythm. Seemingly, there must be a conductor somewhere that directs their cooperation.

The nearest thing to an answer is the pituitary, a small gland suspended from the bottom of the brain (but not part of it). The name of the gland arose from an ancient notion that its function was to secrete phlegm, the Latin word for which is *pituita* (also the source of the word *spit*). Because this notion is false, scientists have renamed the gland the *hypophysis* (from Greek words meaning "growing under"—that is, under the brain), but *pituitary* is still the more common term.

The gland has three parts: the anterior lobe, the posterior lobe, and, in some organisms, a small bridge connecting the two. The anterior lobe is the most important, for it produces at least six hormones (all small-molecule proteins), which seem to act specifically upon other ductless glands. In other words, the anterior pituitary can be viewed as the orchestra leader that keeps the other glands playing in time and in tune. (It is interesting that the pituitary is located just about in the center of the skull, as if deliberately placed in a spot of maximum security.)

One of the pituitary's messengers is the *thyroid-stimulating hormone* (TSH). It stimulates the thyroid on a feedback basis: that is, it causes the thyroid to produce thyroid hormone. The rise in concentration of thyroid hormone in the blood, in turn, inhibits the formation of TSH by the pituitary; the fall of TSH in the blood in its turn reduces the thyroid's production; that stimulates the production of TSH by the pituitary, and so the cycle maintains a balance.

In the same way, the *adrenal-cortical-stimulating hormone*, or *adrenocorticotropic hormone* (ACTH), maintains the level of cortical hormones. If extra ACTH is injected into the body, it will raise the level of these hormones and thus can serve the same purpose as the injection of cortisone itself. ACTH has therefore been used to treat rheumatoid arthritis.

Research into the structure of ACTH has proceeded with vigor because of this tie-in with arthritis. By the early 1950s, its molecular weight had been determined as 20,000, but it was easily broken down into smaller fragments (*corticotropins*), which possessed full activity. One of them, made up of a chain of thirty-nine amino acids, has had its structure worked out completely, and even shorter chains have been found effective.

ACTH has the ability of influencing the skin pigmentation of animals, and even humans are affected. In diseases involving overproduction of ACTH, human skin darkens. It is known that in lower animals, particularly the amphibians, special skin-darkening hormones exist. A hormone of this sort was finally detected among the pituitary products in the human being in 1955. It is called *melanocyte-stimulating hormone* (melanocytes being the cells that produce skin pigment) and is usually abbreviated MSH.

The molecule of MSH has been largely worked out; it is interesting to note that MSH and ACTH share a seven amino-acid sequence in common. The indication that structure is allied to function (as, indeed, it must be) is unmistakable.

While on the subject of pigmentation, it might be well to mention the pineal gland, a conical body attached, like the pituitary, to the base of the brain and so named because it is shaped like a pine cone. The pineal gland has seemed glandular in nature, but no hormone could be located until the late 1950s. Then the discoverers of MSH, working with 200,000 beef pineals, finally isolated a tiny quantity of substance that, on injection, lightened the skin of a tadpole. The hormone, named *melatonin,* does not, however, appear to have any effect on human melanocytes.

The list of pituitary hormones is not yet complete. A couple of pituitary hormones, ICSH (*interstitial cell-stimulating hormone*) and FSH (*follicle-stimulating hormone*) control the growth of tissues involved in reproduction. There is also the *lactogenic hormone,* which stimulates milk production.

Lactogenic hormone stimulates other postpregnancy activities. Young female rats injected with the hormone busy themselves with nest building even though they have not given birth. On the other hand, mice whose pituitaries have been removed shortly before giving birth to young exhibit little interest in the baby mice. The newspapers at once termed lactogenic hormone the "mother-love hormone."

These pituitary hormones, associated with sexual tissues, are lumped together as the *gonadotropins.* Another substance of this type is produced by the *placenta* (the organ that serves to transfer nourishment from the mother's blood to the blood of the developing infant and to transfer wastes in the opposite direction). The placental hormone is called *human chorionic gonadotropin* and is abbreviated HCG. As early as two to four weeks after the beginning of pregnancy, HCG is produced in appreciable quantities and makes its appearance in the urine. When extracts of the urine of a pregnant woman are injected into mice, frogs, or rabbits, recognizable effects can be detected. Pregnancy can be determined in this way at a very early stage.

The most spectacular of the anterior pituitary hormones is the *somatotropic hormone* (STH), more popularly known as the *growth hormone*. Its effect is general, stimulating growth of the whole body. A child who cannot produce a sufficient supply of the hormone will become a dwarf; one who produces too much will turn into a circus giant. If the disorder that results in an oversupply of the growth hormone does not occur until after the person has matured (that is, when the bones have been fully formed and hardened), only the extremities —such as the hands, feet, and chin—grow grotesquely large—a condition known as *acromegaly* (Greek for "large extremities"). It is this growth hormone that Li (who first determined its structure in 1966) synthesized in 1970.

THE ROLE OF THE BRAIN

Hormones act slowly. They have to be secreted, carried by the blood to some target organ, and build up to some appropriate concentration. Nerve action is very rapid. Both slow control and fast control are needed by the body under various conditions, and to have both systems in action is more efficient than to have either alone. It is not likely that the two systems are entirely independent.

The pituitary, which is a kind of master gland, is suspiciously close to the brain, almost a part of it. The part of the brain to which the pituitary is attached by a narrow stalk is the *hypothalamus;* and from the 1920s, it was suspected that there was some kind of connection.

In 1945, the British biochemist, Geoffrey W. Harris, suggested that the cells of the hypothalamus produced hormones that could be taken by the bloodstream directly to the pituitary. These hormones were detected and termed *releasing factors.* Each particular releasing factor will bring about the production by the anterior pituitary of one of its hormones.

In this way, the nervous system can, to an extent, control the hormone system.

The brain, as a matter of fact, seems increasingly to be not merely a "switchboard" of nerve cells in superintricate arrangement, but is a highly specialized chemical factory that may turn out to be just as intricate.

The brain, for instance, contains certain receptors that receive nerve impulses to which it ordinarily responds by producing the sensation of pain. Anesthetics such as morphine and cocaine attach themselves to the receptors and blank out the pain.

Sometimes people, under the stress of strong emotion, do not feel pain when ordinarily they would. Some natural chemical must block the pain receptors on those occasions. In 1975, such chemicals were found and isolated from the brains of animals at a number of laboratories. They are peptides, short chains of amino acids, the shortest (*enkephalins*) made up of five amino acids only, while longer ones are *endorphins*.

It may well be that the brain creates, fleetingly, large numbers of different peptides each of which modifies brain action in some way—easily produced, easily broken down. To understand the brain, it is likely that it will have to be studied intimately both chemically and electrically.

THE PROSTAGLANDINS

Before leaving the hormones, I should mention a group that have recently become prominent that are built up of neither amino acids nor a steroid nucleus.

In the 1930s, the Swedish physiologist Ulf Svante von Euler isolated a fat-soluble substance from the prostate gland, which, in small quantities, lowered blood pressure and caused certain smooth muscles to contract. (Von Euler was the son of Nobel Laureate Euler-Chelpin and went on to win a share of the 1970 Nobel Prize for physiology and medicine for his work on nerve transmission.) Von Euler called the substance *prostaglandin* because of its source.

It turned out to be not one substance but many. At least fourteen prostaglandins are known. Their structure has been worked out, and they are found all to be related to polyunsaturated fatty acids. It may be because of the need to form prostaglandins that the body, which cannot manufacture these fatty acids, requires them in the diet. They all have similar effects on blood pressure and smooth muscle but to different degrees, and their functions are not yet entirely elucidated.

HORMONE ACTION

How do hormones work?

It seems certain that the hormones do not act as enzymes. At least, no hormone has been found to catalyze a specific reaction directly. The next alternative is to suppose that a hormone, if not itself an enzyme, acts upon an enzyme: that it either promotes or inhibits an enzyme's activity. Insulin, the most thoroughly investigated of all the hormones, does seem to be definitely connected with an enzyme called *glucokinase,* which is essential for the conversion of glucose to glycogen. This enzyme is inhibited by extracts from the anterior pituitary and the adrenal cortex, and insulin can nullify that inhibition. Thus, insulin in the blood may serve to activate the enzyme and so speed up the conversion of glucose to glycogen. That would help to explain how insulin lowers the glucose concentration in the blood.

Yet the presence or the absence of insulin affects metabolism at so many points that it is hard to see how this one action could bring about all the abnormalities that exist in the body chemistry of a diabetic. (The same is true for other hormones.) Some biochemists have therefore tended to look for grosser and more wholesale effects.

There is the suggestion that insulin somehow acts as an agent to get glucose into the cell. On this theory, a diabetic has a high glucose level in his blood for the simple reason that the sugar cannot get into his cells and therefore he cannot use it. (In explaining the insatiable appetite of a diabetic, Mayer, as

I have already mentioned, suggested that glucose in the blood has difficulty in entering the cells of the appestat.)

If insulin assists glucose in entering the cell, then it must act on the cell membrane in some way. How? Cell membranes are composed of protein and fatty substances. We can speculate that insulin, as a protein molecule, may somehow change the arrangement of amino-acid side chains in the protein of the membrane and thus open doors for glucose (and possibly many other substances).

If we are willing to be satisfied with generalities of this kind, we can go on to suppose that the other hormones also act on the cell membranes, each in its own fashion because each has its own specific amino-acid arrangement. Similarly, steroid hormones, as fatty substances, may act on the fatty molecules of the membrane, either opening or closing the door to certain substances. Clearly, by helping a given material to enter the cell or preventing it from doing so, a hormone could exert a drastic effect on what goes on in the cell. It could supply one enzyme with plenty of substrate to work on and deprive another of material, thus controlling what the cell produces. Assuming that a single hormone may decide the entrance or nonentrance of several different substances, we can see how the presence or the absence of a hormone could profoundly influence metabolism, as in fact it does in the case of insulin.

The picture I have drawn is attractive but vague. Biochemists would much prefer to know the exact reactions that take place at the cell membrane under the influence of a hormone. The beginning of such knowledge came with the discovery in 1960 of a nucleotide like adenylic acid except that the phosphate group was attached to two different places in the sugar molecule. Its discoverers, Earl Wilbur Sutherland, Jr., and Theodore W. Rall, called it *cyclic AMP*. It was "cyclic" because the doubly attached phosphate-group formed a circle of atoms, and AMP stood for "*a*denine *m*ono*p*hosphate," an alternate name for adenylic acid. Sutherland received the 1971 Nobel Prize for physiology and medicine for this work.

Once discovered, cyclic AMP was found to be widely spread in tissue, and to have a pronounced effect on the activity of many different enzymes and cell processes. Cyclic AMP is produced from the universally occurring ATP by means of an enzyme named *adenyl cyclase*, which is located at the surface of cells. There may be several such enzymes, each geared for activity in the presence of a particular hormone. In other words, the surface activity of hormones serves to activate an adenyl cyclase that leads to the production of cyclic AMP, which alters the enzyme activity within the cell, producing many changes.

Undoubtedly, the details are enormously complex, and compounds other than cyclic AMP may be involved (possibly the prostaglandins)—but it is a beginning.

Death

The advances made by modern medicine in the battle against infection, against cancer, against nutritional disorders, have increased the probability that any given individual will live long enough to experience old age. Half the people born in this generation can be expected to reach the age of seventy (barring a nuclear war or some other prime catastrophe).

The rarity of survival to old age in earlier eras no doubt accounts in part for the extravagant respect paid to longevity in those times. The *Iliad,* for instance, makes much of "old" Priam and "old" Nestor. Nestor is described as having survived three generations of men; but at a time when the average length of life could not have been more than twenty to twenty-five, Nestor need not have been older than seventy to have survived three generations. That is old, yes, but not extraordinary by present standards. Because Nestor's antiquity made such an impression on people in Homer's time, later mythologists supposed that he must have been something like two hundred years old.

To take another example at random, Shakespeare's *Richard II* opens with the rolling words: "Old John of Gaunt, time-honored Lancaster." John's own contemporaries, according to the chroniclers of the time, also considered him an old man. It comes as a slight shock to realize that John of Gaunt lived only to the age of fifty-nine. An interesting example from our own history is that of Abraham Lincoln. Whether because of his beard, or his sad, lined face, or songs of the time that referred to him as "Father Abraham," most people think of him as an old man at the time of his death. One could only wish that he had lived to be one. He was assassinated at the age of fifty-six.

All this is not to say that really old age was unknown in the days before modern medicine. In ancient Greece, Sophocles, the playwright, lived to be ninety; and Isocrates, the orator, to ninety-eight. Flavius Cassiodorus of fifth-century Rome died at ninety-five. Enrico Dandolo, the twelfth-century doge of Venice, lived to be ninety-seven. Titian, the Renaissance painter, survived to ninety-nine. In the era of Louis XV, the Duc de Richelieu, grandnephew of the famous cardinal, lived ninety-two years; and the French writer Bernard Le Bovier de Fontenelle managed to arrive at just a month short of one hundred years.

This emphasizes the point that although the average life expectancy in medically advanced societies has risen greatly, the maximum life span has not. We expect very few individuals, even today, to attain or exceed the lifetime of an Isocrates or a Fontenelle. Nor do we expect modern nonagenarians to be able to participate in the business of life with any greater vigor. Sophocles was writing great plays in his nineties, and Isocrates was composing great orations. Titian painted to the last year of his life. Dandolo was the indomitable leader of a Venetian war against the Byzantine Empire at the age of ninety-six. (Among comparably vigorous oldsters of our day, the best example I can think of is George Bernard Shaw, who lived to ninety-four, and the

English mathematician and philosopher Bertrand Russell, who was still active in his ninety-eighth year, when he died.)

Although a far larger proportion of our population reaches the age of sixty than ever before, beyond that age life expectancy has improved very little over the past. The Metropolitan Life Insurance Company estimates that the life expectancy of a sixty-year-old American male in 1931 was just about the same as it was a century and a half earlier—that is, 14.3 years against the estimated earlier figure of 14.8. For the average American woman, the corresponding figures were 15.8 and 16.1. Since 1931, the advent of antibiotics has raised the expectancy at sixty for both sexes by two and a half years. But on the whole, despite all that medicine and science have done, old age overtakes a person at about the same rate and in the same way as it always has. We have not yet found a way to stave off the gradual weakening and eventual breakdown of the human machine.

ATHEROSCLEROSIS

As in other forms of machinery, it is the moving parts that go first. The circulatory system—the pulsing heart and arteries—is the human's Achilles' heel in the long run. Our progress in conquering premature death has raised disorders of this system to the rank of the number-one killer. Circulatory diseases are responsible for just over half the deaths in the United States; and of these diseases, a single one, atherosclerosis, accounts for one death out of four.

Atherosclerosis (from Greek words meaning "mealy hardness") is characterized by grainlike fatty deposits along the inner surface of the arteries, which force the heart to work harder to drive blood through the vessels at a normal pace. The blood pressure rises, and the consequent increase in strain on the small blood vessels may burst them. If this happens in the brain (a particularly vulnerable area), one has a cerebral hemorrhage, or *stroke.* Sometimes the bursting of a vessel is so minor that it occasions only a trifling and temporary discomfort or even goes unnoticed, but a massive collapse of vessels will bring on paralysis or a quick death.

The *coronary arteries* which supply oxygen to the heart itself, are particularly susceptible to atherosclerotic narrowing. The resulting oxygen starvation of the heart gives rise to the agonizing pain of *angina pectoris,* and, eventually though not necessarily quickly, to death.

The roughening and narrowing of the arteries introduces another hazard. Because of the increased friction of the blood scraping along the roughened inner surface of the vessels, blood clots are more likely to form, and the narrowing of the vessels heightens the chances that a clot will completely block the blood flow. In the coronary artery, feeding the heart muscle itself, a block (*coronary thrombosis*) can produce almost instant death.

Just what causes the formation of deposits on the artery wall is a matter of much debate among medical scientists. Cholesterol certainly seems to be involved, but how it is involved is still far from clear. The plasma of human blood contains *lipoproteins,* which consist of cholesterol and other fatty substances bound to certain proteins. Some of the fractions making up lipoprotein maintain a constant concentration in the blood—in health and in disease, before and after eating, and so on. Others fluctuate, rising after meals. Still others are particularly high in obese individuals. One fraction, rich in cholesterol, is particularly high in overweight people and in those with atherosclerosis.

Atherosclerosis tends to go along with a high blood-fat content, and so does obesity. Overweight people are more susceptible to atherosclerosis than are thin people. Diabetics also have high blood-fat levels and are more susceptible to atherosclerosis than are normal individuals. And, to round out the picture, the incidence of diabetes among the stout is considerably higher than among the thin.

It is thus no accident that those who live to a great age are often scrawny, little fellows. Large, fat men may be jolly, but they do not keep the sexton waiting unduly, as a rule. (Of course, there are always exceptions, and one can point to men such as Winston Churchill and Herbert Hoover, who passed their ninetieth birthdays but were never noted for leanness.)

The key question, at the moment, is whether atherosclerosis can be fostered or prevented by the diet. Animal fats—such as those in milk, eggs, and butter —are particularly high in cholesterol; plant fats lack it altogether. Moreover, the fatty acids of plant fats are mainly of the unsaturated type, which has been reported to counter the deposition of cholesterol. Investigations of these matters seemed to show conclusively, in 1984, that cholesterol in the diet is involved in atherosclerosis and people have been flocking to *low-cholesterol diets*, in the hope of staving off thickening of the artery walls.

Of course, the cholesterol in the blood is not necessarily derived from the cholesterol of the diet. The body can and does make its own cholesterol with great ease, and even though you live on a diet that is completely free of cholesterol, you will still have a generous supply of cholesterol in your blood lipoproteins. It therefore seems reasonable to suppose that what matters is not the mere presence of cholesterol but the individual's tendency to deposit it where it will do the most harm. It may be that there is a hereditary tendency to manufacture excessive amounts of cholesterol. Biochemists are seeking drugs that will inhibit cholesterol formation, in the hope that such drugs may forestall the development of atherosclerosis in those who are susceptible to the disease.

Meanwhile *coronary bypass surgery*—the use of other blood vessels from a patient's body to attach to the coronary arteries in such a way that blood flows freely through the bypass around the atherosclerotic region and supplies the heart with an ample blood supply—has become very common since its introduction in 1969, and very successful. It does not seem to lengthen one's overall life expectancy, but it makes one's final years free of crippling anginal pain, and (as they know who have experienced it) that is a great deal.

OLD AGE

But even those who escape atherosclerosis grow old. Old age is a disease of universal incidence. Nothing can stop the creeping enfeeblement, the increasing brittleness of the bones, the weakening of the muscles, the stiffening of the joints, the slowing of reflexes, the dimming of sight, the declining agility of the mind. The rate at which this happens is somewhat slower in some people than in others—but, fast or slow, the process is inexorable.

Perhaps we ought not complain too loudly about this. If old age and death must come, they arrive unusually slowly. In general, the life span of mammals correlates with size. The smallest mammal, the shrew, may live one and a half years, and a rat may live four or five. A rabbit may live up to fifteen years, a dog up to eighteen, a pig up to twenty, a horse up to forty, and an elephant up to seventy. To be sure, the smaller the animal the more rapidly it lives— the faster its heartbeat, for instance. A shrew with a heartbeat of 1,000 per minute can be matched against an elephant with a heartbeat of 20 per minute.

In fact, mammals in general seem to live, at best, as long as it takes their hearts to count a billion. To this general rule, human beings themselves are the most astonishing exception. Though considerably smaller than a horse and far smaller than an elephant, the human being can live longer than any mammal can. Even if we discount tales of vast ages from various backwoods where accurate records have not been kept, there are reasonably convincing data for life spans of up to 115 years. The only vertebrates to outdo this record, without question, are certain large, slow-moving tortoises.

A man's heartbeat of about seventy-two per minute is just what is to be expected of a mammal of his size. In seventy years, which is the average life expectancy in the technologically advanced areas of the world, the human heart has beaten 2.5 billion times; at 115 years, it has beaten about 4 billion times. Even our nearest relatives, the great apes, cannot match this, even closely. The gorilla, considerably larger than a man, is in extreme old age at fifty.

There is no question but that the human heart outperforms all other hearts in existence. (The tortoise's heart may last longer but it lives nowhere near as intensely.) Why we should be so long-lived is not known; but as humans, we are far more interested in asking why we do not live still longer.

What is old age, anyway? So far, there are only speculations. Some have suggested that the body's resistance to infection slowly decreases with age (at a rate depending on heredity). Others speculate that *clinkers* of one kind or another accumulate in the cells (again, at a rate that varies from individual to individual). These supposed side products of normal cellular reactions, which the cell can neither destroy nor get rid of, slowly build up in the cell as the years pass, until they eventually interfere with the cell's metabolism so seriously that it ceases to function. When enough cells are put out of action, so the theory goes, the body dies. A variation of this notion holds that the protein molecules themselves become clinkers, because cross links develop between them so that they become stiff and brittle and finally bring the cell machinery grinding to a halt.

If this is so, then "failure" is built into the cell machinery. Carrel's ability to keep a piece of embryonic tissue alive for decades had made it seem that cells themselves might be immortal: it was only the organization into combinations of trillions of individual cells that brought death. The organization failed, not the cells.

Not so, apparently. It is now thought that Carrel may (unwittingly) have introduced fresh cells into his preparation in the process of feeding the tissue. Attempts to work with isolated cells or groups of cells in which the introduction of fresh cells was rigorously excluded seem to show that the cells inevitably age and will not divide more than fifty times all told—presumably through irreversible changes in the key cell components.

And yet there is the extraordinarily long human life span. Can it be that human tissue has developed methods of reversing or inhibiting cellular aging effects, methods that are more efficient than those in any other mammal? Again, birds tend to live markedly longer than mammals of the same size despite the fact that bird metabolism is even more rapid than mammal metabolism—again, superior ability of old age reversal or inhibition.

If old age can be staved off more by some organisms than by others, there seems no reason to suppose that humans cannot learn the method and improve upon it. Might not old age, then, be curable, and might not we develop the ability to enjoy an enormously extended life span—or even immortality?

General optimism in this respect is to be found among some people. Medical miracles in the past would seem to herald unlimited miracles in the future. And if that is so, what a shame to live in a generation that will just miss a cure for cancer, or for arthritis, or for old age!

In the late 1960s, therefore, a movement grew to freeze human bodies at the moment of death, in order that the cellular machinery might remain as intact as possible, until the happy day when whatever it was that marked the death of the frozen individual, could be cured. He or she would then be revived and made healthy, young, and happy.

To be sure, there is no sign at the present moment that any dead body can be restored to life, or that any frozen body—even if alive at the moment of freezing—can be thawed to life. Nor do the proponents of this procedure (*cryonics*) give much attention to the complications that might arise in the flood of dead bodies returned to life. The personal hankering for immortality governs all.

Actually, it makes little sense to freeze intact bodies, even if all possible revival could be done. It is wasteful. Biologists have so far had much more luck with the developing of whole organisms from groups of specialized cells. Skin cells or liver cells, after all, have the same genetic equipment that other cells have, and that the original fertilized ovum had in the first place. The cells are specialized because the various genes are inhibited or activated to varying extents. But might not the genes be deinhibited or deactivated, and might they not then make their cell into the equivalent of a fertilized ovum and develop an organism all over again—the same organism, genetically speaking, as the one of which they had formed part? Surely, this procedure (called *cloning*) offers more hope for a kind of preservation of the genetic pattern (if not the memory and personality). Instead of freezing an entire body, chop off the little toe and freeze that.

But do we really want immortality—either through cryonics, through cloning, or through simple reversal of the aging phenomenon in each individual? There are few human beings who would not eagerly accept an immortality reasonably free of aches, pains, and the effects of age—but suppose we were all immortal?

Clearly, if there were few or no deaths on earth, there would have to be few or no births. It would mean a society without babies. Presumably that is not fatal; a society self-centered enough to cling to immortality would not stop at eliminating babies altogether.

But will that do? It would be a society composed of the same brains, thinking the same thoughts, circling the same ruts in the same way, endlessly. It must be remembered that babies possess not only young brains but *new* brains. Each baby (barring identical multiple births) has genetic equipment unlike that of any human individual who ever lived. Thanks to babies, there are constantly fresh genetic combinations injected into humanity, so that the way is open toward improvement and development.

It would be wise to lower the level of the birth rate, but ought we to wipe it out entirely? It would be pleasant to eliminate the pains and discomforts of old age, but ought we to create a species consisting of the old, the tired, the bored, the same, and never allow for the new and the better?

Perhaps the prospect of immortality is worse than the prospect of death.

Chapter 16

The Species

Varieties of Life

Our knowledge of our own bodies is incomplete without a knowledge of our relationship to the rest of life on the earth.

In primitive cultures, the relationship was often considered to be close indeed. Many tribes regarded certain animals as their ancestors or blood brothers, and made it a crime to kill or eat them, except under certain ritualistic circumstances. This veneration of animals as gods or near-gods is called *totemism* (from an American Indian word), and there are signs of it in cultures that are not so primitive. The animal-headed gods of Egypt were a hangover of totemism, and so, perhaps, is the modern Hindu veneration of cows and monkeys.

On the other hand, Western culture, as exemplified in Greek and Hebrew ideas, very early made a sharp distinction between human beings and the "lower animals." Thus, the Bible emphasizes that Adam was produced by a special act of creation in the image of God, "after our likeness" (Genesis 1:26). Yet the Bible attests, nevertheless, to man's remarkably keen interest in the lower animals. Genesis mentions that Adam, in his idyllic early days in the Garden of Eden, was given the task of naming "every beast of the field, and every fowl of the air."

Offhand, that seems not too difficult a task—something that one could do in perhaps an hour or two. The scriptural chroniclers put "two of every sort" of animal in Noah's Ark, whose dimensions were 450 by 75 by 45 feet (if we take the cubit to be 18 inches). The Greek natural philosophers thought of the living world in similarly limited terms: Aristotle could list only about 500 kinds of animals, and his pupil Theophrastus, the most eminent botanist of ancient Greece, listed only about 500 different plants.

Such a list might make some sense if one thought of an elephant as always an elephant, a camel as just a camel, or a flea as simply a flea. Things began to get a little more complicated when naturalists realized that animals had to be differentiated on the basis of whether they could breed with each other. The Indian elephant could not interbreed with the African elephant; therefore, they had to be considered different *species* of elephant. The Arabian camel (one hump) and the Bactrian camel (two humps) also are separate species. As for the flea, the small biting insects (all resembling the common flea) are divided into 500 different species!

Through the centuries, as naturalists counted new varieties of creatures in the field, in the air, and in the sea, and as new areas of the world came into view through exploration, the number of identified species of animals and plants grew astronomically. By 1800 it had reached 70,000. Today more than 1,500,000 million different species—two-thirds animal and one-third plant— are known, and no biologist supposes the count to be complete.

Even fairly large animals remain to be found in odd corners of the globe. The okapi, a relative of the giraffe and the size of a zebra, became known to biologists only in 1900 when it was finally tracked down in the Congo forests. Even in 1983, a new kind of albatross was recorded on an island in the Indian ocean, and two new kinds of monkey were found in the Amazon jungles.

Undiscovered varieties of organisms are sure to be hidden in the ocean depths where investigation is more difficult. The giant squid, the largest of all invertebrates, was not proved to exist until the 1860s. The coelacanth (see chapter 4) was discovered only in 1938.

As for small animals—insects, worms, and so on—new varieties are discovered every day. A conservative estimate would have it that there are 10 million species of living things existing in the world today. If it is true that some nine-tenths of all the species that have ever lived are now extinct then 100 million species of living things have been found on Earth at some time or other.

CLASSIFICATION

The living world would be exceedingly confusing if we were unable to classify this enormous variety of creatures according to some scheme of relationships. One can begin by grouping together the cat, the tiger, the lion, the panther, the leopard, the jaguar, and other catlike animals in the *cat family;* likewise, the dog, the wolf, the fox, the jackal, and the coyote form a *dog family,* and so on. On the basis of obvious general criteria, one can go on to classify some animals as meat eaters and others as plant eaters. The ancients also set up general classifications based on habitat and so considered all animals that live in the sea to be fishes and all that fly in the air to be birds. But this standard made the whale a fish and the bat a bird. Actually, in a fundamental sense, the whale and the bat are more like each other than the one is like a fish or the other like a bird. Both bear live young. Moreover, the whale has air-breathing lungs, rather than the gills of a fish, and the bat has hair instead of the feathers of a bird. Both are classed with the mammals, which give birth to living babies (instead of laying eggs) and feed them on mother's milk.

One of the earliest attempts to make a systematic classification was that of an Englishman named John Ray (or Wray), who in the seventeenth century classified all the known species of plants (about 18,600), and later the species of animals, according to systems that seemed to him logical. For instance, he divided flowering plants into two main groups, on the basis of whether the seed contained one embryonic leaf or two. The tiny embryonic leaf or pair of leaves had the name *cotyledon*, from the Greek word for a kind of cup (*kotyle*), because it lay in a cuplike hollow in the seed. Ray therefore named the two types respectively *monocotyledonous* and *dicotyledonous*. The classification (similar, by the way, to one set up 2,000 years earlier by Theophrastus) proved so useful that it is still in effect today. The difference between one embryonic leaf and two in itself is unimportant, but there are a number of important ways in which all monocotyledonous plants differ from all dicotyledonous ones. The difference in the embryonic leaves is just a handy tag which is symptomatic of many general differences. (In the same way, the distinction between feathers and hair is minor in itself but is a handy marker for the vast array of differences that separates birds from mammals.)

Although Ray and others contributed some useful ideas, the real founder of the science of classification, or *taxonomy* (from a Greek word meaning "arrangement"), was a Swedish botanist best known by his Latinized name of Carolus Linnaeus, who did the job so well that the main features of his scheme still stand today. Linnaeus set forth his system in 1737 in a book entitled *Systema Naturae*. He grouped species resembling one another into a *genus* (from a Greek word meaning "race" or "sort"), put related genera in turn into an *order*, and grouped similar orders in a *class*. Each species was given a double name, made up of the name of the genus and of the species itself. (This is much like the system in the telephone book, which lists Smith, John; Smith, William; and so on.) Thus the members of the genus of cats are *Felis domesticus* (the pussycat), *Felis leo* (the lion), *Felis tigris* (the tiger), *Felis pardus* (the leopard), and so on. The genus to which the dog belongs includes *Canis familiaris* (the dog), *Canis lupus* (the European gray wolf), *Canis occidentalis* (the American timber wolf), and so on. The two species of camel are *Camelus bactrianus* (the Bactrian camel) and *Camelus dromedarius* (the Arabian camel).

Around 1800, the French naturalist Georges Leopold Cuvier went beyond *classes* and added a more general category called the *phylum* (from a Greek word for "tribe"). A phylum includes all animals with the same general body plane (a concept that was emphasized and made clear by none other than the great German poet Johann Wolfgang von Goethe). For instance, the mammals, birds, reptiles, amphibia, and fishes are placed in one phylum because all have backbones, a maximum of four limbs, and red blood containing hemoglobin. Insects, spiders, lobsters, and centipedes are placed in another phylum; clams, oysters, and mussels in still another; and so on. In the 1820s, the Swiss botanist Augustin Pyramus de Candolle similarly improved Linnaeus's classification of plants. Instead of grouping species together according to external appearance, he laid more weight on internal structure and functioning.

The tree of life now is arranged as I shall describe in the following paragraphs, going from the most general divisions to the more specific.

We start with the *kingdoms*, which for a long time were assumed to be two in number: animals and plants. (The assumption is still made in the popular game of "Twenty Questions," in which everything is classified as "animal, vegetable or mineral.") However, the growing knowledge concerning the microorganisms complicated matters, and the American biologist Robert Harding Whittaker suggested that living organisms be divided into no fewer than five kingdoms.

By Whittaker's system, the plant kingdom and the animal kingdom are confined to multicellular organisms. The plants are characterized by the possession of chlorophyll (so that they are the so-called *green plants*) and the use of photosynthesis. The animals ingest other organisms as food and have digestive systems.

A third kingdom, the *fungi*, are multicellular and resemble plants in some ways but lack chlorophyll. They live on other organisms though they do not ingest them as animals do, but excrete digestive enzymes, digest their food outside the body, then absorb it.

The remaining two kingdoms contain one-celled organisms. *Protista*, a word coined in 1866 by the German biologist Ernst Heinrich Haeckel, includes the eukaryotes: both those that are made of cells resembling those that constitute animals (*protozoa*, such as the amoeba and the paramecium); and those that are cells resembling those that constitute plants (*algae*).

Finally, a kingdom known as *monera* contain the one-celled organisms that are prokaryotes—the bacteria and the blue-green algae. Left out of this scheme are the viruses and viroids which are subcellular and might well form a sixth kingdom.

The plant kingdom, according to one system of classification, is divided into two main phyla—the Bryophyta (the various mosses) and the Tracheophyta (plants with systems of tubes for the circulation of sap), which includes all the species that we ordinarily think of as plants.

This last great phylum is made up of three main classes: the Filicineae, the Gymnospermae, and the Angiospermae. In the first class are the ferns, which reproduce by means of spores. The gymnosperms, forming seeds on the surface of the seed-bearing organs, include the various evergreen cone-bearing trees. The angiosperms, with the seeds enclosed in ovules, make up the vast majority of the familiar plants.

As for the animal kingdom, I shall list only the more important phyla.

The Porifera are animals consisting of colonies of cells within a pore-bearing skeleton; these are the sponges. The individual cells show signs of specialization but retain a certain independence, for when all are separated by straining through a silk cloth, they may aggregate to form a new sponge.

(In general, as the animal phyla grow more specialized, individual cells and tissues grow less "independent." Simple creatures can regrow to entire organisms even though badly mutilated, a process called *regeneration*. More complex ones can regrow limbs. By the time we reach humans, however, the capacity for regeneration has sunk quite low. We can regrow a lost fingernail but not a lost finger.)

The first phylum whose members can be considered truly multicellular animals is the Coelenterata (meaning "hollow gut"). These animals have the basic shape of a cup and consist of two layers of cells—the *ectoderm* ("outer skin") and the *endoderm* ("inner skin"). The most common examples of this phylum are the jellyfish and the sea anemones.

All the rest of the animal phyla have a third layer of cells—the *mesoderm* ("middle skin"). From these three layers, first recognized in 1845 by the German physiologists Johannes Peter Müller and Robert Remak, are formed the many organs of even the most complex animals, including man.

The mesoderm arises during the development of the embryo, and the manner in which it arises divides the animals involved into two *superphyla*. Those in which the mesoderm forms at the junction of the ectoderm and the endoderm make up the Annelid superphylum; those in which the mesoderm arises in the endoderm alone are the Echinoderm superphylum.

Let us consider the Annelid superphylum first. Its simplest phylum is Platyhelminthes (Greek for "flat worms"). This includes not only the parasitic tapeworm but also free-living forms. The flatworms have contractile fibers that can be considered primitive muscles, and they also possess a head, a tail, special reproductive organs, and the beginnings of excretory organs. In addition, the flatworms display bilateral symmetry: that is, they have left and right sides that are mirror images of each other. They move headfirst, and their sense organs and rudimentary nerves are concentrated in the head area, so that the flatworm can be said to possess the first step toward a brain.

Next comes the phylum Nematoda (Greek for "thread worm"), whose most familiar member is the hookworm. These creatures possess a primitive bloodstream—a fluid within the mesoderm that bathes all the cells and conveys food and oxygen to them. This allows the nematodes, in contrast to animals such as the flat tapeworm, to have bulk, for the fluid can bring nourishment to interior cells. The nematodes also possess a gut with two openings, one for the entry of food, the other (the anus) for ejection of wastes.

The next two phyla in this superphylum have hard external *skeletons*—that is, shells (which are found in some of the simpler phyla, too). These two groups are the Brachiopoda, which have calcium carbonate shells on top and bottom and are popularly called *lampshells*, and the Mollusca (Latin for "soft"), whose soft bodies are enclosed in shells originating from the right and left sides instead of the top and bottom. The most familiar molluscs are the clams, oysters, and snails.

A particularly important phylum in the Annelid superphylum is Annelida. These are worms, but with a difference: they are composed of segments, each of which can be looked upon as a kind of organism in itself. Each segment has its own nerves branching off the main nerve stem, its own blood vessels, its own tubules for carrying off wastes, its own muscles, and so on. In the most familiar annelid, the earthworm, the segments are marked off by little constrictions of flesh which look like little rings around the animal; in fact, *Annelida* is from a Latin word meaning "little ring."

Segmentation apparently endows an animal with superior efficiency, for all the most successful species of the animal kingdom, including the human, are segmented. (Of the nonsegmented animals, the most complex and successful is the squid.) If you wonder how the human body is segmented, think of the vertebrae and the ribs; each vertebra of the backbone and each rib represents a separate segment of the body, with its own nerves, muscles, and blood vessels.

The annelids, lacking a skeleton, are soft and relatively defenseless. The phylum Arthropoda ("jointed feet"), however, combines segmentation with a skeleton, the skeleton being as segmented as the rest of the body. The skeleton is not only more maneuverable for being jointed; it is also light and tough, being made of a polysaccharide called *chitin* rather than of heavy, inflexible limestone or calcium carbonate. On the whole, the Arthropoda—which includes lobsters, spiders, centipedes, and insects—is the most successful phylum in existence. At least it contains more species than all the other phyla put together.

This accounts for the main phyla in the Annelid superphylum. The other superphylum, the Echinoderm, contains only two important phyla. One is Echinodermata ("spiny skin"), which includes such creatures as the starfish and the sea urchin. The echinoderms differ from other mesoderm-containing phyla in possessing radial symmetry and having no clearly defined head and tail (though, in early life, echinoderms do show bilateral symmetry, which they lose as they mature).

The second important phylum of the Echinoderm superphylum is important indeed, for it is the one to which human beings themselves belong.

THE VERTEBRATES

The general characteristic that distinguishes the members of this phylum (which embraces the human being, ostrich, snake, frog, mackerel, and a varied host of other animals) is an internal skeleton (figure 16.1). No animal outside this phylum possesses one. The particular mark of such a skeleton is the backbone. In fact, the backbone is so important a feature that, in common parlance, all animals are loosely divided into vertebrates and invertebrates. Actually, there is an in-between group which has a rod of cartilage called a

Figure 16.1. A philogenetic tree, showing evolutionary lines of the verte-brates.

NERVE CHORD
NOTOCHORD   FIN RAYS   SEGMENTAL
                           MUSCLES   TAIL

CIRRI   ENDOSTYLE
   VELUM   GILL SLITS

Figure 16.2. Amphioxus, a primitive, fishlike chordate with a notochord.

notochord ("backcord") in the place of the backbone (figure 16.2). The notochord, first discovered by Von Baer, who had also discovered the mammalian ovum, seems to represent a rudimentary backbone; in fact, it makes its appearance even in mammals during the development of the embryo. So the animals with notochords (various wormlike, sluglike, and mollusclike creatures) are classed with the vertebrates. The whole phylum was named Chordata in 1880, by the English zoologist Francis Maitland Balfour; it is divided into four subphyla, three of which have only a notochord. The fourth, with a true backbone and general internal skeleton, is Vertebrata.

The vertebrates in existence today form two superclasses: the Pisces ("fishes") and the Tetrapoda ("four-footed" animals).

The Pisces group is made up of three classes: (1) the Agnatha ("jawless") fishes, which have true skeletons but no limbs or jaws—the best-known representative, the lamprey, possessing a rasping set of files in a round suckerlike mouth; (2) the Chondrichthyes (*cartilage fish*), with a skeleton of cartilage instead of bone, sharks being the most familiar example; and (3) the Osteichthyes, or *bony fishes*.

The *tetrapods*, or four-footed animals, all of which breathe by means of lungs, make up four classes. The simplest are the Amphibia ("double life")—for example, the frogs and toads. The double life means that in their immature youth (for example, as tadpoles), they have no limbs and breathe by means of gills; then as adults they develop four feet and lungs. The amphibians, like fishes, lay their eggs in the water.

The second class are the Reptilia (from a Latin word meaning "creeping"). They include snakes, lizards, alligators, and turtles. They breathe with lungs from birth, and hatch their eggs (enclosed in a hard shell) on land. The most advanced reptiles have essentially four-chambered hearts, whereas the amphibian's heart has three chambers, and the fish's heart only two.

The final two groups of tetrapods are the Aves (birds) and the Mammalia (mammals). All are warm-blooded: that is, their bodies possess devices that maintain an even internal temperature regardless of the temperature outside (within reasonable limits). Since the internal temperature is usually higher than the external, these animals require insulation. As aids to this end, the birds are equipped with feathers and the mammals with hair, both serving to trap a layer of insulating air next to the skin. The birds lay eggs like those of reptiles. The mammals, of course, bring forth their young already "hatched" and supply them with milk produced by mammary glands (*mammae* in Latin).

In the nineteenth century, zoologists heard reports of a great curiosity so amazing that they refused to believe it. The Australians had found a creature that had hair and produced milk (through mammary glands that lacked nipples), yet laid eggs! Even when the zoologists were shown specimens of the animal (not alive, unfortunately, because it is not easy to keep it alive away from its natural habitat), they were inclined to brand it a clumsy fraud. The beast was a land-and-water animal that looked a good deal like a duck: it had a bill and webbed feet. Eventually the *duckbilled platypus* had to be recognized as a genuine phenomenon and a new kind of mammal. Another egg-laying mammal, the echidua has since been found in Australia and New Guinea. Nor is it only in the laying of eggs that these mammals show themselves to be still close to the reptile. They are only imperfectly warm-blooded; on cold days their internal temperature may drop as much as 10 degrees centigrade.

The mammals are now divided into three subclasses. The egg-laying mammals form the first class, Prototheria (Greek for "first beasts"). The embryo in the egg is actually well developed by the time the egg is laid, and it hatches out not long afterward. The second subclass of mammals, Metatheria ("mid-beasts"), includes the opossums and kangaroos. Their young, though born alive, are in a very undeveloped form and will die in short order unless they manage to reach the mother's protective pouch and stay at the mammary nipples until they are strong enough to move about. These animals are called *marsupials* (from *marsupium,* Latin for "pouch").

Finally, at the top of the mammalian hierarchy, we come to the subclass Eutheria ("true beasts"). Their distinguishing feature is the placenta, a blood-suffused tissue that enables the mother to supply the embryo with food and oxygen and carry off its wastes, so that she can develop the offspring for a long period inside her body (nine months in the case of the human being, two years in the case of elephants and whales). The eutherians are usually referred to as *placental mammals.*

The placental mammals are divided into well over a dozen orders, of which the following are examples:

Insectivora ("insect-eating")—shrews, moles, and others.

Chiroptera ("hand-wings")—the bats.

Carnivora ("meat-eating")—the cat family, the dog family, weasels, bears, seals, and so on, but not including human beings.

Rodentia ("gnawing")—mice, rats, rabbits, squirrels, guinea pigs, beavers, porcupines, and so on.

Edentata ("toothless")—the sloths and armadillos, which have teeth, and anteaters, which do not.

Artiodactyla ("even toes")—hoofed animals with an even number of toes on each foot, such as cattle, sheep, goats, swine, deer, antelopes, camels, giraffes, and so on.

Perissodactyla ("odd toes")—horses, donkeys, zebras, rhinoceroses, and tapirs.

Proboscidea ("long nose")—the elephants, of course.

Odontoceti ("toothed whales")—the sperm whale and others with teeth.

Mysticeti ("mustached whales")—the right whale, the blue whale, and others that filter their small sea food through fringes of whalebone that look like a colossal mustache inside the mouth.

Primates ("first")—humans, apes, monkeys, and some other creatures with which we may be surprised to find ourselves associated.

The primates are characterized by hands and sometimes feet that are equipped for grasping, with opposable thumbs and big toes. The digits are topped with flattened nails rather than with sharp claws or enclosing hoofs. The brain is enlarged, and the sense of vision is more important than the sense of smell. There are many other, less obvious, anatomical criteria.

The primates are divided into nine families. Some have so few primate characteristics that it is hard to think of them as primates, but so they must be classed. One is the family Tupaiidae, which includes the insect-eating tree-shrews! Then there are the lemurs—nocturnal, tree-living creatures with foxlike muzzles and a rather squirrelly appearance, found particularly in Madagascar.

The families closest to humans are, of course, the monkeys and apes. There are three families of monkeys (a word possibly derived from the Latin *homunculus,* meaning "little man").

The two monkey families in the Americas, known as the *New World monkeys,* are the Cebidae (for example, the organ-grinder's monkey) and the Callithricidae (for example, the marmoset). The third, the *Old World* family, are the Cercopithecidae; they include the various baboons.

The apes all belong to one family, called Pongidae. They are native to the Eastern Hemisphere. Their most noticeable outward differences from the monkeys are, of course, their larger size and their lack of tails. The apes fall into four types: the gibbon, smallest, hairiest, longest-armed, and most primitive of the family; the orangutan, larger, but also a tree-dweller like the gibbon; the gorilla, rather larger than a man, mainly ground-dwelling, and a native of Africa; and the chimpanzee, also a dweller in Africa, rather smaller than a man and the most intelligent primate next to humans themselves.

As for our own family, Hominidae, it consists today of only one genus and, as a matter of fact, only one species. Linnaeus named the species *Homo sapiens* ("man the wise"), and no one has dared change the name, despite provocation.

Evolution

It is almost impossible to run down the roster of living things, as I have just done, without ending with a strong impression that there has been a slow development of life from the very simple to the complex. The phyla can be arranged so that each seems to add something to the one before. Within each phylum, the various classes can be arranged likewise; and within each class, the orders.

Furthermore, the species often seem to melt together, as if they were still evolving along their slightly separate roads from common ancestors not very far in the past. Some species are so close together that under special circumstances they will interbreed, as in the case of the horse and the donkey, which, by appropriate cooperation, can produce the mule. Cattle can interbreed with buffaloes, and lions with tigers. There are also intermediate species, so to speak —creatures that link together two larger groups of animals. The cheetah is a cat with a smattering of doggish characteristics, and the hyena is a dog with some cattish characteristics. The platypus is a mammal only halfway removed from a reptile. There is a creature called *peripatus*, which seems half worm, half centipede. The dividing lines become particularly thin when we look at certain animals in their youthful stages. The infant frog seems to be a fish; and there is a primitive chordate called *balanoglossus*, discovered in 1825, which as a youngster is so like a young echinoderm that at first it was so classified.

We can trace practically a re-enactment of the passage through the phyla, even in the development of a human being from the fertilized egg. The study of this development (*embryology*) began in the modern sense with Harvey, the discoverer of the circulation of the blood. In 1759, the German physiologist Kaspar Friedrich Wolff demonstrated that the change in the egg is really a development: that is, specialized tissues grow out of unspecialized precursors by progressive alteration rather than (as many had previously thought) through the mere growth of tiny, already specialized structures existing in the egg to begin with.

In the course of this development, the egg starts as a single cell (a kind of protozoon), then becomes a small colony of cells (as in a sponge), each of which at first is capable of separating and starting life on its own, as happens when identical twins develop. The developing embryo passes through a two-layered stage (like a coelenterate), then adds a third layer (like an echinoderm), and so continues to add complexities in roughly the order that the progressively higher species do. The human embryo has at some stage in its development the notochord of a primitive chordate, later gill pouches reminiscent of a fish, and still later the tail and body hair of a lower mammal.

EARLY THEORIES

From Aristotle on, many men speculated on the possibility that organisms had evolved from one another. But as Christianity grew in power, such speculations were discouraged. The first chapter of Genesis in the Bible stated flatly that each living thing was created "after his kind," and, taken literally, had to mean that the species were "immutable" and had had the same form from the very beginning. Even Linnaeus, who must have been struck by the apparent kinships among living things, insisted firmly on the immutability of species.

The literal story of Creation, strong as its hold was on the human mind, eventually had to yield to the evidence of the *fossils* (from the Latin word meaning "to dig"). As long ago as 1669, the Danish scientist Nicolaus Steno had pointed out that lower layers (*strata*) of rock had to be older than the upper strata. At any reasonable rate of rock formation, it became more and more evident that lower strata had to be *much* older than upper strata. Petrified remnants of once living things were often found buried so deep under layers of rock that they had to be immensely older than the few thousand years that had elapsed since the creation described in the Bible. The fossil evidence also pointed to vast changes in the structure of the earth. As long ago as the sixth century B.C., the Greek philosopher Xenophanes of Colophon had noted fossil sea shells in the mountains and had surmised that those mountains had been under water long ages before.

Believers in the literal words of the Bible could and did maintain that the fossils resembled once-living organisms only through accident, or that they had been created deceitfully by the Devil. Such views were most unconvincing, and a more plausible suggestion was made that the fossils were remnants of creatures drowned in the Flood. Sea shells on mountain tops would certainly be evidence for that theory, since the biblical account of the Deluge states that water covered all the mountains.

But on close inspection, many of the fossil organisms proved to be different from any living species. John Ray, the early classifier, wondered if they might represent extinct species. A Swiss naturalist named Charles Bonnet went farther and, in 1770, suggested that fossils were indeed remnants of extinct species which had been destroyed in ancient geological catastrophes going back to long before the Flood.

It was an English land surveyor named William Smith, however, who laid a scientific foundation for the study of fossils and ancient life (*paleontology*). While working on excavations for a canal in 1791, he was impressed by the fact that the rock through which the canal was being cut was divided into strata, and that each stratum contained its own characteristic fossils. It now became possible to put fossils in a chronological order, depending on their place in the series of successive layers, and to associate each fossil with a particular type of rock stratum which would represent a certain period in geological history.

About 1800, Cuvier (the man who invented the notion of the phylum) classified fossils according to the Linnaean system and extended comparative anatomy into the distant past. Although many fossils represented species and genera not found among living creatures, all fitted neatly into one or another of the known phyla and so made up an integral part of the scheme of life. In 1801, for instance, Cuvier studied a long-fingered fossil of a type first discovered twenty years earlier, and demonstrated it to be the remains of a leathery-winged flying creature like nothing now existing—at least like nothing now existing *exactly*. He was able to show from the bone structure that these *pterodactyls* ("wing-fingers"), as he called them, were nevertheless reptiles, clearly related to the snakes, lizards, alligators, and turtles of today.

Furthermore, the deeper the stratum in which the fossil was to be found, and therefore the older the fossil, the simpler and less highly developed it seemed. Not only that, but fossils sometimes represented intermediate forms connecting two groups of creatures which, as far as living forms were concerned, seemed entirely separate. A particularly startling example, discovered after Cuvier's time, is a very primitive bird called *archaeopteryx* (Greek for "ancient wing"). This now-extinct creature had wings and feathers, but it also had a lizardlike, feather-fringed tail and a beak that contained reptilian teeth! In these and other respects it was clearly midway between a reptile and a bird (figure 16.3).

Cuvier still supposed that terrestrial catastrophes, rather than evolution, had been responsible for the disappearance of the extinct forms of life; but in the 1830s, Charles Lyell's new view of fossils and geological history in his history-making work *The Principles of Geology* won scientific opinion to his side. Some reasonable theory of evolution became a necessity, if any sense at all was to be made of the paleontological evidence.

If animals had evolved from one form to another, what had caused them to do so? This was the main stumbling block in the efforts to explain the

Figure 16.3. Archaeopteryx.

varieties of life. The first to attempt an explanation was the French naturalist Jean Baptiste de Lamarck. In 1809, he published a book, entitled *Zoological Philosophy,* in which he suggested that the environment caused organisms to acquire small changes which were then passed on to their descendants. Lamarck illustrated his idea with the giraffe (a newly discovered sensation of the time). Suppose that a primitive, antelopelike creature that fed on tree leaves ran out of food within easy reach, and had to stretch its neck as far as it could to get more food. By habitual stretching of its neck, tongue, and legs, it would gradually lengthen those appendages. It would then pass on these developed characteristics to its offspring, which in turn would stretch farther and pass on a still longer neck to their descendants, and so on. Little by little, by generation after generation of stretching, the primitive antelope would evolve into a giraffe.

Lamarck's notion of the *inheritance of acquired characteristics* quickly ran afoul of difficulties. How had the giraffe developed its blotched coat, for instance? Surely no action on its part, deliberate or otherwise, could have effected this change. Furthermore, a skeptical experimenter, the German biologist August Friedrich Leopold Weismann, cut off the tails of mice for generation after generation and reported that the last generation grew tails not one whit shorter than the first. (He might have saved himself the trouble by considering the case of the circumcision of Jewish males, which after more than a hundred generations had produced no shriveling of the foreskin.)

By 1883, Weismann had observed that the germ cells, which were eventually to produce sperm or ova, separated from the remainder of the embryo at an early stage and remained relatively unspecialized. From this, and from his experiments with rat tails, Weismann deduced the notion of the *continuity of the germ plasm.* The germ plasm (that is, the protoplasm making up the germ cells) had, he felt, an independent existence, continuous across the generations, with the remainder of the organism but a temporary housing, so to speak, built up and destroyed in each generation. The germ plasm guided the characteristics of the body and was not itself affected by the body. In all this, he was at the extreme opposite to Lamarck and was also wrong, although, on the whole, the actual situation seemed closer to the Weismann view than to that of Lamarck.

Despite its rejection by most biologists, Lamarckism lingered on into the twentieth century and even had a strong but apparently temporary revival in the form of Lysenkoism (hereditary modification of plants by certain treatments) in the Soviet Union. (Trofim Denisovich Lysenko, the exponent of this belief, was powerful under Stalin, retained much influence under Khrushchev, but underwent an eclipse when Khrushchev fell from power in 1964.) Modern geneticists do not exclude the possibility that the action of the environment may bring about certain transmittable changes in simple organisms, but the Lamarckian idea as such was demolished by the discovery of genes and the laws of heredity.

DARWIN'S THEORY

In 1831, a young Englishman named Charles Darwin, a dilettante and sportsman who had spent a more or less idle youth and was restlessly looking for something to do to overcome his boredom, was persuaded by a ship captain and a Cambridge professor to sign on as naturalist on a ship setting off on a five-year voyage around the world. The expedition was to study continental coastlines and make observations of flora and fauna along the way. Darwin, aged twenty-two, made the voyage of the *Beagle* the most important sea voyage in the history of science.

As the ship sailed slowly down the east coast of South America and then up its west coast, Darwin painstakingly collected information on the various forms of plant and animal life. His most striking discovery came in a group of islands in the Pacific, about 650 miles west of Ecuador, called the Galapagos Islands because of giant tortoises living on them (*Galapagos* coming from the Spanish word for "tortoise"). What most attracted Darwin's attention during his five-week stay was the variety of finches on the islands; they are known as *Darwin's finches* to this day. He found the birds divided into at least fourteen different species, distinguished from one another mainly by differences in the size and shape of their bills. These particular species did not exist anywhere else in the world, but they resembled an apparently close relative on the South American mainland.

What accounted for the special character of the finches on these islands? Why did they differ from ordinary finches, and why were they themselves divided into no fewer than fourteen species? Darwin decided that the most reasonable theory was that all of them were descended from the mainland type of finch and had differentiated during long isolation on the islands. The differentiation had resulted from varying methods of obtaining food. Three of the Galapagos species still fed on seeds, as the mainland finch did, but each ate a different kind of seed and varied correspondingly in size, one species being rather large, one medium, and one small. Two other species fed on cacti; most of the others fed on insects.

The problem of the changes in the finches' eating habits and physical characteristics preyed on Darwin's mind for many years. In 1838, he began to get a glimmering of the answer from reading a book that had been published forty years before by an English clergyman named Thomas Robert Malthus. In his *An Essay on the Principle of Population*, Malthus maintained that a population always outgrew its food supply and so eventually was cut back by starvation, disease, or war. It was in this book that Darwin came across the phrase "the struggle for existence," which his theories later made famous. Thinking of his finches, Darwin at once realized that competition for food would act as a mechanism favoring the more efficient individuals. When the finches that had colonized the Galapagos multiplied to the point of outrunning the seed supply, the only survivors would be the stronger birds or those particularly adept at obtaining seeds or those able to get new kinds of food.

A bird that happened to be equipped with slight variations of the finch characteristics, which enabled it to eat bigger seeds or tougher seeds or, better still, insects, would find an untapped food supply. A bird with a slightly thinner and longer bill could reach food that others could not, or one with an unusually massive bill could use otherwise unusable food. Such birds, and their descendants, would gain in numbers at the expense of the original variety of finch. Each of the adaptive types would find and fill a new, unoccupied niche in the environment. On the Galapagos Islands, virtually empty of bird life to begin with, all sorts of niches were there for the taking, with no established competitors to bar the way. On the South American mainland, with all the niches occupied, the ancestral finch did well merely to hold its own. It proliferated into no further species.

Darwin suggested that every generation of animals was composed of an array of individuals varying randomly from the average. Some would be slightly larger; some would possess organs of slightly altered shape; some abilities would be a trifle above or below normal. The differences might be minute, but those whose make-up was even slightly better suited to the environment would tend to live slightly longer and have more offspring. Eventually, an accumulation of favorable characteristics might be coupled with an inability to breed with the original type or other variations of it, and thus a new species would be born.

Darwin called this process *natural selection*. According to his view, the giraffe got its long neck not by stretching but because some giraffes were born with longer necks than their fellows, and the longer the neck, the more chance a giraffe had of reaching food. By natural selection, the long-necked species won out. Natural selection explained the giraffe's blotched coat just as easily: an animal with blotches on its skin would blend against the sun-spotted vegetation and thus have more chance of escaping the attention of a prowling lion.

Darwin's view of the way in which species were formed also made clear why it was often difficult to make clear-cut distinctions between species or between genera. The evolution of species is a continuous process and, of course, takes a very long time. There must be any number of species with members that are even now slowly drifting apart into separate species.

Darwin spent many years collecting evidence and working out his theory. He realized that it would shake the foundations of biology and society's thinking about the place of human beings in the scheme of things, and he wanted to be sure of his ground in every possible respect. Darwin started collecting notes on the subject and thinking about it in 1834, even before he read Malthus; and in 1858, he was still working on a book dealing with the subject. His friends (including Lyell, the geologist) knew what he was working on; several had read his preliminary drafts. They urged him to hurry, lest he be anticipated. Darwin would not (or could not) hurry, and he *was* anticipated.

The man who anticipated him was Alfred Russel Wallace, fourteen years younger than Darwin. Wallace's life paralleled that of Darwin. He, too, went on an around-the-world scientific expedition as a young man. In the East

Indies, he noticed that the plants and animals in the eastern islands were completely different from those in the western islands. A sharp line could be drawn between the two types of life forms: it ran between Borneo and Celebes, for instance, and between the small islands of Bali and Lombok farther to the south. The line is still called *Wallace's line*. (Wallace went on, later in his life, to divide the earth into six large regions, characterized by differing varieties of animals, a division that, with minor modifications, is still considered valid today.)

Now the mammals in the eastern islands and in Australia were distinctly more primitive than those in the western islands and Asia—or, indeed, in the rest of the world. It looked as if Australia and the eastern islands had split off from Asia at some early time when only primitive mammals existed, and the placental mammals had developed later only in Asia. New Zealand must have been isolated even longer, for it lacked mammals altogether and was inhabited by primitive flightless birds, of which the best-known survivor today is the kiwi.

How had the higher mammals in Asia arisen? Wallace first began puzzling over this question in 1855. In 1858 he, too, came across Malthus's book; and from it, he, too, drew the conclusions Darwin had drawn. But Wallace did not spend fourteen years writing his conclusions. Once the idea was clear in his mind, he sat down and wrote a paper on it in two days. He decided to send his manuscripts to some well-known competent biologist for criticism and review, and he chose Charles Darwin.

When Darwin received the manuscript, he was thunderstruck. It expressed his own thoughts in almost his own terms. At once he passed Wallace's paper to other important scientists and offered to collaborate with Wallace on reports summarizing their joint conclusions. Their reports appeared in the *Journal of the Linnaean Society* in 1858.

The next year Darwin's book was finally published. Its full title is *On the Origin of Species by Means of Natural Selection, or the Preservation of Favoured Races in the Struggle for Life*. We know it simply as *The Origin of Species*.

The theory of evolution has been modified and sharpened since Darwin's time, through knowledge of the mechanism of inheritance, of genes, and of mutations (see chapter 13). It was not until 1930, indeed, that the English statistician and geneticist Ronald Aylmer Fisher succeeded in showing that Mendelian genetics provides the necessary mechanism for evolution by natural selection. Only then did evolutionary theory gain its modern guise.

Naturally, advances in other branches of science continued to sharpen and focus the Darwinian concept. An understanding of plate tectonics (see chapter 4) explained a great deal concerning the forces that drive evolution and the manner in which similar species appear in widely separated parts of the earth. The ability to analyze proteins and nucleic acids in detail has made it possible to trace molecular evolution and, to judge from the degree of differences among molecules (as I will describe later in the chapter), the degree of relationship among organisms.

Naturally, in anything so complex as evolutionary development of living organisms over billions of years of time, controversy continues about the details of the mechanism. Thus, in the 1970s, such biologists as Stephen Jay Gould have advanced the notion of *punctuated evolution*. They do not picture evolutionary development as a slow, more or less evenly and continually moving process. Rather, they feel that there are long periods of relative changelessness interspersed by situations in which comparatively sudden and pronounced changes occur (not overnight, but perhaps over a few hundred thousand years, which is fast on the evolutionary scale).

Nevertheless, no reputable biologist feels any doubt about the validity of the evolutionary concept. Darwin's basic point of view has stood firm; and indeed, the evolutionary idea has been extended to every field of science—physical, biological, and social.

OPPOSITION TO THE THEORY

The announcement of the Darwinian theory naturally blew up a storm. At first, a number of scientists held out against the notion. The most important of these was the English zoologist Richard Owen, who was the successor of Cuvier as an expert on fossils and their classification. Owen stooped to rather unmanly depths in his fight against Darwinism. He not only urged others into the fray while remaining hidden himself, but even wrote anonymously against the theory and quoted himself as an authority.

The English naturalist Philip Henry Gosse tried to wriggle out of the dilemma by suggesting that the earth had been created by God complete with fossils to test human faith. To most thinking people, however, the suggestion that God would play juvenile tricks on humankind seemed more blasphemous than anything Darwin had suggested.

Its counterattacks blunted, opposition within the scientific world gradually subsided and, within the generation, nearly disappeared. The opponents outside science, however, carried on the fight much longer and much more intensively. The Fundamentalists (literal interpreters of the Bible) were outraged by the implication that human beings might be mere descendants from an apelike ancestor. Benjamin Disraeli (later to be prime minister of Great Britain) created an immortal phrase by remarking acidly: "The question now placed before society is this, 'Is man an ape or an angel?' I am on the side of the angels." Churchmen, rallying to the angels' defense, carried the attack to Darwin.

Darwin himself was not equipped by temperament to enter violently into the controversy, but he had an able champion in the eminent biologist Thomas Henry Huxley. As "Darwin's bulldog," Huxley fought the battle tirelessly in the lecture halls of England. He won his most telling victory in 1860 in a famous debate with Samuel Wilberforce, a bishop of the Anglican Church, a mathematician, and so accomplished and glib a speaker that he was familiarly known as "Soapy Sam."

Bishop Wilberforce, after apparently having won the audience, turned at last to his solemn, humorless adversary. As the report of the debate quotes him,

Wilberforce "begged to know whether it was through his grandfather or his grandmother that [Huxley] claimed his descent from a monkey."

While the audience roared with glee, Huxley whispered to a neighbor, "The Lord hath delivered him into my hands"; then he rose slowly to his feet and answered: "If, then, the question is put to me, would I rather have a miserable ape for a grandfather, or a man highly endowed by nature and possessing great means and influence, and yet who employs those faculties and that influence for the mere purpose of introducing ridicule into a grave scientific discussion —I unhesitatingly affirm my preference for the ape."

Huxley's smashing return apparently not only crushed Wilberforce but also put the Fundamentalists on the defensive. In fact, so clear was the victory of the Darwinian viewpoint that, when Darwin died in 1882, he was buried, with widespread veneration, in Westminster Abbey, where lie England's greats. In addition, the town of Darwin in northern Australia was named in his honor.

Another powerful proponent of evolutionary ideas was the English philosopher Herbert Spencer, who popularized the phrase *survival of the fittest* and the word *evolution*—a word Darwin himself rarely used. Spencer tried to apply the theory of evolution to the development of human societies (he is considered the founder of the science of sociology). His arguments were invalid, however, for the biological changes involved in evolution are in no way similar to social changes; and, contrary to his intention, they were later misused to support war and racism.

In the United States, a dramatic battle against evolution took place in 1925; it ended with the anti-evolutionists winning the battle and losing the war.

The Tennessee legislature had passed a law forbidding teachers in publicly supported schools of the state from teaching that humans had evolved from lower forms of life. To challenge the law's constitutionality, scientists and educators persuaded a young high-school biology teacher named John Thomas Scopes to tell his class about Darwinism. Scopes was thereupon charged with violating the law and brought to trial in Dayton, Tennessee, where he taught. The world gave fascinated attention to his trial.

The local population and the judge were solidly on the side of anti-evolution. William Jennings Bryan, the famous orator, three times unsuccessful candidate for the Presidency, and outstanding Fundamentalist, served as one of the prosecuting attorneys. Scopes had as his defenders the noted criminal lawyer Clarence Darrow and associated attorneys.

The trial was for the most part disappointing, for the judge refused to allow the defense to place scientists on the stand to testify to the evidence behind the Darwinian theory, and restricted testimony to the question whether Scopes had or had not discussed evolution. But the issues nevertheless emerged in the courtroom when Bryan, over the protests of his fellow prosecutors, volunteered to submit to cross-examination on the Fundamentalist position. Darrow promptly showed that Bryan was ignorant of modern developments in science and had only a stereotyped Sunday-school acquaintance with religion and the Bible.

Scopes was found guilty and fined one hundred dollars. (The conviction was later reversed on technical grounds by the Tennessee Supreme Court.) But the Fundamentalist position (and the State of Tennessee) had stood in so ridiculous a light in the eyes of the educated world that the anti-evolutionists were forced on the defensive and retired into the background for half a century.

But the forces of darkness and ignorance are never permanently defeated; and in the 1970s, the anti-evolutionists returned for a new and even more insidious stand against the scientific view of the universe. They abandoned their earlier (at least forthright) stand on the literal words of the Bible, which had been totally discredited, and pretended to scientific respectability. They spoke vaguely of a "Creator" and were careful not to use the words of the Bible. They then argued that evolutionary theory was full of flaws and could not be true and that, therefore, *creationism* was true.

To demonstrate that evolutionary theory was not true, they did not hesitate to misquote, distort, take out of context, and in other ways violate the biblical injunction against false witness. And even so they proclaimed their own view as true only by default and never, at any time, have presented rational evidence in favor of their creationism, which they solemnly (but ridiculously) call "scientific creationism."

Their demand is that their foolish viewpoint be given "equal time" in the schools, and that any teacher or school textbook that discusses evolution should also discuss "scientific creationism." At this writing, they have won no battles in the courts of the land; but their spokesmen, backed by earnest churchgoers who know no science, and to whom everything outside the Bible is a misty fog of ignorance, bully school boards, libraries, and legislators into censorship and suppression of science.

The results may be sad indeed, for the creationist view that the earth is only a few thousand years old, as is the entire universe—that life was created suddenly with all its species distinct from the start—makes utter nonsense out of astronomy, physics, geology, chemistry, and biology and could create a generation of American youngsters whose minds are shrouded in the darkness of night.

EVIDENCE FOR THE THEORY

One of the arguments of the creationists is that no one has ever seen the forces of evolution at work. That would seem the most nearly irrefutable of their arguments, and yet it, too, is wrong.

In fact, if any confirmation of Darwinism were needed, it has turned up in examples of natural selection that have taken place before our eyes (now that we know what to watch for). A notable example occurred in Darwin's native land.

In England, it seems, the peppered moth exists in two varieties, a light and a dark. In Darwin's time, the white variety was predominant because it was less prominently visible against the light lichen-covered bark of the trees it frequented. It was saved by this *protective coloration*, more often than were the clearly visible, dark variety, from those animals that would feed on it. As England grew more industrialized, however, soot killed the lichen cover and blackened the tree bark. It was then the dark variety that was less visible against the bark and was protected. Therefore, the dark variety became predominant—through the action of natural selection.

In 1952, the British Parliament passed laws designed to clean the air. The quantity of soot declined, the trees regained some of their light lichen covering, and at once the percentage of the light variety of moth began to increase. All this change is quite predictable by evolutionary theory, and it is the mark of a successful theory that it not only explains the present but can predict the future.

The Course of Evolution

A study of the fossil record has enabled paleontologists to divide the history of the earth into a series of *eras*. These were roughed out and named by various nineteenth-century British geologists, including Lyell himself, Adam Sedgwick, and Roderick Impey Murchison. Those named eras start some 600 million years ago with the first unmistakable fossils (when all the phyla except Chordata were already established). The first fossils do not, of course, represent the first life. For the most part, it is only the hard portions of a creature that fossilize, so the clear fossil record contains only animals that possessed shells or bones. Even the simplest and oldest of these creatures are already far advanced and must have a long evolutionary background. One evidence of that assumption is that, in 1965, fossil remains of small clamlike creatures were discovered and seem to be about 720 million years old.

Paleontologists can now do far better. It stands to reason that simple one-celled life must extend much farther back in time than anything with a shell; and indeed, signs of blue-green algae and of bacteria have been found in rocks that were 1 billion years old and more. In 1965, the American paleontologist Elso Sterrenberg Barghoorn discovered minute bacteriumlike objects (*microfossils*) in rocks over 3 billion years old. They are so small, their structure must be studied by electron microscope.

It would seem then that chemical evolution, moving toward the origin of life, began almost as soon as the earth took on its present shape some 4.6 billion years ago. Within a billion years, chemical evolution had reached the stage where systems complicated enough to be called living had formed. At this time, Earth's atmosphere was still reducing and contained no significant quantity of oxygen (see chapter 5). The earliest forms of life must therefore have been adapted to this situation, and their descendants survive today.

In 1970, Carl R. Woese began to study in detail certain bacteria that exist only under circumstances where free oxygen is absent. Some of these reduce carbon dioxide to methane and are therefore called *methanogens* ("methane producers"). Other bacteria engage in other reactions that yield energy and support life but do not involve oxygen. Woese lumps them together as *archaeobacteria* ("old bacteria") and suggests that it might be well to consider them a separate kingdom of life.

Once life was established, however, the nature of the atmosphere began to change—very slowly, at first. About two and a half billion years ago, blue-green algae may have already been in existence, and the process of photosynthesis began the slow change from a nitrogen–carbon-dioxide atmosphere into a nitrogen–oxygen atmosphere. By about a billion years ago, eukaryotes may have been well established, and the one-celled life of the seas must have been quite diversified and included distinctly animal protozoa, which would then have been the most complicated forms of life in existence—monarchs of the world.

For 2 billion years after blue-green algae came into existence, the oxygen content must have been very slowly increasing. As the most recent billion years of Earth's history began to unfold, the oxygen concentration may have been 1 percent or 2 percent of the atmosphere, enough to supply a rich source of energy for animal cells beyond anything that had earlier existed. Evolutionary change spurted in the direction of increased complication; and by 600 million years ago, there could begin the rich fossil record of elaborate organisms.

The earliest rocks with elaborate fossils are said to belong to the *Cambrian age*; and the entire 4-billion-year history of our planet that preceded it has been, until recently, dismissed as the *pre-Cambrian age*. Now that the traces of life have unmistakably been found in it, the more appropriate name *Cryptozoic eon* (Greek for "hidden life") is used, while the last 600 million years make up the *Phanerozoic eon* ("visible life").

The Cryptozoic eon is even divided into two sections: the earlier *Archeozoic era* ("ancient life"), to which the first traces of unicellular life belong; and the later *Proterozoic era* ("early life").

The division between the Cryptozoic eon and the Phanerozoic eon is extraordinarily sharp. At one moment in time, so to speak, there are no fossils at all above the microscopic level; and at the next, there are elaborate organisms of a dozen different basic types. Such a sharp division is called an *unconformity,* and an unconformity leads invariably to speculations about possible catastrophes. It seems there should have been a more gradual appearance of fossils, and what may have happened is that geological events of some extremely harsh variety wiped out the earlier record.

ERAS AND AGES

The broad divisions of the Phanerozoic eon are the *Paleozoic era* ("ancient life"), *the Mesozoic* ("middle life"), and the *Cenozoic* ("new life"). According to modern methods of geological dating, the Paleozoic era covered a span of perhaps 350 million years; the Mesozoic, 150 million years; and the Cenozoic, the last 50 million years of the earth's history.

Each era is in turn subdivided into ages. The Paleozoic begins, as I have stated, with the Cambrian age (named for a location in Wales—actually an ancient tribe that occupied it—where these strata were first uncovered). During the Cambrian period shellfish were the most elaborate form of life. This was the era of the *trilobites*, primitive arthropods of which the modern horseshoe crab is the closest living relative. The horseshoe crab, because it has survived with few evolutionary changes for 200 million years, is an example of what is sometimes rather dramatically called a *living fossil*.

The next age is the *Ordovician* (named for another Welsh tribe). This was the age, between 450 million and 500 million years ago, when the chordates made their first appearance in the form of *graptolites*, small animals living in colonies and now extinct. They are possibly related to the balanoglossus, which, like the graptolites, belongs to the *hemichordata*, the most primitive subphylum of the chordate phylum.

Then came the *Silurian* (named for still another Welsh tribe) and the *Devonian* (from Devonshire). The Devonian age, between 350 million and 400 million years ago, witnessed the rise of fish to dominance in the ocean, a position they still hold. In that age, however, came also the colonization of the dry land by life forms. It is hard to realize, but true, that during five-sixths or more of its history, life was confined to the waters, and the land remained dead and barren. Considering the difficulties represented by the lack of water, by extremes of temperature, by the full force of gravity unmitigated by the buoyancy of water, it must be understood that the spread to land of life forms that evolved to meet the conditions of the ocean was the greatest single victory won by life over the inanimate environment.

The move toward the land probably began when competition for food in the crowded sea drove some organisms into shallow tidal waters, until then unoccupied because the bottom was exposed for hours at a time at low tide. As more and more species crowded into the tidewaters, relief from competition could be attained only by moving farther and farther up the shore, until eventually some mutant organisms were able to establish themselves on dry land.

The first life forms to manage the transition were plants. This took place about 400 million years ago. The pioneers belonged to a now extinct plant group called *psilopsids*—the first multicellular plants. (The name comes from the Greek word for "bare," because the stems were bare of leaves, a sign of the primitive nature of these plants.) More complex plants developed; and by 350 million years ago, the land was covered with forest. Once plant life had begun to grow on dry land, animal life could follow suit. Within a few million years, the land was occupied by arthropods, molluscs, and worms. All these first land animals were small, because heavier animals, without an internal skeleton, would have collapsed under the force of gravity. (In the ocean, of course, buoyancy largely negated gravity, which was not therefore a factor. Even today the largest animals live in the sea.) The first land creatures to gain much mobility were the insects; thanks to their development of wings, they were able to counteract the force of gravity, which held other animals to a slow crawl.

Finally, 100 million years after the first invasion of the land, there came a new invasion by creatures that could afford to be bulky despite gravity because they had a bracing of bone within. The new colonizers from the sea were bony fishes belonging to the subclass Crossopterygii ("fringed fins"). Some of their fellow members had migrated to the uncrowded sea deeps, including the coelacanth, which biologists discovered in 1938 to be still in existence (much to their astonishment).

The fishy invasion of land began as a result of competition for oxygen in brackish stretches of fresh water. With oxygen available in unlimited quantities in the atmosphere, those fish best survived that could most effectively gulp air when the oxygen content of water fell below the survival point. Devices for storing such gulped air had survival value, and fish developed pouches in their alimentary canals in which swallowed air could be kept. These pouches developed into simple lungs in some cases. Descendants of these early fish include the *lungfishes*, a few species of which still exist in Africa and Australia. These live in stagnant water where ordinary fishes would suffocate, and can even survive summer droughts when their habitat dries up. Even fish who live in the sea, where the oxygen supply is no problem, show signs of their descent from the early-lunged creatures, for they still possess air-filled pouches, used not for respiration but for buoyancy.

Some of the lung-possessing fishes, however, carried the matter to the logical extreme and began living, for shorter or longer stretches, out of the water altogether. These crossopterygian species with the strongest fins could do so most successfully for, in the absence of water buoyancy, they had to prop themselves up against the pull of gravity. By the end of the Devonian age, some of the primitive-lunged crossopterygians found themselves standing on the dry land, propped up shakily on four stubby legs.

After the Devonian came the *Carboniferous* ("coal-bearing") age, so named by Lyell because it was the period of the vast, swampy forests that, some 300 million years ago, represented what was perhaps the lushest vegetation in earth's history; eventually, they were buried and became this planet's copious coal beds. This was the age of the amphibians; the crossopterygians by then were spending their entire adult lives on land. Next came the *Permian* age (named for a district in the Urals, for the study of which Murchison made the long trip from England). The first reptiles now made their appearance. They ushered in the Mesozoic era, in which reptiles were to dominate the earth so thoroughly that it has become known as the *age of the reptiles*.

The Mesozoic is divided into three ages—the *Triassic* (it was found in three strata), the *Jurassic* (from the Jura mountains in France), and the *Cretaceous* ("chalk-forming"). In the Triassic arose the *dinosaurs* (Greek for "terrible lizards"). The dinosaurs reached their peak form in the Cretaceous, when *Tyrannosaurus rex* thundered over the land—the largest carnivorous land animal in the history of our planet.

It was during the Jurassic that the earliest mammals and birds developed, each from a separate group of reptiles. For millions of years, these creatures remained inconspicuous and unsuccessful. With the end of the Cretaceous, however, all the dinosaurs vanished in a relatively short period. So did other large reptiles that are not classified with the dinosaurs—ichthyosaurs, plesiosaurs, and the pterosaurs. (The first two were sea reptiles; the third, winged reptiles.) In addition, certain groups of invertebrates, such as the ammonites (related to the still-living chambered nautilus), died out—as did many smaller organisms, down to many types of microscopic organisms in the sea.

According to some estimates, as many as 75 percent of all species then living died during what is sometimes called *the Great Dying* and the end of the Cretaceous. Of the 25 percent that survived, there may have been great individual carnage, and it would not be surprising that 95 percent of all organisms died. Something happened that nearly sterilized the earth—but what?

In 1979, the American paleontologist Walter Alvarez headed a team who were trying to test ancient sedimentation rates, by testing for the concentration of certain metals along the length of a core taken out of rocks in central Italy. One of the metals being tested for, by neutron-activation techniques, happened to be iridium; and, somewhat to his astonishment, Alvarez found a concentration of iridium in a single narrow band that was 25 times as high as the concentrations immediately below or above.

Where could the iridium have come from? Could the sedimentation rate have been unusually high at that point? Or could it have come from some unusually rich iridium source. Meteorites are richer in iridium and certain other metals than the earth's crust is, and that section of the core was rich in the other metals as well. Alvarez suspected that a meteor had fallen, but there was no sign of any ancient crater in the region.

Later investigations, however, showed that the iridium-rich layer occurred in widely separated places on Earth and always in rocks of the same age. It began to look as though a huge meteor could have fallen, and enormous quantities of material had been thrown, by the impact, into the upper atmosphere (including the entire vaporized meteor itself) and they had slowly settled out over the whole earth.

At what time did this happen? The rock from which the iridium-rich material was taken was 65 million years old—precisely the end of the Cretaceous. Many geologists and paleontologists (but not all, by any means) began to look with favor on the suggestion that the dinosaurs and the other organisms that seemed to have come to a sudden end, during the Great Dying at the close of the Cretaceous, had died as a result of the catastrophic impact with the earth of an object perhaps as much as 10 kilometers in diameter—either a small asteroid or the core of a comet.

There may well have been periodic collisions of this sort, each of which may have produced a Great Dying. The one at the end of the Cretaceous is merely the most spectacular of the recent ones and therefore the easiest to document in detail. And, of course, similar events may take place in the future unless humanity's developing space capability eventually makes it possible to destroy threatening objects while they are still in space and before they strike. Indeed, it now appears that Great Dyings take place regularly every 28 million years. In 1984, it was speculated that the sun has a small dim star as a companion and that its approach to perihelion every 28 million years disrupts the Oort cloud of comets (see chapter 3) and sends millions into the inner solar system. A few are bound to strike the Earth.

Such an impact devastates areas in the vicinity at once, but the planetary effect is more the result of the vast quantity of dust lofted into the stratosphere —dust that produced a long, frigid night over the world and put a temporary end to photosynthesis.

In 1983, the astronomer Carl Sagan and the biologist Paul Ehrlich have pointed out that, in the event of a nuclear war, the explosion of as little as 10 percent of the present-day armory of nuclear weapons would send enough matter into the stratosphere to initiate an artificial wintry night that might last long enough to put human life on Earth into serious jeopardy—another Great Dying we certainly cannot afford.

But, in any case, the death of the dominant reptiles at the end of the Cretaceous, whatever the cause, meant that the Cenozoic era that followed became the *age of mammals.* It brought in the world we know.

BIOCHEMICAL CHANGES

The unity of present life is demonstrated in part by the fact that all organisms are composed of proteins built from the same amino acids. The same kind of evidence has recently established our unity with the past as well. The new science of *paleobiochemistry* (the biochemistry of ancient forms of life) was opened in the late 1950s, when it was shown that certain 300-million-year-old fossils contained remnants of proteins consisting of precisely the same amino acids that make up proteins today—glycine, alanine, valine, leucine, glutamic acid, and aspartic acid. Not one of the ancient amino acids differed from present ones. In addition, traces of carbohydrates, cellulose, fats, and porphyrins were located, with (again) nothing that would be unknown or unexpected today.

From our knowledge of biochemistry we can deduce some of the biochemical changes that may have played a part in the evolution of animals.

Let us take the excretion of nitrogenous wastes. Apparently, the simplest way to get rid of nitrogen is to excrete it in the form of the small ammonia molecule (NH_3), which can easily pass through cell membranes into the blood. Ammonia happens to be extremely poisonous; if its concentration in the blood exceeds one part in a million, the organism will die. For a sea animal, this is no great problem; it can discharge the ammonia into the ocean continuously through its gills. But for a land animal, however, ammonia excretion is out of the question. To discharge ammonia as quickly as it is formed would require such an excretion of urine that the animal would quickly be dehydrated and die. Therefore a land organism must produce its nitrogenous wastes in a less toxic form than ammonia. The answer is urea. This substance can be carried in the blood in concentrations up to one part in a thousand without serious danger.

Now fish eliminate nitrogenous wastes as ammonia, and so do tadpoles. But when a tadpole matures to a frog, it begins to eliminate nitrogenous wastes as urea. This change in the chemistry of the organism is every bit as crucial for the changeover from life in the water to life on land as is the visible change from gills to lungs.

Such a biochemical change must have taken place when the crossopterygians invaded the land and became amphibians. Thus, there is every reason to believe that *biochemical* evolution played as great a part in the development of organisms as *morphological evolution* (that is, changes in form and structure).

Another biochemical change was necessary before the great step from amphibian to reptile could be taken. If the embryo in a reptile's egg excreted urea, it would build up to toxic concentrations in the limited quantity of water within the egg. The change that took care of this problem was the formation of uric acid instead of urea. Uric acid (a purine molecule resembling the adenine and guanine that occur in nucleic acids) is insoluble in water; it is therefore precipitated in the form of small granules and thus cannot enter the cells.

In adult life, reptiles continue eliminating nitrogenous wastes as uric acid. They have no urine in the liquid sense. Instead, the uric acid is eliminated as a semisolid mass through the same body opening that serves for the elimination of feces. This single body opening is called the *cloaca* (Latin for "sewer").

Birds and egg-laying mammals, which lay eggs of the reptilian type, preserve the uric-acid mechanism and the cloaca. In fact, the egg-laying mammals are often called *monotremes* (from Greek words meaning "one hole").

Placental mammals, on the other hand, can easily wash away the embryo's nitrogenous wastes, for the embryo is connected, indirectly, to the mother's circulatory system. Mammalian embryos, therefore, manage well with urea. It is transferred to the mother's bloodstream and passes out through the mother's kidneys.

An adult mammal has to excrete substantial amounts of urine to get rid of its urea. Hence, there are two separate openings: an anus to eliminate the indigestible solid residues of food and a urethral opening for the liquid urine.

The account just given of nitrogen excretion demonstrates that, although life is basically a unity, there are systematic minor variations from species to species. Furthermore, these variations seem to be greater as the species considered are farther removed from each other in the evolutionary sense.

Consider, for instance, that antibodies can be built up in animal blood to some foreign protein or proteins as, for example, those in human blood. Such *antisera*, if isolated, will react strongly with human blood, coagulating it, but will not react in this fashion with the blood of other species. (This is the basis of the tests indicating whether bloodstains are of human origin, which sometimes lend drama to murder investigations.) Interestingly, antisera that will react with human blood will respond weakly with chimpanzee blood, while antisera that will react strongly with chicken blood will react weakly with duck blood, and so on. Antibody specificity thus can be used to indicate close relationships among life forms.

Such tests indicate, not surprisingly, the presence of minor differences in the complex protein molecule—differences small enough in closely related species to allow some overlapping in antiserum reactions.

When biochemists developed techniques for determining the precise amino-acid structure of proteins, in the 1950s, this method of arranging species according to protein structure was vastly sharpened.

In 1965, even more detailed studies were reported on the hemoglobin molecules of various types of primates, including humans. Of the two kinds of peptide chains in hemoglobin, one, the *alpha chain*, varied little from primate to primate; the other, the *beta chain*, varied considerably. Between a particular primate and the human species, there were only six differences in the amino acids and the alpha chain, but twenty-three in those of the beta chains. Judging by differences in the hemoglobin molecules, it is believed that human beings diverged from the other apes about 75 million years ago, or just about the time the ancestral horses and donkeys diverged.

Still broader distinctions can be made by comparing molecules of *cytochrome c*, an iron-containing protein molecule made up of about 105 amino acids and found in the cells of every oxygen-breathing species—plant, animal, or bacterial. Through analysis of the cytochrome-c molecules from different species, it was found that the molecules in humans differed from those of the rhesus monkey in only one amino acid in the entire chain. Between the cytochrome-c of a human being and that of a kangaroo, there were ten differences in amino acid; between those of human and a tuna fish, twenty-one differences; between those of a human and a yeast cell, some forty differences.

With the aid of computer analysis, biochemists have decided it takes on the average some 7,000,000 years for a change in one amino-acid residue to establish itself, and estimates can be made of the time in the past when one type of organism diverged from another. It was about 2,500,000,000 years ago, judging from cytochrome-c analysis, that higher organisms diverged from bacteria (that is, it was about that long ago that a living creature was last alive that might be considered a common ancestor of all eukaryotes). Similarly, it was about 1,500,000,000 years ago that plants and animals had a common ancestor, and 1,000,000,000 years ago that insects and vertebrates had a common ancestor. We must understand, then, that evolutionary theory stands not alone on fossils but is supported by a wide variety of geological, biological, and biochemical detail.

RATE OF EVOLUTION

If mutations in the DNA chain, leading to changes in amino-acid pattern, were established by random factors only, it might be supposed that the rate of evolution would continue at an approximately constant rate. Yet there are occasions when evolution seems to progress more rapidly than at others— when there is a sudden flowering of new species—as described in Gould's notion of punctuated evolution, earlier mentioned. It may be that the rate of mutations is greater at some periods in Earth's history than at others, and these more frequent mutations may establish an extraordinary number of new species or render unviable an extraordinary number of old ones. (Or else some of the new species may prove more efficient than the old and compete them to death.)

One environmental factor that encourages the production of mutations is energetic radiation, and Earth is constantly bombarded by energetic radiation from all directions at all times. The atmosphere absorbs most of it, but even the atmosphere is helpless to ward off cosmic radiation. Can it be that cosmic radiation is greater at some period than at others?

A difference can be postulated in each of two different ways. Cosmic radiation is diverted to some extent by Earth's magnetic field. However, the magnetic field varies in intensity, and there are periods, at varying intervals, when it sinks to zero intensity. Bruce Heezen suggested in 1966 that these periods when the magnetic field, in the process of reversal, goes through a time of zero intensity may also be periods when unusual amounts of cosmic radiation reach the surface of the earth, bringing about a jump in mutation rate. This is a sobering thought in view of the fact that the earth seems to be heading toward such a period of zero intensity.

Then, too, what about the occurrence of supernovas in earth's vicinity—close enough to the solar system, that is, to produce a distinct increase in the intensity of bombardment by cosmic rays of the earth's surface? Some astronomers have speculated on that possibility.

The Descent of Man

James Ussher, a seventeenth-century Irish archbishop, dated the creation of man (a term commonly used for human beings of both sexes until the rise of the women's movement in the 1960s) precisely in the year 4004 B.C.

Before Darwin, few people dared to question the Biblical interpretation of early human history. The earliest reasonably definite date to which the events recorded in the Bible can be referred is the reign of Saul, the first king of Israel, who is believed to have become king about 1025 B.C. Bishop Ussher and other biblical scholars who worked back from that date through the chronology of the Bible came to the conclusion that human beings and the universe could not be more than a few thousand years old.

EARLY CIVILIZATIONS
Documented human history, as recorded by Greek historians, was no better, or more ancient than that of the Bible. It began only about 700 B.C. Beyond this hard core of history, dim oral traditions went back to the Trojan War,

about 1200 B.C., and more dimly still to a pre-Greek civilization on the island of Crete under a King Minos. Nothing beyond the writings of historians in known languages, with all the partiality and distortion that might involve— was known to moderns concerning the everyday life of ancient times prior to the eighteenth century. Then, in 1738, the cities of Pompeii and Herculaneum, buried in an eruption of Vesuvius in 79 A.D., began to be excavated. For the first time, historians grew aware of what could be done by digging, and the science of archaeology got its start.

At the beginning of the nineteenth century, archaeologists began to get their first real glimpses of human civilizations that came before the periods described by the Greek and Hebrew historians. In 1799, during General Bonaparte's invasion of Egypt, an officer in his army, named Boussard, discovered an inscribed stone in the town of Rosetta, on one of the mouths of the Nile. The slab of black basalt had three inscriptions, one in Greek, one in an ancient form of Egyptian picture writing called *hieroglyphic* ("sacred writing"), and one in a simplified form of Egyptian writing called *demotic* ("of the people").

The inscription in Greek was a routine decree of the time of Ptolemy V, dated the equivalent of 27 March 196 B.C. Plainly, it must be a translation of the same decree given in the other two languages on the slab (compare the no-smoking signs and other official notices that often appear in three languages in public places, especially airports, today). Archaeologists were overjoyed: at last they had a "pony" with which to decipher the previously undecipherable Egyptian scripts. Important work was done in "cracking the code" by Thomas Young, the man who had earlier established the wave theory of light (see chapter 8), but it fell to the lot of a French student of antiquities, Jean François Champollion, to solve the *Rosetta stone* completely. He ventured the guess that Coptic, a still-remembered language of certain Christian sects in Egypt, could be used as a guide to the ancient Egyptian language. By 1821, he had cracked the hieroglyphs and the demotic script and opened the way to reading all the inscriptions found in the ruins of ancient Egypt.

An almost identical find later broke the undeciphered writing of ancient Mesopotamia. On a high cliff near the ruined village of Behistun in western Iran, scholars found an inscription that had been carved about 520 B.C. at the order of the Persian emperor Darius I. It announced the manner in which he had come to the throne after defeating a usurper; to make sure that everyone could read it, Darius had had it carved in three languages—Persian, Sumerian, and Babylonian. The Sumerian and Babylonian writings were based on pictographs formed as long ago as 3100 B.C. by indenting clay with a stylus; these had developed into a *cuneiform* ("wedge-shaped") script, which remained in use until the first century A.D.

An English army officer, Henry Creswicke Rawlinson, climbed the cliff, transcribed the entire inscription, and, by 1846, after ten years of work, managed to work out a complete translation, using local dialects as his guide where necessary. The deciphering of the cuneiform scripts made it possible to read the history of the ancient civilizations between the Tigris and the Euphrates.

Expedition after expedition was sent to Egypt and Mesopotamia to look for tablets and the remains of the ancient civilizations. In 1854, a Turkish scholar, Hurmuzd Rassam, discovered the remnants of a library of clay tablets in the ruins of Nineveh, the capital of ancient Assyria—a library that had been collected by the last great Assyrian king, Ashurbanipal, about 650 B.C. In 1873, the English Assyriologist George Smith discovered clay tablets giving legendary accounts of a flood so like the story of Noah that it became clear that much of the first part of the book of Genesis was based on Babylonian legend. Presumably, the Jews picked up the legends during their Babylonian captivity in the time of Nebuchadnezzar, a century after the time of Ashurbanipal. In 1877, a French expedition to Iraq uncovered the remains of the culture preceding the Babylonian—that of the aforementioned Sumerians. This finding carried the history of the region back to earliest Egyptian times. And in 1921, remains of a totally unexpected civilization were discovered along the Indus Valley in what is now Pakistan. It had flourished between 2500 and 2000 B.C.

Yet Egypt and Mesopotamia were not quite in the same league with Greece when it came to dramatic finds on the origins of modern Western culture. Perhaps the most exciting moment in the history of archaeology came in 1873 when a German ex-grocer's boy found the most famous of all legendary cities.

Heinrich Schliemann as a boy developed a mania for Homer. Although most historians regarded the *Iliad* as mythology, Schliemann lived and dreamed of the Trojan War. He decided that he must find Troy and, by nearly superhuman exertions, raised himself from grocer's boy to millionaire so that he could finance the quest. In 1868, at the age of forty-six, he set forth. He persuaded the Turkish government to give him permission to dig in Asia Minor; and, following only the meager geographical clues afforded by Homer's accounts, he finally settled upon a mound near the village of Hissarlik. He browbeat the local population into helping him dig into the mound. Excavating in a completely amateurish, destructive and unscientific manner, he began to uncover a series of buried ancient cities, each built on the ruins of the other. And then, at last, success: he uncovered Troy—or at least a city he proclaimed to be Troy. Actually, the particular ruins he named Troy are now known to be far older than Homer's Troy, but Schliemann had proved that Homer's tales are not mere legends.

Inexpressibly excited by his triumph, Schliemann went on to mainland Greece and began to dig at the site of Mycenae, a ruined village which Homer had described as the once powerful city of Agamemnon, leader of the Greeks in the Trojan War. Again, Schliemann uncovered an astounding find—the ruins of a city with gigantic walls, which we now know to date back to 1500 B.C.

Schliemann's successes prompted the British archaeologist Arthur John Evans to start digging on the island of Crete, described in Greek legends as the site of a powerful early civilization under a King Minos. Evans, exploring the island in the 1890s, laid bare a brilliant, lavishly ornamented Minoan civilization that stretched back many centuries before the time of Homer's Greece. Here, too, written tablets were found. They were in two different scripts, one of which, called Linear B, was finally deciphered in the 1950s and shown to be a form of Greek, through a remarkable feat of cryptography and linguistic analysis, by a young English architect named Michael Vestris.

As other early civilizations were uncovered—the Hittites and the Mittanni in Asia Minor, the Indus civilization in India, and so on—it became obvious that the history recorded by Greece's Herodotus and the Hebrews' Old Testament represented comparatively advanced stages of human civilization. The earliest cities were at least several thousand years old, and the prehistoric existence of humans in less civilized modes of life must stretch many thousands of years farther into the past.

THE STONE AGE

Anthropologists find it convenient to divide cultural history into three major periods: the Stone Age, the Bronze Age, and the Iron Age (a division first suggested by the Roman poet and philosopher Lucretius and introduced to modern science by the Danish paleontologist Christian Jurgenson Thomsen in 1834). Before the Stone Age, there may have been a "Bone Age," when pointed horns, chisel-like teeth, and clublike thigh bones served human beings at a time when the working of relatively intractable rock had not yet been perfected.

The Bronze and Iron ages are, of course, very recent; as soon as we delve into the time before written history, we are back in the Stone Age. What we call *civilization* (from the Latin word for "city") began perhaps around 8000 B.C., when humans first turned from hunting to agriculture, learned to domesticate animals, invented pottery and new types of tools, and started to develop permanent communities and a settled way of life. Because the archaeological remains from this period of transition are marked by advanced stone tools formed in new ways, it is called the New Stone Age, or the Neolithic period; and although it antedated the supposed Biblical date of creation, humanity was already old at the time.

This Neolithic Revolution seems to have started in the Near East, at the crossroads of Europe, Asia, and Africa (where later the Bronze and Iron ages also were to originate). From there, it appears, the revolution slowly spread in widening waves to the rest of the world. It did not reach western Europe and India until 3000 B.C., northern Europe and eastern Asia until 2000 B.C., and central Africa and Japan until perhaps 1000 B.C. or later. Southern Africa and Australia remained in the Old Stone Age until the eighteenth and nineteenth centuries. Most of America also was still in the hunting phase when the Europeans arrived in the sixteenth century, although a well-developed civilization, possibly originated by the Mayas, had developed in Central America and Peru as early as the first centuries of the Christian era.

Evidences of humanity's pre-Neolithic cultures began to come to light in Europe at the end of the eighteenth century. In 1797 an Englishman named John Frere dug up in Suffolk some crudely fashioned flint tools too primitive to have been made by Neolithic human beings. They were found thirteen feet underground, which, allowing for the normal rate of sedimentation, testified to great age. In the same stratum with the tools were bones of extinct animals. More and more signs of the great antiquity of tool-making human beings were discovered, notably by two nineteenth-century French archeologists, Jacques Boucher de Perthes and Edouard Armand Lartet. Lartet, for instance, found a mammoth tooth on which some early human being had scratched an excellent drawing of the mammoth, obviously from living models. The mammoth was a hairy species of elephant that disappeared from the earth well before the beginning of the New Stone Age.

Archaeologists launched upon an active search for early stone tools. They found that these could be assigned to a relatively short Middle Stone Age (Mesolithic) and a long Old Stone Age (Paleolithic). The Paleolithic was divided into Lower, Middle, and Upper periods. The earliest objects that could be considered tools (*eoliths,* or "dawn stones") seemed to date back nearly a million years!

What sort of creature had made the Old Stone Age tools? It turned out that Paleolithic human beings, at least in their late stages, was far more than a hunting animal. In 1879, a Spanish nobleman, the Marquis de Sautuola, explored some caves that had been discovered a few years earlier—after having been blocked off by rock slides since prehistoric times—at Altamira in northern Spain near the city of Santander. While he dug into the floor of a cave, his five-year-old daughter, who had come along to watch papa dig, suddenly cried: "Toros! Toros!" ("Bulls! Bulls!"). The father looked up, and there on the walls of the cave were drawings of various animals, in vivid color and vigorous detail.

Anthropologists found it hard to believe that these sophisticated drawings could have been made by primitive people. But some of the pictured animals were plainly extinct types. The French archaeologist Henri Edouard Prosper Breuil found similar art in caves in southern France. All the evidence finally forced archaeologists to agree with Breuil's firmly expressed views and to conclude that the artists must have lived in the late Paleolithic, say about 10,000 B.C.

Something was already known about the physical appearance of these Paleolithic men. In 1868, workmen excavating a roadbed for a railroad had uncovered the skeletons of five human beings in the so-called Cro-Magnon caves in southwest France. The skeletons were unquestionably *Homo sapiens,* yet some of them, and similar skeletons soon found elsewhere, seemed to be up to 35,000 or 40,000 years old, according to the geological evidence. They were given the name *Cro-Magnon man* (figure 16.4). Taller than the average modern man and equipped with a large braincase, a Cro-Magnon man is pictured by artists as a handsome, stalwart fellow, modern enough, it would

Figure 16.4. Reconstructed skulls of (A) Zinjanthropus, (B) Pithecanthropus, (C) Neanderthal, and (D) Cro-Magnon.

certainly appear, to be able to interbreed with present-day human beings.

Human beings, traced thus far back, were not a planet-wide species as they are now. Prior to 20,000 B.C. or so, they were confined to the great "world island" of Africa, Asia, and Europe. It was only later that hunting bands began to migrate across narrow ocean passages into the Americas, Indonesia, and Australia. It was not until 400 B.C., and later, that daring Polynesian navigators crossed wide stretches of the Pacific, without compasses, and in what were little more than canoes, to colonize the islands of the Pacific. Finally, it was not until the twentieth century that a human foot rested on Antarctica.

But if we are to trace the fortunes of prehistoric peoples at a time when they were confined to only part of the earth's land area, there must be some manner of dating events, at least roughly. A variety of ingenious methods have been used.

Archaeologists have, for instance, used tree rings for the purpose, a technique (*dendrochronology*) introduced in 1914 by the American astronomer Andrew Ellicott Douglass. Tree rings are widely separated in wet summers when much new wood is laid down, and closely spaced in dry summers. The pattern over the centuries is quite distinctive. A piece of wood forming part of a primitive abode can have its ring pattern matched with the one place of the scheme where it will fit, and, in this way, its date can be obtained.

A similar system can be applied to layers of sediment or *varves* laid down summer after summer by melting glaciers in such places as Scandinavia. Warm summers will leave thick layers, cool summers thin ones; and again, there is a distinctive pattern. In Sweden, events can be traced back 18,000 years in this way.

An even more startling technique is that developed in 1946 by the American chemist Willard Frank Libby. Libby's work had its origin in the 1939 discovery by the American physicist Serge Korff that cosmic ray bombardment of the atmosphere produced neutrons. Nitrogen reacts with these neutrons, producing radioactive carbon 14 in nine reactions out of every ten, and radioactive hydrogen 3 in the tenth reaction.

As a result, the atmosphere would always contain small traces of carbon 14 (and even smaller traces of hydrogen 3). Libby reasoned that radioactive carbon 14 created in the atmosphere by cosmic rays would enter all living tissue via carbon dioxide, first absorbed by plants and then passed on to animals. As long as a plant or animal lived, it would continue to receive radiocarbon and maintain it at a constant level in its tissues. But when the organism died and ceased to take in carbon, the radiocarbon in its tissues would begin to diminish by radioactive breakdown, at a rate determined by its 5,600-year half-life. Therefore, any piece of preserved bone, any bit of charcoal from an ancient campfire, or organic remains of any kind could be dated by measuring the amount of radiocarbon left. The method is reasonably accurate for objects up to 30,000 years old, and this covers archaeological history from the ancient civilizations back to the beginnings of Cro-Magnon man. For developing this technique of *archaeometry*, Libby was awarded the Nobel Prize for chemistry in 1960.

Cro-Magnon was not the first early man dug up by the archaeologists. In 1857, in the Neanderthal valley of the German Rhineland, a digger discovered part of a skull and some long bones that looked human in the main but only crudely human. The skull had a sharply sloping forehead and very heavy brow ridges. Some archaeologists maintained that they were the remains of a human being whose bones had been deformed by disease; but as the years passed, other such skeletons were found, and a detailed and consistent picture of Neanderthal man was developed. Neanderthal was a short, squat, stooping biped, the men averaging a little taller than five feet, the women somewhat shorter. The skull was roomy enough for a brain nearly as large as a modern human's (figure 16.4). Anthropological artists picture the creature as barrel-chested, hairy, beetle-browed, chinless, and brutish in expression—a picture originated by the French paleontologist Marcellin Boule, who was the first to describe a nearly complete Neanderthal skeleton in 1911. Actually, Neanderthal was probably not as subhuman as pictured. Modern examination of the skeleton described by Boule show it to have belonged to a badly arthritic creature. A normal skeleton gives rise to a far more human image. In fact, give a Neanderthal man a shave and a haircut, dress him in well-fitted clothes, and he could probably walk down New York's Fifth Avenue without getting much notice.

Traces of Neanderthal man were eventually found not only in Europe but also in northern Africa, in Russia and Siberia, in Palestine, and in Iraq. About a hundred different skeletons have now been located at some forty different sites, and human beings of this sort may still have been alive as recently as 30,000 years ago. Skeletal remains somewhat resembling Neanderthal man were discovered in still more widely separated places; these were Rhodesian man, dug up in northern Rhodesia (now Zambia) in southern Africa in 1921, and Solo man, found on the banks of the Solo River in Java in 1931. They were considered separate species of the genus *Homo,* and so the three types were named *Homo neanderthalensis, Homo rhodesiensis,* and *Homo solensis.* But some anthropologists and evolutionists maintain that all three should be placed in the same species as *Homo sapiens,* as *varieties* or *subspecies* of man. There were humans that we call *sapiens* living at the same time as Neanderthal, and intermediate forms have been found which suggest that there may have been interbreeding between them. If Neanderthal and his cousins can be classed as *sapiens,* then our species is perhaps 250,000 years old.

HOMINIDS

Darwin's *Origin of Species* launched a great hunt for our distinctly subhuman ancestors—what the popular press came to call the *missing link* between us and our presumably apelike forerunners. This hunt, in the very nature of things, could not be an easy one. Primates are quite intelligent, and few allow themselves to be trapped in situations that lead to fossilization. It has been estimated that the chance of finding a primate skeleton by random search is only one in a quadrillion.

In the 1880s, a Dutch paleontologist, Marie Eugène François Thomas Dubois, got it into his head that the ancestors of human beings might be found in the East Indies (modern Indonesia), where great apes still flourished (and where he could work conveniently because those islands then belonged to the Netherlands). Surprisingly enough, Dubois, working in Java, the most populous of the Indonesian islands, did turn up a creature somewhere between an ape and a human! After three years of hunting, he found the top of a skull which was larger than an ape's but smaller than any recognized as human. The next year he found a similarly intermediate thighbone. Dubois named his "Java man" *Pithecanthropus erectus* ("erect apeman") (figure 16.4). Half a century later, in the 1930's, another Dutchman, Gustav H. R. von Koenigswald, discovered more bones of *Pithecanthropus,* and they composed a clear picture of a small-brained, very beetling-browed creature with a distant resemblance to Neanderthal.

Meanwhile other diggers had found, in a cave near Peking, skulls, jaws, and teeth of a primitive man they called *Peking man.* Once this discovery was made, it came to be realized that such teeth had been located earlier—in a Peking drugstore, where they were kept for medicinal purposes. The first intact skull was located in December 1929, and Peking man was eventually recognized as markedly similar to Java man. It lived perhaps half a million years ago, used fire, and had tools of bone and stone. Eventually, fragments from forty-five individuals were accumulated, but they disappeared in 1941 during an attempted evacuation of the fossils in the face of the advancing Japanese. In 1949, Chinese archeologists resumed digging, and fragments of forty individuals of both sexes and all ages have now been located.

Peking man was named *Sinanthropus pekinensis* ("China man of Peking"), but closer examination of more of these comparatively small-brained *hominids* ("manlike" creatures) made it seem that it was poor practice to place Peking man and Java man in separate genera. The German-American biologist Ernst Walter Mayr felt it wrong to place them in a separate genus from modern human beings, so that Peking man and Java man are now considered two varieties of the species *Homo erectus*, whose earliest members may have appeared 700,000 years ago.

It is unlikely that humankind originated in Java, despite the existence there of a small-brained hominid. For a while the vast continent of Asia, early inhabited by Peking man, was suspected of being the birthplace of human beings; but as the twentieth century progressed, attention focused more and more firmly on Africa, which, after all, is the continent richest in primate life generally and of the higher primates particularly.

The first significant African finds were made by two English scientists, Raymond Dart and Robert Broom. One spring day in 1924, workers blasting in a limestone quarry near Taungs in South Africa picked up a small skull that looked nearly human. They sent it to Dart, an anatomist working in Johannesburg. Dart immediately identified it as a being between an ape and a human, and called it *Australopithecus africanus* ("southern ape of Africa"). When his paper announcing the find was published in London, anthropologists thought he had blundered, mistaking a chimpanzee for an apeman. But Broom, an ardent fossil hunter who had long been convinced that human beings originated in Africa, rushed to Johannesburg and proclaimed *Australopithecus* the closest thing to a missing link that had yet been discovered.

Through the following decades, Dart, Broom, and several anthropologists searched for and found many more bones and teeth of South African apemen, as well as clubs that they used to kill game, the bones of animals that they killed, and caves in which they lived. *Australopithecus* was a short, small-brained creature with a snoutlike face, in many ways less human than Java man. But *Australopithecus* had more human brows and more human teeth than *Pithecanthropus* and walked erect, used tools, and probably had a primitive form of speech. In short, *Australopithecus* was an African variety of hominid living at least half a million years ago and definitely more primitive than *Homo erectus*.

There were no clear grounds for suspecting priority between the African and the Asian varieties of hominids at first, but the balance swung definitely and massively toward Africa with the work of the Kenya-born Englishman Louis Seymour Bazett Leakey and his wife Mary. With patience and persist-

ence, the Leakeys combed likely areas in eastern Africa for early fossil hominids. The most promising was Olduvai Gorge, in what is now Tanzania, and
there, on 17 July 1959, Mary Leakey crowned a more than quarter-century
search by discovering fragments of a skull that, when pieced together, proved
to encase the smallest brain of any hominid yet discovered. Other features
showed this hominid, however, to be closer to humans than to apes, for it
walked upright and the remains were surrounded by small tools formed out
of pebbles. The Leakeys named their find *Zinjanthropus* ("East African man,"
using the Arabic word for East Africa) (figure 16.4).

Zinjanthropus does not seem to be in the direct line of ancestry of modern
humans. Still older fossils, some 2 million years old, may qualify. These, given
the name of *Homo habilis* ("nimble man"), were 4½-foot-tall creatures who
already had hands with opposable thumbs which were nimble enough (hence,
the name) to make them utterly like humans in this respect.

In 1977, the American archaeologist Donald Johanson discovered a hominid fossil that was perhaps 4 million years old. Enough bones were dug up to
make up about 40 percent of a complete individual. It was a little creature
about three and a half feet tall with slender bones. Its scientific name is
Australopithecus afarensis, but it is popularly known as Lucy.

The most interesting thing about Lucy is that she is completely bipedal—
as much as we are. It would seem that the first important anatomical characteristic that marked off hominids from apes was the development of bipedality
at a time when the hominid brain was no larger than that of a gorilla. It might
be argued, in fact, that the sudden and remarkable expansion of the hominid
brain in the last million years came about as the result of bipedality. The
forelimbs were freed to become delicate hands with which to feel and manipulate various objects, and the flood of information reaching the brain put a
premium on any chance increase, which was then given survival value by the
processes of natural selection.

It may be that Lucy represents the ancestors of two branches of the hominid
line. On one side are various australopithecines, whose brains had a volume
of between 450 and 650 cubic centimeters, and which became extinct about
a million years ago. On the other side are the ancestral hominids, the members
of genus Homo, which passed through *Homo habilis,* then *Homo erectus*
(with a brain capacity of from 800 to 1,100 cubic centimeters), and then
finally *Homo sapiens* (with a brain capacity of from 1,200 to 1,600 cubic
centimeters).

Naturally, if we look beyond Lucy, we find fossils of animals that are too
primitive to be called hominids, and we approach the common ancestor of the
hominids, of which the living members are human beings, and the pongids
(or apes), of which the living members are the chimpanzee, the gorilla, the
orangutan, and several species of gibbon.

There is Ramapithecus, whose upper jaw was located in northern India in
the early 1930s by G. Edward Lewis. The upper jaw was distinctly closer to
the human than is that of any living primate other than ourselves; it was

perhaps 3 million years old. In 1962, Leakey discovered an allied species which isotope studies showed to be 14 million years old.

In 1948, Leakey had discovered a still older fossil (perhaps 25 million years old), which was named Proconsul. (This name, meaning "before Consul" honored Consul, a chimpanzee in the London Zoo.) Proconsul seems to be the common ancestor of the larger great apes, the gorilla, chimpanzee, and orangutan. Farther back, then, there must be a common ancestor of Proconsul and *Ramapithecus* (and of the primitive ape that was ancestral to the smallest modern ape, the gibbon). Such a creature, the first of all the apelike creatures, would date back perhaps 40 million years.

PILTDOWN MAN

For many years anthropologists were greatly puzzled by a fossil that did look like a missing link, but of a curious and incredible kind. In 1911, near a place called Piltdown Common in Sussex, England, workmen building a road found an ancient, broken skull in a gravel bed. The skull came to the attention of a lawyer named Charles Dawson, and he took it to a paleontologist, Arthur Smith Woodward, at the British Museum. The skull was high-browed, with only slight brow ridges; it looked more modern than Neanderthal. Dawson and Woodward went searching in the gravel pit for other parts of the skeleton. One day Dawson, in Woodward's presence, came across a jawbone in about the place where the skull fragments had been found. It had the same reddish-brown hue as the other fragments and therefore appeared to have come from the same head. But the jawbone, in contrast to the human upper skull, was like that of an ape! Equally strange, the teeth in the jaw, though apelike, were ground down, as human teeth are, by chewing.

Woodward decided that this half-ape, half-man might be an early creature with a well-developed brain and a backward jaw. He presented the find to the world as the *Piltdown man*, or *Eoanthropus dawsoni* ("Dawson's dawn man").

Piltdown man became more and more of an anomaly as anthropologists found that, in all other fossil finds that included the jaw, jawbone development did keep pace with skull development. Finally, in the early 1950s, three British scientists—Kenneth Oakley, Wilfrid Le Gros Clark, and Joseph Sidney Weiner—decided to investigate the possibility of fraud. It *was* a fraud. The jawbone, that of a modern ape, had been planted.

The tale of Piltdown man is perhaps the best known and most embarrassing example of scientists being fooled for a long time by an arrant hoax. In hindsight, we can be astonished that scientists were fooled by so clumsy a jape, but hindsight is cheap. We must remember that in 1911 very little was known about hominid evolution. Today, a similar hoax would fool no knowledgeable scientist for a moment.

Another odd story of primate relics had a happier ending. In 1935, von Koenigswald had come across a huge but manlike fossil tooth for sale in a Hong Kong pharmacy. The Chinese pharmacist considered it a "dragon tooth" of valuable medicinal properties. Von Koenigswald ransacked other Chinese pharmacies and had four such molars before the Second World War temporarily ended his activities.

The manlike nature of the teeth made it seem that gigantic human beings, possibly 9 feet high, once roamed the earth. There was a tendency to accept this theory, perhaps, because the Bible says, "There were giants in the earth in those days" (Genesis 6:4).

Between 1956 and 1968, however, four jawbones were discovered into which such teeth would fit. The creature, Gigantopithecus, was seen to be the largest primate ever known to exist, but was distinctly an ape and not a hominid, for all its human-appearing teeth. Very likely it was a gorillalike creature, standing 9 feet tall when upright and weighing 600 pounds. It may have existed contemporaneously with *Homo erectus* and possessed the same feeding habits (hence the similarity in teeth). It has, of course, been extinct for at least a million years and could not possibly have been responsible for that biblical verse.

RACIAL DIFFERENCES

It is important to emphasize that the net result of human evolution has been the production today of a single species: that is, while there may have been a number of species of hominids, one only has survived. All men and women today, regardless of differences in appearances, are *Homo sapiens;* and the difference between blacks and whites is approximately that between horses of different coloring.

Still, ever since the dawn of civilization, human beings have been more or less acutely conscious of racial differences and usually have viewed other races with the emotions generally evoked by strangers, ranging from curiosity to contempt to hatred. But seldom has racism had such tragic and long-persisting results as the modern conflict between white people and black. (White people are often referred to as Caucasians, a term first used, in 1775, by the German anthropologist Johann Friedrich Blumenbach, who was under the mistaken impression that the Caucasus contained the most perfect representatives of the group. Blumenbach also classified blacks as Ethiopians and East Asians as Mongolians, terms that are still sometimes used.)

The racist conflict between white and black, between Caucasian and Ethiopian, so to speak, entered its worst phase in the fifteenth century, when Portuguese expeditions down the west coast of Africa began a profitable business of carrying off black Africans into slavery. As the trade grew and nations built their economies on slave labor, rationalizations to justify the enslavement of blacks were invoked in the name of the Scriptures, of social morality, and even of science.

According to the slaveholders' interpretation of the Bible—an interpretation believed by many people to this day—blacks were descendants of Ham and, as such, an inferior tribe subject to Noah's curse: "a servant of servants shall he be unto his brethren" (Genesis 9:25). Actually, the curse was laid upon Ham's son, Canaan, and on his descendants, the Canaanites, who were reduced to servitude by the Israelites when the latter conquered the land of Canaan. No doubt the words in Genesis 9:25 represent a comment after the fact, written by the Hebrew writers of the Bible to justify the enslavement of the Canaanites. In any case, the point of the matter is that the reference is to the Canaanites only, and the Canaanites were certainly white. It was a twisted interpretation of the Bible that the slaveholders used, with telling effect in centuries past, to defend their subjugation of blacks.

The "scientific" racists of more recent times took their stand on even shakier ground. They argued that black people were inferior to white as obviously representing a lower stage of evolution. Were not a dark skin and wide nose, for instance, reminiscent of the ape? Unfortunately for the "scientific" racists' case, this line of reasoning actually leads to the opposite conclusion. Black people are the least hairy of all human groups; in this respect and in the fact that their hair is crisp and woolly, rather than long and straight, they are farther from the ape than white people are! The same can be said of the black's thick lips; they resemble those of an ape less than do the white's thin lips.

The fact of the matter is that any attempt to rank the various groups of *Homo sapiens* on the evolutionary ladder is to try to do fine work with blunt tools. Humanity consists of but one species, and so far the variations that have developed in response to natural selection are superficial.

The dark skin of dwellers in the earth's tropical and subtropical regions has obvious value in preventing sunburn. The fair skin of northern Europeans is useful to absorb as much ultraviolet radiation as possible from the comparatively feeble sunlight in order that enough vitamin D be formed from the sterols in the skin. The narrowed eyes of the Eskimo and the Mongol have survival value in lands where the glare from snow or desert sands is intense. The high-bridged nose and narrow nasal passages of the European serve to warm the cold air of the northern winter. And so on.

Since the tendency of *Homo sapiens* has been to make our planet one world, no basic differences in the human constitution have developed in the past and are even less likely to develop in the future. Interbreeding is steadily evening out the human inheritance. The American black is one of the best cases in point. Despite social barriers against intermarriage, nearly four-fifths of the black people in the United States, it is estimated, have some white ancestry. By the end of the twentieth century probably there will be no "pure-blooded" black people in North America.

BLOOD GROUPS AND RACE

Anthropologists nevertheless are keenly interested in race, primarily as a guide to the migrations of early human beings. It is not easy to identify specific races. Skin color, for instance, is a poor guide; the Australian aborigine and the African black are both dark in color but are no more closely related to each other than either is to the European. Nor is the shape of the head—*dolichocephalic* (long) versus *brachycephalic* (wide), terms introduced in 1840 by the Swedish anatomist Anders Adolf Retzius—much better despite the classifications of Europeans into subgroups on this basis. The ratio of head length to head width multiplied by 100 (*cephalic index*, or, if skull measurements were substituted, *cranial index*) served to divide Europeans into Nordics, Alpines, and Mediterraneans. The differences, however, from one group to another are small, and the spread within a group is wide. In addition, the shape of the skull is affected by environmental factors such as vitamin deficiencies, the type of cradle in which an infant sleeps, and so on.

But the anthropologists have found an excellent marker for race in blood groups. The Boston University biochemist William Clouser Boyd was prominent in this connection. He pointed out that blood groups are inherited in a simple and known fashion, are unaltered by the environment, and show up in distinctly different distributions in the various races.

The American Indian is a particularly good example. Some tribes are almost entirely O; others are O but with a heavy admixture of A; virtually no Indians have B or AB blood. An American Indian testing as a B or AB is almost certain to possess some European ancestry. The Australian aborigines are likewise high in O and A, with B virtually nonexistent. But they are distinguished from the American Indian in being high in the more recently discovered blood group M and low in blood group N, while the American Indian is high in N and low in M.

In Europe and Asia, where the population is more mixed, the differences between peoples are smaller, yet still distinct. For instance, in London 70 percent of the population has O blood; 26 percent, A; and 5 percent, B. In the city of Kharkov, Russia, on the other hand, the corresponding distribution is 60, 25, and 15. In general, the percentage of B increases as one travels eastward in Europe, reaching a peak of 40 percent in central Asia.

Now the blood-type genes show the not-yet-entirely-erased marks of past migrations. The infiltration of the B gene into Europe may be a dim mark of the invasion by the Huns in the fifth century and by the Mongols in the thirteenth. Similar blood studies in the Far East seem to indicate a comparatively recent infiltration of the A gene into Japan from the southwest and of the B gene into Australia from the north.

A particularly interesting, and unexpected, echo of early human migrations in Europe showed up in Spain. It came out in a study of Rh blood distribution. (The Rh blood groups are so named from the reaction of the blood to antisera developed against the red cells of a rhesus monkey. There are at least eight alleles of the responsible gene; seven are called *Rh positive*, and the eighth, recessive to all the others, is called *Rh negative* because it shows its effect only when a person has received the allele from both parents.) In the United States, about 85 percent of the population is Rh positive; 15 percent, Rh negative. The same proportion holds in most of the European peoples. But, curiously, the Basques of northern Spain stand apart, with something like 60 percent Rh negative to 40 percent Rh positive. And the Basques are also notable in having a language unrelated to any other European language.

The conclusion that can be drawn is that the Basques are a remnant of a prehistoric invasion of Europe by an Rh-negative people. Presumably a later wave of invasions by Rh-positive tribes penned them up in their mountainous refuge in the western corner of the continent, where they remain the only sizable group of survivors of the *early Europeans*. The small residue of Rh-negative genes in the rest of Europe and in the American descendants of the European colonizers may represent a legacy from those early Europeans.

The peoples of Asia, the African blacks, the American Indians, and the Australian aborigines are almost entirely Rh positive.

Humanity's Future

Attempting to foretell the future of the human race is a risky proposition that had better be left to mystics and science-fiction writers (though, to be sure, I am a science-fiction writer myself, among other things). But of one thing we can be fairly sure. Provided there are no worldwide catastrophes—such as a full-scale nuclear war, or a massive attack from outer space, or a pandemic of a deadly new disease—the human population will increase rapidly. It is now five times as large as it was only two centuries ago. Some estimates are that the total number of human beings who have lived over a period of 600,000 years comes to 77,000,000,000. If so, then nearly 6 percent of all the human beings who have ever lived are living at this moment. And the world population is still growing at a tremendous rate.

Since we have no censuses of ancient populations, we must estimate them roughly on the basis of what we know about the conditions of human life. Ecologists have estimated that the pre-agricultural food supply—obtainable by hunting, fishing, collecting wild fruit and nuts, and so on—could not have supported a world population of more than 20,000,000, and, in all likelihood, the actual population during the Paleolithic era was only one-third or one-half of this figure at most. Hence, as late as 6000 B.C., it could not have numbered more than 6 to 10 million people—less than the population of a single present-day city such as Shanghai or Mexico City. (When America was discovered, the food-gathering Indians occupying what is now the United States probably numbered not much more than 250,000—or as if the population of Dayton, Ohio, were spread out across the continent.)

THE POPULATION EXPLOSION

The first big jump in world population came with the Neolithic Revolution and agriculture. The British biologist Julian Sorrell Huxley (grandson of the Huxley who was "Darwin's bulldog") estimates that the population began to increase at a rate that doubled its numbers every 1,700 years or so. By the opening of the Bronze Age, the world population may have been about 25 million; by the beginning of the Iron Age, 70 million; by the start of the Christian era, 170 million, with one-third crowded into the Roman empire, another third into the Chinese empire, and the rest scattered. By 1600, the earth's population totaled perhaps 500 million, considerably less than the present population of India alone.

At that point, the smooth rate of growth ended, and the population began to explode. World explorers opened up some 18 million square miles of almost empty land on new continents to colonization by the Europeans. The eighteenth-century Industrial Revolution accelerated the production of food and of people. Even backward China and India shared in the population explosion. The doubling of the world's population now took place not in a period of nearly two millennia but in less than two centuries. The population expanded from 500,000,000 in 1600 to 900,000,000 in 1800. Since then it has grown at an ever faster rate. By 1900, it had reached 1,600,000,000. In the first seventy years of the twentieth century, it has climbed to 3,600,000,000, despite two world wars.

In 1970 the world population was increasing at the rate of 220,000 each day, or 70,000,000 each year. This was an increase at the rate of 2.0 percent each year (as compared with an estimated increase of only 0.3 percent per year in 1650). At this rate, the population of the earth would double in about thirty-five years; and in some regions, such as Latin America, the doubling would take place in a shorter time.

At the moment, students of the population explosion are leaning strongly toward the Malthusian view, which has been unpopular ever since it was advanced in 1798. As I said earlier, Thomas Robert Malthus maintained, in *An Essay on the Principle of Population*, that population always tends to grow faster than the food supply, with the inevitable result of periodic famines and wars. Despite his predictions, the world population has grown apace without any serious setbacks in the past century and a half. But, for this postponement of catastrophe, we can be grateful, in large measure, that large areas of the earth were still open for the expansion of food production. Now we are running out of tillable new lands. A majority of the world's population is underfed, and we must make mighty efforts to wipe out this chronic undernourishment. To be sure, the sea can be more rationally exploited, and its food yield multiplied. The use of chemical fertilizers must yet be introduced to wide areas. Proper use of pesticides will reduce the loss of food to insect depredation in areas where such loss has not yet been countered. There are also ways of encouraging growth directly. Plant hormones such as *gibberellin* (studied by Japanese biochemists before the Second World War and coming to Western attention in the 1950s) could accelerate plant growth, while small quantities of antibiotics added to animal feed will accelerate animal growth (perhaps by suppressing intestinal bacteria that otherwise compete for the food supply passing through the intestines, and by suppressing mild but debilitating infections). Nevertheless, with new mouths to feed multiplying at their current rate, it will take Herculean efforts merely to keep the world's population up to the present none-too-good mark in which some 300 million children under five, the world over, are undernourished to the point of suffering permanent brain damage.

Even so common (and, till recently, disregarded) a resource as fresh water is beginning to feel the pinch. Fresh water is now being used at the rate of nearly 2 trillion gallons a day the world over; although total rainfall, which at the moment is the main source of fresh water, is 50 times this quantity, only a fraction of the rainfall is easily recoverable. And in the United States, where fresh water is used at a total rate of 350 billion gallons a day at a larger per-capita rate than in the world generally, some 10 percent of the total rainfall is being consumed one way or another.

The result is that the world's lakes and rivers are being quarreled over more intensely than ever. (The quarrels of Syria and Israel over the Jordan, and of Arizona and California over the Colorado River, are cases in point.) Wells are being dug ever deeper; and in many parts of the world, the ground-water level is sinking dangerously. Attempts to conserve fresh water have included the use of cetyl alcohol as a cover for lakes and reservoirs in such regions as Australia, Israel, and East Africa. Cetyl alcohol spreads out into a film one molecule thick, cutting down on water evaporation without polluting the water. (Of course, increasing water pollution by sewage and by industrial wastes is an added strain on the diminishing fresh-water surplus.)

Eventually, it seems, it will be necessary to obtain fresh water from the oceans, which, for the foreseeable future, offer an unlimited supply. The most promising methods of desalting sea water include distillation and freezing. In addition, experiments are proceeding with membranes that will selectively permit water molecules to pass, but not the various ions. Such is the importance of this problem that the Soviet Union and the United States are discussing a joint attack on it, at a time when cooperation between these two competing nations is, in other respects, exceedingly difficult to arrange.

But let us be as optimistic as we can and admit no reasonable limits to human ingenuity. Let us suppose that, by miracles of technology, we raise the productivity of the earth tenfold; suppose that we mine the metals of the ocean, bring up gushers of oil in the Sahara, find coal in Antarctica, harness the energy of sunlight, develop fusion power. Then what? If the rate of increase of the human population continues unchecked at its present rate, all our science and technical invention will still leave us struggling uphill like Sisyphus.

If you are not certain whether to accept this pessimistic appraisal, let us consider the powers of a geometric progression. It has been estimated that the total quantity of living matter on earth is now equal to 2×10^{19} grams. If so, the total mass of humanity in 1970 was about 1/100,000 of the mass of all life.

If the earth's population continues to double every thirty-five years (as it was then doing), by 2570 A.D. it will have increased 100,000-fold. It may prove extremely difficult to increase as a whole the mass of life the earth can support (though one species can always multiply at the expense of others). In that case, by 2570 A.D. the mass of humanity would comprise all of life, and we would be reduced to cannibalism if some people were to continue to survive.

Even if we could imagine artificial production of foodstuffs out of the inorganic world via yeast culture, *hydroponics* (the growth of plants in solutions of chemicals), and so on, no conceivable advance could match the inexorable number increase involved in this doubling-every-thirty-five years. At that rate, by 2600 A.D., it would reach 630,000 billion! Our planet would have standing room only, for there would be only 2½ square feet per person on the entire land surface, including Greenland and Antarctica. In fact, if the human species could be imagined as continuing to multiply further at the same rate, by 3550 A.D. the total mass of human tissue would be equal to the mass of the earth.

If there are people who see a way out in emigration to other planets, they may find food for thought in the fact that, assuming there were 1,000 billion other inhabitable planets in the universe and people could be transported to any of them at will, at the present rate of increase of human numbers every one of those planets would be crowded literally to standing room only by 5000 A.D. By 7000 A.D., the mass of humanity would be equal to the mass of the known universe!

Obviously, the human race cannot increase at the present rate for very long, regardless of what is done with respect to the supply of food, water, minerals, and energy. I do not say "will not" or "dare not" or "should not"; I say quite flatly "cannot."

Indeed, it is not mere numbers that will limit our growth if it continues at a high rate. It is not only that there are more men, women, and children each minute, but that each individual uses (on the average) more of Earth's unrenewable resources, expends more energy, and produces more waste and pollution each minute. Where population doubles every thirty-five years, energy utilization, in 1970, was increasing at such a rate, that, in thirty-five years, it would have increased not twice but sevenfold.

The blind urge to waste and poison faster and faster each year is driving us to destruction even more rapidly, then, than mere multiplication alone. For instance, smoke from burning coal and oil is freely dumped into the air by home and factory, as is the gaseous chemical refuse from industrial plants. Automobiles by the hundreds of millions discharge fumes of gasoline and of its breakdown and oxidation products, to say nothing of carbon-monoxide and lead compounds. Oxides of sulfur and nitrogen (produced either directly or through later oxidation by ultraviolet light from the sun), together with other substances, can corrode metals, weather construction materials, embrittle rubber, damage crops, cause and exacerbate respiratory diseases, and even serve as one of the causes of lung cancer.

When atmospheric conditions are such that the air over a city remains stagnant for a period of time, the pollutants collect, seriously contaminating the air and encouraging the formation of a smoky fog (*smog*) that was first publicized in Los Angeles but had long existed in many cities and now exists in more. At its worst, it can take thousands of lives among those who, out of age or illness, cannot tolerate the added stress placed on their lungs. Such disasters took place in Donora, Pennsylvania, in 1948 and in London in 1952.

The fresh waters of the earth are polluted by chemical wastes, and occasionally one of them will come to dramatic notice. Thus, in 1970, it was found that mercury compounds heedlessly dumped into the world's waters were finding their way into sea organisms in sometimes dangerous quantities. At this rate, far from finding the ocean a richer source of food, we may make a good beginning at poisoning it altogether.

Indiscriminate use of long-lingering pesticides results in their incorporation first into plants, then into animals. Because of the poisoning, some birds find it increasingly difficult to form normal eggshells, so that, in attacking insects, we are bringing perilously close to extinction the peregrine falcon.

Almost every new so-called technological advance, hastened into without due caution by the eagerness to overreach one's competitors and multiply one's profits, can bring about difficulties. Since the Second World War, synthetic detergents have replaced soaps. Important ingredients of those detergents are various phosphates, which washed into the water supply and greatly accelerated the growth of microorganisms that, however, used up the oxygen supply of the waters—thus leading to the death of other sea organisms. These deleterious changes in water habitats (*eutrophication*) are rapidly aging the Great Lakes, for instance—the shallow Lake Erie in particular—and are shortening their natural lives by millions of years. Thus, Lake Erie may become Swamp Erie, while the swampy Everglades may dry up altogether.

Living species are utterly interdependent. There are obvious cases like the interconnection of plants and bees, where the plants are pollinated by the bees and the bees are fed by the plants, and a million other cases less obvious. Every time life is made easier or more difficult for one particular species, dozens of other species are affected—sometimes in hard-to-predict ways. The study of this interconnectability of life, *ecology,* is only now attracting attention, for in many cases human beings, in an effort to achieve some short-term benefit for themselves have so altered the ecological structure as to bring about some long-term difficulty. Clearly we must learn to look far more carefully before we leap.

Even so apparently other-worldly an affair as rocketry must be carefully considered. A single large rocket may inject over 100 tons of exhaust gases into the atmosphere at levels above 60 miles. Such quantities of material could appreciably change the properties of the thin upper atmosphere and lead to hard-to-predict climatic changes. In the 1970s supersonic transport planes (SSTs) traveling through the stratosphere at higher-than-sound velocities were introduced. Those who object to their use cite not only the noise factor involved in sonic booms but also the chance of climate-affecting pollution.

Another factor that makes the increase in numbers even worse is the uneven distribution of human beings over the face of the earth. Everywhere there is a trend toward accumulation within metropolitan areas. In the United States, even while the population goes up and up, certain farming states not only do not share in the explosion but are actually decreasing in population. It is estimated that the urban population of the earth is doubling not every thirty-five years but every eleven years. By 2005 A.D., when the earth's total population will have doubled, the metropolitan population will, at this rate, have increased over ninefold.

This is serious. We are already witnessing a breakdown in the social structure—a breakdown that is most strongly concentrated in just those advanced nations where urbanization is most apparent. Within those nations, it is most concentrated in the cities, especially in their most crowded portions. There is no question but that when living beings are crowded beyond a certain point, many forms of pathological behavior become manifest. This has been found to be true in laboratory experiments on rats, and the newspaper and our own experience should convince us that this is also true for human beings.

It would seem obvious, then, that *if present trends continue. unchanged,* the world's social and technological structure will have broken down well within the next half-century, with incalculable consequences. Human beings, in sheer madness, may even resort to the ultimate catastrophe of thermonuclear warfare.

But *will* present trends continue?

Clearly, changing them will require a massive effort and will mean that we must change long-cherished beliefs. For most of human history, people have lived in a world in which life was brief and many children died while still infants. If the tribal population were not to die out, women had to bear as many babies as they could. For this reason, motherhood was deified, and every trend that might lower the birthrate was stamped out. The status of women was lowered so that they might be nothing but baby-making and baby-rearing machines. Sexual mores were so controlled that only those actions were approved of that led to conception; everything else was considered perverted and sinful.

But now we live in a crowded world. If we are to avoid catastrophe, motherhood must become a privilege sparingly doled out. Our views on sex and on its connection with childbirth must be changed.

Again, the problems of the world—the really serious problems—are global in nature. The dangers posed by overpopulation, overpollution, the disappearance of resources, the risk of nuclear war, affect every nation, and there can be no real solutions unless all nations cooperate. What this means is that a nation can no longer go its own way, heedless of the others; nations can no longer act on the assumption that there is such a thing as a "national security" whereby something good can happen to them if something bad happens to someone else. In short, an effective world government is necessary—one that is federalized to allow the free play of cultural differences and one that (we hope) can guarantee human rights.

Can this sort of thing come to pass?

Perhaps.

In the preceding pages, I have talked of world population and rate of population increase as of 1970. That is because since that date the rate of increase seems to have slowed. Governments have increasingly come to realize the enormous danger of overpopulation and are increasingly aware that *no* problem can be solved as long as the population problem is not. Increasingly, population planning is encouraged, and China (which, with its 1-billion population represents nearly one-quarter of the world's people) is, at the moment, strongly pushing the one-child family.

The result is that the world population increase has declined from 2 percent in 1970 to an estimated 1.6 percent in the early 1980s. To be sure, the world population has increased to 4,500,000,000 by now, so that a 1.6 percent increase represents 72,000,000 additional people each year—if anything, a trifle more than the yearly increment in 1970. We have not moved far enough, in other words; but we are moving in the right direction.

What's more, we are witnessing a steady strengthening of feminism. Women realize the importance of taking an equal role in every facet of living and are increasingly determined to do so. The importance of this development (aside from the simple justice of it) is that women engaged in the work of the world will find other ways of reaching self-fulfillment than in their traditional roles of baby machine and household drudge, and the birth rate is more likely to stay low.

To be sure, the movement in the direction of population control, essential though it would seem to anyone capable of a moment's thought, is not without its opponents. In the United States, an active group opposes not only abortion but also the kind of sex education in schools and the availability of contraceptive devices that would make abortion unnecessary. The only way of legitimately lowering the birth rate, in their view, is by sexual abstinence, something that no sane person would suppose that people can be talked into. This group calls itself the "Right to Life," but a better name for people who do not recognize the dangers of overpopulation would be the "Right to Fatal Stupidity."

Then, too, in 1973, the Arab nations, which control most of the world's oil supply, effected a temporary oil blockade to punish Western nations which, in their view, were supporting Israel. This policy, and the several years thereafter when the price of oil was steadily increased, served to convince the industrial nations of the absolute necessity of energy conservation. If this policy continues—and if to it is added a resolute determination to replace the fossil fuels, as far as possible, with solar power, nuclear fusion, and renewable energy sources—we will have taken a giant step toward survival.

There is also increasing concern over the quality of the environment. In the United States, the administration of Ronald Reagan, which came into power in 1981, put into action many programs that favor business over the humanitarian ideals that had been practiced since the days of Franklin D. Roosevelt's "New Deal" a half-century before. In this, the Reagan administration felt it had the support of the majority of the American people. However, when the Environmental Protection Agency was put into the hands of those who felt that profits for a few were worth the poisoning of many, there arose a howl of protest that forced a reorganization of the body and an admission that the Reagan administration had "misread its mandate."

Nor ought we to underestimate the effect of advancing technology. There is, for instance, the revolution in communications. The proliferation of communications satellites may make it possible in the near future for every person to be within reach of every other person. Underdeveloped nations can leapfrog over the earlier communications networks' necessity of involving large capital investments and move directly into a world in which everyone has a personal television station, so to speak, for receipt and emission of messages.

The world will become so much smaller as to resemble in social structure a kind of neighborhood village. (Indeed, the phrase *global village* has come into use to describe the new situation.) Education can penetrate every corner of the global village with the ubiquity of television. The new generation of every underdeveloped nation may grow up learning about modern agricultural methods, about the proper use of fertilizers and pesticides, and about the techniques of birth control.

There may even be, for the first time in Earth's history, a tendency toward decentralization. With ubiquitous television making all parts of the world equally accessible to business conferences and libraries and cultural programs, there will be less need to conglomerate everything into a large, decaying mass.

Computers and robots (which will be discussed in the next chapter) may also have a salutary effect.

Who knows, then? Catastrophe seems to have the edge, but the race for salvation is perhaps not quite over.

LIVING IN THE SEA

Assuming that the race for salvation is won; that the population levels off and a slow and humane decrease begins to take place; that an effective and sensible world government is instituted, allowing local diversity but not local murder; that the ecological structure is cared for and the earth systematically preserved—what then?

For one thing, humanity will probably continue to extend its range. Beginning as a primitive hominid in east Africa—at first perhaps no more widespread or successful than the modern gorilla—hominids slowly moved outward until by 15,000 years ago *Homo sapiens* had colonized the entire world island (Asia, Africa, and Europe). Human beings then made the leap into the Americas, Australia, and even through the Pacific islands. By the twentieth century, the population remained thin in particularly undesirable areas—such as the Sahara, the Arabian Desert, and Greenland—but no sizable area was utterly uninhabited by humans except for Antarctica. Now scientific stations, at least, are permanently established even on that least habitable of continents.

Where next?

One possible answer is the sea. It was in the sea that life originated and where it still flourishes best in terms of sheer quantity. Every kind of land animal, except for the insects, has tried the experiment of returning to the sea for the sake of its relatively unfailing food supply and for the relative equability of the environment. Among mammals, such examples as the otter, the seal, or the whale, indicate progressive stages of readaptation to a watery environment.

Can we return to the sea, not by the excessively slow alteration of our bodies through evolutionary change, but by the rapid help of technological advance? Encased in the metal walls of submarines and bathyscaphes, human beings have penetrated the ocean to its very deepest floor.

For bare submergence, much less is required. In 1943, the French oceanographer Jacques-Ives Cousteau invented the *aqualung.* This device brings oxygen to a person's lungs from a cylinder of compressed air worn on one's back and makes possible the modern sport of scuba diving (*scuba* is an acryonym for "*s*elf-*c*ontained *u*nderwater-*b*reathing *a*pparatus"). This makes it possible for one to stay underwater for considerable periods in one's skin, so to speak, without being encased in ships or even in enclosed suits.

Cousteau also pioneered in the construction of underwater living quarters in which people could remain submerged for even longer periods. In 1964, for instance, two men lived two days in an air-filled tent 432 feet below sea level. (One was Jon Lindbergh, son of the aviator.) At shallower depths, men have remained underwater for many weeks.

Even more dramatic is the fact that, beginning in 1961, the biologist Johannes A. Kylstra, at the University of Leyden, began to experiment with actual water-breathing in mammals. The lung and the gill act similarly, after all, except that the gill is adapted to work on lower levels of oxygenation. Kylstra made use of a water solution sufficiently like mammalian blood to avoid damaging lung tissue, and then oxygenated it heavily. He found that both mice and dogs could breathe such liquid for extended periods without apparent ill effect.

Hamsters have been kept alive under ordinary water when they were enclosed in a sheet of thin silicone rubber through which oxygen could pass from water to hamster and carbon dioxide from hamster to water. The membrane was virtually an artificial gill. With such advances and still others to be expected, can human beings look forward to a future in which we can remain underwater for indefinite periods and make all the planet's surface—land and sea—their home?

SETTLING IN SPACE

And what of outer space? Need we remain on our home planet, or can we venture to other worlds?

Once the first satellites were launched into orbit in 1957, the thought naturally arose that the dream of space travel, till then celebrated only in science-fiction stories, might become an actuality. It took only three and a half years after the launching of *Sputnik I* for the first step to be taken and only eight years after that first step for human beings to stand on the moon.

The space program has been expensive and has met with growing resistance from scientists who think that too much of it has been public-relations-minded and too little scientific, or who think it obscures other programs of greater scientific importance. It has also met with growing resistance from the general public, which considers it too expensive, particularly in the light of urgent sociological problems on Earth.

Nevertheless, the space program will probably continue, if only at a reduced pace; and if humanity can figure out how to spend less of its energies and resources on the suicidal folly of war, the program may even accelerate. There are plans for the establishment of space stations—in effect, large vehicles in more or less permanent orbit about the earth and capable of housing sizable numbers of men and women for extended periods—so that observations and experiments can be conducted that will presumably be of great value. Shuttle vessels, quite reusable, have been devised, work well, and are the essential preliminary to all this.

It is to be hoped that further trips to the moon will eventually result in the establishment of more or less permanent colonies there that, we may further hope, can exploit lunar resources and become independent of Earth's day-to-day help.

In 1974, the American physicist Gerard Kitchen O'Neill suggested that a full settlement need not be made on the moon, which could be reserved as a mining station alone. Although life began on a planetary surface, it need not confine itself to one. He pointed out that large cylinders, spheres, or doughnuts could be placed in orbit and set to rotating quickly enough to produce a centrifugal effect that would hold people to the inner surface with a kind of pseudogravity.

Such settlements could be built of metal and glass, and the inside lined with soil, all from the moon. The interior could be engineered into an Earthlike environment and could be settled by 10,000 human beings or more, depending on the size. Its orbit could be in the Trojan position with respect to the earth and the moon (so that Earth, moon and settlement would be at the apices of an equilateral triangle).

There are two such positions and dozens of settlements might cluster at each. So far, neither the United States nor the Soviet Union seem to be planning such structures, but the sanguine O'Neill feels that if humanity plunged into such a project wholeheartedly, it would not be long before there were more human beings living in space than on Earth.

O'Neill's settlements, at least at first, are planned for the lunar orbit. But can human beings penetrate beyond the moon?

In theory, there is no reason why they cannot, but flights to the next nearest world on which they can land, Mars (Venus, though closer, is too hot for a manned landing), will require flights not of days, as in the case of the moon, but of months. And for those months, they will have to take a livable environment along with them.

Human beings have already had some experience along these lines in descending into the ocean depths in submarines and vessels such as the bathyscaphe. As on those voyages, they will go into space in a bubble of air enclosed in a strong metal shell, carrying a full supply of the food, water, and other necessities they will require for the journey. But the take-off into space is complicated enormously by the problem of overcoming gravity. In the space ship, a large proportion of weight and volume must be devoted to the engine and fuel, and the possible "payload" of crew and supplies will at first be small indeed.

The food supply will have to be extremely compact: there will be no room for any indigestible constituents. The condensed, artificial food might consist of lactose, a bland vegetable oil, an appropriate mixture of amino acids, vitamins, minerals, and a dash of flavoring, the whole enclosed in a tiny carton made of edible carbohydrate. A carton containing 180 grams of solid food would suffice for one meal. Three such cartons would supply 3,000 calories. To this a gram of water per calorie (2½ to 3 liters per day per person) would have to be added; some of it might be mixed in the food to make it more palatable, increasing the size of the carton. In addition, the ship would have to carry oxygen for breathing in the amount of about 1 liter (1,150 grams) of oxygen in liquid form per day per person.

Thus the daily requirement for each person would be 540 grams of dry food, 2,700 grams of water, and 1,150 grams of oxygen. Total, 4,390 grams, or roughly 9½ pounds. Imagine a trip to the moon, then, taking one week each way and allowing two days on the moon's surface for exploration. Each person on the ship would require about 150 pounds of food, water, and oxygen. This can probably be managed at present levels of technology.

For an expedition to Mars and back, the requirements are vastly greater. Such an expedition might well take two and a half years, allowing for a wait on Mars for a favorable phase of the planetary orbital positions to start the return trip. On the basis I have just described, such a trip would call for about 5 tons of food, water, and oxygen per person. To transport such a supply in a space ship is, under present technological conditions, unthinkable.

The only reasonable solution for a long trip is to make the space ship self-sufficient, in the same sense that the earth, itself a massive "ship" traveling through space, is self-sufficient. The food, water, and air taken along to start with would have to be endlessly reused by recycling the wastes.

Such *closed systems* have already been constructed in theory. The recycling of wastes sounds unpleasant, but this is, after all, the process that maintains life on the earth. Chemical filters on the ship could collect the carbon dioxide and water vapor exhaled by the crew members; urea, salt, and water could be recovered by distillation and other processes from urine and feces; the dry fecal residue could be sterilized of bacteria by ultraviolet light and, along with the carbon dioxide and water, could then be fed to algae growing in tanks. By photosynthesis, the algae would convert the carbon dioxide and nitrogenous compounds of the feces to organic food, plus oxygen, for the crew. The only thing that would be required from outside the system is energy for the various processes, including photosynthesis, and this could be supplied by the sun.

It has been estimated that as little as 250 pounds of algae per person could take care of the crew's food and oxygen needs for an indefinite period. Adding the weight of the necessary processing equipment, the total weight of supplies per man would be perhaps 350 pounds, certainly no more than 1,000 pounds. Studies have also been made with systems in which hydrogen-using bacteria are employed. These do not require light, merely hydrogen which can be obtained through the electrolysis of water. The efficiency of such systems is much higher, according to the report, than that of photosynthesizing organisms.

Aside from supply problems, there is that of prolonged weightlessness. Astronauts have survived half a year of continuous weightlessness without permanent harm, but there have been enough minor disturbances to make prolonged weightlessness a disturbing factor. Fortunately, there are ways to counteract it. A slow rotation of the space vehicle, for instance, could produce the sensation of weight by virtue of the centrifugal force, acting like the force of gravity.

More serious and less easily countered are the hazards of high acceleration and sudden deceleration, which space travelers will inevitably encounter in taking off and landing on rocket flights.

The normal force of gravity at the earth's surface is called 1 g. Weightlessness is 0 g. An acceleration (or deceleration) that doubles the body's weight is 2 g, a force tripling the weight is 3 g, and so on.

The body's position during acceleration makes a big difference. If you are accelerated head first (or decelerated feet first), the blood rushes away from your head. At a high enough acceleration (say 6 g for 5 seconds), this means *blackout*. On the other hand, if you are accelerated feet first (called *negative acceleration*, as opposed to the *positive* headfirst acceleration), the blood rushes to your head. This is more dangerous, because the heightened pressure may burst blood vessels in the eyes or the brain. The investigators of acceleration call it *redout*. An acceleration of 2½ g for 10 seconds is enough to damage some of the vessels.

By far the easiest to tolerate is *transverse acceleration*—that is, with the force applied at right angles to the long axis of the body, as in a sitting position. Men have withstood transverse accelerations as high as 10 g for more than 2 minutes in a centrifuge without losing consciousness.

For shorter periods the tolerances are much higher. Astounding records in sustaining high g decelerations were made by Colonel John Paul Stapp and other volunteers on the sled track of the Holloman Air Force Base in New Mexico. On his famous ride of 10 December 1954, Stapp took a deceleration of 25 g for about a second. His sled was brought to a full stop from a speed of more than 600 miles per hour in just 1.4 seconds. This, it was estimated, amounted to driving an automobile into a brick wall at 120 miles per hour! Of course, Stapp was strapped in the sled in a manner to minimize injury. He suffered only bruises, blisters, and painful eye shocks that produced two black eyes.

An astronaut, on take-off, must absorb (for a short while) as much as 6½ g and, at re-entry, up to 11 g.

Devices such as contour couches, harnesses, and perhaps even immersion in water in a water-filled capsule or space suit will give a sufficient margin of safety against high g forces.

Similar studies and experiments are being made on the radiation hazards, the boredom of long isolation, the strange experience of being in soundless space where night never falls, and other eerie conditions that space fliers will have to endure. All in all, those preparing for humanity's first venture away from the neighborhood of its home planet see no insurmountable obstacles ahead.

The psychological difficulties of long space voyages may not, in fact, prove very serious if we do not persist in thinking of the astronauts as Earthpeople. If they are, of course, then there is an enormous difference between their life on the outside of a huge planet and their stay inside a small spaceship.

What, however, if the explorers are people from space settlements of the type Gerard O'Neill envisions. These settlers will be accustomed to an internal environment, to having their food, drink and air tightly cycled, to having variations in the gravitational effect, to living in a space environment. A spaceship will be a smaller version of the settlement they are used to and have lived in, perhaps, all their lives.

It may be the space settlers then that will be the cutting edge of human exploration in the twenty-first century and thereafter. It will be they, perhaps, who reach the asteroids. There, mining operations will supply a new level of resources for expanded humanity, and many of the asteroids may be hollowed out as natural settlements, many of them considerably larger than any that would be practical in the Earth-moon system.

From the asteroids as a base, human beings can explore the vast reaches of the outer solar system . . . and beyond that lie the stars.

Chapter 17

The Mind

The Nervous System

Physically speaking, we humans are rather unimpressive specimens, as organisms go. We cannot compete in strength with most other animals our size. We walk awkwardly, compared with, say, the cat; we cannot run with the dog or the deer; in vision, hearing, and the sense of smell, we are inferior to a number of other animals. Our skeletons are ill suited to our erect posture: the human is probably the only animal that develops "low back pain" from normal posture and activities. When we think of the evolutionary perfection of other organisms—the beautiful efficiency of the fish for swimming or of the bird for flying, the great fecundity and adaptability of the insects, the perfect simplicity and efficiency of the virus—the human being seems a clumsy and poorly designed creature indeed. As sheer organism, we can scarcely compete with the creatures occupying any specific environmental niche on earth. We have come to dominate the earth only by grace of one rather important specialization—the brain.

NERVE CELLS

A cell is sensitive to a change in its surroundings (*stimulus*) and will react appropriately (*response*). Thus, a protozoon will swim toward a drop of sugar solution deposited in the water near it, or away from a drop of acid. Now this direct, automatic sort of response is fine for a single cell, but it would mean chaos for a collection of cells. Any organism made up of a number of cells must have a system that coordinates their responses. Without such a system, it would be like a city of people completely out of communication with one another and acting at cross purposes. So even the coelenterates, the most primitive multicelled animals, have the beginnings of a nervous system. We can see in them the first nerve cells (*neurons*)—special cells with fibers that extend from the main cell body and put out extremely delicate branches (figure 17.1).

Figure 17.1. A nerve cell.

The functioning of nerve cells is so subtle and complex that even at this simple level we are already a little beyond our depth when it comes to explaining just what happens. In some way not yet understood, a change in the environment acts upon the nerve cell. It may be a change in the concentration of some substance, or in the temperature, or in the amount of light, or in the movement of the water, or it may be an actual touch by some object. Whatever the stimulus, it sets up a tiny *nerve impulse*, an electric current that progresses rapidly along the fiber. When it reaches the end of the fiber, the impulse jumps a tiny gap (*synapse*) to the next nerve cell; and so it is transmitted from cell to cell. (In well-developed nervous systems, a nerve cell may make thousands of synapses with its neighbors.) In the case of a coelenterate, such as a jellyfish, the impulse is communicated throughout the organism. The jellyfish responds by contracting some part or all of its body. If the stimulus is contact with a food particle, the organism engulfs the particle by contraction of its tentacles.

All this is strictly automatic, of course, but since it helps the jellyfish, we like to read purpose into the organism's behavior. Indeed, humans, as creatures who behave in a purposeful, motivated way, naturally tend to attribute purpose even to inanimate nature. Scientists call this attitude *teleological*, and try to avoid such a way of thinking and speaking as much as they can. But in describing the results of evolution, it is so convenient to speak in terms of development toward more efficient ends that even among scientists all but the most fanatical purists occasionally lapse into teleology. (Readers of this book have noticed, of course, that I have sinned often.) Let us, however, try to avoid teleology in considering the development of the nervous system and the brain. Nature did not design the brain; it came about as the result of a long series of evolutionary accidents, so to speak, which happened to produce helpful features that at each stage gave an advantage to organisms possessing them. In the fight for survival, an animal that was more sensitive to changes in the environment than its competitors, and could respond to them faster, would be favored by natural selection. If, for instance, an animal happened to possess some spot on its body that was exceptionally sensitive to light, the advantage would be so great that evolution of eye spots, and eventually of eyes, would follow almost inevitably.

Specialized groups of cells that amount to rudimentary *sense organs* begin to appear in the Platyhelminthes, or flatworms. Furthermore, the flatworms also show the beginnings of a nervous system that avoids sending nerve impulses indiscriminately throughout the body, but instead speeds them to the critical points of response. The development that accomplishes this is a central nerve cord. The flatworms are the first to develop a *central nervous system.*

This is not all. The flatworm's sense organs are localized in its head end, the first part of its body that encounters the environment as it moves along, and so naturally the nerve cord is particularly well developed in the head region. That knob of development is the beginning of a brain.

Gradually the more complex phyla add new features. The sense organs increase in number and sensitivity. The nerve cord and its branches grow more elaborate, developing a widespread system of *afferent* ("carrying to") nerve cells that bring messages to the cord and *efferent* ("carrying away") fibers that transmit messages to the organs of response. The knot of nerve cells at the crossroads in the head becomes more and more complicated. Nerve fibers evolve into forms that can carry the impulses faster. In the squid, the most highly developed of the unsegmented animals, this faster transmission is accomplished by a thickening of the nerve fiber. In the segmented animals, the fiber develops a sheath of fatty material (*myelin*) which is even more effective in speeding the nerve impulse. In human beings, some nerve fibers can transmit the impulse at 100 meters per second (about 225 miles per hour), compared with only about 1/10 mile per hour in some of the invertebrates.

The chordates introduce a radical change in the location of the nerve cord. In them this main nerve trunk (better known as the *spinal chord*) runs along the back instead of along the belly, as in all lower animals. This may seem a step backward—putting the cord in a more exposed position. But the vertebrates have the cord well protected within the bony spinal column. The backbone, though its first function was protecting the nerve cord, produced amazing dividends, for it served as a girder upon which chordates could hang bulk and weight. From the backbone they can extend ribs that enclose the chest, jawbones that carry teeth for chewing, and long bones that form limbs.

BRAIN DEVELOPMENT

The chordate brain develops from three structures that are already present in simple form in the most primitive vertebrates. These structures, at first mere swellings of nerve tissue, are the *forebrain*, the *midbrain*, and the *hindbrain*, a division first noted by the Greek anatomist Erasistratus of Chios about 280 B.C. At the head end of the spinal cord, the cord widens smoothly into the hindbrain section known as the *medulla oblongata*. On the front side of this section in all but the most primitive chordates is a bulge called the *cerebellum* ("little brain"). Forward of this is the midbrain. In the lower vertebrates the midbrain is concerned chiefly with vision and has a pair of *optic lobes*, while the forebrain is concerned with smell and taste and contains *olfactory bulbs*. The forebrain, reading from front to rear, is divided into the olfactory-bulb section, the *cerebrum*, and the *thalamus*, the lower portion of which is the *hypothalamus*. (*Cerebrum* is Latin for "brain"; in humans, at least, the cerebrum is the largest and most important part of the organ.) By removing the cerebrum from animals and observing the results, the French anatomist Marie Jean Pierre Flourens was able to demonstrate in 1824 that it is indeed the cerebrum that is responsible for acts of thought and will. (For the human brain, see figure 17.2.)

It is the roof of the cerebrum, moreover—the cap called the *cerebral cortex*—that is the star of the whole show. In fishes and amphibians, this is merely a smooth covering (called the *pallium*, or "cloak"). In reptiles a patch of new nerve tissue, called the *neopallium* ("new cloak") appears. This is the real

Figure 17.2. The human brain.

PARIETAL LOBE

PINEAL BODY

THALAMUS

OCCIPITAL LOBE

FRONTAL LOBE

TEMPORAL LOBE

HYPOPHYSIS

CEREBELLUM

PONS

MEDULLA OBLONGATA

forerunner of things to come. It will eventually take over the supervision of vision and other sensations. In the reptiles, the clearing house for visual messages has already moved from the midbrain to the forebrain in part; in birds this move is completed. With the first mammals, the neopallium begins to take charge. It spreads virtually over the entire surface of the cerebrum. At first it remains a smooth coat, but as it goes on growing in the higher mammals, it becomes so much larger in area than the surface of the cerebrum that it is bent into folds, or *convolutions*. This folding is responsible for the complexity and capacity of the brain of a higher mammal, notably that of *Homo sapiens*.

More and more, as one follows this line of species development, the cerebrum comes to dominate the brain. The midbrain fades to almost nothing. In the case of the primates, which gain in the sense of sight at the expense of the sense of smell, the olfactory lobes of the forebrain shrink to mere blobs. By this time the cerebrum has expanded over the thalamus and the cerebellum.

Even the early humanlike fossils had considerably larger brains than the most advanced apes. Whereas the brain of the chimpanzee or of the orangutan weighs less than 400 grams (under 14 ounces), and the gorilla, though far larger than a man, has a brain that averages about 540 grams (19 ounces), *Pithecanthropus's* brain apparently weighed about 850 to 1,000 grams (30 to 35 ounces). And these were the *small-brained* hominids. Rhodesian man's brain weighed about 1,300 grams (46 ounces); the brain of Neanderthal and of modern *Homo sapiens* comes to about 1,500 grams (53 ounces or 3.3 pounds). The modern human's mental gain over Neanderthal apparently lies in the fact that a larger proportion of the human brain is concentrated in the foreregions, which apparently control the higher aspects of mental function. Neanderthal was a low-brow whose brain bulged in the rear; a human today, in contrast, is a high-brow whose brain bulges in front.

The hominid brain has about tripled in size in the last 3 million years—a fast increase as evolutionary changes go. But why did that happen? Why the hominids?

One possible reason is that we now know that even very early, small-brained hominids walked erect, exactly as modern humans do. Erect posture long preceded the enlarged brain. Erect posture has two important consequences: first, the eyes are lifted higher above the ground and deliver more information to the brain; second, the forelimbs are *permanently* freed so that they might feel and manipulate the environment. The flood of sense perceptions, long-distance sight, and short-distance touch make an enlarged brain capable of dealing with new material useful, and any individuals whose brains happen to be more efficient (through superior size or organization) are bound to have an advantage over others. Evolutionary procedures would then inevitably produce large-brained hominids. (Or so it seems to us in hindsight.)

THE HUMAN BRAIN

The modern human brain is about 1/50 of the total human body weight. Each gram of brain weight is in charge, so to speak, of 50 grams of body. In comparison, the chimpanzee's brain is about 1/150 the weight of its body, and the gorilla's about 1/500 its body. To be sure, some of the smaller primates have an even higher brain/body ratio than human beings do. (So do the hummingbirds.) A marmoset can have a brain that is 1/18 the weight of its body. However, there the mass of the brain is too small in absolute terms for it to be able to pack into itself the necessary complexity for intelligence on the human scale. In short, what is needed, and what humans have, is a brain that is large both in the absolute sense and in relation to body size.

This is made plain by the fact that two types of mammal have brains that are distinctly larger than the human brain and yet that do not lend those mammals superintelligence. The largest elephants can have brains as massive as 6,000 grams (about 13 pounds) and the largest whales can have brains that reach a mark of 9,000 grams (or nearly 19 pounds). The size of the bodies those brains must deal with is, however, enormous. The elephant's brain, despite its size, is only 1/1,000 the weight of its body, and the brain of a large whale may be only 1/10,000 the weight of its body.

In only one direction, however, do humans have a possible rival. The dolphins and porpoises, small members of the whale family, show possibilities. Some of these are no heavier than a person and yet have brains that are larger (with weights up to 1,700 grams, or 60 ounces) and more extensively convoluted.

It is not safe to conclude from this evidence alone that the dolphin is more intelligent than we are, because there is the question of the internal organization of the brain. The dolphin's brain (like that of Neanderthal man) may be oriented more in the direction of what we might consider lower functions.

The only safe way to tell is to attempt to gauge the intelligence of the dolphin by actual experiment. Some investigators, notably John C. Lilly, seem convinced that dolphin intelligence is indeed comparable to our own, that dolphins and porpoises have a speech pattern as complicated as ours, and that possibly a form of interspecies communication may yet be established.

Even if this is so, there can be no question but that dolphins, however intelligent, lost their opportunity to translate that intelligence into control of the environment when they readapted to sea life. It is impossible to make use of fire under water, and it was the discovery of the use of fire that first marked off hominids from all other organisms. More fundamentally still, rapid locomotion through a medium as viscous as water requires a thoroughly streamlined shape. This has made impossible in the dolphin the development of anything equivalent to the human arm and hand with which the environment can be delicately investigated and manipulated.

Another interesting point is that human beings caught up and surpassed the cetaceans. When hominids were still small-brained, the dolphins were already large-brained, yet the latter could not stop the former. It would seem inconceivable to us today that we would allow the evolutionary development of a large-brained rat, or even a large-brained dog to threaten our position on Earth, but the dolphins, trapped at sea, could do nothing to prevent hominid development to the point where we now can wipe out cetacean life with scarcely an effort, if we wish. (It is to our credit that so many of us do *not* wish and are making every effort to prevent it.)

The dolphins may, in some philosophical manner we do not yet understand, surpass us in some forms of intelligence, but as far as effective control of the environment and the development of technology are concerned, *Homo sapiens* stands without a peer on Earth at present or, as far as we know, in the past. (It goes without saying that the activity of human beings in exercising their intelligence and technological capacity has not necessarily always been for the good of the planet—or for themselves, for that matter.)

INTELLIGENCE TESTING

While considering the difficulty in determining the precise intelligence level of a species such as the dolphin, it might be well to say that no completely satisfactory method exists for measuring the precise intelligence level of individual members of our own species.

In 1904, the French psychologists Alfred Binet and Théodore Simon devised means of testing intelligence by answers given to judiciously chosen questions. Such *intelligence tests* give rise to the expression *intelligence quotient* (or I.Q.), representing the ratio of the mental age, as measured by the test, to the chronological age—this ratio being multiplied by 100 to remove decimals. The public was made aware of the significance of I.Q. chiefly through the work of the American psychologist Lewis Madison Terman.

The trouble is that no test has been devised that is not culturally centered. Simple questions about ploughs might stump an intelligent city boy, and simple questions about escalators might stump an equally intelligent farm boy. Both would puzzle an equally intelligent Australian aborigine, who might nevertheless dispose of questions about boomerangs that would leave us gasping.

What is more, it is difficult for people not to have preconceived notions about who is intelligent and who is not. An investigator is bound to find higher intelligence in culturally similar subjects. Stephen Jay Gould in his book *The Mismeasure of Man* (1981) describes in full detail how I.Q. measurements ever since the First World War have been placed at the service of unconscious or taken-for-granted racism.

The most recent and blatant example is that of the British psychologist Cyril Lodowic Burt, who was educated at Oxford and taught both at Oxford and Cambridge. He studied the I.Q.'s of children and correlated those I.Q.'s with the occupational status of the parents: higher professional, lower professional, clerical, skilled labor, semiskilled labor, unskilled labor.

He found that the I.Q.'s fit those occupations perfectly. The lower the parent was in the social scale, the lower the I.Q. of the child. In other words, people belonged where they were, and those who were better off deserved to be.

Furthermore, Burt found that men had higher I.Q.'s than women, Englishmen than Irishmen, Gentiles than Jews, and so on. He tested identical twins who were separated soon after birth and found that their I.Q.'s were nevertheless very similar—again pointing out the great importance of heredity over environment.

Burt was greatly honored and was knighted before his death in 1971. After his death, however, it was discovered that, quite beyond doubt, he had fabricated his data.

It is not necessary to go into the psychological reasons for this. It is sufficient (to me) that people are so anxious to be considered intelligent that it is next to impossible for them to find figures that would yield opposite results. The whole field of intelligence testing is so involved with emotion and self-love that any results must be approached very cautiously.

Another familiar test is aimed at an aspect of the mind even more subtle and elusive than intelligence. This consists of ink-blot patterns first prepared by a Swiss doctor, Hermann Rorschach, between 1911 and 1921. Subjects are asked to convert these ink blots into images; from the type of image a person builds into such a *Rorschach test,* conclusions concerning his or her personality are drawn. Even at best, however, such conclusions are not likely to be truly conclusive.

THE SPECIALIZATION OF FUNCTIONS

Oddly enough, many of the ancient philosophers almost completely missed the significance of the organ under the human skull. Aristotle considered the brain merely an air-conditioning device, so to speak, designed to cool the overheated blood. In the generation after Aristotle, Herophilus of Chacedon, working at Alexandria, correctly recognized the brain as the seat of intelligence, but, as usual, Aristotle's errors carried more weight than did the correctness of others.

The ancient and medieval thinkers therefore often tended to place the seat of emotions and personality in organs such as the heart, the liver, and the spleen (*vide* the expressions "broken-hearted," "lily-livered," "vents his spleen").

The first modern investigator of the brain was a seventeenth-century English physician and anatomist named Thomas Willis; he traced the nerves that led to the brain. Later, a French anatomist named Felix Vicq d'Azyr and others roughed out the anatomy of the brain itself. But it was the eighteenth-century Swiss physiologist Albrecht von Haller who made the first crucial discovery about the functioning of the nervous system.

Von Haller found that he could make a muscle contract much more easily by stimulating a nerve than by stimulating the muscle itself. Furthermore, this contraction was involuntary; he could even produce it by stimulating a nerve after the organism had died. Von Haller went on to show that the nerves carry sensations. When he cut the nerves attached to specific tissues, these tissues could no longer react. The physiologist concluded that the brain receives sensations by way of nerves and then sends out, again by way of nerves, messages that lead to such responses as muscle contraction. He supposed that the nerves all come to a junction at the center of the brain.

In 1811 the Austrian physician Franz Joseph Gall focused attention on the *gray matter* on the surface of the cerebrum (which is distinguished from the *white matter* in that the latter consists merely of the fibers emerging from the nerve-cell bodies, these fibers being white because of their fatty sheaths). Gall suggested that the nerves do not collect at the center of the brain as von Haller had thought, but that each runs to some definite portion of the gray matter, which he considered the coordinating region of the brain. Gall reasoned that different parts of the cerebral cortex are in charge of collecting sensations from different parts of the body and sending out the messages for responses to specific parts as well.

If a specific part of the cortex is responsible for a specific property of the mind, what is more natural than to suppose that the degree of development of that part would reflect a person's character or mentality? By feeling for bumps on a person's skull, one might find out whether this or that portion of the brain was enlarged and so judge whether a person was particularly generous or particularly depraved or particularly something else. With this reasoning, some of Gall's followers founded the pseudo-science of *phrenology,* which had quite a vogue in the nineteenth century and is not exactly dead even today. (Oddly enough, although Gall and his followers emphasized the high forehead and domed head as a sign of intelligence—a view that still influences people today—Gall himself had an unusually small brain, about 15 percent smaller than the average.)

But the fact that phrenology, as developed by charlatans, is nonsense, does not mean that Gall's original notion of the specialization of functions in particular parts of the cerebral cortex was wrong. Even before specific explorations of the brain were attempted, it was noted that damage to a particular portion of the brain might result in a particular disability. In 1861, the French surgeon Pierre Paul Broca, by assiduous postmortem study of the brain, was able to show that patients with *aphasia* (the inability to speak, or to understand speech) usually possessed physical damage to a particular area of the left cerebrum, an area called *Broca's convolution* as a result.

Then, in 1870, two German scientists, Gustav Fritsch and Eduard Hitzig, began to map the supervisory functions of the brain by stimulating various parts of it and observing what muscles responded. A half-century later, this technique was greatly refined by the Swiss physiologist Walter Rudolf Hess, who was awarded a share of the 1949 Nobel Prize for medicine and physiology in consequence.

It was discovered by such methods that a specific band of the cortex was particularly involved in the stimulation of the various voluntary muscles into movement. This band is therefore called the *motor area.* It seems to bear a generally inverted relationship to the body; the uppermost portions of the motor area, toward the top of the cerebrum, stimulate the lowermost portions of the leg; as one progresses downward in the motor area, the muscles higher in the leg are stimulated, then the muscles of the torso, then those of the arm and hand, and finally those of the neck and hand.

Behind the motor area is another section of the cortex that receives many types of sensation and is therefore called the *sensory area*. As in the case of the motor area, the regions of the sensory area in the cerebral cortex are divided into sections that seem to bear an inverse relation to the body. Sensations from the foot are at the top of the area, followed successively as we go downward with sensations from leg, hip, trunk, neck, arm, hand, fingers, and, lowest of all, the tongue. The sections of the sensory area devoted to lips, tongue, and hand are (as one might expect) larger in proportion to the actual size of those organs than are the sections devoted to other parts of the body.

If, to the motor area and the sensory area, are added those sections of the cerebral cortex primarily devoted to receiving the impressions from the major sense organs, the eye and the ear, there still remains a major portion of the cortex without any clearly assigned and obvious function.

It is this apparent lack of assignment that has given rise to the statement, often encountered, that the human being "uses only one-fifth of his brain." That, of course, is not so; the best we can really say is that one-fifth of the human brain has an obvious function. We might as well suppose that a construction firm engaged in building a skyscraper is using only one-fifth of its employees because that one-fifth is actually engaged in raising steel beams, laying down electric cables, transporting equipment, and such. This assumption would ignore the executives, secretaries, filing clerks, supervisors, and others. Analogously, the major portion of the brain is engaged in what we might call white-collar work, in the assembling of sensory data, in its analysis, in deciding what to ignore, what to act upon, and just how to act upon it. The cerebral cortex has distinct *association areas*—some for sound sensations, some for visual sensations, some for others.

When all these association areas are taken into account, there still remains one area of the cerebrum that has no specific and easily definable function. This is the area just behind the forehead, which is called the *prefrontal lobe*. Its lack of obvious function is such that it is sometimes called the *silent area*. Tumors have made it necessary to remove large areas of the prefrontal lobe without any particular significant effect on the individual; yet surely it is not a useless mass of nerve tissue.

One might even suppose it to be the most important portion of the brain if one considers that, in the development of the human nervous system, there has been a continual piling up of complication at the forward end. The prefrontal lobe might therefore be the brain area most recently evolved and most significantly human.

In the 1930s, it seemed to a Portuguese surgeon, Antonio Egas Moniz, that where a mentally ill patient was at the end of his rope, it might be possible to help by taking the drastic step of severing the prefrontal lobes from the rest of the brain. The patient might then be cut off from a portion of the associations he had built up, which were, apparently, affecting him adversely, and make a fresh and better start with the brain he had left. This operation, *prefrontal lobotomy,* was first carried out in 1935; in a number of cases, it did indeed seem to help. Moniz shared (with W. R. Hess) the Nobel Prize for medicine and physiology in 1949 for his work. Nevertheless, the operation never achieved popularity and is less popular now than ever. Too often, the cure is literally worse than the disease.

The cerebrum is actually divided into two *cerebral hemispheres* connected by a tough bridge of white matter, the *corpus callosum.* In effect, the hemispheres are separate organs, unified in action by the nerve fibers that cross the corpus callosum and act to coordinate the two. Nonetheless, the hemispheres remain potentially independent.

The situation is somewhat analogous to that of our eyes. Our two eyes act as a unit ordinarily, but if one eye is lost, the other can meet our needs. Similarly, the removal of one of the cerebral hemispheres does not make an experimental animal brainless; the remaining hemisphere learns to carry on.

Ordinarily, each hemisphere is largely responsible for a particular side of the body: the left cerebral hemisphere for the right side, the right cerebral hemisphere for the left side. If both hemispheres are left in place and the corpus callosum is cut, coordination is lost, and the two body halves come under more or less independent control. A literal case of twin brains, so to speak, is set up.

Monkeys can be so treated (with further operation upon the optic nerve to make sure that each eye is connected to only one hemisphere), and when this is done, each eye can be separately trained to do particular tasks. A monkey can be trained to select a cross over a circle to indicate, let us say, the presence of food. If only the left eye is kept uncovered during the training period, only the left eye will be useful in this respect. If the right eye is uncovered and the left eye covered, the monkey will have no right-eye memory of its training. It will have to hunt for its food by trial and error. If the two eyes are trained to contradictory tasks and if both are then uncovered, the monkey alternates activities, as the hemispheres politely take their turns.

Naturally, in any such "two in charge" situation, there is always the danger of conflict and confusion. To avoid that, one cerebral hemisphere (almost always the left one in human beings) is dominant, when both are normally connected. Broca's convolution, which controls speech, is in the left hemisphere, for instance. The *gnostic area,* which is an over-all association area, a kind of court of highest appeal, is also in the left hemisphere. Since the left cerebral hemisphere controls the motor activity of the right-hand side of the body, it is not surprising that most people are right-handed (though even left-handed people usually have a dominant left cerebral hemisphere). Where clear-cut dominance is not established between left and right, there may be ambidexterity, rather than a clear right-handedness or left-handedness, along with some speech difficulties and, perhaps, manual clumsiness.

It has become fashionable in recent years to suppose that the two halves of the brain think differently. The left hemisphere, which is clearly in control of speech, would think logically, mathematically, step by step. The right hemisphere would be left with intuition, artistic conception, thinking as a whole.

The cerebrum is not the whole of the brain. There are areas of gray matter embedded below the cerebral cortex. These are called the *basal ganglia;* included is a section called the *thalamus* (see figure 17.2). The thalamus acts as a reception center for various sensations. The more violent of these—such as pain, extreme heat or cold, or rough touch—are filtered out. The milder sensations from the muscles—the gentle touches, the moderate temperatures —are passed on to the sensory area of the cerebral cortex. It is as though mild sensations can be trusted to the cortex, where they can be considered judiciously and where reaction can come after a more or less prolonged interval of consideration. The rough sensations, however, which must be dealt with quickly and for which there is no time for consideration, are handled more or less automatically in the thalamus.

Underneath the thalamus is the *hypothalamus* center for a variety of devices for controlling the body. The body's appestat, mentioned in chapter 15 as controlling the body's appetite, is located there; so is the control of the body's temperature. It is through the hypothalamus, moreover, that the brain exerts at least some influence over the pituitary gland (see chapter 15); this is an indication of the manner in which the nervous controls of the body and the chemical controls (the hormones) can be unified into a master supervisory force.

In 1954, the physiologist James Olds discovered another and rather frightening function of the hypothalamus. It contains a region that, when stimulated, apparently gives rise to a strongly pleasurable sensation. An electrode affixed to the *pleasure center* of a rat, so arranged that it can be stimulated by the animal itself, will be stimulated up to 8,000 times an hour for hours or days at a time, to the exclusion of food, sex, and sleep. Evidently, all the desirable things in life are desirable only insofar as they stimulate the pleasure center. To stimulate it directly makes all else unnecessary.

The hypothalamus also contains an area that has to do with the wake-sleep cycle, since damage to parts of it induces a sleeplike state in animals. The exact mechanism by which the hypothalamus performs its function is uncertain. One theory is that it sends signals to the cortex, which sends signals back in response, in mutually stimulating fashion. With continuing wakefulness, the coordination of the two fails, the oscillations become ragged, and the individual becomes sleepy. A violent stimulus (a loud noise, a persistent shake of the shoulder, or, for that matter, a sudden interruption of a steady noise) will arouse one. In the absence of such stimuli, coordination will be restored eventually between hypothalamus and cortex, and sleep will end spontaneously; or perhaps sleep will become so shallow that a perfectly ordinary stimulus, of which the surroundings are always full, will suffice to wake one.

During sleep, dreams—sensory data more or less divorced from reality—will take place. Dreaming is apparently a universal phenomenon; people who report dreamless sleep are merely failing to remember their dreams. The American physiologist William Dement, studying sleeping subjects in 1952, noticed periods of rapid eye movements that sometimes persisted for minutes (*REM sleep*). During this period, one's breathing, heartbeat, and blood pressure, rose to waking levels. This takes place about a quarter of the sleeping time. A sleeper who was awakened during these periods generally reported having had a dream. Furthermore, a sleeper who was continually disturbed during these periods began to suffer psychological distress; the periods of distress were multiplied during succeeding nights as though to make up for the lost dreaming.

It would seem, then, that dreaming has an important function in the working of the brain. It is suggested that dreaming is a device whereby the brain runs over the events of the day to remove the trivial and repetitious that might otherwise clutter it and reduce its efficiency. Sleep is the natural time for such activity, for the brain is then relieved of many of its waking functions. Failure to accomplish this task (because of interruption) may so clog the brain that clearing attempts must be made during waking periods, producing hallucinations (that is, waking dreams, so to speak) and other unpleasant symptoms. One might naturally wonder if this is not a chief function of sleep, since there is very little physical resting in sleep that cannot be duplicated by quiet wakefulness. REM sleep even occurs in infants who spend half their sleeping time at it and who would seem to lack anything about which to dream. It may be that REM sleep helps the development of the nervous system. (It has been observed in mammals other than humans, too.)

THE SPINAL CORD

Below the cerebrum is the smaller cerebellum (also divided into two *cerebellar hemispheres*) and the *brain stem*, which narrows and leads smoothly into the *spinal cord* extending about 18 inches down the hollow center of the spinal column.

The spinal cord consists of gray matter (at the center) and white matter (on the periphery); to it are attached a series of nerves that are largely concerned with the internal organs—heart, lungs, digestive system, and so on—organs that are more or less under involuntary control.

In general, when the spinal cord is severed, through disease or through injury, that part of the body lying below the severed segment is disconnected, so to speak. It loses sensation and is paralyzed. If the cord is severed in the neck region, death follows, because the chest is paralyzed, and with it the action of the lungs. It is this that makes a broken neck fatal, and hanging a feasible form of quick execution. It is the severed cord, rather than a broken bone, that is fatal.

The entire structure of the *central nervous system*, consisting of cerebrum, cerebellum, brain stem, and spinal cord, is carefully coordinated. The white matter of the spinal cord is made up of bundles of nerve fibers that run up and down the cord, unifying the whole. Those that conduct impulses downward from the brain are the *descending tracts,* and those that conduct them upward to the brain are the *ascending tracts.*

In 1964, research specialists at Cleveland's Metropolitan General Hospital reported the isolation from rhesus monkeys of brains that were then kept independently alive for as long as eighteen hours. This offers the possibility of detailed specific study of the brain's metabolism through a comparison of the nutrient medium entering the blood vessels of the isolated brain and of the same medium leaving it.

The next year they were transplanting dogs' heads to the necks of other dogs, hooking them up to the host's blood supply, and keeping the brains in the transplanted heads alive and working for as long as two days. By 1966, dogs' brains were lowered to temperatures near freezing for six hours and then revived to the point of showing clear indications of normal chemical and electrical activity. Brains are clearly tougher than they might seem to be.

Nerve Action

It is not only the various portions of the central nervous system that are hooked together by nerves, but, clearly, all the body that, in this fashion, is placed under the control of that system. The nerves interlace the muscles, the glands, the skin; they even invade the pulp of the teeth (as we learn to our cost at every toothache).

The nerves themselves were observed in ancient times, but their structure and function were consistently misunderstood. Until modern times, they were felt to be hollow and to function as carriers of a subtle fluid. Rather complicated theories developed by Galen involved three different fluids carried by the veins, the arteries, and the nerves, respectively. The fluid of the nerves, usually referred to as *animal spirits,* was the most rarefied of the three. Galvani's discovery that muscles and nerves could be stimulated by an electric discharge laid the foundation for a series of studies that eventually showed nerve action to be associated with electricity—a subtle fluid, indeed, more subtle than Galen could have imagined.

Specific work on nerve action began in the early nineteenth century with the German physiologist Johannes Peter Müller, who, among other things, showed that sensory nerves always produce their own sensations regardless of the nature of the stimulus. Thus, the optic nerve registers a flash of light, whether stimulated by light itself or by the mechanical pressure of a punch in the eye. (In the latter case, you "see stars.") This emphasizes that our contact with the world is not with reality at all but with specialized stimuli that the brain usually interprets in a useful manner, but can interpret in a non-useful manner.

Study of the nerves was advanced greatly in 1873, when an Italian physiologist, Camillo Golgi, developed a cellular stain involving silver salts that was well adapted to react with nerve cells, making clear their finest details. He was able to show, in this manner, that nerves are composed of separate and distinct cells, and that the processes of one cell might approach very closely to those of another, but that they do not fuse. There remained the tiny gap of the synapse. In this way, Golgi bore out, observationally, the contentions of a German anatomist, Wilhelm von Waldeyer, to the effect that the entire nervous system consists of individual nerve cells or neurons (this contention being termed the *neuron theory*).

Golgi did not, however, himself support the neuron theory. This proved to be the task of the Spanish neurologist Santiago Ramón y Cajal, who, by 1889, using an improved version of Golgi's stain, worked out the connections of the cells in the gray matter of the brain and spinal cord and fully established the neuron theory. Golgi and Ramón y Cajal, although disputing the fine points of their findings, shared the Nobel Prize for medicine and physiology in 1906.

These nerves form two systems: the *sympathetic* and the *parasympathetic*. (The terms date back to semimystical notions of Galen.) Both systems act on almost every internal organ, exerting control by opposing effects. For instance, the sympathetic nerves act to accelerate the heartbeat, the parasympathetic nerves to slow it; the sympathetic nerves slow up secretion of digestive juices, the parasympathetic stimulate such secretions, and so on. Thus, the spinal cord, together with the subcerebral portions of the brain, regulates the workings of the organs in an automatic fashion. This set of involuntary controls was investigated in detail by the British physiologist John Newport Langley in the 1890s, and he named it the *autonomic nervous system.*

REFLEX ACTION

In the 1830s, the English physiologist Marshall Hall had studied another type of behavior which seemed to have voluntary aspects but proved to be really quite involuntary. When you accidentally touch a hot object with your hand, the hand draws away instantly. If the sensation of heat had to go tò the brain, be considered and interpreted there, and evoke the appropriate message to the hand, your hand would be pretty badly scorched by the time it got the message. The unthinking spinal cord disposes of the whole business automatically and much faster. It was Hall who gave the process the name *reflex.*

The reflex is brought about by two or more nerves working in coordination, to form a *reflex arc* (figure 17.3). The simplest possible reflex arc is one consisting of two neurons, a sensory (bringing sensations to a *reflex center* in the central nervous system, usually at some point in the spinal cord) and a motor (carrying instructions for movement from the central nervous system). The two neurons may be connected by one or more *connector neurons.* A particular study of such reflex arcs and of their function in the body was made by the English neurologist Charles Scott Sherrington, who won a share in the 1932 Nobel Prize for medicine and physiology in consequence. It was Sherrington who, in 1897, coined the word *synapse.*

Figure 17.3. The reflex arc.

Reflexes bring about so rapid and certain a response to a particular stimulus that they offer simple methods for checking the general integrity of the nervous system. A familiar example is the *patellar reflex* or, as it is commonly called, the *knee jerk*. When the legs are crossed, a sudden blow below the knee of the upper leg will cause it to make a quick, kicking motion—a fact first brought into medical prominence in 1875 by the German neurologist Carl Friedrich Otto Westphal. The patellar reflex is not important in itself, but its nonappearance can mean some serious disorder involving the portion of the nervous system in which that reflex arc is to be found.

Sometimes damage to a portion of the central nervous system brings about the appearance of an abnormal reflex. If the sole of the foot is scratched, the normal reflex brings the toes together and bent downward. Certain types of damage to the central nervous system will cause the big toe to bend upward in response to this stimulus, and the little toes to spread apart as they bend down. This is the *Babinski reflex*, named for a French neurologist, Joseph François Félix Babinski, who described it in 1896.

In human beings, reflexes are sometimes decidedly subordinate to the conscious will. Thus, you may up your rate of breathing when ordinary reflex action would keep it slow and so on. The lower phyla of animals are much more strictly controlled by their reflexes than human beings are and also have them far more highly developed.

One of the best examples is a spider spinning its web. Here the reflexes produce such an elaborate pattern of behavior that it is difficult to think of it as mere reflex action; instead, it is usually called *instinctive* behavior. (Because the word *instinct* is often misused, biologists prefer the term *innate* behavior.) The spider is born with a nerve-wiring system in which the switches have been preset, so to speak. A particular stimulus sets it off on weaving a web, and each act in the process in turn acts as a stimulus determining the next response.

Looking at the spider's intricate web, built with beautiful precision and effectiveness for the function it will serve, it is almost impossible to believe that the thing has been done without purposeful intelligence. Yet the very fact that the complex task is carried through so perfectly and in exactly the same way every time is itself proof that intelligence has nothing to do with it. Conscious intelligence, with the hesitations and weighings of alternatives that are inherent in deliberate thought, will inevitably give rise to imperfections and variations from one construction to another.

With increasing intelligence, animals tend more and more to shed instincts and inborn skills. Thereby they doubtless lose something of value. A spider can build its amazingly complex web perfectly the first time, although it has never before seen web spinning or even a web. Human beings, on the other hand, are born almost completely unskilled and helpless. A newborn baby can automatically suck on a nipple, wail if hungry, and hold on for dear life if about to fall, but can do very little else. Every parent knows how painfully and with what travail a child comes to learn the simplest forms of suitable behavior. And yet, a spider or an insect, though born with perfection, cannot deviate from it. The spider builds a beautiful web, but if its preordained web should fail, it cannot learn to build another type of web. A child, on the other hand, reaps great benefits from being unfettered by inborn perfection. One may learn slowly and attain only imperfection at best, but one can attain a variety of imperfections of one's own choosing. What human beings have lost in convenience and security, they have gained in an almost limitless flexibility.

Recent work, however, emphasizes the fact that there is not always a clear division between innate and learned behavior not only in the case of human feedback but among lower animals as well. It would seem, on casual observation, for instance, that chicks or ducklings, fresh out of the shell, follow their mothers out of instinct. Closer observation shows that they do not.

The instinct, however, is not to follow their mother but merely to follow something of a characteristic shape or color or faculty of movement. Whatever object provides this sensation at a certain period of early life is followed by the young creature and is thereafter treated as the mother. This may really be the mother; it almost invariably is, in fact, but it need not be! In other words, following is instinctive, but the "mother" that is followed is learned. (Much of the credit for this discovery goes to the remarkable Austrian naturalist Konrad Zacharias Lorenz. Lorenz, during the course of studies now some thirty years old, was followed hither and yon by a gaggle of goslings.)

The establishment of a fixed pattern of behavior in response to a particular stimulus encountered at a particular time of life is called *imprinting*. The specific time at which imprinting takes place is a *critical period*. For chicks, the critical period of *mother imprinting* lies between thirteen and sixteen hours after hatching. For a puppy there is a critical period between three and seven weeks, during which the stimulations it is usually likely to encounter imprint various aspects of what we consider normal doggish behavior.

Imprinting is the most primitive form of learned behavior, one that is so automatic, takes place in so limited a time, and under so general a set of conditions that it is easily mistaken for instinct.

A logical reason for imprinting is that it allows a certain desirable flexibility. If a chick were born with some instinctive ability of distinguishing its true mother so that it might follow only her, and if the true mother were for any reason absent in the chick's first day of life, the little creature would be helpless. As it is, the question of motherhood is left open for just a few hours, and the chick may imprint itself to any hen in the vicinity and thus adopt a foster mother.

ELECTRICAL IMPULSES

As stated earlier, it had been Galvani's experiments just before the opening of the nineteenth century that had first indicated some connection between electricity and the actions of muscle and nerve.

The electrical properties of muscle led to a startling medical application, thanks to the work of the Dutch physiologist Willem Einthoven. In 1903, he developed an extremely delicate galvanometer, one delicate enough to respond to the tiny fluctuations of the electric potential of the beating heart. By 1906, Einthoven was recording the peaks and troughs of this potential (the recording being an *electrocardiogram*) and correlating them with various types of heart disorder.

The more subtle electrical properties of nerve impulses were thought to have been initiated and propagated by chemical changes in the nerve. This was elevated from mere speculation to experimental demonstration by the nineteenth-century German physiologist Emil Du Bois-Reymond, who by means of a delicate galvanometer was able to detect tiny electric currents in stimulated nerves.

With modern electronic instruments, researches into the electrical properties of the nerve have been incredibly refined. By placing tiny electrodes at different spots on a nerve fiber and by detecting electrical changes through an oscilloscope, it is possible to measure a nerve impulse's strength, duration, speed of propagation, and so on. For such work, the American physiologists Joseph Erlanger and Herbert Spencer Gasser were awarded the 1944 Nobel Prize for medicine and physiology.

If you apply small electric pulses of increasing strength to a single nerve cell, up to a certain point there is no response whatever. Then suddenly the cell fires: an impulse is initiated and travels along the fiber. The cell has a threshold: it will not react at all to a stimulus below the threshold; and to any stimulus above the threshold, it will respond only with an impulse of a certain fixed intensity. The response, in other words, is "all or nothing." And the nature of the impulse elicited by the stimulus seems to be the same in all nerves.

How can such a simple yes-no affair, identical everywhere, lead to the complex sensations of sight, for instance, or to the complex finger responses involved in playing a violin? It seems that a nerve, such as the optic nerve, contains a large number of individual fibers, some of which may be *firing* and others not, and where the firing may be in rapid succession or slowly, forming a pattern, possibly a complex one, shifting continuously with changes in the over-all stimulus. (For work in this field, the English physiologist Edgar Douglas Adrian shared, with Sherrington, the 1932 Nobel Prize in medicine and physiology.) Such a changing pattern may be continually *scanned* by the brain and interpreted appropriately. But nothing is known about how the interpretation is made or how the pattern is translated into action such as the contraction of a muscle or secretion by a gland.

The firing of the nerve cell itself apparently depends on the movement of ions across the membrane of the cell. Ordinarily, the inside of the cell has a comparative excess of potassium ions, while outside the cell there is an excess of sodium ions. Somehow the cell holds potassium ions in and keeps sodium ions out so that the concentrations on the two sides of the cell membrane do not equalize. It is now believed that a *sodium pump* of some kind inside the cell keeps pumping out sodium ions as fast as they come in. In any case, there is an electric potential difference of about 1/10 volt across the cell membrane, with the inside negatively charged with respect to the outside. When the nerve cell is stimulated, the potential difference across the membrane collapses, and this represents the firing of the cell. It takes a couple of thousandths of a second for the potential difference to be re-established; and during that interval, the nerve will not react to another stimulus. This is the *refractory period*.

Once the cell fires, the nerve impulse travels down the fiber by a series of firings, each successive section of the fiber exciting the next in turn. The impulse can travel only in the forward direction, because the section that has just fired cannot fire again until after a resting pause.

Research that related, in the fashion just described, nerve action and ion permeability led to the award of the 1963 Nobel Prize for medicine and physiology to two British physiologists, Alan Lloyd Hodgkin and Andrew Fielding Huxley, and to an Australian physiologist, John Carew Eccles.

What happens, though, when the impulse traveling along the length of the nerve fiber comes to a synapse—a gap between one nerve cell and the next? Apparently, the nerve impulse also involves the production of a chemical that can drift across the gap and initiate a nerve impulse in the next nerve cell. In this way, the impulse can travel from cell to cell.

One of the chemicals definitely known to affect the nerves is the hormone adrenalin. It acts upon nerves of the sympathetic system, which slows the activity of the digestive system and accelerates the rate of respiration and the heartbeat. When anger or fear excites the adrenal glands to secrete the hormone, its stimulation of the sympathetic nerves sends a faster surge of blood through the body, carrying more oxygen to the tissues; and by slowing down digestion for the duration, it saves energy during the emergency.

The American psychologists and police officers John Augustus Larsen and Leonard Keeler took advantage of this finding in 1921 to devise a machine to detect the changes in blood pressure, pulse rate, breathing rate, and perspiration brought on by emotion. This device, the *polygraph*, detected the emotional effort involved in telling a lie, which always carries with it the fear of detection in any reasonably normal individual and therefore brings adrenalin into play. While far from infallible, the polygraph has gained great fame as a *lie detector.*

In the normal course, the nerve endings of the sympathetic nervous system themselves secrete a compound very like adrenalin, called *noradrenalin.* This chemical serves to carry the nerve impulses across the synapses, transmitting the message by stimulating the nerve endings on the other side of the gap.

In the early 1920s the English physiologist Henry Dale and the German physiologist Otto Loewi (who were to share the Nobel Prize in physiology and medicine in 1930) studied a chemical that performed this function for most of the nerves other than those of the sympathetic system. The chemical is called *acetylcholine.* It is now believed to be involved not only at the synapses but also in conducting the nerve impulse along the nerve fiber itself. Perhaps acetylcholine acts upon the sodium pump. At any rate, the substance seems to be formed momentarily in the nerve fiber and to be broken down quickly by an enzyme called *cholinesterase.* Anything that inhibits the action of cholinesterase will interfere with this chemical cycle and will stop the transmission of nerve impulses. The deadly substances now known as *nerve gases* are cholinesterase inhibitors. By blocking the conduction of nerve impulses, they can stop the heartbeat and produce death within minutes. The application to warfare is obvious. They can be used, less immorally, as insecticides.

A less drastic interference with cholinesterase is that of local anesthetics, which in this way suspend (temporarily) those nerve impulses associated with pain.

Thanks to the electric currents involved in nerve impulses, it is possible to "read" the brain's activity, in a way, though no one has yet been able to translate fully what the brain waves are saying. In 1929, a German psychiatrist, Hans Berger, reported earlier work in which he applied electrodes to various parts of the head and was able to detect rhythmic waves of electrical activity.

Berger gave the most pronounced rhythm the name of *alpha wave.* In the alpha wave, the potential varies by about 20 microvolts in a frequency of roughly 10 times a second. The alpha wave is clearest and most obvious when the subject is resting with eyes closed. When the eyes are open but viewing featureless illumination, the alpha wave persists. If, however, the ordinary variegated environment is in view, the alpha view vanishes, or is drowned, by other more prominent rhythms. After a while, if nothing visually new is presented, the alpha wave reappears. Typical names for other types of waves are *beta waves, delta waves,* and *theta waves.*

Electroencephalograms ("electrical writings of the brain" or, as abbreviated, EEG) have since been extensively studied and show that each individual has his or her own pattern, varying with excitement and in sleep. Although the electroencephalogram is still far from being a method of "reading thoughts" or tracing the mechanism of the intellect, it does help in the diagnosis of major upsets of brain function, particularly epilepsy. It can also help locate areas of brain damage or brain tumors.

In the 1960s, specially designed computers were called into battle. If a particular small environmental change is applied to a subject, it is presumed that there will be some response in the brain that will be reflected in a small alteration in the EEG pattern at the moment when the change is introduced. The brain will be engaged in many other activities, however, and the small alteration in the EEG will not be noticeable. Notwithstanding, if the process is repeated over and over again, a computer can be programed to average out the EEG pattern and find the consistent difference.

By 1964, the American psychologist Manfred Clynes reported analyses fine enough to be able to tell, by a study of the EEG pattern alone, what color a subject was looking at. The English neurophysiologist William Grey Walter similarly reported a brain-signal pattern that seems characteristic of the learning process. It comes when the subject under study has reason to think he or she is about to be presented with a stimulus that will call for thought or action. Walter calls it the *expectancy wave* and points out that it is absent in children under three and in certain psychotics. The reverse phenomenon, that of bringing about specific actions through direct electrical stimulation of the brain, was also reported in 1965. José Manuel Rodriguez Delgado of Yale, transmitting electrical stimulation by radio signals, caused animals to walk, climb, yawn, sleep, mate, switch emotions, and so on at command. Most spectacularly, a charging bull was made to stop short and trot peacefully away.

Human Behavior

Unlike physical phenomena, such as the motions of planets or the properties of light, the behavior of living things has never been reduced to rigorous natural laws and perhaps never will be. There are many who insist that the study of human behavior cannot become a true science, in the sense of being able to explain or predict behavior in any given situation on the basis of universal natural laws. Yet life is no exception to the rule of natural law, and it can be argued that living behavior would be fully explainable if all the factors were known. The catch lies in that last phrase. It is unlikely that all the factors will ever be known; they are too many and too complex. We need not, however, despair of ever being able to improve our understanding of ourselves. There is ample room for better knowledge of our own mental complexities, and even if we never reach the end of the road, we may yet hope to travel along it quite a way.

Not only is the subject particularly complex, but its study has not been progressing for long. Physics came of age in 1600, and chemistry in 1775, but the much more complex study of *experimental psychology* dates only from 1879, when the German physiologist Wilhelm Wundt set up the first laboratory devoted to the scientific study of human behavior. Wundt interested himself primarily in sensation and in the manner in which humans perceive the details of the universe about them.

At almost the same time, the study of human behavior in one particular application—that involving the individual as an industrial cog—arose. In 1881, the American engineer Frederick Winslow Taylor began measuring the time required to do certain jobs and to work out methods for so organizing the work as to minimize that time. He was the first *efficiency expert* and was (like all efficiency experts who tend to lose sight of values beyond the stop watch) unpopular with the workers.

But as we study human behavior, step by step, either under controlled conditions in a laboratory or empirically in a factory, it does seem that we are tackling a fine machine with blunt tools.

In the simple organisms we can see direct, automatic responses of the kind called *tropisms* (from a Greek word meaning "to turn"). Plants show *phototropism* ("turning toward light"), *hydrotropism* ("turning toward water," in this case by the roots), and *chemotropism* ("turning toward particular chemical substances"). Chemotropism is also characteristic of many animals, from protozoa to ants. Certain moths are known to fly toward a scent as far as 2 miles away. That tropisms are completely automatic is shown by the fact that a phototropic moth will even fly into a candle flame.

The reflexes mentioned earlier in this chapter do not seem to progress far beyond tropisms, and imprinting, also mentioned, represents learning, but in so mechanical a fashion as scarcely to deserve the name. Yet neither reflexes nor imprinting can be regarded as characteristic of the lower animals only; human beings have their share.

CONDITIONED RESPONSES

The human infant from the moment of birth will grasp a finger tightly if it touches his palm and will suck at a nipple if that is put to his lips. The importance of such instincts to keep the infant secure from falling and from starvation is obvious.

It seems almost inevitable that the infant is subject also to imprinting. This is not a fit subject for experimentation, of course, but knowledge can be gained through incidental observations. Children who, at the babbling stage, are not exposed to the sounds of actual speech may not develop the ability to speak later, or do so to an abnormally limited extent. Children brought up in impersonal institutions where they are efficiently fed and their physical needs are amply taken care of, but where they are not fondled, cuddled, and dandled, become sad little specimens indeed. Their mental and physical development is greatly retarded and many die for no other reason apparently than lack of *mothering*—by which may be meant the lack of adequate stimuli to bring about the imprinting of necessary behavior patterns. Similarly, children who are unduly deprived of the stimuli involved in the company of other children during critical periods in childhood develop personalities that may be seriously distorted in one fashion or another.

Of course, one can argue that reflexes and imprinting are a matter of concern only for infancy. When one achieves adulthood, one is then a rational being who responds in more than a mechanical fashion. But does one? To put it another way: Do we possess *free will* (as we like to think)? Or, is our behavior in some respects absolutely determined by the stimulus, as the bull's was in Delgado's experiment I have just described?

One can argue for the existence of free will on philosophical or theological grounds, but I know of no one who has ever found a way to demonstrate it experimentally. To demonstrate *determinism,* the reverse of free will, is not exactly easy either. Attempts in that direction, however, have been made. Most notable were those of the Russian physiologist Ivan Petrovich Pavlov.

Pavlov started with a specific interest in the mechanism of digestion. He showed, in the 1880s, that gastric juice was secreted in the stomach as soon as food was placed on a dog's tongue; the stomach would secrete this juice even if food never reached it. But if the *vagus nerve* (which runs from the medulla oblongata to various parts of the alimentary canal) was cut near the stomach, the secretions stopped. For his work on the physiology of digestion, Pavlov received the Nobel Prize in physiology and medicine in 1904. But like some other Nobel laureates (notably, Ehrlich and Einstein) Pavlov went on to other discoveries that dwarfed the accomplishments for which he actually received the prize.

He decided to investigate the automatic, or reflex, nature of secretions, and he chose the secretion of saliva as a convenient, easy-to-observe example. The sight or odor of food causes a dog (and a human being, for that matter) to salivate. What Pavlov did was to ring a bell every time he placed food before a dog. Eventually, after twenty to forty associations of this sort, the dog salivated when it heard the bell even though no food was present. An association had been built up. The nerve impulse that carried the sound of the bell to the cerebrum had become equivalent to one representing the sight or the odor of food.

In 1903, Pavlov invented the term *conditioned reflex* for this phenomenon; the salivation was a *conditioned response.* Willy-nilly, the dog salivated at the sound of the bell just as it would at the sight of food. Of course, the conditioned response could be wiped out—for instance, by repeatedly denying food to the dog when the bell was rung and subjecting it to a mild electric shock instead. Eventually, the dog would not salivate but instead would wince at the sound of the bell, even though it received no electric shock.

Furthermore, Pavlov was able to force dogs to make subtle decisions by associating food with a circular patch of light and an electric shock with an elliptical patch. The dog could make the distinction, but as the ellipse was made more and more nearly circular, distinction became more difficult. Eventually, the dog, in an agony of indecision, developed what could only be called a *nervous breakdown.*

Conditioning experiments have thus become a powerful tool in psychology. Through them, animals sometimes almost talk to the experimenter. The technique has made it possible to investigate the learning abilities of various animals, their instincts, their visual abilities, their ability to distinguish colors, and so on. Of all the investigations, not the least remarkable are those of the Austrian naturalist Karl von Frisch. Von Frisch trained bees to go to dishes placed in certain locations for their food, and he learned that these foragers soon told the other bees in their hive where the food was located. From his experiments von Frisch learned that the bees could distinguish certain colors —including ultraviolet, but excluding red—which they communicated with one another by means of a dance on the honeycombs; that the nature and vigor of the dance told the direction and distance of the food dish from the hive and even how plentiful or scarce the food supply was; and that the bees were able to tell direction from the polarization of light in the sky. Von Frisch's fascinating discoveries about the language of the bees opened up a whole new field of study of animal behavior.

In theory, all learning can be considered to consist of conditioned responses. In learning to type, for instance, you start by watching the typewriter keyboard and gradually substitute certain automatic movements of the fingers for visual selection of the proper key. Thus the thought k is accompanied by a specific movement of the middle finger of the right hand; the thought *the* causes the first finger of the left hand, the first finger of the right hand, and the second finger of the left hand, to hit certain spots in that order. These responses involve no conscious thought. Eventually a practiced typist has to stop and think to recall where the letters are. I am myself a rapid and completely mechanical typist, and if I am asked where the letter f, say, is located on the keyboard, the only way I can answer (short of looking at the keyboard) is to move my fingers in the air as if typing and try to catch one of them in the act of typing f. Only my fingers know the keyboard; my conscious mind does not.

The same principle may apply to more complex learning, such as reading or playing a violin. Why, after all, does the design CRAYON in black print on this piece of paper automatically evoke a picture (to an English-speaking person) of a pigmented stick of wax and a certain sound that represents a word? You do not need to spell out the letters or search your memory or reason out the possible message contained in the design; from repeated conditioning, you automatically associate the symbol with the thing itself.

In the early decades of this century, the American psychologist John Broadus Watson built a whole theory of human behavior, called *behaviorism,* on the basis of conditioning. Watson went so far as to suggest that people have no deliberate control over the way they behave; it is all determined by conditioning. Although his theory was popular for a time, it never gained wide support among psychologists. In the first place, even if the theory is basically correct—if behavior is dictated solely by conditioning—behaviorism is not very enlightening on those aspects of human behavior that are of most interest to us, such as creative intelligence, artistic ability, and the sense of right and wrong. It would be impossible to identify all the conditioning influences and relate them to the pattern of thought and belief in any measurable way; and something that cannot be measured is not subject to any really scientific study.

In the second place, what does conditioning have to do with a process such as intuition? The mind suddenly puts two previously unrelated thoughts or events together, apparently by sheer chance, and creates an entirely new idea or response.

Cats and dogs, in solving tasks (as in finding out how to work a lever in order to open a door) may do so by a process of trial and error. They may move about randomly and wildly until some motion of theirs trips the lever. If they are set to repeating the task, a dim memory of the successful movement may lead them to it sooner, and then still sooner at the next attempt, until finally they move to the lever at once. The more intelligent the animal, the fewer attempts will be required to graduate from sheer trial and error to purposive useful action.

By the time we reach people, memory is no longer feeble. Your tendency might be to search for a dropped dime by glances randomly directed at the floor, but from past experience you may look in places where you have found the dime before, or look in the direction of the sound, or institute a systematic scanning of the floor. Similarly, if you were in a closed place, you might try to escape by beating and kicking at the walls randomly; but you would also know what a door would look like and would concentrate your efforts on that.

People can, in short, simplify trial and error by calling on years of experience, and transfer it from thought to action. In seeking a solution, you may do nothing, you may merely act in thought. It is this etherealized trial and error we call *reason*, and it is not even entirely restricted to the human species.

Apes, whose patterns of behavior are simpler and more mechanical than ours, show some spontaneous insight, which may be called reason. The German psychologist Wolfgang Köhler, trapped in one of the German colonies in Africa by the advent of the First World War, discovered some striking illustrations of this insight in his famous experiments with chimpanzees. In one case a chimp, after trying in vain to reach bananas with a stick that was too short, suddenly picked up another bamboo stick that the experimenter had left lying handy, joined the two sticks together, and so brought the fruit within reach. In another instance, a chimp piled one box on another to reach bananas hanging overhead. These acts had not been preceded by any training or experience that might have formed the association for the animal; apparently they were sheer flashes of inspiration.

To Köhler, it seemed that learning involved the entire pattern of a process, rather than individual portions of it. He was one of the founders of the Gestalt school of psychology (*Gestalt* being the German word for "pattern").

Chimpanzees and the other great apes are so nearly human in appearance and in some of their behavior that there have not been lacking attempts to bring up young apes with human children in order to see how long they would keep up with the latter. At first, maturing more quickly, young apes forge ahead of their human counterparts. However, once human children learn to speak, the apes fall behind forever. They lack the equivalent of Broca's convolution.

In the wild, however, chimpanzees communicate not only by a small catalogue of sounds but by gesture. It occurred to Beatrice and Allen Gardner at the University of Nevada, in 1966, to try to teach a sign language to a one-and-one-half-year-old female chimpanzee named Washoe. They were amazed at the results. Washoe learned dozens of symbols, used them correctly, and understood them easily.

Other chimpanzees were so taught by others—and young gorillas, too. And, with that, came controversy. Were the apes actually communicating creatively, or were they merely responding mechanically in conditioned-reflex fashion?

Those who taught the apes had many anecdotes of their charges inventing new and creative combinations of symbols, but such things are dismissed as unconvincing by critics, or as uncertain. The controversy will undoubtedly continue.

The power of conditioning has turned out to be greater than had been expected, in fact, even in human beings. For a long time it had been assumed that certain body functions—such as heartbeat, blood pressure, and intestinal contractions—were essentially under the control of the autonomic nervous system and therefore beyond conscious control. There were catches, of course. A man adept at yoga can produce effects on his heartbeat by control of chest muscles, but that is no more significant than stopping the blood flow through a wrist artery by applying thumb pressure. Again, one can make one's heart beat faster by fantasying a state of anxiety, but that is the conscious manipulation of the autonomic nervous system. Is it possible simply to will the heart to beat faster or the blood pressure to rise without extreme manipulation of either the muscles or the mind?

The American psychologist Neal Elgar Miller and his co-workers carried out conditioning experiments, in the early 1960s, where rats were rewarded when they happened to increase their blood pressure for any reason, or when their heartbeat was increased or decreased. Eventually, for the sake of the reward, they learned to perform voluntarily a change effected by the autonomic nervous system—just as they might learn to press a lever, and for the same purpose.

At least one experimental program, using human volunteers (male) who were rewarded by flashes of light revealing photographs of nude girls, demonstrated the volunteers' ability to produce increases or decreases in blood pressure in response. The volunteers did not know what was expected of them in order to produce the flashing light—and the nude—but just found that, as time went along, they caught the desired glimpses more often.

More systematic experimentation showed that if people were made aware, at all times, of some property they are ordinarily unaware of—say, blood pressure, heart rate, or skin temperature—they can, through a voluntary effort (in some fashion not easily defined), change the value. This process is called *biofeedback*.

There were hopes at first that biofeedback might accomplish, more efficiently and easily, some of the claims of the accomplishments of Eastern mystics: that it might control or ameliorate some otherwise intransigent metabolic disorders. These hopes seem to have faded in the last decade or so.

THE BIOLOGICAL CLOCK

There are additional subtleties to the autonomic body controls which had earlier gone unsuspected. Since living organisms are subjected to natural rhythms—the ebb and flow of the tides, the somewhat slower alternation of day and night, the still slower swing of the seasons—it is not surprising that they themselves respond rhythmically. Trees shed their leaves in fall and bud in the spring; humans grow sleepy at night and rouse themselves at dawn.

What did not come to be fully appreciated until lately is the complexity and multiplicity of the rhythmic responses, and their automatic nature, which persists even in the absence of the environmental rhythm.

Thus, the leaves of plants rise and fall in a daylong rhythm to match the coming and going of the sun. This is made apparent by time-lapse photography. Seedlings grown in darkness showed no such cycle, but the potentiality was there. One exposure to light—one only—was enough to convert that potentiality into actuality. The rhythm then began, and it continued even if the light was cut off again. From plant to plant, the exact period of rhythm varied—anywhere from 24 to 26 hours in the absence of light—but it was always about 24 hours, under the regulating effect of the sun. A 20-hour cycle could be established if artificial light were used on a 10-hour-on and 10-hour-off cycle, but as soon as the light was turned off altogether, the about-24-hour rhythm reestablished itself.

This daily rhythm, a kind of *biological clock* that works even in the absence of outside hints, permeates all life. Franz Halberg of the University of Minnesota named it *circadian rhythm,* from the Latin *circa dies,* meaning "about a day."

Human beings are not immune to such rhythms. Men and women have voluntarily lived for months at a time in caves where they separated themselves from any time-telling mechanism and had no idea whether it was night or day outside. They soon lost all track of time and ate and slept rather erratically. However, they also noted their temperature, pulse, blood pressure, and brain waves, and sent these and other measurements to the surface, where observers kept track of them in connection with time. It turned out that, however time-confused the cave dwellers were, their bodily rhythm was not. The rhythm remained stubbornly at a period of about a day, with all measurements rising and falling regularly, through all the stay in the cave.

This is by no means only an abstract matter. In nature, the earth's rotation remains steady, and the alternation of day and night remains constant and beyond human interference—but only if you remain in the same spot on earth or only shift north or south. If you travel east or west for long distances and quite rapidly, however, you change the time of day. You may land in Japan at lunchtime (for Japanese) when your biological clock tells you it is time to go to bed. The jet-age traveler often has difficulty matching his activity to that of the at-home people surrounding him. If he does so—with his pattern of hormone-secretion, for instance, not matching the pattern of his activity—he will be tired and inefficient, suffering from *jet fatigue*, or *jet lag*.

Less dramatically, the ability of an organism to withstand a dose of X rays or various types of medication often depends on the setting of the biological clock. It may well be that medical treatment ought to vary with the time of day or, for maximum effect and minimum side effect, be restricted to one particular time of day.

What keeps the biological clock so well regulated? Suspicion in this respect has fallen upon the pineal gland (see chapter 15). In some reptiles, the pineal gland is particularly well developed and seems to be similar in structure to the eye. In the tuatara, a lizardlike reptile that is the last surviving species of its order and is found only on some small islands off New Zealand, the *pineal eye* is a skin-covered patch on top of its skull, particularly prominent for about six months after birth and definitely sensitive to light.

The pineal gland does not "see" in the ordinary sense of the word, but may produce some chemical that rises and falls in rhythmic response to the coming and going of light. It thus may regulate the biological clock and do so even after light ceases to be periodic (having learned its chemical lesson by a kind of conditioning).

But then how does the pineal gland work in mammals, where it is no longer located just under the skin at the top of the head but is buried deep in the center of the brain? Can there be something more penetrating than light— something that is rhythmic in the same sense? There are speculations that cosmic rays might be the answer. These have a circadian rhythm of their own, thanks to Earth's magnetic field and the solar wind, and perhaps this force is the external regulator.

Even if the external regulator is found, is the internal biological clock something that can be identified? Is there some chemical reaction in the body that rises and falls in a circadian rhythm and that controls all the other rhythms? Is there some "master reaction" that we can tab as *the* biological clock? If so, it has not yet been found.

PROBING HUMAN BEHAVIOR

It does not seem likely, however, that we are ever going to pin anything as complex as life into complete determinism. It is easy to be deterministic about something like the replication of nucleic acids, and yet environmental factors introduce errors that result in mutations and evolution. Nor is it conceivable that the course of evolution can be predicted in detail.

More fundamentally, we know from quantum mechanics that there are indeterminacies and uncertainties inherent in the behavior of objects; and that the lighter the objects are and the less massive, the greater the indeterminacies. The behavior of electrons is, in some ways, unpredictable, and there are arguments to the effect that certain properties of electrons cannot be known until they are measured. It may even be that the state of the universe is, in a certain subtle way, defined at each instant of time by the observations and measurements made by human beings. (This is called the *anthropic principle* from the Greek word for "human being.")

It is easy to see that there may be times when the course of human behavior or a human decision (or even perhaps those of lower animals) may rest upon the indeterminate motion of an electron somewhere in the body. This would, in principle, wreck determinism, but it would not establish free will either. It would, instead, introduce a random factor, which may well be harder to understand than either.

But not necessarily harder to handle. Random factors can be allowed for if there are enough such events. Individual gas molecules move about in random fashion; but in any ordinary quantity of gas, there are so many molecules that the randomness cancels out, and the *gas laws* will apply with great precision to such properties as temperature, pressure, and volume.

We have not come to this yet, however, and there have, instead, been attempts to attack human behavior by methods that are themselves highly intuitive and as difficult to handle as the behavior they attempt to deal with.

These methods can be traced back nearly two centuries to an Austrian physician, Franz Anton Mesmer, who became the sensation of Europe for his experiments with a powerful tool for probing human behavior. He used magnets at first, and then his hands only, obtaining his effects by what he called *animal magnetism* (soon renamed *mesmerism*): he would put a patient into a trance and pronounce the patient cured of his illness. Mesmer may well have produced some cures (since some disorders can be treated by suggestion) and gained many ardent followers, including the Marquis de Lafayette, fresh from his American triumph. However, Mesmer, an ardent astrologer and all-round mystic, was investigated skeptically but fairly by a committee, which included Lavoisier and Benjamin Franklin, and was then denounced as a fake and eventually retired in disgrace.

Nevertheless, he had started something. In the 1850s a British surgeon named James Braid revived *hypnotism* (he was the first to use this term in place of *mesmerism*) as a medical device, and other physicians also took it up. Among them was a Viennese doctor named Josef Breuer, who in the 1880s began to use hypnosis specifically for mental and emotional disorders.

Hypnotism (Greek for "putting to sleep") had been known, of course, since ancient times and had often been used by mystics. But Breuer and others now began to interpret its effects as evidence of the existence of an *unconscious* level of the mind. Motivations of which the individual was unaware were buried there, and they could be brought to light by hypnosis. It was tempting to suppose that these motivations were suppressed from the conscious mind because they were associated with shame or guilt, and that they might account for useless, irrational, or even vicious behavior.

Breuer set out to employ hypnosis to probe the hidden causes of hysteria and other behavior disorders. Working with him was a pupil named Sigmund

Freud. For a number of years, they treated patients together, putting the patients under light hypnosis and encouraging them to speak. They found that the patients' venting of experiences or impulses buried in the unconscious often acted as a cathartic, relieving their symptoms after they awoke from the hypnosis.

Freud came to the conclusion that practically all of the suppressed memories and motivations were sexual in origin. Sexual impulses tabooed by society and the child's parents were driven underground, but still strove for expression and generated intense conflicts which were the more damaging for being unrecognized and unadmitted.

In 1894, after breaking with Breuer because the latter disagreed with his concentration on the sexual factor, Freud went on alone to develop his ideas about the causes and treatment of mental disturbances. He dropped hypnosis and urged his patients to babble in a virtually random manner—to say anything that came into their minds. As the patient came to feel that the physician was listening sympathetically without any moral censure, slowly—sometimes very slowly—the individual began to unburden himself, to remember things long repressed and forgotten. Freud called this slow analysis of the *psyche* (Greek for "soul" or "mind") *psychoanalysis*.

Freud's involvement with the sexual symbolism of dreams and his description of infantile wishes to substitute for the parent of the same sex in the marital bed (the *Oedipus complex* in the case of boys, and the *Electra complex* in girls—named for characters in Greek mythology) horrified some and fascinated others. In the 1920s, after the dislocations of the First World War and amid the further dislocations of Prohibition in America and changing mores in many parts of the world, Freud's views struck a sympathetic note, and psychoanalysis attained the status almost of a popular fad.

Nearly a century after its beginnings, however, psychoanalysis still remains an art rather than a science. Rigorously controlled experiments, such as those conducted in physics and the other "hard" sciences, are, of course, exceedingly difficult in psychiatry. The practitioners must base their conclusions largely on intuition or subjective judgment. *Psychiatry* (of which psychoanalysis is only one of the techniques) has undoubtedly helped many patients, but it has produced no spectacular cures and has not notably reduced the incidence of mental disease. Nor has it developed any all-embracing and generally accepted theory, comparable to the germ theory of infectious disease. In fact, there are almost as many schools of psychiatry as there are psychiatrists.

Serious mental illness takes various forms, ranging from chronic depression to a complete withdrawal from reality into a world in which some, at least, of the details do not correspond to the way most of us see things. This form of psychosis is usually called *schizophrenia*, a term introduced by the Swiss psychiatrist Eugen Bleuler. The word covers such a multitude of disorders that it can no longer be described as a specific disease. About 60 percent of all the chronic patients in our mental hospitals are diagnosed as schizophrenics.

Until recently, drastic treatments, such as prefrontal lobotomy, or shock therapy using electricity or insulin (the latter technique introduced in 1933 by the Austrian psychiatrist Manfred Sakel), were all that could be offered. Psychiatry and psychoanalysis have been of little avail, except occasionally in the early stages when a physician is still able to communicate with the patient. But some recent discoveries concerning drugs and the chemistry of the brain (*neurochemistry*) have introduced an encouraging note.

Even the ancients knew that certain plant juices could induce hallucinations (fantasies of vision, hearing, and so on) and others could bring on happy states. The Delphic priestesses of ancient Greece chewed some plant before they pronounced their cryptic oracles. Indian tribes of the southwestern United States have made a religious ritual of chewing peyote or mescal buttons (which produce hallucinations in color). Perhaps the most dramatic case was that of the Moslem sect in a mountain stronghold in Iran who used *hashish*, the juice of hemp leaves, more familiarly known to us as *marijuana*. The drug, taken in their religious ceremonies, gave the communicants the illusion that they caught glimpses of the paradise to which their souls would go after death, and they would obey any command of their leader, called the Old Man of the Mountains, to receive this key to heaven. His commands took the form of ordering them to kill enemy rulers and hostile Moslem government officials, and thus gave rise to the word *assassin*, from *hashishin* ("a user of hashish"). The sect terrorized the region throughout the twelfth century, until the Mongol invaders in 1226 swarmed into the mountains and killed every last assassin.

The modern counterpart of the euphoric herbs of earlier times (aside from alcohol) is the group of drugs known as the *tranquilizers*. As a matter of fact, one of the tranquilizers had been known in India as long ago as 1000 B.C. in the form of a plant called *Rauwolfia serpentinum*. It was from the dried roots of this plant that American chemists in 1952 extracted *reserpine*, the first of the currently popular tranquilizing drugs. Several substances with similar effects but simpler chemical structure have since been synthesized.

The tranquilizers are sedatives, but with a difference: they reduce anxiety without appreciably depressing other mental activity. Nevertheless, they do tend to make people sleepy, and they may have other undesirable effects. They were at once found to be immensely helpful in relieving and quieting mental patients, including some schizophrenics. The tranquilizers are not cures for any mental illness, but they suppress certain symptoms that stand in the way of adequate treatment. By reducing the hostilities and rages of patients, and by quieting their fears and anxieties, they reduce the necessity for drastic physical restraints, make it easier for psychiatrists to establish contacts with patients, and increase a patient's chances of release from the hospital.

But where the tranquilizers had their runaway boom was among the public at large, which apparently seized upon them as a panacea to banish all cares.

DRUG USE

Reserpine turns out to have a tantalizing resemblance to an important substance in the brain. A portion of its complex molecule is rather similar to the substance called *serotonin*. Serotonin was discovered in the blood in 1948, and it has greatly intrigued physiologists ever since. It was found to be present in the hypothalamus region of the human brain and proved to be widespread in the brain and nerve tissues of other animals, including invertebrates.

What is more, various other substances that affect the central nervous system have turned out to resemble serotonin closely. One of them is a compound in toad venom called *bufotenin*. Another is mescaline, the active drug in mescal buttons. Most dramatic of all is a substance named *lysergic acid diethylamide* (popularly known as LSD). In 1943, a Swiss chemist named Albert Hofmann happened to absorb some of this compound in the laboratory and was overcome by strange sensations. Indeed, what he seemed to perceive by way of his senses in no way matched what we would take to be the objective reality of the environment. He suffered what we call *hallucinations*, and LSD is an example of what we now call a *hallucinogen*.

Those who take pleasure in the sensations they experience when under the influence of a hallucinogen refer to this as *mind expansion*—apparently indicating that they sense, or think they sense, more of the universe than they would under ordinary conditions. But then, so do drunks once they bring themselves to the stage of *delirium tremens*. The comparison is not as unkind as it may seem, for investigations have shown that a small dose of LSD, in some cases, can produce many of the symptoms of schizophrenia!

What can all this mean? Well, serotonin (which is structurally like the amino acid tryptophan) can be broken down by means of an enzyme called *amine oxidase*, which occurs in brain cells. Suppose that this enzyme is taken out of action by a competitive substance with a structure like serotonin's—lysergic acid, for example. With the breakdown enzyme removed, serotonin will accumulate in the brain cells, and its level may rise too high. This will upset the serotonin balance in the brain and may bring on the schizophrenic state.

Is it possible that schizophrenia arises from some naturally induced upset of this sort? The manner in which a tendency to schizophrenia is inherited certainly makes it appear that some metabolic disorder (one, moreover, that is gene-controlled) is involved. In 1962, it was found that with a certain course of treatment, the urine of schizophrenics often contained a substance absent from the urine of nonschizophrenics. The substance eventually turned out to be a chemical called *dimethoxyphenylethylamine*, with a structure that lies somewhere between adrenalin and mescaline. In other words, certain schizophrenics seem, through some metabolic error, to form their own hallucinogens and to be, in effect, on a permanent drug-high.

Not everyone reacts identically to a given dose of one drug or another. Obviously, however, it is dangerous to play with the chemical mechanism of the brain. To become a mental cripple is a price surely too high for any amount of "mind-expanding" fun. Nevertheless, the reaction of society to drug use—particularly to that of marijuana, which has not yet been definitely shown to be as harmful as other hallucinogens—tends to be overstrenuous. Many of those who inveigh against the use of drugs of one sort or another are themselves thoroughly addicted to the use of alcohol or tobacco, both of which, in the mass, are responsible for much harm both to the individual and to society. Hypocrisy of this sort tends to decrease the credibility of much of the antidrug movement.

MEMORY

Neurochemistry also offers a hope for understanding that elusive mental property known as *memory*. There are, it seems, two varieties of memory: short-term and long-term. If you look up a phone number, it is not difficult to remember it until you have dialed; it is then automatically forgotten and, in all probability, will never be recalled again. A telephone number you use frequently, however, enters the long-term memory category. Even after a lapse of months, you can dredge it up.

Yet even of what we would consider long-term memory items, much is lost. We forget a great deal and even, alas, forget much of vital importance (as every student facing an examination is woefully aware). Yet is it forgotten? Has it really vanished, or is it simply so well stored that it is difficult to recall—buried, so to speak, under too many extraneous items?

The tapping of such hidden memories has become an almost literal tap. The American-born surgeon Wilder Graves Penfield at McGill University in Montreal, while operating on a patient's brain, accidentally touched a particular spot that caused the patient to hear music. That happened over and over again. The patient could be made to relive an experience in full, while remaining quite conscious of the present. Proper stimulation can apparently reel off memories with great accuracy. The area involved is called the *interpretative cortex*. It may be that the accidental tapping of this portion of the cortex gives rise to the phenomenon of *déjà vu* (the feeling that something has happened before) and other manifestations of *extrasensory perception*.

But if memory is so detailed, how can the brain find room for it all? It is estimated that, in a lifetime, a brain can store 1,000,000,000,000,000 (a million billion) units of information. To store so much, the units of storage must be of molecular size. There would be room for nothing more.

Suspicion is currently falling on ribonucleic acid (RNA) in which the nerve cell, surprisingly enough, is richer than almost any other type of cell in the body. This is surprising because RNA is involved in the synthesis of protein (see chapter 13) and is therefore usually found in particularly high quantity in those tissues producing large quantities of protein either because they are actively growing or because they are producing copious quantities of protein-rich secretions. The nerve cell falls into neither classification.

A Swedish neurologist, Holger Hyden, developed techniques that could separate single cells from the brain and then analyze them for RNA content.

He took to subjecting rats to conditions where they were forced to learn new skills—that of balancing on a wire for long periods of time, for instance. By 1959, he had discovered that the brain cells of rats that were forced to learn increased their RNA content up to 12 percent higher than that of the brain cells of rats allowed to go their normal way.

The RNA molecule is so very large and complex that, if each unit of stored memory is marked off by an RNA molecule of distinctive pattern, we need not worry about capacity. So many different RNA patterns are available that even a number such as a million billion is insignificant in comparison.

But ought one to consider RNA by itself? RNA molecules are formed according to the pattern of DNA molecules in the chromosomes. Is it that each of us carries a vast supply of potential memories—a *memory bank,* so to speak—in the DNA molecules we were born with, called upon and activated by actual events with appropriate modifications?

And is RNA the end? The chief function of RNA is to form specific protein molecules. Is it the protein, rather than the RNA, that is truly related to the memory function?

One way of testing this hypothesis is to make use of a drug called *puromycin,* which interferes with protein formation by way of RNA. The American man-and-wife team Louis Barkhouse Flexner and Josepha Barbara Flexner conditioned mice to solve a maze, then immediately injected puromycin. The mice forgot what they had learned. The RNA molecule was still there, but the key protein molecule could not be formed. Using puromycin, the Flexners showed that while short-term memory could be erased in this way in rats, long-term memory could not. The proteins for the latter had presumably already been formed.

And yet it may be that memory is more subtle and is not to be fully explained on the simple molecular level. There are indications that patterns of neural activity may be involved, too. Much yet remains to do.

Automatons

It is only very recently, however, that the full resources of science have been turned upon the effort to analyze the functioning of living tissues and organs, in order that the manner in which they perform—worked out hit-and-miss over billions of years of evolution—might be imitated in man-made machines. This study is called *bionics,* a term—suggested by *"bio*logical electronics"* but much broader in scope—coined by the American engineer Jack Steele in 1960.

As one example of what bionics might do, consider the structure of dolphin skin. Dolphins swim at speeds that would require 2.6 horsepower if the water about them were as turbulent as it would be about a vessel of the same size.

For some reason, water flows past the dolphin without turbulence, and therefore little power is consumed overcoming water resistance. Apparently this happens because of the nature of dolphin skin. If we can reproduce that effect in vessel walls, the speed of an ocean liner could be increased and its fuel consumption decreased—simultaneously.

Then, too, the American biophysicist Jerome Lettvin studied the frog's retina in detail by inserting tiny platinum electrodes into its optic nerve. It turned out that the retina did not merely transmit a mélange of light and dark dots to the brain and leave it to the brain to do all the interpretation. Rather, there were five different types of cells in the retina, each designed for a particular job. One cell reacted to edges—that is, to sudden changes in the nature of illumination, as at the edge of a tree marked off against the sky. A second reacted to dark curved objects (the insects eaten by the frog). A third reacted to anything moving rapidly (a dangerous creature that might better be avoided). A fourth reacted to dimming light; and a fifth, to the watery blue of a pond. In other words, the retinal message went to the brain already analyzed to a considerable degree. If man-made sensors made use of the tricks of the frog's retina, they could be made far more sensitive and versatile than they now are.

If, however, we are to build a machine that will imitate some living device, the most attractive possibility is the imitation of that unique device that interests us most profoundly—the human brain.

The human mind is not a "mere" machine; it is safe enough to say that. On the other hand, even the human mind, which is certainly the most complex object or phenomenon we know of, has certain aspects that remind us of machines in certain ways. And the resemblances can be important.

Thus, if we analyze what it is that makes a human mind different from other minds (to say nothing of different from mindless objects), one thought that might strike us is that, more than any other object, living or nonliving, the human mind is a self-regulating system. It is capable of controlling not only itself but also its environment. It copes with changes in the environment, not by yielding but by reacting according to its own desires and standards. Let us see how close a machine can come to this ability.

About the simplest form of self-regulating mechanical device is the controlled valve. Crude versions were devised as early as 50 A.D. by Hero of Alexandria, who used one in a device to dispense liquid automatically. A very elementary version of a safety valve is exemplified in a pressure cooker invented by Denis Papin in 1679. To keep the lid on against the steam pressure, he placed a weight on it, but he used a weight light enough so that the lid could fly off before the pressure rose to the point where the pot would explode. The present-day household pressure cooker or steam boiler has more sophisticated devices for this purpose (such as a plug that will melt when the temperature gets too high); but the principle is the same.

FEEDBACK

Of course, this is a "one shot" sort of regulation. But it is easy to think of examples of continuous regulation. A primitive type was a device patented in 1745 by an Englishman, Edmund Lee, to keep a windmill facing squarely to the wind. He devised a *fantail* with small vanes that caught the wind whenever the wind shifted direction; the turning of these vanes operated a set of gears that rotated the windmill itself so that its main vanes were again head on to the wind in the new quarter. In that position, the fantail vanes remained motionless; they turned only when the windmill was not facing the wind.

But the archetype of modern mechanical self-regulators is the *governor* invented by James Watt for his steam engine (figure 17.4). To keep the steam

Figure 17.4. Watt's governor.

output of his engine steady, Watt conceived a device consisting of a vertical shaft with two weights attached to it laterally by hinged rods, allowing the weights to move up and down. The pressure of the steam whirled the shaft. When the steam pressure rose, the shaft whirled faster, and the centrifugal force drove the weights upward. In moving up, they partly closed a valve, choking off the flow of steam. As the steam pressure fell, the shaft whirled less rapidly, gravity pulled the weights down, and the valve opened. Thus, the governor kept the shaft speed, and hence the power delivered, at a uniform level. Each departure from that level set in train a series of events that corrected the deviation. This is called *feedback:* the error itself continually sends back information and serves as the measure of the correction required.

A very familiar example of a feedback device is the *thermostat,* first used in crude form by the Dutch inventor Cornelis Drebble in the early seventeenth century. A more sophisticated version, still used today, was invented in principle by a Scottish chemist named Andrew Ure in 1830. Its essential component consists of two strips of different metals laid against each other and soldered together. Since the two metals expand and contract at different rates with changes in temperature, the strip bends. The thermostat is set, say, at 70° F. When the room temperature falls below that, the thermocouple bends in such a fashion as to make a contact that closes an electric circuit and turns on the heating system. When the temperature rises above 70° F, the thermocouple bends back enough to break the contact. Thus, the heater regulates its own operation through feedback.

It is feedback that similarly controls the workings of the human body. To take one example of many, the glucose level in the blood is controlled by the insulin-producing pancreas, just as the temperature of a house is controlled by the heater. And just as the working of the heater is regulated by the departure of the temperature from the norm, so the secretion of insulin is regulated by the departure of the glucose concentration from the norm. A too-high glucose level turns on the insulin, just as a too-low temperature turns on the heater. Likewise, as a thermostat can be turned up to higher temperature, so an internal change in the body, such as the secretion of adrenalin, can raise the operation of the human body to a new norm, so to speak.

Self-regulation by living organisms to maintain a constant norm was named *homeostasis* by the American physiologist Walter Bradford Cannon, who was a leader in investigation of the phenomenon in the first decades of the twentieth century.

The feedback process in living systems is essentially the same as in machines and ordinarily is not given a special name. The use of biofeedback for cases where voluntary control of autonomic nerve functions is sought is an artificial distinction for convenience.

Most systems, living and nonliving, lag a little in their response to feedback. For instance, after a heater has been turned off, it continues for a time to emit its residual heat; conversely, when it is turned on, it takes a little time to heat up. Therefore, the room temperature does not hold to 70° F but oscillates around that level; it is always overshooting the mark on one side or the other. This phenomenon, called *hunting*, was first studied in the 1830s by George Airy, the Astronomer Royal of England, in connection with devices he had designed to turn telescopes automatically with the motion of the earth.

Hunting is characteristic of most living processes, from control of the glucose level in the blood to conscious behavior. When you reach to pick up an object, the motion of your hand is not a single movement but a series of movements continually adjusted in both speed and direction, with the muscles correcting departures from the proper line of motion, those departures being judged by the eye. The corrections are so automatic that you are not aware of them. But watch an infant, not yet practiced in visual feedback, try to pick up something: the child overshoots and undershoots because the muscular corrections are not precise enough. And victims of nerve damage that interferes with the ability to utilize visual feedback go into pathetic oscillations, or wild hunting, whenever they attempt a coordinated muscular movement.

The normal, practiced hand goes smoothly to its target and stops at the right moment because the control center looks ahead and makes corrections in advance. Thus, when you drive a car around a corner you begin to release the steering wheel before you have completed the turn, so that the wheels will be straight by the time you have rounded the corner. In other words, the correction is applied in time to avoid overshooting the mark to any significant degree.

It is the chief role of the cerebellum, evidently, to take care of this adjustment of motion by feedback. It looks into the future and predicts the position of the arm a few instants ahead, organizing motion accordingly. It keeps the large muscles of the torso in constantly varying tensions to keep you in balance and upright if you are standing. It is hard work to stand and "do nothing"; we all know how tiring just standing can be.

Now this principle can be applied to a machine. Matters can be arranged so that, as the system approaches the desired condition, the shrinking margin between its actual state and the desired state will automatically shut off the corrective force before it overshoots. In 1868, a French engineer, Léon Farcot, used this principle to invent an automatic control for a steam-operated ship's rudder. As the rudder approached the desired position, his device automatically closed down the steam valve; by the time the rudder reached the specified position, the steam pressure had been shut off. When the rudder moved away from this position, its motion opened the appropriate valve so that it was pushed back. Farcot called his device a *servomechanism*, and in a sense it ushered in the era of *automation* (a term introduced in 1951 by the American engineer John Diebold).

EARLY AUTOMATION

The invention of mechanical devices that imitated human foresight and judgment, no matter how crudely, was enough to set off the imagination of some into considering the possibility of some device that could imitate human actions more or less completely—an *automaton.* Myths and legends are full of them.

To translate the accomplishments of gods and magicians into those of mere men required the gradual development of clocks during the Middle Ages. As clocks advanced in complexity, *clockwork,* the use of intricately related wheels that cause a device to perform certain motions in the right order and at appropriate times, made it possible to consider the manufacture of objects that mimick the actions associated with life more closely than ever.

The eighteenth century began a kind of golden age of automatons. Automatic toy soldiers were constructed for the French dauphin; an Indian ruler had a six-foot mechanical tiger.

Such royal conveniences, however, were outstripped by commercial ventures. In 1738, a Frenchman, Jacques de Vaucanson, constructed a mechanical duck of copper that could quack, bathe, drink water, eat grain, seem to digest and then excrete it. People paid to see the duck, and it earned money for its owners for decades but no longer survives.

A later automaton does survive in a Swiss museum at Neuchâtel. It was constructed in 1774 by Pierre Jacquet-Droz and is an automatic scribe. It is in the shape of a boy who dips his pen in an inkwell and writes a letter.

To be sure, such automatons are completely inflexible. They can *only* follow the motions dictated by the clockwork.

Nevertheless, it was not long before the principles of automatism were made flexible and turned to useful labor rather than mere show.

The first great example was an invention of a French weaver, Joseph Marie Jacquard. In 1801, he devised the *Jacquard loom*.

In such a loom, needles ordinarily move through holes set in a block of wood and there engage the threads in such a way as to produce the weaving interconnections. Suppose, though, that a punched card is interposed between needles and holes. Holes in the card here and there allow needles to pass through and enter the wood as before. In places where needles are *not* punched through the card, the needles are stopped. Thus, some interconnections are made, and some are not.

If there are different punched cards, with different arrangements of holes, and if these are inserted into the machine in a particular order, then the changes in the stitches that are allowed or not allowed can produce a pattern. By appropriate adjustment of the cards, any pattern, in principle, can be formed quite automatically. In modern terms, we would say that the card arrangement serves to *program* the loom, which then does something, of its own apparent accord, that could be mistaken for artistic creativity.

The most important aspect of the Jacquard loom was that it accomplished its amazing successes (by 1812, there were 11,000 of these looms in France; and once the Napoleonic wars were over, they spread to Great Britain) by a simple yes-no dichotomy. Either a hole existed in a special place or it did not, and the pattern yes-no-yes-yes-no and so on over the face of the card was all that was necessary.

Ever since, more and more complicated devices designed to mimic human thought have made use of ever more subtle methods of dealing with yes-no patterns. It might seem totally ridiculous to expect to get complicated, human-seeming results, from a simple yes-no pattern; but actually the mathematical basis for it had been demonstrated in the seventeenth century, after thousands of years of attempts to mechanize arithmetical calculations and to find aids (increasingly subtle) for the otherwise unaided operation of the human mind.

ARITHMETICAL CALCULATIONS

The first tools for the purpose must have been human fingers. Mathematics began when human beings used their own fingers to represent numbers and combinations of numbers. It is no accident that the word *digit* stands both for a finger (or toe) and for a numerical integer.

From that, another step leads to the use of other objects in place of fingers —small pebbles, perhaps. There are more pebbles than fingers, and intermediate results can be preserved for future reference in the course of solving the problem. Again, it is no accident that the word *calculate* comes from the Latin word for "pebble."

1870

+549
=2419

+2801
=5220

+1193
=6413

Figure 17.5. Adding with an abacus. Each counter below the bar counts 1; each counter above the bar counts 5. A counter registers when it is pushed to the bar. Thus in the top setting here, the right-hand column reads 0; the one to the left of that reads 7 or (5 + 2); the next left reads 8 or (5 + 3); and the next left reads 1: the number shown, then, is 1870. When 549 is added to this, the right column becomes 9 or (9 + 0); the next addition (4 + 7) becomes 1 with 1 to carry, which means that one counter is pushed up in the next column; the third addition is 9 + 5, or 4 with 1 to carry; and the fourth addition is 1 + 1 or 2: the addition gives 2419, as the abacus shows. The simple maneuver of carrying 1 by pushing up a counter in the next column makes it possible to calculate very rapidly; a skilled operator can add faster than an adding machine can, as was shown by an actual test in 1946.

Pebbles or beads lined up in slots or strung on wires, formed the *abacus*, the first really versatile mathematical tool (figure 17.5). With this device, it became easy to represent units, tens, hundreds, thousands, and so on. By manipulating the pebbles, or counters, of an abacus, one could quickly carry through an addition such as 576 + 289. Furthermore, any instrument that can add can also multiply, for multiplication is only repeated addition. And multiplication makes raising to a power possible, because this is only repeated multiplication (for example, 4^5 is shorthand for $4 \times 4 \times 4 \times 4 \times 4$). Finally, running the instrument backward, so to speak, makes possible the operations of subtraction, division, and extracting a root.

The abacus can be considered the second *digital computer*. (The first, of course, was the fingers.)

For thousands of years the abacus remained the most advanced form of calculating tool. It actually dropped out of use in the West after the end of the Roman Empire and was reintroduced by Pope Sylvester II about 1000 A.D., probably from Moorish Spain, where its use had lingered. It was greeted on its return as an Eastern novelty, its Western ancestry forgotten.

The abacus was not replaced until a numerical notation was introduced that imitated the workings of the abacus. (This notation, the one familiar to us nowadays as *Arabic numerals,* was originated in India some time about 800 A.D., was picked up by the Arabs, and finally introduced to the West about 1200 A.D. by the Italian mathematician Leonardo of Pisa.)

In the new notation, the nine different pebbles in the units row of the abacus were represented by nine different symbols, and those same nine symbols were used for the tens row, hundreds row, and thousands row. Counters differing only in position were replaced by symbols differing only in position, so that in the written number 222, for instance, the first 2 represents 200, the second 20, and third represents two itself; that is, $200 + 20 + 2 = 222$.

This "positional notation" was made possible by recognition of an all-important fact which the ancient users of the abacus had overlooked. Although there are only nine counters in each row of the abacus, there are actually ten possible arrangements. Besides using any number of counters from one to nine in a row, it is also possible to use *no* counter—that is, to leave the place at the counting position empty. This escaped all the great Greek mathematicians and was not recognized until the ninth century, when some unnamed Hindu thought of representing the tenth alternative by a special symbol which the Arabs called "sifr" ("empty") and which has come down to us, in consequence, as "cipher" or, in more corrupt form, "zero." The importance of the zero is recorded in the fact that the manipulation of numbers is still sometimes called "ciphering," and that to solve any hard problem is to "decipher" it.

Another powerful tool grew out of the use of the exponents to express powers of numbers. To express 100 as 10^2, 1,000 as 10^3, 100,000 as 10^5, and so on, is a great convenience in several respects; not only does it simplify the writing of large numbers but it reduces multiplication and division to simple addition or subtraction of the exponents (e.g., $10^2 \times 10^3 = 10^5$) and makes raising to a power or extraction of a root a simple matter of multiplying or dividing exponents (e.g., the cube root of 1,000,000 is $10^{6/3} = 10^2$). Now this is all very well, but very few numbers can be put into simple exponential form. What could be done with a number such as 111? The answer to that question led to the tables of logarithms.

The first to deal with this problem was the seventeenth-century Scottish mathematician John Napier. Obviously, expressing a number such as 111 as a power of 10 involves assigning a fractional exponent to 10 (the exponent is between 2 and 3). In more general terms, the exponent will be fractional whenever the number in question is not a multiple of the base number. Napier worked out a method of calculating the fractional exponents of numbers, and he named these exponents *logarithms.* Shortly afterward, the English mathematician Henry Briggs simplified the technique and worked out logarithms with 10 as the base. The *Briggsian logarithms* are less convenient in calculus, but they are the more popular for ordinary computations.

All nonintegral exponents are irrational: that is, they cannot be expressed in the form of an ordinary fraction. They can be expressed only as an indefinitely long decimal lacking a repeating pattern. Such a decimal can be calculated, however, to as many places as necessary for the desired precision.

For instance, let us say we wish to multiply 111 by 254. The Briggsian logarithm of 111 to five decimal places is 2.04532, and for 254 it is 2.40483. Adding these logarithms, we get $10^{2.04532} \times 10^{2.40483} = 10^{4.45015}$. That number is approximately 28,194, the actual product of 111×254. If we want to get still closer accuracy, we can use the logarithms to six or more decimal places.

Tables of logarithms simplified computation enormously. In 1622 an English mathematician named William Oughtred made things still easier by devising a *slide rule.* Two rulers are marked with a *logarithmic scale,* in which the distances between numbers get shorter as the numbers get larger: for example, the first division holds the numbers from 1 to 10; the second division, of the same length, holds the numbers from 10 to 100; the third from 100 to 1,000; and so on. By sliding one rule along the other to an appropriate position, one can read off the result of an operation involving multiplication or division. The slide rule makes computations as easy as addition and subtraction on the abacus; though in both cases, to be sure, one must be skilled in the use of the instrument.

CALCULATING MACHINES

The first step toward a truly automatic calculating machine was taken in 1642 by the French mathematician Blaise Pascal. He invented an adding machine that did away with the need to move the counters separately in each row of the abacus. His machine consisted of a set of wheels connected by gears. When the first wheel—the units wheel—was turned ten notches to its 0 mark, the second wheel turned one notch to the number 1, so that the two wheels together showed the number 10. When the tens wheel reached its 0, the third wheel turned a notch, showing 100, and so on. (The principle is the same as that of the mileage indicator in an automobile.) Pascal is supposed to have had more than fifty such machines constructed; at least five are still in existence.

Pascal's device could add and subtract. In 1674, the German mathematician Gottfried Wilhelm von Leibnitz went a step further and arranged the wheels and gears so that multiplication and division were as automatic and easy as addition and subtraction. In 1850, a United States inventor named D. D. Parmalee patented an important advance which added greatly to the calculator's convenience: in place of moving the wheels by hand, he introduced a set of keys—pushing down a marked key with the finger turned the wheels to the correct number. This is the mechanism of what is now familiar to us as the old-fashioned cash register.

Leibnitz, however, went on to do something more. Perhaps as a result of his efforts to mechanize calculation, he thought of its ultimate simplification by inventing the binary system.

Human beings usually use a ten-based system (*decinary*), in which ten different digits (0, 1, 2, 3, 4, 5, 6, 7, 8, 9) are used to represent, in different amounts and combinations, all conceivable numbers. In some cultures, other bases are used (there are five-based systems, twenty-based systems, twelve-based systems, sixty-based systems and so on) but the ten-based is by far the most popular. It undoubtedly arose out of the fact that we happen to have evolved with ten fingers on our two hands.

Leibnitz saw that *any* number could be used as a base and that, in many ways, the simplest to operate mechanically would be a two-based system (*binary*).

The binary notation uses only two digits: 0 and 1. It expresses all numbers in terms of powers of 2. Thus, the number one is 2^0, the number two is 2^1, three is $2^1 + 2^0$, four is 2^2, and so on. As in the decimal system, the power is indicated by the position of the symbol. For instance, the number four is represented by 100, read thus: $(1 \times 2^2) + (0 \times 2^1) + (0 \times 2^0)$, or $4 + 0 + 0 = 4$ in the decimal system.

As an illustration, let us consider the number 6,413. In the decimal system it can be written $(6 \times 10^3) + (4 \times 10^2) + (1 \times 10^1) + (3 \times 10^0)$; remember that any number to the zero power equals 1. Now in the binary system we add numbers in powers of 2, instead of powers of 10, to compose a number. The highest power of 2 that leaves us short of 6,413 is 12; 2^{12} is 4,096. If we now add 2^{11}, or 2,048, we have 6,144, which is 269 short of 6,413. Next, 2^8 adds 256 more, leaving 13; we can then add 2^3, or 8, leaving 5; then 2^2, or 4, leaving 1; and 2^0 is 1. Thus we might write the number 6,413 as $(1 \times 2^{12}) + (1 \times 2^{11}) + (1 \times 2^8) + (1 \times 2^3) + (1 \times 2^2) + (1 \times 2^0)$. But, as in the decimal system, each digit in a number, reading from the left, must represent the next smaller power. Just as in the decimal system we represent the additions of the third, second, first, and zero powers of 10 in stating the number 6,413, so in the binary system we must represent the additions of the powers of 2 from 12 down to 0. In the form of a table this would read:

$$
\begin{aligned}
1 \times 2^{12} &= 4{,}096 \\
1 \times 2^{11} &= 2{,}048 \\
0 \times 2^{10} &= 0 \\
0 \times 2^{9} &= 0 \\
1 \times 2^{8} &= 256 \\
0 \times 2^{7} &= 0 \\
0 \times 2^{6} &= 0 \\
0 \times 2^{5} &= 0 \\
0 \times 2^{4} &= 0 \\
1 \times 2^{3} &= 8 \\
1 \times 2^{2} &= 4 \\
0 \times 2^{1} &= 0 \\
1 \times 2^{0} &= 1 \\
\hline
& 6{,}413
\end{aligned}
$$

Taking the successive multipliers in the column at the left (as we take 6, 4, 1, and 3 as the successive multipliers in the decimal system), we write the number in the binary system as 1100100001101.

This looks pretty cumbersome. It takes 13 digits to write the number 6,413, whereas in the decimal system we need only four. But for a computing machine the system is just about the simplest imaginable. Since there are only two different digits, any operation can be carried out in terms of yes-and-no.

Presumably something as simple as the presence or the absence of a needle in a Jacquard loom can somehow mimic the yes and the no respectively, or the 1 and the 0. With the proper ingenious combinations, one can have the combinations so adjusted as to have $0 + 0 = 0$, $0 + 1 = 1$, $0 \times 0 = 0$; $0 \times 1 = 0$ and $1 \times 1 = 1$. Once such combinations are possible, we can imagine all arithmetical calculations performable on something like a Jacquard loom.

Nor are just ordinary calculations conceivably possible. The system can be widened to include logical statements that we do not often think of as representing arithmetic.

In 1936, the English mathematician Alan Mathison Turing showed that any problem could be solved mechanically if it could be expressed in the form of a finite number of manipulations that could be performed by the machine.

In 1938, an American mathematician and engineer, Claude Elwood Shannon, pointed out in his master's thesis that deductive logic, in a form known as *Boolean algebra,* could be handled by means of the binary system. Boolean algebra refers to a system of *symbolic logic* suggested in 1854 by the English mathematician George Boole in a book entitled *An Investigation of the Laws of Thought.* Boole observed that the types of statement employed in deductive logic could be represented by mathematical symbols, and he went on to show how such symbols could be manipulated according to fixed rules to yield appropriate conclusions.

To take a very simple example, consider the following statement: "Both A and B are true." We are to determine the truth or falsity of this statement by a strictly logical exercise, assuming that we know whether A and B, respectively, are true or false. To handle the problem in binary terms, as Shannon suggested, let 0 represent "false" and 1 represent "true." If A and B are both false, then the statement "Both A and B are true" is false. In other words, 0 and 0 yield 0. If A is true but B is false (or vice versa), then the statement again is false: that is, 1 and 0 (or 0 and 1) yield 0. If A is true and B is true, then the statement "Both A and B are true" is true. Symbolically, 1 and 1 yield 1.

Now these three alternatives correspond to the three possible multiplications in the binary system—namely: $0 \times 0 = 0$, $1 \times 0 = 0$, and $1 \times 1 = 1$. Thus the problem in logic posed by the statement "Both A and B are true" can be manipulated by multiplication. A device (properly programed) therefore can handle this logical problem as easily, and in the same way, as it handles ordinary calculations.

In the case of the statement "Either A or B is true," the problem is handled by addition instead of by multiplication. If neither A nor B is true, then this statement is false. In other words, $0 + 0 = 0$. If A is true and B false, or vice versa, the statement is true; in these cases $1 + 0 = 1$ and $0 + 1 = 1$. If both A and B are true, the statement is certainly true, and $1 + 1 = 10$. (The significant digit in the 10 is the 1; the fact that it is moved over one position is immaterial. In the binary system, 10 represents $(1 \times 2^1) + (0 \times 2^0)$, which is equivalent to 2 in the decimal system.)

Boolean algebra has become important in the engineering of communications and forms part of what is now known as *information theory*.

Artificial Intelligence

The first person who really saw the potentialities of the punch cards of the Jacquard loom was an English mathematician, Charles Babbage. In 1823, he began to design and build a device he called a Difference Engine, and then, in 1836, a more complicated Analytical Engine, but completed neither.

His notions were, in theory, completely correct. He planned to have arithmetical operations carried out automatically by the use of punch cards and then to have the results either printed out or punched out on blank cards. He also planned to give the machine a memory by enabling it to store cards, which had been properly punched out, and then making use of them at later times when called upon to do so.

The engine's physical movements were to be performed by rods, cylinders, gear racks, and geared wheels cut in accordance with the ten-digit decimal system. Bells would tell attendants to feed in certain cards, and louder bells would tell them whether they had inserted a wrong card.

Unfortunately, Babbage, a hot-tempered and eccentric person, periodically tore his machines apart to rebuild them in more complex fashion as new ideas came to him, and he inevitably ran out of money.

Even more important was the fact that the mechanical wheels and levers and gears on which he had to depend were simply not up to the demands he put upon them. The Babbage machines required technology more subtle and responsive than those that sufficed for a Pascal machine, and such a technology had not yet arrived.

For these reasons, Babbage's work petered out and was forgotten for a century. When calculating machines of the Babbage type were eventually constructed successfully, it was because his principles were independently rediscovered.

ELECTRONIC COMPUTERS

A more successful application of punch cards to the task of calculation arose out of the demands of the United States census. The American Constitution directs a census every ten years and a statistical survey of the nation's population and economy proved invaluable. In fact, every ten years, not only did the population and wealth of the nation increase, but the statistical detail demanded increased as well. The result was that, increasingly, it took enormous time to work out all the statistics. By the 1880s, it began to seem that the 1880 census might not be truly complete till the 1890 census was nearly due.

It was then that Herman Hollerith, a statistician with the Census Bureau, worked out a way of recording statistics by a system of the mechanical formation of holes in appropriate positions in cards. The cards themselves were nonconductors, but electrical currents could pass along contacts made through the holes; and in this way, counting and other operations could be carried through automatically by electrical currents—an important, and even crucial, advance on Babbage's purely mechanical devices. Electricity, you see, was up to the job.

Hollerith's electromechanical tabulating machine was successfully used in the U.S. Censuses of 1890 and 1900. The 1890 census of 65 million people took two and a half years to tabulate even with the Hollerith device. By 1900, he had improved his machines, however, so that cards could be automatically fed through brushes for reading, and the new and larger 1900 census had its count completed in a little over one and a half years.

Hollerith founded a firm that later became International Business Machines (IBM). The new company, and Remington Rand, under the leadership of Hollerith's assistant, John Powers, steadily improved the system of electromechanical computations over the next thirty years.

They had to.

The world economy, with advancing industrialization, was steadily becoming more complex; and, increasingly, the only way to run the world successfully was to know more and more about the details of the statistics involved, of numbers, of information. The world was becoming an *information society*, and it would collapse under its own weight if humanity did not learn to collect, understand, and respond to the information quickly enough.

It was this sort of unforgiving pressure, of *having* to handle increasing quantities of information, that drove society forward toward the invention of successively more subtle, variegated, and capacious computing devices throughout the twentieth century.

Electromechanical machines became faster and were used through the Second World War, but their speed and reliability was limited as long as they depended on moving parts like switching relays and on electromagnets that controlled counting wheels.

In 1925, the American electrical engineer Vannevar Bush and his colleagues constructed a machine capable of solving differential equations. It could do what Babbage had hoped to do with his machine, and was the first successful instrument that we would today call a *computer*. It was electromechanical.

Also electromechanical, but even more impressive, was a machine designed in 1937 by Howard Aiken, of Harvard, working with IBM. The machine, the IBM Automatic Sequence Controlled Calculator, known at Harvard as Mark I, was completed in 1944 and was intended for scientific applications. It could perform mathematical operations involving up to twenty-three decimal places. In other words, two eleven-digit numbers could be multiplied, correctly, in three seconds. It was electromechanical; and since it dealt primarily with the manipulation of numbers, it is the first modern *digital computer.* (Bush's device solved problems by converting numbers into lengths, as a slide rule does; and because it used *analogous quantities,* not numbers themselves, it was an *analog computer.*)

For complete success, however, the switches in such computers had to be electronic. Mechanical interruption and reinstatement of electric currents, while far superior to wheels and gears, was still clumsy and slow, to say nothing of unreliable. In electronic devices, such as radio tubes, the electron flow could be manipulated far more delicately, accurately, and speedily, and it was this which was the next step.

The first large electronic computer, containing 19,000 vacuum tubes, was built at the University of Pennsylvania by John Presper Eckert and John William Mauchly during the Second World War. It was called ENIAC, for *E*lectronic *N*umerical *I*ntegrator *a*nd *C*omputer. ENIAC ceased operation in 1955 and was dismantled in 1957, a hopelessly outmoded dotard at twelve years of age, but it left behind an amazingly numerous and sophisticated progeny. Whereas ENIAC weighed 30 tons and took up 1,500 square feet of floor space, the equivalent computer thirty years later—using switching units far smaller, faster, and more reliable than the old vacuum tubes—could be built into an object the size of a refrigerator.

So fast was progress that by 1948, small electronic computers were being produced in quantity; within five years, 2,000 were in use; by 1961, the number was 10,000. By 1970, the number had passed the 100,000 mark, and that was scarcely a beginning.

The reason for the rapid advance was that although electronics was the answer, the vacuum tube was not. It was large, fragile, and required a great deal of energy. In 1948, the transistor (see chapter 9) was invented; and thanks to such solid-state devices, electronic control could be carried through sturdily, compactly, and with trivial expenditure of energy.

Computers shrank and grew cheap even as they increased their capacity and versatility enormously. In the generation after the invention of the transistor, new ways were found, in rapid succession, to squeeze ever more information capacity and memory into smaller and smaller bits of solid-state devices. In the 1970s, the *microchip* came into its own—a tiny bit of silicon on which numbers of circuits were etched under a microscope.

The result was that computers became affordable to private individuals of no great wealth. It may be that the 1980s will see the proliferation of *home computers* as the 1950s saw the proliferation of home television sets.

The computers that came into use after the Second World War already seemed to be "thinking machines" to the general public, so that both scientists and laypeople began to think of the possibilities, and consequences, of *artificial intelligence,* a term first used in 1956 by an M.I.T. computer engineer, John McCarthy.

How much more so when, in just forty years, computers have become giants without which our way of life would collapse. Space exploration would be impossible without computers. The space shuttle could not fly without them. Our war machine would collapse into Second World War weaponry without them. No industry of any size, scarcely any office, could continue as presently constituted without them. The government (including particularly the Internal Revenue Service) would become even more helpless than it ordinarily is without them.

And consequently new uses are being worked out for them. Aside from solving problems, doing graphics, storing and retrieving data, and so on, they can be bent to trivial tasks. Some can be programed to play chess with near-master ability, while some can be used for games of all kinds that by the 1980s had caught the imagination of the younger public to the tune of billions of dollars. Computer engineers are laboring to improve the ability of computers to translate from one language to another, and to give them the ability to read, to hear, and speak.

ROBOTS

The question arises, inevitably, is there anything computers can, in the end, not do? Are they not, inevitably, going to do anything we can imagine? For instance, can a computer of the proper sort somehow be inserted into a structure resembling the human body, so that we can finally have true automata—not the toys of the seventeenth century, but artificial human beings with a substantial fraction of the abilities of human beings.

Such matters were considered quite seriously by science-fiction writers even before the first modern computers were built. In 1920, a Czech playwright, Karel Capek, wrote *R.U.R.*, a play in which automata are mass-produced by an Englishman named Rossum. The automata are meant to do the world's work and to make a better life for human beings; but in the end they rebel, wipe out humanity, and start a new race of intelligent life themselves.

Rossum comes from a Czech word, *rozum*, meaning "reason"; and *R.U.R.* stands for "Rossum's Universal Robots," where *robot* is a Czech word for "worker," with the implication of involuntary servitude, so that it might be translated as "serf" or "slave." The popularity of the play threw the old term automaton out of use. *Robot* has replaced it in every language, so that now a robot is commonly thought of as any artificial device (often pictured in at least vaguely human form) that will perform functions ordinarily thought to be appropriate for human beings.

On the whole, though, science fiction writers did not treat robots realistically but used them as cautionary objects, as villains or heroes designed to point up the human condition.

In 1939, however, Isaac Asimov,* only nineteen at the time, tiring of robots that were either unrealistically wicked or unrealistically noble, began to devote some of the science-fiction stories he was publishing to robots that were viewed

*Yes, the author of this book.

merely as machines and built, as all machines are, with some rational attempt at adequate safeguards. Throughout the 1940s, he published stories of this sort; and in 1950, nine of them were collected into a book entitled *I, Robot.*

Asimov's safeguards were formalized as the "Three Laws of Robotics." The phrase was first used in a story published in March 1942, and that was the very first known use of the word *robotics,* the now-accepted term for the science and technology of the design, construction, maintenance and use of robots.

The three rules are:

1. A robot may not injure a human being, or, through inaction, allow a human being to come to harm.

2. A robot must obey the orders given it by human beings except where such orders would conflict with the First Law.

3. A robot must protect its own existence as long as such protection does not conflict with the First or Second Law.

What Asimov did was, of course, purely speculative and could, at best, only serve as a source of inspiration. The real work was being done by scientists in the field.

Partly, this was being done through the pressures of the Second World War. The application of electronics made it possible to endow weapons with a sensitivity and swiftness of response even beyond the capabilities of a living organism. Furthermore, radio extended their sphere of action over a considerable distance. The German buzz bomb of the war was essentially a flying *servomechanism,* and it introduced the possibility not only of guided missiles but also of self-operated or remotely operated vehicles of all sorts, from subway trains to space ships. Because the military establishments had the keenest interest in these devices, and the most abundant supply of funds, servomechanisms have reached perhaps their highest development in aiming-and-firing mechanisms for guns and rockets. These systems can detect a swiftly moving target hundreds of miles away, instantly calculate its course (taking into account the target's speed of motion, the wind, the temperatures of the various layers of air, and numerous other conditions), and hit the target with pinpoint accuracy, all without any human guidance.

Automation found an ardent theoretician and advocate in the mathematician Norbert Wiener, who worked on such targeting problems. In the 1940s, he and his group at the Massachusetts Institute of Technology worked out some of the fundamental mathematical relationships governing the handling of feedback. He named this branch of study *cybernetics,* from the Greek word for "helmsman," which seems appropriate, since the first use of servomechanisms was in connection with a helmsman. (Cybernetics also harks back to Watt's centrifugal governor, for *governor* comes from the Latin word for "helmsman.")

This was the first important book to be devoted entirely to theory of computer control, and cybernetic principles made it possible to build, if not a robot, then at least systems that utilized these principles to mimic the behavior of simple animals.

The British neurologist William Grey Walter, for instance, built a device in the 1950s that explores and reacts to its surroundings. His turtlelike object, which he calls a *testudo* (Latin for "tortoise"), has a photoelectric cell for an eye, a sensing device to detect touch, and two motors—one to move forward or backward, and the other to turn around. In the dark, it crawls about, circling in a wide arc. When it touches an obstacle, it backs off a bit, turns slightly and moves forward again; it will do this until it gets around the obstacle. When its photoelectric eye sees a light, the turning motor shuts off and the testudo advances straight toward the light. But its phototropism is under control; as it gets close to the light, the increase in brightness causes it to back away, so that it avoids the mistake of the moth. When its batteries run down, however, the now "hungry" testudo can crawl close enough to the light to make contact with a recharger placed near the light bulb. Once recharged, the testudo is again sensitive enough to back away from the bright area around the light.

And yet neither can we entirely underplay the influence of inspiration. In the early 1950s, a Columbia undergraduate, Joseph F. Engelberger, read Asimov's *I, Robot* and was, as a result, infected with a life-long enthusiasm for work with robots.

In 1956, Engelberger met George C. Devol, Jr., who, two years before, had obtained the first patent for an industrial robot. He called its control and computer memory system *universal automation*—or *unimation,* for short.

Together, Engelberger and Devol founded Unimation, Inc., and Devol then developed thirty to forty related patents.

None of these were really practical, because the robots could not really do their work unless they were computerized; and computers were too bulky and expensive to make robots competitive enough for any tasks. It was only with the development of the microchip that the robot designs of Unimation became attractive in the marketplace. Unimation quickly became the most important and most profitable robotics firm in the world.

With that began the era of the *industrial robot.* The industrial robot does not have the appearance of the classical robot; there is nothing obviously humanoid about it. It is essentially a computerized arm, which can perform simple operations with great precision and which possesses, because of its computerization, a certain flexibility.

Industrial robots have found their greatest use so far on assembly lines (particularly those in Japan along which automobiles are assembled). For the first time, we have machines that are complex enough and "talented" enough to do jobs that until now required human judgment—but so little human judgment that the human brain, caught in the necessity of doing a repetitious and stultifying job does not reach anything near its potential and is probably damaged as a result.

It is clearly useful to have machines do jobs that are insufficient for the human brain (though too much for anything short of robots) and thus leave human beings the possibility of devoting themselves to more creative labors that will stretch and expand their minds.

Already, however, the use of industrial robots is showing uncomfortable side effects in the short term. Human workers are being replaced. We are probably headed for a painful transition period during which society will be faced with the problem of taking care of the new unemployed; of re-educating or retraining them to do other work; or, where that is impossible, of finding some useful occupation they can do; or, where all else fails, of simply supporting them.

Presumably, as time passes, a new generation educated to be part of a computerized, robotized society will come into being, and matters will improve.

And yet technology will continue to advance. There is a strong push in favor of developing robots with greater abilities, with more flexibility, with the ability to "see," "speak," "hear." What's more, *home robots* are being developed—robots of more humanoid appearance which can be useful about the house and do some of the functions classically assigned to human servants. (Joseph Engelberger has a prototype of such a device which he hopes before long to introduce into his home: something that will be capable of accepting coats, passing out drinks, and performing other simple tasks. He calls it Isaac.)

Can we help but wonder whether computers and robots may not eventually replace *any* human ability? Whether they may not replace human beings by rendering them obsolete? Whether artificial intelligence, of our own creation, is not fated to be our replacement as dominant entities on the planet?

One might be fatalistic about this. If it is inevitable, then it is inevitable. Besides, the human record is not a good one, and we are in the process, perhaps, of destroying ourselves (along with much of life) in any case. Perhaps it is not computer replacement we should fear, but the possibility that it will not come along quickly enough.

We might even feel triumphant about it. What achievement could be grander than the creation of an object that surpasses the creator? How could we consummate the victory of intelligence more gloriously than by passing on our heritage, in triumph, to a greater intelligence—of our own making?

But let us be practical. Is there really danger of replacement?

In the first place, we must ask whether intelligence is a one-dimensional variant, or whether there may not be qualitatively different kinds of intelligence, even very many different kinds. If dolphins have intelligence similar to ours, for instance, it seems nevertheless to be of so different a nature from our own that we have not yet succeeded in establishing communication across the species line. Computers may, in the end, differ from us qualitatively also. It would certainly not be surprising if that were so.

After all, the human brain, built of nucleic acid and protein against a watery background, has been the product of the development of three and a half billion years of biological evolution, based on the random effects of mutation, natural selection, and other influences, and driven forward by the necessity of survival.

The computer, on the other hand, built of electronic switches and electric current against a semiconductor background, has been the product of the development of forty years of human design, based on the careful foresight and ingenuity of human beings, and driven forward by the necessity of serving its human users.

When two intelligences are so different in structure, history, development, and purpose, it would certainly not be surprising if their intelligences were widely different in nature as well.

From the very start, for instance, computers were capable of solving complex problems involving arithmetical operations upon numbers, of doing so with far greater speed than any human being could, and with far less chance of error. If arithmetical skill is the measure of intelligence, then computers have been more intelligent than human beings all along.

But it may be that arithmetical skill and other similar talents are not at all what the human brain is primarily designed for—that such things, not being our metier, we naturally do very poorly.

It may be that the measure of human intelligence involves such subtle qualities as insight, intuition, fantasy, imagination, creativity—the ability to view a problem as a whole and guess the answer by the "feel" of the situation. If that is so, then human beings are very intelligent, and computers are very unintelligent indeed. Nor can we see right now how this deficiency in computers can be easily remedied, since human beings cannot program a computer to be intuitive or creative for the very good reason that we do not know what we ourselves do when we exercise these qualities.

Might we someday learn how to program computers into a display of human intelligence of this sort?

Conceivably; but in that case we might choose not do so out of a natural reluctance to be replaced. Besides, what would be the point of duplicating human intelligence—of building a computer that might glow with a faint humanity—when we can so easily form the real thing by ordinary biological processes? It would be much like training human beings from infancy to perform "mathematical marvels" similar to those a computer can do. Why, when the cheapest calculating device will do it for us?

It would surely pay us to continue to develop two intelligences that were differently specialized, so that different functions could be performed with the highest efficiency. We might even imagine numerous classes of computers with different types of intelligence. And, by the use of genetic engineering methods (and the help of computers), we might even develop varieties of human brains displaying different species of human intelligences.

With intelligences of different species and genera, there is the possibility at least of a symbiotic relationship, in which all will cooperate to learn how best to understand the laws of nature and how most benignly we might cooperate with them. Certainly, the cooperation will do better than any intelligence variety on its own.

Viewed in this fashion, the robot/computer will not replace us but will serve us as our friend and ally in the march toward the glorious future—if we do not destroy ourselves before the march can begin.

Appendix

Mathematics
in Science

Gravitation

As I explained in chapter 1, Galileo initiated science in its modern sense
by introducing the concept of reasoning back from observation and experiment to basic principles. In doing so, he also introduced the essential technique of measuring natural phenomena accurately and abandoned the practice
of merely describing them in general terms. In short, he turned from the
qualitative description of the universe by the Greek thinkers to a quantitative
description.

Although science depends so much on mathematical relationships and
manipulations, and could not exist in the Galilean sense without it, I have
nevertheless written this book nonmathematically, and have done so deliberately. Mathematics, after all, is a highly specialized tool. To have discussed the
developments in science in mathematical terms would have required a prohibi-
tive amount of space, as well as a sophisticated knowledge of mathematics on
the part of the reader. But in this appendix, I would like to present an example
or two of the way in which simple mathematics has been fruitfully applied to
science. How better to begin than with Galileo himself?

THE FIRST LAW OF MOTION

Galileo (like Leonardo da Vinci nearly a century earlier) suspected that falling objects steadily increase their velocity as they fall. He set out to measure exactly by how much and in what manner the velocity increases.

The measurement was anything but easy for Galileo, with the tools he had at his disposal in 1600. To measure a velocity requires the measurement of time. We speak of velocities of 60 miles *an hour,* of 13 feet *a second.* But there were no clocks in Galileo's time that could do more than strike the hour at approximately equal intervals.

Galileo resorted to a crude water clock. He let water trickle slowly from a small spout, assuming, hopefully, that it dripped at a constant rate. This water he caught in a cup; and, by the weight of water caught during the interval in which an event took place, Galileo measured the elapsed time. (He also used his pulse beat for the purpose on occasion.)

One difficulty was, however, that a falling object dropped so rapidly that Galileo could not collect enough water, in the interval of falling, to weigh accurately. What he did, then, was to *dilute* the pull of gravity by having a brass ball roll down a groove in an inclined plane. The more nearly horizontal the plane, the more slowly the ball moved. Thus Galileo was able to study falling bodies in whatever degree of *slow motion* he pleased.

Galileo found that a ball rolling on a perfectly horizontal plane moves at constant speed. (This supposes a lack of friction, a condition that could be assumed within the limits of Galileo's crude measurements.) Now a body moving on a horizontal track is moving at right angles to the force of gravity. Under such conditions, the body's velocity is not affected by gravity either way. A ball resting on a horizontal plane remains at rest, as anyone can observe. A ball set to moving on a horizontal plane moves at a constant velocity, as Galileo observed.

Mathematically, then, it can be stated that the velocity v of a body, *in the absence of any external force,* is constant k, or:

$$v = k.$$

If k is equal to any number other than zero, the ball is moving at constant velocity. If k is equal to zero, the ball is at rest; thus, rest is a "special case" of constant velocity.

Nearly a century later, when Newton systemized the discoveries of Galileo in connection with falling bodies, this finding became the First Law of Motion (also called the *principle of inertia*). This law can be stated: Every body persists in a state of rest or of uniform motion in a straight line unless compelled by external force to change that state.

When a ball rolls down an inclined plane, however, it is under the continuous pull of gravity. Its velocity then, Galileo found, is not constant but increases with time. Galileo's measurements showed that the velocity increases in proportion to the lapse of time t.

In other words, when a body is under the action of constant external force, its velocity, starting at rest, can be expressed as:

$$v = kt.$$

What is the value of k?

That, it was easy to find by experiment, depends on the slope of the inclined plane. The more nearly vertical the plane, the more quickly the rolling ball gains velocity and the higher the value of k. The maximum gain in speed comes when the plane is vertical—in other words, when the ball drops freely under the undiluted pull of gravity. The symbol g (for "gravity") is used where the undiluted force of gravity is acting, so that the velocity of a ball in free fall, starting from rest, was:

$$v = gt.$$

Let us consider the inclined plane in more detail. In the diagram:

the length of the inclined plane is AB, while its height at the upper end is AC. The ratio of AC to AB is the sine of the angle x, usually abbreviated as "sin x."

The value of this ratio—that is, of sin x—can be obtained approximately by constructing triangles with particular angles and actually measuring the height and length involved. Or it can be calculated by mathematical techniques to any degree of precision, and the results can be embodied in a table. By using such a table, we can find, for instance, that sin 10° is approximately equal to 0.17365, that sin 45° is approximately equal to 0.70711, and so on.

There are two important special cases. Suppose that the "inclined" plane is precisely horizontal. Angle x is then zero, and as the height of the inclined plane is zero, the ratio of its height to its length is also zero. In other words, sin 0° = 0. When the "inclined" plane is precisely vertical, the angle it forms with the ground is a right angle, or 90°. Its height is then exactly equal to its length, so that the ratio of one to the other is just 1. Consequently, sin 90° = 1.

Now let us return to the equation showing that the velocity of a ball rolling down an inclined plane is proportional to time:

$$v = kt.$$

It can be shown by experiment that the value of k changes with the sine of the angle so that:

$$k = k' \sin x$$

(where k' is used to indicate a constant that is different from k).

(As a matter of fact, the role of the sine in connection with the inclined plane was worked out somewhat before Galileo's time by Simon Stevinus, who also performed the famous experiment of dropping different masses from a height—an experiment traditionally, but wrongly, ascribed to Galileo. Still, if Galileo was not the very first to experiment and measure, he was the first to impress the scientific world, indelibly, with the necessity to experiment and measure, and that is glory enough.)

In the case of a completely vertical inclined plane, sin x becomes sin 90°, which is 1, so that in free fall

$$k = k'.$$

It follows that k' is the value of k in free fall under the undiluted pull of gravity, which we have already agreed to symbolize as g. We can substitute g for k' and, for any inclined plane:

$$k = g \sin x.$$

The equation for the velocity of a body rolling down an inclined plane is, therefore:

$$v = (g \sin x) \, t.$$

On a horizontal plane with sin $x = 0° = 0$, the equation for velocity becomes:

$$v = 0.$$

This is another way of saying that a ball on a horizontal plane, starting from rest, will remain motionless regardless of the passage of time. An object at rest tends to remain at rest, and so on. That is part of the First Law of Motion, and it follows from the inclined plane equation of velocity.

Suppose that a ball does not start from rest but has an initial motion before it begins to fall. Suppose, in other words, you have a ball moving along a horizontal plane at 5 feet per second, and it suddenly finds itself at the upper end of an inclined plane and starts rolling downward.

Experiment shows that its velocity thereafter is 5 feet per second greater, at every moment, than it would have been if it had started rolling down the plane from rest. In other words, the equation for the motion of a ball down an inclined plane can be expressed more completely as follows:

$$v = (g \sin x) \, t + V$$

where V is the original starting velocity. If an object starts at rest, then V is equal to 0 and the equation becomes as we had it before:

$$v = (g \sin x) \, t.$$

If we next consider an object with some initial velocity on a horizontal plane, so that angle x is 0°, the equation becomes:

$$v = (g \sin 0°) + V$$

or, since sin 0° is 0:

$$v = V.$$

Thus the velocity of such an object remains its initial velocity, regardless of the lapse of time. That is the rest of the First Law of Motion, again derived from observed motion on an inclined plane.

The rate at which velocity changes is called *acceleration*. If, for instance, the velocity (in feet per second) of a ball rolling down an inclined plane is, at the end of successive seconds, 4, 8, 12, 16 . . . then the acceleration is 4 feet per second per second.

In a free fall, if we use the equation:

$$v = gt,$$

each second of fall brings an increase in velocity of g feet per second. Therefore, g represents the acceleration due to gravity.

The value of g can be determined from inclined-plane experiments. By transposing the inclined-plane equation, we get:

$$g = v \, / \, (t \sin x).$$

Since v, t, and x can all be measured, g can be calculated, and it turns out to be equal to 32 feet per second per second at the earth's surface. In free fall under normal gravity at earth's surface, then, the velocity of fall is related to time thus:

$$v = 32t.$$

This is the solution to Galileo's original problem—namely, determining the rate of fall of a falling body and the manner in which that rate changes.

The next question is: How far does a body fall in a given time? From the equation relating the velocity to time, it is possible to relate distance to time by the process in calculus called *integration*. It is not necessary to go into that, however, because the equation can be worked out by experiment; and, in essence, Galileo did this.

He found that a ball rolling down an inclined plane covers a distance proportional to the square of the time. In other words, doubling the time increases the distance fourfold; tripling it increases the distance ninefold; and so on.

For a freely falling body, the equation relating distance d and time is:

$$d = \tfrac{1}{2}gt^2$$

or, since g is equal to 32:

$$d = 16t^2.$$

Next, suppose that instead of dropping from rest, an object is thrown horizontally from a position high in the air. Its motion would then be a compound of two motions—a horizontal one and a vertical one.

The horizontal motion, involving no force other than the single original impulse (if we disregard wind, air resistance, and so on), is one of constant velocity, in accordance with the First Law of Motion, and the distance the object covers horizontally is proportional to the time elapsed. The vertical motion, however, covers a distance, as I have just explained, that is proportional to the square of the time elapsed. Prior to Galileo, it had been vaguely believed that a projectile such as a cannon ball travels in a straight line until the impulse that drives it is somehow exhausted, after which it falls straight down. Galileo, however, made the great advance of *combining* the two motions.

The combination of these two motions (proportional to time horizontally, and proportional to the square of the time vertically) produces a curve called a *parabola*. If a body is thrown, not horizontally, but upward or downward, the curve of motion is still a parabola.

Such curves of motion, or trajectories, apply, of course, to a projectile such as a cannon ball. The mathematical analysis of trajectories, stemming from Galileo's work, made it possible to calculate where a cannon ball would fall when fired with a given propulsive force and a given angle of elevation of the cannon. Although people had been throwing objects for fun, to get food, to attack, and to defend, for uncounted thousands of years, it was only due to Galileo that for the first time, thanks to experiment and measurement, there was a science of *ballistics*. As it happened, then, the very first achievement of modern experimental science proved to have a direct and immediate military application.

It also had an important application in theory. The mathematical analysis of combinations of more than one motion answered several objections to the Copernican theory. It showed that an object thrown upward will not be left behind by the moving earth, since the object will have two motions: one imparted to it by the impulse of throwing, and one that it shares along with the moving earth. This analysis also made it reasonable to expect the earth to have two motions at once: rotation about its axis and revolution about the sun —a situation that some of the non-Copernicans insisted was unthinkable.

THE SECOND AND THIRD LAWS

Isaac Newton extended the Galilean concepts of motion to the heavens and showed that the same set of laws of motion apply to the heavens and the earth alike.

He began by considering that the moon might be falling toward the earth in response to the earth's gravity but never struck the earth's surface because of the horizontal component of its motion. A projectile fired horizontally, as I said, follows a parabolically curved path downward to intersection with the earth's surface. But the earth's surface curves downward, too, since the earth is a sphere. A projectile given a sufficiently rapid horizontal motion might curve downward no faster than the earth's surface and would therefore eternally circle the earth.

Now the moon's elliptical motion around the earth can be split into horizontal and vertical components. The vertical component is such that, in the space of a second, the moon falls a trifle more than 1/20 inch toward the earth. In that time, it also moves about 3,300 feet in the horizontal direction, just far enough to compensate for the fall and carry it around the earth's curvature.

The question was whether this 1/20-inch fall of the moon is caused by the same gravitational attraction that causes an apple, falling from a tree, to drop 16 feet in the first second of its fall.

Newton visualized the earth's gravitational force as spreading out in all directions like a vast, expanding sphere. The surface area A of a sphere is proportional to the square of its radius r:

$$A = 4\pi r^2.$$

He therefore reasoned that the gravitational force, spreading out over the spherical area, must weaken as the square of the radius. The intensity of light and of sound weakens as the square of the distance from the source. Why not the force of gravity as well?

The distance from the earth's center to an apple on its surface is roughly 4,000 miles. The distance from the earth's center to the moon is roughly 240,000 miles. Since the distance to the moon was 60 times greater than to the apple, the force of the earth's gravity at the moon must be 60^2, or 3,600, times weaker than at the apple. Divide 16 feet by 3,600, and you come out with roughly 1/20 of an inch. It seemed clear to Newton that the moon does indeed move in the grip of the earth's gravity.

Newton was persuaded further to consider *mass* in relation to gravity. Ordinarily, we measure mass as weight. But weight is only the result of the attraction of the earth's gravitational force. If there were no gravity, an object would be weightless; nevertheless, it would still contain the same amount of matter. Mass, therefore, is independent of weight and should be capable of measurement by a means not involving weight.

Suppose you tried to pull an object on a perfectly frictionless surface in a direction horizontal to the earth's surface, so that there was no resistance from gravity. It would take effort to set the body in motion and to accelerate its motion, because of the body's inertia.

If you measured the applied force accurately—say, by pulling on a spring balance attached to the object—you would see that the force f required to bring about a given acceleration a would be directly proportional to the mass m. If you doubled the mass, it would take double the force. For a given mass, the force required would be directly proportional to the acceleration desired. Mathematically, this is expressed in the equation:

$$f = ma.$$

The equation is known as Newton's Second Law of Motion.

Now, as Galileo had found, the pull of the earth's gravity accelerates all bodies, heavy or light, at precisely the same rate. (Air resistance may slow the fall of very light bodies; but in a vacuum, a feather will fall as rapidly as a lump of lead, as can easily be demonstrated.) If the Second Law of Motion is to hold, one must conclude that the earth's gravitational pull on a heavy body must be greater than on a light body, in order to produce the same acceleration. To accelerate a mass that is eight times as great as another, for instance, takes eight times as much force. It follows that the earth's gravitational pull on any body must be exactly proportional to the mass of that body. (That, in fact, is why mass on the earth's surface can be measured quite accurately as weight.)

Newton evolved a Third Law of Motion, too: "For every action there is an equal and opposite reaction." This law applies to force. In other words, if the earth pulls at the moon with a certain force, then the moon pulls on the earth with an equal force. If the moon were suddenly doubled in mass, the earth's gravitational force upon it would also be doubled, in accordance with the Second Law; of course, the moon's gravitational force on the earth would then have to be doubled in accordance with the Third Law.

Similarly, if it were the earth rather than the moon that doubled in mass, it would be the moon's gravitational force on the earth that would double, according to the Second Law, and the earth's gravitational force on the moon that would double, in accordance with the Third.

If both the earth and the moon were to double in mass, there would be a doubled doubling, each body doubling its gravitational force twice, for a fourfold increase all told.

Newton could only conclude, by this sort of reasoning, that the gravitational force between any two bodies in the universe was directly proportional to the product of the masses of the bodies. And, of course, as he had decided earlier, it is inversely proportional to the square of the distance (center to center) between the bodies. This is Newton's Law of Universal Gravitation.

If we let f represent the gravitational force, m_1 and m_2 the masses of the two bodies concerned, and d the distance between them, then the law can be stated:

$$f = \frac{Gm_1m_2}{d^2}.$$

G is the *gravitational constant,* the determination of which made it possible to "weigh the earth" (see chapter 4). It was Newton's surmise that G has a fixed value throughout the universe. As time went on, it was found that new planets, undiscovered in Newton's time, temper their motions to the requirements of Newton's law; even double stars incredibly far away dance in time to Newton's analysis of the universe.

All this came from the new quantitative view of the universe pioneered by Galileo. As you see, much of the mathematics involved was really very simple. Those parts of it I have quoted here are high-school algebra.

In fact, all that was needed to introduce one of the greatest intellectual revolutions of all time was:

1. A simple set of observations any high-school student of physics might make with a little guidance.

2. A simple set of mathematical generalizations at high school level.

3. The transcendent genius of Galileo and Newton, who had the insight and originality to make these observations and generalizations for the first time.

Relativity

The laws of motion as worked out by Galileo and Newton depended on the assumption that such a thing as absolute motion exists—that is, motion with reference to something at rest. But everything that we know of in the universe is in motion: the earth, the sun, our galaxy, the systems of galaxies. Where in the universe, then, can we find absolute rest against which to measure absolute motion?

THE MICHELSON-MORLEY EXPERIMENT

It was this line of thought that led to the Michelson-Morley experiment, which in turn led to a scientific revolution as great, in some respects, as that initiated by Galileo (see chapter 8). Here, too, the basic mathematics is rather simple.

The experiment was an attempt to detect the absolute motion of the earth against an *ether* that was supposed to fill all space and to be at rest. The reasoning behind the experiment was as follows.

Suppose that a beam of light is sent out in the direction in which the earth is traveling through the ether; and that at a certain distance in that direction, there is a fixed mirror which reflects the light back to the source. Let us symbolize the velocity of light as c, the velocity of the earth through the ether as v, and the distance of the mirror as d. The light starts with the velocity $c + v$: its own velocity plus the earth's velocity. (It is traveling with a tail wind, so to speak.) The time it takes to reach the mirror is d divided by $(c + v)$.

On the return trip, however, the situation is reversed. The reflected light now is bucking the head wind of the earth's velocity, and its net velocity is $c - v$. The time it takes to return to the source is d divided by $(c - v)$.

The total time for the round trip is:

$$\frac{d}{c + v} + \frac{d}{c - v}.$$

Combining the terms algebraically, we get:

$$\frac{d(c - v) + d(c + v)}{(c + v)(c - v)} = \frac{dc - dv + dc + dv}{c^2 - v^2} = \frac{2dc}{c^2 - v^2}.$$

Now suppose that the light-beam is sent out to a mirror at the same distance in a direction at right angles to the earth's motion through the ether.

The beam of light is aimed from S (the source) to M (the mirror) over the distance d. However, during the time it takes the light to reach

direction
of earth's
motion

the mirror, the earth's motion has carried the mirror from M to M', so that the actual path traveled by the light beam is from S to M'. This distance we call x, and the distance from M to M' we call y (see diagram above).

While the light is moving the distance x at its velocity c, the mirror is moving the distance y at the velocity of the earth's motion v. Since both the light and the mirror arrive at M' simultaneously, the distances traveled must be exactly proportional to the respective velocities. Therefore:

$$\frac{y}{x} = \frac{v}{c}$$

or:

$$y = \frac{vx}{c}.$$

Now we can solve for the value of x by use of the Pythagorean theorem, which states that the sum of the squares of the sides of a right triangle is equal to the square of the hypotenuse. In the right triangle SMM', then, substituting vx/c for y:

$$x^2 = d^2 + \left(\frac{vx}{c}\right)^2$$

$$x^2 - \left(\frac{vx}{c}\right)^2 = d^2$$

$$x^2 - \frac{v^2 x^2}{c^2} = d^2$$

$$\frac{c^2 x^2 - v^2 x^2}{c^2} = d^2$$

$$(c^2 - v^2)x^2 = d^2c^2$$

$$x^2 = \frac{d^2c^2}{c^2 - v^2}$$

$$x = \frac{dc}{\sqrt{c^2 - v^2}}.$$

The light is reflected from the mirror at M' to the source, which meanwhile has traveled on to S'. Since the distance $S'S''$ is equal to SS', the distance $M'S''$ is equal to x. The total path traveled by the light beam is therefore $2x$, or $2dc/\sqrt{c^2 - v^2}$.

The time taken by the light beam to cover this distance at its velocity c is:

$$\frac{2dc}{\sqrt{c^2 - v^2}} \div c = \frac{2d}{\sqrt{c^2 - v^2}}.$$

How does this compare with the time that light takes for the round trip in the direction of the earth's motion? Let us divide the time in the parallel case $(2dc/(c^2 - v^2)$ by the time in the perpendicular case $(2d/\sqrt{c^2 - v^2})$:

$$\frac{2dc}{c^2 - v^2} \div \frac{2d}{\sqrt{c^2 - v^2}} = \frac{2dc}{c^2 - v^2} \times \frac{\sqrt{c^2 - v^2}}{2d} = \frac{c\sqrt{c^2 - v^2}}{c^2 - v^2}.$$

Now any number divided by its square root gives the same square root as a quotient, that is, $x/\sqrt{x} = \sqrt{x}$. Conversely, $\sqrt{x}/x = 1/\sqrt{x}$. So the last equation simplifies to:

$$\frac{c}{\sqrt{c^2 - v^2}}.$$

This expression can be further simplified if we multiply both the numerator and the denominator by $\sqrt{1/c^2}$ (which is equal to $1/c$):

$$\frac{c\sqrt{1/c^2}}{\sqrt{c^2 - v^2}\sqrt{1/c^2}} = \frac{c/c}{\sqrt{c^2/c^2 - v^2/c^2}} = \frac{1}{\sqrt{1 - v^2/c^2}}.$$

And there you are. That is the ratio of the time that light should take to travel in the direction of the earth's motion as compared with the time it should take in the direction perpendicular to the earth's motion. For any value of v greater than zero, the expression $1/\sqrt{1 - v^2/c^2}$ is greater than 1. Therefore, if the earth is moving through a motionless ether, it should take longer for light to travel in the direction of the earth's motion than in the perpendicular direction. (In fact, the parallel motion should take the maximum time and the perpendicular motion the minimum time.)

Michelson and Morley set up their experiment to try to detect the directional difference in the travel time of light. By trying their beam of light in all directions, and measuring the time of return by their incredibly delicate interferometer, they felt they ought to get differences in apparent velocity. The direction in which they found the velocity of light to be at a minimum should be parallel to the earth's absolute motion, and the direction in which the velocity would be at a maximum should be perpendicular to the earth's motion. From the difference in velocity, the amount (as well as the direction) of the earth's absolute motion could be calculated.

They found no differences at all in the velocity of light with changing direction! To put it another way, the velocity of light was always equal to c, regardless of the motion of the source—a clear contradiction of the Newtonian laws of motion. In attempting to measure the absolute motion of the earth, Michelson and Morley had thus managed to cast doubt not only on the existence of the ether, but on the whole concept of absolute rest and absolute motion, and upon the very basis of the Newtonian system of the universe.

THE FITZGERALD EQUATION

The Irish physicist G. F. FitzGerald conceived a way to save the situation. He suggested that all objects decrease in length in the direction in which they are moving by an amount equal to $\sqrt{1 - v^2/c^2}$. Thus:

$$L' = L \sqrt{1 - v^2/c^2}$$

where L' is the length of a moving body in the direction of its motion and L is what the length would be if it were at rest.

The foreshortening fraction $\sqrt{1 - v_2/c^2}$, FitzGerald showed, would just cancel the ratio $1 \sqrt{- v^2/c^2}$, which related the maximum and minimum velocities of light in the Michelson-Morley experiment. The ratio would become unity, and the velocity of light would seem to our foreshortened instruments and sense organs to be equal in all directions, regardless of the movement of the source of light through the ether.

Under ordinary conditions, the amount of foreshortening is very small. Even if a body were moving at one-tenth the velocity of light, or 18,628 miles per second, its length would be foreshortened only slightly, according to the FitzGerald equation. Taking the velocity of light as 1, the equation says:

$$L' = L\sqrt{\left(1 - \frac{0.1}{1}\right)^2}$$
$$L' = L\sqrt{1 - 0.01}$$
$$L' = L\sqrt{0.99}.$$

Thus L' turns out to be approximately equal to $0.995L$, a foreshortening of about half of 1 percent.

For moving bodies, velocities such as this occur only in the realm of the subatomic particles. The foreshortening of an airplane traveling at 2,000 miles per hour is infinitesimal, as you can calculate for yourself.

At what velocity will an object be foreshortened to half its rest-length? With L' equal to one-half L, the FitzGerald equation is:

$$L/2 = L\sqrt{1 - v^2/c^2}$$

or, dividing by L:

$$\tfrac{1}{2} = \sqrt{1 - v^2/c^2}.$$

Squaring both sides of the equation:

$$\tfrac{1}{4} = 1 - v^2/c^2$$
$$v^2/c^2 = \tfrac{3}{4}$$
$$v = \sqrt{3c/4} = 0.866c.$$

Since the velocity of light in a vacuum is 186,282 miles per second, the velocity at which an object is foreshortened to half its length is 0.866 times 186,282, or roughly 161,300 miles per second.

If a body moves at the speed of light, so that v equals c, the FitzGerald equation becomes:

$$L' = L\sqrt{1 - c^2/c^2} = L\sqrt{0} = 0.$$

At the speed of light, then, length in the direction of motion becomes zero. It would seem, therefore, that no velocity faster than that of light is possible.

THE LORENTZ EQUATION

In the decade after FitzGerald had advanced his equation, the electron was discovered, and scientists began to examine the properties of tiny charged particles. Lorentz worked out a theory that the mass of a particle with a given charge is inversely proportional to its radius. In other words, the smaller the volume into which a particle crowds its charge, the greater its mass.

Now if a particle is foreshortened because of its motion, its radius in the direction of motion is reduced in accordance with the FitzGerald equation. Substituting the symbols R and R' for L and L', we write the equation:

$$R' = R\sqrt{1 - v^2/c^2}$$
$$R'/R = \sqrt{1 - v^2/c^2}.$$

The mass of a particle is inversely proportional to its radius. Therefore:

$$\frac{R'}{R} = \frac{M}{M'}$$

where M is the mass of the particle at rest and M' is its mass when in motion.

Substituting M/M' for R'/R in the preceding equation, we have:

$$\frac{M}{M'} = \sqrt{1 - v^2/c^2}$$

$$M' = \frac{M}{\sqrt{1 - v^2/c^2}}.$$

The Lorentz equation can be handled just as the FitzGerald equation was. It shows, for instance, that for a particle moving at a velocity of 18,628 miles per second (one-tenth the speed of light), the mass M' would appear to be 0.5 percent higher than the rest-mass M. At a velocity of 161,300 miles per second, the apparent mass of the particle would be twice the rest-mass.

Finally, for a particle moving at a velocity equal to that of light, so that v is equal to c, the Lorentz equation becomes:

$$M' = \frac{M}{\sqrt{1 - c^2/c^2}} = \frac{M}{0}.$$

Now as the denominator of any fraction with a fixed numerator becomes smaller and smaller (*approaches zero*), the value of the fraction itself becomes larger and larger without limit. In other words, from the equation preceding, it would seem that the mass of any object traveling at a velocity approaching that of light becomes infinitely large. Again, the velocity of light would seem to be the maximum possible.

All this led Einstein to recast the laws of motion and of gravitation. He considered a universe, in other words, in which the results of the Michelson-Morley experiments were to be expected.

Yet even so we are not quite through. Please note that the Lorentz equation assumes some value for M that is greater than zero. This is true for most of the particles with which we are familiar and for all bodies, from atoms to stars, that are made up of such particles. There are, however, neutrinos and antineutrinos for which M, the mass at rest, or *rest-mass*, is equal to zero. This is also true of photons.

Such particles travel at the speed of light in a vacuum, provided they are indeed in a vacuum. The moment they are formed they begin to move at such a velocity without any measurable period of acceleration.

We might wonder how it is possible to speak of the rest-mass of a photon or a neutrino, if they are never at rest but can only exist while traveling (in the absence of interfering matter) at a constant speed of 186,280 miles per second. The physicists Olexa-Myron Bilaniuk and Ennackal Chandy George Sudarshan have therefore suggested that M be spoken of as *proper mass*. For a particle with mass greater than zero, the proper mass is equal to the mass measured when the particle is at rest relative to the instruments and observer making the measurement. For a particle with mass equal to zero, the proper mass is obtained by indirect reasoning. Bilaniuk and Sudarshan also suggest that all particles with a proper mass of zero be called *luxons* (from the Latin word for "light") because they travel at light-speed, while particles with a proper mass greater than zero be called *tardyons* because they travel at less than light-speed, or at *subluminal velocities*.

In 1962, Bilaniuk and Sudarshan began to speculate on the consequences of faster-than-light velocities (*superluminal velocities*). Any particle traveling with faster-than-light velocities would have an imaginary mass. That is, the mass would be some ordinary value multiplied by the square root of -1.

Suppose, for instance, a particle were going at twice the speed of light, so that in the Lorentz equation $v = 2c$. In that case:

$$M' = \frac{M}{\sqrt{1 - (2c)^2/c^2}} = \frac{M}{\sqrt{1 - 4c^2/c^2}} = \frac{M}{\sqrt{-3}}.$$

This works out to the fact that its mass while in motion would be some proper mass (M) divided by $\sqrt{-3}$. But $\sqrt{-3}$ is equal to $3 \times \sqrt{-1}$ and therefore to $1.74 \sqrt{-1}$. The proper mass M is therefore equal to $M' \times 1.74 \times \sqrt{-1}$. Since any quantity that includes $\sqrt{-1}$ is called imaginary, we conclude that particles at superluminal velocities must have imaginary proper masses.

Ordinary particles in our ordinary universe always have masses that are zero or positive. An imaginary mass can have no imaginable significance in our universe. Does this mean that faster-than-light particles cannot exist?

Not necessarily. Allowing the existence of imaginary proper masses, we can make such faster-than-light particles fit all the equations of Einstein's Special Theory of Relativity. Such particles, however, display an apparently paradoxical property: the more slowly they go, the more energy they contain. This is the precise reverse of the situation in our universe and is perhaps the significance of the imaginary mass. A particle with an imaginary mass speeds up when it meets resistance and slows down when it is pushed ahead by a force. As its energy declines, it moves faster and faster, until when it has zero energy it is moving at infinite speed. As its energy increases, it moves slower and slower until, as its energy approaches the infinite, it slows down to approach the speed of light.

Such faster-than-light particles have been given the name of *tachyons* from the Greek word for "speed," by the American physicist Gerald Feinberg.

We may imagine, then, the existence of two kinds of universes. One, our own, is the *tardyon-universe*, in which all particles go at subluminal velocities and may accelerate to nearly the speed of light as their energy increases. The other is the *tachyon-universe*, in which all particles go at superluminal velocities and may decelerate to nearly the speed of light as their energy increases. Between is the infinitely narrow *luxon wall* in which there are particles that go at exactly luminal velocities. The luxon wall can be considered as being held by both universes in common.

If a tachyon is energetic enough and therefore moving slowly enough, it might have sufficient energy and remain in one spot for a long enough period of time to give off a detectable burst of photons. (Tachyons would leave a wake of photons even in a vacuum as a kind of Cerenkov radiation.) Scientists are watching for those bursts, but the chance of happening to have an instrument in just the precise place where one of those (possibly very infrequent) bursts appears for a trillionth of a second or less, is not very great.

There are those physicists who maintain that "anything that is not forbidden is compulsory." In other words, any phenomenon that does not actually break a conservation law *must* at some time or another take place; or, if tachyons do not actually violate special relativity, they *must* exist. Nevertheless, even physicists most convinced of this as a kind of necessary "neatness" about the universe, would be rather pleased (and perhaps relieved) to obtain some evidence for the non-forbidden tachyons. So far, they have not been able to.

EINSTEIN'S EQUATION

One consequence of the Lorentz equation was worked out by Einstein to produce what has become perhaps the most famous scientific equation of all time.

The Lorentz equation can be written in the form:

$$M' = M (1 - v^2/c^2)^{-\frac{1}{2}}$$

since in algebraic notation $1/\sqrt{x}$ can be written $x^{-\frac{1}{2}}$. This puts the equation into a form that can be expanded (that is, converted into a series of terms) by a formula discovered by, of all people, Newton. The formula is the binomial theorem.

The number of terms into which the Lorentz equation can be expanded is infinite, but since each term is smaller than the one before, if you take only the first two terms you are approximately correct, the sum of all the remaining terms being small enough to be neglected. The expansion becomes:

$$(1 - v^2/c^2)^{-\frac{1}{2}} = 1 + \frac{\frac{1}{2}v^2}{c^2} \ldots$$

Substituting that in the Lorentz equation, we get:

$$M' = M\left(1 + \frac{\frac{1}{2}v^2}{c^2}\right) = M + \frac{\frac{1}{2}Mv^2}{c^2}.$$

Now, in classical physics, the expression $\frac{1}{2}Mv^2$ represents the energy of a moving body. If we let the symbol e stand for energy, the equation above becomes:

$$M' = M + e/c^2$$

or:

$$M' - M = e/c^2.$$

The increase in mass due to motion $(M' - M)$ can be represented as m, so:

$$m = e/c^2$$

or:

$$e = MC^2.$$

It was this equation that for the first time indicated mass to be a form of energy. Einstein went on to show that the equation applies to all mass, not merely to the increase in mass due to motion.

Here again, most of the mathematics involved is only at the high-school level. Yet it presented the world with the beginnings of a view of the universe greater and broader even than that of Newton, and also pointed the way to concrete consequences. It pointed the way, for instance, to the nuclear reactor and the atom bomb.

Bibliography

A guide to science would be incomplete without a guide to more reading. I am setting down here a brief selection of books. The list is miscellaneous and does not pretend to be a comprehensive collection of the best modern books about science, but I have read most or all of each of them myself and can highly recommend all of them, even my own.

General

ASIMOV, ISAAC. *A Choice of Catastrophes*. New York: Simon & Schuster, 1979.

ASIMOV, ISAAC. *Asimov's Biographical Encyclopedia of Science and Technology* (2nd rev. ed.). New York: Doubleday, 1982.

ASIMOV, ISAAC. *Exploring the Earth and the Cosmos*. New York: Crown Publishers, 1982.

ASIMOV, ISAAC. *Measure of the Universe*. New York: Harper & Row, 1983.

ASIMOV, ISAAC. *Understanding Physics* (1-vol. ed.). New York: Walker, 1984.

CABLE, E. J., *et al*. *The Physical Sciences*. New York: Prentice-Hall, 1959.

GAMOW, GEORGE. *Matter, Earth, and Sky*. New York: Prentice-Hall, 1958.

HUTCHINGS, EDWARD, JR., ed. *Frontiers in Science*. New York: Basic Books, 1958.

SAGAN, CARL. *Cosmos*. New York: Random House, 1980.

SHAPLEY, HARLOW; RAPPORT, SAMUEL; and WRIGHT, HELEN, eds. *A Treasury of Science* (4th ed.). New York: Harper, 1958.

SLABAUGH, W. H.; and BUTLER, A. B. *College Physical Science*. New York: Prentice-Hall, 1958.

WATSON, JANE WERNER. *The World of Science*. New York: Simon & Schuster, 1958.

Chapter 1   What Is Science?

BERNAL, J. D. *Science in History*. New York: Hawthorn Books, 1965.

CLAGETT, MARSHALL. *Greek Science in Antiquity*. New York: Abelard-Schuman, 1955.

CROMBIE, A. C. *Medieval and Early Modern Science* (2 vols.). New York: Doubleday, 1959.

DAMPIER, SIR WILLIAM CECIL. *A History of Science.* New York: Cambridge University Press, 1958.

DREYER, J. L. E. *A History of Astronomy from Thales to Kepler.* New York: Dover Publications, 1953.

FORBES, R. J.; and DIJKSTERHUIS, E. J. *A History of Science and Technology* (2 vols.). Baltimore: Penguin Books, 1963.

RONIN, COLIN A. *Science: Its History and Development among the World's Cultures.* New York: Facts on File Publications, 1982.

TATON, R., ed. *History of Science* (4 vols.). New York: Basic Books, 1963–66.

Chapter 2 The Universe

ABELL, GEORGE O. *Exploration of the Universe* (4th ed.). Philadelphia: Saunders College Publishing, 1982.

ASIMOV, ISAAC. *The Collapsing Universe.* New York: Walker, 1977.

ASIMOV, ISAAC. *The Universe* (new rev. ed.). New York: Walker, 1980.

BURBIDGE, G.; and BURBIDGE, M. *Quasi-Stellar Objects.* San Francisco: W. H. Freeman, 1967.

FLAMMARION, G. C., *et al. The Flammarion Book of Astronomy.* New York: Simon & Schuster, 1964.

GOLDSMITH, DONALD. *The Universe,* Menlo Park, Calif.: W. A. Benjamin, 1976.

HOYLE, FRED. *Astronomy.* New York: Doubleday, 1962.

KIPPENHAHN, RUDOLF. *100 Billion Suns.* New York: Basic Books, 1983.

LEY, WILLY. *Watchers of the Skies.* New York: Viking Press, 1966.

MCLAUGHLIN, DEAN B. *Introduction to Astronomy.* Boston: Houghton Mifflin, 1961.

MITTON, SIMON, ed.-in-chief. *The Cambridge Encyclopaedia of Astronomy.* New York: Crown, 1977.

SHKLOVSKII, I. S.; and SAGAN, CARL. *Intelligent Life in the Universe.* San Francisco: Holden-Day, 1966.

SMITH, F. GRAHAM. *Radio Astronomy.* Baltimore: Penguin Books, 1960.

STRUVE, OTTO; and ZEBERGS, VELTA. *Astronomy of the 20th Century.* New York: Macmillan, 1962.

Chapter 3 The Solar System

BEATTY, J. KELLY; O'LEARY, BRIAN; and CHAIKIN, ANDREW, eds. *The New Solar System.* Cambridge, Mass.: Sky Publishing, and Cambridge, England: Cambridge University Press, 1981.

RYAN, PETER; and PESEK, LUDEK. *Solar System.* New York: Viking Press, 1978.

Chapter 4 The Earth

ADAMS, FRANK DAWSON. *The Birth and Development of the Geological Sciences.* New York: Dover Publications, 1938.
ASIMOV, ISAAC. *The Ends of the Earth.* New York: Weybright & Talley, 1975.
ASIMOV, ISAAC. *Exploring the Earth and the Cosmos.* New York: Crown Publishers, 1982.
BURTON, MAURICE. *Life in the Deep.* New York: Roy Publishers, 1958.
GAMOW, GEORGE. *A Planet Called Earth.* New York: Viking Press, 1963.
GILLULY, J.; WATERS, A. C.; and WOODFORD, A. O. *Principles of Geology.* San Francisco: W. H. Freeman, 1958.
JACKSON, DONALD DALE. *Underground Worlds.* Alexandria, Va.: Time-Life Books, 1982.
KUENEN, P. H. *Realms of Water.* New York: John Wiley, 1963.
MASON, BRIAN. *Principles of Geochemistry.* New York: John Wiley, 1958.
MOORE, RUTH. *The Earth We Live On.* New York: Alfred A. Knopf, 1956.
SCIENTIFIC AMERICAN, eds. *The Planet Earth.* New York: Simon & Schuster, 1957.
SMITH, DAVID G., ed.-in-chief. *The Cambridge Encyclopaedia of Earth Sciences.* New York: Crown, 1981.
SULLIVAN, WALTER. *Continents in Motion.* New York: McGraw-Hill, 1974.
TIME-LIFE BOOKS, eds. *Volcano.* Alexandria, Va.: Time-Life Books, 1982.

Chapter 5 The Atmosphere

BATES, D. R., ed. *The Earth and Its Atmosphere.* New York: Basic Books, 1957.
GLASSTONE, SAMUEL. *Sourcebook on the Space Sciences.* New York: Van Nostrand, 1965.
LEY, WILLY. *Rockets, Missiles, and Space Travel.* New York: Viking Press, 1957.
LOEBSACK, THEO. *Our Atmosphere.* New York: New American Library, 1961.
NEWELL, HOMER E., JR. *Window in the Sky.* New York: McGraw-Hill, 1959.
NININGER, H. H. *Out of the Sky.* New York: Dover Publications, 1952.
ORR, CLYDE, JR. *Between Earth and Space.* New York: Collier Books, 1961.
YOUNG, LOUISE B. *Earth's Aura.* New York: Alfred A. Knopf, 1977.

Chapter 6 The Elements

ALEXANDER, W.; and STREET, A. *Metals in the Service of Man.* New York: Penguin Books, 1954.
ASIMOV, ISAAC. *A Short History of Chemistry.* New York: Doubleday, 1965.
ASIMOV, ISAAC. *The Noble Gases.* New York: Basic Books, 1966.

DAVIS, HELEN MILES. *The Chemical Elements.* Boston: Ballantine Books, 1959.

HOLDEN, ALAN; and SINGER, PHYLIS. *Crystals and Crystal Growing.* New York: Doubleday, 1960.

IHDE, AARON J. *The Development of Modern Chemistry.* New York: Harper & Row, 1964.

LEICESTER, HENRY M. *The Historical Background of Chemistry.* New York: John Wiley, 1956.

PAULING, LINUS. *College Chemistry* (3rd ed.). San Francisco: W. H. Freeman, 1964.

PRYDE, LUCY T. *Environmental Chemistry: An Introduction.* Menlo Park, Calif.: Cummings Publishing, 1973.

WEEKS, MARY E.; and LEICESTER, H. M. *Discovery of the Elements* (7th ed.). Easton, Pa.: Journal of Chemical Education, 1968.

Chapter 7   The Particles

ALFREN, HANNES. *Worlds Antiworlds.* San Francisco: W. H. Freeman, 1966.

ASIMOV, ISAAC. *The Neutrino.* New York: Doubleday, 1966.

FEINBERG, GERALD. *What Is the World Made Of?* Garden City, N.Y.: Anchor Press/Doubleday, 1977.

FORD, KENNETH W. *The World of Elementary Particles.* New York: Blaisdell Publishing, 1963.

FRIEDLANDER, G.; KENNEDY, J. W.; and MILLER, J. M. *Nuclear and Radiochemistry* (2nd ed.). New York: John Wiley, 1964.

GARDNER, MARTIN. *The Ambidextrous Universe* (2nd rev. ed.). New York: Charles Scribner's, 1979.

GLASSTONE, SAMUEL. *Sourcebook on Atomic Energy* (3rd ed.). Princeton: Van Nostrand, 1967.

HUGHES, DONALD J. *The Neutron Story.* New York: Doubleday, 1959.

MASSEY, SIR HARRIE. *The New Age in Physics.* New York: Harper, 1960.

PARK, DAVID. *Contemporary Physics.* New York: Harcourt, Brace & World, 1964.

WEINBERG, STEVEN. *The Discovery of Subatomic Particles.* New York: Scientific Library, 1983.

Chapter 8   The Waves

BENT, H. A. *The Second Law.* New York: Oxford University Press, 1965.

BERGMANN, P. G. *The Riddle of Gravitation.* New York: Charles Scribner's, 1968.

BLACK, N. H.; and LITTLE, E. P. *An Introductory Course in College Physics.* New York: Macmillan, 1957.

FREEMAN, IRA M. *Physics Made Simple.* New York: Made Simple Books, 1954.

GARDNER, MARTIN. *Relativity for the Million.* New York: Macmillan, 1962.

HOFFMAN, BANESH. *The Strange Story of the Quantum.* New York: Dover Publications, 1959.

Rouse, Robert S.; and Smith, Robert O. *Energy: Resource, Slave Pollutant.* New York: Macmillan, 1975.

Schwartz, Jacob T. *Relativity in Illustrations.* New York: New York University Press, 1962.

Shamos, Morris H. *Great Experiments in Physics.* New York: Henry Holt, 1959.

Chapter 9 The Machine

Bitter, Francis. *Magnets.* New York: Doubleday, 1959.

Clarke, Donald, ed. *The Encyclopedia of How It Works.* New York: A & W Publishers, 1977.

De Camp, L. Sprague. *The Ancient Engineers.* New York: Doubleday, 1963.

Kock, W. E. *Lasers and Holography.* New York: Doubleday, 1969.

Larsen, Egon. *Transport.* New York: Roy Publishers, 1959.

Lee, E. W. *Magnetism.* Baltimore: Penguin Books, 1963.

Lengyel, Bela A. *Lasers.* New York: John Wiley, 1962.

Neal, Harry Edward. *Communication.* New York: Julius Messner, 1960.

Pierce, John R. *Electrons, Waves and Messages.* New York: Doubleday, 1956.

Pierce, John R. *Symbols, Signals and Noise.* New York: Harper, 1961.

Singer, Charles; Holmyard, E. J.; and Hall, A. R., eds. *A History of Technology* (5 vols.). New York: Oxford University Press, 1954.

Taylor, F. Sherwood. *A History of Industrial Chemistry.* New York: Abelard-Schuman, 1957.

Upton, Monroe. *Electronics for Everyone* (2nd rev. ed.). New York: New American Library, 1959.

Usher, Abbott Payson. *A History of Mechanical Inventions.* Boston: Beacon Press, 1959.

Warschauer, Douglas M. *Semiconductors and Transistors.* New York: McGraw-Hill, 1959.

Chapter 10 The Reactor

Alexander, Peter. *Atomic Radiation and Life.* New York: Penguin Books, 1957.

Bishop, Amasa S. *Project Sherwood.* Reading, Mass: Addison-Wesley, 1958.

Fowler, John M. *Fallout: A Study of Superbombs, Strontium 90, and Survival.* New York: Basic Books, 1960.

Jukes, John. *Man-Made Sun.* New York: Abelard-Schuman, 1959.

Jungk, Robert. *Brighter Than a Thousand Suns.* New York: Harcourt, Brace, 1958.

Purcell, John. *The Best-Kept Secret.* New York: Vanguard Press, 1963.

Riedman, Sarah R. *Men and Women behind the Atom.* New York: Abelard-Schuman, 1958.

Scientific American, eds. *Atomic Power.* New York: Simon & Schuster, 1955.

Wilson, Robert R.; and Littauer, R. *Accelerators.* New York: Doubleday, 1960.

Chapter 11 The Molecule

FIESER, L. F.; and FIESER, M. *Organic Chemistry.* Boston: D. C. Heath, 1956.

GIBBS, F. W. *Organic Chemistry Today.* Baltimore: Penguin Books, 1961.

HUTTON, KENNETH. *Chemistry.* New York: Penguin Books, 1957.

PAULING, LINUS. *The Nature of the Chemical Bond* (3rd ed.). Ithaca, N.Y.: Cornell University Press, 1960.

PAULING, LINUS; and HAYWARD, R. *The Architecture of Molecules.* San Francisco: W. H. Freeman, 1964.

Chapter 12 The Proteins

ASIMOV, ISAAC. *Photosynthesis.* New York: Basic Books, 1969.

BALDWIN, ERNEST. *Dynamic Aspects of Biochemistry* (5th ed.). New York: Cambridge University Press, 1967.

BALDWIN, ERNEST. *The Nature of Biochemistry.* New York: Cambridge University Press, 1962.

HARPER, HAROLD A. *Review of Physiological Chemistry* (8th ed.). Los Altos, Calif.: Lange Medical Publications, 1961.

KAMEN, MARTIN D. *Isotopic Tracers in Biology.* New York: Academic Press, 1957.

KARLSON, P. *Introduction to Modern Biochemistry.* New York: Academic Press, 1963.

LEHNINGER, ALBERT L. *Biochemistry* (2nd ed.). New York: Worth Publishers, 1975.

LEHNINGER, ALBERT L. *Bioenergetics.* New York: Benjamin Company, 1965.

Chapter 13 The Cell

ANFINSEN, CHRISTIAN B. *The Molecular Basis of Evolution.* New York: John Wiley, 1959.

ASIMOV, ISAAC. *Extraterrestrial Civilizations.* New York: Crown Publishers, 1979.

ASIMOV, ISAAC. *The Genetic Code.* New York: Orion Press, 1962.

ASIMOV, ISAAC. *A Short History of Biology.* New York: Doubleday, 1964.

BUTLER, J. A. V. *Inside the Living Cell.* New York: Basic Books, 1959.

CARR, DONALD E. *The Deadly Feast of Life.* Garden City, N.Y.: Doubleday, 1971.

FEINBERG, GERALD; and SHAPIRO, ROBERT. *Life Beyond Earth.* New York: William Morrow, 1980.

HARTMAN, P. E.; and SUSKIND, S. R. *Gene Action.* Englewood Cliffs, N.J.: Prentice-Hall, 1965.

HUGHES, ARTHUR. *A History of Cytology.* New York: Abelard-Schuman, 1959.

NEEL, J. V.; and SCHULL, W. J. *Human Heredity.* Chicago: University of Chicago Press, 1954.

OPARIN, A. I. *The Origin of Life on the Earth.* New York: Academic Press, 1957.

SINGER, CHARLES. *A Short History of Anatomy and Physiology from the Greeks to Harvey*. New York: Dover Publications, 1957.

SULLIVAN, WALTER. *We Are Not Alone*. New York: McGraw-Hill, 1964.

TAYLOR, GORDON R. *The Science of Life*. New York: McGraw-Hill, 1963.

WALKER, KENNETH. *Human Physiology*. New York: Penguin Books, 1956.

Chapter 14 The Microorganism

BURNET, F. M. *Viruses and Man* (2nd ed.). Baltimore: Penguin Books, 1955.

CURTIS, HELENA. *The Viruses*. Garden City, N.Y.: Natural History Press, 1965.

DE KRUIF, PAUL. *Microbe Hunters*. New York: Harcourt, Brace, 1932.

DUBOS, RENÉ. *Louis Pasteur*. Boston: Little, Brown, 1950.

LUDOVICI, L. J. *The World of the Microscope*. New York: G. P. Putnam, 1959.

MCGRADY, PAT. *The Savage Cell*. New York: Basic Books, 1964.

RIEDMAN, SARAH R. *Shots without Guns*. Chicago: Rand, McNally, 1960.

SINGER, CHARLES; and UNDERWOOD, E. ASHWORTH. *A Short History of Medicine* (2nd ed.). New York: Oxford University Press, 1962.

SMITH, KENNETH M. *Beyond the Microscope*. Baltimore: Penguin Books, 1957.

WILLIAMS, GREER. *Virus Hunters*. New York: Alfred A. Knopf, 1959.

ZINSSER, HANS. *Rats, Lice and History*. Boston: Little, Brown, 1935.

Chapter 15 The Body

ASIMOV, ISAAC. *The Human Body*. Boston: Houghton Mifflin, 1963.

CARLSON, ANTON J.; and JOHNSON, VICTOR. *The Machinery of the Body*. Chicago: University of Chicago Press, 1953.

CHANEY, MARGARET S. *Nutrition*. Boston: Houghton Mifflin, 1954.

MCCOLLUM, ELMER VERNER. *A History of Nutrition*. Boston: Houghton Mifflin, 1957.

SMITH, ANTHONY. *The Body*. London: George Allen & Unwin, 1968.

TANNAHILL, REAY. *Food in History*. New York: Stein & Day, 1973.

WILLIAMS, ROGER J. *Nutrition in a Nutshell*. New York: Doubleday, 1962.

WILLIAMS, SUE RODWELL. *Essentials of Nutrition and Diet Therapy*. St. Louis: C. V. Mosby, 1974.

Chapter 16 The Species

ASIMOV, ISAAC. *Wellsprings of Life*. New York: Abelard-Schuman, 1960.

BOULE, M.; and VALLOIS, H. V. *Fossil Men*. New York: Dryden Press, 1957.

CALVIN, MELVIN. *Chemical Evolution.* New York: Oxford University Press, 1969.

CAMPBELL, BERNARD. *Human Evolution* (2nd ed.). Chicago: Aldine Publishing, 1974.

CARRINGTON, RICHARD. *A Biography of the Sea.* New York: Basic Books, 1960.

DARWIN, FRANCIS, ed. *The Life and Letters of Charles Darwin* (2 vols.). New York: Basic Books, 1959.

DE BELL, G. *The Environmental Handbook.* New York: Ballantine Books, 1970.

EHRLICH, PAUL; and EHRLICH, ANN. *Extinction.* New York: Random House, 1981.

HANRAHAN, JAMES S.; and BUSHNELL, DAVID. *Space Biology.* New York: Basic Books, 1960.

HARRISON, R. J. *Man, the Peculiar Animal.* New York: Penguin Books, 1958.

HEPPENHEIMER, T. A. *Colonies in Space.* Harrisburg, Pa.: Stackpole Books, 1977.

HOWELLS, WILLIAM. *Mankind in the Making.* New York: Doubleday, 1959.

HUXLEY, T. H. *Man's Place in Nature.* Ann Arbor: University of Michigan Press, 1959.

LEWONTIN, RICHARD. *Human Diversity.* New York: Scientific American Library, 1982.

MEDAWAR, P. B. *The Future of Man.* New York: Basic Books, 1960.

MILNE, L. J.; and MILNE, M. J. *The Biotic World and Man.* New York: Prentice-Hall, 1958.

MONTAGU, ASHLEY. *The Science of Man.* New York: Odyssey Press, 1964.

MOORE, RUTH. *Man, Time, and Fossils* (2nd ed.). New York: Alfred A. Knopf, 1963.

O'NEILL, GERARD K. *2081.* New York: Simon & Schuster, 1981.

ROMER, A. S. *Man and the Vertebrates* (2 vols.). New York: Penguin Books, 1954.

ROSTAND, JEAN. *Can Man Be Modified?* New York: Basic Books, 1959.

SAX, KARL. *Standing Room Only.* Boston: Beacon Press, 1955.

SIMPSON, GEORGE G.; PITTENDRIGH, C. S.; and TIFFANY, L. H. *Life: An Introduction to College Biology* (2nd ed.). New York: Harcourt, Brace, 1965.

TAYLOR, GORDON RATTRAY. *The Doomsday Book.* New York: World Publishing, 1970.

TINBERGEN, NIKO. *Curious Naturalists.* New York: Basic Books, 1960.

UBBELOHDE, A. R. *Man and Energy.* Baltimore: Penguin Books, 1963.

Chapter 17 The Mind

ASIMOV, ISAAC. *The Human Brain.* Boston: Houghton Mifflin, 1964.

BERKELEY, EDMUND C. *Symbolic Logic and Intelligent Machines.* New York: Reinhold Publishing, 1959.

EVANS, CHRISTOPHER. *The Making of the Micro, A History of the Computer.* New York: Van Nostrand Reinhold, 1981.

FACKLAM, MARGERY; and FACKLAM, HOWARD. *The Brain.* New York: Harcourt, Brace, Jovanovich, 1982.

GARDNER, HOWARD. *Frames of Mind.* New York: Basic Books, 1983.

GOULD, STEPHEN JAY. *The Mismeasure of Man.* New York: W. W. Norton, 1981.

JONES, ERNEST. *The Life and Work of Sigmund Freud* (3 vols.). New York: Basic Books, 1957.

LASSEK, A. M. *The Human Brain.* Springfield, Ill.: Charles C Thomas, 1957.

MENNINGER, KARL. *Theory of Psychoanalytic Technique.* New York: Basic Books, 1958.

MURPHY, GARDNER. *Human Potentialities.* New York: Basic Books, 1958.

RAWCLIFFE, D. H. *Illusions and Delusions of the Supernatural and Occult.* New York: Dover Publications, 1959.

SANDERS, DONALD H. *Computers Today.* New York: McGraw-Hill, 1983.

SCIENTIFIC AMERICAN, eds. *Automatic Control.* New York: Simon & Schuster, 1955.

SCOTT, JOHN PAUL. *Animal Behavior.* Chicago: University of Chicago Press, 1957.

THOMPSON, CLARA; and MULLAHY, PATRICK. *Psychoanalysis: Evolution and Development.* New York: Grove Press, 1950.

Appendix   Mathematics in Science

COURANT, RICHARD; and ROBBINS, HERBERT. *What Is Mathematics?* New York: Oxford University Press, 1941.

DANTZIG, TOBIAS. *Number, the Language of Science.* New York: Macmillan, 1954.

FELIX, LUCIENNE. *The Modern Aspect of Mathematics.* New York: Basic Books, 1960.

FREUND, JOHN E. *A Modern Introduction to Mathematics.* New York: Prentice-Hall, 1956.

KLINE, MORRIS. *Mathematics and the Physical World.* New York: Thomas Y. Crowell, 1959.

KLINE, MORRIS. *Mathematics in Western Culture.* New York: Oxford University Press, 1953.

NEWMAN, JAMES R. *The World of Mathematics* (4 vols.). New York: Simon & Schuster, 1956.

STEIN, SHERMAN K. *Mathematics, the Man-Made Universe.* San Francisco: W. H. Freeman, 1963.

VALENS, EVANS G. *The Number of Things.* New York: Dutton, 1964.

Name Index

Abbe, Ernst Karl, 372, 608
Abderhalden, Emil, 511
Abel, Frederick Augustus, 494
Abelson, Philip, 247, 596
Acheson, Edward Goodrich, 274
Adam, 696
Adams, John Couch, 127
Adams, Leason Heberling, 156
Adams, Walter Sydney, 49, 360
Addison, Thomas, 684
Adrian, Edgar Doublas, 768, 769
Agamemnon, 726
Agassiz, Jean Louis Rodolphe, 183, 184
Aiken, Howard, 796
Airy, George Biddell, 607, 787
Albert (Prince), 478
Alburger, David Elmer, 54
Alcmaeon, 551
Alder, Kurt, 481
Alfven, Hannes, 84
Alhazen, 606
Allbutt, Thomas Clifford, 364
Alvarez, Luis Walter, 302, 328
Alvarez, Walter, 720
Amagat, Emile Hilaire, 273
Ambartsumian, Victor Amazaspovich, 65
Amontons, Guillaume, 363
Ampère, André Marie, 209, 390
Amundsen, Roald, 182
Anderson, Carl David, 299, 300, 322
Andrews, Thomas, 266
Angstrom, Anders Jonas, 59, 343
Apian, Peter, 135, 137
Appert, François, 657

Appleton, Edward Victor, 205
Aquinas, Thomas, 11, 12, 592
Archimedes, 10, 96
Archytas, 194
Aristarchus, 11, 20, 22, 27, 85
Aristotle, 7, 10, 11, 45, 46, 89, 96, 135, 138, 147, 192, 232, 234, 697, 707, 758
Armstrong, Edwin Howard, 414
Armstrong, Neil Alden, 97
Arnon, Daniel Israel, 547
Arp, Halton C., 68
Arrhenius, Svante August, 204, 205, 389, 594
Artsimovich, Lev Andreevich, 459
Aschheim, Selman, 683
Ashurbanipal, 726
Asimov, Isaac, 450, 798–800
Astbury, William Thomas, 514
Aston, Francis William, 291, 292, 369–71
Atlas, 19
Aumann, Harmut H., 602
Avery, Oswald Theodore, 582
Avogadro, Amedeo, 235

Baade, Walter, 34, 35, 64, 132, 134
Babbage, Charles, 795, 796
Babinet, Jacques, 344
Babinski, François Félix, 766
Bacon, Francis, 12, 157
Bacon, Roger, 12, 382, 607
Baekeland, Leo Hendrik, 497, 498, 513

Baer, Karl Ernst von, 556, 703
Baeyer, Adolf von, 479, 497, 548
Baffin, William 178
Baird, John Logie, 414
Baker, B. L., 599
Baker, C. P. 433
Balfour, Francis Maitland, 703
Balmer, Johann Jakob, 60
Bang, Olaf, 648
Banting, Frederick Grant, 680
Bardeen, John, 417, 418
Barghoorn, Elso Sterrenberg, 716
Barkhausen, Heinrich, 208, 214
Barkla, Charles Glover, 240
Barnard, Christiaan, 642
Barnard, Edward Emerson, 113
Barnes, D. E., 308
Barringer, Daniel Moreau, 223
Bartlett, Neil, 254
Barton, Otis, 178
Basov, Nicolai Gennediyevich, 421
Bateson, William, 564
Baum, William Alvin, 116
Baumann, Eugen, 675
Bawden, Frederick Charles, 628
Bay, Zoltan Lajos, 100
Bayliss, William Maddock, 530, 677, 680
Beadle, George Wells, 572–5
Becker, Herbert, 296
Beckwith, Jonathan, 568
Becquerel, Alexandre Edmond, 58, 242
Becquerel, Antoine Henri, 242–4, 285
Beebe, Charles William, 177
Beer, Wilhelm, 106
Beguyer de Chancourtois, A. E., 237
Behring, Emil von, 636, 637
Beijerinck, Martinus Willem, 627
Bell, Alexander Graham, 394
Bell, Jocelyn, 71
Bell, Peter, M., 276
Bellingshausen, Fabian Gottlieb, 180
Benda, C., 538
Beneden, Eduard van, 558
Benedict, Francis Gano, 682
Bennett, Floyd, 180
Benz, Karl, 405
Berg, Otto, 242, 245
Berg, Paul, 591
Berger, Hans, 770
Bergius, Friedrich, 428
Bergmann, Max, 510, 513
Berliner, Emile, 395

Bernard, Claude, 680
Bernoulli, Daniel, 358
Berthelot, Pierre Eugène Marcelin, 477
Berzelius, Jöns Jacob, 220, 235, 464–6, 492, 507, 525
Bessel, Friedrich Wilhelm, 25, 49
Bessemer, Henry, 278, 279
Best, Charles Herbert, 680
Bethe, Hans Albrecht, 39, 442
Biermann, Ludwig Franz, 215
Bigeleisen, Jacob, 188
Bijvoet, Johannes Martin, 473
Bilaniuk, Olexa-Myon, 816, 817
Billroth, Theodor, 608
Binet, Alfred, 757
Biot, Jean Baptiste, 220, 469, 470
Birkeland, Olaf Kristian, 214
Bittner, John Joseph, 649
Black, Joseph, 200, 364, 365
Blackett, Patrick M. S., 295, 371
Blaiberg, Philip, 642
Blanchard, Jean Pierre, 195
Blanqui, Jérôme Adolphe, 385
Blériot, Louis, 407
Bleuler, Eugen, 780
Bloch, Felix, 516
Bloch, Konrad Emil, 543
Bloembergen, Nicolaas, 420
Blout, Elkan Rogers, 516
Blumenbach, Johann Friedrich, 735
Bohr, Niels H. D., 96, 257, 313, 317, 356, 374, 375, 377, 435, 437
Bolt, C. T., 74
Bolton, John C., 63
Boltzmann, Ludwig, 227, 367
Bonaparte, Napoleon, 725
Bond, George Phillips, 60
Bond, William Cranch, 120
Bondi, Hermann, 43
Bonnett, Charles, 707
Boole, George, 794
Bordet, Jules, 637
Borman, Frank, 97
Born, Max, 375
Borrel, Amédée, 648
Bosch, Karl, 496
Bose, Satyendranath, 310
Bothe, Walter, 296
Bouchardat, Gustave, 491
Boucher de Perthes, Jacques, 728
Boule, Marcelin, 730
Boussard, 725
Boussingault, Jean Baptiste, 724, 725

Bouvard, Alexis, 126
Boveri, Theodor 647
Bovet, Daniel, 642
Bowen, Edward George, 222
Bowen, Ira Sprague, 202
Boyd, William Clouser, 651, 737
Boyle, Robert, 13, 193, 233, 235, 363
Boylston, Zabdiel, 634
Braconnot, Henri, 489, 508
Bradley, James, 345
Brady, Matthew, 400
Bragg, William Henry, 241
Bragg, William Lawrence, 241
Brahe, Tycho, 14, 46, 47, 51, 96, 104, 135
Braid, James, 779
Brand, Erwin, 519
Brandenberger, Jacques Edwin, 502
Branley, Edouard, 410
Brattain, Walter Houser, 417, 418
Braun, Karl Ferdinand, 411, 417
Braun, Werner von, 93
Brearley, Harry, 280
Breuer, Josef, 779–80
Breuil, Henri Edouard Prosper, 728
Bridgman, Percy Williams, 273, 275, 276
Briggs, Henry, 791
Briggs, Robert William, 561
Brill, Rudolf, 514
Broca, Pierre Paul, 759
Broglie, Louis Victor de, 372
Broom, Robert, 732
Brown, Robert (astronomer), 118
Brown, Robert (botanist), 236, 557
Brunhes, Bernard, 213
Bruno, Giordano, 234
Brush, Charles Francis, 404
Bryan, William Jennings, 714
Buchner, Eduard, 528
Buffon, Georges Louis Leclerc de, 81
Buist, John Brown, 628
Bullen, Keith Edward, 155
Bunsens, Robert Wilhelm, 59
Burbank, Luther, 563
Burke, Bernard, 219
Burnet, MacFarlane, 642
Burt, Cyril Lodowic, 571, 757, 758
Bury, C. R., 257
Bush, Vannevar, 796, 797
Butenandt, Adolf, 683
Byrd, Richard Evelyn, 180, 182

Cabrera, Blas, 350
Cagniard de la Tour, Charles, 527
Cailletet, Louis Paul, 266
Callinicus, 382
Calvin, Melvin, 548, 550
Canaan, 735
Candolle, Augustin Pyramus de, 699
Cannizzaro, Stanislao, 236, 237
Cannon, Annie J., 59
Cannon, Walter Bradford, 787
Canton, John, 213
Capek, Karel, 798
Capp, Al, 438
Capra, J. Donald, 640
Cardano, Geronimo, 13
Carlson, Chester, 389
Carnot, Nicolas Léonard Sadi, 365, 384
Carothers, Wallace Hume, 503
Carpenter, Roland L., 101
Carrel, Alexis, 637, 638, 694
Carrier, Willis Haviland, 266
Carrington, Richard Christopher, 214
Carson, Rachel Louise, 619
Cartier, Jacques, 658
Caspersson, Torbjörn, 581
Cassini, Jacques, 141
Cassini, Jean Dominique, 23, 87, 105, 114, 115, 120, 122, 123, 221
Cassiodorus, Flavius, 690
Castle, William Bosworth, 674
Cavendish, Henry, 144, 145, 147, 200, 201, 233
Caventou, Joseph Bienaimé, 546
Cayley, George, 407
Celsius, Anders, 213, 364
Chadwick, James, 296, 297, 432
Chaffee, Roger, 97
Chain, Ernst Boris, 616–18
Chamberlain, Owen, 309
Chamberlin, Thomas Chrowder, 83
Champollion, Jean François, 725
Chance, Britton, 531
Chandrasekhar, Subrahmanyan, 56
Chapman, Sydney, 197, 202, 214
Chardonnet, Hilaire B. de, 501–2
Chargaff, Erwin, 582
Charles II (King), 14
Charles, Jacques Alexandre César, 262
Charpentier, Johann von, 183
Chen Ning Yang, 336, 337
Cherenkov, Paul Alekseyevich, 353

Chevreul, Michel Eugène, 679
Chien Shiung Wu, 337
Cho Hao Li, 523
Christison, Robert, 482
Christofilos, Nicholas, 217, 308
Christy, James W., 129
Churchill, Winston, 692
Clairault, Alexis Claude, 142
Clark, Alvan, 49
Clark, Wilfrid LeGros, 734
Clarke, Arthur C., 198
Claude, Georges, 399
Clausius, Rudolf Julius Emmanuel, 367
Clementi, Enrico, 488
Clerk, Dugald, 404
Clynes, Manfred, 771
Coblentz, William Weber, 108, 517
Cockcroft, John Douglas, 96, 305
Code, Arthur Dodd, 116
Cohen, B., 662
Cohn, Edwin Joseph, 673
Cohn, Ferdinand Julius, 608
Coleridge, Samuel Taylor, 180
Collins, Samuel Cornette, 269
Colt, Samual, 405
Columbus, Christopher, 20, 178, 212, 246, 247, 491, 614
Compton, Arthur Holly, 299, 356, 374, 441
Conant, James Bryant, 440
Congreve, William, 92
Cook, Captain James, 180, 658
Cook, Newell C., 284
Coolidge, William David, 398
Cooper, Peter, 386
Copernicus, Nicolaus, 11, 21, 22, 88, 96, 99, 552
Copp, D. Harold, 681
Corbet, Ruth Elizabeth, 662
Corey, Robert Brainard, 514
Cori, Carl Ferdinand, 535
Cori, Gerty Theresa, 535
Coriolis, Gaspard Gustave de, 139
Correns, Karl Erich, 563
Corson, Dale Raymond, 246
Coryell, Charles DuBois, 246
Coster, Dirk, 241
Cotton, Gardner Quincy, 483
Coulomb, Charles Augustin de, 388
Couper, Archibald Scott, 467
Cousteau, Jacques-Ives, 209, 746
Cowan, Clyde Lorrain, 319
Crafts, James Mason, 480

Craig, Lyman Creighton, 681
Cranston, John Arnold, 244
Crewe, Albert Victor, 237, 374
Crick, Francis Harry Compton, 583, 604
Cromwell, Townsend, 164
Cronin, James Watson, 338
Crookes, William, 249, 289, 292, 397
Cross, Charles Frederick, 502
Cullen, William, 265
Curie, Jacques, 170, 243
Curie, Marie Sklodowska, 96, 243, 244, 247, 285, 296, 368, 369, 448, 453, 647
Curie, Pierre, 96, 170, 208, 243, 244, 247, 285, 296, 453
Curtis, Heber Doust, 33, 47
Cuvier, Georges Léopold, 698, 707, 708, 713
Cyrano de Bergerac, 91

Daguerre, Louis Jacques Mande, 60, 400
Daimler, Gottlieb, 405
Dale, Henry, 770
Dalton, John, 96, 235, 250
Dam, Carl Peter Henrik, 664
Dandolo, Enrico, 690
Dante Alighieri, 139
Danysz, Marian, 327, 328
Darby, Abraham, 278, 426
D'Arrest, Heinrich Ludwig, 127
Darrow, Clarence, 714
Dart, Raymond, 732
Darwin, Charles Robert, 571, 710–15, 724, 731, 738
Darwin, George Howard, 159, 160
Daubrée, Gabriel Auguste, 155
Davaine, Casimir Joseph, 612
Da Vinci, Leonardo, 552, 554, 791
Davis, Marguerite, 660
Davis, Raymond R., 320, 325
Davisson, Clinton Joseph, 372
Davy, Humphry, 234, 255, 281, 365, 389, 482
Dawson, Charles, 734
Debierne, André Louis, 244
De Broglie, Louis Victor, 375, 476
Debye, Peter Joseph Wilhelm, 271, 475
DeForest, Lee, 413
Delgado, José Manuel Rodriguez, 771

Delporte, Eugène, 134
Dement, William, 763
Democritus, 234, 250
Dempster, Arthur Jeffrey, 292
Desaguliers, John Théophile, 387
Descartes, René, 342
Destriau, Georges, 399
Deutsch, Martin, 300
Devol, Jr., George C., 800
De Vries, Hugo, 563
Dewar, James, 267, 494
D'Hérelle, Félix Hubert, 632
Dicke, Robert Henry, 44
Diebold, John, 788
Diels, Otto, 481
Diener, T. O., 633
Diesel, Rudolf, 406
Dietz, Robert S., 174
Dirac, Paul Adrien Maurice, 299, 300,
    304, 305, 310, 349, 375
Disraeli, Benjamin, 713
Döbereiner, Johann Wolfgang 525
Dobzhansky, Theodosius, 568
Doering, William von Eggers, 482
Doisy, Edward, 664, 683
Dole, Stephen H., 602–3
Dollfus, Audouin, 123
Dollond, John, 58
Domagk, Gerhard, 615, 669, 683
Donath, William Frederick, 661
Donn, William L., 188
Doppler, Christian Johann, 40
Dorfman, D. D., 571
Dorn, Friedrich Ernst, 244
Doty, Paul Mead, 516
Douglass, Andrew Ellicott, 729
Down, John Landgon Haydon, 559
Drake, Edwin Laurentine, 427
Drake, Frank Donald, 219, 604
Drebbel, Cornelis, 177, 786
Drummond, Jack Cecil, 660
Dubois, Marie E. F. T., 731
Du Bois-Reymond, Emil, 768
Dubos, René Jules, 616
Du Fay, Charles F. de C., 386
Duggar, Benjamin Minge, 617
Dumas, Jean Baptiste André, 466,
    518
Dunning John Ray, 437
Du Pouy, Gaston, 629
Dutrochet, René Joachim Henri,
    555
Dutton, Clarence Edward, 157
Dyce, Rolf Buchanan, 100

Eastman, George, 400
Eccles, John Carew, 769
Eckert, John Presper, 797
Eddington, Arthur Stanley, 39, 51,
    96, 360
Eddy, John A., 87
Edelman, Gerald Maurice, 640
Edison, Thomas Alva, 393–5
    397, 398, 402, 404, 411
Edman, Pehr, 522
Edward VII (King), 565
Ehrlich, Paul, 613, 614, 637, 639, 773
Ehrlich, Paul R., 721
Eijkman, Christiaan, 659, 660
Einstein, Albert, 7, 40, 42, 50, 96, 236,
    247, 300, 306, 310, 316, 338, 340,
    355–62, 369–71, 374, 377, 420, 436,
    773, 816–19
Einthoven, Willem, 768
Elford, William Joseph, 627
Elizabeth I (Queen), 208
Ellerman, Wilhelm, 648
Elliot, James, L., 126
Elsasser, Walter Maurice, 210, 218
Elvehjem, Conrad Arnold, 667
Emiliani, Cesare, 189
Empedocles, 232
Enders, John Franklin, 638
Engelberger, Joseph F., 800, 801
Epicurus, 234
Erasistratus, 551, 754
Eratosthenes, 20, 21
Erlanger, Joseph, 768
Euclid, 9, 12
Euler, Ulf Svante von, 688
Euler-Chelpin, Hans K. A. S. von,
    666, 688
Evans, Arthur John, 726, 727
Evans, Herbert McLean, 663
Eve, 4
Evenson, Kenneth M., 347
Even, Harold Irving, 76
Ewing, William Maurice, 156, 172,
    188
Eysenck, Hans J., 571

Fabricius, David, 46
Fabricius, Hieronymus, 553
Fabry, Charles, 203
Fahrenheit, Gabriel Daniel, 363
Faraday, Michael, 209, 210, 249, 264,
    266, 348, 350, 389–92, 473

Farcot, Léon, 788
Feinberg, Gerald, 817
Feokstistov, Konstantin P., 97
Ferdinand II (Grand Duke), 363
Fermi, Enrico, 96, 246, 247, 310, 312, 317, 335, 433, 440
Fernel, Jean, 552
Fessenden, Reginald Aubrey, 411
Feulgen, Robert, 587, 613
Feynman, Richard Phillips, 377
Fick, Adolf Eugen, 607
Field, Cyrus West, 168
Findlay, George William Marshall, 647
Finsen, Niels Ryberg, 647
Fischer, Emil, 472, 489, 510, 511, 513, 523
Fischer, Hans, 486, 487, 547
Fisher, Ronald Aylmer, 712
Fitch, John, 385
Fitch, Val Logsden, 338
FitzGerald, George F., 352, 353, 814, 815
Fizeau, Armand H. L., 40, 345-7
Flamsteed, John, 125
Fleming, Alexander, 616, 617
Fleming, John Ambrose, 411
Flemming, Walther, 557, 613
Flerov, Georgii Nicolaevich, 248, 314
Flexner, Josepha Barbara, 784
Flexner, Louis Barkhouse, 784
Florey, Howard Walter, 616, 617
Flourens, Marie Jean Pierre, 754
Folkers, Karl August, 673
Fontenelle, Bernard le Bovier de, 690
Forbes Jr., Edward, 168
Ford, Henry, 405, 406
Foucault, Jean Bernard Léon, 139, 140, 212, 346, 347
Fourier, Jean Baptiste, Joseph, 365
Fowler, Ralph Howard, 49
Fox, Sidney Walter, 598
Fracastoro, Girolamo, 135, 137
Fraenkel-Conrat, Heinz, 630, 631
Francis Joseph I (Emperor), 233
Franck, James, 375, 444
Frank, Ilya Mikhailovich, 354
Frankland, Edward, 467
Franklin, Benjamin, 163, 195, 386-8, 607, 779
Franklin, John, 179
Franklin, Kenneth Linn, 219
Franklin, Rosalind Elsie, 583
Fraunhofer, Joseph von, 58, 344

Frere, John, 728
Fresnel, Augustin Jean, 343, 344
Freud, Sigmund, 779, 780
Fridericia, L. S., 669
Friedel, Charles, 480
Friedman, Herbert, 69, 70
Friedmann, Alexander Alexandrovich, 43
Frisch, Karl von, 774
Frisch, Otto Robert, 435
Fritsch, Gustav, 759
Fruton, Joseph Stewart, 510
Fuchs, Vivian Ernest, 183
Fukui, Saburo, 294
Fulton, Robert, 385
Funk, Casimir, 660, 667

Gabor, Dennis, 424
Gadolin, Johan, 255, 256
Gagarin, Yuri Alexeyevich, 96
Gahn, Johann Gottlieb, 671
Galen, 552-4, 635, 764, 765
Galileo Galilei, 11, 12, 14, 23, 26, 85, 87, 89, 90, 94, 99, 111, 113, 119, 120, 124, 139, 142, 146, 192, 344, 345, 362, 363, 553, 607, 803, 804, 806-11
Gall, Franz Joseph, 759
Galle, Johann Gottfried, 127
Galois, Evariste, 96
Galton, Francis, 571
Galvani, Luigi, 389, 764, 768
Gama, Vasco da, 658
Gamow, George, 43, 586
Ganswindt, Hermann, 92
Gardner, Allen, 776
Gardner, Beatrice, 776
Garrod, Archibald Edward, 577
Gassendi, Pierre, 213, 234
Gasser, Herbert Spencer, 768
Gatling, Richard, 494
Gauss, Johann Karl Friedrich, 131
Gautier, Marthe, 559
Gay-Lussac, Joseph Louis, 195
Geiger, Hans, 215
Geissler, Heinrich, 249
Gell-Mann, Murray, 329, 331
Germer, Lester Halbert, 372
Gernsback, Hugo, 96
Ghiorso, Albert, 247, 248
Giauque, William Francis, 272, 370

Gibbs, Josiah Willard, 368
Gilbert, Joseph Henry, 655
Gilbert, William, 208, 386, 553
Gill, David, 60
Gillett, Fred, 602
Glaser, Donald Arthur, 294
Glashow, Sheldon Lee, 338
Glendenin, L. E., 246
Glenn, John Herschel, 96, 97
Goddard, Robert Hutchings, 92, 93
Godwin, Francis, 91
Goeppert-Mayer, Maria, 314
Goethe, Johann Wolfgang von, 698
Gold, Thomas, 43, 72
Goldberger, Joseph, 667
Goldmark, Peter, 395
Goldstein, Eugen, 249, 287
Goldstein, Richard M., 101
Golgi, Camillo, 765
Goodpasture, Ernest William, 638
Goodricke, John, 41, 46
Goodyear, Charles, 491, 504
Gordon, John B., 561
Gosse, Philip Henry, 713
Goudsmit, Samuel Abraham, 310
Gould, Stephen Jay, 713, 757
Goward, F. K., 308
Graaf, Regnier de, 554
Graebe, Karl, 479
Graff, Samuel, 649
Graham, Thomas, 512
Gram, Hans Christian Joachim, 616
Green, Howard, 568
Greenstein, Jesse Leonard, 66, 67
Grew, Nehemiah, 554
Griffith, Fred, 582
Grigniard, Victor, 480
Grijns, Gerrit, 659
Grissom, Virgil I., 97
Grosseteste, Robert, 606
Groth, Wilhelm, 596
Grove, William, 391
Guericke, Otto von, 383
Guettard, Jean Etienne, 153
Guillaume, Charles Edouard, 141
Gutenberg, Beno, 155, 156
Gutenberg, Johann, 382
Guth, Alan, 340
Guyot, Arnold Henry, 171

Haber, Fritz, 496
Hadfield, Robert Abbott, 279

Hahn, Otto, 244, 248, 434, 435
Halberg, Franz, 777
Haldane, John Burdon Sanderson, 577
Hale, George Ellery, 35, 87, 214
Hales, Stephen, 545
Hall, Angelina Stickney, 105, 110
Hall, Asaph, 105, 107, 110
Hall, Charles Martin, 281
Hall, Marshall, 765
Haller, Albrecht von, 758, 759
Halley, Edmund, 24, 38, 135, 136
Halsted, William Stewart, 612
Ham, 735
Ham, Johann, 556
Hampson, William, 266
Harden, Arthur, 534, 535, 666, 667
Hare, Robert, 396, 456
Harkins, William Draper, 442
Harris, Geoffrey W., 687
Harteck, Paul, 303
Hartmann, Johannes Franz, 75
Hartwig, Ernst, 47
Harvey, William, 553, 554, 592, 706
Hata, Sahachiro, 614
Hauksbee, Francis, 298
Hawking, Stephen, 75
Haworth, Walter Norman, 669
Hays, J. D., 187
Hazard, Cyril, 66
Heaviside, Oliver, 205
Heezen, Bruce Charles, 171
Hefner-Alteneck, Friedrich von, 392
Heisenberg, Werner, 297, 321, 375, 376
Helmholtz, Hermann Ludwig Ferdinand von, 38, 82, 366, 401, 442
Hench, Philip Showalter, 684, 685
Hencke, Karl L., 131
Henderson, Thomas, 25
Henry, Joseph, 392, 393
Henseleit, K., 539
Heracleides, 99, 139
Herbig, George, 53
Hercules, 19
Hero, 382, 785
Herodotus, 727
Héroult, Paul Louis Toussaint, 281
Herophilus, 551, 758
Herrick, James Bryan, 574
Herschel, John, 400

Herschel, William, 26, 27, 30–32, 60, 61, 86, 105, 124, 125, 130, 224, 400
Hershey, Alfred Day, 632
Hertz, Gustav Ludwig, 375
Hertz, Heinrich Rudolf, 61, 62, 96, 375, 410
Hertzsprung, Ejnar, 30, 51, 96
Herzog, R. A., 514
Hess, Harry Hammond, 158, 171, 174
Hess, Victor Francis, 298–300
Hess, Walter Rudolf, 759, 761
Hevesy, Georg von, 241, 540
Hewish, Anthony, 71
Heyl, Paul R., 145
Hill, Archibald Vivian, 535
Hillary, Edmund Percival, 183
Hillier, James, 374
Hipparchus, 20, 21, 23, 28, 46, 90, 140, 220
Hippocrates, 610, 611, 635
Hitler, Adolf, 444
Hitzig, Eduard, 759
Hoagland, Mahlon Bush, 587
Hodgkin, Alan Lloyd, 769
Hodgkin, Dorothy Crowfoot, 524, 674
Hodson, G. W., 599
Hofman, Albert, 782
Hofmann, August Wilhelm, 478–80
Hofmeister, Wilhelm F. B., 558
Hofstadter, Robert, 329
Hohenheim, Theophrastus B. von, 232
Hollerith, Herman, 796
Holley, Robert William, 587, 588
Holloway, M. G., 433
Holm, E., 669
Holmes, Harry Nicholls, 662
Holmes, Oliver Wendell, 283, 483
Homer, 132, 277, 690, 726, 727
Hong Yee Chiu, 66, 318
Hooft, Gerard 't, 349
Hooke, Robert, 115, 192, 554, 607
Hooker, John B., 34
Hoover, Herbert, 692
Hopkins, Frederick G., 656, 660, 662
Hoppe-Seyler, Felix, 578, 579
Houssay, Bernardo Alberto, 681
Hoyle, Fred, 43, 53–6, 83, 85, 218, 596
Hubble, Edwin Powell, 34, 36, 42, 43, 47

Hudson, Henry, 178
Huggins, William 41, 202
Hughes, Vernon Willard, 322
Humason, Milton La Salle, 42
Humboldt, Alexander von, 86, 163
Hunter, William, 613
Hutton, James, 37, 153
Huxley, Andrew Fielding, 769
Huxley, Julian Sorrell, 738
Huxley, Thomas Henry, 713, 714, 738
Huygens, Christian, 33, 105, 120, 121, 142, 342, 347
Hyatt, John Wesley, 497
Hyden, Holger, 783

Ichikawa, K., 645
Ikhnaton, 85
Imbrie, John, 187
Ingenhousz, Jan, 545
Ingram, Vernon Martin, 576
Isaacs, Alick, 643
Ishtar, 7
Isis, 7
Isocrates, 690
Ivanovski, Dmitri Iosifovich, 627

Jacob, François, 586, 588
Jacquard, Joseph Marie, 789
Jacquet-Droz, Pierre, 789
James, A. T., 519
Jansen, Barend Coenraad Petrus, 661
Jansenn, Pierre Jules César, 59
Janssen, Zacharias, 527
Jansky, Karl, 62, 64
Javan, Ali, 422
Jeans, James Hopwood, 83, 354
Jefferson, Thomas, 220
Jeffreys, Harold, 83
Jeffries, John, 195
Jeffries, Zay, 298
Jenner, Edward, 634, 635
Jensen, Johannes Hans Daniel, 315
Johannsen, Wilhelm Ludwig, 562
Johanson, Donald, 733
John of Gaunt, 690
Joliot-Curie, Frédéric, 296, 300–302, 304, 439, 540
Joliot-Curie, Irène, 296, 300–302, 304, 434, 448, 540, 647
Joule, James Prescott, 267, 365, 366

Joyce, James, 331
Junkers, Hugo, 407

Kamen, Martin David, 542, 548, 549
Kant, Immanuel, 33, 82
Kapitza, Peter Leonidovich, 271
Kapteyn, Jacobus Cornelis, 27, 30
Karrer, Paul, 669
Kasper, Jerome V. V., 422
Katchalski, E., 513
Kauffman, Walter, 353
Keeler, James Edward, 41
Keeler, Leonard, 769
Keesom, A. P., 270
Keesom, Willem Hendrik, 270
Keilin, David, 531, 537, 672
Kekulé von Stradonitz, Friedrich A., 467, 471, 473–5, 480
Kelvin, William Thomson, Lord, 38, 39, 82, 262–4, 267, 366, 394
Kendall, Edward Calvin, 678, 684, 685
Kendrew, John Cowdery, 523
Kennedy, Joseph William, 247
Kennelly, Arthur Edwin, 205
Kepler, Johannes, 14, 22, 47, 51, 96, 105, 111
Kerst, Donald William, 307
Kettering, Charles Franklin, 405
Key, Francis Scott, 92
Khorana, Har Gobind, 589
Khrushchev, Nikita, 709
Kiliani, Heinrich, 489
King, Charles Glen, 662
King, Thomas J., 561
Kipling, Rudyard, 495
Kipping, Frederic Stanley, 501
Kircher, Athanasius, 193
Kirchhoff, Gottlieb Sigismund, 489, 525
Kirchhoff, Gustav Robert, 59
Kirkwood, Daniel, 131
Klaproth, Martin Heinrich, 255
Klebs, Edwin, 613
Kleist, Ewald Georg von, 387
Knoll, Max, 374
Knoop, Franz, 539, 540
Koch, Robert, 613, 636
Kocher, Emil Theodor, 675
Koenigswald, Gustav H. R. von, 731, 734
Köhler, Wolfgang, 775

Kohman, Truman Paul, 292
Kohoutek, Lajos, 136
Kolbe, Adolphe Wilhelm Hermann, 465, 477
Koller, Carl, 482
Komarov, Vladimir M., 97
Korff, Serge, 730
Kornberg, Arthur, 585
Kossel, Albrecht, 579
Kowal, Charles T., 113, 132, 133
Kozyrev, Nikolai Alexandrovich, 224
Krebs, Hans Adolf, 537, 539
Kruse, Walther, 627
Kuhn, Richard, 668, 683
Kühne, Wilhelm 528
Kuiper, Gerard P., 108, 122, 125, 128
Kurchatov, Igor Vasilievich, 248
Kurti, Nicholas, 272
Kylstra, Johannes, A., 746

Laënnec, René T. H., 611
Lafayette, Marquis de, 779
Lagrange, Joseph Louis, 123
Lamarck, Jean Baptiste de, 709
Lamont, Johann von, 86
Lampland, Carl Otto, 108
Land, Edwin Herbert, 401, 469
Landau, Lev Davidovich, 50
Landsteiner, Karl, 569, 570, 639, 640
Langerhans, Paul, 680
Langevin, Paul, 170
Langley, John Newport, 765
Langley, Samuel Pierpont, 407
Langmuir, Irving, 197, 251, 268, 398, 458, 526, 530
Langsdorf, Alexander, 294
Lankard, John R. 423
Laplace, Pierre Simon de, 82, 83, 120, 121
Larsen, John Augustus, 769
Lartet, Edouard Armand, 728
Lassell, William, 125
Laue, Max Theodore Felix von, 240
Laveran, Charles Louis, 623
Lavoisier, Antoine Laurent, 200, 233, 234, 248, 274, 369, 525, 779
Lawes, John Bennett, 655
Lawrence, Ernest O., 245, 248, 305, 306
Lazear, Jesse William, 625
Leakey, Louis Seymour Burnett, 732–4

Leakey, Mary, 732, 733
Leavitt, Henrietta Swan, 28–30, 32, 35
Lebedev, Peter Nicolaevich, 594
Le Bel, Joseph Achille, 471, 472
LeCanu, L. R., 485
Lecoq de Boisbaudran, Paul Emile, 240
Lederberg, Joshua, 610, 628
Lee, Edmund, 786
Leeuwenhoek, Anton van, 527, 556, 607, 608
Lehmann, Inge, 156
Leibnitz, Gottfried Wilhelm von, 792, 793
Leith, Emmet N., 424
Lejeune, Jérôme Jean, 559
Lemaître, Georges, 43
Lemonnier, Pierre Charles, 125
Lenard, Philipp, 355
Lenoir, Etienne, 404
Leonardo of Pisa, 791
Leonov, Aleksei A., 97
Leopold (Prince), 565
Lerner, Richard A., 640
Lettvin, Jerome, 785
Leucippus, 234
Levene, Phoebus Aaron Theodore, 579, 580
Leverrier, Urbain Jean Joseph, 127, 360
Lewis, G. Edward, 733
Lewis, Gilbert Newton, 251
Ley, Willy, 96
Libby, Willard Frank, 730
Liebig, Justus von, 465, 466, 477, 478, 508, 518, 654, 671, 672
Lilienthal, Otto, 407
Lilly, John C., 170, 756
Lincoln, Abraham, 690
Lind, James, 658
Lindbergh, Charles Augustus, 408, 637
Lindbergh, Jon, 746
Linde, Karl von, 266
Linnaeus, Carolus, 698, 699, 705, 707
Lipmann, Fritz Albert, 535–8
Lippershey, Hans, 607
Lippmann, Gabriel, 401
Lister, Joseph, 612
Lister, Joseph Jackson, 608
Locke, Richard Adams, 94
Lockyer, Norman, 59
Lodge, Oliver Joseph, 410

Loewi, Otto, 770
Löffler, Friedrich August Johannes, 627
Lohmann, Karl Heinrich Adolf, 669
Lomonosov, Mikhail Vasilievich, 96
London, Heinz, 272
Long, Crawford Williamson, 528
Lorentz, Hendrik Antoon, 96, 352, 353, 357, 815
Lorenz, Konrad Zacharias, 767
Lorin, René, 408
Louis XV (King), 690
Love, Augustus Edward Hough, 156
Lovell, Bernard, 63
Lowell, Percival, 107, 117, 128, 129
Lucian of Samosata, 91
Lucretius, 234, 727
Lunge, George, 526, 530
Luther, Martin, 382
Lwoff, André Michael, 588
Lyell, Charles, 37, 708, 711, 716, 719
Lynen, Feodor, 538, 543
Lyons, Harold, 419
Lyot, Bernard Ferdinand, 69
Lysenko, Trofim Denisovich, 709
Lyttleton, R. A., 83

Mach, Ernst, 409
McCarthy, John, 797
McCarty, Maclyn, 582
McClintock, Barbara, 567
McCollum, Elmer Vernon, 660, 663
McCoy, Herbert Newby, 290
Macintosh, Charles, 491
Mackenzie, Kenneth Ross, 246
Macleod, Colin Munro, 582
Macleod, John James Rickard, 680
McMillan, Edwin Mattison, 247, 306
Macquer, Pierre Joseph, 507
Maffei, Paolo, 36
Magellan, Ferdinand, 29, 446, 658
Magendie, François, 655
Maiman, Theodore Harold, 421
Malpighi, Marcello, 554, 607
Malthus, Thomas Robert, 710–12, 739
Malus, Etienne Louis, 469
Mann, Thaddeus Robert Rudolph, 531, 672
Manson, Patrick, 623
Marconi, Guglielmo, 205, 411
Marcus Aurelius, 611

Marduk, 7
Marine, David, 675
Mariner, Ruth, 597
Marinsky, J. A., 246
Marius, Simon, 33, 112
Martin, Archer John Porter, 517–19
Matthei, J. Heinrich, 589
Mauchley, John William, 797
Maunder, Edward Walter, 87, 108, 109
Maurice (Prince), 607
Maury, Antonia C., 59
Maury, Matthew Fontaine, 154, 168, 171
Maxim, Hiram Stevens, 494
Maxwell, James Clerk, 61, 82, 96, 121, 210, 227, 348, 349, 357, 367, 392, 394, 401, 410, 594
Mayer, Cornell H., 101
Mayer, Jean, 682, 688
Mayer, Julius Robert, 366, 545
Maynard, J. Parkers, 496
Mayr, Ernst Walter, 732
Mechnikov, Ilya Ilitch, 639
Medawar, Peter, 642
Meissner, Walther, 270
Meister, Joseph, 636
Meitner, Lise, 244, 317, 434, 435
Melotte, P. J., 113
Ménard, Louis Nicolas, 496
Mendel, Gregor Johann, 561–4, 566, 571
Mendel, Lafayette Benedict, 662
Mendeleev, Dmitri Ivanovich, 96, 237, 240, 241, 247, 249, 256
Menghini, Vincenzo Antonio, 671
Menten, Maud Lenora, 530
Menzel, Donald Howard, 96, 116
Mering, Joseph von, 680
Merrifield, Robert Bruce, 523
Mersenne, Marin, 142
Mesmer, Franz Anton, 779
Messier, Charles, 33
Meyer, Julius Lothar, 237
Meyer, Viktor, 472
Meyerhof, Otto Fritz, 535
Michaelis, Leonor, 530
Michell, John, 147
Michelson, Albert Abraham, 344, 347, 350–2, 813, 814
Midgely, Thomas, 265, 406, 500
Miescher, Friedrich, 578, 579
Milankovich, Milutin, 187
Miller, Jacques F. A. P., 642

Miller, Neal Elgar, 776
Miller, Stanley Lloyd, 595–7
Millikan, Robert Andrews, 299
Milne, John, 147
Minkowski, Hermann, 358
Minkowski, Oscar, 680
Minkowski, Rudolf, 64
Minos, 725, 727
Minot, George Richards, 673
Mirsky, Alfred Ezra, 581
Miyamoto, Shotaro, 294
Mohl, Hugo von, 555
Mohorovicic, Andrija, 156
Mohs, Friedrich, 274
Moissan, Ferdinand F. H., 274, 275
Molchanoff, Pyotr A., 196
Moller, Christian, 298
Mondino de Luzzi, 552
Moniz, Antonio Egas, 760, 761
Monod, Jacques Lucien, 586, 582
Montagu, Lady Mary Wortley, 634
Montgolfier, Jacques Etienne, 195
Montgolfier, Joseph Michel, 195
Moore, Stanford, 517
Moore, Thomas, 662
Morgan, Thomas Hunt, 564, 566–8
Morgan, William Wilson, 76
Morley, Edward Williams, 351, 813, 814
Mörner, K. A. H., 509
Morse, Samuel Finley Breese, 392
Morton, William Thomas Green, 483
Morveau, Louis Bernard, Guyton de, 274
Mosander, Carl Gustav, 255, 256
Moseley, Henry G. J., 96, 241, 250
Moses, 426
Mössbauer, Rudolf Ludwig, 361, 362
Moulton, Forest Ray, 83
Mueller, Erwin Wilhelm, 237
Mulder, Gerardus Johannes, 507, 508
Muller, Hermann Joseph, 568
Müller, Johannes Peter, 700, 764
Müller, Otto Frederik, 168, 608
Müller, Paul, 618
Murchison, Roderick Impey, 716
Murdock, William, 396, 404
Murphy, William Parry, 673
Musschenbroek, Pieter van, 387

Nägeli, Karl Wilhelm von, 563
Nansen, Fridtjof, 180

Napier, John, 791
Napoleon I (Emperor), 396, 657
Napoleon III (Emperor), 278, 281
Nathans, Daniel, 591
Natta, Giulio, 499
Nebuchadnezzar, 726
Ne'eman, Yuval, 329
Nernst, Hermann Walther, 96, 270, 272
Nestor, 690
Neumann, John von, 376
Newcomen, Thomas, 383, 384
Newlands, John Alexander Reina, 237
Newton, Isaac, 14, 57, 58, 62, 79, 81, 85, 91, 135, 140, 142, 144, 146, 233, 235, 342, 343, 348, 351, 357-9, 366, 469, 592, 804, 808-11, 819
Nicholas II (Tsar), 565
Nicholas of Cusa, 24
Nicholson, Seth B., 113
Nicol, William, 469
Nicolet, Marcel, 204
Nicolle, Charles, 625
Niepce, Joseph Nicéphore, 400
Nightingale, Florence, 658
Nilson, Lars Fredrik, 240
Ninninger, Harvey Harlow, 225
Nirenberg, Marshall Warren, 589
Noah, 426, 697, 726, 735
Nobel, Alfred Bernhard, 495
Noddack, Walter, 242, 245
Norgay, Tenzing, 183
Norman, Robert, 212
Northrop, John Howard, 529, 531, 627

Oakley, Kenneth, 734
Oatley, C. W., 630
Oberth, Hermann, 93
Ochoa, Severo, 585, 589
Odin, 7
Oersted, Hans Christian, 209, 210, 281
Ohdake, S., 660, 661
Ohm, Georg Simon, 410
Olbers, Heinrich W. M., 130, 131
Oldham, Richard Dixon, 155
Olds, James, 762
Olds, Ransom E., 405
Oliphant, Marcus Lawrence Elwin, 303

Oliver, Bernard, 604
O'Neill, Gerard Kitchen, 747, 750
Onnes, Heike Kamerlingh, 268, 269
Oort, Jan, 31, 77, 136
Oparin, Aleksandr Ivanovich, 592
Oppenheimer, J. Robert, 70, 73, 440
Oro, Juan, 596
Ostwald, Wilhelm, 236, 368, 496
Otis, Elisha Graves, 279
Otto, Nikolaus, 404
Oughtred, William, 792
Owen, Robert, 713

Palade, George Emil, 587
Palmer, Nathaniel B., 182
Palmieri, Luigi, 147
Pandora, 4
Paneth, Friedrich Adolf, 540
Papin, Denis, 383, 785
Paracelsus, 232, 233, 614
Parker, Eugene Norman, 215
Parkes, Alexander, 496, 497
Parmalee, D. D., 792
Parsons, Charles Algernon, 385
Pascal, Blaise, 194, 792
Pasierb, Elaine, 325
Pasteur, Louis, 96, 470, 472, 527, 593-5, 610-13, 616, 626, 635-7, 639, 648, 659
Pauli, Wolfgang, 251, 317, 319
Pauling, Linus, 254, 455, 476, 514, 575, 583, 640, 665
Pavlov, Ivan Petrovich, 773
Payen, Anselme, 528
Payer, Julius, 179
Peary, Robert Edwin, 180
Pedro II (Emperor), 394
Pekelharing, Cornelis Adrianis, 660
Pelletier, Pierre Joseph, 546
Penfield, Wilder Graves, 783
Penzias, Arno Allan, 44
Peregrinus, Peter, 207, 209
Perey, Marguerite, 246
Pericles, 611
Perier, Florin, 194
Perkin, William Henry, 478-80, 482, 557, 613
Perkins, Jacob, 265
Perl, Marin L., 324
Perrier, Carlo, 245, 246
Perrin, Jean, 236, 237
Perrine, Charles Dillon, 113
Persoz, Jean François, 528
Perutz, Max Ferdinand, 523

Pestka, Sydney, 644
Peter I (Tsar), 607
Petri, Julius Richard, 613
Petrjak, K. A., 314
Pettengill, Gordon H., 100
Petterson, Hans, 222
Pfann, William Gardner, 418
Pfeffer, Wilhelm, 513
Phipps, James, 635
Piazzi, Giuseppe, 130, 131
Piccard, Auguste, 178, 196
Piccard, Jacques, 178
Piccard, Jean Félix, 196
Pickering, Edward Charles, 59
Pickering, William Henry, 122
Pictet, Raoul, 266
Pierce, John Robinson, 198, 417
Pipkin, Marvin, 398
Pirie, Norman Wingate, 628
Pitt-Rivers, Rosalind, 678
Planck, Max K. E. L., 354–6, 371, 374
Planté, Gaston, 404
Plass, Gilbert N., 190
Plato, 9, 149, 234
Platt, Hugh, 426
Pliny, 152, 507
Pliny the Younger, 152
Plotinus, 11
Pniewski, Jerzy, 327, 328
Pogson, Norman Robert, 28
Polo, Marco, 426
Polyakov, Alexander, 349, 350
Pomeranchuk, Isaak Yakovievich, 272
Ponnamperuma, Cyril, 596, 597
Pontecorvo, Bruno, 319
Pope, William Jackson, 472
Popov, Alexander Stepanovich, 96
Porta, Giambattista della, 400
Posidonius, 20, 167
Poulsen, Valdemar, 396
Powell, Cecil Frank, 326
Powell, George, 182
Powers, John, 796
Praxagoras, 553
Prebus, Albert F., 374
Pregl, Fritz, 466, 518
Priam, 690
Priestley, Joseph, 200, 491, 545
Prochorov, Aleksandr Mikhailovich, 421
Proust, Joseph Louis, 235
Prout, William, 289, 292, 654
Ptolemy, 10, 11, 21, 23, 96, 606

Ptolemy V (King), 725
Pupin, Michael Idvorsky, 395
Purcell, Edward Mills, 76, 516
Purkinje, Jan Evangelista, 555
Pythagoras, 9, 98, 138

Rabi, Isidor Isaac, 310, 441
Rabinowitch, Eugene I., 546
Rall, Theodore W., 689
Raman, Chandrasekhara Venkata, 516
Ramon, Gaston, 637
Ramon y Cajal, Santiago, 765
Ramsay, William, 201
Rankine, William John Macquorn, 366
Ranson, Stephen Walter, 682
Rasmussen, Howard, 681
Rasputin, Gregory, 565
Rassam, Hurmuzd, 726
Rawlinson, Henry Creswicke, 725
Ray, John, 698, 707
Rayleigh, John W. Strutt, Lord, 156, 201, 354
Reagan, Ronald, 744, 745
Réaumur, René A. F. de, 278, 527, 528
Reber, Grote, 63, 64
Redi, Francesco, 592, 593
Reed, Walter, 624, 627
Regiomontanus, 135
Regnault, Henri Victor, 200
Reichstein, Tadus, 664, 665, 684, 685
Reines, Frederick, 319, 325, 326
Reinmuth, Karl, 134
Remak, Robert, 700
Retzius, Anders Adolf, 737
Riccioli, Giovanni Battista, 96
Richards, Theodore William, 236
Richardson, Owen Willans, 411
Richelieu, Duc de, 690
Richer, Jean, 23
Richet, Charles Robert, 641
Richter, Burton, 332
Richter, Charles Francis, 149
Ricketts, Howard Taylor, 626
Ride, Sally, 97
Ringer, Sidney, 671
Rittenberg, David, 542, 543
Ritter, Johann Wilhelm, 61
Robbins, Frederick Chapman, 638
Robinson, Robert, 481, 482

Roche, Edouard, 121
Roemer, Olaus, 345
Roentgen, Wilhelm Konrad, 62, 96, 240, 646
Roget, Peter Mark, 401
Roosevelt, Franklin Delano, 436, 638, 745
Rorschach, Hermann, 758
Rose, William Cumming, 510, 656
Rosen, James M., 222
Rosenheim, Otto, 663
Ross, James Clark, 168, 182
Ross, Ronald, 623
Ross, William Horace, 290
Rosse, William Parsons, Lord, 33, 50, 385
Rossi, Bruno, 70
Rous, Francis Peyton, 648, 649
Rowland, Henry Augustus, 344
Ruark, Arthur Edward, 300
Ruben, Samuel, 542, 548, 549
Rumford, Benjamin Thompson, (Count), 365
Ruska, Ernst, 374
Russell, Bertrand, 691
Russell, Henry Norris, 51, 83
Rutherford, Daniel, 20
Rutherford, Ernest, 39, 62, 248, 285–90, 292, 293, 295, 304, 305
Ruzicka, Leopold, 480
Ryle, Martin, 66

Sabatier, Paul, 480
Sabin, Albert Bruce, 639
Sabine, Edward, 86
Sachs, Julius von, 546
Sagan, Carl, 109, 597, 601, 603, 721
Sager, Ruth, 581
Sainte-Claire Deville, Henri E., 281
Sakel, Manfred, 781
Salam, Abdus, 338
Salk, Jonas Edward, 638
Sandage, Allen, 66, 68
Sanger, Frederick, 520–22
Sarrett, Lewis H., 684
Saul, 724
Saussure, Nicolas Théodore de, 545
Sautuola, Marquis de, 728
Savery, Thomas, 383
Savitch, P., 434
Schäberle, John Martin, 50
Schaefer, Vincent Joseph, 197

Schaffer, Frederick I, 628
Scheele, Carl Wilhelm, 234, 477
Scheiner, Christoph, 85
Schiaparelli, Giovanni Virginio, 99, 107
Schirra, Walter M., 97
Schleiden, Matthia Jakob, 555
Schlieman, Heinrich, 726
Schmidt, Bernard, 60
Schmidt, Maarten, 66, 67
Schoenheimer, Rudolf, 541–3
Schönbein, Christian Friedrich, 494, 495
Schrödinger, Erwin, 375, 476
Schrötter, Anton, Ritter von, 397
Schulze, Max Johann Sigismund, 555
Schuster, P., 669
Schwabe, Heinrich Samuel, 86
Schwann, Theodor, 528, 555
Schweigger, Johann S. C., 209
Schwerdt, Carlton E., 628
Schwinger, Julian, 377
Scopes, John Thomas, 714, 715
Scott, K. J., 663
Scott, Robert Falcon, 182
Seaborg, Glenn Theodore, 240, 261
Secchi, Petro Angelo, 59, 60
Sedgwick, Adam, 716
Seebeck, Thomas Johann, 428
Segrè, Emilio Gino, 245, 246, 309
Semmelweiss, Ignaz Philipp, 611, 612
Senderens, Jean Baptiste, 480
Seneca, 46
Seyfert, Carl, 68
Shackleton, Ernest, 182
Shackleton, N. J., 187
Shakespeare, William, 690
Shannon, Claude Elwood, 794
Shapiro, Irwin Ira, 101
Shapley, Harlow, 30–35
Sharpey-Schafer, Edward Albert, 680
Shaw, George Bernard, 690
Shawlow, Arthur L., 423
Shemin, David, 543, 544
Sherrington, Charles Scott, 765, 769
Shimamura, T., 660
Shklovskii, I. S., 77
Shockley, William Bradford, 417, 418, 571
Siebe, Augustus, 177
Siebold, Karl Theodor Ernst, 608
Siegbahn, Karl Manne George, 241
Siemens, Karl Wilhelm, 279
Sikorsky, Igor Ivan, 407

Silliman, Benjamin, 220
Simon, Théodore, 757
Simpson, James Young, 483
Sinsheimer, Robert Louis, 585
Slipher, Vesto Melvin, 42, 202
Slye, Maude, 649
Smith, Ernest Lester, 673
Smith, George, 726
Smith, Hamilton Othniel, 591
Smith, J. L. B., 176
Smith, William (explorer), 180
Smith, William (geologist), 707
Snell, Willebrord, 342
Sobel, Henry W., 325
Sobrero, Ascani, 495
Socrates, 481
Soddy, Frederick, 244, 290
Solomon, 37
Sommerfeld, Arnold Johannes
    Wilhelm, 374
Sophocles, 690
Sorokin, Peter, 423
Spallanzani, Lazzaro, 169, 593
Speck, Richard, 566
Spedding, Frank Harold, 260, 437
Spencer, Herbert, 714
Sperry, Elmer Ambrose, 212
Spiegelman, Sol, 585, 631
Spinrad, Hyron, 116
Spitzer, Lyman, 83, 197, 459
Stalin, Josef V., 709
Stanley, Francis Edgar, 404
Stanley, Wendell Meredith, 627, 628
Stapp, John Paul, 749, 750
Stark, Johannes, 287
Starling, Ernest Henry, 677, 680
Stas, Jean Servais, 236
Staudinger, Hermann, 492
Steele, Jack, 784
Stefan, Josef, 354
Stein, William Howard, 517
Steinmetz, Charles Proteus, 305, 394
Stelluti, Francesco, 607
Stenhouse, John, 526
Steno, Nicolaus, 707
Stephenson, George, 385
Stern, Otto, 310
Stevens, Edward, 528
Stevinus, Simon, 806
Stokes, George Gabriel, 61
Stoney, George Johnstone, 250
Strassman, Fritz, 434
Struve, Friedrich Wilhelm von, 26,
    120

Sturgeon, William, 391
Sturtevant, Alfred Henry, 567
Sudarshan, Ennackal, C. G., 816, 817
Suess, Edward, 175
Sumner, James Batcheller, 529, 627
Sutherland Jr., Earl Wilbur, 689
Sutton, Walter Stanborough, 564
Suzuki, Umetaro, 660
Svedberg, Theodor, 512
Swallow, John Crossley, 166
Swammerdam, Jan, 554, 608
Swan, Joseph, 398
Swift, Jonathan, 110
Sylvester II (Pope), 790
Synge, Richard L. M. 517–19
Szent-Györgyi, Albert, 662
Szilard, Leo, 436

Tacke, Ida, 242, 245
Tainter, Charles Sumner, 395
Takaki, Kanehiro, 659
Takamine, Jokichi, 677
Talbot, William Henry Fox, 400
Tamm, Igor Yevgenevich, 354
Tartaglia, Niccolo, 13
Tatum, Edward Lawrie, 572–4, 610
Tayler, Gordon, 126
Taylor, Elizabeth, 618
Taylor, Frederick Winslow, 772
Teisserenc de Bort, Léon Philippe,
    195, 201, 204
Teller, Edward, 436, 455
Tempel, Ernst Wilhelm, 115
Tennant, Smithson, 274
Tereshkova, Valentina V., 97
Terman, Lewis Madison, 757
Tesla, Nikola, 393, 394
Thales, 7, 207, 386
Theiler, Max, 638
Theophrastus, 426, 697
Theorell, Hugo, 668
Thiele, Johannes, 474
Thilorier, C. S. A., 264–6
Thomas, Sidney Gilchrist, 279
Thomsen, Christian Jurgenson, 727
Thomson, Charles Wyville, 168, 169
Thomson, George Paget, 372
Thomson, Joseph John, 249, 250, 285,
    287, 291
Thomson, Robert William, 504
Thor, 6
Ting, Samuel Chao Chung, 332

Tiselius, Arne Wilhelm Kaurin, 575
Tissandier, Gaston, 195
Titian, 552, 690
Titov, Gherman Stepanovich, 96
Todd, Alexander Robertus, 580
Tombaugh, Clyde William, 128
Tomonaga, Sin-itiro, 377
Torricelli, Evangelista, 192
Townes, Charles Hard, 420, 421
Trefouël, Jacques, 615
Trevithick, Richard, 385
Ts'ai Lun, 382
Tschermak von Seysenegg, Erich, 563
Tsiolkovsky, Konstantin E., 92, 93, 96
Tsung Dao Lee, 336, 337
Tswett, Mikhail Semenovich, 260
Tuppy, Hans, 521
Ture, Merle A., 57
Turing, Alan Mathison, 794
Turpin, Raymond, 559
Twort, Frederick William, 632
Tyndall, John, 513

Uhlenbeck, George Eugene, 310
Upatnieks, Juris, 424
Urbain, Georges, 241
Ure, Andrew, 786
Urey, Harold Clayton, 162, 188, 230, 303, 541, 595
Ussher, James, 37, 724

Van Allen, James Alfred, 216
Van Calcar, Jan Stevenzoon, 552
Van de Graaf, Robert Jemison, 305
Van de Hulst, Hendrik Christoffel, 76
Van de Kamp, Peter, 601
Van der Waals, Johannes Diderick, 266
Van Helmont, Jan Baptista, 199, 200, 262, 545
Van Maanen, Adriaan, 33, 34
Van't Hoff, Jacobus Hendricus, 471, 472
Vaucanson, Jacques de, 788
Veksler, Vladimir Iosifovich, 306
Venetz, Ignatz, 183
Verne, Jules, 92, 96, 385
Vesalius, Andreas, 552, 553
Vestris, Michael, 727

Vicq d'Azyr, Felix, 758
Victoria (Queen), 385, 395, 478, 483, 565
Vigneaud, Vincent du, 522
Villard, Paul Ulrich, 285
Viviani, Vincenzo, 192
Virchow, Rudolf, 556, 645
Vogel, Hermann Carl, 42, 46
Volta, Alessandro, 389, 390
Voltaire, 110, 146
Vonnegut, Bernard, 197
Voorhis, Arthur D., 164

Wahl, Arthur Charles, 247
Waksman, Selman Abraham, 617
Wald, George, 670
Waldeyer, Wilhelm von, 557, 765
Wall, William, 387
Wallace, Alfred Russel, 711, 712
Wallach, Otto, 480
Wallich, George C., 168
Walsh, Don, 178
Walter, William Grey, 771, 800
Walton, Ernest Thomas Sinton, 305
Warburg, Otto Heinrich, 536, 537, 668
Washington, George, 281
Wasserman, August von, 637
Watson, James Dewey, 583
Watson, John Broadus, 774
Watson-Watt, Robert Alexander, 205, 347
Watt, James, 383–5, 786
Weber, Josep i, 316
Webster, T. A., 663
Weddell, James, 182
Wegener, Alfred Lothar, 157, 158, 173, 174, 180
Weinberg, Robert A., 648
Weinberg, Steven, 338, 340
Weiner, Joseph Sidney, 734
Weismann, August Friedrich Leopold, 709
Weiss, Pierre, 208
Weizsäcker, Carl Friedrich von, 84
Weller, Thomas Huckle, 638
Welles, Orson, 107
Wells, Herbert George, 107
Wells, Horace, 483
Wells, John West, 160
Welsbach, Karl Auer, Baron von, 396
Wendelin, Godefroy, 22

Werner, Abraham Gottlob, 153
Werner, Alfred, 472
Westall, R. G., 544
Westinghouse, George, 394
Westphal, Carl Friedrich Otto, 766
Weyprecht, Carl, 179
Weyssenhoff, H. von, 596
Wharton, Leonard, 488
Wheatstone, Charles, 392
Wheeler, John Archibald, 73
Whewell, William, 204
Whipple, Fred Lawrence, 136, 221
Whipple, George Hoyt, 673
White, Edward H., 97
Whitney, Eli, 405
Whittaker, Robert Harding, 699
Whittle, Frank, 409
Wickramasinghe, Chandra, 596
Wieland, Heinrich, 536, 537, 542, 650
Wien, Wilhelm, 354
Wiener, Norbert, 96, 799
Wigner, Eugene, 314, 336, 436, 440
Wilberforce, Samuel, 713, 714
Wildt, Rupert, 116, 228
Wilhelm II (Kaiser), 505
Wilkes, Charles, 182
Wilkins, George Hubert, 182
Wilkins, Maurice Hugh Frederick, 583
Willcock, Edith Gertrude, 656
Williams, Carroll Milton, 619
Williams, Robert Runnels, 661, 664
Williams, Robley Cook, 629, 630
Willis, Thomas, 758
Willstätter, Richard, 260, 484, 528, 529, 547
Wilm, Alfred, 282
Wilska, Alvar P., 629
Wilson, Alexander, 85, 195
Wilson, Charles Thomson Rees, 293
Wilson, Robert Woodrow, 44
Windaus, Adolf, 663
Winkler, Clemens Alexander, 240
Wislicenus, Johannes, 470

Witt, Gustav, 133
Woese, Carl R., 716
Wöhler, Friedrich, 464–6, 477
Wolf, Maximilian F. J. C., 132
Wolf, Rudolf, 86
Wolff, Kaspar Friedrich, 706
Wolfgang, Richard Leopold, 487
Wollaston, William Hyde, 509
Woodward, Arthur Smith, 734
Woodward, Robert Burns, 482, 513, 514, 547
Woolfson, M. M., 83
Wright, Orville, 407
Wright, Seth, 563, 568
Wright, Wilbur, 407
Wunderlich, Karl August, 363
Wundt, Wilhelm, 771
Wyckoff, Ralph Walter Graystone, 629
Wynn-Williams, Charles Eryl, 293

Xenophanes, 707
Yagi, Kunio, 531
Yamagiwa, K., 645
Yarmolinsky, Michael, 621
Yeager, Charles Elwood, 409
Yegorov, Boris G., 97
Young, John W., 97
Young, Thomas, 343, 725
Young, William John, 534, 535
Yukawa, Hideki, 322, 326, 377

Zelikoff, Murray, 202
Zeus, 6
Ziegler, Karl, 498, 499, 505, 525
Zinn, Walter Henry, 447
Zondek, Bernhard, 683
Zsigmondy, Richard Adolf, 513
Zwicky, Fritz, 47, 70
Zworykin, Vladimir Kosma, 414, 415

Subject Index

Abacus, 790, 791
Aberration, light, 345
A, B, O blood groups, 737
A-bombs, 441
Abortion, 744
Abrasives, 274–6
Absolute magnitude, 28
Absolute motion, 350
Absolute rest, 350
Absolute space, 350
Absolute zero, 263, 272
Absorption spectrum, 517
Acceleration, 807; space-travel and, 749, 750
Acetic acid, 233, 465, 477
Acetic anhydride, 480
Acetylation, 538
Acetylcholine, 770
Acetylene, 467, 468
Achaeans, 278
Achilles, 132
Achromatic lenses, 58
Achromatic microscopes, 608
Achromycin, 617
Acid rain, 429
Acquired characteristics, inheritance of, 709
Acquired immune deficiency syndrome (AIDS), 644
Acrilan, 504
Acromegaly, 687
Acrylonitrile, 503
ACTH, 685, 686
Actinides, 238, 239, 261
Actinium, 244, 261; breakdown of, 290

Action at a distance, 348
Activated carbon, 526
Actomyosin, 524
Adamantine substance, 416
Addison's disease, 684
Adenine, 579
Adenoma, 645
Adenosine triphosphate (ATP), 536
Adenyl cyclase, 689
Adonis, 134
Adrenal glands, 677, 683
Adrenalin, 677, 782; nerves and, 769
Adrenocorticotropic hormone (ACTH), 533, 685, 686
Adsorption, 526
Aegina, 277
Aether (see also Ether) 89, 232
Afferent nerves, 753
Agammaglobulinemia, 643
Agar, 613
Agnatha, 703
AIDS, 644
Air (see Atmosphere, Earth's)
Air conditioning, 204, 265, 266
Airglow, 201, 202, 214
Air, liquid, 266, 267
Airplanes, 407–10
Air pressure, 193, 194
Air pump, 383
Alabamine, 245
Alanine-transfer-RNA, 587, 588
Albino, 572
Albumen, 507
Alchemy, 233
Alcoholism, 485
Aldosterone, 685

Algae, 609, 699
Algol, 41, 46
Aliphatic compounds, 474
Alizarin, 479
Alkaline earths, 255
Alkaloid, 481, 482
Alleles, 562
Allergies, 641
Allotropes, 273
Alloying, 283, 284
Alloys, 269
Alloy steel, 279, 280
Allyltoluidine, 478
Alnico, 280
Alpha-amino acids, 508
Alpha Centauri, 28, absolute
    magnitude of, 28; distance of, 25, 26;
    possible habitable planets of, 603;
    spectral class of, 59
Alpha-interferon, 644
Alpha particles, 285–8; danger of,
    449; detection of, 292; energy of,
    317; nuclear change and, 290, 291
Alpha rays, 285
Alpha wave, 770
Alphonsus, 224
Alpine Europeans, 737
Alps, 183
Alternating current, 393, 394
Aluminum, 281, 282;
    superconductivity of, 269
Aluminum chloride, 480
Aluminum oxide, 260
Amalthea, 113; size of, 118
Amber, 386
American Indians, 737
Americium, 247
Amine group, 508
Amine oxidase, 782
Amino acids, 502, 508; abiogenic
    formation of, 595, 596;
    chromatographic separation of,
    517; essential, 655–7; fossils and,
    721; optical activity of, 515;
    symbols of, 519
Amino groups, 502
Ammonia, 265, 508; atmosphere and,
    228, 229; electron configuration
    of, 252; excretion of, 721; explosives
    and, 496; interstellar, 78;
    molecular vibration of, 419;
    structural formula of, 467
Ammonites, 720
Ammonium cyanate, 465

Amor, 134
Ampere, 209, 390
Amphibians, 703, 719
AM radio, 411
Analog computer, 797
Anaphylactic shock, 641
Anatomy, 551–3
Ancestors, common, 723
Androgens, 683
Andromeda galaxy, 34; distance of,
    36; local group and, 36; nova in,
    47; radial velocity of, 42; spiral
    arms of, 76
Andromeda Nebula, 33, 114
Androsterone, 683
Anemia, 671; pernicious, 673
Anesthetics, 483
Angels, 347
Angina pectoris, 691
Angiospermae, 699
Angular momentum, conservation of,
    82, 83
Aniline, 478
Animalcules, 608
Animal electricity, 389
Animal kingdom, 699
Animal magnetism, 779
Animal spirits, 764
Animal viruses, 628
Annelida, 701
Annelid superphylum, 700
Anopheles, 623, 624
Antarctica, 181; fossils in, 158; ice cap
    of, 180, 184; meteorites in, 225,
    226; temperatures in, 183
Antarctic Circle, 180
Antarctic convergence, 165
Antarctic peninsula, 182
Antares, 51
Antenna, radio, 411
Anthrax, 613
Anthropic principle, 778, 779
Antibiotics, 616, 617, 639ff.
Antibiotic spectrum, 617
Antibodies, 639ff.
Antideuteron, 311
Antielectron, 299, 300
Antiferromagnetism, 208
Antigens, 639
Antihelium-3, 311
Antihistamines, 642, 643
Anti-knock compounds, 406
Antimatter, 310, 311
Antimuonium, 323

Blacktongue, 667
Blast furnace, 279
Blasting oil, 435
Blink comparator, 128
Blood, 671; circulation of, 553–5
Blood poisoning, 643
Blood pressure, 691
Blood transfusion, 569, 570
Blood types, 569, 570; races and, 737, 738
Blue-green algae, 599, 600, 609, 622, 717
Blue stellar objects, 68
Bombs, nuclear, 440–44
Bonds, chemical, 467
Bone Age, 727
Bones, 671
Bony fish, 703
Boolean algebra, 794
Borazon, 276
Boron, 276; plants and, 672; steel and, 284
Boron nitride, 276
Bose-Einstein statistics, 310
Bosons, 310
Boyle's law, 278
B-quark, 332
Brachycephalic skull, 736
Brain, 551, 756ff.; evolution of, 754, 755; function of parts of, 759, 760; hormones and, 687; storage capacity of, 4
Brain–body ratio, 756
Brain stem, 763
Bread, 526
Bread mold, 572
Breeder reactors, 448
Bremsstrahlung, 308, 646
Brent Crater, 224
Briggsian logarithms, 791, 792
Brine, 427
Broad-spectrum antibiotics, 617
Broca's convolution, 759
Brome-grass mosaic virus, 629
Bronze, 277
Bronze Age, 277, 727; population during, 738, 739
Brownian motion, 236
Bryophyta, 699
BSOs, 68
Bubble chamber, 294
Bufotenin, 782
Bulldog, 168
Buna rubber, 506

Butadiene, 475, 505, 506
Butane, 467, 468
Buzz-bombs, 409, 799
B-vitamin complex, 661, 664, 665, 669

Calciferol, 663
Calcitonin, 681
Calcium, 234, 255; parathormone and 681
Calcium carbonate, 700
Calcium oxide, 255, 282, 396
Calcium phosphate, 615
Calcium rigor, 615
Calculating machines, 792, 795
Calculation, 789
Californium, 247
Callisto, 112; surface of, 117
Caloric, 365
Caloris, 103
Cambrian age, 717, 718
Camphor, 496
Canals, Martian, 107–9
Canaries current, 163
Cancer, 644ff.; skin, 203
Candide, 146
Canned food, 657, 658
Cannibal galaxy, 65
Capacitor, 387
Capella, 28
Capillaries, 554
Carbohydrates, 489–91, 654
Carbon, 471; allotropes on, 273–6; electron shells and, 252, high pressure and, 273–6
Carbon 11, 548
Carbon 12, 370
Carbon 13, 541
Carbon 14, 730; half-life of, 450; sunspot cycle and, 87; tracer activity of 542ff., 548
Carbonaceous chondrites, 223
Carbon compounds, asymmetric, 472, 473
Carbon dioxide, 672; atmosphere and, 230, 231; discovery of, 200; electron configuration of, 252; fossil fuels and, 430; greenhouse effect and, 189, 190; liquid, 264; Mars's atmosphere and, 108; photosynthesis and, 546, 548, 549; solid, 265, 266; sublimation point

Carbon dioxide – *cont.*
of, 267; Venus's atmosphere and, 102
Carbonic anhydrase, 672
Carboniferous age, 719
Carbon monoxide, 201; liquid, 266; poisonous nature of, 530
Carbon tetrachloride, 499
Carborundum, 274–6
Carboxyl groups, 502, 508
Carcinogens, 645, 646, 648, 650
Carcinomas, 645
Carnivores, 653, 704
Carotene, 662
Cartilage, 701, 703
Casein, 507, 508, 656
Cash register, 792
Cassini's division, 120
Cast iron, 278
Catalase, 529
Catalysis, 525ff.
Catalysts, 498, 499, 525; protein, 529; specificity of, 526
Catastrophism, 708
Cathode, 249
Cathode rays, 249
Caucasian race, 735
Causality, 377
Cell division, 557, 558
Cellophane, 502
Cells, 554ff.; differentiation of, 561; division of, 557, 558; dyes and, 557; hybrid, 648; origin of, 598; size of, 556, 626
Cells, electric, 389, 391
Cells, photovoltaic, 431, 432
Celluloid, 497
Cellulose, 489, 524; decomposition of, 621; modification of, 494, 496, 497, 501, 502; structure of, 490
Cellulose acetate, 497, 502
Cellulose nitrate, 501
Celsius scale, 364
Celtium, 241
Cenozoic era, 717
Census, American, 795, 796
Centigrade scale, 364
Central nervous system, 753, 763, 764
Centrifuge, 511, 512
Cephalic index, 737
Cepheids, 27–30; stellar populations and, 35
Cerebellar hemispheres, 763

Cerebellum, 552, 754; feedback and, 788
Cerebral cortex, 754
Cerebral hemispheres, 761
Cerebrum, 552, 752
Cerenkov counters, 354
Cerenkov radiation, 354
Ceres, 131
Ceria, 255
Cerium, 256, 361
Cesium, 59; rocketry and, 268
Cetane, 407
Cetane number, 407
Cetyl alcohol, 740
Chain reaction, nuclear, 435
Challenger (I), 169, 171
Challenger (II), 175
Challenger Deep, 175
Chambered nautilus, 720
Chandrasekhar's limit, 56
Characteristic X-rays, 240, 241
Charcoal, 278, 526
Charge conjugation, 337
Charged-current reaction, 325
Charge, electric: fractional, 331; mass and, 315
Charles's law, 262
Charmed particles, 332
Charon, 129, 130
Cheetah, 706
Chemical accelerator, 487, 488
Chemical atomic weights, 370
Chemical compound, 235
Chemical evolution, 595–8, 716
Chemical fertilizers, 546
Chemical lasers, 422, 423
Chemical symbols, 465
Chemical thermodynamics, 368
Chemosynthetic bacteria, 550
Chemotherapy, 613ff.
Chemotropism, 772
Chewing gum, 493
Chick embryo, 638
Chickenpox, 634
Chicle, 493
Chimneys, oceanic, 176
Chimpanzees, 776
Chiron, 133
Chiroptera, 704
Chitin, 701
Chloral hydrate, 484
Chloramphenicol, 617, 621
Chlorella, 548
Chlorine, 234; isotopes of, 292;

liquid, 264; nuclear structure of, 297
Chlorophyll, 546, 547, 599
Chloroplasts, 546, 547
Chloroprene, 506
Cholera, 613, 635
Cholesterol, 663; atherosclerosis and, 691, 692; synthesis in body of, 542, 543
Cholinesterase, 770
Chondrichthyes, 703
Chordates, 703; appearance of, 718
Christofilos effect, 217
Chromatic aberration, 58
Chromatin, 557
Chromatography, 260, 261
Chromium, 672; steel and, 279; synthetic diamonds and, 276
Chromosome maps, 567
Chromosomes, 557ff.; crossing over, 566, 567; DNA and, 581; genes and, 564
Chubb crater, 224
Chymotrypsinogen, 532
Cigarette lighter, 397
Cigarettes, 399, 646
Cilia, 608
Circadian rhythms, 777
Circle of fire, 173
Circulation, blood, 553–5
Circumcision, 709
Citric acid cycle, 537
Citrulline, 574
Civilization, 727
Class, 698
Classical physics, 355
Clermont, 385
Clinical thermometer, 364
Cloaca, 722
Clock, 142
Clock paradox, 358, 359
Clockwork, 788
Clones, 560, 561, 694, 695
Cloud chamber, 293, 294
Cloud seeding, 197
Coagulation, blood, 664
Coal, 274, 426; formation of, 719; oil from, 428
Coal gas, 396
Coalsack, 76
Coal tar, 478
Cobalt, 167; ferromagnetism of, 208; life and, 672–4; steel and, 279; vitamin B12 and, 673

Cobalt 60, 650
Cocaine, 481–4; anesthetic qualities of, 482; synthesis of, 484
Cocci, 608, 609
Cod-liver oil, 665
Codon, 586
Codon dictionary, 589, 590
Coelacanth, 176
Coelenterata, 700
Coenzyme A, 538
Coenzymes, 666–9
Coherent light, 422
Coincidence counters, 296
Coins, 277
Coke, 278, 426
Collapsed stars, 49, 50
Cold, common, 627
Collagen, 524
Collodion, 496, 500, 501
Colloids, 512ff.
Color (quark), 332
Color-blindness, 565, 566
Color photography, 401
Color television, 261, 415
Coma, 137
Cometary cloud, 136, 721
Comet Encke, 136, 223
Comet Halley, 136
Comet Kohoutek, 136
Comets, 45, 46, 135–7, 596; meteors and, 221; orbits of, 135, 136; structure of, 136; tails of, 135, 137, 215
Communication satellites, 198
Comparative anatomy, 554
Compass: gyroscopic, 212; magnetic, 207
Competitive inhibition, 531, 620
Complement, 637
Complementarity, principle of, 356
Compound A, 684, 685
Compound, chemical, 235
Compound E, 684
Compound microscope, 527
Compression ratio, 406
Compton effect, 374
Computers, 795ff.; chemistry and, 488, 674
Concorde, 410
Condensation, 488
Condenser, electric, 387
Conditioned responses, 773ff.
Conductors, electric, 387
Cones, retinal, 670

Coniine, 481
Conjugated double bond, 475
Conjugation, 610
Connector neurons, 765
Conservation laws, 335ff.
Contact lenses, 607
Continental drift, 159–9, 173
Continental shelves, 185
Continuous creation, 43
Convolutions, brain, 755
Coordination theory, 472
Copper, 277; alloys of, 277;
    beryllium and, 284; life and, 672;
    nodules and, 167
Copper sulfate, 682
Coptic language, 725
Cordite, 494
Core, Earth's, 155, 156
Coriolis effect, 139, 163, 164
Corona (solar), 69, 70, 214, 215;
    sunspots and, 87
Coronary arteries, 691
Coronary bypass surgery, 692
Coronary thrombosis, 691
Coronium, 70
Corpus callosum, 761
Corrosion, 280
Cortex: adrenal, 683; cerebral, 754;
    interpretative, 783
Cortical hormones, 683–6
Corticotropins, 786
Cortisone, 684
Cosmic particles, 299
Cosmic rays, 215, 298, 299, 311–13,
    730; evolution and, 724
Cosmonauts, 96, 97
Cosmotron, 308
Cotyledon, 698
Coulomb, 389
Coumarin, 479
Cowpox, 635
Cowpox virus, 629
Cows, 490
CP conservation, 337, 338
CPT symmetry, 338
C-quark, 332
Crab Nebula, 50, 51; pulsar in, 71, 72;
    radio waves from 64; X-rays
    from, 70
Craters: fossil, 224; lunar, 224, 225
Creation, Biblical date of, 37
Creationism, 715
Cretaceous age, 719
Cretan civilization, 149, 725–7

Critical temperature, 266
Cro-Magnon man, 728
Cromwell current, 164
Crossing over, 566, 567
Crossopterygii, 719
Cross-section, nuclear, 433
Cryogenics, 271
Cryolite, 281
Cryonics, 694
Cryostat, 269
Cryptozoic eon, 717
Crystalloids, 512
Crystals, 240, 241; asymmetric, 470
Crystal sets, 411
Culture media, 613
Cuprous oxide, 682
Curie, 453
Curie temperature, 208
Curiosity, 3ff.
Curium, 247
Currents: electric, 389ff.; ocean,
    163–6
Cyanocobalamin, 664, 673, 674
Cybernetics, 799
Cyclic-AMP, 689
Cyclone, 140
Cyclonite, 495
Cyclotron, 306, 307
Cygnus X-1, 74
Cysteine, 509
Cystine, 509
Cytochrome C, 723
Cytochrome oxidase, 537
Cytochromes, 536, 547, 723
Cytosine, 579

Dacron, 504
Daguerrotypes, 60, 400
D-amino acids, 598, 616, 620
Dark nebulae, 76
Darwin's finches, 710
Day, lengthening of, 160
DDT, 618, 619, 625
Death rate, 611
Decay bacteria, 621
Decinary system, 792, 793
Declination, magnetic, 212
Decomposers, 621
Deduction, 9
Deficiency diseases, 658–60
Degenerate gas, 50
De Humani Corporis Fabrica, 552

Dehydrogenases, 536, 667
Deimos, 106; size of, 110
Delayed neutrons, 439
Delirium tremens, 485, 782
Delta-aminolevulinic acid, 544
Delta Cephei, 27, 28
Delta wave, 770
De Magnete, 208
De Motus Cordis, 554
Denaturation, protein, 515, 516
Dendrochronology, 729
Deoxyribonucleic acid (DNA), 580
Deoxyribose, 579, 580
Desalination, 167
Desiccant, 526
Detergents, synthetic, 742
Determinism, 773
Detonators, 495
Deuterium, 303
Deuteron, 303
Devonian age, 718
Dewar flasks, 267
Dextrorotation, 470
Diabetes mellitus, 679, 680, 692
Diacetylmorphine, 484
Dialysis, 513, 666
Diamagnetism, 270
Diamond, 273–6; synthetic, 274–6
Diastase, 528
Diatomaceous earth, 495
Diatoms, 495
Dibenzanthracene, 645
Dichlorodiphenyltrichloroethane (DDT), 618, 619
Dicotyledonous plants, 698
Diels-Alder reaction, 481
Diesel engine, 406
Diesel fuel, 406, 407
Diet, 652ff.
Dietetics, 658
Diethyl ether, 266, 483
Differentiation, cell, 561
Diffraction grating, 240, 343, 344
Diffusion, 512
Diffusion cloud chamber, 294
Digestion, 527, 528, 653
Digit, 789
Digital computer, 790, 797
Diisopropylfluorophosphate group, 532
Dimethoxyphenylethylamine, 782
Dimethyl ether, 464; boiling point of, 476; structural formula of, 468
Dinitrophenyl group, 520

Dinosaurs, 718; disappearance of, 720
Dinucleotide, 597
Diode, 412
Dione, 123
Dipeptides, 521
Diphosphopyridine nucleotide, 667
Diphtheria, 613, 636, 637
Dip needle, 208
Dipole moment, 475, 476
Direct current, 393
Disaccharide, 492
Disease: deficiency, 658–60; germ theory of, 610; transmission of, 612ff.; virus, 627
Disulfide link, 515
Diving suit, 177
DNA, 580ff., chromosomes and, 581; enzyme formation and, 586ff.; replication of, 584, 585; structure of, 582, 583, 584; viruses and, 628, 631–2
DNA ligase, 591
Dogs, 773
Dog whistle, 170
Dolichocephalic skull, 736
Dolphins, 756, 757; swimming technique of, 784, 785
Domains, magnetic, 208, 280
Doppler effect, 40
Doppler-Fizeau effect, 41, 287
Dorians, 278
Down's syndrome, 559
D-quark, 331
Dreams, 762, 763
Drug addiction, 483
Dry ice, 265
D-series, 473
Duckbilled platypus, 704, 706
Ductless glands, 677, 678
Duralumin, 282
Dwarf, 687
Dyes, 478, 479; cells and, 557
Dynamic electricity, 389
Dynamic State of Body Constituents, The, 541
Dynamite, 495
Dynamo, 391
Dynel, 504
Dysprosium, 208

Early Bird, 199
Early Europeans, 738

Earth, 138ff.; age of, 37–9; ancient temperatures of, 188, 189; atmosphere of, 91; central temperature of, 155; circumnavigation of, 20; density of, 106, 145; formation of, 161; inner core of, 156; interior of, 153ff.; internal pressure of, 273; liquid core of, 155, 156; magnetic field of, 86, 210ff.; magnetic poles of, 210, 212; mantle of, 156, 157; mass of, 144, 145; oblateness of, 115, 140, 141; orbital changes of, 187; original atmosphere of, 230; rotation of, 139, 140; shape of, 138, 140–43; size of, 20; surface speed of, 139

Earth-grazers, 133–5
Earthlight, 89
Earthquake belts, 150, 151
Earthquake waves, 153, 154
Earthquakes, 146–9; energy of, 148; plate tectonics and, 173
Earth resource satellites, 199
Earths (minerals), 255
Earthworm, 701
Ebonite, 493
Echidna, 704
Echinodermata, 701
Echinoderm superphylum, 700
Echo I, 198
Echo-location, 169
Eclipse: lunar, 89; solar, 59, 90
Eclipsing binary stars, 41, 42
Ecology, 742
Ectoderm, 700
Edentata, 704
Edison effect, 411
EEG, 770, 771
Efferent nerves, 753
Efficiency expert, 772
Egg cells, 556; chromosomes in, 558, 559
Eggs, 722
Eightfold way, 329
Einsteinium, 247
Einstein shift, 50, 360
Elastomers, 506
Electra complex, 780
Electric arc, 397
Electric battery, 389, 390
Electric chairs, 393
Electric cells, 389, 391
Electric charge, 388; conservation of, 302, 328; fractional, 331; mass and, 315

Electric current, 389ff.
Electric generator, 391, 392
Electricity, 386ff.; living tissue and, 768; magnetism and, 209; solutions and, 204, 205
Electric light, 397–400
Electric motor, 392, 393
Electric potential, 387
Electric relays, 392
Electric storm, 213
Electric transformer, 393
Electrocardiogram, 768
Electroencephalogram (EEG), 770, 771
Electroluminescence, 399, 400
Electrolysis, 389
Electromagnetic interaction, 61, 320, 334
Electromagnetic radiation, 210, 349
Electromagnetism, 209, 210
Electromagnets, 391, 392; superconductivity and, 270
Electromotive force, 387, 390
Electron gun, 415
Electronics, 410ff.
Electron microscope, 372, 373; viruses and, 629
Electron-positron pair, 371
Electrons, 287, 288, 315, 386, 387; atomic sharing of, 475; discovery of, 250; nucleus and, 289, 302; periodic table and, 250, 251; spin of, 295, 296; wave nature of, 372, 375
Electron shells, 251ff.; transitional elements and, 257ff.
Electron synchrotron, 307, 308
Electron volt, 305
Electrophoresis, 575
Electroscope, 298
Electrostatic generator, 305
Electrostriction, 170
Electroweak field, 338
Elementary particles (cellular), 538
Elements, chemical, 232ff.; atomic numbers of, 241; electron shells and, 251ff.; families of, 253; Greek, 232; radioactive, 242ff., spectral lines and, 59; superconductivity of, 269; super-heavy, 248, 249; supernova formation of, 56; symbols of, 465; transitional,

257ff.; transmutation of, 295; transuranium, 246

Elephants, 682; brains of, 756

Elevators, 279

Embryo, 706

Embryology, 706

Empirical formulae, 465

Enantiomorphs, 470

Enceladus, 123

Endocrine glands, 677, 678

Endoderm, 700

Endorphins, 687

Energy, 366ff., 425ff.; attomic, 432; conservation of, 38, 317, 366; mass and, 40, 369; uses of, 379ff.

Engine, 383

ENIAC, 797

Enkephalins, 687

Entropy, 270, 367

Environmental quality, 744, 745

Enzymes, 528f.; active regions of, 531, 532; body chemistry and, 534; coenzymes and, 666–9; DNA replication and, 585; efficiency of, 529; genes and, 574; nucleic acids and, 586ff; protein nature of, 528, 529; specificity of, 530; surface catalytic activity of, 530

Enzyme-substrate complex, 530, 531

Eoanthropus dawsoni, 734

Epilepsy, 610

Epinephrine, 677

Equatorial bulge, 140

Equatorial diameter, 115

Erect posture, 733; brain and, 755

Erg, 369

Ergosterol, 663

Erie, Lake, 742

Eros, 133; parallax of, 23

Escape velocity, 226, 227

Escherischia coli, 644

Essential amino acids, 655, 656

Essential fatty acids, 657

Estrogens, 683

Estrone, 683

Eta Carinae, 48

Ether (see also Aether), 191, luminiferous, 348, 811

Ethiopian race, 735

Ethyl alcohol, 464; boiling point of, 476; consumption of, 485; structural formula of, 468; synthetic, 477; yeast and, 527

Ethylene, 498; structural formula of, 467, 468

Ethylene bromide, 406

Ethyl gasoline, 406

Eugenics, 570, 571

Eukaryotes, 590, 591, 609, 717

Europa, 112; possible life on, 601; surface of, 117, 118

Europium oxide, 261

Eutheria, 704

Eutrophication, 742

Evaporation, 264

Evening star, 98

Everest, Mount, 183

Everglades, 742

Evolution, 706ff.; biochemical aspects of, 721–3; chemical, 595–8; course of, 716ff.; human, 724; natural selection and, 711; opposition to, 713, 714; rate of, 723, 724

Exchange forces, 321

Exclusion principle, 251

Exons, 590

Exosphere, 197

Exotic atoms, 326

Expectancy wave, 771

Experimental psychology, 771

Experimentation, 12

Exploding galaxies, 65

Explorer I, 94, 216

Explorer III, 216

Explorer IV, 216, 217

Explorer XIV, 216

Explorer XVI, 221

Explosives, 494–6

Exponential notation, 791

External skeletons, 700, 701

Extra-galactic nebulae, 34

Extrasensory perception, 783

Extraterrestrial intelligence, 604, 605

Fahrenheit scale, 363

Fallout, radioactive, 452ff.

Fantail, 786

Farad, 390

Faraday, 390

Fast pulsar, 73

Fats, 654

Fat-soluble A, 660

Fatty acid oxidation cycle, 538

Fatty acids, 538, 657

Faults, geologic, 173
Feedback, 680, 681, 786–8
Fermentation, 527
Ferments, 528
Fermi-Dirac statistics, 310
Fermions, 310
Fermium, 247
Ferns, 699
Ferrites, 280
Ferromagnetism, 208
Fibers, protein, 524
Fibroin, 514
Field ion microscope, 237
Fields, 333; magnetic, 209
Filicineae, 699
Film, photographic, 400, 401
Filtrable virus, 627
Fine arts, 5
Finnegans Wake, 331
Firmament, 19
First-generation stars, 56
Fish, 718
Fission bombs, 440–2
Fission (nuclear), 314; discovery of, 434, 435
Fission products, 449
FitzGerald equation, 352, 814, 815
Flagellae, 608
Flare, solar, 214
Flatworms, 700
Flavors, 324, 332
Fleas, 697
Flint and steel, 460
Flints, lighter, 261
Flood, Biblical, 707
Florentium, 245
Fluorescence, 242, 243
Fluorescent lighting, 399
Fluoridation, 675, 676
Fluorine, 499; electrons and, 252; inert gases and, 254, 255; liquid, 267
Fluorine molecules, 252
Fluorocarbons, 265, 499
FM radio, 414
Folic acid, 620, 664
Follicles, ovarian, 554
Follicle-stimulating hormone, 686
Food, 652ff.
Force, 809, 810
Forebrain, 754
Forests, 425
Formaldehyde, 497, 548, 549; interstellar, 78

Formulae, chemical, 466–8
Fossil craters, 224
Fossil fuels, 426ff.; atmospheric pollution by, 429, 430
Fossils, 707
Foucault pendulum, 139
Francium, 246
Fraunhofer lines, 58
Free energy, 367
Free radicals, 455
Free will, 773
Freon, 203, 204, 265, 499, 500
Frequency, 61, 62
Frequency modulation (FM), 414
Fresh water, 167, 740
Friedel-Crafts reaction, 480
Frogs, 703; retina of, 785
Frosted light bulbs, 398
Fructose, 490
Fructose diphosphate, 534
Fruit flies, 564ff.
Fuel cells, 391
Fuels, 425ff.; fossil, 426ff.
Full moon, 88
Fulminates, 466
Fungi, 608, 699
Fusion bomb, 442–4
Fusion, controlled, 456, 457–9

Galactose, 490
Galactosemia, 577
Galapagos Islands, 710
Galaxies, 34–7; clusters of, 36; cores of, 65; hydrogen and, 77; radial velocities of, 42; radio waves from, 64, 65
Galaxy (Milky Way): center of, 30, 31; dust in, 31; mass of, 31; number of stars in, 31; rotation of, 30–31; size of, 26, 30, 31; spiral arms of, 76
Galilean satellites, 111
Galileo's lamp, 142
Gallium, 240
Galvanometer, 209
Gamma globulin, 643
Gamma rays, 62, 243, 285; Mössbauer effect and, 361, 362
Ganymede, 112; surface of, 117
Gas chromatography, 516
Gases, 200, 427; cooling of, 262; degenerate, 50; interstellar, 75;

kinetic theory of, 367; liquefaction of, 262ff.
Gas lighting, 396
Gas masks, 526
Gasoline, 406
Gastric juice, 528
Gatling gun, 494
Gegenschein, 221
Geiger counters, 215
Geissler tube, 249
Gelatin, 508, 655, 656
General anesthetic, 483
Generalization, 12
General relativity, 359–62
Generator, electric, 391, 392
Genes, 562ff.; carbon 14 and, 450; enzymes and, 574; isolation of, 568; locus of, 567; number of, 567; shifting of, 567
Genetic code, 586, 589, 590
Genetic engineering, 568, 622
Genetic load, 568, 569
Genetics, 564
Genus, 698
Geocentric parallax, 22
Geoid, 142
Geometry, 8
Germanium, 240, 416
Germ-free animals, 664
Germ plasma, continuity of, 709
Germs, 608
Germ theory of disease, 610
Giants, 687, 735
Giant squid, 697
Gibberellin, 739
Gigantopithecus, 735
Glaciers, 183ff.; movement of, 183; retreat of, 190
Glands, 677
Glass, 500; scientific instruments and, 363; ultraviolet and, 61
Gliders, 407
Global village, 745
Globular clusters, 30; galaxies and, 34; stellar evolution and, 53
Globulin, 507
Glucagon, 681
Glucose, 489, 490; blood content of, 682, 683; insulin and, 680; photosynthesis and, 547, 548; structural formula of, 490; urine and, 679, 681, 682
Glucose dehydrogenase, 682
Glucose tolerance test, 682

Glueballs, 333
Gluon, 332, 333
Glyceraldehyde, 472, 473
Glycerol, 477
Glyceryl phosphate, 549
Glycine, 508, 509
Glycoprotein, 675
Gnostic area, 761
Goiter, 675, 679
Gold, 233, 277; ocean and, 166
Gonadotropins, 686
Gondwana, 158
Governor, 786
Graham's law, 512
Gramicidin, 616, 620
Gram-negative bacteria, 617
Gram-positive bacteria, 616
Grana, 547
Grandfather clock, 142
Grand unified theories, 339, 340
Graphite, 273–5; nuclear reactors and, 439
Graptolites, 718
Gravimeter, 142
Gravitational interaction, 321
Gravitational constant, 810
Gravitational lens, 361
Gravitational waves, 315, 316
Graviton, 316, 321
Gravity: distance and, 809; relativity and, 359–60
Gray matter, 759
Great Dying, 720
Great Global Rift, 172, 173, 213
Great Red Spot, 115, 117
Great Winter, 187
Greek fire, 382
Greek history, 724–5
Greenhouse effect, 189, 231
Greenland, 88; ice-cap of, 180
Green plants, 699
Growth hormone, 687; synthesis of, 523
Guacharos, 169
Guanine, 579
Gulf Stream, 163–5
Guinea pigs, 662
Guncotton, 494
Gunpowder, 382, 494
Gutenberg discontinuity, 155
Gutta percha, 491, 492
Guyots, 171
Gymnospermae, 699
Gyroscope, 212

Haber process, 496, 622
Habitable Planets for Man, 602
Hadrons, 329
Hafnium, 241, 256;
    superconductivity of, 269
Hahnium, 248
Hale telescope, 35
Half-life, 449
Hall, 110
Hall-Héroult process, 281
Hallucinations, 781, 782
Hallucinogens, 782
Handedness, 336, 337; human, 762
Haptens, 640
Harden-Young ester, 534
Hard magnets, 280
Hard rubber, 493
Hard water, 260
Hashish, 781
Hawaii, 171
H-bomb, 443
Head shape, 736, 737
Heart, 553
Heart transplant, 642
Heat, 362ff.; conduction of, 270, 271;
    flow of, 365; temperature and,
    364, 368; vibration and, 365; work
    and, 365, 366
Heavy water, 303; nuclear reactors
    and, 439
Helicopter, 408
Heliometer, 25
Heliosphere, 204
Helium, 201, 255; alpha particles and,
    286, 287; atmosphere and, 204,
    228, 229; discovery of, 59;
    electrons and, 252; heat
    conduction of, 270, 271; isotopes
    of, 303, 304; Jupiter and, 116;
    liquid, 269; nuclear structure of,
    297; solid, 270; superfluidity of,
    270, 271
Helium I, 271
Helium II, 271
Helium 3, 272, 273, 303, 304
Helium 4, 272, 273
Hematoxylin, 557
Heme, 485
Hemichordata, 718
Hemlock, 481
Hemoglobin, 485, 511, 530;
    abnormal, 575-7; evolution and, 723;
    shape of, 523
Hemoglobin A, 575-7

Hemoglobin C, 576
Hemoglobin F, 575
Hemoglobin S, 575-7
Hemolytic streptococcus, 615
Hemophilia, 565
Hepatoma, 645
Heptane, 406
Herbivores, 653
Herculaneum, 152, 725
Hermes, 134
Heroin, 484
Hertzian waves, 410
Hertzsprung-Russell diagram, 51
Hidalgo, 132
Hieroglyphics, 725
High-energy phosphate bond, 536
High-fidelity sound, 395
High polymers, 492
Hindbrain, 754
Histamine, 642
Histidine, 656
Histones, 578
Hittites, 277
Hofmann degradation, 480
Hofmann's violets, 479
Holography, 425
Home computers, 797
Homeostasis, 787
Home robots, 801
Hominids, 705, 731ff.; brain-size of,
    733, 755
Homo habilis, 733
Homo erectus, 732, 733, 735
Homo neanderthalensis, 731
Homo rhodesiensis, 731
Homo sapiens, 731, 733, 735, 736
Homo solensis, 731
Hood, Mount, 152
Hoof and mouth disease virus, 629
Hooker telescope, 34
Hookworm, 700
Hormones, 671ff.; mechanism of,
    688, 689
Horsehead Nebula, 76
Horseshoe crab, 718
Hospital staph, 618
Hot spots, oceanic, 176
Hubble's law, 42
Hudson Bay, 178
Human chorionic gonadotropin, 686
Human evolution, 724ff.
Humanists, 11
Human temperature, 363, 364
Humboldt current, 163

Hummingbirds, 756
Humulin, 591
Hunting, 787, 788
Hurricane, 140; tracking of, 198
Hybrid cells, 648
Hydrocarbons, 406
Hydrochloric acid, 233, 234
Hydrogen, 200; atmospheres and, 204, 228, 229; atomic, 251, 268; atomic weights and, 235, 236; balloons and, 195; catalysis and, 525, 526; galaxies and, 77; ionized, 76; isotopes of, 303, 457; Jupiter and, 116; liquid, 267, 294; quantum theory and, 374, 375; song of, 76; Sun and, 59
Hydrogen 1, 77
Hydrogen 2, 77, 303; tracer activity of, 541ff.
Hydrogen 3, 730
Hydrogen bomb, 247, 442-4
Hydrogen bond, 514
Hydrogen cyanide, 529
Hydrogen fusion, 371
Hydrogen molecule, 227
Hydrogen peroxide, 529, 531
Hydrolysis, 489
Hydrophobia, 626, 636
Hydrophone, 175
Hydroponics, 741
Hydrotropism, 772
Hydroxyl group, 77, 494
Hyena, 706
Hyperion, 123
Hypernucleus, 328
Hyperons, 328
Hypervitaminoses, 665
Hyperbitaminosis A, 666
Hypnotism, 779, 780
Hypophysis, 685
Hypothalamus, 687, 754, 762

Iapetus, 122, 123
IBM, 796
Icarus, 134
Ice-VII, 273
Ice ages, 184ff.; cause of, 186ff.
Icebergs, 167, 180
Iceland spar, 469
Ice Patrol, 180
Ices, 117
Ichthyosaurs, 720

Iconoscope, 45
Iliad, 132, 220, 277, 726
Illinium, 245
Image orthicon, 415
Imaginary quantities, 817
Immortality, 695
Immunity, 634ff.
Imp-I, 216
Imprinting, 767, 768
Inclusion bodies, 626
Index of refraction, 352
Indigo, 479
Induction, 12; electrical, 210
Induction coil, 405
Indus civilization, 726
Industrial microbiology, 622
Industrial Revolution, 385
Industrial robots, 800
Inert gases, 254, 255
Inertia, 804
Infantile paralysis, 638
Inflationary universe, 340
Influenza, 641
Information theory, 795
Infrared radiation, 61
Infrared spectrophotometer, 518
Infrasonic sound, 169
Inheritance, Mendelian, 561-3
Ink, 382
Innate behavior, 766
Inoculation, smallpox, 634
Inorganic compounds, 464
Insect hormones, 678
Insectivora, 704
Insects, 619, 701, 718, 719
Instinctive behavior, 766
Insulators, electric, 387
Insulin, 679-81, 781; amino acid sequence of, 520-22; isolation of, 680; synthesis of, 523; three-dimensional shape of, 524; varieties of, 533
Integrated circuits, 419
Intelligence quotient (IQ), 757, 758
Intelligence tests, 757, 758
Intercontinental ballistic missiles, 444
Interferometer, 350, 351; laser, 423
Interferon, 643, 644
Interglacial periods, 185
Intermediary metabolism, 535
Internal-combustion engine, 402ff.
International Business Machines, 796
International Geophysical Year, 182, 183

International Polar Year, 182
Interpretative cortex, 783
Interstellar gas, 75
Interstitial cell-stimulating hormone, 686
Intrinsic factor, 674, 675
Introns, 590
Invertebrates, 701
Io, 112, volcanoes on, 118
Iodination, 675
Iodine, 237; thyroid gland and, 674; tincture of, 612
Iodized salt, 675
Ion-exchange resins, 260
Ionic drive, 268
Ionosphere, 205–7, 214
Ions, 204
IQ, 757, 758
Iridium, 720
I, Robot, 800
Iron, 277ff., 671; catalytic properties of, 529; enzymes and, 470; hemoglobin and, 485, 511; magnetism and, 208, 280; smelting of, 278
Iron Age, 727; population during, 739
Iron-deficiency anemia, 671
Iron meteorites, 155, 222
Ishtar Terra, 103
Island universes, 33, 34
Islets of Langerhans, 680, 681
Isobars, 298
Isoclinic lines, 212
Isocyanates, 466
Isodynamic lines, 212
Isogonic lines, 212
Isoleucine, 656
Isomers, 466, 468
Iso-octane, 406
Isoprene, 491, 505
Isostasy, 157
Isotactic polymers, 499
Isotones, 298
Isotopes, 290–92
Ivory, 497

Jackbeans, 529
Jacquard loom, 789
Janus, 123
Jaundice, 684
Java man, 731, 732
Jazz Singer, The, 402

Jellyfish, 700
Jet lag, 778
Jet planes, 408, 409
Jet stream, 196
Joule-Thomson effect, 267
Juno, 131
Journal of the Linnaean Society, 712
Jupiter, 111ff.; atmosphere of, 116, 117; brightness of, 111; central pressure of, 273; central temperature of, 151; density of, 112; diameter of, 111; distance of, 111; interior of, 117; magnetic field of, 219; mass of, 111; minor satellites of, 113, 114, 118, 132; oblateness of, 114, 115; possible life on, 600, 601; radio waves from, 64; ring of, 120; rotation of, 115; satellites of, 111–14, 345; surface markings on, 115; surface temperature of, 116
Jupiter-probes, 116–18
Jurassic age, 719

Kaaba, 220
Kanalstrahlen, 287
Karyotype, 559
Katusha, 92
Kayons, 327
K-capture, 302
Kennelly-Heaviside layer, 205
Keratin, 524
Kerosene, 396, 427
Kidney transplant, 641
Kieselguhr, 495
Kilauea, 152
Killer satellites, 199
Kilocycles, 62
Kilogram, standard, 141
Kinetic energy, 161, 366
Kinetic theory of gases, 367
Kingdoms, 699
Kirkwood gaps, 132
Kites, 194, 195
K-mesons, 327; breakdown of, 336
Knee jerk, 766
Kodak, 401
Korean war, 409
Krakatoa, 149
Krebiozen, 651
Krebs cycle, 536
Krebs-Henseleit cycle, 539

Krypton, 201; electric lights and, 398
Krypton 86, 344
Krypton tetrafluoride, 255
Kurchatovium, 248
Kwashiorkor, 656

Labrador current, 164
Lactic acid, 535; conversion of CO_2 of, 536
Lactogenic hormone, 686
Lactose, 490
Laetrile, 651
Laki, 152, 153
Lalande 21185, 601
Lamarckism, 709
Lambda particle, 328
Lamellae, 547
Lampreys, 703
Lamp shells, 700
Land camera, 401
Land, colonization of, 718, 719
Lanthanides, 238, 239, 261
Lanthanum, 256, 261
Lasers, 421–3; communication and, 423, 424
Laudanum, 484
Laughing gas, 201, 483
Laurasia, 158
Laurentian highlands, 175
Lawrencium, 248
Laws of Thought, An Investigation of the, 794
Lead: isotopes of, 291; radioactive elements and, 244, 290, 291; tracer activity of, 540
Lead azide, 495
Leaded gasoline, 406
Leaven, 526
Lectins, 651
Left brain/right brain, 761, 762
Lemurs, 705
Lenin, 446
Lenses, 607
Leptons, 315
Lepton number, conservation of, 328
Leucine, 508, 656
Leukemia, 448, 449, 645, 647
Levorotation, 382
Leyden jar, 387
Lice, 625
Lichens, 715, 716
Lie detector, 770
Life; deep-sea, 168; origin of, 592ff.

Life expectancy, 691
Lifespans, animal, 693
Light, 341ff.; aberration of, 345, artificial, 396ff.; coherent, 422; electric, 397–400; frequency of, 61; gravitation and, 360, 361; photosynthesis and, 545; polarized, 469; pressure of, 594; refraction of, 58, 341, 342; spectral line shift of, 40, 41; spectrum of, 57, 342; speed of, 344–7; unpolarized, 469; wavelengths of, 61, 343, 344; wave nature of, 58, 342ff.
Lightbulbs, 267
Lighter fluid, 397
Lightning, 168
Lightning, 387, 388; artificial, 305
Lightning rod, 388
Light spectrum, 40
Light-year, 25
Li'l Abner, 438
Lime juice, 659
Limelight, 396
Limestone, 360; ocean and, 166
Limonite, 110
Linac, 309
Linear accelerator, 305, 306
Lines of force, magnetic, 209, 348
Lipoproteins, 691
Liquid-drop model, 313
Lisbon, 146
Lithium, 202; electrons and, 252; hydrogen bomb and, 443
Lithium fluoride, 252, 253
Little America, 182
Little ice age, 88
Liver, 673
Living fossil, 718
Local anesthetic, 483
Local Group, 36
Locomotive, 385, 386
Lodestones, 207, 208
Logarithmic scale, 792
Logarithms, 791, 792
Long Beach, 446
Long-distance interactions, 322
Longitudinal waves, 348
Long-playing record, 395
Long-term memory, 783
Lorentz-FitzGerald equations, 720, 724
Lorentz's equation, 815, 816, 883
Love waves, 156
LSD, 782

L-series, 473
Lucite, 500
Lucy, 733
Luminiferous ether, 348
Luminosities, stellar, 28
Luna IX, 95
Luna X, 95, 217
Lunar Orbiters, 95
Lung cancer, 646
Lungfishes, 719
Lunik I, 95, 215, 218
Lunik II, 95, 215, 218
Lunik III, 95, 96
Lupus vulgaris, 647
Lutetium, 241
Luxons, 816
Lydia, 277
Lymph nodes, 677
Lysenkoism, 709
Lysergic acid diethylamide (LSD), 782
Lysine, 656
Lysozyme, 532, 616

M-82, 77
M-87, 34
Machines, 379ff.
Mach number, 409ff.
Mackintoshes, 491
Maffei *1* and *2*, 36
Magellanic Clouds, 28, 29; distance of, 32
Magenta, 479
Magic numbers, 314, 315
Magma, 173
Magnesium, 200, 234, 255, 282; chlorophyll and, 547; tissues and, 671
Magnesium chloride, 282
Magnesium hydroxide, 282
Magnesium nitride, 200
Magnesium oxide, 255
Magnetic axis, 212
Magnetic bottle, 458
Magnetic compass, 207
Magnetic declination, 212
Magnetic dip, 208
Magnetic domains, 208
Magnetic field (Earth's), 209; evolution and, 724; reversals of, 213; sunspot cycle and, 86
Magnetic lines of force, 209
Magnetic monopoles, 349

Magnetic permeability, 208
Magnetic poles, 207
Magnetic storm, 213
Magnetism: electricity and, 209; superconductivity and, 270
Magnetopause, 216
Magnetosphere, 216
Magnets, 207ff., 280
Magnitude, stellar, 28
Main sequence, 51
Malaria, 478, 614, 623, 624; sickle-cell trait and, 575
Malignant tumors, 645
Malnutrition, 654
Malonic acid, 530, 531
Mammals, *703*–5; appearance of, 719
Mammary glands, 704
Mammoths, 728
Manganese, 672; steel and, 279
Manganese dioxide, 529
Manganese nodules, 166
Manhattan Engineer District, 436
Man in the Moone, 91
Mantle, Earth's 156, 157
Mare Cognitum, 95
Maria, lunar, 90, 95, 224
Marianas Trench, 175
Marijuana, 781
Mariner 2, 101, 215, 219, 221
Mariner 4, 108
Mariner 6, 108
Mariner 7, 108
Mariner 9, 109
Mariner *10*, 103, 219
Mark I, 796
Marmosets, 756
Mars, 104ff., atmosphere of, 105, 108, 109; axis of, 105; brightness of, 104; canals on, 107–9; density of, 106; diameter of, 104; distance from Earth of, 104; distance from Sun of, 104; ice caps of, 105; life on, 110; magnetic field of, 219; maps of, 106ff.; markings on, 105–8; mass of, 106; meteorites and, 226; oblateness of, 115; parallax of, 23; revolution of, 105; rotation of, 105; satellites of, 105; surface of, 109; surface chemistry of; 110, surface gravity of, 106; surface temperature of, 108, 109
Marsupials, 704
Mascons, 224
Masers, 420, 421

Mass, energy and, 40, 369–71; gravity and, 809, 810; speed and, 815–19
Mass defect, 370
Mass-energy, conservation of, 338, 369–71
Mass-luminosity law, 52
Mass production, 405
Mass spectrograph, 291
Masurium, 245
Matches, 397
Matrix machanics, 375, 376
Matter waves, 374
Maunder minima, 87
Mauveine, 479
Maxim gun, 494, 495
Maxwell-Boltzmann law, 227
Maxwell Montes, 103
Maxwell's equations, 349
Measles, 634
Mechanical energy, 366
Mechanical equivalent of heat, 366
Mechanical heart, 637
Mediterranean Europeans, 737
Mediterranean Sea, 166
Medulla, adrenal, 683
Medulla oblongata, 754
Megacycles, 62
Megavitamin therapy, 665
Meiosis, 558
Meissner effect, 270
Melanin, 572
Melanocyte-stimulating hormone, 686
Melatonin, 686
Meltdown, nuclear, 457
Memory, 783
Mendelevium, 247
Mendelian inheritance, 561–3
Mercaptan group, 509
Mercuric chloride, 529
Mercuric oxide, 200
Mercury (element), 192, 193, 200, 232; pollution and, 742; superconductivity of, 269; thermometers and, 363, 364
Mercury (planet), 98ff.; brightness of, 99; density of, 106; diameter of, 99; distance of, 98; magnetic field of, 219; microwave radiation from, 100; perihelion advance of, 360; phases of, 99; revolution of, 99; rotation of, 99, 100; Sun and, 98; surface of, 103
Mercury fulminate, 495

Mercury probes, 103
Mescal 781
Mescaline, 782
Mesmerism, 779
Mesoderm, 700
Mesolithic age, 728
Mesons, 322, 326
Mesosphere, 197
Mesotrons, 322
Mesozoic era, 189, 717, 719
Messenger-RNA, 587
Metabolic intermediate, 534
Metabolism, 534ff.; inborn errors of, 577; rate of, 679
Metalliding, 284
Metals, 276
Metatheria, 704
Meteor, 171
Meteorites, 155, 220, 222ff.; age of, 226; amino acids and, 596
Meteorology, 192
Meteors, 194, 220ff.
Meter, standard, 141
Methane, 201; atmosphere and, 228, 230; electron configuration of, 252; structural formula of, 467; Titan's atmosphere and, 122
Methanogens, 717
Methionine, 656
Methyl alcohol, 467
Methylamine, 467
Methylcholanthrene, 650
Methyl iodide, 480
Methyl methacrylate, 500
Methyl rubber, 505
Metric system, 141
Michaelis-Menten equation, 530
Michelson-Morley experiment, 351, 352, 811ff.
Microbe, 608
Microbiology, 610
Microchips, 419, 797
Microfossils, 716
Micrometers, 61, 221, 222
Microorganisms, 554, 606ff.; size of, 625, 626; spontaneous generation and, 593
Microphone, 395
Microscopes, 372, 527, 607, 608
Microspheres, 598
Microwaves, 100
Mid-Atlantic Range, 171, 172
Midbrain, 754
Middle Stone Age, 728

Mid-Oceanic range, 172
Milankovich cycle, 187, 188
Milk, 660
Milk factor, 649
Milky Way, 26, 27; dark nebulae in, 76
Mimas, 123
Mineral acids, 233
Minerals, essential, 670ff.
Miniaturization, 415ff.; computers and, 797
Mini-black holes, 75
Mira, 46
Miranda, 125
Mischmetal, 261, 397
Mismeasure of Man, The, 757
Missing mass, 325, 326
Mitochondria, 538, 546, 599; DNA and, 581
Mitosis, 557, 558
M, N blood groups, 737
Moderators, neutron, 438, 439
Modern physics, 355
Mohorovicic discontinuity, 156
Mohs scale, 274
Molecular biology, 463
Molecular still, 503
Molecules, 235, 463ff.; interstellar, 78
Moles, 645
Mollusca, 700, 701
Molybdenum, 245; enzymes and, 672; steel and, 279
Monera, 699
Mongolian race, 735
Mongolism, 559
Mongols, 739
Monkeys, 705
Monocotyledonous plants, 698
Monomer, 492
Monoplanes, 407
Monopropellants, 268
Monosaccharides, 492
Monotremes, 722
Monsters, 563
Month, 88
Moon, 88ff., craters on, 96; density of, 106; diameter of, 90; distance of, 20, 21, 90; eclipses of, 89; flights to, 91ff.; glass on, 161; human beings on, 97, 98; life on, 94, 98; magnetic field of, 218; magnitude of, 28; meteorites and, 226; microwave reflection by, 100; orbit of, 808, 809; origin of, 159; other

side of, 95; phases of, 88; photograph of, 60; revolution of, 88; rotation of, 88; surface of, 90, 95; tides and, 160
Moon probe, 95ff., 216
Moon rocks, 97, 98, 161
Morning star, 98
Morphine, 484; structure of, 482
Morse code, 392
Mosquitos, 623-5
Mössbauer effect, 361, 362
Mosses, 699
Mothering, 772
Moths, 772
Motion, laws of, 91, 804ff.
Motion pictures, 401, 402
Motion, retrograde, 101, 104
Motor, 392, 393
Motor area, 759, 760
Mules, 706
Multiplication factor, 439
Mu-meson, 322
Mumps, 634
Muonium, 323
Muons, 322-4
Muscle, 535; electrical properties of, 768; minerals and, 671; nerve stimulation and, 758, 759
Musk, 480
Mustard gas, 647
Mutagens, 648
Mutations, 563ff.; cancer and, 647, 648; DNA replication and, 584; X-rays and, 568, 569, 573
Mutual annihilation, 300, 371
Mycenae, 726
Myelin, 753
Myoglobin: amino acid sequence of, 522; three-dimensional shape of, 523
Mysticeti, 705
Myths, 6, 37

Naphtha, 427
Natural gas, 427
Natural selection, 711
Nature of the Chemical Bond, The, 476
Nautilus (I), 385
Nautilus (II), 446
Navigation satellites, 144

Neanderthal man, 730, 731; brain of, 755
Nebulae, 32–4; dark, 76; planetary, 51
Nebular hypothesis, 82
Nebulium, 202
Negatives, photographic, 400
Nematoda, 700
Neodymium, 261
Neolithic Revolution, 727; population and, 738
Neon, 201, 255; atmosphere and, 229; isotopes of, 291
Neon lights, 399
Neopallium, 754
Neoprene, 506
Neptune, 126–8; diameter of, 127; discovery of, 127; distance of, 127; mass of, 127; revolution of, 127; satellites of, 127–8
Neptunism, 153
Neptunium, 247
Nereid, 128, 132
Nerve action, 764ff.
Nerve cells, 751, 752
Nerve gases, 770
Nerve impulse, 752; speed of, 753
Nerves: electrical impulses in, 768, 769; fatty sheath of, 753; muscle contraction and, 758, 759
Neurochemistry, 781
Neurons, 752
Neuron theory, 765
Neurospora crassa, 572–4, 652, 653
Neutral-current reaction, 325
Neutrino, 317ff.; detection of, 319; mass of, 324–6; muon and, 323; oscillations of, 325
Neutron-activation analysis, 304
Neutrons, 70, 246, 247, 296–8, 730; beta particles and, 301, 302; bombardment with, 433; mass of, 370
Neutron star, 70, 72
New moon, 89
New Stone Age, 727
New York Sun, 94
Niacin, 668
Niacinamide, 668
Nickel, 480; ferromagnetism of, 208; nodules and, 167; steel and, 279
Nicol prism, 469
Nicotinamide, 666, 667
Nicotine, 666

Nicotinic acid, 666–8
Night-blindness, 670
Nimbus I, 230
Nineveh, 726
Niobium, 269
Nitrate group, 494, 495
Nitrates, 621, 622
Nitric acid, 233, 494
Nitric oxide, 202
Nitrifying bacteria, 622
Nitrocellulose, 494
Nitrogen, 276; atmosphere and, 200, 201; atomic, 202; electric lights and, 398; electron shells and, 252; explosives and, 496; lightbulbs and, 267; liquid, 266; Mars's atmosphere and, 109; nuclear spin of, 296, 297; organic compounds and, 466; proteins and, 508; Titan's atmosphere and, 122; transmutation of, 295; Venus's atmosphere and, 102
Nitrogen 15, 541, 543
Nitrogen balance, 655
Nitrogen excretion, 721, 722
Nitrogen-fixing bacteria, 622
Nitrogen molecules, 252
Nitrogen mustards, 647
Nitroglycerine, 494, 495
Nitro group, 495
Nitrous oxide, 201; anesthetic properties of, 483
Nobel Prizes, 495
Nobelium, 248
Noble gases, 255
Non-polar compounds, 476
Noradrenalin, 770
Nordic Europeans, 737
North Atlantic drift, 163, 164
North equatorial current, 163
North Pacific drift, 164
North Pole, 178, 180
Northwest passage, 179
Notochord, 721
Nova Aquilae, 47
Novae, 47, 48; white dwarfs and, 54, 55
Novocaine, 484
Nuclear bombs, 440–42; testing of, 453ff.
Nuclear excursion, 456
Nuclear fission, 318; controlled, 456, 457–9
Nuclear fusion, 442–4, 456, 457–9

Nuclear interaction, 321
Nuclear magnetic resonance, 516
Nuclear proliferation, 441, 442
Nuclear reaction, 301
Nuclear reactors, 446–8
Nuclear test ban, 455, 456
Nuclear transplantation, 561
Nuclear winter, 721
Nucleases, 632
Nucleic acids, 578ff.
Nuclein, 578ff.
Nucleolus, 581
Nucleonics, 298
Nucleons, 298
Nucleoproteins, 580
Nucleotides, 580; abiogenic synthesis of, 596, 597
Nucleus, atomic, 288ff.; binding energy of, 370; structure of, 289ff., 297, 298, 313–15, 320ff.
Nucleus, cell, 557ff.
Nuclide, 292
Nylon, 503

Oak Ridge, 438
Oberon, 125
Oblateness, 114, 115
Oblate spheroid, 140
Ocean, 162ff.; area of, 162; currents in, 163–6; deep-sea life in, 168, 175–7; depth of, 167ff.; desalting, 740; deuterium in, 457; food from, 166; formation of, 229; hot spots in, 176; human beings in, 177, 178, 746; minerals in, 166; mixing of, 165; salt in, 38; trenches of, 171; volume of, 162
Odontoceti, 705
Odyssey, 277
Oedipus complex, 780
Ohm, 390
Ohm's law, 390
Oil, 426
Oil blockade, 744
Oil lamps, 396
Oil well, 427
Okapi, 697
Old age, 690ff.
Old Stone Age, 728
Olduvai Gorge, 733
Olfactory lobes, 754
Olivine, 156

Olympus Mons, 109
Omega Centauri, 30
Omega minus, 329
Omicron Ceti, 46
Omnivores, 653
Oncogene, 648, 649, 651
Open-hearth method, 279
Opsin, 670
Optical activity, 469–73
Optical fibers, 424
Optical masers, 421
Optical pulsar, 73
Optic lobes, 754
Order, 698
Ordovician age, 718
Organelles, 587
Organic chemical analysis, 465, 466
Organic compounds, 464ff.; synthesis of, 464, 465
Organic glass, 500
Organic lasers, 423
Organ transplants, 641, 642
Origin of Life, The, 592
Origin of Species, The, 712, 731
Orion arm, 76
Orion Nebula, 33, 53
Orlon, 504
Ornithine, 539, 574, 620
Ornithine cycle, 539
Orthotolidine, 682
Osmotic pressure, 513
Osteichthyes, 703
Ovarian follicles, 554
Ovaries, 683
Overweight, 654, 655, 692
Ovum, 556; size of, 625
Oxidative phosphorylation, 537, 538, 547
Oxygen, 295; allotropes of, 273, 274; atmosphere and, 200, 201; atomic, 202; atomic weights and, 235, 236; catalysis and, 525, 526; electron shells and, 252; isotopes of, 370; liquid, 266, 267; nuclear structure of, 297; ocean content of, 165; photosynthesis of, 545
Oxygen, 18, 541, 548
Oxygen molecules, 252; speed of, 227
Oxyhydrogen torch, 456
Oxytocin, 522
Oyster shells, 282
Ozone, 203, liquid, 267
Ozonosphere, 203

Pacific Ocean, 165
Packing fraction, 370, 371
Pain, 483
Pair production, 300
Paleobiochemistry, 721
Paleolithic Age, 728
Paleontology, 707
Paleozoic era, 717
Pallas, 131
Pallium, 754
Pancreas, 677
Pangaea, 157; breakup of, 174, 175
Pantothenic acid, 664
Papain, 533
Paper, 382
Paper chromatography, 518, 519
Paper electrophoresis, 576
Papillomas, 649
Para-aminobenzoic acid, 620
Parabola, 808
Parachute, 195
Paraffin, 498
Parallax, 22, 23
Paramagnetism, 208, 271, 272
Paramecium, 625
Parasympathetic nerves, 765
Parathormone, 681
Parathyroid glands, 681
Paricutin, 152
Parity, conservation of, 335ff.
Parrot fever, 617
Parsec, 28
Particle accelerators, 304ff.
Particles: detection of, 292–4;
    subatomic, 285ff, 330; sub-
    baryonic, 329; virtual, 377
Particle spin, 310
Pasiphae, 114
Pasteurization, 527
Patellar reflex, 766
Patroclus, 132
Pea plants, 562, 563
Pegasus I, 221
Peking man, 731, 732
Pelée, Mount, 153
Pellagra, 660, 667
Pellagra-preventive factor, 668
Pendulum, 142
Penicillin, 616, 617, 620
Penicillinase, 618
Penicillium notatum, 616
Peppered moth, 715, 716
Pepsin, 528; crystallization of, 529
Peptide link, 510

Peptides, 510, 511; paper
    chromatography of, 521
Perigee, 90
Perihelion, 103
Periodic table, 237–40; electrons and,
    250, 251
Peripatus, 706
Perissodactyla, 705
Perkin reaction, 480
Permeability, magnetic, 208
Permian age, 719
Pernicious anemia, 673
Peroxidase, 531, 682
Perseus arm, 76
Perspiration, 267
Pertechnates, 280
Perturbations, 126
Peru current, 163
Pesticides, 618, 619; danger of, 742
Peter I Island, 180
Petri dishes, 613
Petroleum, 426
Peyote, 781
Phages, 632
Phagocytes, 639
Phanerozoic eon, 717
Pharmacology, 614
Phase rule, 368
Phases, lunar, 88
Phenol, 612
Phenylalanine, 656
Phenylketonuria, 577
Philadelphia chromosome, 560
Philosophy, 8
Phobos, 106; size of, 110
Phoebe, 122, 132
Phonography, 395
Phosphor, 399
Phosphorus, 274; high pressure and,
    273; isotopes of, 301; metabolism
    and, 534; proteins and, 508; steel
    and, 279; tracer activity of, 540
Photoelectric effect, 250, 355, 356
Photography, 60, 400–402
Photolysis, 549
Photon, 315, 316, 318, 321, 356
Photosynthesis, 544ff., 653; scale of,
    546
Phototropism, 772
Photovoltaic cell, 431, 432
Phrenology, 759
Phylum, 698
Physical atomic weights, 370
Physical chemistry, 496

Physical Geography of the Sea, 163
Piezoelectricity, 170, 243
Pile, atomic, 440
Piltdown man, 734
Pi meson, 326
Pinch effect, 458
Pineal eye, 778
Pineal gland, 686; biological clock and, 778
Pinhole camera, 400
Pion, 326
Pioneer I, 216
Pioneer III, 216
Pioneer X, 116, 219
Pioneer XI, 116, 118, 219
Pionic atom, 326
Pisces, 703
Pitch, 426, 427
Pitchblende, 243
Pithecanthropus erectus, 731
Pituitary gland, 685–7
Placenta, 686
Placental mammals, 704, 705
Plague, 625
Plagues, 611
Planck's constant, 355
Planetary nebulae, 51; formation of, 54
Planetary systems, extra-solar, 601, 602
Planetesimal hypothesis, 83
Planets, 20; orbits of, 22
Plankton, 203
Plant hormones, 678
Plant kingdom, 699
Plant viruses, 697
Plasma, 458, 459
Plasma torch, 459
Plasmids, 610, 618
Plasmochin, 614
Plasmodium, 623
Plasticizer, 496
Plastic magnets, 280
Plastics, 496–500
Plate tectonics, 173–5
Plating, 284
Platinum, 526
Platinum black, 529
Platinum hexafluoride, 254
Platyhelminthes, 700
Pleasure center, 762
Pleistocene epoch, 184
Pleistocene glaciation, 184
Plesiosaurs, 720

Plexiglas, 500
Pluto, 128–30; diameter of, 129; discovery of, 128; distance of, 128; orbit of, 128; rotation of 128; satellite of, 128, 129
Plutonism, 153
Plutonium, 247
Plutonium 239, 440
Plutonium 244, 450
Pneumococci, 582
Pneumonia, 615
Poisons, 529
Polar diameter, 115
Polarized light, 469
Polar molecule, 475
Polaroid, 469
Poles, magnetic, 207
Pole star, 28
Poliomyelitis, 638
Poliomyelitis virus, 629
Pollution, 741, 742
Polonium, 244
Polyacrylic plastics, 500, 501
Polyester fiber, 504
Polyethylene, 498
Polyglycine, 514
Polygraph, 769
Polymers, 492, 493; properties of, 493
Polyneuritis, 659
Polypeptides, 513, 514; X-rays and, 514
Polysaccharides, 492
Polysulfide rubber, 506
Polyterpenes, 492
Polythene, 498
Polytyrosine, 514
Polyunsaturated fatty acids, 657
Polyuridylic acid, 589
Pompeii, 152, 725
Pongids, 733
Population explosion, 739
Population, human, 738, 741; limits of, 741
Populations, stellar, 35
Porifera, 700
Porphin, 486
Porphobilinogen, 544
Porphyrins, 486, 487; body synthesis of, 543, 544
Porpoises, 169
Positional notation, 791
Positive rays, 287
Positives, photographic, 400
Positron, 300, 301

Positronium, 300
Potassium, 671
Potassium 40, 450
Potassium cyanide, 529
Potassium inhibition, 671
Potassium uranyl sulfate, 242, 243
Potential energy, 366
Pre-Cambrian age, 717
Precession of the equinoxes, 140
Prefrontal lobe, 760, 761
Prefrontal lobotomy, 761
Pregnancy, 684, 687
Pressure cooker, 785
Pressures, high, 273ff.
Primary cosmic particles, 299
Primates, 703; brain–body ratio in, 756
Prime mover, 394
Principia Mathematica, 91
Principle of Population, An Essay on the, 710, 739
Principles of Geology, The, 708
Printing, 382
Proboscidea, 705
Procaine, 484
Proconsul, 734
Procyon, 24; companion of, 49, 50; spectral class of, 59
Procyon B, 50
Programming, 789
Project Argus, 217
Project Cyclops, 604
Project Ozma 604
Project West Ford, 218
Prokaryotes, 589, 609, 699
Promethium, 246, 261
Prontosil, 615
Propane, 476
Proper mass, 816
Proper motion, 24
Propylene, 499
Prostaglandins, 688, 689
Protactinium, 244, 290
Protective coloration, 715, 716
Protein-bound iodine, 679
Protein catalysts, 529
Protein deficiency, 654–6
Proteins, 502, 507ff., 654; amino acids from, 517; chromosomes and, 578; digestion of, 510; helical structure of, 514; molecular weight of, 511ff; purification of, 513; synthetic, 522, 523; three-dimensional shape of, 523, 524
Protein structure, evolution of, 723

Proterozoic era, 717
Protista, 699
Protonosphere, 204
Protons, 287, 288; decay of, 339; mass of, 370; spin of, 295, 296; stability of, 328
Proton synchrotron, 308
Proto-oncogene, 648
Protoplasm, 555, 556
Protoporphyrin IX, 486, 487
Protoporphyrins, 486, 487
Prototheria, 704
Protozoa, 608, 699, 717; disease and, 623, 624
Provitamin, 662
Pseudopods, 608
Psilopsids, 718
Psittacosis, 617
Psychiatry, 780
Psychoanalysis, 780
Pterodactyl, 708
Pterosaurs, 720
Publication, scientific, 13
Puerperal fever, 611, 612
Pulsars, 71; neutron stars as, 72; period change in, 72
Pump, suction, 191, 192
Punched cards, 789, 796
Punctuated evolution, 713
Purines, 579
Puromycin, 621, 784
P waves, 153, 154
Pyridine, 667
Pyridoxine, 664
Pyrimidines, 579
Pyroxylin, 496
Pyrrole, 486

Quadrupole moments, 313
Quantum chromodynamics, 332
Quantum electrodynamics, 332
Quantum mechanics, 375
Quantum numbers, 251
Quantum theory, 354ff.; atomic structure of, 374, 375; molecules and, 476
Quarks, 329ff.
Quartz, 61
Quasars, 66; distance of, 67; galaxies and, 69, gravitational lens and, 361
Quasi-stellar radio sources, 66

Queen Mary, 169
Quinine, 478, 614, 615; structural
    formula of, 482
Quintessence, 191

Rabies, 636
Racemic acid, 470
Races, human, 735–8; evolution of,
    736
Rad, 455
Radar, 63, 347
Radial velocity, 40
Radiation, cancer and, 646
Radiation sickness, 539
Radicals, organic, 466
Radio, 410ff.
Radioactive isotopes, 540ff.
Radioactivity, 39, 161, 162, 243ff.,
    285; artificial, 301; dangers of,
    448ff.; energy from, 368, 369
Radio astronomy, 62ff.
Radiolead, 290
Radio sources, 63
Radio sun, 64
Radiotelegraphy, 411
Radio telescope, 62, 63
Radiothorium, 290
Radio-wave background, 43, 44
Radio waves, 61, 410; atmosphere
    and, 205
Radium, 244
Radon, 244, 254
Railroad, 386
Rain, 167, 222, 740
Rainbow, 57
Rainier, Mount, 152
Rainmaking, 197
Raman effect, 516
Ramapithecus, 733, 735
Randomness, 779
Ranger 7, 95
Ranger 8, 95
Ranger 9, 95
Rare earth metals, 256
Rare earths, 255, 256;
    chromatography and, 261
Rayleigh waves, 156
Rayon, 502
Reason, 7
Receptors, brain, 687
Recombinant-DNA, 591, 592
Recurrent novae, 55

Red blood corpuscles, 557, 608; size
    of, 626
Red dwarfs, 50
Red giants, 51, 54
Red Sea, 174
Red shift, 41
Reference, frame of, 357
Reflecting telescope, 58
Reflex arc, 765, 766
Reflex center, 765
Reflexes, 765–8; conditioned, 773
Refracting telescopes, 58
Refraction, light, 58, 341, 342
Refractory period, 769
Refrigeration, 204, 265, 657
Regeneration, 700
Regulator genes, 588
Relativity, theory of, 40, 42, 356–62
Relays, electric, 392
Releasing factors, 687
REM sleep, 763
Renal threshold, 681
Reptiles, 719
Reptilia, 703
Reserpine, 781
Resins, 497
Resistance, electrical, 390
Resonance particles, 329
Resonance stabilization, 476
Responses, conditioned, 773ff.
Rest-mass, 315, 816
Restriction enzymes, 591
Retina, 670
Retinene, 670
Retrograde motion, 101, 104
Rh blood groups, 737, 738
Rhea Mons, 103
Rhenium, 242, 245
Rheumatoid arthritis, 643, 684
Rhodesian man, 731
Rhodopsin, 670
Ribitol, 668
Riboflavin, 668, 669
Ribonuclease, 532; amino acid
    sequence of, 522; synthesis of,
    523
Ribonucleic acid (RNA), 580
Ribose, 579, 580
Ribosomes, 587
Ribulose diphosphate, 549
Rice, 659, 660
Richter scale, 149
Rickets, 660
Rickettsia, 626

Right brain, left brain, 761, 762
Rime of the Ancient Mariner, The, 180
Ringer's solution, 671
Rings, Saturn's, 119–21, 124; structure of, 121
Rings, Uranus's, 126
RNA, 580ff.; enzyme formation and, 586ff.; memory and, 783, 784; replication of, 585; viruses and, 628, 631–3
Robotics, three laws of, 799
Robots, 798ff.; future of, 801, 802
Roche limit, 121
Rocket fuel, 267, 268
Rockets, 91ff.; dangers of, 742; liquid-fuel, 92; warfare and, 91, 92
Rocky Mountain spotted fever, 626
Rodentia, 704
Rods, retinal 670
Roentgen, 455
Roentgen rays, 646
Rorschach test, 758
Rosetta stone, 725
Ross Ice Shelf, 182
Ross Sea, 182
Royal Society, 13
Rubber, 491ff., 504; synthetic, 505
Rubidium, 59
R.U.R., 798
Rutherfordium, 248

Sabatier-Senderens reduction, 480
Sabin vaccine, 639
Safety glass, 500
Safety matches, 397
Sagittarius arm, 76
Sahara Desert, 186
Saint Helens, Mount, 152
Saint Pierre, 153
Salk vaccine, 638, 639
Salt, 470
Salt lick, 671
Salt, table, 671
Salvarsan, 614
San Andreas fault, 173
Sand, 166
S Andromedae, 48
Saran, 502
Sarcomas, 645
Satellite galaxies, 36

Satellites, 111; artificial, 93ff., 198, 199, 215ff.
Saturn, 119ff.; axis of rotation of, 120; brightness of, 119; central pressure of, 273; diameter of, 119; density of, 119; distance of, 119; magnetic field of, 219; mass of, 119; oblateness of, 119; radio waves from, 64; revolution of, 119; rings of, 82, 119–21, 124; rotation of, 119; satellites of, 121–4
Sauerkraut, 658
Savannah, 446
Scalers, 293
Scandium, 238
Scanning electron microscope, 237, 374, 630
Sceptical Chymist, The, 233
Schmidt camera, 60
Schizophrenia, 780, 782
Science fiction, 91, 96, 101, 798, 799
Scintillation counter, 293
Scopes trial, 714, 715
Screw propeller, 383
Scuba diving, 177, 746
Scurvy, 658, 659, 660, 662
Sea anemones, 700
Sea-floor spreading, 173, 174, 213
Seamounts, 171
Sea spout, 140
Secondary cosmic particles, 299
Secretin, 677
Sedatives, 484
Sedoheptulose phosphate, 549
Seismograph, 147
Selenium, 672
Self-regulating systems, 785ff.
Self-starter, 405
Semiconductors, 416, 417
Semi-magic numbers, 314
Semipermeable membranes, 513
Sense organs, 753
Sensory area, 760
Septicemia, 643
Sequenator, 522
Serine, 532
Serotonin, 782
Serum albumin, 519
Servomechanism, 788
Settlements, space, 747
7-Dehydrocholesterol, 663
Sex hormones, 683; cancer and, 649, 650
Sex-linked disease, 565

Sexual repression, 780
Sexual reproduction, 560
Seyfert galaxies, 69
Shadow zone, 156
Shale, 427
Sharks, 703
Shellfish, 718
Shell model, 314
Shell numbers, 314
Shells, 700, 701
Shock therapy, 780, 781
Shooting stars, 220
Short-term memory, 783
Sickle-cell anemia, 574, 575
Sickle-cell trait, 575
Side-chain, amino acid, 509
Sidereal month, 88
Sigma particle, 328
Silanes, 500
Silent Spring, 619
Silica gel, 526
Silicon, 416, 432, 500; isotopes of, 301
Silicon carbide, 274, 275
Silicone resins, 501
Silicones, 501
Silk, 502; artificial, 502
Silurian age, 718
Silver, 277; ocean and, 166
Silver chloride, 400
Simple microscope, 527
Sinanthropus pekinesis, 731
Singularity, 73
Sinope, 114
Sinuses, 552
Sirius: absolute magnitude of, 28;
    companion of, 49; magnitude of,
    28; proper motion of, 24; radial
    velocity of, 41; spectral class of,
    59
Sirius B, 49
61 Cygni, 25; magnitude of, 28;
    possible planets of, 601
Skin cancer, 647
Skin diving, 177
Sleep, 762, 763
Sleeping pills, 485
Sleeping sickness, 614
Slide rule, 792, 797
Slow viruses, 633
Small Magellanic Cloud, 29
Smallpox, 614, 634, 635
Smog, 429, 741, 742
Smokeless powder, 494
Smoking, 646

Snorkels, 177
Sodium, 202, 203
Sodium acetate, 480
Sodium chloride, 202, 671
Sodium perxenate, 255
Sodium pump, 769
Sodium thiosulfate, 400
Soft landing, 95
Soft magnets, 280
Soft water, 260
Soil, microorganisms in, 616
Solar battery, 431
Solar energy, 431, 432
Solar flare, 214
Solar system: atmospheres in, 200;
    craters on worlds of, 225; escape
    velocity from bodies of, 227; origin
    of, 79ff.; size of, 23, 24
Solar wind, 137, 215
Solid-state electronics, 417
Solo man, 731
Soluble RNA, 587
Somatotropic hormone, 687
Sonar, 170, 171
Sonic boom 409
Sound, 40; transmission of, 193
Sound barrier, 409
Southequatorial current, 164
South Pole, 182
South Shetland Islands, 180
Space, human beings in, 746–50
Spaceships, 748, 749
Space-time, 358
Space walk, 97
Spanish-American war, 494
Spark chamber, 294
Special relativity, 356–9, 817
Species, 690ff.; classification of, 697,
    698; extinct, 707ff.; intermediate,
    706; naming, 698; number of, 696,
    697; similar, 706
Specific impulse, 267
Spectacles, 607
Spectral classes, 59
Spectral lines, 58; atomic structure
    and, 60
Spectrophotometer, 517
Spectroscope, 59
Spectroscopic binary stars, 42
Spectrum, light, 57, 342
Spermatozoa, 556
Sperm cells, 556; chromosomes in,
    558, 559
Spiders, 766, 767

Spin, particle, 295, 296, 310; conservation of, 315
Spinal column, 753
Spinal cord, 753, 763, 764; reflex action and, 765
Spine, 552
Spiral galaxies, 35
Spiral nebulae, 33
Spirilla, 608, 609
Spirochetes, 614
Sponges, 700
Spontaneous fission, 314
Spontaneous generation, 592–4
Spray cans, 203, 204
Sputnik I, 93, 747
Sputnik III, 216
Spy satellites, 199
S-quark, 331
Stainless steel, 280
Staphcyllin, 618
Staphylococcus, 616
Starch, 260, 489; catalysis and, 525; structure of, 490
Starch chromatography, 517
Starfish, 701
Starquake, 72
Stars, 24ff.; absolute magnitude of, 28; apparent magnitude of, 28; binary, 42; birth of, 53; brightness of, 28; changes in, 46; clusters of, 30; collapse of, 49–51; collisions of, 83; distance of, 25, 26; dwarf, 50; evolution of, 51ff.; first-generation, 56; giant, 51; internal temperature of, 51, 52; lifetime of, 53; luminosity of, 28; neutrino production and, 46, 47; new, 46, 47; number of, 26, 601; parallax of, 28; photographs of, 60; populations of, 35; proper motion of, 24; radial velocity of, 41; variable, 27
Star Spangled Banner, The, 92
Static electricity, 386–8
Steady-state universe, 43
Steam digester, 383
Steam engine, 383–6; self-regulation of, 786
Steam locomotive, 385, 386
Steamship, 385
Steam turbine, 385
Steel, 277ff.; formation of, 278, 279
Stefan's law, 354
Stellarator, 459

Sterility, 663
Sterilization, 612
Stereophonic sound, 395
Steroid hormones, 683–5
Steroids, 650
Stethoscope, 611
Stickney, 110
Stimulus and response, 751, 752
Stone Age, 727ff.; population during, 738
Stonehenge, 89
Stony meteorites, 155, 222
Storage battery, 404, 405
Storage rings, 309
Strange particles, 331, 332
Strata, 707
Stratosphere, 195
Streamer chamber, 294, 295
Streptomyces, 617, 621
Streptomycin, 617, 620, 621
Stroke, 691
Strong-focusing synchrotron, 308
Strong interaction, 334
Strontium 90, 451, 453–5
Strontium units, 453
Structural formulae, 467
Strychnine: structure of, 482; synthesis of, 482
Subatomic particles, 285ff.; table of, 330
Sub-baryonic particles, 329
Sublimation, 265
Subliminal velocities, 816
Submarines, 177, 385; nuclear, 446
Substrate, 530, 531
Succinic acid, 530, 531
Succinic acid dehydrogenase, 530
Sucrose, 490
Sugar: ethyl alcohol from, 534; lactic acid from, 535
Sulfadiazine, 615
Sulfanilamide, 615
Sulfapyridine, 615
Sulfathiazole, 615
Sulfur, 232; proteins and, 508; rubber and, 491
Sulfur dioxide, 265
Sulfuric acid, 233, 494; catalysis and, 525; formula of, 465; Venus's atmosphere and, 102
Sumerians, 726
Sun, 85ff.; absolute magnitude of, 28; corona of, 69, 70; diameter of, 85; distance of, 21–3, 85; eclipses of,

Sun – *cont.*
59; elements in, 59; energy source
of, 38–40; galactic orbit of, 31;
galactic position of, 30; infrared
radiation of, 61; magnetic field
of, 84, 211; magnitude of, 28; mass
of, 85; motion of, 26; neutrinos
and, 319, 320; photographs of, 60;
possible companion body of,
721; rotation of, 41; spectral class
of, 59; X-rays from, 69
Sunlamp, 399
Sunspots, 41, 85, 86, 211, 212, 214;
cycle of, 86, 87
Superclasses, 703
Superconductivity, 269, 270
Superfluidity, 270, 271
Superheavy elements, 248, 249
Superheterodyne receiver, 414
Superluminal velocity, 68, 817
Supernovae, 48, 83; formation of, 55;
star collapse and, 73
Superphyla, 700
Supersonic planes, 203, 409, 410
Surface catalysis, 526
Surveyor *1*, 95
Surveyor *5*, 95
Svedberg, 512
S waves, 153, 154
Swift, 110
Symbiosis, 491
Symbolic logic, 794
Sympathetic nerves, 765
Synapse, 752
Synchrocyclotron, 306
Syncom II, 198, 199
Syncom III, 198, 199
Synodic month, 88
Synthetics, 477
Syphilis, 614
Systema Naturae, 698

Table salt, 202
Tachyons, 817
Tachyon-universe, 817
Tadpoles, 703
Talking pictures, 403
Tambora, Mount, 153
Tanks, 440
Tape recording, 396
Tardyons, 816
Tardyon-universe, 817

Tartaric acid, 470
Tau electron, 324
Tau meson, 337
Tauon, 324
Taxonomy, 698
Technetium, 246; superconductivity
of, 269; uranium fission and, 435
Technology, 379ff.
Tectonic plates, 173
Teflon, 499
Tektites, 225
Telegraph Plateau, 164, 168
Telegraphy, 392
Telemetering, 196, 197
Telephone, 394, 395; radio and, 414
Telescopes, 57, 58, 607
Television, 198, 206, 207, 414, 415;
color, 261
Tellurium, 237, 244
Telstar I, 198
Temperature, 362–4; heat and, 364,
368
Temperature scales, 263
Terminal amino acids, 520
Termites, 490
Terpene, 492
Terramycin, 617
Terylene, 504
Testes, 683
Testudo, 800
Tetanus, 636, 637
Tethys, 122–4
Tetrachloroethylene, 320
Tetracyclines, 617
Tetraethyl lead, 406
Tetrafluoroethylene, 499
Tetrahedral carbon atom, 471
Tetrapoda, 703
Tevatron, 309
Thalamus, 754, 762
Thera, 149
Thermal neutrons, 433, 438
Thermal pollution, 457
Thermistors, 417
Thermodynamics, 365–8, 384; laws
of, 272, 366, 367, 544
Thermoelectricity, 428, 429
Thermonuclear reactions, 442
Thermometer, 363, 364
Thermoplastics, 498
Thermosetting plastics, 498
Thermosphere, 197
Thermostat, 786, 787
Theta meson, 337

Theta waves, 770
Thiamine, 661, 664
Thiokol, 506, 509, 529
34 Tauri, 125
Thorium, 290; isotopes of, 291
Thorium *282*, 448
Thorium X, 289, 290
Thermos bottle, 267
3C273, 66, 67
Three Mile Island, 447
Threonine, 656
Thresher, 452
Thymus gland, 642
Thymus nucleic acid, 579
Thyroid cancer, 650
Thyroid gland, 678, 679, 681, 685
Thyroid-stimulating hormone, 685
Thyroxine, 678
Ticks, 626
Tidal hypothesis, 83
Tidal wave, 146
Tides, lunar, 160
Time, 358, 359
Tin, 277; superconductivity of, 269
Tires, 504; automobile, 407
Tiros I, 198
Tiros II, 198
Tiros III, 198
Tissue culture, 637
Titan: atmosphere of, 122; possible
    life on, 601
Titania, 125
Titanic, 180
Titanium, 283; superconductivity of,
    269
Toads, 703
Tobacco mosaic disease, 627, 629
Tobacco mosaic virus, 522
Tocopherol, 663
Tokamak, 459
Tooth decay, 675, 676
Tornado, 140
Torricellian vacuum, 193
Totemism, 696
Toxoid, 637
T-quark, 332
Trace-element deserts; 672
Trace minerals, 671ff.
Tracers, 539ff.
Tracheophyta, 699
Tranquilizers, 781
Transactinides, 262
Transduction, 628
Transfer-RNA, 587, 588, 621

Transformer, electric, 210, 393
Transitional elements, 257ff.
Transistors, 417–19, 797
Transit 4A, 451
Transit 4B, 451
Transuranium elements, 246
Transverse waves, 348
Tree-shrews, 705
Trenches, ocean, 171
Triassic age, 719
Trieste, 178
Trigonometry, 23
Triiodothyronine, 679
Trilobites, 718
Trinitrotoluene, 495
Triode, 413
Triphosphopyridine nucleotide, 667
Tristan da Cunha, 171
Tritium, 303; tracer activity of, 542
Triton (nucleus), 303
Triton (satellite), 127
Trojan asteroids, 132
Trojan situation, 123, 124
Trojan war, 724, 725
Tropisms, 772
Tropopause, 195
Troposphere, 195
Troy, 726
Trypanosomes, 614
Trypan red, 614
Trypsin, 530, 532; amino acids of, 522
Tryptophan, 655, 656, 782
Tsunami, 146
Tuatara, 778
Tube, radio, 411, 412
Tuberculosis, 613
Tumor, 645
Tumor viruses, 648, 649
Tungsten, 283; electric lights and,
    398; steel and, 279
Tunguska event, 223
Tinicates, 672
Turbidity currents, 172
Turbinia, 385
Turbo-jet engine, 408
Turboprop planes, 408
*Twenty Thousand Leagues under the
    Sea,* 385
Tycho, 225
Tyndall effect, 513
Typhoon, 140
Typhus fever, 617, 625
Tyrannosaurus Rex, 719
Tyrian purple, 479

Tyrocidin, 616, 620
Tyrosine, 508, 677
Tyrothricin, 616

U-bomb, 443
Ultracentrifuge, 512
Ultramicroscope, 513
Ultrasonic sound, 169, 170
Ultraviolet radiation, 61, 646;
    atmosphere and, 230; cancer and,
    647; ozone and, 203, 204
Umbriel, 125
Uncertainty principle, 376
Unconformity, 717
Undernutrition, 654
Underwater breathing, 746
Unified field theory, 338–40
Uniformitarian principle, 37
Universal gravitation, law of, 810
Universe, 19ff.; age of, 44; average
    temperature of, 44; expansion of,
    42–5; heat-death of, 367;
    inflationary, 340; origin of, 43;
    possible life in, 601ff.
Unpolarized light, 469
Unsaturated fatty acids, 657
Upper atmosphere, 196
U-quark, 331
Uracil, 579
Uranium, 246; breakdown of, 289–
    91; isotopes of, 292, 437ff., 512;
    neutron bombardment of, 433ff.;
    nuclear bomb and, 437ff.; ocean
    and, 166; superconductivity of, 269
Uranium 233, 448
Uranium 235, 292; half-life of, 450;
    purification of, 437, 438
Uranium 238, 292, 437; breakdown
    energy of, 371; breeder reactors
    and, 447, 448; half-life of, 449;
    nuclear bombs and, 443
Uranium fission, 371
Uranium hexafluoride, 438, 499
Uranium X (breakdown product),
    289
Uranium X (hypothetical element),
    246, 433
Uranus, 124–6; axis of rotation of,
    125; diameter of, 124; discovery
    of, 124; distance of, 124; mass of,
    124; orbit of, 126; rings of, 126;
    rotation of, 125; satellites of, 125

Urbanization, 743
Urea, 465; excretion of, 721, 722;
    proteins and, 534, 535
Urea cycle, 539
Urease, 529
Urethra, 722
Uric acid, 722
Urine: diabetes and, 679, 681, 682;
    pregnancy and, 683

V-1, 409
V-2, 93
Vaccinations, 635
Vaccines, 635, 636
Vaccinia, 635
Vacuum, 191, 193, 249; food
    preservation and, 657; uses of,
    383
Vacuum tube, 797
Valence, 466, 467
Valine, 656
Valles Marineris, 109
Valve, 412
Valves, vein, 553
Vanadium, 279; tunicates and, 672
Vanadium pentoxide, 525
Van Allen radiation belts, 216
Vanguard I, 143, 432
Vanguard II, 143
Vapors, 262
Varves, 729, 730
Vaseline, 474
Vega, 602; distance of, 26
Vegetarianism, 655
Veil Nebula, 51
Veins, 553
Venera probes, 102
Ventricles, 553
Venus, 98ff.; atmosphere of, 102;
    brightness of, 99; cloud layer of,
    99, 100; density of, 106; diameter
    of, 99; distance of, 98; magnetic
    field of, 219; magnitude of, 28;
    microwave reflection by 100;
    phases of, 99; radio waves from, 64;
    revolution of, 100; rotation of,
    101; sun and, 98; surface of, 102,
    103; surface temperature of, 101
Venus probes, 101–3
Vertebrates, 701ff.; colonization of
    land by, 719
Vesta, 131

Vesuvius, Mount, 149, 152, 725
Videocassettes, 415
Videotape, 415
Viking *1*, 109
Viking *2*, 109
Vinyl acetate, 504
Vinyl chloride, 500, 504
Vinyl cyanide, 503
Vinyl resin, 500, 502
Violet shift, 41
Virginium, 245
Viroids, 633
Virtual particles, 377
Virus diseases, 627
Viruses, 623, 626ff.; attenuation of,
    636; cancer and, 648, 649;
    crystalline, 627, 628; culture of,
    637–9; nucleic acid in, 628, 631–
    3, shape and size of, 629–30
Viscose, 502
Visual purple, 670
Vitalism, 464
Vital principle, 493
Vitamin A, 660–62, 665, 669, 670
Vitamin B, 660, 661, 664, 665
Vitamin B_1, 661
Vitamin B_2, 668
Vitamin B_{12}, 673, 674
Vitamin C, 660–62
Vitamin D, 661, 663, 665
Vitamin E, 661, 663, 664
Vitamine, 660
Vitamin K, 661, 664
Vitamins, 66off.; fortification with,
    665; synthetic, 664, 665
Volcanoes, 149–53, 189; Io and, 118;
    Mars and, 109
Volt, 390
Voltage multiplier, 305
Voltaic pile, 389
Voltaire, 110
Voyager *1*, 116, 122, 123
Voyager *2*, 116, 122
Voyage to the Moon, 91
Vulcanization, 491

Wallace's line, 712
Warburg flask, 536, 537
Warburg's yellow coenzyme, 668
Warm-bloodedness, 703
War of 1812, 92
War of the Worlds, 107

Warts, 645
Wash-and-wear fabrics, 504
Washington Monument, 281
Wasserman test, 637
Water: atmosphere and, 228, 230;
    boiling point of, 476; electron
    configuration of, 252; empirical
    formula of, 465; fresh, 167; high
    pressures and, 273; human use of,
    740; interstellar, 78;
    photodissociation of, 230;
    photolysis of, 549; photosynthesis
    and, 545, 548; purification of, 260;
    structural formula of, 467
Water-soluble B, 660
Water vapor, 262
Wavelength, light, 61
Wave mechanics, 375, 376
Weak interaction, 334, 335
Weather forecasting, 192, 198, 215
Weather modification, 197
Weather satellites, 198
Weddell Sea, 182
Weightlessness, 749
Welsbach mantle, 396
West-wind drift, 164
Whale oil, 396
Whales, 698; brains of, 756
Whip, 409
Whiskers, crystal, 383
Whistlers, 214
White blood cells, 639
White dwarfs, 49, 50; Einstein shift
    and, 360; formation of, 54; novae
    and, 54, 55
White matter, 759
Windmill, 786
Wireless telegraphy, 411
Wireless telephony, 414
Wirephotos, 414
Withdrawal symptoms, 483
Woad, 479
Work: heat and, 365, 366
Worlds in the Making, 594
World War I, 414
World War II, 92, 93, 260, 347, 409,
    435, 436, 625
Wood, 405; digestion of, 490
W-particle, 335, 338
Writing, ancient, 725
Wrought iron, 278

X-15 plane, 409
X chromosome, 564–6
Xerophthalmia, 660
Xi particle, 328
X-ray diffraction, 240; nucleic acids and, 583; proteins and, 523; vitamin B_{12} and, 674
X-rays, 62, 240–42, 372; black holes and, 74, 75; cancer and, 646; dangers of, 448, 449; diffraction of, 240; fluorescence and, 242; mutations and, 568, 569, 573; nature of, 240, 241; particle characteristics of, 374; polypeptides and, 514
X-ray sources, 69ff.
Xenic acid, 255
Xenon, 201, 254
Xenon difluoride, 255
Xenon fluoroplatinate, 254
Xenon oxytetrafluoride, 255
Xenon trioxide, 255
Xerography, 389

Y chromosome, 564–6
Yeast, 527, 534, 535, 622, 660
Yeast nucleic acid, 579
Yellow fever, 624, 625, 638
Yellow fever virus, 626, 629
Yttria, 255
Yttrium, 255
Yttrium oxide, 261

Zein, 656
Zeolites, 260
Zeppelins, 282
Zero, 791
Zero-point energy, 376
Zinc, 672
Zinc sulfide, 292
Zinjanthropus, 733
Zirconium, 256
Zodiacal light, 221
Zone refining, 418
Zoological Philosophy, 709
Z° particle, 338
Zurich sunspot number, 86